Advances in Intelligent Systems and Computing

Volume 1268

The series "Advances in Intelligent Systems and Computing" contains publications on theory, applications, and design methods of Intelligent Systems and Intelligent Computing. Virtually all disciplines such as engineering, natural sciences, computer and information science, ICT, economics, business, e-commerce, environment, healthcare, life science are covered. The list of topics spans all the areas of modern intelligent systems and computing such as: computational intelligence, soft computing including neural networks, fuzzy systems, evolutionary computing and the fusion of these paradigms, social intelligence, ambient intelligence, computational neuroscience, artificial life, virtual worlds and society, cognitive science and systems, Perception and Vision, DNA and immune based systems, self-organizing and adaptive systems, e-Learning and teaching, human-centered and human-centric computing, recommender systems, intelligent control, robotics and mechatronics including human-machine teaming, knowledge-based paradigms, learning paradigms, machine ethics, intelligent data analysis, knowledge management, intelligent agents, intelligent decision making and support, intelligent network security, trust management, interactive entertainment, Web intelligence and multimedia.

The publications within "Advances in Intelligent Systems and Computing" are primarily proceedings of important conferences, symposia and congresses. They cover significant recent developments in the field, both of a foundational and applicable character. An important characteristic feature of the series is the short publication time and world-wide distribution. This permits a rapid and broad dissemination of research results.

**** Indexing: The books of this series are submitted to ISI Proceedings, EI-Compendex, DBLP, SCOPUS, Google Scholar and Springerlink ****

More information about this series at http://www.springer.com/series/11156

Álvaro Herrero · Carlos Cambra ·
Daniel Urda · Javier Sedano ·
Héctor Quintián · Emilio Corchado
Editors

15th International Conference on Soft Computing Models in Industrial and Environmental Applications (SOCO 2020)

Springer

International Conference on
SOCO
Soft Computing Models in Industrial and Environmental Applications

Editors
Álvaro Herrero
Grupo de Inteligencia Computacional
Aplicada (GICAP), Departamento
de Ingeniería Informática, Escuela
Politécnica Superior
Universidad de Burgos
Burgos, Spain

Carlos Cambra
Grupo de Inteligencia Computacional
Aplicada (GICAP), Departamento
de Ingeniería Informática, Escuela
Politécnica Superior
Universidad de Burgos
Burgos, Spain

Daniel Urda
Grupo de Inteligencia Computacional
Aplicada (GICAP), Departamento
de Ingeniería Informática, Escuela
Politécnica Superior
Universidad de Burgos
Burgos, Spain

Javier Sedano
Technological Institute of Castilla y León
Burgos, Spain

Emilio Corchado
University of Salamanca
Salamanca, Spain

Héctor Quintián
Department of Industrial Engineering
University of A Coruña
La Coruña, Spain

ISSN 2194-5357 ISSN 2194-5365 (electronic)
Advances in Intelligent Systems and Computing
ISBN 978-3-030-57801-5 ISBN 978-3-030-57802-2 (eBook)
https://doi.org/10.1007/978-3-030-57802-2

This Springer imprint is published by the registered company Springer Nature Switzerland AG
The registered company address is: Gewerbestrasse 11, 6330 Cham, Switzerland

Preface

This volume of Advances in Intelligent and Soft Computing contains accepted papers presented at SOCO 2020 conference held in the beautiful and historic city of Burgos (Spain), in September 2020.

Soft computing represents a collection or set of computational techniques in machine learning, computer science, and some engineering disciplines, which investigate, simulate, and analyze very complex issues and phenomena.

After a through peer-review process, the SOCO 2020 International Program Committee selected 83 papers which are published in these conference proceedings and represents an acceptance rate of 35%. Due to the COVID-19 outbreak, the SOCO 2020 edition was blended, combining on-site and on-line participation. In this relevant edition, a special emphasis was put on the organization of special sessions. Eleven special sessions were organized related to relevant topics such as: soft computing applications in precision agriculture, manufacturing and management systems, management of industrial and environmental enterprises, logistics and transportation systems, robotics and autonomous vehicles, computer vision, laser-based sensing and measurement and other topics such as forecasting industrial time series, IoT, big data and cyberphysical systems, nonlinear dynamical systems and fluid dynamics, modeling and control systems.

The selection of papers was extremely rigorous in order to maintain the high quality of SOCO conference editions and we would like to thank the members of the Program Committees for their hard work in the reviewing process. This is a crucial process to the creation of a high standard conference and the SOCO conference would not exist without their help.

SOCO 2020 has teamed up with "Neurocomputing" (Elsevier) "Logic Journal of the IGPL" (Oxford University Press) and "Cybernetics and Systems: An International Journal" (Taylor and Francis) for a suite of special issues including selected papers from SOCO 2020.

Particular thanks go as well to the conference main sponsors Startup Ole and the IEEE Systems, Man, and Cybernetics Society-Spanish, Portuguese, French, and Italian Chapters, who jointly contributed in an active and constructive manner to the success of this initiative.

We would like to thank all the special session organizers, contributing authors, as well as the members of the Program Committees and the Local Organizing Committee for their hard and highly valuable work. Their work has helped to contribute to the success of the SOCO 2020 event.

September 2020

Álvaro Herrero
Carlos Cambra
Daniel Urda
Javier Sedano
Héctor Quintián
Emilio Corchado

Soco 2020 Organization

General Chair

Emilio Corchado	University of Salamanca, Spain

General Co-chair

Álvaro Herrero	University of Burgos, Spain

International Advisory Committee

Ashraf Saad	Georgia Southern University, USA
Amy Neustein	Linguistic Technology Systems, USA
Ajith Abraham	Machine Intelligence Research Labs-MIR Labs, Europe
Jon G. Hall	The Open University, UK
Paulo Novais	University of Minho, Portugal
Amparo Alonso Betanzos	President Spanish Association for Artificial Intelligence (AEPIA), Spain
Michael Gabbay	Kings College London, UK
Aditya Ghose	University of Wollongong, Australia
Saeid Nahavandi	Deakin University, Australia
Henri Pierreval	LIMOS UMR CNRS 6158 IFMA, France

Program Committee Chairs

Emilio Corchado	University of Salamanca, Spain
Álvaro Herrero	University of Burgos, Spain
Javier Sedano	Technological Institute of Castilla y León, Spain
Héctor Quintián	University of A Coruña, Spain

Program Committee

Agostino Marcello Mangini	Politecnico di Bari, Italy
Agustina Bouchet	UNMDP, Argentina
Akemi Galvez-Tomida	University of Cantabria, Spain
Albeto Herreros López	University of Valladolid, Spain
Alfredo Jimenez	KEDGE Business School, Spain
Álvaro Herrero	University of Burgos, Spain
Anca Draghici	Polyethnic University of Timisoara, Romania
Andreea Vescan	Babes-Bolyai University, Romania
Andres Iglesias Prieto	University of Cantabria, Spain
Angel Arroyo	University of Burgos, Spain
Angelo Costa	University of Minho, Portugal
Anna Bartkowiak	University of Wroclaw, Poland
Anna Burduk	Wrocław University of Technology, Poland
Anton Koval	Luleå University of Technology, Sweden
Antonio Caamaño	Rey Juan Carlos University, Spain
Antonio Bahamonde	University of Oviedo, Spain
Bogdan Okreša Đurić	University of Zagreb, Croatia
Bruno Baruque	University of Burgos, Spain
Camelia Serban	Babes-Bolyai University, Romania
Camelia-M. Pintea	Technical University of Cluj-Napoca, Romania
Carlos Cambra	University of Burgos, Spain
Carlos Casanova	Polytechnic University of Madrid, Spain
Carlos Pereira	ISEC, Portugal
Carmen Benavides	University of León, Spain
Cosmin Sabo	Technical University of Cluj-Napoca, Romania
Damian Krenczyk	Silesian University of Technology, Poland
Daniel Urda	University of Burgos, Spain
Daniela Perdukova	Technical University of Kosice, Slovakia
David Alvarez Leon	University of León, Spain
David Camacho	Autonomous University of Madrid, Spain
David Griol	University Carlos III de Madrid, Spain
Eduardo Solteiro Pires	UTAD University, Portugal
Eleni Mangina	University College Dublin, Ireland
Eloy Irigoyen	University of the Basque Country, Spain
Enrique De La Cal Marín	University of Oviedo, Spain
Enrique Onieva	University of Deusto, Spain
Esteban Jove	University of A Coruña, Spain
Eva Volna	University of Ostrava, Czechia
Fernando Sanchez Lasheras	University of Oviedo, Spain
Florentino Fdez-Riverola	University of Vigo, Spain
Francisco Martínez-Álvarez	Pablo de Olavide University, Spain
Francisco Zayas Gato	University of A Coruña, Spain
Gabriel Villarrubia	University of Salamanca, Spain

Grzegorz Ćwikła	Silesian University of Technology, Poland
Grzegorz J. Nalepa	AGH University, Poland
Héctor Quintián	University of A Coruña, Spain
Henri Pierreval	LIMOS-IFMA, France
Humberto Bustince	University of Navarra, Spain
Iñigo Lecuona Mugica	Mondragon University, Spain
Ioana Zelina	Technical University of Cluj-Napoca, Romania
Isaias Garcia	University of León, Spain
Iwona Pisz	Opole University, Poland
Javier Sanchis Saez	Polytechnic University of Valencia, Spain
Jaume Jordán	Polytechnic University of Valencia, Spain
Javier del Ser	Fundación Tecnalia Research & Innovation, Spain
Javier Palanca	Polytechnic University of Valencia, Spain
Jesus Ariel Carrasco-Ochoa	INAOE, Mexico
Jesús D. Santos	University of Oviedo, Spain
Jiri Pospichal	University of Ss. Cyril and Methodius, Slovakia
Jorge Barbosa	ISEC - Instituto Superior de Engenharia de Coimbra, Portugal
Jorge García-Gutiérrez	University of Seville, Spain
Jose Alfredo Ferreira Costa	Federal University, UFRN, Brazil
Jose Dorronsoro	Autonomous University of Madrid, Spain
José Francisco Torres Maldonado	Pablo de Olavide University, Spain
José Gámez	University of Castilla-La Mancha, Spain
José Luis Calvo-Rolle	University of A Coruña, Spain
José-Luis Casteleiro-Roca	University of A Coruña, Spain
Jose M. Molina	University Carlos III de Madrid, Spain
Jose Manuel Gonzalez-Cava	University of La Laguna, Spain
Jose Manuel López-Guede	University of the Basque Country, Spain
José Ramón Villar	University of Oviedo, Spain
José Valente de Oliveira	University of Algarve, Portugal
Juan Albino Mendez	University of La Laguna, Spain
Juan Gomez Romero	University of Granada, Spain
Juan M. Alberola	Polytechnic University of Valencia, Spain
Julio César Puche Regaliza	University of Burgos, Spain
Khalid Raza	Jamia Millia Islamia, India
Krzysztof Kalinowski	Silesian University of Technology, Poland
Lidia Sánchez-González	Universidad de León, Spain
Luis Paulo Reis	University of Porto, Portugal
M. Chadli	University of Paris-Saclay, France
Maciej Grzenda	Warsaw University of Technology, Poland
Manuel Castejón-Limas	Universidad de Leon, Spain
Manuel Mejia-Lavalle	Cenidet, Mexico
Marcin Iwanowski	Warsaw University of Technology, Poland

Marcin Paprzycki	Polish Academy of Sciences, Poland
Maria Luisa Sanchez	University of Oviedo, Spain
Maria Tomas Rodriguez	The City University of London, UK
Marius Balas	Aurel Vlaicu University of Arad, Romania
Matilde Santos	Complutense University of Madrid, Spain
Mehmet Emin Aydin	University of the West of England, UK
Michael O'Grady	University College Dublin, Ireland
Michal Wozniak	Wroclaw University of Technology, Poland
Michele Roccotelli	Politecnico di Bari, Italy
Mihaela I. Chidean	Rey Juan Carlos University, Spain
Mitiche Lahcene	Laboratoire de Recherche Modélisation Simulation et Optimisation des Systèmes Complexes Réels, Algeria
Nayat Sánchez-Pi	Inria, France
Oscar Castillo	Tijuana Institute of Technology, Mexico
Ovidiu Cosma	Technical University Cluj Napoca, Romania
Pablo Chamoso	University of Salamanca, Spain
Paul Eric Dossou	ICAM, France
Paulo Moura Oliveira	UTAD University, Portugal
Paulo Novais	University of Minho, Portugal
Pedro Antonio Gutierrez	University of Cordoba, Spain
Petr Dolezel	University of Pardubice, Czechia
Petrica Pop	Technical University of Cluj-Napoca, Romania
Ravinesh C. Deo	University of Southern Queensland, Australia
Reggie Davidrajuh	University of Stavanger, Norway
Ricardo Aler	University Carlos III, Spain
Richard Duro	University of A Coruña, Spain
Robert Burduk	Wroclaw University of Science and Technology, Poland
Rosangela Ballini	UNICAMP, Brazil
Sancho Salcedo-Sanz	Universidad de Alcalá, Spain
Sara Rodríguez	University of Salamanca, Spain
Sebastian Saniuk	University of Zielona Gora, Poland
Sebastián Ventura	University of Cordoba, Spain
Stefano Pizzuti	Energy New Technologies and Sustainable Economic Development Agency (ENEA), Italy
Sung-Bae Cho	Yonsei University, South Korea
Tzung-Pei Hong	National University of Kaohsiung, Taiwan
Valeriu Manuel Ionescu	University of Pitesti, Romania
Vicente Matellan	University of Leon, Spain
Vicente Julian	Polytechnic University of Valencia, Spain
Wei-Chiang Hong	Jiangsu Normal University, Taiwan
Wilfried Elmenreich	Alpen-Adria-Universität Klagenfurt, Austria
Zita Vale	Polytechnic of Porto, Portugal

Special Sessions

Contributions of Soft Computing to Precision Agriculture

Special Session Organizers

Petr Dolezel	University of Pardubice, Czech Republic
Daniel Honc	University of Pardubice, Czech Republic
Bruno Baruque	University of Burgos, Spain
Jan Mares	University of Chemistry and Technology Prague, Czech Republic

Program Committee

Daniel Honc	University of Pardubice, Czechia
Dominik Stursa	University of Pardubice, Czechia
Eva Volna	University of Ostrava, Czechia
Francisco Martínez-Álvarez	Pablo de Olavide University, Spain
Isabel Sofia Sousa Brito	Polytechnic Institute of Beja, Portugal
Jan Mares	UCT Prague, Czechia
Jan Merta	University of Pardubice, Czechia
Jaroslav Marek	University of Pardubice, Czechia
Laura Melgar-García	Pablo de Olavide University, Spain
Maria Teresa Godinho	Polytechnic Institute of Beja, Portugal
Martin Kotyrba	University of Ostrava, Czechia
Pavel Hrncirik	University of Chemistry and Technology Prague, Czechia
Pavel Skrabanek	Brno University of Technology, Czechia
Santiago Porras Alfonso	Universidad de Burgos, Spain

Soft Computing Methods in Manufacturing and Management Systems

Special Session Organizers

Damian Krenczyk	Silesian University of Technology, Poland
Bożena Skołud	Silesian University of Technology, Poland
Anna Burduk	Wroclaw University of Science and Technology, Poland
Krzysztof Kalinowski	Silesian University of Technology, Poland
Grzegorz Cwikla	Silesian University of Technology, Poland
Marek Placzek	Silesian University of Technology, Poland

Program Committee

Arkadiusz Gola	Lublin University of Technology, Poland
Bozena Skolud	Silesian University of Technology, Poland
Cezary Grabowik	Silesian Technical University, Poland
Dumitru Nedelcu	Gheorghe Asachi Technical University of Iasi, Romania
Franjo Jovic	University of Osijek, Croatia
Grzegorz Ćwikła	Silesian University of Technology, Poland
Ivan Kuric	University of Zilina, Slovakia
Iwona Pisz	Opole University, Poland
Karol Velisek	Slovak University of Technology in Bratislava, Slovakia
Kyratsis Panagiotis	University of Western Macedonia, Greece
Laszlo Dudas	University of Miskolc, Hungary
Marek Płaczek	Silesian University of Technology, Poland
Reggie Davidrajuh	University of Stavanger, Norway
Sebastian Saniuk	University of Zielona Gora, Poland
Wojciech Bozejko	Wroclaw University of Technology, Poland

Soft Computing Applications for the Management of Industrial and Environmental Enterprises

Special Session Organizers

Secil Bayraktar	TBS Business School, France
Alfredo Jiménez	KEDGE Business School, France
Álvaro Herrero	University of Burgos, Spain

Program Committee

Cristina Pérez	University Rey Juan Carlos, Spain
David Griol	University of Granada, Spain
Jose Luis Calvo-Rolle	University of A Coruña, Spain
José Ramón Villar	University of Oviedo, Spain
Julio César Puche Regaliza	University of Burgos, Spain
Manuel Grana	University of the Basque Country, Spain
Montserrat Jimenez Partearroyo	University Rey Juan Carlos, Spain
Pablo Chamoso	University of Salamanca, Spain
Pedro Antonio Gutierrez	University of Cordoba, Spain

Optimization, Modeling and Control by Soft Computing Techniques

Special Session Organizers

Eloy Irigoyen Gordo	University of the Basque Country, Spain
Matilde Santos Peñas	Complutense University of Madrid, Spain
José Luis Calvo Rolle	University of A Coruña, Spain
Mikel Larrea Sukia	University of the Basque Country, Spain
Ahmed Al-Jumaily	Auckland University of Technology, New Zealand

Program Committee

Agustin Jimenez	Polytechnic University of Madrid, Spain
Anna Burduk	Wrocław University of Technology, Poland
Antonio Javier Barragán	University of Huelva, Spain
Antonio Robles Alvarez	University of Oviedo, Spain
Antonio Sala	Polytechnic University of Valencia, Spain
Emilio Jimenez	University of La Rioja, Spain
Fernando Artaza	University of the Basque Country, Spain
Fernando Castaño Romero	Polytechnic University of Madrid, Spain
Fernando Matia	Polytechnic University of Madrid, Spain
Graciliano Marichal	University of La Laguna, Spain
Hilario López	University of Oviedo, Spain
Javier Muguerza	University of the Basque Country, Spain
Jesus Lozano	University of Extremadura, Spain
Jesús M. Zamarreño	University of Valladolid, Spain
Joaquim Melendez	University of Girona, Spain
Jorge Luis Madrid	CSIC, Spain
Jose Basilio Galvan	University of Navarra, Spain
José Luis Casteleiro-Roca	University of A Coruña, Spain
Jose Manuel Lopez-Guede	University of the Basque Country, Spain
Jose-Luis Diez	Polytechnic University of Valencia, Spain
Joseba Quevedo	Polytechnic University of Catalonia, Spain
Joshué Pérez-Rastelli	Tecnalia, Spain
Juan Albino Mendez Perez	University of Laguna, Spain
Juan José Valera	University of the Basque Country, Spain
Juan Pérez Oria	University of Cantabria, Spain
Luciano Alonso	University of Cantabria, Spain
Luis Magdalena	Polytechnic University of Madrid, Spain
Maria Fuente	University of Valladolid, Spain
María José Pérez-Ilzarbe	University of Navarra, Spain
Oscar Barambones	University of the Basque Country, Spain
Petr Dolezel	University of Pardubice, Czechia

Raquel Martinez Rodriguez	University of the Basque Country, Spain
Vicente Gomez-Garay	University of the Basque Country, Spain
Xabier Basogain Olabe	University of the Basque Country, Spain

Soft Computing and Machine Learning in Nonlinear Dynamical Systems and Fluid Dynamics: New Methods and Applications

Special Session Organizers

Soledad Le Clainche	Polytechnic University of Madrid, Madrid
José Miguel Pérez	Polytechnic University of Madrid, Madrid
David Gutiérrez Avilés	Pablo de Olavide University, Spain
Ricardo Vinuesa	KTH Royal Institute of Technology, Sweden

Program Committee

Soledad Le Clainche	Polytechnic University of Madrid, Madrid
José Miguel Pérez	Polytechnic University of Madrid, Madrid
David Gutiérrez Avilés	Pablo de Olavide University, Spain
Ricardo Vinuesa	KTH Royal Institute of Technology, Sweden

Soft Computing Techniques and Applications in Logistics and Transportation Systems

Special Session Organizers

Dragan Simić	University of Novi Sad, Serbia
Petrica Pop	Technical University of Cluj-Napoca, Romania
José Ramón Villar	University of Oviedo, Spain
Cosmin Sabo	Technical University of Cluj-Napoca, Romania
Javier Díez	University of León, Spain
Vladimir Ilin	University of Novi Sad, Serbia

Program Committee

Cosmin Sabo	Technical University of Cluj-Napoca, Romania
Dragan Simić	University of Novi Sad, Serbia
Javier Díez González	University of León, Spain
José R. Villar	University of Oviedo, Spain
Petrica Pop	Technical University of Cluj-Napoca, Romania
Vladimir Ilin	University of Novi Sad, Serbia

Soft Computing and Machine Learning in IoT, Big Data, and Cyberphysical Systems

Special Session Organizers

José Ramón Villar	University of Oviedo, Spain
Nashwa El-Bendary	Arab Academy for Science, Technology & Maritime Transport, Egypt
Qing Tan	Athabasca University, Canada

Program Committee

Alberto Cano	Virginia Commonwealth University, USA
Antony Bagnall	University of East Anglia, UK
Ashraf Darwish	Helwan University, Egypt
Bartosz Krawczyk	VCU College of Engineering, USA
Beatriz de la Iglesia	University of East Anglia, UK
Dragan Simic	University of Novi Sad, Faculty of Technical Sciences, Serbia
Dunwei Wen	Athabasca University, Canada
Enrique de la Cal	University of Oviedo, Spain
Harris Wang	Athabasca University, Canada
Irene Díaz	University of Oviedo, Spain
Jairo Cugliari	Université Paris-Sud XI, France
Kadry Ezzat	Higher Technological Institute, Egypt
Lamia Nabil Mahdy	Higher Technological Institute, Egypt
Larbi Esmahi	Athabasca University, Canada
Nashwa El-Bendary	Arab Academy for Science, Technology, and Maritime Transport, Egypt
Noelia Rico	University of Oviedo, Spain
Oscar Lin	Athabasca University, Canada
Qing Tan	Athabasca University, Canada
Sung-Bae Cho	Yonsei University, South Korea
Xiaokun Zhang	Athabasca University, Canada
Yu-Lin Jeng	Southern Taiwan University of Science and Technology, Taiwan
Yueh-Ming Huang	National Cheng Kung University, Taiwan

Soft Computing Applied to Robotics and Autonomous Vehicles

Special Session Organizers

J. Enrique Sierra García	ASTI Mobile Robotics, Spain
Matilde Santos Peñas	Complutense University of Madrid, Spain

Ioannis Mariolis	Centre for Research and Technology Hellas, Greece
Carlos Cambra Baseca	University of Burgos, Spain

Program Committee

Enrique Onieva	University of Deusto, Spain
Felipe Espinosa	University of Alcalá, Spain
Joshué Pérez-Rastelli	Tecnalia, Spain
Juan Manuel López Guede	University of the Basque Country, Spain
Miguel A. Olivares-Mendez	University of Luxembourg, Luxembourg

Soft Computing for Forecasting Industrial Time Series

Special Session Organizers

Alicja Krzemień	Central Mining Institute, Poland
Fernando Sánchez Lasheras	University of Oviedo, Spain
Gregorio Fidalgo Valverde	University of Oviedo, Spain
Pedro Riesgo Fernández	University of Oviedo, Spain

Program Committee

Alicja Krzemień	Central Mining Institute, Poland
Fernando Sánchez Lasheras	University of Oviedo, Spain
Gregorio Fidalgo Valverde	University of Oviedo, Spain
Javier García	University of Oviedo, Spain
Pedro Riesgo Fernández	University of Oviedo, Spain

Machine Learning in Computer Vision

Special Session Organizers

Jose Garcia Rodriguez	University of Alicante, Spain
Alexandra Psarrou	University of Westminster, UK
Eldon Caldwel	University of Costa Rica, Costa Rica
Jorge Azorin Lopez	University of Alicante, Spain
Andres Fuster Guillo	University of Alicante, Spain
Enrique Dominguez	University of Malaga, Spain

Program Committee

Alexandra Psarrou	University of Westminster, UK
Andres Fuster Guillo	University of Alicante, Spain
Eldon Caldwel	University of Costa Rica, Costa Rica
Enrique Dominguez	University of Malaga, Spain

Jorge Azorin Lopez	University of Alicante, Spain
Jose Garcia Rodriguez	University of Alicante, Spain
Marcelo Salva	University of Alicante, Spain

Computational Intelligence for Laser-Based Sensing and Measurement

Special Session Organizers

Manuel Graña	University of the Basque Country, Spain
Leyre Torre	University of the Basque Country, Spain
Jose Manuel Lopez-Guede	University of the Basque Country, Spain
Anna Kamińska-Chuchmała	Wroclaw University of Science and Technology, Poland
Marina Aguilar	University of the Basque Country, Spain

Program Committee

Anna Kamińska-Chuchmała	Wroclaw University of Science and Technology, Poland
Javier Barandiaran	Vicomtech
Jose Manuel Lopez-Guede	University of the Basque Country, Spain
Leyre Torre	University of the Basque Country, Spain
Manuel Graña	University of the Basque Country, Spain
Marcos Alonso	University of the Basque Country, Spain
Marina Aguilar	University of the Basque Country, Spain

Organising Committee Chairs

Álvaro Herrero	University of Burgos, Spain
Javier Sedano	ITCL, Spain
Carlos Cambra	University of Burgos, Spain
Daniel Urda	University of Burgos, Spain

Organising Committee

Emilio Corchado	University of Salamanca, Spain
Héctor Quintián	University of A Coruña, Spain
Carlos Alonso de Armiño	University of Burgos, Spain
Ángel Arroyo	University of Burgos, Spain
Bruno Baruque	University of Burgos, Spain
Nuño Basurto	University of Burgos, Spain

Pedro Burgos	University of Burgos, Spain
David Caubilla	University of Burgos, Spain
Leticia Curiel	University of Burgos, Spain
Raquel Redondo	University of Burgos, Spain
Jesús Enrique Sierra	University of Burgos, Spain
Belén Vaquerizo	University of Burgos, Spain
Juan Vicente Martín	University of Burgos, Spain

Contents

Special Session: Soft Computing Techniques and Applications in Logistics and Transportation Systems

Special Session: Soft Computing and Machine Learning in IoT, Big Data and Cyber Physical Systems

Special Session: Computational Intelligence for Laser-Based Sensing and Measurement

Soft Computing Applications

Advanced Oversampling for Improved Detection of Software Anomalies in a Robot

Nuño Basurto[1], Michał Woźniak[2], Carlos Cambra[1], and Álvaro Herrero[1](✉)

[1] Grupo de Inteligencia Computacional Aplicada (GICAP), Departamento de Ingeniería Informática, Escuela Politécnica Superior, Universidad de Burgos, Av. Cantabria s/n, 09006 Burgos, Spain
{nbasurto,ccbaseca,ahcosio}@ubu.es

[2] Department of Systems and Computer Networks, Wroclaw University of Science and Technology, Wybrzeze Wyspianskiego 27, 50-370 Wroclaw, Poland
michal.wozniak@pwr.edu.pl

Abstract. Anomaly detection has been a challenging topic for decades and it still is open to new contributions nowadays. More specifically, the detection of anomalies (not only hardware ones but also those affecting the software) suffers from many problems when monitoring cyber-physical systems. One such usual problem is the much fewer data samples of anomalies than those available for the normal functioning of systems. This class-imbalance problem is addressed in the present paper and a novel strategy for oversampling the minority class is applied to an open dataset containing information about the performance of a component-based robot. The proposed strategy mainly consists on selecting the instances to be oversampled according to different criteria instead of randomly oversampling. Obtained results demonstrate that the proposed strategy improves predictive performance, especially when the SVM (*Support Vector Machine*) is used as classifier.

Keywords: Oversampling · Imbalanced data · Anomaly detection · Pattern classification · Support Vector Machines · Robot

1 Introduction

Up to now, both unsupervised [7], supervised, and reinforcement [8] learning models have been previously applied to robotics. A variety of problems have been addressed so far, such as control [21,22] and communications [5] to enumerate only a few. In the case of anomaly detection, most of the works have been focused on the hardware anomaly detection [10], while software anomalies have been scarcely addressed. This work deals with the detection of performance anomalies in a robot software system. A data collection including actions undertaken by

Á. Herrero et al. (Eds.): SOCO 2020, AISC 1268, pp. 3–12, 2021.
https://doi.org/10.1007/978-3-030-57802-2_1

a robot (some of them suffering from an abnormal behavior) has been gathered for this purpose. The dataset is publicly available (see its detailed description in Sect. 3.1) [17,18].

One can find several solutions to this problem. Initially, Wienke et al. [19] employed methods inspired by *Support Vector Machines* (SVM) as One-Class SVM.

The later work [20] includes the analysis of the individual components of the robot to discover the potential changes that can take place in the use of resources. Such an analysis makes it possible to predict how changes in the operation of one component may affect the others.

One-class classification, together with data balancing techniques, has been previously discussed. In [4], the authors analyzed the effect of class imbalance in several datasets. They applied the Tomek-Links undersampling technique together with six different models, comprising Naïve Bayes (NB) and SVM, among others. Finally, the authors proposed a classification model, based on the SVM classifier, that improves the classification results from the minority class without getting worse performance on the majority class.

He and Garcia [6] proposed *dataspace weighting* by assigning different weights to instances from different classes. As a result, classes have the same total weight, with a positive impact on the classification rate. On the other hand, Cerqueira et al. [2] used the *Synthetic Minority Over-sampling Technique* (SMOTE) [3] to get a class-balanced distribution of data that improved the classification performance. The aim of such classification was carrying out predictive maintenance (that is, detecting anomalies) on the air pressure system of heavy trucks. More recently, another study [14] applied SMOTE for anomaly detection in an assembly line. Data was processed (to remove outliers) with DBSCAN, and then SMOTE was applied for data balancing. Finally, Random Forest (RF) was used to predict anomaly. RF is also applied in [1] to detect and classify failures of a vehicle fleet. Additionally, a parameter tuning framework is proposed to overcome the class imbalance problem. Similarly, Luo et al. [11] considered the task of imbalanced data and its implications in anomaly detection. To solve it, they generated new synthetic data samples using a technique called *Synthetic*, which is an extended version of SMOTE. They have used some standard classifiers, such as Decision Trees, Logistic Regression, SVM, and Naive Bayes so that they can verify the universality of the algorithms. Rather than proposing the application of one balancing method, such as SMOTE, our work proposes a novel strategy for selecting data instances to be oversampled. Such a strategy is based on Euclidean distance and the k-NN algorithm trying to improves the oversampling by promoting key examples. Its effect is validated when applying different well-known oversampling methods. Taking into account previous work on this same dataset, the paper is intended to improve the classification results previously obtained by using Support Vector Machines (SVM).

The rest of this article is organized as follows: the applied algorithms for oversampling and the used metrics are described in Sect. 2 while the setup of experiments, the dataset under analysis, and the obtained results are described in Sect. 3. Finally, the conclusions of the present study, as well as proposals for future work, are stated in Sect. 4.

2 Data Oversampling Proposal

As data preprocessing algorithms, well-known oversampling methods try to get an equal number of instances per class. These methods generate new data instances from the minority class. Present paper contributes to the initial step of selecting the data instances to perform oversampling. The 5-near neighbors taxonomy is used to create the map in which the data instances are distributed [12].

Each instance from minority class is assigned into four types according to the following criteria:

- *Safe*: at least 4 of the nearest neighbors belong to the minority class.
- *Borderline*: 2 or 3 of the nearest neighbors belong to the minority class.
- *Rare*: only 1 of the nearest neighbors belong to the minority class.
- *Outlier*: none of the nearest neighbors belong to the minority class.

We propose to oversample only selected types of minority class samples. Then we aim to find the best fraction of them by analyzing all possible settings (i.e., if a given type should or not be preprocessed). Let us define a binary code, where each of its positions refers to a kind of minority examples. "1" means that a given type should be oversampled, while "0" stands for leaving this type of minority class as it is [15]. An example is shown in Fig. 1.

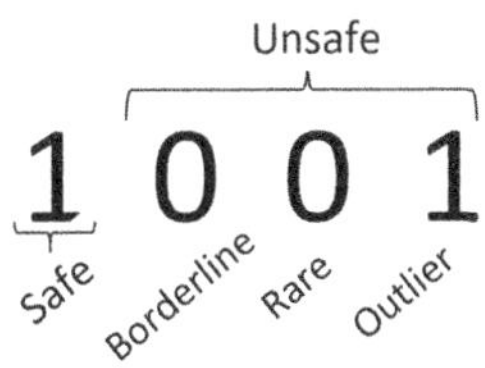

Fig. 1. Sample binary code for oversampling, which means that "Safe" and "Outlier" data should be preprocessed.

Each one of the instances of the minority class are classified in the four types exposed above, in Fig. 2 the criterion used for it is graphically depicted.

The "Outlier" instances are the most isolated ones completely surrounded by the majority class, the "Rares" are surrounded by the majority class but have an instance of the minority class within their neighborhood. "Borderline" are those that are in between the majority and minority instances, with two or three of the latter. Finally, the "Safe" ones are those that have an immense majority of minority class instances in their neighborhood, that is, four or five minority instances. It should be noted that this study is based on a neighbourhood made up of a total of five elements.

Among the many binary combinations that are generated by the previously-explained method, the best one must be selected. In order to do that, the kNN classifier is applied to each one of the different combinations (instances from

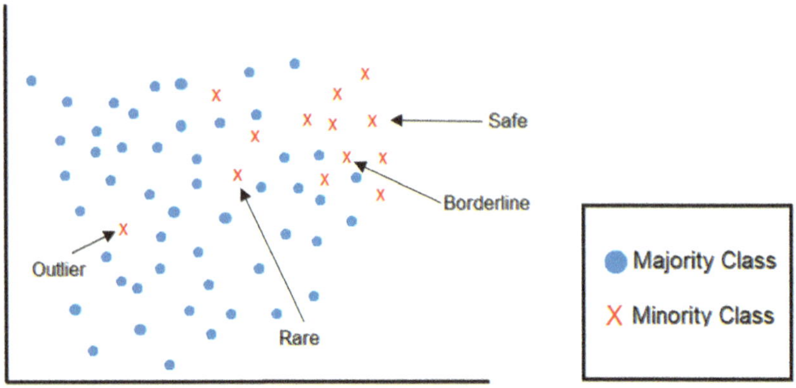

Fig. 2. Graphic example of the selection of the instances for each of the types. With a neighborhood formed by five elements (k = 5).

types taking the 1 value in the binary code of the combination) in order to maximize the value of the g-mean metric. This classification algorithm is applied with a value of the k parameter equal to 5 and the instances are distributed as follows: 75% of the data is selected for training and 25% for testing.

After instances grouping, oversampling itself is carried out. In this work, the selection strategy is combined with the well-known SMOTE oversampling algorithm [3]. To compare obtained results, different metrics are calculated after classifying the oversampled dataset, as described in the following subsection. They are calculated on the basis of the confusion matrix, using four basic indicators:

- True Positives (TP) – how many anomalies (minority class instances) are properly classified.
- True Negatives (TN) – how many normal examples (majority class instances) are correctly predicted.
- False Positives (FP) – how many normal data are classified as an anomaly.
- False Negatives (FN) – how many anomalies are assigned to the normal examples.

Based on these indicators, some standard metrics are calculated: *Accuracy*, *Precision*, False Positive Rate (FPR), and *Recall*. Furthermore, the following advanced ones are also used in this work:

F_1 Score. In order to find a new measure between Precision and Recall which maximizes, taking into account the difficulty to improve both metrics, this one is used given its expression of the harmonic mean between both, as it can be observed in the following formula:

$$F_1 = 2 * \frac{Precision * Recall}{Precision + Recall} \tag{1}$$

ROC Curve. ROC is a visual tool for finding the balance point between the TPR and FPR indicators. The larger the area under the curve (AUC), the better. AUC is recognized as a good indicator to assess the model to distinguish between classes, and it was the most representative metric used by the authors of the dataset.

g-Mean. The geometric mean (g-mean) [9] relates to a point in the ROC curve. It is used in present research as it maximizes the accuracy for both the majority and the minority classes while also taking into account a balance among them, as defined by:

$$g - mean = \sqrt{Precision * Recall} \tag{2}$$

3 Experiments and Results

This section presents the analyzed dataset as well as the obtained results in the different executions and the experimental setup.

3.1 Dataset

The dataset used in this research is publicly available [18]. It includes the anomalies in robotic systems and its details can be found in [17]. The observations were recorded from a robotic system during the RoboCup@Home competition. The analyzed robot consisted of several components, meaning that different manufacturers can make them, but a middleware interconnects all of them. In the analyzed robot, the event-based RBS Middleware [16] has been used.

The relationship among components and anomalies is not one-to-one as sometimes an anomaly affects more than one component or a component may have different anomalies that affect it. This scenario has been chosen to carry out present research, where the "state machine" component is analyzed. This component is crucial as it centralizes the control of the system state, based on the proposal of Siepmann and Wachsmuth [13]. It is also in charge of connecting the rest of the system components. There are three anomalies linked to this component: btlAngleAlgo, bonsaiParticipantLeak, and bonsaiTalkTimeout. They are explained in more detail in Table 1.

These anomalies were induced in the robot, being activated through the RSB middleware. As a result, the precise moment they were produced and the lasting time are known.

The analyzed dataset consists of 71 trials, in which the experiment is reproduced in the same order. However, anomalies are induced only in some of them and at different times. To select the most significant datasets, those trials in which there is a higher amount of induced anomalies have been selected. As a result, trial no. 45 has been selected for btlAngleAlgo anomaly, trial no. 24 for bonsaiParticipanLeak anomaly, and trial no. 18 for bonsaiTalkTimeout anomaly. The number of both normal and anomalous instances in each one of these datasets is shown in Table 2.

Table 1. Anomalies affecting the statemachine component, analyzed in the present study.

Name	Description
btlAngleAlgo	During the tracking of people a mathematical error is added
bonsaiParticipantLeak	Participants are not properly eliminated
bonsaiTalkTimeout	The RSB scope is incorrectly set

Table 2. Figures about the class distribution of data samples in each one of the analyzed anomalies.

Anomaly	Normal samples	Anomalous samples	Total
btlAngleAlgo	553 (74,8%)	186 (25.2%)	739
bonsaiParticipantLeak	522 (76.5%)	160 (23.5%)	682
bonsaiTalkTimeout	554 (77,6%)	160 (22.4%)	714

3.2 Obtained Results

This section presents the results obtained when analyzing each one of the anomalies described in the previous section. For a fair comparison, the calculated values for the different metrics are shown. SMOTE and ROS algorithms have been applied for subsequent classification by SVM to validate the effect of the proposed strategy comprehensively. According to previous SVM experiments on this same dataset, similar values have been chosen for the SVM parameters in order to make a fair comparison of the data. That is: cost = 10, gamma = 0.1, and sigmoid kernel function. Additionally, classification results when no oversampling technique is applied are also shown (denoted as "None").

Results are validated using the 10-fold cross-validation technique, while only 75% of the data have been used for oversampling and the remaining 25% for testing.

For comparison purposes, results obtained by traditional oversampling techniques are shown in Table 3. In these results, we may observe that although accuracy is penalized, all the other metrics and especially those recommended for imbalanced datasets are greatly improved by oversampling.

More precisely, AUC and g-mean values are improved when applying both ROS and SMOTE. ROS obtains the highest AUC value for 2 of the anomalies (btlAngleAlgo and bonsaiTalkTimeout) while SMOTE obtains it in the case of bonsaiParticipantLeak anomaly. Similarly, the highest g-mean values are obtained by ROS (SMOTE obtains the same value for the bonsaiParticipantLeak anomaly). When applying oversampling, the highest AUC and g-mean values are obtained for the btlAngleAlgo anomaly (the least imbalance one) and ROS algorithm, while the lowest ones are obtained for the bonsaiTalkTimeout anomaly (the most imbalance one) and SMOTE algorithm. On the other hand, the obtained metric values greatly vary among anomalies; as an example, the

Table 3. Obtained results according to different metrics for the three anomalies by any kind of strategy, traditional ROS and SMOTE algorithms and the proposed algorithm.

		btlAngleAlgo	bonsaiParticipantLeak	bonsaiTalkTimeout
None	Accuracy	0.7049	0.7329	0.6185
	Precision	0.3829	0.4572	0.1301
	Recall	0.3784	0.4397	0.1323
	F_1 score	0.3767	0.4461	0.1296
	AUC	0.5935	0.6375	0.4432
	g-mean	0.3807	0.4484	0.1312
ROS	Accuracy	0.6951	0.6771	0.4775
	Precision	0.7136	0.6947	0.4691
	Recall	0.4395	0.4065	0.2074
	F_1 Score	0.5426	0.5112	0.2857
	AUC	0.7013	0.6821	0.5295
	g-mean	0.5600	0.5314	0.3119
SMOTE	Accuracy	0.6880	0.6865	0.4652
	Precision	0.6287	0.7123	0.4116
	Recall	0.4244	0.3965	0.1802
	F_1 score	0.5056	0.5082	0.2494
	AUC	0.6676	0.6954	0.5175
	g-mean	0.5165	0.5314	0.2723
Proposed	Accuracy	0.7391	0.6412	0.6011
	Precision	0.7292	0.7083	0.5238
	Recall	0.5000	0.4198	0.3014
	F_1 score	0.5932	0.5271	0.3826
	AUC	0.7359	0.6615	0.5744
	g-mean	0.6038	0.5453	0.3973

g-mean value obtained by ROS is 0.56 for the btlAngleAlgo anomaly while it is 0.3119 for the bonsaiTalkTimeout one.

Results obtained by the proposed oversampling strategy are also detailed in Table 3 (at the bottom). It can be observed that the best overall results are obtained for the btlAngleAlgo anomaly, taking into account all the given metrics. For bonsaiTalkTimeout the results are much worse, especially in the case of the g-mean metric whose value is very low, penalized by a very low recall value. All in all, both AUC and g-mean values are greatly improved by applying the proposed data selection strategy. It outperforms not only the original SMOTE results but also the ROS ones; the highest values of AUC and g-means metrics are obtained for all anomalies when applying the proposed strategy. The only exception is the AUC metric in the case of bonsaiParticipantLeak anomaly; AUC for SMOTE is 0.6954 while for the proposed algorithm is 0.6615.

It is worth noting that during the experiments, the best binary codes (out of 16) for the data selection associated with each one of the anomalies have been:

- btlAngleAlgo: 1 1 0 0.
- bonsaiParticipantLeak: 0 1 1 0.
- bonsaiTalkTimeout: 0 0 1 0.

This means that "Outlier" elements have never been oversampled. For two of the anomalies, the "Rare" and "Borderline" groups have been oversampled and the "Safe" group has been selected for oversampling only once. It should be noted that it means a big difference on the subsequent application of the SMOTE algorithm; although the instances are chosen at random, they are only taken from the selected types.

To ease comparison, the obtained results are depicted in a radar chart by the anomalies in Fig. 3.

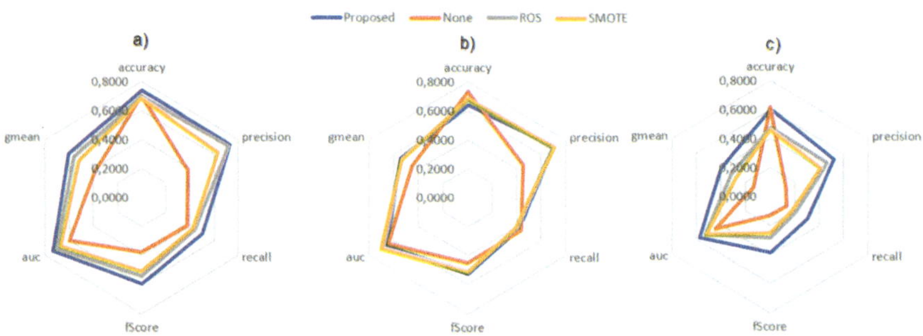

Fig. 3. Radar chart of classification metrics for each anomaly: a) btlAngleAlgo, b) bonsaiParticipantLeak, and c) bonsaiTalkTimeout.

From that figure, some conclusions can be obtained, similar to the ones derived from previous results (in Table 3). Best values for most of the metrics are obtained by the proposed strategy for the btlAngleAlgo and bonsaiTalkTimeout anomalies. Furthermore, thanks to the novel method, higher values are obtained for both recall (TPR) and precision metrics. On the other hand, similar rates are obtained in the case of the bonsaiParticipantLeak for most of the metrics. For all the oversampling alternatives (and without any of them), worst results are obtained for the bonsaiTalkTimeout anomaly (the most imbalance one), but it is worth noting that the proposed method behaves as well as the original SMOTE, and it outperforms other techniques. The similar behavior of SMOTE and the proposed techniques could be caused by the fact that probably the rare objects were taken to oversample and this fraction of the objects dominate the minority class population.

4 Conclusions and Future Work

The paper presents a method of selective oversampling of minority class instances for the problem of anomaly detection. Experimental studies have confirmed the usability of our proposal and encourage us to continue working on this method. The formulated proposition significantly improves classification results, according to class-balance metrics and what is essential, it can outperform other methods for prediction quality of both minority and majority class.

In the future, authors will focus on using the selective oversampling scheme for multi-class problems, mainly since the analyzed dataset (presented in the publication) contains many anomalies, which form several minority classes. Additionally, the further steps in this ongoing research will consist of analyzing some other data sources as well as benchmarking the effect of the oversampling strategy when applying some different classifiers (in addition to SVM).

References

1. Bergmeir, P., Nitsche, C., Nonnast, J., Bargende, M.: Classifying component failures of a hybrid electric vehicle fleet based on load spectrum data. Neural Comput. Appl. **27**(8), 2289–2304 (2016)
2. Cerqueira, V., Pinto, F., Sá, C., Soares, C.: Combining boosted trees with metafeature engineering for predictive maintenance. In: Boström, H., Knobbe, A., Soares, C., Papapetrou, P. (eds.) Advances in Intelligent Data Analysis XV, pp. 393–397. Springer International Publishing, Cham (2016)
3. Chawla, N.V., Bowyer, K.W., Hall, L.O., Kegelmeyer, W.P.: SMOTE: synthetic minority over-sampling technique. J. Artif. Intel. Res. **16**, 321–357 (2002)
4. Devi, D., Biswas, S.K., Purkayastha, B.: Learning in presence of class imbalance and class overlapping by using one-class SVM and undersampling technique. Connection Sci. **31**(2), 105–142 (2019)
5. Alsamhi, S.H., Ma, O., Ansari, M.S.: Survey on artificial intelligence based techniques for emerging robotic communication. Telecommun. Syst. **72**(3), 483–503 (2019). https://doi.org/10.1007/s11235-019-00561-z
6. He, H., Garcia, E.A.: Learning from imbalanced data. IEEE Trans. Knowl. Data Eng. **21**(9), 1263–1284 (2009)
7. Jayaratne, M., de Silva, D., Alahakoon, D.: Unsupervised machine learning based scalable fusion for active perception. IEEE Trans. Auto. Sci. Eng. **16**(4), 1653–1663 (2019). https://doi.org/10.1109/TASE.2019.2910508
8. Kober, J., Bagnell, J.A., Peters, J.: Reinforcement learning in robotics: a survey. Int. J. Robot. Res. **32**(11), 1238–1274 (2013). https://doi.org/10.1177/0278364913495721
9. Kubat, M., Matwin, S., et al.: Addressing the curse of imbalanced training sets: one-sided selection. In: ICML, Nashville, USA, vol. 97, pp. 179–186 (1997)
10. Lu, H., Li, Y., Mu, S., Wang, D., Kim, H., Serikawa, S.: Motor anomaly detection for unmanned aerial vehicles using reinforcement learning. IEEE Internet Things J. **5**(4), 2315–2322 (2018). https://doi.org/10.1109/JIOT.2017.2737479
11. Luo, M., Wang, K., Cai, Z., Liu, A., Li, Y., Cheang, C.F.: Using imbalanced triangle synthetic data for machine learning anomaly detection. Comput. Mater. Continua **58**(1), 15–26 (2019)

12. Napierala, K., Stefanowski, J.: Types of minority class examples and their influence on learning classifiers from imbalanced data. J. Intel. Inf. Syst. **46**(3), 563–597 (2016)
13. Siepmann, F., Wachsmuth, S.: A modeling framework for reusable social behavior. In: De silva, R., reidsma, D. (eds.) Work in Progress Workshop Proceedings ICSR, pp. 93–96 (2011)
14. Syafrudin, M., Fitriyani, N.L., Alfian, G., Rhee, J.: An affordable fast early warning system for edge computing in assembly line. Appl. Sci. **9**(1), 84–102 (2018)
15. Sáez, J.A., Krawczyk, B., Woźniak, M.: Analyzing the oversampling of different classes and types of examples in multi-class imbalanced datasets. Pattern Recogn. **57**, 164–178 (2016). http://www.sciencedirect.com/science/article/pii/S0031320316001072
16. Wienke, J., Wrede, S.: A middleware for collaborative research in experimental robotics. In: 2011 IEEE/SICE International Symposium on System Integration (SII), pp. 1183–1190, December 2011
17. Wienke, J., Meyer zu Borgsen, S., Wrede, S.: A data set for fault detection research on component-based robotic systems. In: Alboul, L., Damian, D., Aitken, J.M. (eds.) Towards Autonomous Robotic Systems, pp. 339–350. Springer International Publishing, Cham (2016)
18. Wienke, J., Wrede, S.: A fault detection data set for performance bugs in component-based robotic systems (2016)
19. Wienke, J., Wrede, S.: Autonomous fault detection for performance bugs in component-based robotic systems. In: 2016 IEEE/RSJ International Conference on Intelligent Robots and Systems (IROS), pp. 3291–3297. IEEE (2016)
20. Wienke, J., Wrede, S.: Continuous regression testing for component resource utilization. In: IEEE International Conference on Simulation, Modeling, and Programming for Autonomous Robots (SIMPAR), pp. 273–280. IEEE (2016)
21. Xiao, B., Yin, S.: Exponential tracking control of robotic manipulators with uncertain dynamics and kinematics. IEEE Trans. Ind. Inf. **15**(2), 689–698 (2019)
22. Zhao, D., Ni, W., Zhu, Q.: A framework of neural networks based consensus control for multiple robotic manipulators. Neurocomputing **140**, 8–18 (2014). https://doi.org/10.1016/j.neucom.2014.03.041

A Preliminary Study for Automatic Activity Labelling on an Elder People ADL Dataset

Enrique de la Cal[1](✉), Mirko Fáñez[2], Alvaro DaSilva[2], Jose Ramón Villar[1], Javier Sedano[2], and Victor Suárez[3]

[1] Computer Science Department, University of Oviedo, Oviedo, Spain
{delacal,villarjose}@uniovi.es

[2] Instituto Tecnológico de Castilla y León, Pol. Ind. Villalonquejar, 09001 Burgos, Spain
mirko@mirkoo.es, ada@ubu.es, javier.sedano@itcl.es

[3] Control and Automatica Department, EPI, University of Oviedo, Gijón, Spain
vmsuarez@uniovi.es

Abstract. One consequence of the aging population is an increase in life expectancy implying greater healthcare needs as well as a serious healthy aging program. So healthy aging is one of the main challenges in the first world nowadays, and as much as possible devices, software and technological solutions applied to measure and improve the quality of life of the elder people are necessary.

Recently, we presented a first prototype of an activity monitoring kit, and this study includes the analysis of the dataset gathered after six months of use. Since the wearable devices employed in this monitoring kit have not the automatic activity recognition service available, current work proposes several techniques to label automatically the Time Series (TS) obtained in the experiment. Thus, a new device with the same sensors as the old one plus the automatic activity recognition service available will be used to obtain a new labelled dataset, that will be used to learn a new model using semi-supervised learning to tag the not-labelled dataset.

Keywords: ADL automatic identification · Falls in elderly · Wearable sensors · Fall detection · Human Activity Recognition

1 Introduction and Motivation

The old-age dependency ratio (people aged 65 and above relative to those aged 15 to 64) in the EU is projected to increase by 21.6% points, from 29.6% in 2016 to 51.2% in 2070. This implies that the EU would go from having 3.3 working-age people for every person aged over 65 years to only two working-age people [2]. One consequence of the ageing population is an increase in life

Á. Herrero et al. (Eds.): SOCO 2020, AISC 1268, pp. 13–21, 2021.
https://doi.org/10.1007/978-3-030-57802-2_2

expectancy implying greater healthcare needs [3]. Thus, non-invasive tools to monitor and analyse Elderly health and activity are required. The most common non-invasive and easy-to-use tools to measure the activity in elderly are the wearable devices. Static daily living activities like standing, sitting and lying are simple to detect, whilst dynamic ones such as walking, running, jumping, are more difficult to recognise. Concerning this, two classification techniques have been regularly used: the supervised and unsupervised activity classification approaches. Traditionally the supervised activity classification approaches have not been considered in real free-living environments where external factors can affect negatively their performances. Moreover, the collection of sufficient amounts of labelled data for a representative set of free-living activities may be sometimes difficult to achieve and computationally expensive. On the other hand, unsupervised machine learning represents the second approach used in Human Activity Recognition (HAR). In this case, labelled data is not required which can overcome the aforementioned limitations of the supervised techniques [5]. There are two categories of unsupervised recognition models: Static [4] and temporal classification approaches [6].

In previous work a first prototype of an autonomous, low-cost and easy-to-use elderly activity monitoring kit was presented [1]. This kit includes a set of 12 smart-bands (2 TICWATCH E2 and 10 SAMSUNG Gear Fit2) with 3DACC, GYROSCOPE and HR Sensors, as well as other components (miniPC and 4G router) to store and access the data remotely. This prototype was deployed in a nursery house in June 2019, and it has been gathering data for 6 months.

While the prototype presented [1] had the main goal of gathering data to obtain a dataset with real falls, current work will be focused on the analysis of the levels of daily life activity excluding Falls. It is worth to state that the obtained data has been split in two datasets: i) one labelled dataset obtained from two participants wearing TICWATCH smartwatch, and ii) other no-labelled dataset gathered from the 10 participants wearing a SAMSUNG smart-band. Thus, a simple automatic labelling technique based on semi-supervised learning is proposed to label the SAMSUNG dataset using a classification model obtained from the TICWATCH dataset. Furthermore, in order to contrast the labelling results, a statistical study involving several movement features like AOM, ACC and SMA was included.

This work is structured as follows: next section includes the design issues of the semi-supervised proposal presented here while the experimentation and the discussion of the results are coped in Sect. 3. Finally, conclusions and future work is included in Sect. 4.

2 The Proposal

The main goal of our proposal is to analyse and characterize the daily levels of activity of the unlabelled data collected for 6 months from a group of participants using the activity monitoring kit presented in [1].

As, this first prototype of monitoring kit used a model of smart-band (OLD-DEVICES) with the automatic activity identification service not available, two units of these smart-bands were replaced by other two new models of smartwatch (NEWDEVICES) with this capability activated. The OLDDEVICES use the following sensors: a 3D Accelerometer, a gyroscope and a heart rate sensor, whilst the NEWDEVICES have the same sensors plus the automatic activity identification service activated. The NEWDEVICES have been collecting data for the last 2 months of the experiment and they will replace all the OLDDEVICES in the next release of the monitoring kit.

Hence, the idea is to learn a model of activity level labelling using the NEWDEVICES and use semi-supervised learning to apply these models to the OLDDEVICES dataset in order to label the activity of the participants.

Consequently, it's proposed a method based on the following steps: i) OLDDEVICES and NEWDEVICES datasets clean and pre-processing, and ii) Design and perform an automatic segmentation algorithm taking as input the NEWDEVICES dataset, and deploy the models on the OLDDEVICES dataset.

2.1 OLDDEVICES Dataset Clean and Pre-processing

The big volume of data obtained for 6 months needs to be pre-processed and cleaned since some days either the participants did not wear the monitoring device or several OLDDEVICES ran out of battery quickly because of an operating system failure. Thus, several statistics have been considered to remove the waste data:

- Mean ($MEAN_p$) and Standard Deviation ($STDN_p$) of the number of hours recorded by day for participant p.
- Hours per Day Threshold (HDT_p): Minimum number of recorded hours to consider a day valid for the participant p. $HDT_p = 0.7 * MEAN_p$
- Percentage of Recorded Days ($PoRD_p$): The percentage of recorded days out of the 6 months, for the participant p.
- Valid Percentage of Recorded Days ($VPoRD_p$): The percentage of recorded days out of the 6 months, for the participant p with a number of hours overpassing HDT_p.

Therefore, all the days with a number of recorded hours under HDT_p, as well as all the data of those participants with a $VPoRD_p$ under 30%, will be removed. This subsection will be covered later on the Numerical Results section (see Sect. 3.2).

2.2 Automatic Segmentation of Activities of the OLDDEVICES Dataset

This study proposes to label the activity level of the OLDDEVICES dataset segmenting the TSs in high and low activity periods.

We have decided to define an algorithm based on the HR sensor to segment the TSs in high and low activity. Therefore, a simple algorithm based on thresholds is defined:

1. Select the TS windows on the NEWDEVICES dataset that has been automatically labelled by the Android Activity Recognition API (using sliding windows of 10 s) as ON_FOOT series (walking or running and labelled as HIGH) and STILL series (no activity or low activity and labelled as LOW).
2. Calculate the mean HR on both types of TSs, grouped by participant ($ONFOOT\_HR_p$ and $STILL\_HR_p$).
3. Calculate the mean HR on both types of TSs (not by participant, ONFOOT_MEAN and STILL_MEAN).

Accordingly, the activity level threshold is calculated as:

$$ONFOOT\_TH = ONFOOT\_MEAN - 0.7 * (ONFOOT\_MEAN - STILL\_MEAN) \quad (1)$$

So the TS windows with HR lower than ONFOOT_TH is considered Low-Activity; High-Activity otherwise.

Finally, the ONFOOT_TH will be deployed on the OLDDEVICES dataset to segment the TS windows, correlating the type of activity with different well-known features on HAR related with the intensity of movement like Simple Moving Averages (SMA), Amount Of Movement (AOM) and Magnitude of Acceleration (MAG).

3 Numerical Results

3.1 Materials and Methods

The Devices: concerning the specific brand and model of the OLDDEVICES and NEWDEVICES referred above, we can say that for the experiments included in this section we have considered the smart-band SAMSUNG Gear Fit 2 as the OLDDEVICES model, and the smartwatch TICWATCH E2 as the NEWDEVICES model (see Fig. 1).

The Participants: When the first prototype of monitoring kit was presented, it was defined a very strict protocol of participant inclusion and exclusion supervised by an expert gerontologist [1]. As a product of this protocol a group of 10 people with ages between 76 and 98 was recruited.

The Timeline of the Experiment: The experiment lasted six months between June 2019 and November 2019, collecting data from the 10 participants using the SAMSUNG devices, and from the 10th October to 22th November two SAMSUNG devices were replaced by other two TICWATCH ones for participants #1 and #2.

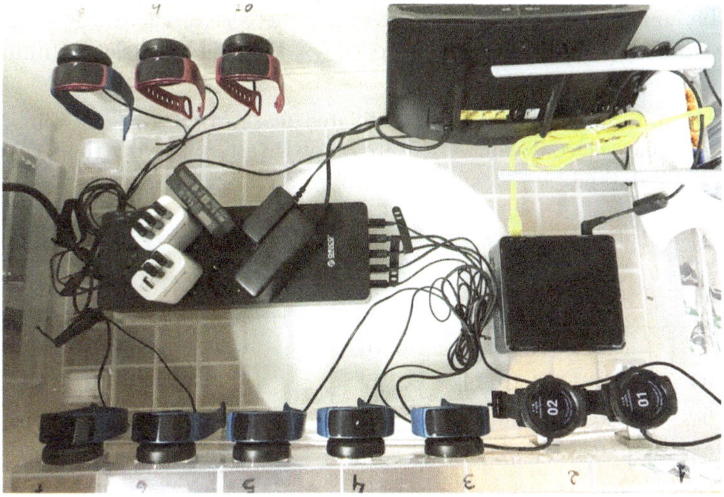

Fig. 1. Monitoring kit release 0.0 with SAMSUNG devices #1 and #2 replaced by TICWATCH devices.

The Methods: The first stage is the clean and pre-processing of the OLDDEVICES and NEWDEVICES datasets. After this stage, the OLDDEVICES dataset will be analysed by performing a semi-supervised learning using the HR threshold-based (HRT) models learned with the NEWDEVICES dataset to label the OLDDEVICES dataset. Hence, this section comprise the following steps: i) both datasets will be cleaned and pre-processed, ii) the HRT will be estimated on the NEWDEVICES dataset in order to segment the datasets in High-Activity and Low-Activity TSs and iii) finally the OLDDEVICES dataset will be characterized analyzing the segmentation based on the HRT.

Facilities and Running Time: The experiments were carried out on a 2.4 GHz Intel Core i9 with 32 GB of RAM MACOSX Laptop. With this configuration, the most time-consuming R script was the OLDDEVICES TSs Characterization based on the HRT values learned with the NEWDEVICES dataset (the last step), which took 6 h to complete (not using R parallel execution).

3.2 Dataset Clean and Pre-processing

Due to erratic battery behaviour for the 6 months that data was being recorded with the OLDDEVICES, the dataset was not very homogeneous among all participants. This leads to a data consolidation process. Table 1 shows data statistics previous to the consolidation, which will be used to perform this pre-processing stage. In the light of these results, we must establish as a valid day threshold the mean recorded hours per day (HDT, calculated as mentioned in Sect. 2.1), and use it to calculate the percentage of valid recorded days per participant (VPoRD). There are 2 participants (#2 and #9) that has low VPoRD, which will not be considered in this study, keeping the rest of them (although not all have the same number of valid recorded days, there is a consistent minimum).

Table 1. Results for both the un-consolidated features (Registered hours per day, HDT, PoRD) and the consolidated feature VPoRD (after applying the HDT), for the OLDDEVICES dataset.

PartID	Regist. hours/day		HDT (hours)	PoRD (%)	VPoRD (%)
	Mean	Std			
01	7.1173	1.6168	4.9821	38.70	38.70
02	6.9631	0.3994	4.8742	3.22	**3.22**
03	7.4811	1.1911	5.2368	40.32	37.09
04	7.0347	2.2072	4.9242	33.87	32.25
05	6.9846	0.0712	4.8892	32.25	32.25
06	6.6304	0.7328	4.6413	90.32	87.09
07	8.4405	2.2326	5.9084	37.09	35.48
08	7.0163	0.9142	4.9114	90.32	87.09
09	6.4032	1.6624	4.4822	17.74	**16.12**
10	6.8757	0.9580	4.8130	87.09	80.64
Mean	7.0947	1.1986	4.9663	47.0920	44.9930
Std	0.5520	0.7259	0.3864	31.1535	29.6479

3.3 Automatic Segmentation of Activities

Table 2 shows the ONFOOT_TH and STILL_TH statistics computed using the data of the two participants belonging to the NEWDEVICES dataset. The ONFOOT_TH HR threshold is calculated based on the Eq. 1:

$$ONFOOT\_TH = 69.7261 \tag{2}$$

Once the ONFOOT_TH is calculated and deployed on the OLDDEVICES dataset, the TSs of this dataset are classified as High-Activity and Low-Activity TSs. In order to characterize these two activity levels, the well-known HAR features: Acceleration Magnitude, SMA and AOM were calculated. Figure 2 includes the boxplots of these features arranged in two columns: first column includes the High activity boxplots while second column includes the Low activity ones. The results show that AOM and Acceleration Magnitude standard deviation are decisive in this characterization, since its values are totally correlated with the activity level. The mean (red dashed line) as well as the deviation of the AOM High Activity TS (c–i) is clearly higher than the corresponding for Low Activity TSs (c–ii). Regarding the other features, just the deviation of the ACCMAG is clearly correlated with the level of activity: the deviation of the Low Activity TSs (a–ii) is significantly lower than the High Activity TSs (a–i). Table 3 shows the numerical values corresponding to the boxplots included in Fig. 2.

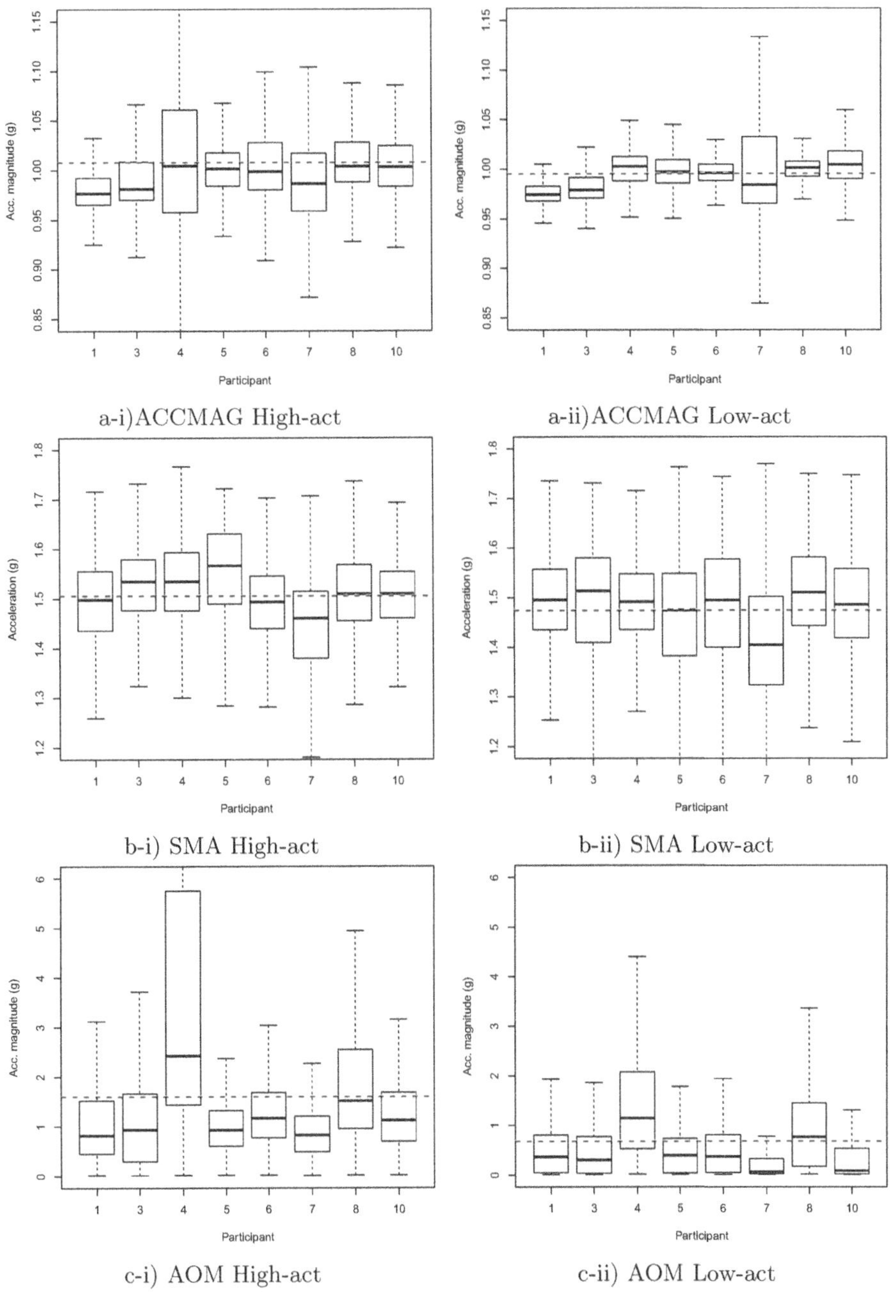

Fig. 2. Boxplots for the 8 participants considered in this study, segmented by activity level: High-Act and Low-Act. The red dashed line is the mean value for all the participants.

Table 2. HR segmentation results for both participants in the NEWDEVICES dataset.

PartID	ONFOOT		STILL	
	Mean	Std	Mean	Std
1	88.5301	22.77587	56.74898	16.42893
2	89.5713	24.79433	66.13932	11.46430
Mean	89.0507	23.7851	61.4442	13.9466
Std	0.7362	1.4273	6.6400	3.5105

Table 3. Results for OLDDEVICES dataset, after segmentation of data in High-Activity and Low-Activity, using ONFOOT_TH HR threshold learned with the NEWDEVICES dataset.

pID	High-Activity TS						Low-Activity TS					
	Acc. magn.		SMA		AOM		Acc. magn.		SMA		AOM	
	Mean	Std	Mean	Std	Mean	Std	Mean	Std	Mean	Std	Mean	Std
01	0.9839	0.1187	1.4919	0.0928	1.2328	1.3764	0.9774	0.0564	1.4938	0.0961	0.6203	0.8378
03	0.9932	0.1177	1.5112	0.1212	1.1845	1.2105	0.9839	0.0540	1.4798	0.1515	0.5096	0.6556
04	1.0352	0.3105	1.5337	0.0877	3.5741	2.6082	1.0092	0.1444	1.4885	0.0966	1.6632	1.7859
05	1.0044	0.0816	1.5475	0.1088	1.0244	0.7686	0.9984	0.0484	1.4551	0.1330	0.4875	0.5956
06	1.0092	0.1235	1.4929	0.0836	1.3898	1.0554	0.9977	0.0513	1.4786	0.1303	0.5607	0.7205
07	0.9911	0.0930	1.4446	0.1071	0.9736	0.8440	0.9935	0.0395	1.4190	0.1604	0.2255	0.3427
08	1.0386	0.2831	1.5287	0.1551	2.2347	2.1323	1.0046	0.0970	1.5075	0.1073	1.1643	1.5425
10	1.0115	0.1159	1.5041	0.0841	1.2896	0.9036	1.0042	0.0363	1.4846	0.1109	0.3411	0.5377
Mean	1.0084	0.1555	1.5068	0.1050	1.6129	1.3624	0.9961	0.0659	1.4759	0.1233	0.6965	0.8773
Std	0.0200	0.0887	0.0320	0.0243	0.8840	0.6648	0.0108	0.0367	0.0274	0.0244	0.4787	0.5104

4 Conclusion and Future Work

This study presents a method to characterize the activity levels of a real not-labelled ADL dataset based on a semi-supervised automatic labelling technique using the HR sensor. The automatic segmentation model has been learned taking as input the automatically labelled dataset gathered from the TICWATCH E2 smartwatches. This model has been deployed on a not-labelled long-term dataset collected using SAMSUNG Gear Fit 2 smart-bands. The results obtained state that the automatic segmentation model based on HR classify quite coherently the TS of the SAMSUNG dataset in High and Low levels of activity. In addition, the AOM feature and the standard deviation of the Acceleration magnitude show a high correlation with the level of activity, verifying that the classification is quite good.

Considering that the baseline of this study was a Faller Monitoring kit [1] that has been collecting data for 6 months without a valid fall, we think that the experiment has been successful. So, next release of the monitoring kit will comprise TICWATCH smartwatches instead of SAMSUNG Gear Fit 2 smart-bands, since the first ones are more robust and stable.

Acknowledgement. This research has been funded partially by Spanish Ministry of Economy, Industry and Competitiveness (MINECO) under grant TIN2017-84804-R and by Foundation for the Promotion of Applied Scientific Research and Technology in Asturias, under grant FC-GRUPIN-IDI2018000226.

References

1. de la Cal, E., DaSilva, A., Fáñez, M., Villar, J., Sedano, J., Suárez, V.: An autonomous fallers monitoring kit: release 0.0. In: Proceedings of the 19th International Conference on Intelligent Systems Design and Applications (2019)
2. European Commission: 2018 Ageing Report: Policy challenges for ageing societies (2020). Accessed 12 Feb 2020. https://ec.europa.eu/info/news/economy-finance/policy-implications-ageing-examined-new-report-2018-may-25_en
3. King, R.C., Villeneuve, E., White, R.J., Sherratt, R.S., Holderbaum, W., Harwin, W.S.: Application of data fusion techniques and technologies for wearable health monitoring. Med. Eng. Phys. **42**, 1–12 (2017)
4. Quante, M., Kaplan, E.R., Rueschman, M., Cailler, M., Buxton, O.M., Redline, S.: Practical considerations in using accelerometers to assess physical activity, sedentary behavior, and sleep. Sleep Health **1**(4), 275–284 (2015)
5. Trabelsi, D., Mohammed, S., Amirat, Y., Oukhellou, L.: Activity recognition using body mounted sensors: an unsupervised learning based approach. In: The 2012 International Joint Conference on Neural Networks (IJCNN). pp. 1–7. IEEE (2012)
6. Trabelsi, D., Mohammed, S., Chamroukhi, F., Oukhellou, L., Amirat, Y.: An unsupervised approach for automatic activity recognition based on hidden markov model regression. IEEE Trans. Auto. Sci. Eng. **10**(3), 829–835 (2013)

How Noisy and Missing Context Influences Predictions in a Practical Context-Aware Data Mining System

Anca Avram(✉), Oliviu Matei, Camelia-M. Pintea, Petrica C. Pop, and Carmen Ana Anton

Technical University of Cluj-Napoca, North University Center at Baia Mare, Baia-Mare, Romania
{anca.avram,dr.camelia.pintea}@ieee.org, oliviu.matei@holisun.com, {petrica.pop,carmen.anton}@cunbm.utcluj.ro

Abstract. The focus of this research is finding out how different levels of context noise and missing data, affect the overall prediction results in a Context-Aware Data Mining (CADM) system for predicting soil moisture. Experiments were performed using more machine learning algorithms and varying the levels of noise and missing context data in realistic scenarios. The results show that context with missing data has a higher impact on the predictions than noise. Results comparable to the clean context baseline are obtained when the 20% threshold of noise and missing data is not exceeded.

Keywords: Context-aware data mining · Time series forecasting · Moisture prediction

1 Introduction

According to Kotu and Deshpande [15] data mining, "in simple terms, is finding useful patterns in the data". The main value brought by data mining is that the patterns discovered can then be transformed in actionable knowledge that can be used to bring improvements in the process that generated the data. Context-Aware Data Mining (CADM) is a variation of the classical data mining method, that integrates context in the process [18].

Previous research [3] has proven advantages in using CADM approach when predicting the value of the soil moisture, in a given location. Since knowing in advance this value is a very valuable information for farmers, that helps them organize their activity, the current research wants to extend existing research in this area and analyze the impact of realistic scenarios like context with noise or missing context data. More than that, it aims of being a proof of concept in evaluating the context based on these two criteria.

Á. Herrero et al. (Eds.): SOCO 2020, AISC 1268, pp. 22–32, 2021.
https://doi.org/10.1007/978-3-030-57802-2_3

Section 1.1 presents a brief overview on CADM and context quality. Section 1.2 presents the current status of the CADM related research. It continues with the experimental methodology in Sect. 2, while Sect. 3 presents the numerical results obtained. The overall conclusions and possible future work are depicted in Sect. 4.

1.1 CADM and Context Quality

As Dey [8] stated, by context we understand "any information that can be used to characterize the situation of an entity". CADM respects the same steps as classical data mining, but comes with an extra step of integrating context data in the process. Lee et al. [16], defined some steps generally applicable to context aware systems: (1) context acquisition; (2) storage of context; (3) abstraction; (4) usage.

Currently, in the industry, when discussing about quality of data, the main focus is on the following dimensions: completeness - data meets the expectations; consistency - data is the same stored and registered in all systems; conformity - data follows the same set of standards agreed in all systems; accuracy - data correctly reflects the reality; integrity - data is valid across all existing relationships and is traceable at any point; timeliness - the degree to which data represent reality from the required point in time [10].

Witten et al. [25] identified some important questions that should be asked when performing data mining: is the collected data useful in terms of what one wants to achieve? Also, is there availability of the data? The research [25] focuses on two main factors affecting the quality of the context in real-life scenarios: noise and missing data. Noise affects the accuracy of data and can be caused by different external issues influencing the measurements like, for example, low battery levels. Missing data affects the completeness of the context and can be caused by various factors, starting from human error, sensors not working or problems in communication.

1.2 Related Work

Lee and Chang [16] stated that a context-aware system is one that is capable of actively and autonomously adapting its operation using contextual information in order to provide the most appropriate functionality to the consumers. Kotte et al. [14] identified the capture and use of context data as a major step in a CADM system. Choosing the context can often be subjective, depending on the overall experience of the ones performing the analysis.

In [21] Scholze et al. validated using context awareness is a reliable option to create a holistic solution for (self-)optimization of discrete flexible manufacturing systems. Vajirkar et al. [22] proposed a CADM framework to test the suitability of different context factor, applicable in the medical field.

As Marakas specifies in [17], quality of data is a very important detail, that could influence dramatically the results when data mining.

Starting from the quality of data premise, we wanted to know how the context quality, in simulated real-life scenarios, would affect the forecasting in a CADM system for predicting soil moisture.

Context-awareness is a research subject starting from 1999 [23,24]. Still the focus of current research on context is mainly on capturing and using context data for obtaining actionable knowledge [20] than on analyzing the quality of the context. Bellavista et al. [5] performed a survey on quality of context for context-aware services. After analyzing different parameters, they defined the quality of context based on context data validity, precision and up-to-dateness.

In previous research [2,3] was validated that using context data when predicting soil moisture, positively influences the forecast results. Avram et al. [4] performed independent experiments on the influence of noise and missing context data in the CADM process. The conclusion was that, taken separately, missing context data has a higher influence on the prediction results than context with noise. The current research extends the work from [4] starting from the premise that most of the time context is affected by more external factors, hence it most often would be the subject of both incomplete and noisy data.

2 Experimental Methodology

The purpose of this research is to simulate real-life scenarios for predicting the soil moisture in a context-aware system. These scenarios focus on the quality of the context and the way noise and missing data in context would influence the overall prediction results.

2.1 Data Sources

The experiment is performed starting from real data, collected from several weather stations in the Transylvania plain, a Romanian areal; information on Romanian geographic data instances are provided in [7]. Table 1 presents the locations and intervals chosen as main subject for the research.

One stream of data provides information on date, precipitations (if available), soil temperatures at three different depths (10 cm, 30 cm and 50 cm) and soil moisture. The moisture of the soil is the attribute that we want to predict.

The context modeled and analyzed in this research is around the daily air temperatures. One could argue that air temperatures could be considered just as an extra attribute when creating the process for predicting the soil moisture.

Table 1. Locations studied and time intervals

Location	From date	To date
Location 1	06/01/2011	06/01/2015
Location 2	05/01/2011	02/01/2014
Location 3	12/01/2011	06/01/2013

It is a fine line between what could be considered regular data attributes and what to consider context data. In this research, we needed a simple scenario, in order to provide a proof of concept on how noise and missing data influence the prediction results in a context-aware data mining system. Another reason why air temperature was considered context, is that it is an information that can be modeled separately and also the source for this can easier be changed without influencing the rest of the process, but only the context related part - for example using other sensors or weather web sites as source for the air temperatures.

2.2 Methods

One of the first steps when performing data mining on data is preprocessing the data, that involves cleaning it and preparing it for further analysis. This step implies smoothing noisy data, identifying or removing outliers, and resolving inconsistencies [6]. Still, the noise cannot be completely removed and the missing values could make a difference in the outcome of the predictions.

The main purpose of this research is to analyze how the quality of the context is influencing the prediction results, in some realistic life-like scenarios, that involve noise and missing context data. Since in the preprocessing phase some of the noise is already eliminated, for this research we considered three levels of noise: Low, Medium and High. For each of these levels we varied the percent of affected data from 0% to 30% and the level of missing data from 0% to 40%. Having more than 30% noise and 40% of the context data missing could lead to re-evaluation of the entire process on whether or not the context would bring any value to the system.

Table 2 presents an overview of all the tests performed for each of the locations chosen. The first line in the table is actually the baseline CADM - the "ideal" situation when noise and missing data did not affect the context and will serve as reference point in the analysis of the results.

To give a value for the three chosen types of noise, the average standard deviation of the air temperature for the three locations was considered. The Low value for the noise was then considered as being 10% of the standard deviation (1.862), the Medium value 40% of standard deviation (7.45), while the High value was computed as 90% of the standard deviation (16.76).

2.3 Machine Learning Algorithms and Measurements

After some preliminary tests that analyzed the performance of more machine learning algorithms on the selected data, the chosen algorithms to be analyzed in this research were: Deep Learning (DL), Gradient Boosted Trees (GBT) and Decision Trees (DT). Statistical analysis is further performed to analyze the performance of the already mentioned algorithms; the following statistical measures were considered to be collected after each test:

- Root Mean Squared Error (RMSE) - the standard deviation of the residuals (prediction errors). It is calculated by finding the square root of the mean/average of the square of all errors [13].

Table 2. Test scenarios covered in the experiment

Noise type	Noise percent	Missing percent	Noise type	Noise percent	Missing percent	Noise type	Noise percent	Missing percent
No	0	0	No	0	0	No	0	0
Low	10	10	Medium	10	10	High	10	10
Low	10	20	Medium	10	20	High	10	20
Low	10	30	Medium	10	30	High	10	30
Low	10	40	Medium	10	40	High	10	40
Low	20	10	Medium	20	10	High	20	10
Low	20	20	Medium	20	20	High	20	20
Low	20	30	Medium	20	30	High	20	30
Low	20	40	Medium	20	40	High	20	40
Low	30	10	Medium	30	10	High	30	10
Low	30	20	Medium	30	20	High	30	20
Low	30	30	Medium	30	30	High	30	30
Low	30	40	Medium	30	40	High	30	40

- Absolute Error (AE) - the average absolute deviation of the prediction from the actual value. This value is used for Mean Absolute Error which is very common measure of forecast error in time series analysis [12].
- Relative Error (RE) - the average of the absolute deviation of the prediction from the actual value divided by actual value [1].
- Spearman Rho ρ - computes the rank correlation between the actual and predicted values [9].

2.4 Tools and Setup Parameters

The tool used to create and run all the proposed test scenarios for each location was Rapid Miner. It provides an integrated environment that offers support for all steps in the data mining process, and helps in modeling complex processes. As Hofmann and Klinkenberg specify [11], due to the flexibility and broad functionality for all steps of the data mining process, Rapid Miner becomes a tool of

Table 3. The parameters used for machine learning algorithms

Algorithm:	GBT	Algorithm:	DT	Algorithm:	DL
Number of trees:	150	Maximal depth:	25	Activation:	Rectifier
Maximal depth:	7	Minimal gain:	0.01	Epochs:	10
Learning rate:	0.1	Minimal leaf size:	2		
Number of bins:	20				

choice. One of the features that is offered by Rapid Miner is the possibility to obtain the best possible combination of parameters in a tested scenario, using the Optimize Parameter operator. Table 3 depicts the parameter setup that was used in the experiments, after the optimize step was performed.

3 Experimental Results

We made experiments on the test scenarios presented in Table 2 for each selected location and algorithm. This resulted in 39 tests for each location, hence 117 tests for one algorithm. Table 4 is an example of the results obtained for a location, for the deep learning algorithm, Low noise affecting 10% of the context and missing data affecting from 10% to 40% of it. The average results in each situation were obtained and further analyzed per each algorithm.

Table 4. Example of results for one location

Location	Criterion	Value	Standard deviation	Noise type	Noise	Missing
Location 1	RMSE	0.046467442	0.006972224	Low	Noise 10%	Missing 10%
Location 1	AE	0.038036865	0.005141079	Low	Noise 10%	Missing 10%
Location 1	RE	0.207048586	0.031709525	Low	Noise 10%	Missing 10%
Location 1	RMSE	0.046705010	0.007191877	Low	Noise 10%	Missing 20%
Location 1	AE	0.036331590	0.006297689	Low	Noise 10%	Missing 20%
Location 1	RE	0.202366312	0.028628403	Low	Noise 10%	Missing 20%
Location 1	RMSE	0.039730800	0.006355081	Low	Noise 10%	Missing 30%
Location 1	AE	0.030320045	0.004282694	Low	Noise 10%	Missing 30%
Location 1	RE	0.169016364	0.032538821	Low	Noise 10%	Missing 30%
Location 1	RMSE	0.040649792	0.008426164	Low	Noise 10%	Missing 40%
Location 1	AE	0.031052963	0.006252677	Low	Noise 10%	Missing 40%
Location 1	RE	0.169260977	0.031731461	Low	Noise 10%	Missing 40%

Table 5 presents the overall computed values for the Spearman Rho coefficient for each algorithm and tested scenario. From this perspective it can be concluded that GBT gives the best results with a coefficient close to 0.9 for the low noise and less than 20% missing data.

For each studied algorithm, the RE average results for the three locations are presented in Figs. 1, 2 and 3. RMSE and AE follow basically the same pattern, but in a different value range and are not presented in the graphs.

3.1 Results of Dirty Context Impact for DT Algorithm

Figure 1 shows the merged results for the tested scenarios when using the decision tree algorithm. Several conclusions related to the evolution of RE, correlated with the statistical Spearman Rho correlation coefficient (Table 5) when the noise and missing context data vary follows.

Table 5. Overall Spearman statistical correlation values per tested scenarios.

		DT			DL			GBT		
Missing (%)	Noise affected (%)	Low noise	Med. noise	High noise	Low noise	Med. noise	High noise	Low noise	Med. noise	High noise
0	0	0.60998	0.60998	0.60998	0.77812	0.77812	0.77812	0.89757	0.89754	0.89757
10	10	0.61570	0.57938	0.56361	0.77460	0.74814	0.75094	0.89147	0.87805	0.87674
10	20	0.63911	0.64203	0.56851	0.73760	0.75263	0.74078	0.88718	0.88364	0.88012
10	30	0.60505	0.53467	0.55704	0.76821	0.74506	0.76513	0.88443	0.87421	0.88179
20	10	0.57717	0.58827	0.58567	0.76383	0.72430	0.72952	0.88809	0.88057	0.87442
20	20	0.57907	0.62339	0.57697	0.75102	0.73428	0.76268	0.87622	0.87838	0.88148
20	30	0.58180	0.59830	0.51809	0.70237	0.73753	0.74686	0.87607	0.87538	0.87380
30	10	0.58490	0.56349	0.54537	0.75624	0.72330	0.74869	0.88425	0.87271	0.87772
30	20	0.54493	0.57113	0.51552	0.74450	0.75014	0.75988	0.86612	0.87919	0.87630
30	30	0.54131	0.53097	0.59364	0.73427	0.74947	0.74776	0.87230	0.86835	0.87201
40	10	0.60500	0.60493	0.56158	0.73676	0.74308	0.72391	0.88020	0.87894	0.87381
40	20	0.52482	0.52629	0.57271	0.73125	0.72876	0.76718	0.88160	0.87439	0.87110
40	30	0.51533	0.55594	0.53725	0.74476	0.72555	0.77035	0.87884	0.86603	0.87707

- The best results are obtained when the context is not affected by noise and missing data.
- Competitive results can also be obtained if the noise is low and affects only about 10%–20% of the data, while missing data is no higher than 20%.
- Medium and high level of noise affecting context data, combined with more than 10% missing data are scenarios in which RE increases in average with almost 25%, so depending on the situation, the use of context might need re-evaluation.

3.2 Results of Dirty Context Impact for DL Algorithm

Figure 2 presents a summary for the results obtained for the deep learning algorithm. Even though there seem to be a lot of spikes in the results for the deep learning algorithm, it is very important to notice the fact that the results for RE in the tested scenarios vary between 2.87% and 3.4%, making the DL algorithm a very good option when the data context is affected by noise and missing data. The results obtained for the Spearman Rho correlation coefficient show that for deep learning algorithm is higher than 0.7.

3.3 Results of Dirty Context Impact for GBT Algorithm

Figure 3 presents the summary of the results obtained for the gradient boosted tree algorithm. Follows the main observations based on the GBT algorithm experiments.

- The best results are obtained for the “clean” context scenario and when noise is low and missing data affects no more than 10% of the context.
- the higher the percentage of missing data, the higher the values for RE.
- There is no significant distinction in the way GBT handles dirty context when the level of noise is low or medium.

Fig. 1. Decision tree results - relative error (%)

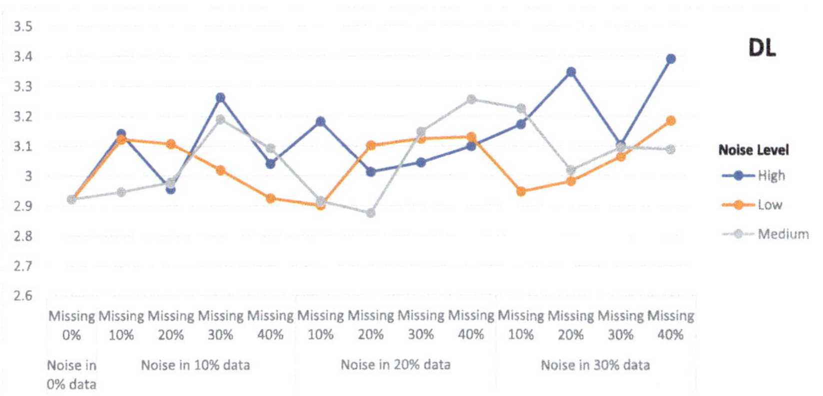

Fig. 2. Deep learning results - relative error (%)

Fig. 3. Gradient boosted tree results - relative error (%)

4 Conclusions

This article presents a case study that has as main objective finding what would be the level of noise and missing context data that would bring positive results in a CADM scenario. The tested scenarios covered three level of noise, affecting different percentage of the context data and several levels of missing data, trying to simulate possible real-life situations.

The main conclusions based on the analysis performed are as follows.

- Deep learning algorithm produces the lowest relative error from the algorithms chosen, making it a good option when forecasting time series in a CADM system.
- From the perspective of the RE results and Spearman Rho coefficient, the best algorithm is GBT, followed by DL and DT.
- A 10% noise and 10% missing data scenario produces comparable results with the "clean" baseline scenario, no matter the level of noise.
- A 10% missing data in the context provides higher variations than 10% noise add to the context, meaning that noise is better handled by missing data.
- Results comparable to the clean context baseline are obtained when the 20% threshold of both noise and missing data is not exceeded.

Future research will be conducted to identify the impact of context with outliers, that are observation points very distant from other observations [19], in a CADM process.

Another research will be conducted on improving the methods that were used to run all the test scenarios in Rapid Miner, based on an example set created a priori and share this with the research community.

References

1. Abramowitz, M., Stegun, I.A.: Handbook of mathematical functions with formulas, graphs, and mathematical table. In: US Department of Commerce. National Bureau of Standards Applied Mathematics Series, vol. 55 (1965)
2. Anton, C.A., et al.: Performance analysis of collaborative data mining vs context aware data mining in a practical scenario for predicting air humidity. In: Proceedings of the Computational Methods in Systems and Software, CoMeSySo 2019, pp. 31–40. Springer, Cham (2019)
3. Avram, A., et al.: Context-aware data mining vs classical data mining: case study on predicting soil moisture. Adv. Intell. Syst. Comput. **950**, 199–208 (2019)
4. Avram, A., Matei, O., Pintea, C.-M., Pop, P.: Context quality impact in context-aware data mining for predicting soil moisture. Cybern. Syst. Taylor & Francis, 1–17 (2020). https://doi.org/10.1080/01969722.2020.1798642
5. Bellavista, P., Corradi, A., Fanelli, M., Foschini, L.: A survey of context data distribution for mobile ubiquitous systems. ACM Comput. Surv. **44**(4), 24 (2012)
6. Chakrabarti, S., et al.: Data Mining: Know it All. Morgan Kaufmann, Massachusetts (2008)
7. Crişan, G.C., Pintea, C.-M., Palade, V.: Emergency management using geographic information systems: application to the first romanian traveling salesman problem instance. Knowl. Inf. Syst. **50**(1), 265–285 (2017)
8. Dey, A.K.: Understanding and using context. Pers. Ubiquit. Comput. **5**(1), 4–7 (2001)
9. Dodge, Y.: Spearman rank correlation coefficient. In: The Concise Encyclopedia of Statistics, pp. 502–505. Springer, New York (2008)
10. Han, J., Kamber, M., Pei, J.: Data Mining: Concepts and Techniques. Morgan Kaufmann Series in Data Management Systems, pp. 230–240 (2006)
11. Hofmann, M., Klinkenberg, R.: RapidMiner: Data Mining use Cases and Business Analytics Applications. CRC Press, Boca Raton (2016)
12. Hyndman, R.J., Athanasopoulos, G.: Forecasting: Principles and Practice. OTexts Melbourne, Australia (2018)
13. Hyndman, R.J., Koehler, A.B.: Another look at measures of forecast accuracy. Int. J. Forecast. **22**(4), 679–688 (2006)
14. Kotte, O., Elorriaga, A., Stokic, D., Scholze, S.: Context sensitive solution for collaborative decision making on quality assurence in software development processes. In: Intelligent Decision Technologies: KES-IDT 2013, vol. 255, pp. 130–139. IOS Press (2013)
15. Kotu, V., Deshpande, B.: Predictive Analytics and Data Mining: Concepts and Practice with RapidMiner. Morgan Kaufmann, San Francisco (2014)
16. Lee, S., Chang, J., Lee, S.-G.: Survey and trend analysis of context-aware systems. Inf. Int. Interdisc. J. **14**(2), 527–548 (2011)
17. Marakas, G.M.: Modern Data Warehousing, Mining, and Visualization: Core Concepts. Prentice Hall, Upper Saddle River (2003)
18. Matei, O., et al.: Context-aware data mining: embedding external data sources in a machine learning process. In: de Martínez Pisón, F., Urraca, R., Quintián, H., Corchado, E. (eds.) International Conference on Hybrid Artificial Intelligence Systems, pp. 415–426. Springer, Cham (2017)
19. Ramaswamy, S., Rastogi, R., Shim, K.: Efficient algorithms for mining outliers from large data sets. In: ACM Sigmod Record, vol. 29(2), pp. 427–438. ACM (2000)

20. Scholze, S., Barata, J.: Context awareness for flexible manufacturing systems using cyber physical approaches. In: Camarinha-Matos, L.M., Falcão, A.J., Vafaei, N., Najdi, S. (eds.) Conference on Computing, Electrical and Industrial Systems, pp. 107–115. Springer, Cham (2016)
21. Scholze, S., Barata, J., Stokic, D.: Holistic context-sensitivity for run-time optimization of flexible manufacturing systems. Sensors **17**(3), 455 (2017)
22. Vajirkar, P., Singh, S., Lee, Y.: Context-aware data mining framework for wireless medical application. In: Mařík, V., Retschitzegger, W., Štěpánková, O. (eds.) International Conference on Database and Expert Systems Applications, pp. 381–391. Springer, Cham (2003)
23. Voida, S., Mynatt, E.D., MacIntyre, B., Corso, G.M.: Integrating virtual and physical context to support knowledge workers. IEEE Pervasive Comput. **1**(3), 73–79 (2002)
24. Weiser, M., Gold, R., Brown, J.S.: The origins of ubiquitous computing research at parc in the late 1980s. IBM Syst. J. **38**(4), 693–696 (1999)
25. Witten, I.H., Frank, E., Hall, M.A.: Data mining: Practical Machine Learning Tools and Techniques. Morgan Kaufmann Series in Data Management Aystems, vol. 104, p. 113. Morgan Kaufmann, Los Altos (2005)

Small-Wind Turbine Power Generation Prediction from Atmospheric Variables Based on Intelligent Techniques

Bruno Baruque[1(✉)], Esteban Jove[2], Santiago Porras[3], and José Luis Calvo-Rolle[2]

[1] Departmento de Ingeniería Informática, University of Burgos, Avd. de Cantabria, s/n, 09006 Burgos, Spain
bbaruque@ubu.es

[2] Departamento de Ingeniería Industrial, University of A Coruña, Avda. 19 de febrero s/n, 15495 Ferrol, A Coruña, Spain

[3] Departamento de Economía Aplicada, University of Burgos, Plaza Infanta Doña Elena, s/n, 09001 Burgos, Spain

Abstract. The present research work deals the model creation obtaining for power generation prediction of a small-wind turbine, based on the atmospheric variables of its location. For testing purposes, a real dataset has been obtained of a bio-climate house located in Sotavento Experimental Wind Farm in the north of Spain. A deep study of the system and atmospheric variables has been performed. Then, some different regression techniques have been tested for accomplishing prediction, obtaining excellent results.

Keywords: Small-wind turbine · Atmospheric measurements · Regression tree · SVR · MLP

1 Introduction

The increasing concern about the climate change has led to the promotion of clean energies that avoid the harmful consequences of fossil fuels use. To reduce greenhouse gases emissions, international, national and regional governments have made significant investments in policies to promote the use of renewable energies [24]. These policies must help develop a sustainable energy generation system, that could be able to mitigate the climate change [5,20].

Although many developed countries started to focus their efforts in increasing the renewable electric power, it only represent a 15% of the global electric production in 2007, being the hydroelectric power the most significant [18]. In 2012, this percentage was raised to a 22% [20]. This increasing trend is especially consequence of the wind energy development. According to [25], the installed wind power increased from 17 GW in 2000 to 514 GW in 2017. A recent work [1,12] estimates that in 2030, only the wind energy would represent the 22.6% of the energy generation [10].

Á. Herrero et al. (Eds.): SOCO 2020, AISC 1268, pp. 33–43, 2021.
https://doi.org/10.1007/978-3-030-57802-2_4

The wind turbine design is crucial to determine the power generated as well the losses [16]. Focusing on the power generated, turbines can be designed to produce from kilowatts to megawatts, depending on the application [16]. Nowadays, according to the axis direction, two different turbine configurations are considered: Vertical Axis Wind Turbines (VAWT) and Horizontal Axis Wind Turbines (HAWT) [7,14–17]. Since the VAWT systems must be placed near the ground, they tend to produce less power for the same size. This configuration presents the advantage of producing electricity with low wind speed, which means less generation cuts.

These installations take advantage of the air masses movements produced mainly by the differential solar heating of the atmosphere [12]. The wind speed is a key factor in the energy generation, since the power is proportional to the cube of the wind speed. This parameter can change from year to year, with the season, on a daily basis, or even in seconds, known as turbulence [12]. As in many different fields, such as medicine or industry [3,6,13,22,26], an accurate prediction of the energy produce through a wind turbine can play a significant role to implement a generation system.

The present work deals with the power generation prediction of a wind turbine placed in a bioclimatic house. The prediction is carried from an original dataset of 50,834 samples registered during one year.

The rest of the document is structured as follows. Section 2 describes briefly the case of study. In Sect. 3, a study of the characteristics of the measured atmospheric is presented in order to determine the most important ones. Section 4 the techniques applied to achieve the energy prediction are explained. Section 5 details the experiments and achieved results and finally, the conclusions and future works are exposed in Sect. 6.

2 Case of Study: Bioclimatic House of Sotavento

The bioclimatic house of Sotavento is a sustainable building funded by the Sotavento Galicia Foundation. This foundation aims to make use of different renewable energy sources, and disseminate their positive consequences. This facility is placed in the borders of the provinces of A Coruña and Lugo, in Galicia. It is located in 43° 21' North, 7° 52' West, at a height of 640 m and is 30 km away from the sea.

The electric power is generated basically from the following systems:

- **Photovoltaic system.** Twenty two photovoltaic modules with a total power of 2,7 kW.
- **Wind turbine system.** A low power generator capable of generating 1,5 kW.
- **Power network.** In charge of supplying electricity when photovoltaic and wind energies do not satisfy the demand.

In addition, the following systems are used to supply Hot Domestic Water.

- **Solar thermal system.** Eight panels to absorb solar radiation and transfer it to an accumulator.
- **Biomas system.** This system has a boiler with configurable power, from 7 kW to 20 kW, with a yield of pellets of 90%.
- **Geothermal system.** A one hundred meters horizontal collector supplies heat from the ground.

2.1 Wind Turbine

This study focuses on the estimation of the wind turbine power generation. A detailed description of this system is presented in this section.

The wind turbine is a BORNAY INCLIN 1.500 model, whose blades are made of fiberglass and carbon fiber. It has a three phase synchronous generator with neodymium permanent magnets. The alternating current is generated with variables frequencies and voltages, depending on the wind speed. Hence, a rectifying stage converts this electric energy into direct current and, then, an inverting system is in charge of obtaining a alternating current waveform suitable for the network (230 Vrms and 50 Hz). This process is shown in Fig. 1.

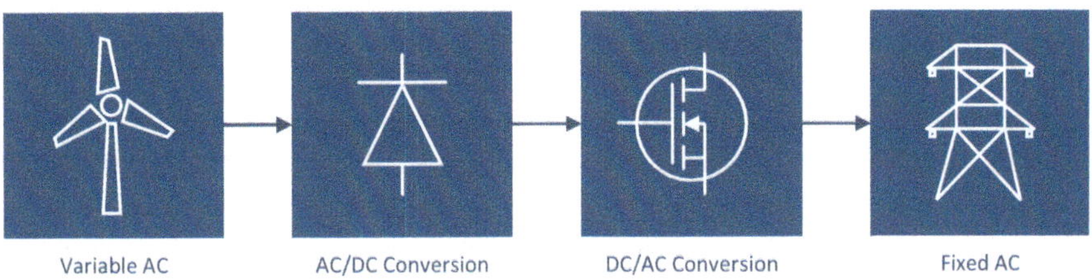

Fig. 1. Energy conversion from turbine to network

Some additional features are summarized next:

- Number of blades: 2.
- Diameter: 2,86 m.
- Nominal power: 1500 W.
- Nominal voltage: 120 Vrms.
- Starter wind: 3,5 ms/s.
- Wind for nominal power: 12 m/s.

2.2 Dataset Description

During the system operation, the following variables are monitored with a 10 min sample rate, recorded along a period of one year.

- **Wind variables**
  - Wind speed at the top of the turbine, at 10 m and its standard deviation.
  - Wind direction at the top of the turbine, at 10 m and its standard deviation. The wind gusts at 10 m are also registered.
- **Atmospherical variables**
  - Temperature at 1,5 m, 0,1 m and ground temperature at –0,1 m. The rain temperature at 1,5 m is also measured.
  - Solar information: sun hours and global radiation.
  - Pressure: atmospheric pressure and atmospheric reduced pressure.
  - Others: rain and relative humidity at 1,5 m.
- **Electric variables**
  - Voltage, current, energy and power.

3 Atmospheric Variables Study

3.1 Pre-process

Data was pre-processed as follows: First, a process of matching and cleaning was carried out in the original data sources. Atmospheric data and electrical output data were collected in different data systems, so the data was matched using the timestamp assigned to each reading in each source. This means there is not a perfect match between both, but they are close enough in time to consider them obtained in the same time instant. After that, a selection of data was performed, obtaining data for a complete year (from 1st April 2017 to 31st March 2018). This was done to avoid having some big time gaps in data, since the original source had some problems registering data for a period of several weeks in February-March of 2017. In addition, some data samples were removed from the dataset as at that time instant, either atmospheric or output of the system data were missing. At the end, only samples with both data type were kept in the dataset. After the process of matching and cleaning, we obtained a dataset of 50,834 samples, with 24 dimensions (19 corresponding to atmospheric conditions and 5 to the output of the generator). Before running the experiments, a standard score (or Z-score) normalization was performed in order to make values of all data dimensions more similar between them, regardless their measure units.

3.2 Feature Reduction

The dataset contains a total of 19 variables. As they can be considered as too many to evaluate their influence in the power generation, firstly a Principal Component Analysis (PCA) is performed in order to reduce the data dimensionality. This is a widespread technique that has been applied in problems quite similar to the one under presented here [8,19].

In Table 1 the KMO test shows a value of 0.701, and a value of 0.000 in Barlett's sphericity test significance. These values indicate that the data is adequate to perform a PCA analysis.

Table 1. KMO and Bartlett's test

Kaiser-Meyer-Olkin measure of sampling adequacy		0.701
Bartlett's test of sphericity	Aproxx. Chi-square	1709315.434
	df	171
	Sig	0.000

Table 2. Total variance explained

Comp.	Extraction sums of squared loadings			Rotation sums of squared loadings		
	Total	%Variance	%Cumm.	Total	%Variance	%Cumm.
1	5,522	29,063	29,063	4,428	23,306	23,306
2	3,791	19,950	49,013	3,177	16,722	40,028
3	2,708	14,251	63,264	3,027	15,931	55,959
4	1,656	8,714	71,978	2,549	13,418	69,377
5	1,499	7,891	79,868	1,993	10,491	79,868

Table 2 shows the quantity of explained variance. As the objective is to reduce the dimensionality of the data, the components with a total eigenvalue higher than 1 are those selected for consequent steps. Therefore a total of 5 components are finally considered. They summarize the 19 variables in consistent way, accumulating a 79.868% of the total variance.

A further analysis can be performed studying the Rotated component matrix, as can be observed in Table 3.

The objective of this is to analyze how the original variables contribute to the new five principal components and if this new latent variables can be renamed in a significant way. To perform this, it is necessary to study the value and sign of the score of each variable in each factor, seeing where a greater contribution is made. This is indicated by a high value and the direction of that contribution is indicated by the sign.

In the first component we find a higher value of 5 variables, which are related to the strength of the wind, thus it can be concluded that this first component summarizes the force of the wind. In the second one, the higher values correspond with the variables that measure the temperatures, so we can name this component as temperature. In third one, we can find three high values, two of them with positive value, corresponding to the solar radiation and the sun hours, and other with a negative sing corresponding to the humidity. This means that the more value of the solar influence, the lower the humidity. We can name this factor as solar power. The fourth factor gather the variables related to the wind direction and the last one the pressure variables, so we can denote these factors as wind direction and atmospheric pressure.

There are two variables that do not have a significant influence in the PCA result, which are "Typical deviation of the wind direction" and "Rain".

Table 3. Rotated component matrix

	Component				
	1	2	3	4	5
Gusts speed at 10 m	**,963**	−, 099	−, 039	, 001	−, 136
Wind speed at 10 m	**,926**	−, 098	−, 099	−, 044	−, 128
Wind speed standard deviation at 10 m	**,907**	−, 066	, 123	, 063	−, 123
Energy	**,893**	−, 165	−, 111	, 047	−, 077
Wind speed at top	**,883**	−, 042	, 017	, 014	−, 121
Rain	, 250	−, 032	−, 144	, 141	−, 224
Rain temperature at 1.5 m	−, 077	**,948**	−, 011	, 157	, 025
Ground temperature at 0.1 m	−, 218	**,913**	, 069	−, 051	, 037
Temperature at 1.5 m	−, 110	**,849**	, 448	−, 005	, 014
Temperature at 0.1 m	−, 064	**,768**	, 608	, 039	, 014
Global radiation	, 025	, 197	**,901**	, 005	, 019
Sun hours	−, 016	, 109	**,882**	−, 069	, 058
Relative humidity at 1, 5 m	, 059	−, 032	**−,739**	, 248	−, 004
Wind direction standard deviation at 10 m	−, 040	, 091	, 496	, 174	−, 021
Wind direction at 10 m	, 061	, 039	, 021	**,925**	−, 086
Gusts direction at 10 m	, 065	, 034	, 014	**,907**	−, 098
Wind direction at top	−, 030	, 035	−, 082	**,836**	−, 049
Atmospheric reduced pressure	−, 212	−, 056	−, 047	−, 098	**,965**
Atmospheric pressure	−, 230	, 112	, 034	−, 096	**,956**

Analyzing Tables 2 and 3 together, we can conclude that the atmospheric conditions are summarized in 5 factors, being the most important the wind speed, followed by the temperature, solar power, wind direction and atmospheric pressure.

4 Used Techniques

Classification and Regression Trees. The algorithms known as classification and regression trees (CART) encompass a wide family of techniques and variants, within this study three of them are used in particular, the simple regression tree and two ensemble models such as the bagging tree and the gradient boosted tree.

Simple Regression Tree [4] is one of the most popular and straightforward regression techniques. The basic idea behind is the recursive partition of the data in small groups to find a simple model to fit them. This method tends to be highly unstable and a poor predictor. However, by applying ensemble techniques we can improve the performance of the algorithm.

Combining the regression tree with the bagging ensemble technique, obtains as result the bagging trees meta-algorithm. This model constructs several classification trees using bootstrap sampling of the training data and then combines their predictions to produce a final one [11].

A more advance technique is to include the use of boosting mechanisms. In particular, LSBoost trees [9] have been used in this study. This method uses a least square improvement to fit the regression set in order to minimize the mean square error.

Support Vector Regression. Inspired by support vector machines, the support vector regression (SVR) has been proven to be an effective tool in real-value function estimation. As a supervised-learning approach, SVR trains using a symmetrical loss function, which equally penalizes high and low misestimates [2].

Multilayer Perceptron. Multilayer Perceptron (MLP) is one of the well-know neural networks configurations, based in a layers structure and a activation function, it allows to perform classification and regression functions quickly and with acceptable results although with less precision than other more advanced techniques [21].

5 Experiments and Results

The main objective of this initial study is to check if the atmospherical conditions and the generator power output measures can be modelled in a way that an automated machine learning system could extrapolate the power outputs from atmospherical conditions. A second step would be to try to predict future outputs given present conditions. As a preliminary study, these tests have been completed without taking into account the temporary dimension of the data; in order to be able to use simpler regression models.

After completing the PCA analysis described in Sect. 3, the 19 atmospheric dimensions were grouped and combined into only 5 dimensions. As output, the power extracted from the wind turbine is considered as the most interesting variable.

5.1 Quality Measures

In order to obtain a reference values to asses the quality of the results obtained, four of the most widely used measures for data regression [23] were employed:

- Mean Absolute Error (MAE), which tries to convey the mean of the errors obtained between predictions and real data; in the same units used to express the predicted value.
- Root Mean Squared Error (RMSE), which express the error between predicted and real values, putting more emphasis in penalizing few highest errors over many small errors.
- Coefficient of Determination (R^2), which expresses the closeness of the predicted values and the real data. Contrary to the other, this is not an error measure. As a general rule, a perfect fit would have the value of 1.

5.2 Baseline Tests

To have a framework to which compare the results of our test, the initial analysis consisted in calculating what would be the error if we used previous data readings of the output to predict the current output of the system. From this point on, all experiments refer to the power obtained (W) as the output that intends to be predicted. We calculated measures for a temporal difference of one, three and six hours, which gave the results presented in Table 4.

Table 4. Errors calculated by trying to predict a future value of W with the value that was registered exactly a given time lapse before (comparing one, three and six hours).

	MAE	RMSE	R^2
1 h	0.229	0.576	0.668
3 h	0.32	0.767	0.411
6 h	0.413	0.962	0.075

As expected, the furthest in time the prediction performed is, the higher are the errors obtained by the prediction, compared with the actual values. These values will be used as acceptance thresholds for the prediction models presented in the regression tests.

5.3 Regression Tests

The tests included in this part are preliminary tests used to check if the atmospheric data obtained can be used to accurately predict the power generation of the wind turbine and to verify to what extent this can yield satisfactory results. In this series of tests, the time component has been completely ignored. Even if atmospheric conditions are obviously dependent on the season and period time of the year, these study aims at the partial objective of assessing if the relationship between atmospheric inputs and future outputs can be considered strong enough to be captured by simpler automated learning algorithms.

The test are performed using a 10 fold cross-validation scheme in which samples form all along the studied period are included in each of the partitions. The inputs to the models are the 5 variables obtained by the PCA (see Sect. 3), while the outputs are the corresponding power (W) values obtained as output of the turbine system in a period of one, three and six hours in the future. Results are included in Tables 5a, 5b and 6.

As conclusions extracted form the presented results, we can highlight that comparing with rather close in time predictions (one hour), automated learning models do not offer a clear advantage over a prediction with the previous reading: only the Bagged Trees algorithms obtains a slightly lower error than the threshold prediction. Conversely, when time gaps are higher, most of the tested regression algorithms are capable of generalize better and obtain more adjusted predictions than the previous time instants. So, in three hours predictions and specially in six hours predictions almost all regressors, with the exception of the SVM are able to outperform the baseline prediction errors.

Table 5. Errors calculated by trying to predict the power output of the turbine in different future time instant by using atmospheric conditions

	MAE	RMSE	R^2
SRTree	0.253	0.620	0.612
BaggedT	0.217	0.483	0.764
BoostedT	0.286	0.581	0.660
SVR	0.383	0.819	0.328
MLP	0.265	0.541	0.705

(a) Errors calculated by trying to predict the power output of the turbine in one hour period by using the current atmospheric conditions.

	MAE	RMSE	R^2
SRTree	0.277	0.677	0.540
BaggedT	0.252	0.540	0.707
BoostedT	0.340	0.650	0.576
SVR	0.400	0.890	0.206
MLP	0.314	0.615	0.619

(b) Errors calculated by trying to predict the power output of the turbine in three hours time by using current atmospheric conditions

Table 6. Errors calculated by trying to predict the power output of the turbine in six hours time by using current atmospheric conditions

	MAE	RMSE	R^2
SRTree	0.304	0.743	0.445
BaggedT	0.282	0.589	0.652
BoostedT	0.402	0.737	0.454
SVR	0.415	0.964	0.060
MLP	0.366	0.696	0.512

As a second conclusion, we can highlight that simpler models are more suited to this particular task of regression without time dependencies: the best performing model in all experiments seems to be the Bagged Tree, ahead of more complex ensemble versions of it, such as the Gradient Boosted Trees. Also, comparing only simple models good performing model seems to be both the Simple Regression Tree, and a more advanced model such as the Multi-Layer Perceptron. The SVR seems to be the worst suited to this task in these experiments.

6 Conclusions and Future Work

In this contribution we have presented the bio-climatic house of Sotavento and described its wind power generation system. By modeling the atmospheric conditions that were presented in the area, the experimental study performed proves that the prediction of the power generated by this renewable energy system is an attainable result. This would be quite beneficial in smart grid scenarios, in which the optimization of power consumption could include the previsions of power generated in the different subsystems of the house.

As lines of developing future work to improve further results obtained in this problem, we have identified two main approaches. From the results obtained to this moment, it can be observed that using ensembles of simple learners yield the best results for this regression task. A straightforward method to improve results would be to study the application of several other variants of ensemble models and compare the results to determine if there is a explicit pattern or characteristics of the ensemble models that has a clear influence in the results.

Another line of work would be to include the time component of the dataset in the analyses performed. There are some models both in the statistics and artificial neural network areas that are specially designed to take sequential data and relationships of precedence of samples into account. The use of those models for this kind of prediction task offers a potential improvement of results over the ones presented in this contribution.

References

1. Aláiz-Moretón, H., Castejón-Limas, M., Casteleiro-Roca, J.L., Jove, E., Fernández Robles, L., Calvo-Rolle, J.L.: A fault detection system for a geothermal heat exchanger sensor based on intelligent techniques. Sensors **19**(12), 2740 (2019)
2. Awad, M., Khanna, R.: Support Vector Regression, pp. 67–80. Apress, Berkeley (2015). https://doi.org/10.1007/978-1-4302-5990-9_4
3. Baruque, B., Porras, S., Jove, E., Calvo-Rolle, J.L.: Geothermal heat exchanger energy prediction based on time series and monitoring sensors optimization. Energy **171**, 49–60 (2019)
4. Breiman, L.: Classification and Regression Trees. Routledge, Abingdon (2017)
5. Casteleiro-Roca, J.L., Gómez-González, J.F., Calvo-Rolle, J.L., Jove, E., Quintián, H., Gonzalez Diaz, B., Mendez Perez, J.A.: Short-term energy demand forecast in hotels using hybrid intelligent modeling. Sensors **19**(11), 2485 (2019)
6. Casteleiro-Roca, J.L., Jove, E., Sánchez-Lasheras, F., Méndez-Pérez, J.A., Calvo-Rolle, J.L., de Cos Juez, F.J.: Power cell SOC modelling for intelligent virtual sensor implementation. J. Sens. **2017**, 1–10 (2017)
7. Cecilia, A., Costa-Castelló, R.: High gain observer with dynamic dead zone to estimate liquid water saturation in pem fuel cells. Revista Iberoamericana de Automática e Informática Ind. **17**(2), 169–180 (2020)
8. De Giorgi, M.G., Congedo, P.M., Malvoni, M.: Photovoltaic power forecasting using statistical methods: impact of weather data. IET Sci. Measur. Technol. **8**(3), 90–97 (2014)
9. Friedman, J.H.: Greedy function approximation: a gradient boosting machine. Ann. Stat. **29**, 1189–1232 (2001)
10. Gomes, I.L.R., Melicio, R., Mendes, V.M.F., Pousinho, H.M.I.: Wind power with energy storage arbitrage in day-ahead market by a stochastic MILP approach. Logic J. IGPL **28**(4), 570–582 (2019). https://doi.org/10.1093/jigpal/jzz054
11. Hothorn, T., Lausen, B.: Bundling classifiers by bagging trees. Comput. Stat. Data Anal. **49**(4), 1068–1078 (2005)

12. Infield, D., Freris, L.: Renewable Energy in Power Systems. Wiley, Hoboken (2020)
13. Jove, E., Blanco-Rodríguez, P., Casteleiro-Roca, J.L., Moreno-Arboleda, J., López-Vázquez, J.A., de Cos Juez, F.J., Calvo-Rolle, J.L.: Attempts prediction by missing data imputation in engineering degree. In: International Joint Conference SOCO'17-CISIS'17-ICEUTE'17, Proceeding, León, Spain, September 6–8, 2017, pp. 167–176. Springer, Heidelberg (2017)
14. Jove, E., Casteleiro-Roca, J.L., Quintián, H., Méndez-Pérez, J.A., Calvo-Rolle, J.L.: A new approach for system malfunctioning over an industrial system control loop based on unsupervised techniques. In: Graña, M., López-Guede, J.M., Etxaniz, O., Herrero, Á., Sáez, J.A., Quintián, H., Corchado, E. (eds.) International Joint Conference SOCO'18-CISIS'18-ICEUTE'18, pp. 415–425. Springer International Publishing, Cham (2018)
15. Jove, E., Casteleiro-Roca, J.L., Quintián, H., Méndez-Pérez, J.A., Calvo-Rolle, J.L.: Anomaly detection based on intelligent techniques over a bicomponent production plant used on wind generator blades manufacturing. Revista Iberoamericana de Automática e Informática Ind. **17**(1), 84–93 (2020)
16. Kumar, Y., Ringenberg, J., Depuru, S.S., Devabhaktuni, V.K., Lee, J.W., Nikolaidis, E., Andersen, B., Afjeh, A.: Wind energy: trends and enabling technologies. Renew. Sustain. Energ. Rev. **53**, 209–224 (2016)
17. Luis Casteleiro-Roca, J., Quintián, H., Luis Calvo-Rolle, J., Méndez-Pérez, J.A., Javier Perez-Castelo, F., Corchado, E.: Lithium iron phosphate power cell fault detection system based on hybrid intelligent system. Logic J. IGPL **28**(1), 71–82 (2020). https://doi.org/10.1093/jigpal/jzz072
18. Lund, H.: Renewable energy strategies for sustainable development. Energy **32**(6), 912–919 (2007)
19. Malvoni, M., De Giorgi, M.G., Congedo, P.M.: Forecasting of PV power generation using weather input data preprocessing techniques. Energ. Procedia **126**, 651–658 (2017)
20. Owusu, P.A., Asumadu-Sarkodie, S.: A review of renewable energy sources, sustainability issues and climate change mitigation. Cogent Eng. **3**(1), 1167990 (2016)
21. Pal, S.K., Mitra, S.: Multilayer perceptron, fuzzy sets, classification. IEEE Trans. Neural Netw. **3**(5), 683–697 (1992)
22. Quintián, H., Corchado, E.: Beta scale invariant map. Eng. Appl. Artif. Intell. **59**, 218–235 (2017). http://www.sciencedirect.com/science/article/pii/S0952197617300015
23. Shcherbakov, M.V., et al.: A survey of forecast error measures. World Appl. Sci. J. **24**(2013), 171–176 (2013)
24. Simón, X., Copena, D.: Eolic energy and rural development: an analysis forgalicia. Span. J. Rural Dev. **3**(3), 13–27 (2012)
25. Sorknæs, P., Djørup, S.R., Lund, H., Thellufsen, J.Z.: Quantifying the influence of wind power and photovoltaic on future electricity market prices. Energ. Convers. Manag. **180**, 312–324 (2019)
26. Tomás-Rodríguez, M., Santos, M.: Modelling and control of floatingoffshore wind turbines. Revista Iberoamericana de Automática eInformática Ind. **16**(4), 381–390 (2019)

Supported Decision-Making by Explainable Predictions of Ship Trajectories

Nadia Burkart[1(✉)], Marco F. Huber[2,3], and Mathias Anneken[4]

[1] Fraunhofer IOSB, Karlsruhe, Germany
nadia.burkart@iosb.fraunhofer.de
[2] Institute of Industrial Manufacturing and Management IFF, University of Stuttgart, Stuttgart, Germany
[3] Center for Cyber Cognitive Intelligence (CCI), Fraunhofer IPA, Stuttgart, Germany
[4] Vision and Fusion Laboratory (IES), Karlsruhe Institute of Technology, Karlsruhe, Germany

Abstract. Machine Learning and Deep Learning models make accurate predictions based on a specifically trained task. For instance, models that classify ship vessel types based on their trajectory and other features. This can support human experts while they try to obtain information on the ships, *e.g.*, to control illegal fishing. Besides the support in predicting a certain ship type, there is a need to explain the decision-making behind the classification. For example, which features contributed the most to the classification of the ship type. This paper introduces existing explanation approaches to the task of ship classification. The underlying model is based on a Residual Neural Network. The model was trained on an AIS data set. Further, we illustrate the explainability approaches by means of an explanatory case study and conduct a first experiment with a human expert.

Keywords: Machine learning · Black box · Explainability · Interpretability

1 Introduction

Rapid progress in Machine Learning (ML) and Deep Learning (DL) pave the way in industrial applications, *e.g.*, in the automotive or health care industry. In this century the key objective of ML and DL has changed to solving real world problems. DL and ML algorithms achieve accurate results, but the main drawback is that they lack explainability and thereby, human understanding and further trust.

A model by itself consists of an algorithm that finds the relationship and patterns based on the given data. In most cases, the industry uses less complex machine learning algorithms such as linear models, small tree-based models or knowledge-based approaches because they are considered to be explainable.

Á. Herrero et al. (Eds.): SOCO 2020, AISC 1268, pp. 44–54, 2021.
https://doi.org/10.1007/978-3-030-57802-2_5

However, this often results to a lack of performance. On the other hand complex algorithms like Deep Neural Networks (DNNs) achieve a better performance but the models lack explainability. The availability of large data sets, *e.g.*, in the field of real time vessel data tracking and high computational power is leading to the development of classification models in the maritime domain, *e.g.*, ship vessel classification models. There are research approaches in the field of vessel classification, *e.g.*, using Convolutional Neural Networks (CNN) [1]. If highly complex models like CNNs need to be employed, one way to obtain explainability is to use model-agnostic approaches, which can easily be applied on various types of models. The lack of explainability of DNNs has always been a limiting factor to the application on more sensitive domains that demand explainability, *e.g.*, in health care.

In this paper, different explainability approaches are applied on a Residual Neural Network (ResNet)—a special type of CNN—that was trained for the classification of ship types based on the vessels' trajectory and other features. Our focus in this work does not lie on the training steps of the black box model but on the explainability approaches and the results. Section 2 gives more theoretical background on the model, introduces the field of explainability and the applied methods. Section 3 illustrates a case study where the output of the explainability methods is discussed. Section 4 introduces the set up on a first experiment with a human expert. Section 5 concludes and gives an outlook on the future work.

2 Related Work

In this chapter we describe the prediction model that is used for the ship classification. The main focus of this work does not lie on the training of the model but on the explainability aspects of the model. Furthermore, we will introduce the applied approaches.

2.1 Residual Neural Network for Ship Vessel Classification

In this work, a ResNet [2] as given in [3] was used. The neural network was trained on a set of the Automatic Identification System (AIS) data. AIS data serves to prevent collision between ships, for coastal States to obtain information on the ships and their cargoes and to control illegal fishing [4]. The data was collected from 2018-07-24 until 2018-11-15 worldwide resulting in 2,144,317,101 raw dynamic data points and 209,536 unique vessels. Per sequence we had 360 samples. The features that were generated and used for the training of the residual network are the trajectory, time difference, distance to coast, distance to harbor, speed and course. The *Trajectory* results from position information provided by a Global Navigation Satellite System (GNSS) over time. It is transformed and normalized, because raw GNSS data points are not easily comparable. The *Time Difference* is an additional feature that was added, which is the difference between sending the AIS signals. *Distance to coast* was also considered. This feature approximates the distance to the closest point on the coastline, whereas the *Distance to harbor* feature approximates the distance to the closest harbor. In addition to the above features, *Course* and *Speed* are also included in every sample, which is required to send by all ship vessels in a certain interval. The ship types are reduced to 5 major types. The possible target classes are: Cargo-Tanker, Fishing, Pleasure craft, Passenger and Tug. The behaviour of *Cargo* and *Tanker* vessels is very similar because of this they were combined.

Traditional neural network approaches feed each layer into the next layer whereas in a ResNet, some layers feed into the next layer and also into the layers two to three steps away. In general, the accuracy of DNNs increases with the increasing number of the layers. ResNet are thereby extremly powerful, thanks to their skip Connection. This makes it possible to build very deep models. However, there is a limit to the number of layers added which result in an improvement of the accuracy. This is because of problems like vanishing gradients and the curse of dimensionality.

Let us consider a ResNet with input x and the task is to learn the true distribution y. The difference (or the residual) between this is noted as

$$F(x) = y - x\,. \tag{1}$$

Rearranging it, we get

$$y = F(x) + x\,. \tag{2}$$

The layers are actually trying to learn the residual, $F(x)$, since we have an identity function due to x. Hence, the name residual block. For the ResNet, the introduced shortcuts or skip connection are identity mappings. Instead of only using the outputs of one layer directly as an input for the next one, they are additionally used as an input for layers two or three steps ahead. To allow skipping one layer, the output of a layer is computed according to

$$y = F(x, \{W_i\}) + W_s \cdot x\,, \tag{3}$$

with W_i as the weight matrices, and W_s as a linear projection in order to match the dimensions. For equal dimensions, W_s can be set to the identity matrix.

The trained model for the classification of ship types has 21 layers. After the initial convolutions and max pooling three blocks with skip connections follow. At the end a global averaging pooling layer and a densely connected layer are responsible for the actual classification. Details on the chosen network structure are given in [3].

2.2 Explainability Approaches

Explainability approaches give insights on the models decision process. Doshi-Velez et al. [5] defined interpretability or explainability as the *ability to explain or to present in understandable terms to a human.* In general, explaining neural networks can be divided into *Saliency methods* and *Feature attribution methods*

[6]. Saliency methods are good at illustrating the inner workings of the network regarding the region of interest and also the weights, but fail to give a complete explanation about which feature is the most important for the model. Feature attribution methods work directly on a subset of the entire dataset to find the explanatory power of each input variable with respect to the target variable. We focus on feature attribution methods. In the following, the applied model-agnostic explainability approaches are described.

SHAP: Shapley values have their origin in coalition game theory and were proposed by Lloyd Shapley [7] in order to assign each player of a coalition game a contribution which it has to the overall outcome of the coalition game. Lundberg et al. [8] proposed the model-agnostic SHAP framework inspired by Shapley values and showed how other explainability methods are approximations of SHAP. The basic building block, the Shapley values, are defined as

$$\phi_i = \frac{1}{|N|!} \sum_{R \in \mathcal{R}} [v(P_i^R \cup i) - v(P_i^R)] \tag{4}$$

where ϕ_i is the Shapley value for player i, N is the set of player (features), P_i^R the set of player with order R, $v(P_i^R)$ the contribution of set of player with order R, $v(P_i^R \cup i)$ the contribution of set of player with order R and player i, and $\mathcal{R}$ the set of possible orders. The Shapley value is the average feature value contribution across all possible combinations of feature values [9].

This game theory measure is adapted for interpreting the target model, where each feature acts as a *contributor* and attempts to predict a task, which is a *game*. The *reward* is the prediction subtracted the result from the explanation model. SHAP belongs to the class of feature attribution methods where the explanation is expressed as a linear function of features. Instead of the original feature, SHAP replaces each feature x_i with the binary variable z_i' that represents whether x_i is present or not, resulting in

$$g(z') = \phi_0 + \sum_{i=1}^{M} \phi_i z_i' = \text{Bias} + \sum \textit{Contribution of each feature} \,. \tag{5}$$

Here, 5, $g(z')$ is a local surrogate model of the original model $f(x)$. ϕ_i illustrates how the presence of feature i contributes to the final output. It helps to interpret the original model by providing the contribution of each feature.

Model Class Reliance (MCR): Permutation importance was introduced by Breiman [10] for random forests. Fisher et al. [11] propose the concept of Model Class Reliance (MCR). MCR is model-agnostic and estimates the feature importance for any black-box model. The importance is calculated by measuring how the score decreases in the absence of a feature. For example, the score can be the accuracy or F1. To achieve this, one can eliminate a feature from the dataset, retrain the model and review the score again. However, this step is computationally expensive, as it requires retraining the estimator for each particular feature. In addition, it demonstrates what may be essential in a dataset, not what is essential in a trained model being educated. In order to avoid the computation problem by retraining the estimator, one can remove a feature only from the test dataset instead of the train part, and evaluate the score without using the feature. As the model expects the feature to be present, this is not necessarily feasible. Instead of eliminating a feature, one can replace it with random values. Thereby, the feature column is still available but no longer contains useful information.

Local Model-Agnostic Explanations: LIME is a model-agnostic approach [12] that can be applied to any classification or regression model. The idea is to perturb the data input and review how this affects the output of the model. This allows us to see what the model focuses on and uses to give its predictions. In order to attain a replication of the model's behaviour locally, LIME solves the optimization problem

$$\xi(x) = \text{argmin}_{g \in G} \; L(f, g, \pi_x) \; + \; \Omega(g), \tag{6}$$

where f is the original predictor, x are the original features, g is the interpretable model and π_x as proximity measure between x and a perturbed instance x' in order to define locality around x. Basically it weights x' depending upon their distance from x. $L(f, g, \pi_x)$ is the measure of unfaithfulness of g in approximating f in the locality defined by π. This is termed as locality-aware loss in the original paper [12]. $\Omega(g)$ is the measure of model complexity of explanation g. The interpretable model can be for example a decision tree with the depth of four.

Submodular-Pick LIME: SP-LIME relies on the sub-modular optimization problem [12]. The algorithm selects a sequence of instances and their corresponding predictions which are reflective of the results of the entire model. These selections are conducted in such a manner that input features that explain more different instances have higher weights, which are used for the explanations.

2.3 User Studies in the Field of Explainable Machine Learning

Schmidt et al. [15] describe a quantitative measure of confidence in ML decisions and carried out an experiment. Two approaches were examined in their experiment, COVAR, a glass-box method, and LIME [12]. They found out that COVAR yielded more interpretable explanations. Thereby they highlighted the usefulness of simple methods. Lage et al. [14] conducted a user study to investigate what makes explanations interpretable for humans. They systematically variated the properties of an explanation to measure the effect on the performance for several tasks. The tasks were to simulate the system's response by validating a suggested response, and an counterfactual reasoning. The findings included that counterfactual questions had significant lower accuracies across the experiments. Poursabzi et al. [13] conducted a user study to measure trust. They measured the confidence by determining the difference between the prediction of the model and the prediction of the participant. As application case they used the forecast of real estate prices. Participants received different information about the model and were asked to make their own forecast of the property price.

3 Explainable Ship Trajectory Classifications

In this chapter, the four explainable approaches are illustrated by means of a case study. We will take one instance and look at the results that each explanation approach generates. Finally, the results are compared and discussed. We chose a particular instance from the class of *Cargo-Tanker* to generate the explanations.

The output of the SHAP explanation is illustrated in Fig. 1. We can observe that the predicted value for *Cargo-Tanker* is 0.48 and the corresponding base value is 0.46. The base value is the average model output over the training data set. The features with fractional values are the amount of contributions towards that class by pushing the model output higher (red) or lower (blue) from the base value.

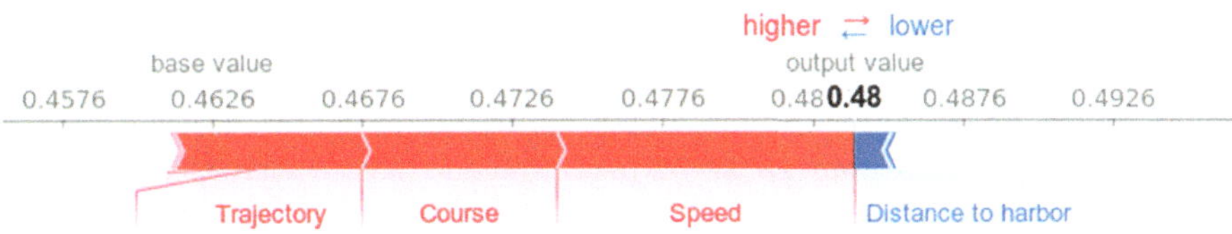

Fig. 1. SHAP explanation for the single prediction *Cargo-Tanker*

If we subtract the length of the blue bars from the length of the red bars, it equals the distance from the base value to the output. The biggest impact origins from the feature *Speed* where its visual size indicates the magnitude of the feature's effect. The evaluation of several explanations for the prediction class *Fishing* shows, that the top important features towards this class are *Distance to coast* and *Distance to harbor*. This means, that the *Distance to coast* and *Distance to harbor* features are more dominant in contributing towards the prediction class *Fishing* than any other classes. A reasonable explanation for this is the generally closer distance to the coast for fishing vessels during their travel compared to other ship types.

Figure 2 illustrates the output of the MCR for the test data set. According to Fig. 2, *Course* and *Speed* are the top features of the model to predict the class *Cargo-Tanker*.

The output of LIME is a single explanation, representing the contribution of each feature to the prediction of an instance. This provides local interpretability and it also allows to determine which feature changes will most likely have the biggest impact on the prediction. The explanation illustrated in Fig. 3 is for the single instance from the class *Cargo-Tanker*. The class *Cargo-Tanker* was predicted with the prediction probability of 1. Figure 3 illustrates the five most

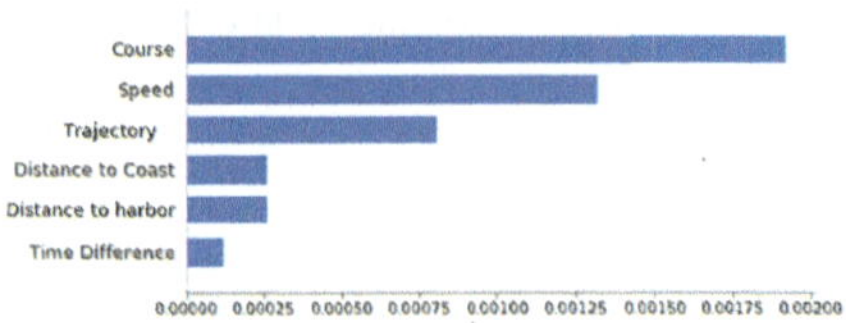

Fig. 2. MCR for *Cargo-Tanker*

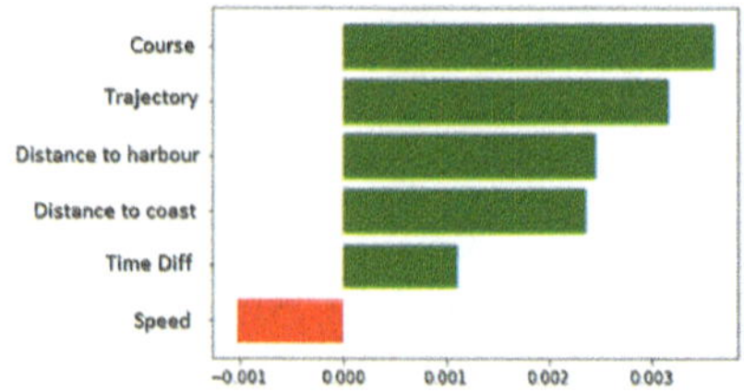

Fig. 3. LIME for *Cargo-Tanker*

important features contributing to the prediction. The features in green support the class *Cargo-Tanker* and features in red are against the class *Cargo-Tanker*. Thereby, we can assume that the feature *Speed* is not contributing to the prediction *Cargo-Tanker*. A typical *Cargo-Tanker* proceeds at 15–20 knots. The floating-point numbers represent the relative contribution of these features.

As we already described, sp-LIME (see Fig. 4) presents a set of representative explanations to illustrate a global understanding of the model. We added the trajectories of the sample instance to gather the decision process of the model more detailed. Thereby, we need to compare the test instance with the generated explanations and instances from sp-LIME and try to grasp the decision process of the model. For the *Cargo-Tanker* explanation the most decisive features were *Course* and *Speed*.

For the local approach SHAP and the global approaches MCR and sp-LIME the most important features were *Speed* and *Course*. For the local approach LIME the *Course* and *Trajectory* were the most important features. *Speed* was contributing against the decision of *Cargo-Tanker*.

4 First Experimental Results

In this section we describe the scenario that was used to evaluate the explainability approaches by a human expert (Navy Officer).

4.1 Methodology

Our experiment was designed to examine how the expert adapts his first prediction if he would get support from an AI assistance system. Four different tasks were designed. The four treatments (T1, T2, T3, T4) showed the expert different approaches of an explanation. The goal of the experiment was to gain insights about what type of explanation the expert favours the most and if the expert would adjust his first prediction or not.

4.2 Experimental Design

The experiment was performed via an online questionnaire. The participant was told basic facts about the corresponding model (ResNet) and how the model can

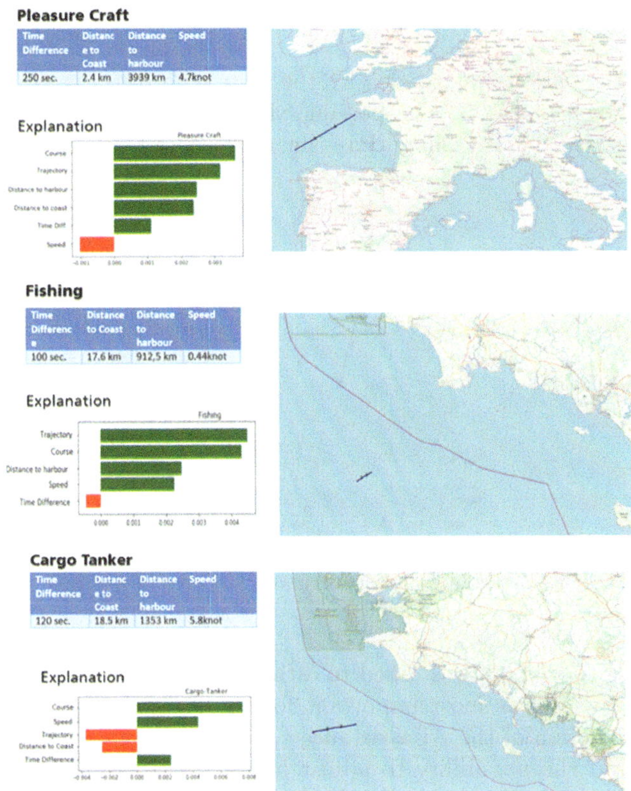

Fig. 4. sp-LIME explanations for example instances *Cargo-Tanker*, *Pleasure Craft* and *Fishing*

be used for producing a prediction (e.g., ship types) when presented with a set of predictor values (e.g., speed, distance to coast). Moreover, it was explained that the used model is sometimes treated as a black box, meaning their prediction techniques are opaque and we cannot say with certainty how the prediction was derived from the model. The participant was also introduced to the basics of the explanations and how they could help in the decision making process. For each task the expert needed to predict the ship type according to the illustrated ship trajectory and the other features. After the initial estimation of the expert, the expert got support by the AI assistance system that predicted a certain ship type. Moreover, the expert got an explanation that explains the prediction. The advice of the AI assistance was either based on SHAP (T1), MCR (T2), LIME (T3), or on sp-LIME (T4).

4.3 Task

The task of the expert was to classify the ship type (see Fig. 5) solely on the information given in Fig. 5. The possible ship-types the expert had to choose from were Cargo-Tanker, Fishing, Passenger, Pleasure Craft and Tug.

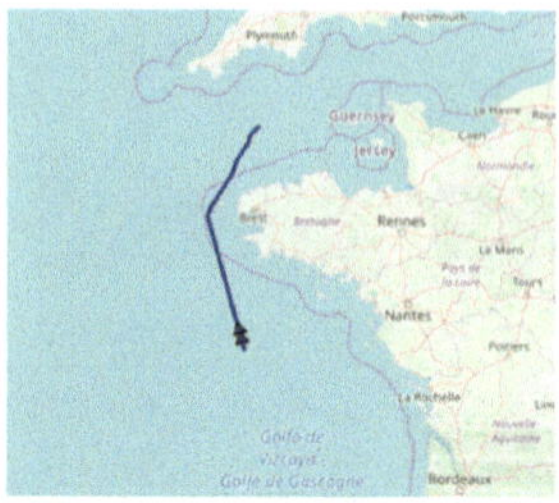

Time Difference	Distance to Coast	Distance to harbour	Speed
80 sec.	20 km	2450 km	5.8 knot

Fig. 5. Example trajectory and features of a ship

Afterwards the expert was asked to estimate the ship type and the features that were considered as most important for the decision. After this, we illustrated the prediction of the AI assistance supplied with an explanation to him. The expert had the possibility to adjust his first estimation based on the AI recommendation. Further, we asked if the explanation was clear to understand and if it was helpful or not for each explanation approach. At the end of the experiment, we asked to fill out a post-experimental questionnaire.

4.4 Results

In this section, we will discuss the results of the experiment. Table 1 illustrates for each task the first estimation of the expert, the AI prediction and the second estimation. The ground truth (real result of the classification) was also the AI's prediction (see Table 1). In the first task the expert changed his second estimation completely, neither according to first estimation nor to the AI prediction. In Task 2 the expert chose two ship types but the AI's prediction confirmed the expert's first estimation and the expert decided to choose *Fishing* only. Task 3 showed that the expert choose again two ship types and the AI supported one of his decisions, but the expert maintained the two ship types as second estimation. It shows that the expert adjusted his estimation completely according to the AI's prediction in task 4. We further asked the expert what he liked about the explanation approaches. The illustration of the influence of the different features and thereby to show their impact was considered positive. The definition of the features was considered negatively by the expert. For example it was not clear enough described if the feature *Distance to harbour* would be the distance to

Table 1. Estimates by the expert and the AI system.

Task	Expert estimation 1	AI prediction	Expert estimation 2	Ground truth
T1	Tug	Cargo-Tanker	Pleasure craft	Cargo-Tanker
T2	Fishing or pleasure craft	Fishing	Fishing	Fishing
T3	Tug or pleasure craft	Pleasure craft	Tug or pleasure craft	Pleasure craft
T4	Cargo Tanker	Pleasure craft	Pleasure craft	Pleasure craft

the port of the destination or any port. This means that the feature description needs to be very clear with some examples. The most favoured explanation approach by the expert was sp-LIME because it gave him an idea about the global environment, the vessel data and the values of the assessment. This is also the explanation approach where the expert adjusted his decision because more of the entire decision process could be grasped. We also asked the expert how we could further improve the explanations. One suggestion was to combine the trajectory and the course. The trajectory can also include changes in the course and the speed which would indicate a smaller vessel. Also an important point that was mentioned for a final assessment was to visualize the sea chart with typical shipping lanes. Bigger ships are normally on these routes, fishers and pleasure crafts are not. The expert indicated that he was undecided (4) which could be related to the unclear feature definitions. The expert also affirmed that in general he would trust an advice from an AI assistance system that is equipped with an explanation more than solely the classification result.

5 Conclusion

In this paper, we applied four explanation approaches on a ResNet for ship vessel classification. In order to apply ML and DL applications in more areas, the approaches need to be comprehensible for humans. We conducted a first experiment in order to evaluate four explanation approaches by a human expert. The overall findings were that the visualization of the feature importance was considered helpful and the explanation approach (sp-LIME) where more of the decision process could be grasped was favoured. On the downside the definition of the parameters was not clear enough to the expert. It is important during the feature engineering step to pick and build features that are very intuitive to understand and to state their meaning very clear to the expert. Our future work will be to improve the experimental design and to conduct a user study with more participants. Further, we want to integrate the experts knowledge into knowledge graphs and combine them with explainability approaches.

References

1. Gundogdu, E., Solmaz, B., Ycesoy, V., Koç, A.: Marvel: A large-scale image dataset for maritime vessels. In: Lai, S.H., Lepetit, V., Nishino, K., Sato, Y. (eds.) Asian Conference on Computer Vision, pp. 165–180. Springer, Cham (2016)

2. He, K., Zhang, X., Ren, S., Sun, J.: Deep residual learning for image recognition. In: Proceedings of the IEEE Conference on Computer Vision and Pattern Recognition, pp. 770–778 (2016)
3. Anneken, M., Strenger, M., Robert, S., Beyerer J.: Classification of Maritime Vessels using Convolutional Neural Networks. UR-AI 2020, accepted for publication (2020)
4. Tetreault, B.J.: Use of the Automatic Identification System (AIS) for maritime domain awareness (MDA). In: Proceedings of OCEANS 2005 MTS/IEEE, pp. 1590–1594. IEEE, September 2005
5. Doshi-Velez, F., Kim, B.: Towards a rigorous science of interpretable machine learning. arXiv preprint arXiv:1702.08608) (2017
6. Denadai, E.P.: Model Interpretability of Deep Neural Networks (2020). http://www.ncbi.nlm.nih.gov
7. Shapley, L.S.: A value for n-person games. Contrib. Theory Games **2**(28), 307–317 (1953)
8. Lundberg, S.M., Lee, S.I.: A unified approach to interpreting model predictions. In: Advances in Neural Information Processing Systems, pp. 4765–4774 (2017)
9. Molnar, C.: Interpretable machine learning. Lulu.com (2019)
10. Breiman, L.: Random forests. Mach. Learn. **45**(1), 5–32 (2001)
11. Fisher, A., Rudin, C., Dominici, F.: Model class reliance: variable importance measures for any machine learning model class, from the "rashomon" perspective. arXiv preprint arXiv:1801.01489, p. 68 (2018)
12. Ribeiro, M. T., Singh, S., Guestrin, C.: "Why should i trust you?" explaining the predictions of any classifier. In: Proceedings of the 22nd ACM SIGKDD International Conference on Knowledge Discovery and Data Mining, pp. 1135–1144 (2016)
13. Poursabzi-Sangdeh, F., Goldstein, D.G., Hofman, J.M., Vaughan, J.W., Wallach, H.: Manipulating and measuring model interpretability. arXiv preprint arXiv:1802.07810 (2018)
14. Lage, I., Chen, E., He, J., Narayanan, M., Kim, B., Gershman, S., Doshi-Velez, F.: An evaluation of the human-interpretability of explanation. arXiv preprint arXiv:1902.00006 (2019)
15. Schmidt, P., Biessmann, F.: Quantifying interpretability and trust in machine learning systems. arXiv preprint arXiv:1901.08558 (2019)

A Natural Language Processing Approach to Represent Maps from Their Description in Natural Language

Silvia Barbero[1], David Griol[2(✉)], and Zoraida Callejas[2]

[1] Universidad Carlos III de Madrid, Avda. de la Universidad, 30, Leganes, Spain
100316961@alumnos.uc3m.es

[2] University of Granada, Periodista Daniel Saucedo Aranda sn, Granada, Spain
{dgriol,zoraida}@ugr.es

Abstract. With the re-emergence of role playing games, interactive adventures, fantasy novels and tabletop games, the storytelling industry has a renewed interest to create engaging stories that require an interactive world-building process, in which the scenario where the story occurs is constructed, establishing the different regions, cultures and people that inhabit that land. This process usually relies on the creation of a map to locate themselves while the story develops. The main objective of this paper is to describe an approach to interpret a textual description of a map written in natural language and extract the main features and elements characterizing that map in order to produce a visual representation of the information provided by a user.

Keywords: Natural language processing · Image descriptions · Maps · World-building

1 Introduction

Nowadays there are multiple business areas that have relation with the increasingly popular entertainment sector, such as fantasy books, role-playing games, video-games, movies or tabletop games [2]. Designers and storytellers follow different processes to create worlds in which their stories can take place, a process known as world-building [3,5,9].

It is popular among role-playing gamers to create their own worlds, investing a lot of time defining the different elements that characterize them and spending countless hours sketching and drawing the maps that the rest of players will explore. Not all people that like these kind of games or designing these environments have the time or skills needed to create a visual representation of these maps or environments. Professionals dedicated to this sector could also benefit from a tool that would help them to obtain rather easily different map representations so that they can build upon them or use them to brainstorm different ideas or variations of the same environment.

Á. Herrero et al. (Eds.): SOCO 2020, AISC 1268, pp. 55–64, 2021.
https://doi.org/10.1007/978-3-030-57802-2_6

The main motivation of our research presented is to facilitate the virtual world-building process by providing a system that is able to obtain a written description of a map and transform it into a graphical representation of the same element. As it can be observed in Fig. 1, the architecture of the system consists of two main modules: the first one takes the written description provided by the user in natural language and extract the relevant information to create the map. The second module takes this information as input, representing it as an image. This division allows to isolate the natural language processing and computer graphics processes in different tasks that are connected to understand the description provided by the user and represent this description by means of a map.

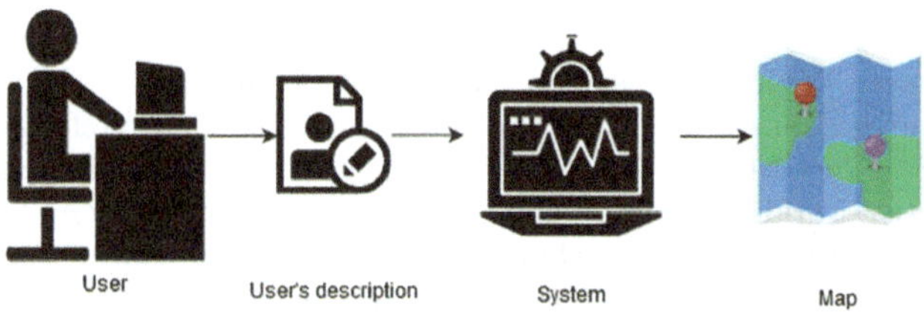

Fig. 1. System overview

This paper is focused on the design of the first stage of the system architecture, providing a proposal to generate a natural language interpretation module that allows the extraction of the different features of the description to make them understandable for the computer to them generate the image.

To achieve this goal, a corpus of users' descriptions of randomly generated geographical maps with some political elements, such as countries and their borders, has been generated. With the provided parsed data, we have learned a model consisting of several machine learning classifiers that can be trained using supervised methods, and match the different concepts and words extracted from the text to a set of predefined map elements that are formatted into a JSON file that can be later used to graphically represent the map.

2 State of the Art

Natural Language Processing (NLP) techniques provide several ways of obtaining relevant information from words or sentences with different levels of detail, which are usually used to process and parse the information into a computer readable format. Tokenization is a process in which a text is broken down into smaller pieces than can be passed onto the next steps of language analysis. These pieces can be defined by different parameters such as words or numbers, and it takes into account language specific aspects such as English contractions for negation or possession indicator and word units like city or country names formed by more than one word in order to decide which is the smaller unit taking into account the parameter defined to specify how they are obtained. Another process usually applied to words is their lemmatization, that is, obtaining the lemma from a word, stripping it from its tense, plural form, third person indicator or other modifications that can be applied to the lexeme of words [6].

Once the tokens of a text have been obtained, there are several possible analysis that can be performed on them. One of those is bag of words, also called vector space model, a method that disregards the words order or structure and focuses on obtaining the appearances of the words. Latent semantic analysis (LSA) or Latent Semantic Indexing (LSI) works much like bag of words, as it does not focus on the structure or semantic, but rather in the meaning of the words. It uses a matrix to represent a document, where the rows denote words and the columns documents. Every cell states the frequency of the word in that specific document, which can be used to obtain relationships between words that might usually appear with other common terms [6].

There are some types of analysis that need to preserve word order in order to obtain or provide information extracted from the test. One of those methods is Part of Speech (POS) tagging, which obtains the syntactic role of every word within a sentence, labeling them and returning the result to the user. This kind of methods are rule-based or stochastic, and the algorithms used to recognize word roles can be trained using artificial intelligence classification methods over a labeled dataset. Once they are trained, they can classify a sequence of words using Hidden Markov models (HMM) or use the same classification algorithms that has already been trained [1,6].

Named entity recognition (NER) is focused on data that can be recognized as a unit (e.g., persons, cities, dates, locations) identifying the named entity and its type. There are several tools that perform this kind of analysis with different levels of detail, such as NLTK[1], CoreNLP[2] or Meaningcloud[3]. Relation extraction on the other hand focuses on obtaining the connections between different entities. These relationships can be used later to identify tasks or causality between actions [4,7].

Considering relationships among the different elements of a sentence, semantic role labeling is a technique which focuses on obtaining relationships between verbs and its arguments, focusing more on the semantic meaning of a sentence rather than its syntactic structure, obtaining its meaning instead of the roles each word performs. Although the goal of semantic role labeling is not to obtain the syntactic structure of the sentence, it needs to perform this type of analysis first in order to identify the elements needed for the semantic analysis, e.g., identifying the verb and its arguments.

Finally, to maintain context information, language models preserve words sequences either using grammars or n-grams. Usually, small n-grams are used, limiting the window to the previous or two previous words, which are called

[1] https://www.nltk.org.
[2] https://stanfordnlp.github.io/CoreNLP/.
[3] https://www.meaningcloud.com/.

bigrams and trigrams respectively. Using more words usually results in a less accurate prediction and may produce worse results [10,11].

In the context of the problem and goal for this paper, several of the described analysis have been combined to obtain the maximum information possible from the description provided by the user, detecting the key information pieces that can better contribute to obtaining the most accurate representation of this description.

3 Data Corpus

As far as we are concerned, there are no corpus regarding maps descriptions. In order to generate a corpus for this domain, 25 people were recruited to write descriptions considering different elements that appeared in a map. They were given a collection of 75 images containing randomly generated maps using Azgaar's Fantasy Map Generator[4]. Two iterations were made over the maps' descriptions. In the first one, the writers were completely free to choose a style and the information they wanted to describe from the maps they were given. After those texts were reviewed, the common points among them were extracted in order to choose the information that seemed more relevant for the description to represent it graphically.

With those main areas in mind, a guideline was written focusing on the map elements that needed to be described and the writers were provided with it, so that the old descriptions could be adapted to this new version, and new descriptions could be written following those directions. The texts provided by those people were compared between each other to look for differences in the writing style, vocabulary and structures used to describe the maps, so that they could be taken into account and enrich the corpus with different variations. A total of 40 descriptions were acquired in order to test the system, check what kind of information is more useful to the machine learning classifiers and check the results and outputs obtained afterward from them. Figure 2 shows the representation of a map and the corresponding description provided by one of the writers who participated in the corpus acquisition.

4 Developed System

As described in the introduction section, the main objective of the module described in this paper is to take a map description and identify the most significant elements and those words which contribute the most to the description, and that will represent the basis from which the map will be depicted.

Users must provide a description of a world and the elements composing and characterizing it. This description is the input of the NLP module, which extracts the set of characteristics using different techniques and practical tools. This information is processed using a set of machine learning classifiers, which

[4] https://azgaar.github.io/Fantasy-Map-Generator/.

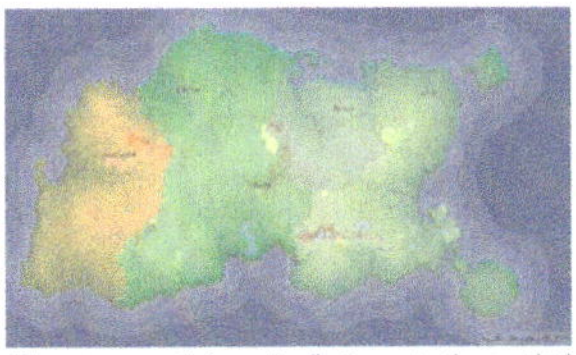

This world is occupied by only a really big continent, with a few islands scattered along its coast. The continent has a trapezoid shape, close to that of a rectangle. It contains five countries.

Schopoldia occupies the continent from northwest to southwest, taking all the west. It is one of the biggest countries, with a kind of rectangular shape. It has borders with Oittmaa in the northeast and with Yongeom from east to southeast. It has half of a big mountain range made up by small mountains, which goes from northwest to east-northeast, continuing then to Yongeom. It also has some smaller mountains spread in its land. The most abundant river is formed by several tributaries it collects along its way, starting in the East of the country, near the border with Yongeom, and going west to finally disembogue into the sea. The next most abundant river follows a similar path and it's located only a few kilometres north from the first one. In the south there are some smaller and shorter rivers that disembogue into the coast. There are also some small lakes, all of which are circular shaped, near the longest rivers. Its coast is pretty rugged, with three main gulfs: the first one located in the north, with an elongated shape and not quite big, the second one in the west, with a really prominent circular shape and delimited by a big elongated cape, which cuts into the sea for several kilometres. The third and final gulf is located in the south, it goes on for several kilometres wide but just touches the land, with a softened circular shape.

Oittmaa is located which half of its country in the northwest and the other half in the north. Its shape is that of a rounded square, and it has borders to Schopoldia in the west, Yongeom in the south and Eschia in the east. This country's rivers are really long and abundant, most of which come together as tributaries of an even bigger river...

Eschia is located mostly in the north, but it also takes part of the northeast. It's the smallest country and it has borders with Oittmaa in the west, Yongeom in the southwest and Aflia from south to northeast. It has a really long river, with several tributaries, that goes from the south to the northwest, where it disembogues into a gulf...

Fig. 2. Image of a map and extract of the description provided in the corpus

split the problem into smaller classification tasks, each of them focusing on each aspect that needs to be obtained from the description. Instead of passing all the data through the different machine learning classifiers, the output from each one is considered to decide which classifier must process the data next, compartmentalizing the classification task and reducing the dimensionality of the problem.

Figure 3 shows the structure of the developed NLP module. Once the text from the description is acquired, the NLP analysis can begin. For this step, the CoreNLP, NLTK and Meaningcloud tools were selected. CoreNLP performs the more general analysis, obtaining the lemma, POS and NER for each word. Meaningcloud enriches the information provided by CoreNLP by performing its own analysis over the text, returning three different results in separate lists: concept, entity and quantity. Concept and entity are quite similar but focus over two different aspects. The entity list provides a list of the different words that can be recognized as individuals, such as country names. The concept list on the other hand groups all the words that can be classified as belonging to a specific field, such as the words river or mountain, or the cardinal points. The quantity analysis recognizes not only the numbers that appear in the text, but also the element to which they refer, providing as an output the numeric value that word represents and the unit of the quantity that appears. Finally, NLTK performs text pre-processing, so it is used to delete all the stop words and other irrelevant information to the analysis and data classification such as punctuation, and it also obtains the raw frequencies for n-grams. In this specific case, different assessments have been completed using single words as input for the machine learning algorithms, bigrams and trigrams, so the raw frequencies for each bigram and trigram that can appear over any of the texts are obtained.

Finally, the results obtained by each individual tool must be gathered and joined to present a unified version of the results for each word and provide an easy way to access this information, parsing all this information into a JSON file, as well as storing it internally in the system, so that it can be used and

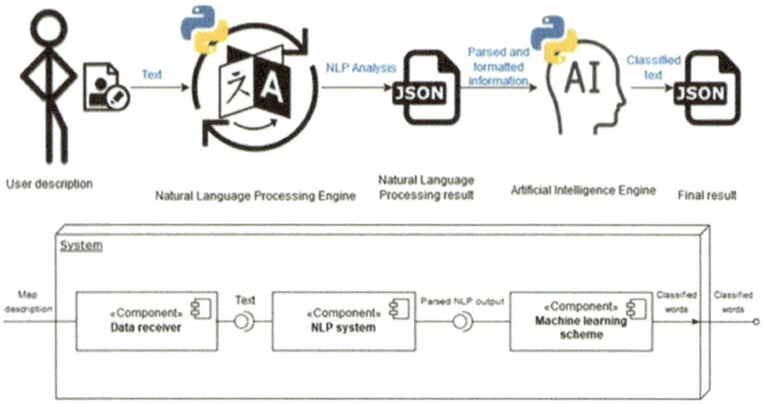

Fig. 3. System overview of the NLP module

accessed at any time by the artificial intelligence engine. This engine carries a more complex series of tasks. First of all, it accesses the information obtained from the NLP analysis and parses the results into a readable format so that it can be used by the different machine learning classifiers. Once the information can be accessed directly by any of the machine learning classifiers, the classification process obtains the classes for every classifier in the scheme for every word that is received. Once all the classes have been obtained, the results are then parsed into a list, where every word has as an output an array with the values acquired for every class from the scheme, and that constitutes the final output of the system.

There are fourteen proposed machine learning classifiers to obtain the different classes for each word in this multilabel problem. These classifiers follow a linear process. Once a word is discarded by any of the classifiers, that word does not continue the classification process. The first classifier discerns if the information of each word refers to the world, a continent or a country. Depending on the output obtained on this classifier, the words can then go on to the second classifier - if the word has been classified as world, the fourth classifier - if the word is considered related to a continent, or the fifth - if the word represents information of a country.

The second classifier obtains the information characterizing the world, coming from the class world obtained from the first classifier, so the outputs it provides are size, external islands or number of continents. The size and number of continents classes are quite self-explainable, but the external islands class is aimed to obtain the information of any islands that are present in the world but not belonging to any of the countries established.

The third classifier obtains the name, size, shape and location of the different elements in the map. The fourth classifier is focused on obtaining information regarding the continent's characteristics, coming from the output continent on the first classifier, and it only has two possible outputs: general information and number of countries.

The fifth classifier processes the country class in the first classifier and decides to which element of the country's aspects does that information refer to, classifying the words into general information, borders, river, lake, mountain or coast. The sixth classifier comes from the borders class in the fifth classifier, and obtains the location of the neighboring countries with respect to the country whose information is being obtained. To do so, it classifies the information considering the different cardinal points, them being North, Northeast, East, Southeast, South, Southwest, West and Northwest.

The seventh classifier is focused on obtaining the rivers information. The eighth classifier constitutes the second part for obtaining the rivers' information (name, start, finish, volume, length and tributaries of the rivers that are individually mentioned on the description). The ninth classifier is focused on obtaining information regarding the lakes. The tenth classifier is pretty similar to both the river's and the lake's ones, but focused on obtaining the information about mountains (name, start, finish, size and mountain size). The eleventh classifier, in a similar way to the eight, obtains specific information about the mountain ranges formations, classifying it into name, size, mountain size, start and finish.

The twelfth classifier obtains general information regarding the coast, dividing the words into presence or not, type and geographic accident (information related to different elements that appear in the coast).

The thirteenth classifier is focused on obtaining this kind of information, dividing the outputs for the words into general information, type, number and other.

Finally, the fourteenth classifier discerns what kind of coast element the words are referring to, contemplating the following: gape, gulf, bay, inner sea, peninsula and island.

5 Experiments

From the different studies and techniques described in the state of the art section, several algorithms stand out among the supervised machine learning options used in similar classification tasks: neural networks (multilayer perceptrons), support vector machines, decision trees and Naive Bayes. As the machine learning tool that was finally selected to process this information is Scikit-learn [8], only soft machine learning techniques were used to classify the words. The training sets were also divided following two different techniques: first dividing the dataset into a training and a test group (dumping 70% of the data into the training set and the remaining 30% into the test set), and second using ten-fold cross validation.

Two metrics were chosen to evaluate the results obtained by these algorithms: accuracy and Hamming loss. The accuracy obtains the percentage of correct predictions. The Hamming loss obtains the average Hamming distance, which calculates the difference between two words or sentences, averaged over all the examples. Hamming loss is only computed for the train/test split, as with cross validation the average of all the folds must be computed, so the accuracy is more representative for this training method.

The formulas that they used to calculate these measures are:

$$accuracy\,(y, y') = \frac{1}{n\,samples} \sum_{i=0}^{n\,samples-1} 1\,(y'_i = y_i) \tag{1}$$

$$Hamming\ loss\,(y, y') = \frac{1}{n\,labels} \sum_{j=0}^{n\,labels-1} 1\,(y'_j \neq y_j) \tag{2}$$

where y' stands for the predicted label for the provided input and y is the actual value for that input. The variable N denotes the number of samples in the corpus. The variable L (labels) denotes the number of classes defined for each classifier.

After analyzing the results provided by the four supervised machine learning techniques for the set of classifiers described in the previous section, the technique that provided the best results for each classification problem was decision trees. Table 1 shows the results obtained for these classifiers.

With regard the comparison of decision trees with the remainder classifiers, for the first classification task, the Naive Bayes classifier provided a mean accuracy of 0.95. For the second classification task, support vector machines and the MLP provided accuracies near to 0.99. Support vector machines also provided an accuracy of 0.81 and 0.97 for the third and fourth classification tasks. The rest of classifiers obtained a maximum of 0.53 for the fifth classification task (support vector machines) and 0.88 for the sixth classification task (Naive Bayes). Support vector machines provided an accuracy of 0.88 and 0.92 for the seventh and eighth classification tasks. An accuracy of 0.98 was provided by the Naive Bayes classifier for the ninth task. Support vector machines provided an accuracy of 0.95 and 0.98 for the tenth and eleventh classification tasks. The MLP classifier provided an accuracy of 0.87 for the twelfth classification task. Support vector machines provided an accuracy of 0.90 and 0.99 for the thirteenth and fourteenth classification tasks.

Delving into the different configurations for the decision trees that are to be used to classify the map information, the ones that include only the word to be classified in the input are the eighth, ninth, eleventh, thirteenth and fourteenth classifiers. Among them, only the thirteenth and fourteenth use a train/test split to train the machine learning classifier, the rest of them sticking to cross validation. Then, the first, fifth, sixth, seventh, and tenth classifiers use bigrams in their inputs, as they achieve the best results providing extra information for the classification algorithm. From them, the first and sixth use a train/test split to train the algorithm. Finally, the remaining classifiers - second, third, fourth and twelfth - use not only the word to be classified, but also the two previous ones and the classes for all the corresponding classifiers. All of them except for the twelfth use cross validation to train the decision tree, the remaining one using a train/test split.

Table 1. Results obtained for the different classification tasks using decision trees

n-grams	Classifier	Training method	Accuracy	Hamming loss	Classifier	Training method	Accuracy	Hamming loss
Monogram		Train/test	0.89	0.11		Train/test	0.95	0.05
		Cross-val.	0.90			Cross-val.	0.96	
Bigram	Classifier 1	Train/test	0.99	0.01	Classifier 8	Train/test	0.95	0.05
		Cross-val.	0.99			Cross-val.	0.96	
Trigram		Train/test	0.98	0.02		Train/test	0.95	0.05
		Cross-val.	0.99			Cross-val.	0.96	
Monogram		Train/test	0.99	0.01		Train/test	0.98	0.02
		Cross-val.	0.99			Cross-val.	0.99	
Bigram	Classifier 2	Train/test	0.99	0.02	Classifier 9	Train/test	0.99	0.01
		Cross-val.	0.99			Cross-val.	0.99	
Trigram		Train/test	0.99	0.01		Train/test	0.98	0.02
		Cross-val.	0.99			Cross-val.	0.98	
Monogram		Train/test	0.76	0.24		Train/test	0.97	0.03
		Cross-val.	0.75			Cross-val.	0.97	
Bigram	Classifier 3	Train/test	0.79	0.22	Classifier 10	Train/test	0.98	0.02
		Cross-val.	0.81			Cross-val.	0.98	
Trigram		Train/test	0.78	0.22		Train/test	0.97	0.03
		Cross-val.	0.81			Cross-val.	0.98	
Monogram		Train/test	0.98	0.02		Train/test	0.99	0.01
		Cross-val.	0.99			Cross-val.	0.99	
Bigram	Classifier 4	Train/test	0.99	0.01	Classifier 11	Train/test	0.99	0.01
		Cross-val.	0.99			Cross-val.	0.99	
Trigram		Train/test	0.98	0.02		Train/test	0.99	0.01
		Cross-val.	0.99			Cross-val.	0.99	
Monogram		Train/test	0.59	0.41		Train/test	0.96	0.04
		Cross-val.	0.58			Cross-val.	0.97	
Bigram	Classifier 5	Train/test	0.92	0.08	Classifier 12	Train/test	0.98	0.02
		Cross-val.	0.93			Cross-val.	0.97	
Trigram		Train/test	0.92	0.08		Train/test	0.97	0.03
		Cross-val.	0.92			Cross-val.	0.97	
Monogram		Train/test	0.97	0.03		Train/test	0.98	0.02
		Cross-val.	0.96			Cross-val.	0.97	
Bigram	Classifier 6	Train/test	0.88		Classifier 13	Train/test	0.98	0.02
		Cross-val.	0.98	0.02		Cross-val.	0.97	
Trigram		Train/test	0.97			Train/test	0.98	0.02
		Cross-val.	0.97	0.03		Cross-val.	0.97	
Monogram		Train/test	0.97	0.03		Train/test	1.00	0.00
		Cross-val.	0.98			Cross-val.	1.00	
Bigram	Classifier 7	Train/test	0.98	0.02	Classifier 14	Train/test	1.00	0.00
		Cross-val.	0.98			Cross-val.	1.00	
Trigram		Train/test	0.98	0.02		Train/test	0.99	0.01
		Cross-val.	0.98			Cross-val.	0.99	

6 Conclusions and Future Work

The research presented in this paper is inspired by the world-building process that most creators of contents follow when designing and constructing new worlds, as most of them start with the definition of the environment in which they are going to take place. There are several resources available to create maps, but they are generally oriented to providing a randomly generated map that users can then tweak and adapt to their needs. In our proposal, the main objective is to take directly the user's description in the form of a written text and provide the interpretation made by the computer using NLP and machine learning techniques.

After testing the system with a data corpus of 40 descriptions of maps, the results are promising and the next steps would involve the extension of this initial corpus to test the system with a much larger set of examples, involving a wider variety and more diversity in the styles, vocabulary and ways the different texts are written. The extended corpus with the set of descriptions and detailed instructions will be uploaded to the GitHub repository hosting service.

As future work, we will also develop the second phase in which we will take the machine-friendly representation of the map and use it to graphically represent the map with the information that has been interpreted by the computer.

References

1. Das, S., Dutta, A., Medina, G., Minjares-Kyle, L., Elgart, Z.: Extracting patterns from Twitter to promote biking. IATSS Res. **43**(1), 51–59 (2019)
2. Hergenrader, T.: Dense worlds, deep characters: role-playing games, world building, and creative writing. In: Proceedings for the Games, Learning and Society 10.0 Conference, Pittsburgh, USA, pp. 118–124 (2004)
3. Hergenrader, T.: Collaborative Worldbuilding for Writers and Gamers. Bloomsbury Academic, London (2019)
4. Ji, G., Bilmes, J.: Dialog act tagging using graphical models. In: Proceedings of ICASSP 2005, Philadelphia, USA, pp. 33–36 (2005)
5. Jokela, M.: Constructing Music Culture - a study in creativity through worldbuilding. Ph.D. thesis, Gothenburg University (2013)
6. McTear, M.F., Callejas, Z., Griol, D.: The Conversational Interface: Talking to Smart Devices. Springer, Cham (2016)
7. Pandita, R., Xiao, X., Zhong, H., Xie, T., Oney, S., Paradkar, A.: Inferring method specifications from natural language API descriptions. In: Proceedings of ICSE 2012, Zurich, Switzerland, pp. 815–825 (2012)
8. Pedregosa, F., Varoquaux, G., Gramfort, A., Michel, V., Thirion, B., Grisel, O., Blondel, M., Muller, A., Nothman, J., Louppe, G., Prettenhofer, P., Weiss, R., Dubourg, V., Vanderplas, J., Passos, A., Cournapeau, D., Brucher, M., Perrot, M., Duchesnay, E.: Scikit-learn: machine learning in Python. J. Mach. Learn. Res. **12**, 2825–2830 (2011)
9. von Stackelberg, P., McDowell, A.: What in the world? Storyworlds, science fiction, and futures studies. J. Futur. Stud. **20**(2), 25–46 (2015)
10. Wang, X., McCallum, A., Wei, X.: Topical N-grams: phrase and topic discovery, with an application to information retrieval. In: Proceedings of ICDM 2007, Omaha, USA, pp. 697–702 (2007)
11. Zhou, D., He, Y.: Discriminative training of the hidden vector state model for semantic parsing. IEEE Trans. Knowl. Data Eng. **21**(1), 66–77 (2009)

Evolutionary Computation

A Novel Formulation for the Energy Storage Scheduling Problem in Solar Self-consumption Systems

Icíar Lloréns[1,2(✉)], Ricardo Alonso[2], Sergio Gil-López[2], Sandra Riaño[2], and Javier Del Ser[2,3]

[1] École Polytechnique Fédérale de Lausanne EPFL, 1015 Lausanne, Switzerland
iciarllorens@gmail.com
[2] TECNALIA, 48160 Derio, Spain
{ricardo.alonso,sergio.gil,sandra.riano}@tecnalia.com
[3] University of the Basque Country UPV/EHU, 48013 Bilbao, Spain
javier.delser@ehu.eus

Abstract. Energy storage systems are key components to increase photovoltaic (PV) self-consumption profitability. Indeed, they allow for the intermittency dampening of the PV production so as to adequately cover end-users' consumption. Given that in most grid-connected PV systems electricity prices are variable, an informed battery scheduling can significantly decrease energy costs. Moreover, energy storage systems can cover consumption peaks to enable contracted power reduction and hence additional savings in electricity bill. This work elaborates on a scalable and flexible optimization system based on production and load forecasting as a Model Predictive Control (MPC) for battery scheduling that aims at minimizing energy costs for consumers. The system provides a 24-hour-ahead battery plan that reduces purchase cost from grid, extends the battery lifetime and guarantees purchases below the maximum contracted power. The formulated problem is solved by means of a MINLP solver and several evolutionary algorithms. Results obtained by these optimization algorithms over real data are promising in terms of cost savings within Spanish electricity market, particularly when compared to the results rendered by other methods from the state of the art. We end by outlying several research directions rooted on the findings reported in this study.

Keywords: Solar energy · Renewable energy integration · Model Predictive Control (MPC) · MINLP optimization · Evolutionary algorithms

1 Introduction

In the context of climate change prevention, the EU has set itself targets for reducing its greenhouse gas emissions progressively up to 2050. These targets

Á. Herrero et al. (Eds.): SOCO 2020, AISC 1268, pp. 67–78, 2021.
https://doi.org/10.1007/978-3-030-57802-2_7

are defined to put the EU on the way to achieve the transformation towards a low-carbon economy. The accomplishment of these goals hinge on reducing the green-house gas emissions, increasing the share of energy coming from renewable sources, and improving the energy efficiency of energy-consuming assets [2]. In this context, PV technology is one of the fastest growing renewable energy technologies [1] due to its clean nature, high availability and ease of installation for consumers [10,20]. Two modalities of PV production can be found in practice: high-scale production in large industrial plants or user-based production in small installations aimed primarily at self-consumption. In what follows, the present work focuses on the latter case.

One of the major drawbacks of PV energy supply in residential buildings derives from the fact that, in general, the production of energy does not match the electricity consumption. Consequently, users have no choice but to sell the surplus of PV energy even though energy can be demanded (and bought) later. This ultimately leads to economic losses to the end-user as the investment in the PV system cannot be entirely capitalized. Moreover, the purpose of self-consumption is not met at its full potential. Energy storage is an attractive solution for this issue, since it compensates for the intermittency of PV production by storing energy during generation, and by releasing it when the demand is high. The use of a battery consequently increases self-consumption – and hence decreases the purchase of energy from fossil fuels. However, it can also increase the cost of electricity for the consumer if the economical investment for purchasing and deploying the battery is never returned, particularly when its usage shortens its lifetime extensively [20].

As a consequence, a wide assortment of methods for battery scheduling aiming to reduce energy costs have been proposed. Deterministic [4] and global (linear [7,15,19], nonlinear [5] and evolutionary [23]) problem-solving methods have been developed in prior work to optimize the battery schedule in micro-grids. They aim to minimize the energy bought from the grid when prices are at their highest. However, these methods do not take into account the battery degradation due to extensive use, which can make the optimized schedule useless due to the shorter battery lifetime and the earlier need for replacing it on site.

The scenario tackled in this work is located in Spain, where approximately half of the electrical bill is the fixed term corresponding to the maximum contracted power. It describes the maximum power a consumer is allowed to buy at each time step. Reducing the maximum contracted power has the most impact on electricity bill savings. A solution taking into account this element was given in [17], which relied on an evolutionary solver operating on the battery schedule. However, the problem in this work does not consider battery costs either.

To address this issue, the study in [14] proposed a novel NLP approach that integrates the battery cost in the cost function, but neglects the effect of contracted power term. Later, the SUNSET system proposed in [12] tackled all three of these issues, yet by using greedy rules rather than optimization algorithms. This manuscript covers this research niche with the following novel aspects with respect to the state of the art:

- We propose a novel formulation of the battery scheduling problem aimed at minimizing the energy costs for the end-user, considering the electrical grid price variation, the storage system cost, and the maximum contracted power. As per the current Spanish legislation, we assume that electrical grid tariffs are independent of the PV installation.
- We design and validate two problem-solving methods, a branch-and-bound algorithm and several evolutionary algorithms, incorporating repair methods aimed at ensuring the fulfilment of the set of imposed constraints.
- We make the battery scheduling problem scalable and flexible. Firstly, our solution allows for the incorporation of production and load data from multiple end-users, making it capable of scheduling storage systems in connected residential neighborhoods. Secondly, it allows weighting or deleting the various concepts contributing to the total energy cost. Therefore, the problem formulation accommodates regulatory differences existing among countries.

The rest of this manuscript is organized as follows: Sect. 2 is devoted to the mathematical formulation of the problem under study. Next, Sect. 3 proposes several ways of solving the problem. Section 4 presents our results by comparing them with a traditional real-time storage strategy, the SUNSET scheme contributed in [12], and a household without any storage system. Finally, concluding remarks and future work are outlined in Sect. 5.

2 Problem Formulation

Assuming N intra-daily time slots of length Δ_t (in hours, i.e. $N = \frac{24}{\Delta_t}$) we contextualize the battery scheduling problem under analysis in a microgrid with PV power production represented by $\mathbf{P}_{PV} = [P_{PV}^n]_{n=1}^N \in \mathbb{R}^N$, with $P_{PV}^n \geq 0$ $\forall n$. Let $\mathbf{P}_S = [P_S^n]_{n=1}^N \in \mathbb{R}^N$ denote the energy at the input or output of the storage system, which can be charged ($P_S^n < 0$) or discharged ($P_S^n > 0$) at time slot n. Following with the same notation, $\mathbf{P}_L = [P_L^n]_{n=1}^N \in \mathbb{R}^N \geq 0$ represents the aggregate load in the microgrid, and $\mathbf{P}_E = [P_E^n]_{n=1}^N \in \mathbb{R}^N$ is the electrical grid power. Regarding the latter, $P_E^n > 0$ corresponds to the case when energy is retrieved (*bought*) from the energy grid, while $P_E^n < 0$ indicates that energy is injected (*sold*) from the microgrid to the main grid. At each time step $n \in \{1, \ldots, N\}$ the energy balance should be zero, this is:

$$P_{PV}^n - P_L^n + P_S^n + P_E^n = 0 \quad \forall n \in \{1, \ldots, N\}. \tag{1}$$

Bearing this in mind, the main goal of this work is to minimize the energy cost C^n [€] for a user, which can be modeled for each time step n as:

$$\mathrm{C}^n = \mathrm{C}_{\text{purchase}}^n - \mathrm{C}_{\text{sale}}^n + \mathrm{C}_{\text{battery}}^n, \tag{2}$$

where superindex n denotes that the cost is measured at time slot n. Since the purchase of energy from the grid occurs when $P_E^n > 0$, the variable energy purchase price $\mathbf{R}_{\text{buy}} = [R_{\text{buy}}^n]_{n=1}^N$ [€/kWh] is applied only to the positive part of the grid power, resulting in $C^n_{\text{purchase}} = \max\{0, P^n_E\ \Delta_t\} \cdot R^n_{\text{buy}}$. Similarly, the sale price R_{sell} is only applied to the negative part of $\mathbf{P}_E$, namely, $C^n_{\text{sale}} = -\min\{0, P^n_E\ \Delta_t\} \cdot R^n_{\text{sell}}$.

We proceed by formulating a cost term representing the battery usage. Such cost depends on the initial investment CAPEX [€/kWh], the operational expenses OPEX [€/(kWh · year)], the battery's maximum capacity $cap_{\max}$ [kWh], the number of cycles at 80% Depth of Discharge (DOD) (or 20% State of Charge) $n_{\text{cycles},80\%}$ and the battery's $lifetime = \frac{n_{\text{cycles},80\%}}{n_{\text{cycles/yr}}}$ [years]. Furthermore, we subtract from this cost the term pow_{saved} [€/year], which describes the savings due to reducing the maximum contracted power granted by the use of the battery. This yields the battery's price per kWh:

$$\text{Price}_{\text{batt}} = \frac{\text{CAPEX} \cdot cap_{\max} + lifetime\ (\text{OPEX} \cdot cap_{\max} - pow_{\text{saved}})}{cap_{\max} \cdot 0.8 \cdot n_{\text{cycles}}}. \tag{3}$$

However, if the battery usage is little to none, a cost related to the initial investment and battery lifetime must be applied. This cost, denoted as $\text{Price}^0_{\text{batt}}$, is a daily cost whose sum over the number of days in $lifetime$ is equal to the price of the battery:

$$\text{Price}^0_{\text{batt}} = \frac{1}{365}\left(\frac{\text{CAPEX}}{lifetime} + \text{OPEX} - \frac{pow_{\text{saved}}}{cap_{\max}}\right). \tag{4}$$

Finally, the rightful usage of the battery is modeled. In this regard, n_{cycles} is given for a constant charging rate of 0.1C. Since the battery under consideration is Li-ion, its number of cycles is assumed to decrease linearly by 15% for a charging rate of 1C [22]. This additional term imposes to apply all battery costs to the charging regime, thereby giving rise to the sought overall battery cost:

$$C_{\text{battery}} = \max\left\{\text{Price}^0_{\text{batt}}, \sum_{n=1}^{N} -\text{Price}_{\text{batt}} \min\{0, P^n_S \Delta_t\}\left(0.85 - 0.15\frac{\min\{0, P^n_S \Delta_t\}}{cap_{\max}}\right)\right\}.$$

The total daily energy cost is the sum of C^n from Eq. (2) over all time steps $n \in \{1, \ldots, N\}$. To cast the objective as a function of the variable to be optimized ($\mathbf{P}_S$), we replace $\mathbf{P}_E$ using Eq. (1), from where the battery scheduling problem under study can be formulated as:

$$\begin{aligned}
\underset{P_S}{\text{minimize}} \quad & \sum_{n=1}^{N} \max\{0, (P^n_L - P^n_{PV} - P^n_S)\Delta_t\} R^n_{\text{buy}} \\
& + \min\{0, (P^n_L - P^n_{PV} - P^n_S)\Delta_t\} R^n_{\text{sell}} + C_{\text{battery}} && (5)\\
\text{subject to} \quad & P^n_S \geq \left(P^n_L - P^n_{PV} - \frac{pow_{\max}}{\Delta_t}\right)\frac{1}{\Psi_S} \text{ if } P^n_L - P^n_{PV} - \frac{pow_{\max}}{\Delta_t} < 0, && (6)\\
& P^n_S \geq \left(P^n_L - P^n_{PV} - \frac{pow_{\max}}{\Delta_t}\right)\Psi_S \text{ if } P^n_L - P^n_{PV} - \frac{pow_{\max}}{\Delta_t} \geq 0, && (7)\\
& \frac{1}{\Psi_S}\left(\frac{cap_{\max}}{\Delta_t}\Delta_{\text{SOC}} - \sum_{k=1}^{n-1} P^k_S\right) \leq P^n_S \leq -\Psi_S \sum_{k=1}^{n-1} P^k_S, && (8)\\
& P^N_S = -\sum_{n=1}^{N-1} P^n_S, && (9)
\end{aligned}$$

where $\Delta_{\text{SOC}} = \text{SOC}_{\max} - \text{SOC}_{\min}$, Ψ_S denotes the (dis)charging efficiency of the battery. The constraints in (6) and (7) compel the battery to cover the consumption peaks that would otherwise require buying an amount of energy from the grid that surpasses the maximum contracted power $pow_{\max}$. Expression (8) takes into account the previous charges and discharges to prevent the battery from overcharging or from withdrawing nonexistent energy. Finally, Eq. (9) sets a boundary condition by forcing the battery's SOC at the end of the day to match the initial SOC.

3 Optimization Methods Under Consideration

The objective function described in Eq. (5) is nonlinear and non-convex. Therefore, linear or convex programming methods are not applicable to our case [11]. Furthermore, our function being piecewise-defined, it is not differentiable. This prevents us from using traditional gradient-based NLP methods [6]. We have explored two approaches to tackle our problem without modifying the formulated objective: Mixed-Integer Non-Linear Programming (MINLP) heuristics and evolutionary meta-heuristics.

3.1 MINLP Heuristics

A popular heuristic approach for piecewise-defined nonlinear problems is MINLP, which offers multiple strategies to solve problems defined over continuous and integer variables. We select the open-source BONMIN solver [3], which addresses general mixed-integer nonlinearly constrained problems. In particular, we utilize the BONMIN B-BB variant, which consists of a simple branch-and-bound algorithm based on solving a continuous nonlinear program at each node of the search tree, and branching on variable [16]. The NLP solver used to optimize each program is IPOPT [21], an interior-point method that iteratively approaches the optimal solution from the interior of the continuous variable's feasible set. No reparation phase is needed to ensure that constraints (6) to (9) are met.

In the case where the objective is non-convex, BONMIN is heuristic [3]. To tackle that, we resort to a multistart strategy. The optimization is run multiple times, each with a different initialization seed. The initialization is drawn uniformly at random between the variable bounds. Then, the best solution is chosen. Such an approach minimizes the chances of falling in a local minimum when producing the solution. Both BONMIN and IPOPT have stopping criteria related to error tolerances. In the worst-case scenario the cumulative maximum number of iterations in BONMIN used to process nodes is set to $72 \cdot 10^3$ in order to match the number of evaluations of the evolutionary algorithms described in Sect. 3.2. This problem has been implemented by using the Pyomo optimization framework [8,9].

3.2 Evolutionary Algorithms

We have also developed several Evolutionary Algorithms tailored to efficiently solve our problem, especially in regards to the set of imposed constraints: a Genetic Algorithm (GA), a Differential Evolution (DE) algorithm and Evolution Strategies (ES). For all these techniques, the fitness value corresponds to the daily energy cost as per Eq. (5), which has low variance across individuals. To tackle this, we used sigma-scaling in all three algorithms [18]. We set a population size of 120 individuals and run every search for a total of 600 generations.

The genetic algorithm utilizes roulette-wheel selection, Gaussian mutation and uniform crossover. The crossover and mutation rates are set to 0.99 and 0.1, respectively. For Differential Evolution, we set the recombination rate to 0.5 and the mutation rate to 0.5. Finally, for the evolution strategies, we implemented a $(\mu + \lambda)$ self-adaptive ES, where we have a population size of μ and λ children are produced therefrom at each generation. The next generation is composed of the best individuals chosen among children and parents [18]. Children are created from two parents, and each individual has an unlimited life expectancy. Both μ and λ were set to 120. A fine-grained parametric tuning was performed off-line using grid-search, which is not discussed hereafter due to the lack of space.

As for constraint handling, three main approaches can be adopted: death penalty, reparation and fitness penalization [13,18]. Since in our case constraints are very stringent, reparation is the only strategy producing solutions within the feasible domain. Besides clipping variable values beyond the allowed range boundaries, the designed repair method also accounts for constraints (6) and (7) by enforcing that the battery load is enough to cover grid consumption peaks. This non-stochastic repair strategy comes along with a penalty in the exploration capability of the meta-heuristics. The initial population needs to be composed of feasible individuals. Consequently, the diversity of the population is greatly reduced from the very start of the search process. Moreover, new individuals that do not comply with the constraints are repaired immediately. Such a strategy for constraint handling, albeit effective, produces early convergence.

4 Experimental Setup, Results and Discussion

In order to quantify and compare the performance of the considered solvers, we have performed numerical simulations over PV generation and consumption data of a single-family house provided within the program under grant agreement no. 691768. We used measured data to test our algorithms in order to decouple forecasting errors and algorithmic results. Should load and production forecasting models be designed and deployed in place, their natural regression error could eventually impact on the quality of the results, thereby making conclusions drawn from the comparison benchmark meaningless. Selected purchase electricity tariff corresponds to 2.0DHA access tariff in Spain for the year 2018. The value of parameters used in our experiments are given in Table 1.

We study the performances of several battery planning strategies: 1) a real-time (RT) strategy in which the grid scheduling is driven only by the current consumption status of the grid; 2) the SUNSET system proposed in [12], which relies on greedy rules; and 3) the optimization-based approach proposed in this paper. We also compare the performance of these strategies with the cost of a system with PV panels, but no storage system whatsoever.

Table 1. Parameter values of the cost function used in the experiments discussed in this work.

Parameter	Value	Parameter	Value	Parameter	Value
SOC_{min}	0.1	Ψ_s	0.95	$n_{cycles/yr}$	300 yrs^{-1}
SOC_{max}	1	CAPEX	250 €/kWh	$n_{cycles,80\%}$	3000
cap_{max}	10 kWh	OPEX	5 €/kWh yr	$Price_{batt}$	9.69 c/kWh
pow_{max}	2.8 kWh	pow_{saved}	67.34 €/yr	$Price^{0}_{batt}$	6.37 c/day

Before proceeding with the discussion of the results rendered by the above strategies, we run a preliminary test to shed light on the statistical stability of the considered solvers. For this purpose we run GA and BONMIN 100 times – each with different seed – to optimize the battery plan for a day chosen randomly. Cost values of the solutions obtained with GA have a standard deviation of $7.6 \cdot 10^{-3}$, and minimum/maximum values of 1.819 and 1.856 respectively. Such a small variation is symptomatic of the high number of constraints requiring the reparation phase to vastly reduce the population's variability. The GA is hence dominated by the reparation phase instead of stochastic search operators. However, some variation is still present. When examining the results of BONMIN, the standard deviation of the 100 cost values is 0, and the only yielded value is 1.814. Despite its multiple random initializations, BONMIN is able to converge to the same solution, which is a better optimum than any GA solution.

We now evaluate the described 5 strategies in terms of self-consumption rate and total yearly cost. Since the battery price has been decreased by the savings in contracted power – Eqs. (3) and (4) –, we must add back such value when the considered strategy fails to keep the electricity bought from the grid below pow_{max}. Table 2 shows the simulation results over 365 days.

As we can see from Table 2, the RT and SUNSET strategies are the ones that favor the most self-consumption. Indeed, RT charges and discharges the battery regardless of external considerations such as battery lifetime or peak-covering and the SUNSET rules are designed to maximise self-consumption. Our strategies sacrifice self-consumption in favor of a lower energy cost. We can also observe that yearly costs of the solvers proposed in this work, as well as SUNSET, are below the yearly cost without a battery as opposed to the RT strategy mainly due to savings related to peak-shaving. Under RT, acquiring a storage system is not cost-effective according to Eq. (5). As was anticipated by the preliminary statistical stability study, BONMIN achieves a slightly lower

Table 2. Annual cost balance for all strategies (best results highlighted in bold).

	Self-consumption (%)	Days > pow_{max}	Total cost (€)
No battery	29.0	36 (9.8%)	221.17
Real-time	**45.5**	28 (7.7%)	341.14
SUNSET [12]	**45.5**	**0**	178.64
ES	34.5	**0**	162.67
GA	31.3	**0**	159.75
DE	37.8	**0**	140.77
BONMIN	39.9	**0**	**138.46**

yearly price than the evolutionary algorithms under consideration. Lastly, our proposed solutions guarantee no purchase of energy above $pow_{\max}$, reproducing SUNSET's peak-shaving characteristics.

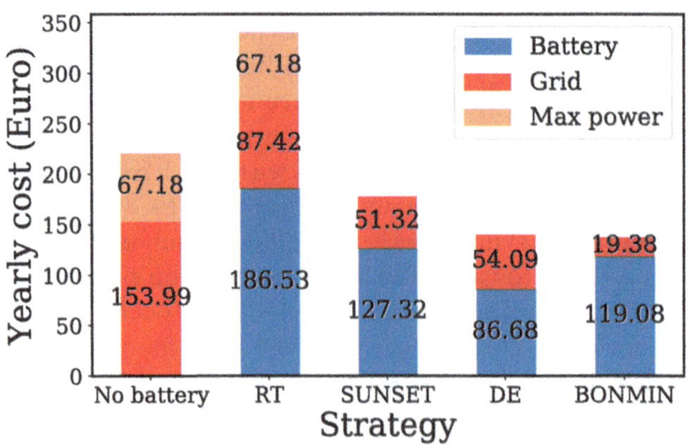

Fig. 1. Yearly cost breakdown for each strategy.

Our discussion follows in Fig. 1, which depicts the breakdown of the yearly costs for all strategies. Such yearly cost decomposes in three terms: $C_{year} = C_{\text{grid}} + C_{\text{battery}} + C_{pow_{max}}$, with $C_{\text{grid}} = C_{\text{purchase}} - C_{\text{sale}}$. In regards to evolutionary algorithms we restrict our attention on DE, since it performed best in the previously discussed benchmark described in Table 2. We find that the main energy price decrease is due to maximum contracted power savings in SUNSET, DE and BONMIN. Moreover, we observe that DE and BONMIN follow two different approaches for reducing the energy price: while BONMIN uses the battery more extensively to reduce the price of the energy bought from the grid, DE focuses more on battery degradation. The differences in the two approaches are likely to be explained by the constraint handling, since DE brings individuals that fall outside the feasible region to the edges of such region, whereas BONMIN always keeps the solution within the feasible region.

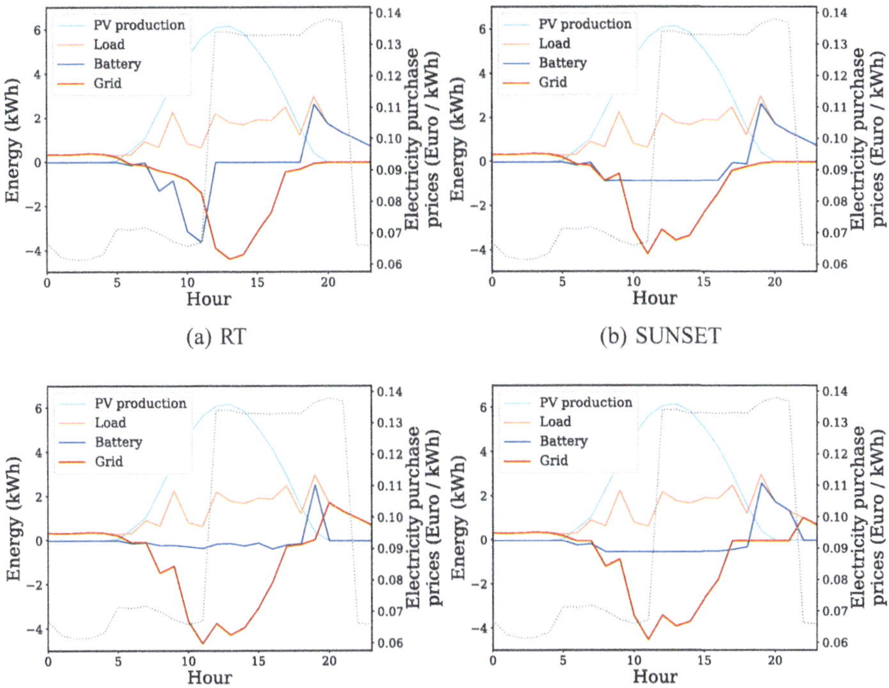

(a) RT

(b) SUNSET

(c) DE

(d) BONMIN

Fig. 2. Energy management in a PV surplus day with high-peak loads.

We end our analysis of the results with Fig. 2, where we observe the energy management for all strategies in a day where the load exceeds the maximum contracted power and there is a PV surplus. SUNSET, DE and BONMIN charge the battery smoothly and supply energy when the load is higher than pow_{max}. BONMIN and SUNSET cover all of the hours when the electricity prices are at the highest, unlike DE. BONMIN tends to charge and use the battery less than SUNSET. Indeed, while SUNSET covers the last hours of the day with battery power, BONMIN satisfies the load by purchasing power from the grid. This can be explained by electricity prices during those hours being lower than the battery's equivalent price. Even though SUNSET and our optimization methods seem to produce similar results, our formulation, especially when solved with BONMIN, is better suited for optimizing costs.

5 Concluding Remarks and Future Research Lines

In this manuscript we have elaborated on a novel formulation for the battery scheduling problem, which takes into account three main aspects influencing the energy cost: 1) suitably-timed energy purchase from the electrical grid; 2) appropriate charging behaviour to increase battery life; and 3) a restriction to avoid surpassing the maximum contracted power. The formulated problem is solved by using several methods: a MINLP solver relying on a branch-and-bound heuristic combined with an interior-point method, and a set of evolutionary algorithms (GA, DE and ES) hybridized with a repair method to ensure that the constraints of the formulated problem are met. This implementation is tariff dependent, and can communicate through simulation whether the acquisition of a storage system is profitable. Experimental results with real data have confirmed that solving the proposed problem formulation reduces effectively the energy costs for the end-user.

In view of the promising results observed in the presented work, we foresee several future research directions aimed either at algorithmic considerations regarding the solving of our problem or at formulation considerations in order to generalize our problem. Such lines can be summarized as follows:

- The problem formulation considers so far a single end-user. An interest generalization of the proposed solution would target networks of multiple end-users in order to model residential neighborhoods where consumers can share the produced PV energy.
- In order to decouple results from the influence of forecasting errors, this work has been tested on measured data rather than on forecasts. The latter would make more practical sense as the intent of use of the proposed approach is for battery scheduling. We could include forecasting error estimations and on-line corrections in the formulation to cope with the potential prediction errors.
- Another research line to pursue is to consider the more general case in which electrical grid tariffs depend on the PV installation.
- Finally, an additional option to explore is the use of Reinforcement Learning, where an agent optimizes the battery scheduling given the PV and load forecasts. This solution would override the need for heuristics.

Acknowledgments. The work herein described has received funding from the EU's Horizon 2020 research and innovation program under grant agreement No 691768. Javier Del Ser receives funding support from the Consolidated Research Group MATHMODE (IT1294-19) granted by the Department of Education of the Basque Government.

References

1. European renewable energy council (2005). erec.org/renewableenergy/photovoltaics.html
2. European commission: Climate strategies & targets (2019). ec.europa.eu/clima/policies/strategiesen
3. Bonami, P., Biegler, L.T., Conn, A.R., Cornuéjols, G., Grossmann, I.E., Laird, C.D., Lee, J., Lodi, A., Margot, F., Sawaya, N., Wächter, A.: An algorithmic framework for convex mixed integer nonlinear programs. Discrete Optim. **5**(2), 186–204 (2008)
4. Colas, F., Lu, D., Lazarov, V., François, B., Kanchev, H.: Energy managementand power planning of a microgrid with a PV-based active generator for smart grid applications. IEEE Trans. Ind. Electron **58**(10), 4583–4592 (2011)
5. Fan, H., Yuan, Q., Cheng, H.: Multi-objective stochastic optimal operation of a grid-connected microgrid considering an energy storage system. Appl. Sci. **8**, 2560 (2018)
6. Gould, N.I.M., Leyffer, S.: An Introduction to Algorithms for Nonlinear Optimization, pp. 109–197. Springer, Heidelberg (2003)
7. Hanna, R., Kleissl, J., Nottrott, A., Ferry, M.: Energy dispatch schedule optimization for demand charge reduction using a photovoltaic-battery storage system with solar forecasting. Sol. Energy **103**, 269–287 (2014)
8. Hart, W., Watson, J.P., Woodruff, D., Watson, J.P.: Pyomo: modeling and solving mathematical programs in Python. Math. Program. Comput. **3**, 219–260 (2011)
9. Hart, W.E., Laird, C.D., Watson, J.P., Woodruff, D.L., Hackebeil, G.A., Nicholson, B.L., Siirola, J.D.: Pyomo–optimization modeling in Python. Springer International Publishing (2017)
10. Kwon, J., Nam, K., Know, B.: Photovoltaic power conditioning system with line connection. IEEE Trans. Ind. Electron. **53**(5), 1048–1054 (2006)
11. Luenberger, D.G., Ye, Y.: Linear and Nonlinear Programming, 3rd edn. Springer (2008)
12. Manjarres, D., Alonso, R., Gil-Lopez, S., Landa-Torres, I.: Solar energy forecasting and optimization system for efficient renewable energy integration. In: Woon, W.L., Aung, Z., Kramer, O., Madnick, S. (eds.) Data Analytics for Renewable Energy Integration: Informing the Generation and Distribution of Renewable Energy, pp. 1–12. Springer International Publishing (2017)
13. Michalewicz, Z., Dasgupta, D., Riche, R.G.L., Schoenauer, M.: Evolutionary algorithms for constrained engineering problems. Comput. Ind. Eng. **30**(4), 851–870 (1996)
14. Michiorri, A., Bossavy, A., Kariniotakis, G., Girard, R.: Impact of PV forecasts uncertainty in batteries management in microgrids. In: IEEE Grenoble Conference, pp. 1–6 (2013)
15. Nottrott, A., Kleissl, J., Washom, B.: Energy dispatch schedule optimization and cost benefit analysis for grid-connected, photovoltaic-battery storage systems. Renewable Energy **55**, 230–240 (2013)
16. Gupta, O.K.: Branch and bound experiments in convex nonlinear integer programming. Manage. Sci. **31**, 1533–1546 (1985)
17. Salcedo-Sanz, S., Camacho-Gómez, C., Mallol-Poyato, R., Jiménez-Fernández, S., Del Ser, J.: A novel coral reefs optimization algorithm with substrate layers for optimal battery scheduling optimization in micro-grids. Soft Comput. **20**(11), 4287–4300 (2016)

18. Simon, D.: Evolutionary Optimization Algorithms: Biologically Inspired and Population-Based Approaches to Computer Intelligence. Wiley, Hoboken (2013)
19. Tziovani, L., Kolios, P., Hadjidemetriou, L., Kyriakides, E.: Energy scheduling in non-residential buildings integrating battery storage and renewable solutions, pp. 1–6 (2018)
20. Vieira, F.M., Moura, P.S., de Almeida, A.T.: Energy storage system for self-consumption of photovoltaic energy in residential zero energy buildings. Renewable Energy **103**, 308–320 (2017)
21. Wächter, A., Biegler, L.T.: On the implementation of an interior-point filter line-search algorithm for large-scale nonlinear programming. Math. Program. **106**(1), 25–57 (2006)
22. Wikner, E., Thiringer, T.: Extending battery lifetime by avoiding high SOC. Appl. Sci. **8**(10), 1825 (2018)
23. Yoon, Y., Kim, Y.H.: Charge scheduling of an energy storage system under time-of-use pricing and a demand charge. Sci. World J. **2014**, 9 (2014)

A Behavioural Study of the Crossover Operator in Diploid Genetic Algorithms

Adrian Petrovan[1(✉)], Oliviu Matei[1], and Rudolf Erdei[2]

[1] Technical University of Cluj-Napoca, North University Center of Baia Mare, Baia Mare, Romania
adrian.petrovan@cunbm.utcluj.ro
[2] Holisun SRL, Baia Mare, Romania

Abstract. The article presents an analysis of seven crossover operators for continuous spaces applied for Diploid Genetic algorithms (DGA). Unlike the classical ones, called in genetic "haploid" in which an individual is synonym with a chromosome, the individuals of DGA carry two chromosomes, which brings in intrinsic diversity in the population. The impact of the recombination operators is analyzed and compared, turning out that BLX operators yields the best results and uniform and arithmetic crossover the worst. With respect to specificity, the uniform crossover and two point crossover have the lowest standard deviation of the results.

Keywords: Diploid genetic algorithms · Crossover · Real-coded genetic algorithms

1 Introduction

Genetic algorithms (GAs) are adaptive heuristic search techniques, based on the principles of genetics and natural selection, inspired from the theory of natural evolution developed by Charles Darwin based on the "survival of the fittest". These algorithms were introduced in practice by Holland, and the mechanism is similar to the biological process of evolution. The process has a feature that only species that are better adapted to the environment are able to survive and evolve over generations, while those less adapted do not survive and eventually disappear, as a result of natural selection. In other words, GAs have the ability to deliver a "good-enough" solution "fast-enough", making them very attractive in solving optimization problems [14].

Usually, the population of genetic algorithms (AGs) consists of haploid individuals, that is, individuals with a single chromosome. As a result, genetic operators involved in solving a problem using genetic algorithms use the only one chromosome as an informational entity. It is also known that if a GA lose the population diversity it will stuck into local optimum and by default its genetic operators like crossover become ineffective [19]. Some solutions have been used to

Á. Herrero et al. (Eds.): SOCO 2020, AISC 1268, pp. 79–88, 2021.
https://doi.org/10.1007/978-3-030-57802-2_8

preserve population diversity like high mutation or random immigration of individuals into the population [3]. Another idea of maintaining population diversity is the use of multiploid (polyploid chromosomes) individuals. The idea of using diploid representation (individual comprising of two chromosomes) in GAs is not new and it is proved it has advantages in problems using real-coded chromosome representation [14]. The idea is that even the phenotype of a diploid individual tends to converge to a local optimum, its different genotypes will preserve the population diversity.

Yukiko and Nobue [19] described a diploid genetic algorithm (DGA) for preserving the population diversity using the idea of meiosis to convert the genotype to phenotype. Lieckens et al. [10] introduce a very simple diploid GA that limits the GA mainly in its discrete time, non-overlapping populations setup and its representation of genotypes and as well provided formal methods to be used to study finite population models of diploid genetic algorithms and Bull [2] presents a new variant of evolutionary algorithm that harnesses the haploid-diploid cycle present in eukaryotic organisms. In [17] Yang investigates the effect of the cardinality of genotypic representation and the existence of uncertainty in the dominance scheme for DGAs in dynamic environments.

Other much practical approaches on the use of DGA are presented in [1]. The authors present the performances of DGA in relation to the greedy approach on the dynamic traveling salesman person problem. Pop et al. [15] used successfully DGAs in order to solve the generalized traveling salesman problem, the generalized minimum spanning tree problem and the family traveling salesman problem. Dulebenets [5] propose a Diploid Evolutionary Algorithm (DEA) that can assist the cross-docking operators with the design of cost-efficient truck schedules, that can facilitate the flow of different products within the cross-docking facilities, ensure the "just-in-time" deliveries within supply chains, and improve sustainability of the supply chain operations.

Approaches regarding the effect of the crossover operator on the behaviour of genetic algorithms (GAs) have been made only in the case of haploid genetic algorithms. In this sense, the works of Herrera et al. [8,9] presents experimental results of the application of different crossover operators.

The aim of this article is to explore the behavior of diploid genetic algorithms when using different cross operators and by testing them on some major benchmark functions used for performance evaluation of genetic algorithms. The rest of the paper is organized as follows: in Sect. 2 we briefly describe the diploid genetic algorithms and in Sect. 3 we describe the proposed genetic crossover operators used in this study. Section 4 details the experiments performed and analyses the obtained results. Finally, the conclusions and future research directions are presented in the last section.

2 The Diploid Genetic Algorithms

In the case of diploid genetic algorithms (DGA), individuals consist of two coupled haploid chromosomes. This type of representation is called the diploid (bichromosomal). Also, each individual carries additional information called phenotype, information necessary for the selection of the individual who will participate in the crossover process for the formation of new offspring. In this way, the diploid representation of superior life forms is mimicked [12]. The superiority of this representation lies in the fact that each person carries twice as much information as compared to the classical haploid approach, thus a greater diversity being ensured in terms of potential workable solutions [7]. The final values of the phenotype of an individual are decided by the dominance schemes that play an important role in the algorithm's performance. It is particularly important to design and experiment with a good dominance scheme to guarantee the performance of the diploid GA as compared to the haploid one.

One of the most important dominance schemes described in the literature have been proposed by Ng and Wong [13], in which the dominant allele will always be part of the phenotype. If there is a conflict between two dominant or two recessive alleles, the selected one is strictly random. Another elaborated approach to the dominance scheme is offered by Yang and Yao [18] and is called the dominance learning scheme in which a dominant probability vector is defined. Within this scheme each element has a dominance probability which represents the probability that a genotypic allele can be expressed within the phenotype.

3 Crossover Operators

Several crossover operators for real numbers, as described by Herrera et al. [8], have been take into account and adapted for diploid representation.

Further we will assume that the two chromosomes to undergo recombination are $C_1 = (c_1^1, ..., c_n^1)$ and $C_2 = (c_1^2, ..., c_n^2)$.

Simple crossover (one cut) (SX). A position $i \in 1, 2, ..., n-1$ is chosen randomly and two offspring chromosomes are created as:

$$O_1 = (c_1^1, ..., c_i^1, c_{i+1}^2, ...c_n^2) \tag{1}$$
$$O_2 = (c_1^2, ..., c_i^2, c_{i+1}^1, ...c_n^1) \tag{2}$$

Two point crossover (TPX). The two point crossover [11] derives from the simple crossover, but uses two cutting points $i, j \in 1, 2, ..., n-1$ with $i < j$. The offspring chromosomes are:

$$O_1 = (c_1^1, ..., c_i^1, c_{i+1}^2, ...c_j^2, c_{j+1}^1, ..., c_n^1) \tag{3}$$
$$O_2 = (c_1^2, ..., c_i^2, c_{i+1}^1, ...c_j^1, c_{j+1}^2, ..., c_n^2) \tag{4}$$

Uniform crossover (UX). In the case of uniform crossover [11], each gene is selected at random from one of the corresponding genes of the chromosomes C_1 or C_2. The two offspring $O_k = (o_1^k, ..., o_n^k)$, $k = 1, 2$ are built from genes as:

$$o_i^k = \begin{cases} c_i^1, if u = 0 \\ c_i^2, if u = 1 \end{cases} \tag{5}$$

where u is a random generated number that can have a value of zero or one.

Arithmetic crossover (AX). The arithmetic crossover produces two chromosomes, with the associated genes [11]:

$$o_i^1 = \lambda \cdot c_i^1 + (1 - \lambda) \cdot c_i^2 \quad (6)$$
$$o_i^2 = \lambda \cdot c_i^2 + (1 - \lambda) \cdot c_i^1 \quad (7)$$

where $\lambda \in [0, 1]$ is randomly generated.

BLX-α crossover (BLX-). The two offspring chromosomes which are created after BLX-α [6] crossover contains genes o_i^k which are uniformly randoms number in the range $[c_i^{min} - I\alpha, c_i^{max} + I\alpha]$, where $c_i^{min} = min\{c_i^1, c_i^2\}$, $c_i^{max} = max\{c_i^1, c_i^2\}$ and $I - c_{max} - c_{min}$.

Max-Min Arithmetic crossover (MMAX). Max-Min Arithmetic crossover [8] combines two parent chromosomes and generates 4 offspring chromosomes (the first two are like in the case of arithmetic crossover), as:

$$o_i^1 = \lambda c_i^1 + (1 - \lambda) c_i^2; o_i^2 = \lambda c_i^2 + (1 - \lambda) c_i^1; o_i^3 = min\{c_i^1, c_i^2\}; o_i^4 = max\{c_i^1, c_i^2\} \quad (8)$$

Linear crossover (LX). The linear crossover [16] creates 3 offspring chromosomes, according to the formulae:

$$o_i^1 = \frac{1}{2} c_i^1 + \frac{1}{2} c_i^2; o_i^2 = \frac{3}{2} c_i^1 - \frac{1}{2} c_i^2; o_i^3 = -\frac{1}{2} c_i^1 + \frac{3}{2} c_i^2 \quad (9)$$

From the point of view of a taxonomy, the selected crossover operators are part of four major groups of classification [8], as follows: simple (SX), two point (TPX) and uniform (UX) crossover are part of discrete crossover operator group (DCO); arithmetical (AX) and linear (LX) are part of aggregation based crossover operator group (ABCO) group; BLX-α is part of the neighborhood based crossover operator group (NBCO) and max-min-arithmetic (MMAX) is from the group of hybrid crossover operator (HCO).

In the case of diploid parents in DGA, the crossover operator work as follows: each chromosome of each parent are crossed with the chromosomes of the other parent, thus resulting in six distinct offspring (see Fig. 1).

Applied to the chosen crossover operators and the principles of forming offspring in DGA presented in Fig. 1 some clarifications are needed: for crossover operators with two offspring there are created by this technique twelve offspring from which six are selected to form the new population, and for the crossover with three offspring created (LX) and four offspring created (MMAX) the principles remain the same.

4 Experimental Study

The purpose of our experiments was the study of the behaviour of diploid genetic algorithm on several benchmark functions under the conditions of major modifications of the crossover operator. Also the influence of a specific crossover on

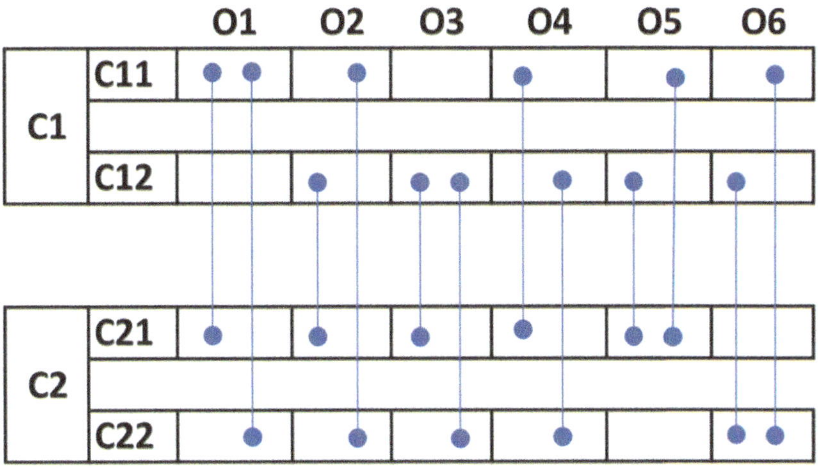

Fig. 1. Offspring creation method in DGA.

the convergence of the solution is analysed. That is the reason why the results are not compared with the results of other techniques reported in other articles.

Our algorithms have been implemented on Java 8 and we have performed 30 independent tests for each considered benchmark function. The experiments have been conducted on a machine with CPU Intel Core i5, 2.4 GHz, 8 GB RAM, running JDK 8.

The two developed algorithms, have been tested on the following benchmark functions described in what it follows:

1. **Sphere function** is a smooth uni-modal function defined as follows:

$$f(\mathbf{x}) = f(x_1, x_2, ..., x_n) = \sum_{i=1}^{n} x_i^2 \tag{10}$$

   where $-5.12 \leq x_i \leq 5.12$.
2. **Ackley function** is a continuous, non-convex multi-modal function defined as follows:

$$f(\mathbf{x}) = -a.exp\Big(-b\sqrt{\frac{1}{n}\sum_{i=1}^{n} x_i^2}\Big) - exp\Big(\frac{1}{n}\sum_{i=1}^{n} cos(cx_i)\Big) + a + exp(1) \tag{11}$$

   where usually $-32.768 \leq x_i \leq 32.768$ and $a = 20$, $b = 0.2$ and $c = 2\pi$.
3. **Griewank function** is a non-linear, multi-modal function defined as follows:

$$f(\mathbf{x}) = f(x_1, ..., x_n) = 1 + \sum_{i=1}^{n} \frac{x_i^2}{4000} - \prod_{i=1}^{n} cos\Big(\frac{x_i}{\sqrt{i}}\Big) \tag{12}$$

   where usually $-600 \leq x_i \leq 600$.

4. **Rastrigin function** which is a non-linear, multi-modal function containing millions of local optima and defined as follows:

$$f(\mathbf{x}) = f(x_1, x_2, ..., x_n) = 20A + \sum_{i=1}^{n} \left(x_i^2 - 10cos(2\pi x_i)\right) \tag{13}$$

where usually $A = 10$ and $-5.12 \leq x_i \leq 5.12$.
5. **Schwefel function** which is a non-linear, multi-modal function defined as follows:

$$f(\mathbf{x}) = f(x_1, x_2, ..., x_n) = 10V - \sum_{i=1}^{n} x_i sin(\sqrt{|x_i|}) \tag{14}$$

where usually $V = 4189, 829101$ and $-500 \leq x_i \leq 500$.

All the considered test functions are dealing with minimization problems. For more information regarding the properties of the considered benchmark functions and some other functions used in order to test the performance of GAs we refer to [4].

5 Experimental Results

The presented results correspond to three sets of experiments on the above presented test functions: Sphere, Ackley, Griewank, Rastrigin and Schwefel. The experiments make use of the same mutation rate $p_m = 0.5\%$ over 200 epochs. The results are summarized in Tables 1, 2, 3, 4 and 5.

Each benchmark has been run 30 times on each function, crossover and setup and the best (columns `Best`), the mean value (column `Mean`), respectively the standard deviation (column `StdDev`) have been computed. Please note that the optimum of each function is zero, therefore the lower the value, the better.

Table 1. The experimental results for Sphere function

Crossover	**Pop. size: 1000; Genes: 25**			**Pop. size: 2000; Genes: 25**			**Pop. size: 1000; Genes: 50**		
	Best	Mean	StdDev	Best	Mean	StdDev	Best	Mean	StdDev
SX	2.44E−01	2.44E−01	2.96E−15	1.16E−01	1.16E−01	2.05E−15	2.75E+00	2.75E+00	2.64E−14
TPX	1.03E−01	1.03E−01	7.49E−16	3.84E−02	3.84E−02	9.25E−16	1.13E+00	1.13E+00	9.86E−15
UX	1.15E−02	1.15E−02	1.53E−16	5.82E−03	5.82E−03	1.43E−16	6.65E−02	6.65E−02	7.97E−16
AX	9.61E−04	9.61E−04	7.54E−18	6.35E−05	6.35E−05	1.12E−18	6.66E−02	6.66E−02	5.33E−16
BLX-0	1.83E−08	1.83E−08	5.68E−15	5.58E−12	5.58E−12	4.94E−17	1.20E−04	1.20E−04	3.52E−09
BLX-0.3	1.43E−22	2.66E−22	3.11E−23	6.46E−23	1.31E−22	1.61E−23	2.78E−14	4.03E−14	2.62E−15
BLX-0.5	2.96E−15	5.86E−15	7.50E−16	2.14E−15	4.70E−15	5.97E−16	1.37E−08	2.22E−08	1.99E−09
MMAX	1.82E−04	1.82E−04	1.10E−15	1.13E−05	1.13E−05	1.82E−15	2.40E−02	2.40E−02	3.25E−13
LX	1.33E−05	1.53E−05	3.74E−07	2.23E−06	3.45E−06	2.46E−07	5.33E−03	5.49E−03	2.99E−05

Table 2. The experimental results for Ackley function

Crossover	Pop. size: 1000; Genes: 25			Pop. size: 2000; Genes: 25			Pop. size: 1000; Genes: 50		
	Best	Mean	StdDev	Best	Mean	StdDev	Best	Mean	StdDev
SX	3.83E+00	3.83E+00	5.28E−14	2.78E+00	2.78E+00	4.15E−14	7.78E+00	7.78E+00	6.87E−14
TPX	3.22E+00	3.22E+00	3.06E−14	2.10E+00	2.10E+00	4.67E−14	5.27E+00	5.27E+00	3.62E−14
UX	1.36E+00	1.36E+00	1.03E−14	5.63E−01	5.63E−01	6.83E−15	1.77E+00	1.77E+00	1.52E−14
AX	2.07E+00	2.07E+00	3.15E−10	9.30E−02	9.30E−02	1.93E−15	2.47E+00	2.47E+00	7.40E−15
BLX-0	1.03E−03	1.03E−03	1.44E−10	1.64E−05	1.64E−05	5.88E−11	7.57E−02	1.14E−01	4.89E−09
BLX-0.3	3.27E−11	4.42E−11	2.75E−12	2.18E−11	3.14E−11	1.97E−12	2.01E−07	3.53E−07	1.29E−08
BLX-0.5	1.44E−07	1.94E−07	1.28E−08	1.25E−07	1.79E−07	1.19E−08	3.06E−04	3.82E−04	1.80E−05
MMAX	1.10E−01	1.10E−01	1.62E−15	3.56E−02	3.56E−02	7.07E−13	1.15E+00	1.15E+00	6.71E−08
LX	8.23E−02	8.39E−02	2.99E−04	1.92E−02	2.08E−02	2.31E−04	4.42E−01	4.48E−01	1.24E−03

Table 3. The experimental results for Griewank function

Crossover	Pop. size: 1000; Genes: 25			Pop. size: 2000; Genes: 25			Pop. size: 1000; Genes: 50		
	Best	Mean	StdDev	Best	Mean	StdDev	Best	Mean	StdDev
SX	1.24E+00	1.24E+00	4.44E−15	4.37E−01	4.37E−01	6.74E−15	9.62E+00	9.62E+00	7.16E−14
TPX	5.76E−01	5.76E−01	5.17E−15	1.30E−01	1.30E−01	3.59E−15	3.27E+00	3.27E+00	3.53E−14
UX	2.98E−01	2.98E−01	3.86E−15	3.56E−02	3.56E−02	6.99E−16	1.33E+00	1.33E+00	1.74E−14
AX	5.66E−03	5.66E−03	5.13E−17	1.78E−04	1.78E−04	3.79E−18	2.61E−01	2.61E−01	2.13E−15
BLX-0	5.77E−08	5.77E−08	1.97E−14	1.02E−12	1.02E−12	1.50E−17	2.09E−03	2.09E−03	1.21E−10
BLX-0.3	3.64E−22	7.16E−22	8.82E−23	1.82E−22	3.94E−22	4.83E−23	5.04E−14	7.20E−14	5.21E−15
BLX-0.5	1.05E−14	1.88E−14	2.28E−15	7.43E−15	1.51E−14	1.91E−15	4.71E−08	7.64E−08	6.76E−09
MMAX	1.05E−03	1.05E−03	1.22E−17	7.95E−05	7.95E−05	1.54E−14	6.19E−02	6.19E−02	2.84E−13
LX	7.83E−05	1.11E−04	3.29E−06	3.14E−05	4.42E−05	1.91E−06	2.21E−02	2.28E−02	1.18E−04

Table 4. The experimental results for Rastrigin function

Crossover	Pop. size: 1000; Genes: 25			Pop. size: 2000; Genes: 25			Pop. size: 1000; Genes: 50		
	Best	Mean	StdDev	Best	Mean	StdDev	Best	Mean	StdDev
SX	1.27E+01	1.27E+01	9.77E−14	1.92E+00	1.15E+00	2.83E−14	3.73E+01	3.73E+01	3.13E−13
TPX	5.76E+00	5.76E+00	6.24E−14	1.77E+00	1.77E+00	5.10E−14	3.20E+01	3.20E+01	2.99E−13
UX	1.75E+00	1.75E+00	8.44E−15	6.16E−01	6.16E−01	1.51E−14	6.76E+00	6.76E+00	3.68E−14
AX	3.23E+01	3.23E+01	1.23E−08	6.68E+00	6.68E+00	7.51E−05	4.41E+01	4.41E+01	6.32E−09
BLX-0	1.64E−02	1.64E−02	1.43E−09	7.79E−−04	7.79E−04	7.73E−10	2.10E+00	2.10E+00	1.55E−07
BLX-0.3	0.00E+00	0.00E+00	0.00E+00	0.00E+00	0.00E+00	0.00E+00	9.04E−10	1.40E−09	9.98E−11
BLX-0.5	2.49E−01	2.49E−01	5.45E−10	1.89E−12	4.98E−12	6.88E−13	6.67E−01	6.69E−01	5.53E−04
MMAX	1.40E−01	1.40E−01	7.49E−14	4.00E−02	4.00E−02	3.32E−13	2.74E+00	2.74E+00	1.09E−11
LX	5.62E−02	6.03E−02	7.11E−04	4.19E−04	5.64E−04	2.32E−05	1.57E+01	1.57E+01	7.34E−04

The first benchmark uses 1000 individuals with 25 genes (meaning also that the dimension of the function is 25) and the results are reported in column `Pop: 1000, genes: 25`. The second benchmark uses 2000 individuals with 25 genes each (column `Pop: 2000, genes: 25`). The last benchmark uses 1000 individuals with 50 genes each (meaning that the dimension of the functions is 50) and its results are reported in column `Pop: 1000, genes: 50`. The reason for choosing these running parameters of these algorithms consisted in following

Table 5. The experimental results for Schwefel function

Crossover	Pop. size: 1000; Genes: 25			Pop. size: 2000; Genes: 25			Pop. size: 1000; Genes: 50		
	Best	Mean	StdDev	Best	Mean	StdDev	Best	Mean	StdDev
SX	4.75E+01	4.75E+01	6.42E−13	2.39E+01	2.39E+01	5.90E−13	5.38E+02	5.38E+02	6.79E−12
TPX	2.28E+01	2.28E+01	1.34E−13	1.08E+01	1.08E+01	6.47E−14	2.20E+02	2.20E+02	2.83E−12
UX	9.69E+00	9.69E+00	0.00E+00	2.62E+00	2.62E+00	0.00E+00	4.30E+01	4.30E+01	2.47E−13
AX	4.81E+03	4.81E+03	4.19E-06	4.49E+03	4.49E+03	6.69E−04	1.28E+04	1.28E+04	2.72E−06
BLX-0	5.96E+03	5.96E+03	2.10E−04	5.01E+03	5.26E+03	2.14E−13	1.15E+04	1.16E+04	1.89E−04
BLX-0.3	5.12E+03	5.24E+05	2.79E−12	4.30E+03	4.63E+05	2.79E−12	9.90E+03	1.02E+06	2.79E−12
BLX-0.5	3.79E+03	3.88E+05	1.80E−08	3.19E+03	3.42E+05	1.80E−08	7.33E+03	7.54E+05	1.80E−08
MMAX	9.86E−01	9.86E−01	1.66E−13	1.01E−01	1.01E−01	2.48E−11	2.66E+01	2.66E+01	3.30E−10
LX	1.64E−01	1.77E−01	4.30E−04	3.53E−02	3.56E−02	1.42E−11	4.25E−01	4.27E−01	2.68E−13

the variations of the mean values obtained in case of doubling the population size or in case of doubling the number of variables of the test functions.

In most cases, the algorithms reach the same optimum per crossover type, that is why the best the worst and the mean values are more or less identical for each crossover, respectively the standard deviation is very close to zero. Therefore, in this setup, no further improvement could be brought without significant change of the evolutionary parameters.

Comparing the crossover operators, AX yields the worst results independent on the function or the genetic parameters. On the other side, BLX-0.3 works the best, followed closely by BLX-0.5. Regarding the specificity of the results, LX causes the least specificity, as it has the highest standard deviation in all the cases, followed by BLX-05. The advantage of the BLX method is that it uses an initial exploration of the parameters field followed by an exploitation phase to improve resolution. The highest specificity is reached by the algorithms using UX and TPX. It is also worth noting that MMAX (the hybrid recombination method) has yielded much better results than the aggregation recombination method AX.

6 Conclusions and Future Work

In this paper we made a comparative study of how is affected the convergence of the diploid genetic algorithms (DGA) due to seven different recombination operators. The experiments have been done on five benchmark functions. It is proved that DGA allow good convergence due to the intrinsic diversity of the population. The high convergence is proven by the very low standard deviations reported in Tables 1, 2, 3, 4 and 5. In most of the tests performed, a significant improvement of the average results obtained by changing the crossing technique is observed, in some situations the results obtained are very close to the global optimum of the function

The obtained results allow us to affirm once again that, in the light of the obtained results, diploid genetic algorithms can be used to solve complex optimization problems with very large number of variables. Future research will focus on the application of the results obtained in solving certain complex problems and by comparing the results obtained with those confirmed in the specialized literature.

References

1. Bhasin, H., Behal, G., Aggarwal, N., Saini, R.K., Choudhary, S.: On the applicability of diploid genetic algorithms in dynamic environments. Soft Comput. **20**(9), 3403–3410 (2016). https://doi.org/10.1007/s00500-015-1803-5
2. Bull, L.: Haploid-diploid evolutionary algorithms: the Baldwin effect and recombination nature's way. In: AISB (2017)
3. Cobb, H.G., Grefenstette, J.J.: Genetic algorithms for tracking changing environments. Technical report, Naval Research Lab Washington DC (1993)
4. Digalakis, J., Margaritis, K.: On benchmarking functions for genetic algorithms. Int. J. Comput. Math. **77**(4), 481–506 (2001). https://doi.org/10.1080/00207160108805080
5. Dulebenets, M.A.: A diploid evolutionary algorithm for sustainable truck scheduling at a cross-docking facility. Sustainability **10**(5), 1333 (2018)
6. Eshelman, L.J., Schaffer, J.D.: Real-coded genetic algorithms and interval-schemata. In: Foundations of Genetic Algorithms, vol. 2, pp. 187–202. Elsevier, Amsterdam (1993)
7. Goldberg, D., Smith, R.: Nonstationary function optimization using genetic algorithms with dominance and diploidy. In: Proceedings of Second International Conference on Genetic Algorithms and their Application, pp. 59–68 (1987)
8. Herrera, F., Lozano, M., Sánchez, A.M.: A taxonomy for the crossover operator for real-coded genetic algorithms: an experimental study. Int. J. Intell. Syst. **18**(3), 309–338 (2003)
9. Herrera, F., Lozano, M., Sánchez, A.M.: Hybrid crossover operators for real-coded genetic algorithms: an experimental study. Soft Comput. **9**(4), 280–298 (2005)
10. Liekens, A., Eikelder, H., Hilbers, P.: Modeling and simulating diploid simple genetic algorithms. In: Proceedings Foundations of Genetic Algorithms VII. FOGA VII, pp. 151–168 (2003)
11. Michalewicz, Z.: Genetic Algorithms + Data Structures = Evolution Programs. Springer Science & Business Media, Heidelberg (2013)
12. Mitchell, M.: An Introduction to Genetic Algorithms. MIT Press, Cambridge (1998)
13. Ng, K.P., Wong, K.C.: A new diploid scheme and dominance change mechanism for non-stationary function optimization. In: Proceedings of the 6th International Conference on Genetic Algorithms, pp. 159–166. Morgan Kaufmann Publishers Inc., San Francisco (1995). http://dl.acm.org/citation.cfm?id=645514.657904
14. Petrovan, A., Pop-Sitar, P., Matei, O.: Haploid versus diploid genetic algorithms. a comparative study. In: International Conference on Hybrid Artificial Intelligence Systems, pp. 193–205. Springer (2019)
15. Pop, P., Matei, O., Pintea, C.: A two-level diploid genetic based algorithm for solving the family traveling salesman problem. In: Proceedings of the Genetic and Evolutionary Computation Conference. GECCO 2018, pp. 340–346. ACM, New York (2018). https://doi.org/10.1145/3205455.3205545
16. Schlierkamp-Voosen, D., Mühlenbein, H.: Strategy adaptation by competing subpopulations. In: International Conference on Parallel Problem Solving from Nature, pp. 199–208. Springer (1994)

17. Yang, S.: On the design of diploid genetic algorithms for problem optimization in dynamic environments. In: 2006 IEEE International Conference on Evolutionary Computation, pp. 1362–1369. IEEE (2006)
18. Yang, S., Yao, X.: Experimental study on population-based incremental learning algorithms for dynamic optimization problems. Soft Comput. **9**(11), 815–834 (2005). https://doi.org/10.1007/s00500-004-0422-3
19. Yukiko, Y., Nobue, A.: A diploid genetic algorithm for preserving population diversity—Pseudo-Meiosis GA. In: International Conference on Parallel Problem Solving from Nature, pp. 36–45. Springer (1994)

Parallel Differential Evolution with Variable Population Size for Global Optimization

Iztok Fister[1,3(✉)], Andres Iglesias[2,3], Akemi Galvez[2,3], Dušan Fister[4], and Iztok Fister Jr.[1]

[1] Faculty of Electrical Engineering and Computer Science, University of Maribor, Koroška cesta 46, 2000 Maribor, Slovenia
iztok.fister@um.si

[2] Toho University, 2-2-1 Miyama, Funabashi 274-8510, Japan

[3] University of Cantabria, Avenida de los Castros, s/n, 39005 Santander, Spain

[4] Faculty of Economics and Business, University of Maribor, Razlagova 14, 2000 Maribor, Slovenia

Abstract. The results of evolutionary algorithms depends on population diversity that normally decreases by increasing the selection pressure from generation to generation. Usually, this can lead evolution process to get stuck in local optima. The study is focused on mechanisms to avoid this undesired phenomenon by introducing parallel differential evolution that decompose a monolithic population into more variable-sized subpopulations, which evolve independently of each other. The proposed parallel algorithm operates with individuals having some characteristics of agents, e.g., they act autonomously by selecting actions, with which they affect the state of environment. This incorporates two additional mechanisms: aging, and adaptive population growth, which direct the decision-making by individuals. The proposed parallel differential evolution was applied to the CEC'18 benchmark function suite, while the produced results were compared with some traditional stochastic nature-inspired population-based and state-of-the-art algorithms.

Keywords: Differential evolution · Variable population size · Aging mechanism · Autonomous agent

1 Introduction

Usually, researchers in Evolutionary Algorithm (EA) community are confronted with the question: How to maintain a diversity of population in the conditions of open-ended evolution, where EA must operate continuously without any breaks [8]. Unfortunately, losing the population diversity normally leads to a premature convergence. A lot of approaches have been proposed for avoiding this phenomenon, such as, for instance by Črepinšek et al. [13], and by Fister et al. [5]. The novel step in mastering the arisen problem in open-ended

Á. Herrero et al. (Eds.): SOCO 2020, AISC 1268, pp. 89–99, 2021.
https://doi.org/10.1007/978-3-030-57802-2_9

evolution and artificial life, obviously, is the development of Parallel Evolutionary Algorithms (PEA) [7] and Evolutionary MultiAgent Systems (EMAS) [3]. PEAs decompose the monolithic population in EAs into more sub-populations, where selection and reproduction are limited to individuals inhabiting one region and a migration operator is used to move selected individual from one region to another. On the other hand, agents in EMAS, representing solution of the problem to be solved, are distributed across islands (i.e., sub-populations) and can interact directly only with their local environment. Due to suffering a lack of global knowledge, their decision-making process is limited only on local information [3].

The purpose of this study is to confront with the problem of premature convergence in Parallel Differential Evolution (PDE) decomposing the monolithic population into more sub-populations (i.e., islands) capable of exploring different regions of the search space independently of each other. Moreover, individuals in the PDE have some characteristics of agents, because they act autonomously in deciding, which action to select in order to affect the state of environment (i.e., problem). Consequently, also some elements of the classical DE algorithm are either eliminated (e.g., survivor selection), or redefined in new way (e.g., variation operators).

PDE incorporates two new mechanisms: aging, and adaptive population growth. The former influences on the size of island by action death and controls decreasing the number of individuals with regard to the feedback obtained from the last generation, similar as the latter that directs individuals by deciding, whether to apply reproduction with growing of the population or clone actions keeping the population size intact. Because both mechanisms use global information, appearing on the population level, the individuals in the proposed PDE are not pure agents.

As a result, the proposed Parallel Variable sized DE for global optimization (gPVaDE) was developed and applied to the CEC-18 benchmark function suite. The results of comparison with the classical EAs, like DE [11], and its self-adaptive variants jDE [2] and SaDE [10], and the state-of-the-art algorithms, like LShade [12] and jSO [1], showed the potential of the proposed algorithm, and encourage us to continue with the research in the same direction also in the future.

2 Mutation Strategies in Differential Evolution

DE was introduced by Storn and Price in 1995 [11] and work with real-valued vectors. Although these vectors in the original DE undergo operations of variation operators, such as mutation, crossover, and selection, we borrow only DE mutation strategies in our study. These strategies describes the way, in which the operations of mutation and crossover are conducted within DE.

In the basic mutation, two solutions are selected randomly, and their scaled difference is added to the third solution, as follows:

$$\mathbf{u}_i^{(t)} = \mathbf{x}_{r0}^{(t)} + F \cdot (\mathbf{x}_{r1}^{(t)} - \mathbf{x}_{r2}^{(t)}), \quad \text{for } i = 1, \ldots, Np, \tag{1}$$

where $F \in [0.1, 1.0]$ denotes the scaling factor that scales the rate of modification, while Np represents the population size and $r0$, $r1$, $r2$ are randomly selected values in the interval $1, \ldots, Np$.

The mentioned mutation strategy is capable for exploring a search space. When the exploitation of the search space is needed, the following mutation strategy is more appropriately:

$$\mathbf{u}_i^{(t)} = \mathbf{x}_{best}^{(t)} + F \cdot (\mathbf{x}_{r1}^{(t)} - \mathbf{x}_{r2}^{(t)}), \quad \text{for } i = 1, \ldots, Np, \tag{2}$$

where $\mathbf{x}_{best}^{(t)}$ is the current best individual, and $r1$, $r2$ are randomly selected values in the interval $1, \ldots, Np$. Let us emphasize that a balancing between exploration and exploitation can be achieved by mixing both strategies [13].

In our study, we employ a binomial crossover [11]. This crossover is uniform in the sense that each parameter, regardless of its location in the trial vector, has the same probability of inheriting its value from a given vector. Thus, the trial vector is built from parameter values copied from either the mutant vector generated by Eq. (1) or parent at the same index position laid i-th vector. Mathematically, this crossover can be expressed as follows:

$$w_{i,j}^{(t)} = \begin{cases} u_{i,j}^{(t)} & \text{rand}_j(0,1) \leq CR \vee j = j_{rand}, \\ x_{i,j}^{(t)} & \text{otherwise}, \end{cases} \tag{3}$$

where $CR \in [0.0, 1.0]$ controls the fraction of parameters that are copied to the trial solution. The condition $j = j_{rand}$ ensures that the trial vector differs from the original solution $\mathbf{x}_i^{(t)}$ in at least one element.

3 Proposed gPVaDE for Global Optimization

The proposed gPVaDE consists of more islands connected in ring topology that evolve in parallel. Each island hosts individuals located in some positions on a grid. They are capable of changing these positions randomly in each generation. On the other hand, these individuals can perform following autonomous actions: reproduction, clone, death, migration, and rebirth. Reproduction generates the trial solution using the exploration DE-mutation strategy (Eq. (1)), and keep it within the island beside its parent to evolve in the next generation. With new individual, the reproduction contribute to the island growth. Clone that generates the trial solution using the exploitation DE-mutation strategy (Eq. (2)) selects the better between parent and trial solutions for the next generation, and therefore keeps the island size unchanged. Death eliminate the individual from the island. Migration enable some individuals in one island to move in the other island according to a migration probability p_m. The action rebirth is launched, when the diversity in the island is lost.

Additionally, two mechanisms are incorporated within the algorithm, i.e., aging and adaptive population growth. The former introduces the concept of individual's age replacing the classical selection and changes this with more natural paradigm asserting, when people are old enough, they must die. The latter directs individual by making decision, whether replacement or clone should be performed in particular situation. In summary, the aging mechanism takes case about reducing the island size, while the adaptive population growth enriches the island with new individuals.

3.1 Aging Mechanism

An aging mechanism presents one of the more popular concepts of adapting the population size during the evolutionary cycle in the EA community, and was used in the Genetic Algorithm with varying population size (GAVaPS) [9]. This mechanism introduced the concept of an individual's "age", which counts the number of generations the individual stays "alive".

The aging mechanism operates as follows: Each individual in a population lives the number of generations (ages) determined by its parameter *lifetime*. This parameter depends on the fitness of the corresponding individual, i.e., the higher the fitness of an individual, the higher a *lifetime* granted to it. Mathematically, the *lifetime* is defined as:

$$lifetime = \begin{cases} MinLT + K \cdot \frac{f_i - MinFit}{AvgFit - MinFit}, & \text{if } AvgFit \geq f_i, \\ \frac{1}{2}(MinLT + MaxLT) + K \cdot \frac{f_i - AvgFit}{MaxFit - AvgFit}, & \text{if } AvgFit < f_i, \end{cases} \tag{4}$$

where *MinLT* and *MaxLT* denotes the minimum and maximum available *lifetime* values, respectively, *AvgFit*, *MinFit*, and *MaxFit* are average, minimum, and maximum values of fitness in the current population, while the coefficient is expressed as $K = \frac{1}{2}(MaxLT - MinLT)$.

3.2 Adaptive Population Growth

The adaptive population growth implements so-called Non-Linear population Size Reduction (NLSR) mechanism, where the population size is adapted following the population dynamics. In population dynamics, measure of the uncertainty in population size is expressed as:

$$\Delta H^{(t+1)} = \log \frac{2 \cdot S^{(t)}}{N^{(t+1)}}, \tag{5}$$

where ΔH denotes a change in the evolutionary entropy [4], $S^{(t)}$ is the number of positive variations in the last population, and $N^{(t+1)}$ is the effective population size in the next evolutionary cycle. Entropy influences increasing/decreasing the current population size regarding the following relations:

$$\Delta_{\max} = \begin{cases} -\text{rand}(1, 2 \cdot (R^{(t+1)} - N^{(t+1)})), & \text{if } \Delta H^{(t+1)} > 0, \\ +\text{rand}(1, 2 \cdot (R^{(t+1)} - N^{(t+1)})), & \text{if } \Delta H^{(t+1)} < 0, \\ 0, & \text{if } \Delta H^{(t+1)} = 0, \end{cases} \tag{6}$$

where $\Delta_{\max}$ denotes a modification in the population size, and $R^{(t+1)}$ is a decreased, linear, reference function that reduce the population size according to:

$$R^{(t+1)} = \left(1 - \frac{t+1}{t_{\max}}\right) \cdot (MAX\_VAL - MIN\_VAL) + (t+1), \quad (7)$$

where t is the generation number, $t_{\max}$ the maximum number of generation, MAX_VAL and MIN_VAL are the maximum and minimum population sizes, respectively.

3.3 Design and Implementation of gPVaDE

Each individual in gPVaDE has its own time of birth and time of death, and lives as long as permitted by its quality of behavior. It is defined as a tuple:

$$X_i = \langle \mathbf{x}_i, f_i, M_i, age_i, lt_i, act_i \rangle, \quad \text{for } i = 1, \ldots, Np, \quad (8)$$

where $\mathbf{x}_i$ denotes vector with elements $\{x_{i,j}\}$ for $i = 1, \ldots, D$, f_i is a fitness function of the problem in question, M_i location of the individual within the island, age_i current age of agent, lt_i calculated lifetime, and act_i action to be performed by the individual.

Three main algorithms need to be implemented for covering the proposed three level program architecture of the gPVaDE, i.e.: algorithm's, island's, and individual's. The first algorithm takes care about: creation, termination and parallel evolving of islands, and synchronization between them. The second one provides global functions for individuals, like: positioning on random locations within grid, aging management, and adaptive population growth. The last one addresses tasks with which an individual is confronted, as: decision-making, and executing actions.

Due to the paper limitation, only the algorithm Evolve coping the second level is illustrated here (Algorithm 1). As can be seen, the Evolve algorithm launched in each generation is divided into two parts: In the first part, the evolving of the particular individuals is performed, and the potential emigrants are collected, while in the second, at first, the target island is selected between two island local neighbors in ring randomly and then migration is performed.

4 Experiments and Results

The goal of this study can be condensed into three hypotheses, whose evidence can be shown that gPVaDE: (1) is capable of solving the global optimization problems introduced by the CEC-18 benchmark function suite, (2) can improve the behavior of classical linear population size reduction, and (3) achieves the results comparable with the results of some traditional EAs. During the experiments, the gPVaDE applied the parameter setup as illustrated in Table 1. Let us emphasize that the initial number of individuals resident on the island depends on the number of islands (normally $100/n$, where 100 is the maximum initial population size). On the other hand, all configurations use the same number of fitness function evaluations that serves also as a termination condition for the proposed algorithm, and $F = 0.5$ and $CR = 0.9$.

Algorithm 1. The EVOLVE algorithm.

```
procedure EVOLVE(island)
    emigrants = ∅;
    island.CALCLIFETIMES;                                   ▷ update aging data
    M = island.GENERATETOPOLOGY;                ▷ set new topology for individuals
    Δ_max = island.EVOLUTIONENTROPY;          ▷ calculate the population dynamics
    for all individual ∈ island do
        individual.DECISION_MAKING(island);              ▷ decision-making process
        individual.DO_ACTION(island);                      ▷ execution of actions
        emigrant = emigrant ∪ COLLECTEMIGRANTS(island);           ▷ tag migration
    end for
    while emigrant ≠ ∅ do                       ▷ performing migration actions
        emg_item = HEAD(emigrant);
        emigrant = TAIL(emigrant);
        target_island = RAND(predecessor, successor);
        DO_MIGRATE(target_island, island,emg_item);             ▷ launch migration
    end while
end procedure
```

Table 1. Parameter setup of the proposed gPVaDE.

Parameter	Abbreviation	Value/Interval
Dimension of the problem	D	10
Island size	Np	$[10, 100]$
Number of islands	n	$[1, 10]$
Number of fitness function evaluations	$nFEs$	$1,000 \times D$
Probability of migration	p_m	0.001
Life time	*lifetime*	$[1, 24]$

The results obtained by the algorithms were evaluated according to five standard statistical measures: *Best, Worst, Mean, Median,* and *StDev* values. Friedman's non-parametric statistical test [6] was conducted in order to estimate the quality of the results obtained by various nature-inspired algorithms for global optimization. This test is a two-way analysis of variances by ranks, where the null hypothesis is stated assuming that medians between the ranks of all algorithms are equal. The second step is performed only if a null hypothesis of a Friedman test is rejected. In this step, the post-hoc tests are conducted using the calculated ranks. Indeed, a Wilcoxon two paired non-parametric test was applied in our study as a post-hoc test after determining the control method (i.e., the algorithm with the lowest rank) by using the Friedman test, while the Nemenyi post-hoc test is used for graphical presentation of the results. Both post-hoc tests were conducted using a significance level of 0.05.

The CEC'18 test suite consists of 30 benchmark functions that are divided into four classes: (1) unimodal functions (1–3), (2) simple multimodal functions

(4–10), (3) hybrid functions (11–20), and (4) composition functions (21–30). Unimodal functions have a single global optimum and no local optima. Unimodal functions in this suite are non-separable and rotated. Multi-modal functions are either separable or non-separable. In addition, they are also rotated and/or shifted. To develop the hybrid functions, the variables are divided randomly into some sub-components and then different basic functions are used for different sub-components. Composition functions consist of a sum of two or more basic functions. In this suite, hybrid functions are used as the basic functions to construct composition functions. The characteristics of these hybrid and composition functions depend on the characteristics of the basic functions. The functions of dimensions D = 10 were used in our experiments due to a limitation of the paper length, while the search range of the problem variables was limited to $x_{i,j} \in [-100, 100]$.

4.1 Results

Influence of the Adaptive Population Growth. The purpose of the experiment was to establish a behavior of the adaptive population growth introducing the NLSR mechanism, and to compare this with the famous Linear population Size Reduction (LSR) used in many popular stochastic nature-inspired population-based algorithms. Actually, while the LSR is capable of uniform decreasing of population size with maturing the search process, the population size can also be increased by the NLSR depending on the number of positive individual's variations.

In order to show the advantage of NLSR, the six configurations of gPVaDE with the number of islands varying from one to six (denoted as gPVaDE-c1 to gPVaDE-c6) using the LSR, were compared with their counterparts using the NLSR. The results of the tests are depicted in Table 2, where the achievements obtained by gPVaDE using different features are compared in the sense of the Wilcoxon 2-paired non-parametric statistical test. From the table, it can be seen that, in summary, the gPVaDE using NLSR were significantly better than their counterparts using LSR even tree times, i.e., in most, especially, if we assume that monolithic population is not relevant.

In the next test, the behavior of the NLSR feature was established, where multimodal function f_4 of dimension $D = 10$ was taken into consideration. The function was optimized by the proposed gPVaDE algorithm with three different

Table 2. Comparison between two different models of population size reduction in various configuration of gPVaDE ($D = 10$).

Method	gPVaDE						Summary
	c1	c2	c3	c4	c5	c6	
LSR	+	−	−	−	+	=	2
NLSR	−	+	+	−	+	=	3

gEMAS configurations, i.e., using one (gPVaDE-c1), using two (gPVaDE-c1), and using three islands (gPVaDE-c3). Actually, one typical run of the function optimization was selected for each algorithm in question, where no optimal solution was found. In line with this, the behavior of these can be monitored during the whole run. The result of the test is illustrated in Fig. 1 that is divided into six diagrams. As can be seen from Fig. 1, the population size oscillates around the reference line representing the traditional LSR. Typically, the population size increasing is followed by size reduction. The increasing is launched either by the decision-making process imposing by the reproduction, while reducing the population size is caused by the aging mechanism. Sometimes, this mechanism can introduce such a high selection pressure that can eliminate the major part of individuals from the island in only one cycle. Indeed, imposing the reproduction in the next cycle causes replacing the vacant places in the island. On the other hand, the diversity of island can be lost, when its size becomes lower. In this case, the new genetic material can be supplied using action rebirth.

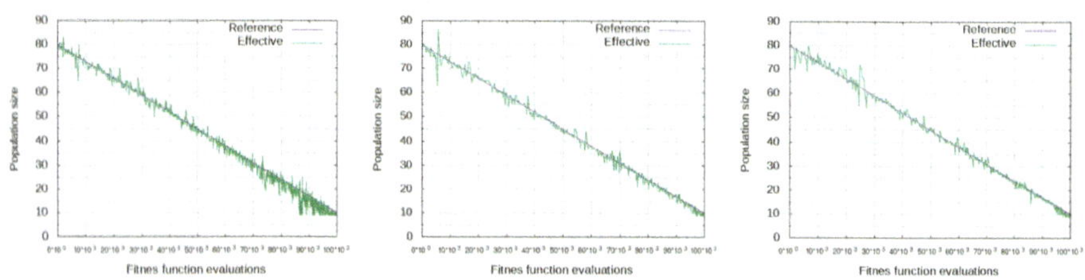

a: gPVaDE-c1 - island 0. b: gPVaDE-c2 - island 1. c: gPVaDE-c2 - island 2.

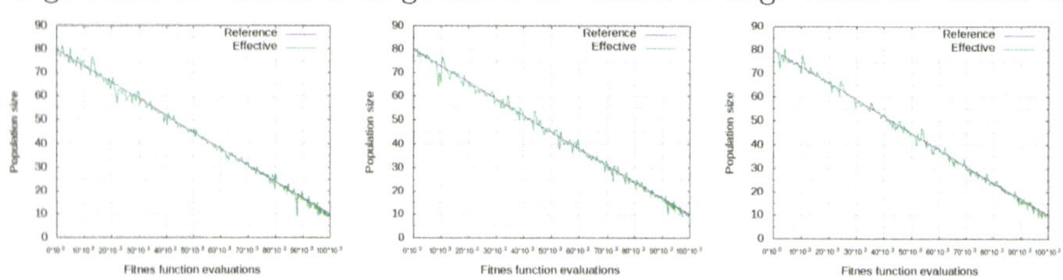

d: gPVaDE-c3 - island 1. e: gPVaDE-c3 - island 2. f: gPVaDE-c3 - island 3.

Fig. 1. Influence of the adaptive population growth on function f_4 of dimension $D = 10$.

Comparative Analysis. The goal of this test was to show that the results of the proposed gPVaDE algorithms are comparable with the results of the traditional stochastic nature-inspired population-based algorithms, like DE, jDE, and SaDE, although these do not achieve those obtained by the state-of-the-art algorithms, like jSO and LShade, at the moment. Thus, even nine configurations of gPVaDE algorithms were taken into consideration with varying the number of islands from one to nine, denoted as gPVaDE-c1 to gPVaDE-c9.

Algorithm	Fri.	Nemenyi		Wilcoxon	
		CD	S.	p-value	S.
gPVaDE-c1	4.83	[4.02,5.64]	†	> 0.05	†
gPVaDE-c2	4.26	[3.45,5.07]	†	> 0.05	†
gPVaDE-c3	3.77	[2.96,4.58]	†	> 0.05	†
gPVaDE-c4	4.42	[3.61,5.23]	†	> 0.05	†
gPVaDE-c5	4.69	[3.88,5.5]	†	> 0.05	†
gPVaDE-c6	5.42	[4.61,6.23]	†	> 0.05	†
gPVaDE-c7	5.43	[4.62,6.24]	†	$\gg 0.05$	†
gPVaDE-c8	6.61	[5.80,7.42]	†	$\gg 0.05$	†
gPVaDE-c9	6.71	[5.90,7.52]	†	$\gg 0.05$	†
DE	3.78	[2.97,4.59]	†	> 0.05	†
jDE	4.72	[3.91,5.53]	†	> 0.05	†
SaDE	4.15	[3.34,4.96]	†	> 0.05	†
jSO	1.12	[0.31,1.93]		$\ll 0.05$	
LShade	1.00	[0.19,1.81]	‡	∞	‡

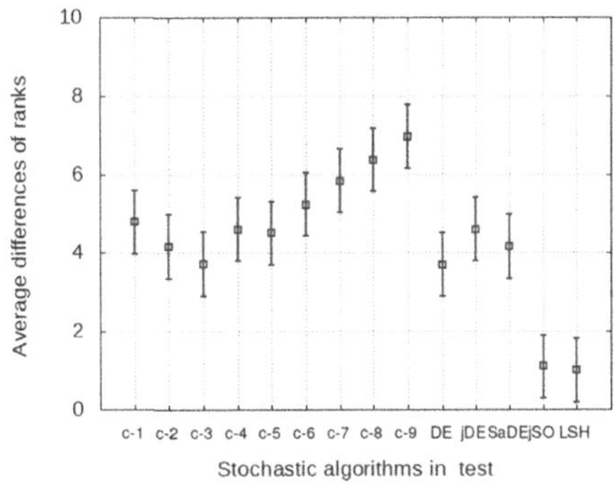

a: $D = 10$, $p_m = 0.001$.

Fig. 2. The results of comparative analysis using Nemenyi post-hoc statistical test.

The results obtained by the particular algorithms were compared using the Friedman non-parametric statistical tests, and refined by a Nemenyi and Wilxocon post-hoc statistical tests. These are depicted in Fig. 2 that is divided into two parts, where the first presents the results in numerical and second in graphical form. As can be seen from the figure, the result quality of the proposed gPVaDE algorithms depends on the number of islands. It turns out that the smaller number of islands is better than the higher. However, the gPVaDE using monolithic population is distinguished as not the preferable configuration.

In summary, the more islands in an algorithm demand more small-sized populations. This is very inefficient for the search process due to suffering a lack of population diversity. On the other hand, the algorithm using a monolithic population maintains the higher population diversity, but suffers a lack of selection pressure. As a result, the proper bias between the population diversity and selection pressure ensure the optimal results for the configuration. In our case, the reasonable number of agents must be higher than or equal to two, but lower than or equal than six.

5 Conclusion

The question how to maintain a diversity of population has been followed researchers in evolutionary community from the same beginning. In the study, this problem is addressed by introducing the parallel gPVaDE algorithm, where the individuals are capable of executing some actions, with which they modify their programming environment and thus have more autonomy as those in the classical EAs. Additionally, the algorithm incorporates two mechanisms: aging, and adaptive population growth. While the former controls the action death, the latter navigates between actions reproduction and clone implemented as different DE mutation strategies.

The proposed gPVaDE was applied for solving the CEC'18 benchmark function suite representing a test-bed for global optimization. During huge experimental work, the influence of the new adaptive population growth mechanism was discovered in detail. Finally, the results produced by various gPVaDE configurations were compared with some traditional stochastic nature-inspired population-based algorithms, like DE, jDE, and SaDE, and some state-of-the-art algorithms, like jSO, and L-Shade. From the comparative analysis, it can be seen that the results of the proposed gPVaDE are comparable with those of the traditional algorithms in question, while needing some improvements to become more competitive also with the state-of-the-art ones.

In the future, we would like to build the gPVaDE with pure agents capable of decision-making at the individual's level. Applying the algorithm for solving the other hard optimization problems seem to be a very challenging for the future.

Acknowledgment. Iztok Fister thanks the financial support from the Slovenian Research Agency (Research Core Funding No. P2-0042 - Digital twin). Iztok Fister Jr. thanks the financial support from the Slovenian Research Agency (Research Core Funding No. P2-0057). Andres Iglesias and Akemi Galvez thank the Computer Science National Program of the Spanish Research Agency and European Funds, Project #TIN2017-89275-R. (AEI/FEDER, UE), and the PDE-GIR project of the European Union's Horizon 2020 programme, Marie Sklodowska-Curie Actions grant agreement #778035. Dušan Fister thanks the financial support from the Slovenian Research Agency (Research Core Funding No. P5-0027).

References

1. Brest, J., Maučec, M.S., Bošković, B.: Single objective real-parameter optimization: algorithm jSO. In: 2017 IEEE Congress on Evolutionary Computation (CEC), pp. 1311–1318, June 2017. https://doi.org/10.1109/CEC.2017.7969456
2. Brest, J., Greiner, S., Bošković, B., Mernik, M., Žumer, V.: Self-adapting control parameters in differential evolution: a comparative study on numerical benchmark problems. IEEE Trans. Evol. Comput. **10**(6), 646–657 (2006). https://doi.org/10.1109/TEVC.2006.872133
3. Byrski, A., Drezewski, R., Siwik, L., Kisiel-Dorohinicki, M.: Evolutionary multi-agent systems. Knowl. Eng. Rev. **30**(2), 171–186 (2015). https://doi.org/10.1017/S0269888914000289
4. Demetrius, L., Legendre, S., Harremöes, P.: Evolutionary entropy: a predictor of body size, metabolic rate and maximal life span. Bull. Math. Biol. **71**(4), 800–818 (2009). https://doi.org/10.1007/s11538-008-9382-6
5. Fister, I., Iglesias, A., Galvez, A., Del Ser, J., Osaba, E., Fister Jr., I., Perc, M., Slavinec, M.: Novelty search for global optimization. Appl. Math. Comput. **347**, 865–881 (2019)
6. Friedman, M.: A comparison of alternative tests of significance for the problem of m rankings. Ann. Math. Statist. **11**(1), 86–92 (1940). https://doi.org/10.1214/aoms/1177731944

7. Luque, G., Alba, E.: Parallel Genetic Algorithms: Theory and Real World Applications. Springer Publishing Company, Incorporated, New York (2013)
8. Lynch, M.: The evolution of genetic networks by non-adaptive processes. Nat. Rev. Genet. **8**, 803–813 (2007). https://doi.org/10.1038/nrg2192
9. Michalewicz, Z.: Genetic Algorithms + Data Structures = Evolution Programs, 2nd edn. Springer Verlag, Berlin (1996)
10. Qin, A.K., Suganthan, P.N.: Self-adaptive differential evolution algorithm for numerical optimization. In: 2005 IEEE Congress on Evolutionary Computation, vol. 2, pp. 1785–1791, September 2005. https://doi.org/10.1109/CEC.2005.1554904
11. Storn, R., Price, K.: Differential evolution–a simple and efficient heuristic for global optimization over continuous spaces. J. Global Optim. **11**(4), 341–359 (1997). https://doi.org/10.1023/A:1008202821328
12. Tanabe, R., Fukunaga, A.S.: Improving the search performance of shade using linear population size reduction. In: 2014 IEEE Congress on Evolutionary Computation (CEC), pp. 1658–1665, July 2014. https://doi.org/10.1109/CEC.2014.6900380
13. Črepinšek, M., Liu, S.H., Mernik, M.: Exploration and exploitation in evolutionary algorithms: a survey. ACM Comput. Surv. **45**(3), 35:1–35:33 (2013). https://doi.org/10.1145/2480741.2480752

A Preliminary Many Objective Approach for Extracting Fuzzy Emerging Patterns

Angel Miguel Garcia-Vico, Cristobal J. Carmona(✉), Pedro Gonzalez, and Maria Jose del Jesus

Interuniversity Andalusian Institute on Data Science and Computation Intelligence, University of Jaén, 23071 Jaén, Spain
{agvico,ccarmona,pglez,mjjesus}@ujaen.es

Abstract. A preliminary many objective algorithm for extracting fuzzy emerging patterns is presented in this contribution. The proposed algorithm employs fuzzy logic together with an evolutionary algorithm. The aim is to expand the complex search space that we have in emerging pattern mining.

The experimental study presented in this paper faces this new proposal regarding an ensemble of one of the most used algorithms within supervised descriptive rule discovery. Results presents a set of patterns with a major interpretability and precision for the new proposal which could be interesting for experts in real-world applications.

Keywords: Many objective evolutionary algorithm · Emerging pattern mining · Fuzzy patterns

1 Introduction

Emerging pattern mining (EPM) is a data mining task that tries to find discriminative patterns whose support increases significantly from one class, or dataset, to another. EPM is halfway prediction and description because it describes a problem by discovering some relationships on the data by means of a target variable, typically used in classification. In fact, EPM belongs to the supervised descriptive rule discovery framework [5].

The quality of an emerging pattern (EP) can be determined by a wide range of quality measures [17]. In fact, there is no consensus in the literature about the most relevant quality measures to analyse the goodness of a supervised descriptive rule algorithm, but rather the quality is based on three fundamentally axis: interpretability of the sets of extracted patterns, balance between generality and reliability, and interest of the emerging patterns.

In this contribution, we present a preliminary approach for extracting emerging patterns through a many objective algorithm, the ManyObjective-EFEP algorithm. The proposal is based on soft computing techniques, in particular, it is an evolutionary fuzzy system (EFS) [22], an hybridization of fuzzy logic [28] and evolutionary algorithms [21]. The former allows us the obtaining of fuzzy

Á. Herrero et al. (Eds.): SOCO 2020, AISC 1268, pp. 100–110, 2021.
https://doi.org/10.1007/978-3-030-57802-2_10

emerging patterns which facilitate the analysis and understanding by the experts; the latter is an evolutionary algorithm based on NSGA-III [9] that allows us the use of a wide number of quality measures within the evolutionary search process without degrading its performance.

The paper is organized as follows: Sect. 2 presents the main concepts and properties of the EPM. In Sect. 3 the main characteristics of the EFSs are shown. Section 4 presents the ManyObjective-EFEP algorithm. Section 5 presents the experimental study carried out to determine the quality of the proposed method. Finally, the conclusions extracted from this work are depicted in Sect. 6.

2 Emerging Pattern Mining

EPM was defined such as the search for patterns whose support increase significantly from one dataset (D_1) to another (D_2) [20]. Specifically, D_1 contains examples for one class and D_2 examples for the remaining classes. A pattern is emerging if the growth rate (GR) is greater than a threshold $\rho > 1$ and it is defined as:

$$GR(x) = \begin{cases} 0, & IF\ Supp_{D_1}(x) = Supp_{D_2}(x) = 0, \\ \infty, & IF\ Supp_{D_1}(x) \neq 0 \wedge Supp_{D_2}(x) = 0, \\ \frac{Supp_{D_1}(x)}{Supp_{D_2}(x)}, & another\ case \end{cases} \quad (1)$$

EPs are usually represented by means of conjunctions of attribute-value pairs, or attribute-value pairs in disjunctive normal form (DNF), which represents the discriminative characteristics they want to describe. For the determination of D_1 and D_2, these patterns are usually labeled with the class or the dataset they try to describe. Generally, these patterns can be represented as rules in the following form [5]:

$$P : Cond \rightarrow Class \quad (2)$$

where *Cond* represents the condition of the pattern and *Class* is the value of the class.

The analysis of the descriptive behaviour of a pattern is key in EPM. For this purpose, a contingency table is usually calculated. In this contingency table, the number of examples covered or not covered by the patterns which belong or do not belong to the class of the pattern is calculated. An example is shown in Table 1.

By means of this table, several quality measures can be used from the EPM for the determination of a wide range of aspects. The most widely used quality measures in EPM are outlined in Table 2 [20].

Table 1. Contingency table of a pattern.

	Class	No class
Covered	p	n
Not covered	$\overline{p}$	$\overline{n}$
	P	N

Table 2. Quality measures used in EPM for the determination of the quality of a pattern.

Name	Abbreviation	Formula
Number of patterns	nP	–
Number of variables	nV	–
Confidence [13]	Conf	$\frac{p}{p+n}$
Weighted Relative Accuracy [5]	WRAcc	$\frac{p+n}{P+N}\left(\frac{p}{p+n} - \frac{P}{P+N}\right)$
Growth Rate [11]	GR	$\frac{p \cdot N}{P \cdot n}$
True Positive Rate [25]	TPR	$\frac{p}{P}$
False Positive Rate [16]	FPR	$\frac{n}{N}$

3 Evolutionary Fuzzy Systems for Extracting Emerging Patterns

A fuzzy system [28] augmented with a learning process based on evolutionary algorithms [12] is defined as evolutionary fuzzy systems (EFSs) as can be observed in [22]. In this definition two concepts are presented: fuzzy systems and evolutionary algorithms. The former are usually considered in the form of fuzzy-rule based systems (FRBSs), which are composed of "IF-THEN" rules where both the antecedent and consequent can contain fuzzy logic statements. Fuzzy systems are based on fuzzy logic [28], which already allow us to consider uncertainty, and also to represent the continuous variables in a manner which is close to human reasoning. In this way interpretable fuzzy rules consider continuous variables as linguistic ones, where values are represented through fuzzy linguistic labels (LLs) in fuzzy sets [24]. These fuzzy sets facilitate the application to real-world problems because the representation of continuous variables is very close to human reasoning, e.g. a variable such as *Age* could be represented with three linguistic labels such as *Small*, *Normal* and *Tall* making it possible to achieve better analysis.

On the other hand, evolutionary algorithms are stochastic algorithms for optimizing and searching. These algorithms were introduced by Holland [23]. Different computational models can be found within these types of algorithms such as genetic algorithms [21,23], evolution strategies [27], evolutionary programming [15] and genetic programming [26], amongst others. The evolutionary algorithms imitate the principles of natural evolution to address optimization and learning problems. They are well suited to perform the EPM task due to their ability to reflect the interaction of variables in a rule-learning process also providing great flexibility in the representation [14].

EPM is a supervised descriptive rule discovery task that can be seen as an approximation problem in which the objective is the learning of the parameters of the model. In this task, the search space can be very complex and the search strategy used becomes a key factor. The use of evolutionary fuzzy systems is very well suited to this task because these types of algorithms perform a global search in the space in a suitable way, as can be observed in the real-world problems solved in the literature. For example, in Bioinformatics [1,7], Medicine [4], E-commerce [6] or Industry [2], amongst others.

4 ManyObjective-EFEP: ManyObjective Evolutionary Algorithm for Extracting Fuzzy Emerging Patterns

Throughout the literature, a wide number of quality measures have been presented both to guide the search process in order to find the best EPs and to measure the quality of these patterns, as can be observed in [17,18]. In fact, as we have presented in our previous review [20], the main purpose of an EPM algorithm is to find a good trade-off between generality, reliability and interest. This could lead us to employ a wide number of quality measures in the search process.

The main proposal of the ManyObjective-EFEP algorithm is to extract emerging fuzzy and/or crisp patterns, depending on the type of variables the problem contains, with a good trade-off between reliability and descriptive capacity through the use of a wide number of objectives in the evolutionary process. Specifically, this algorithm is based on the NSGA-III algorithm [9] where the main difference with respect to NSGA-II is that former uses a set of reference points to maintain the diversity of the Pareto points during the search. This results in a very even distribution of Pareto points across the objective space, even when the number of objectives is large.

ManyObjective-EFEP uses a "chromosome = rule" approach where only the antecedent is represented. In this way, an execution for each value of the class is performed in order to extract knowledge for all the classes. The algorithm is able to extract patterns following a DNF representation because it is the best one for the extraction of descriptive EPs [19]. DNF patterns are codified by means of a bit-vector genotype whose length is equal to the total number of features. The number of features is determined by the number of possible categories for nominal variables, while for numeric variables it is the number of LLs used. A fuzzy emerging pattern and its representation can be observed in Fig. 1. Note that the class must be fixed for a value beforehand. Therefore, it is necessary to execute the algorithm for each value of the class.

$$\begin{array}{c} Genotype \\ \left|\begin{array}{c} X_1 \\ 1\ \emptyset\ 1 \end{array}\right|\left|\begin{array}{c} X_2 \\ 1\ 1\ 1 \end{array}\right|\left|\begin{array}{c} X_3 \\ 1\ \emptyset\ \emptyset\ \emptyset \end{array}\right|\left|\begin{array}{c} X_4 \\ \emptyset\ \emptyset\ \emptyset \end{array}\right| \\ \Downarrow \end{array}$$

$$Phenotype\ IF(X_1 = (Low \vee High)) \wedge (X_3 = Arts)\ THEN\ (Class = Positive)$$

Fig. 1. Representation of a fuzzy DNF pattern with continuous and categorical variables in ManyObjective-EFEP.

In the final stage, the algorithm obtains a set of patterns for each value of the class where the repeated patterns are deleted. The operating scheme of Manyobjective-EFEP algorithms can be seen in Fig. 2.

```
BEGIN
    Create P_0 and reference points
    REPEAT
        Q_t ← ∅
        Generate (Q_t) through genetic operators on P_t
        R_t ← Join(P_t,Q_t)
        Non-dominationed-sort(R_t) based on five objectives
        Associate with reference points
        Apply niche preservation and save in P_{t+1}
        t ← t + 1
    WHILE (num-eval < Max-eval)
    RETURN F_1 without repeated
END
```

Fig. 2. The ManyObjective-EFEP algorithm.

5 Experimental Study

This section presents a summary about the experimental framework in Sect. 5.1, results of the experimental study and a complete analysis of the results are outlined in Sect. 5.2.

5.1 Experimental Framework

The experimental framework used for the evaluation of ManyObjective-EFEP is presented below:

- Algorithms and parameters. The ManyObjective-EFEP algorithm is compared in this paper with an adaptation of the well-known NSGA-II algorithm

Table 3. Algorithms and their parameters used in this experimental study.

Parameters
Population length = 51
Number of labels = 3
Number of evaluations = 10000
Crossover probability = 0.6
Mutation probability = 0.1
Objectives = TPR, FPR, WRAcc, Conf, Strength

[8]. Both algorithms are presented in the jMetal framework[1]. The parameters chosen for both algorithms are identical in order to perform a fair comparison, and they are summarized in Table 3.

- Quality measures in the search process. The main difference between both algorithms is considered with respect to the search process of the evolutionary algorithm. Specifically, for the NSGA-II algorithm we employ an ensemble of algorithms based on the seven possible combinations of the objectives considered in Table 3. In this way, we obtain seven versions of the NSGA-II where all extracted rules for each version are joined and repeated rules are deleted. On the other hand, the ManyObjective-EFEP is executed only once with the five objectives.
- Datasets. The study with datasets from the UCI repository [10] were employed for comparing the quality of the proposed method. They are presented in Table 4. For each data set, it is shown its name and its number of instances, attributes (the number of Real/Integer/Nominal attributes in the data) and classes (number of possible values of the output variable). In addition, the table shows if the corresponding data set has missing values or not (for data sets with missing values the table shows the number of instances without missing values, and the total number of instances between brackets).
- Experiment evaluation. As EPM tries to describe the underlying phenomena in data, an evaluation becomes necessary of the patterns extracted using unseen data. Therefore, this experimental study follows a five-fold stratified cross-validation schema in order to avoid as much as possible bias when creating the training-test partitions.
- Analysis of the quality. The quality measures analyzed in this study were presented in Table 2. These measures are key for the determination of the quality of the patterns extracted regarding the different aspects of EPM. In addition, the number of patterns (nP) and the average number of variables (nV) are analysed in order to determine the model complexity. It is important to remark that the value shown for GR represents the percentage of patterns whose GR in test is greater than one. This is because the domain of GR is $[0, \infty]$, so the average cannot be computed properly.

[1] http://jmetal.github.io/jMetal/.

Table 4. Datasets employed in this experimental study.

Name	♯ Attributes	(R/I/N)	♯ Examples	♯ Classes
appendicitis	7	(7/0/0)	106	2
Australian	14	(3/5/6)	690	2
automobile	25	(15/0/10)	150 (205)	6
bands	19	(13/6/0)	365 (539)	2
breast	9	(0/0/9)	277 (286)	2
car	6	(0/0/6)	1728	4
chess	36	(0/0/36)	3196	2
cleveland	13	(13/0/0)	297 (303)	5
coil2000	85	(0/85/0)	9822	2
contraceptive	9	(0/9/0)	1473	3
crx	15	(3/3/9)	653 (690)	2
dermatology	34	(0/34/0)	358 (366)	6
flare	11	(0/0/11)	1066	6
German	20	(0/7/13)	1000	2
glass	9	(9/0/0)	214	7
heart	13	(1/12/0)	270	2
hepatitis	19	(2/17/0)	80 (155)	2
housevotes	16	(0/0/16)	232 (435)	2
led7digit	7	(7/0/0)	500	10
letter	16	(0/16/0)	20000	26
lymphography	18	(0/3/15)	148	4
magic	10	(10/0/0)	19020	2
mammographic	5	(0/5/0)	830 (961)	2
marketing	13	(0/13/0)	6876 (8993)	9
monk2	6	(0/6/0)	432	2
nursery	8	(0/0/8)	12690	5
pageBlocks	10	(4/6/0)	5472	5
penbased	16	(0/16/0)	10992	10
pima	8	(8/0/0)	768	2
post-operative	8	(0/0/8)	87 (90)	3
ring	20	(20/0/0)	7400	2
saheart	9	(5/3/1)	462	2
satimage	36	(0/36/0)	6435	7
segment	19	(19/0/0)	2310	7
shuttle	9	(0/9/0)	58000	7
thyroid	21	(6/15/0)	7200	3
tictactoe	9	(0/0/9)	958	2
twonorm	20	(20/0/0)	7400	2
vehicle	18	(0/18/0)	846	4
vowel	13	(10/3/0)	990	11
wine	13	(13/0/0)	178	3
winequalityRed	11	(11/0/0)	1599	11
winequalityWhite	11	(11/0/0)	489	8
wisconsin	9	(0/9/0)	683 (699)	2
yeast	8	(8/0/0)	1484	10
zoo	16	(0/0/16)	101	7

5.2 Analysis of the Results Obtained

Due to the extension of the results obtained in this experimental study, the complete results are presented in a website[2]. In addition, the average results of the study are presented in Table 5.

Table 5. Average results extracted from the NSGA-II ensemble and ManyObjective-EFEP methods.

Algorithm	nP	nV	$WRACC$	$CONF$	GR	TPR	FPR
NSGA-II *Ensemble*	157.74	9.65	**0.538**	0.286	**0.426**	**0.270**	0.116
ManyObjective-EFEP	**45.65**	**9.91**	0.523	**0.380**	**0.426**	0.091	**0.015**

The results are analysed based on the three important axis for the supervised descriptive rule discovery tasks [3]:

- *Interpretability*: The ensemble of the different versions of NSGA-II algorithm obtains an elevated number of patterns, three times upper than the algorithm presented in this contribution. Throughout the literature, we are aware about the complexity to incorporate more than three objectives within the evolutionary process because the number of patterns grows very high. In this way, the new approximation keeps a number of pattern more reduced that is more relevant within supervised descriptive rule discovery. On the other hand, there is no difference in the number of variables of the patterns extracted with values very similar. However, it is important to note a high complexity in the knowledge extracted where expert would need to analyse results with a high number of rules and variables which would complicate the understanding of the problem.
- *Tradeoff between generality and reliability*: The generality is measured through the TPR where the percentage of examples covered for the class are calculated. In this way, the algorithm NSGA-II *Ensemble* obtains a value more interesting with a value three times upper than the ManyObjective-EFEP algorithm. However, the reliability of the patterns extracted is far below. In fact, the ratio between TPR and FPR (false positive rate) in the ManyObjective-EFEP algorithm is about six times upper so more precise patterns are extracted for this algorithm. Therefore, the values of confidence are higher in this one. On the other hand, the value in the GR is similar in both algorithms, i.e., the percentage of fuzzy emerging patterns is similar in both algorithms.
- *Interest*: This concept within supervised descriptive rule discovery is calculated through the $WRACC$ quality measure which is key as can be observed in [5]. The interest values obtained by the NSGA-II *Ensemble* algorithm in

[2] https://simidat.ujaen.es/papers/ManyObjectiveEFEP/.

this experimental study are very close to those obtained by ManyObjective-EFEP. This value is determined by the coverage of the rule that, as we have seen previously, is superior in the first algorithm.

6 Conclusions

This contribution presents a first approximation of a many objective algorithm for extracting fuzzy emerging patterns. The ManyObjective-EFEP algorithm combines soft-computing techniques such as fuzzy logic and the NSGA-III evolutionary algorithm. The complexity of the search process with the inclusion of a wide number of objectives in the evolutionary process is analysed in this study, where good results in reliability with interest are obtained but with a low values in generality. However, it is interesting to see how the number of patterns is reduced with respect to an ensemble approach.

As future work, we will study and continue with the analysis of the use of many objective evolutionary algorithms for EPM, because it is a complex space, and the tradeoff among a wide number of quality measures is desired.

Acknowledgement. This study was funded by the FPI 2016 Scholarship reference BES-2016-077738 (FEDER Founds).

References

1. Carmona, C.J., Chrysostomou, C., Seker, H., del Jesus, M.J.: Fuzzy rules for describing subgroups from influenza a virus using a multi-objective evolutionary algorithm. Appl. Soft Comput. **13**(8), 3439–3448 (2013)
2. Carmona, C.J., González, P., García-Domingo, B., del Jesus, M.J., Aguilera, J.: MEFES: an evolutionary proposal for the detection of exceptions in subgroup discovery. An application to concentrating photovoltaic technology. Knowl.-Based Syst. **54**, 73–85 (2013)
3. Carmona, C.J., González, P., del Jesus, M.J., Herrera, F.: Overview on evolutionary subgroup discovery: analysis of the suitability and potential of the search performed by evolutionary algorithms. WIREs Data Min. Knowl. Disc. **4**(2), 87–103 (2014)
4. Carmona, C.J., González, P., del Jesus, M.J., Navío, M., Jiménez, L.: Evolutionary fuzzy rule extraction for subgroup discovery in a psychiatric emergency department. Soft Comput. **15**(12), 2435–2448 (2011)
5. Carmona, C.J., del Jesus, M.J., Herrera, F.: A unifying analysis for the supervised descriptive rule discovery via the weighted relative accuracy. Knowl.-Based Syst. **139**, 89–100 (2018)
6. Carmona, C.J., Ramírez-Gallego, S., Torres, F., Bernal, E., del Jesus, M.J., García, S.: Web usage mining to improve the design of an e-commerce website: OrOliveSur.com. Expert Syst. Appl. **39**, 11243–11249 (2012)
7. Carmona, C.J., Ruiz-Rodado, V., del Jesus, M.J., Weber, A., Grootveld, M., González, P., Elizondo, D.: A fuzzy genetic programming-based algorithm for subgroup discovery and the application to one problem of pathogenesis of acute sore throat conditions in humans. Inf. Sci. **298**, 180–197 (2015)

8. Deb, K.: Multi-Objective Optimization using Evolutionary Algorithms. John Willey & Sons, Hoboken (2001)
9. Deb, K., Jain, H.: An evolutionary many-objective optimization algorithm using reference-point-based nondominated sorting approach, part I: solving problems with box constraints. IEEE Trans. Evol. Comput. **18**(4), 577–601 (2014)
10. Dheeru, D., Karra Taniskidou, E.: UCI machine learning repository (2017). http://archive.ics.uci.edu/ml
11. Dong, G., Li, J.: Efficient mining of emerging patterns: discovering trends and differences. In: Proceedings of the Fifth ACM SIGKDD International Conference on Knowledge Discovery and Data Mining, New York, NY, USA, pp. 43–52. ACM (1999)
12. Eiben, A.E., Smith, J.E.: Introduction to Evolutionary Computation. Springer, Berlin (2003)
13. Fayyad, U.M., Piatetsky-Shapiro, G., Smyth, P.: From data mining to knowledge discovery: an overview. In: Advances in knowledge discovery and data mining, AAAI/MIT Press, Menlo Park, CA, USA, pp. 1–34 (1996)
14. Fernández, A., García, S., Luengo, J., Bernadó-Mansilla, E., Herrera, F.: Genetics-based machine learning for rule induction: state of the art, taxonomy, and comparative study. IEEE Trans. Evol. Comput. **14**(6), 913–941 (2010)
15. Fogel, D.B.: Evolutionary Computation - Toward a New Philosophy of Machine Intelligence. IEEE Press, New York (1995)
16. Gamberger, D., Lavrac, N.: Expert-guided subgroup discovery: methodology and application. J. Artif. Intell. Res. **17**, 501–527 (2002)
17. García-Borroto, M., Loyola-Gonzalez, O., Martínez-Trinidad, J.F., Carrasco-Ochoa, J.A.: Comparing Quality Measures for Contrast Pattern Classifiers, pp. 311–318. Springer, Berlin Heidelberg (2013)
18. García-Borroto, M., Loyola-González, O., Martínez-Trinidad, J.F., Carrasco-Ochoa, J.A.: Evaluation of quality measures for contrast patterns by using unseen objects. Expert Syst. Appl. **83**, 104–113 (2017)
19. García-Vico, A.M., Carmona, C.J., González, P., del Jesus, M.J.: MOEA-EFEP: multi-objective evolutionary algorithm for extracting fuzzy emerging patterns. IEEE Trans. Fuzzy Syst. **26**(5), 2861–2872 (2018)
20. García-Vico, A.M., Carmona, C.J., Martín, D., García-Borroto, M., del Jesus, M.J.: An overview of emerging pattern mining in supervised descriptive rule discovery: taxonomy, empirical study, trends and prospects. WIREs: Data Min. Knowl. Disc. **8**(1), e1231 (2018)
21. Goldberg, D.E.: Genetic Algorithms in search, optimization and machine learning. Addison-Wesley Longman Publishing Co., Inc. (1989)
22. Herrera, F.: Genetic fuzzy systems: taxomony, current research trends and prospects. Evol. Intell. **1**, 27–46 (2008)
23. Holland, J.H.: Adaptation in Natural and Artificial Systems, 2nd edn. University of Michigan Press, Ann Arbor (1975)
24. Hüllermeier, E.: Fuzzy sets in machine learning and data mining. Appl. Soft Comput. **11**(2), 1493–1505 (2011)
25. Kloesgen, W.: Explora: a multipattern and multistrategy discovery assistant. Advances in Knowledge Discovery and Data Mining, pp. 249–271. American Association for Artificial Intelligence, Menlo Park, CA, USA (1996)
26. Koza, J.R.: Genetic Programming: On the Programming of Computers by Means of Natural Selection. MIT Press, Cambridge (1992)

27. Schwefel, H.P.: Evolution and Optimum Seeking. Sixth-generation Computer Technology Series, Wiley (1995)
28. Zadeh, L.A.: The concept of a linguistic variable and its applications to approximate reasoning. Parts I, II, III. Inf. Sci. **8-9**, 43–80, 199–249, 301–357 (1975)

Artificial Neural Networks

A Smart Crutch Tip for Monitoring the Activities of Daily Living Based on a Novel Neural-Network Intelligent Classifier

Asier Brull[1(✉)], Asier Zubizarreta[1], Itziar Cabanes[1], Jon Torres-Unda[2], and Ana Rodriguez-Larrad[2]

[1] Faculty of Engineering of Bilbao, University of the Basque Country UPV/EHU, Bilbao, Spain
asier.brull@ehu.eus
[2] Department of Physiology, Faculty of Medicine and Nursing, University of the Basque Country UPV/EHU, Bilbao, Spain

Abstract. The determination of the objectives of gait rehabilitation therapies is usually based on partial data provided by clinical tests carried out in specific scenarios and the subjective perception of both the therapist and the patient. However, recent studies have shown that individualization is mandatory to maximize the effect of the therapy on the patient. This requires monitoring the Activities of Daily Living of the patient using objective indicators and measurements, which can be achieved using instrumented devices or wearable sensors. In this work, a smart crutch tip is proposed, which integrates a novel neural-network based intelligent Activities of Daily Living classifier. Based on the sensors integrated on the tip, the classifier is able to detect four typical activities (walking, standing still, going up stairs and going down stairs). In order to design the classifier, data from a group of 13 volunteers is used and different network architectures are evaluated in order to consider the most computationally efficient design, obtaining a success rate of 95%.

Keywords: Rehabilitation · Gait monitoring · Artificial neural network · Daily life activities classifier

1 Introduction

Patients suffering from a lower-limb impairment due to a trauma or a neurological disorder have their autonomy and quality of life decreased, which ultimately

Supported by the University of the Basque Country UPV/EHU under grant number PIF18/067 and project number project GIU19/45 (GV/EJ IT1381-19) and by the Ministerio de Ciencia e Innovación (MCI) under grant number DPI2017-82694-R (AEI/FEDER, UE).

Á. Herrero et al. (Eds.): SOCO 2020, AISC 1268, pp. 113–122, 2021.
https://doi.org/10.1007/978-3-030-57802-2_11

results in a lower health status [1,2]. Gait rehabilitation becomes an important tool to avoid this and recover the impaired limb functions.

It has been demonstrated that therapies adapted to the particular state of the patient [3] allow to increase the effectiveness of the rehabilitation process. However, this requires a proper characterization of the functional performance of the patient and its recovery process.

Traditionally, the assessment of patient recovery is carried out using different standardized clinical tests and scales [4,5] and the subjective perception of the therapist. However, these approaches only provide data on specific moments and in controlled clinical environments, neglecting intermediate patient status evolutions which could be of interest. In particular, the characterization of the Activities of Daily Living (ADL) related to the movement of each patient, i.e., their routines, are very interesting, as they can provide an objective measurement for the aforementioned purpose.

Several approaches have been proposed in recent years to perform gait monitoring. The use of wearable sensors, in particular IMU (Inertial Measurement Units) based ones, is one of the most popular ones [6,7]. These sensors are usually small and light, and are attached to different parts of the lower-limbs in order to monitor their motion. Other approaches have considered integrating sensors in assistive devices such as canes or crutches [8,9], allowing to measure not only the motion, but also the interaction force, in a less invasive way.

The monitoring devices provide a wide set of data from their sensors that can be used to identify the different ADL the patient performs during the day. While it is possible to make ADL classifiers by using unprocessed data, these generally require complex algorithms with high computational cost, such as Convolutional Neural Networks [10] or Multi-Layer Perceptron (MLP) neural networks with many hidden layers [11]. Hence, in order to reduce complexity, most approaches first reduce the dimension of the data by defining a set of indicators (usually based on statistical, phase or frequency analysis) [12–14], and then design a classifier based on the set of indicators defined. Traditionally, MLP artificial neural networks are used for this purpose, whose topology varies depending on the set of indicators selected for each case [12,13,15].

In this work a smart crutch tip is presented, which integrates a novel neural-network based intelligent Activities of Daily Living classifier. The smart tip integrates a high precision IMU and a force sensor. A set of indicators are proposed to process these sensor data, and use them as an input to a neural-network based classifier that is able to detect four basic activities (walking, standing still, going up stairs and going down stairs). The success rate of the proposed approach is analyzed to define the most computationally efficient approach.

The rest of the paper is structured as follows. In Sect. 2 the smart tip design, the integrated sensors and its acquisition system are detailed. Section 3 details the dataset used to develop the classifier and the selected indicators. Section 4 shows the procedure followed to design the neural-network based classifier and the obtained results. In addition, an analysis is carried out and an optimized structure is also proposed to develop the classifier. Finally, the most important ideas are summarized in Sect. 5.

2 Smart Tip

In order to monitor gait, a smart tip [16] has been developed which can be attached to any commercial crutch or cane (see Fig. 1). The tip has been manufactured using light aluminum and integrates a series of sensors that allow monitoring both motion and interaction force on the assistive device (crutch or cane).

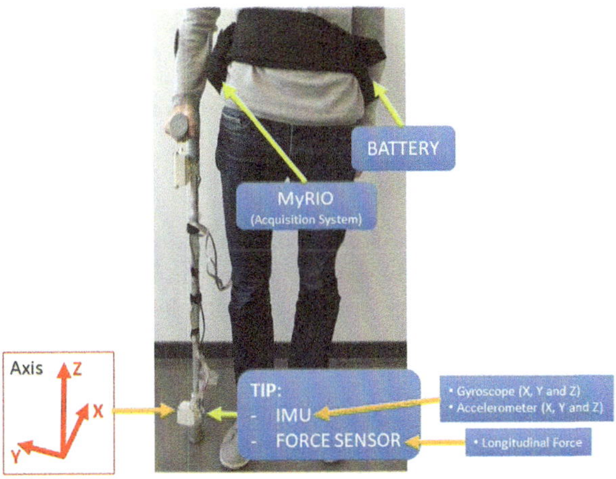

Fig. 1. Crutch, elements that compose it and Reference Axis of the crutch.

The acquisition system and power source of the tip are located externally to reduce the mass of the tip. As seen in Fig. 1 a belt is used to hold both the battery and a National Instrument's myRIO acquisition device.

The latter captures information from the two sensors integrated in the tip: A MPU-6000 IMU (Inertial Measurement Unit), which provides information of the acceleration and angular velocity on the local x, y and z axes; and a HBM C9C piezoelectric force sensor, which measures the load applied by the patient on the assistive device, up to 1200N. The required signal processing electronics are integrated within the tip, while the capture rate is 50 Hz.

3 Database Generation

The development of an ADL classifier requires defining a proper database that considers the different activities to be identified. In the scope of this work, four basic activities are considered: walking, standing still, going up stairs and going down stairs. Next, the procedure used to develop the required database is detailed.

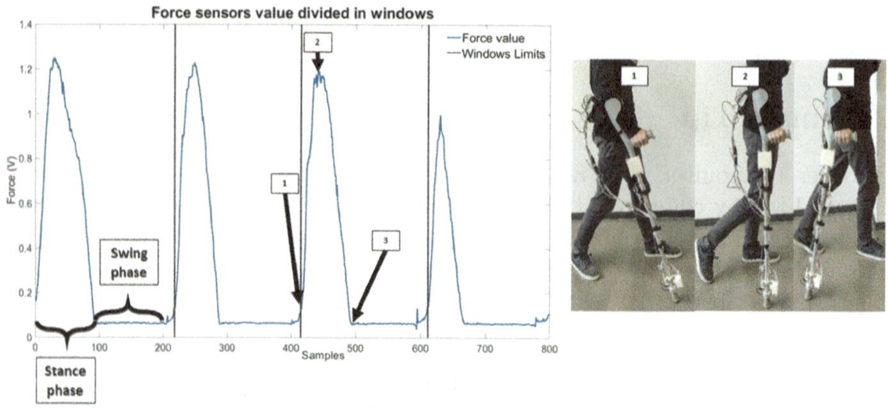

Fig. 2. Phases of use of technical support aid system.

3.1 Experimental Test Definition

Three main tests have been developed to capture data for the aforementioned four scenarios. The first test is based on walking in a straight line for 27 m at a constant, normal speed. The acceleration and deceleration phases are neglected. The second one consists on walking up and down a set of 11 stairs. Finally, the last test requires to stand still for 5 s.

13 individuals (women and men) with heights between 151 cm–187 cm were asked to perform these tests twice using a crutch in which the smart tip was attached. The data was recorded at 50 Hz.

The recorded data was then segmented into windows considering the *crutch cycle*, which can be derived from the force sensor measurement (see Fig. 2). This cycle has two phases: the stance phase (in which the crutch is in contact with the ground) and the swing phase (in which the crutch moves in the air).

Note that in the case of the standing still case, no cycle exist. Hence, a virtual standing still step has been defined when the user of the smart Tip does not apply a force for 3.7 s (the mean of one step).

Finally, each segment was tagged with the identified ADL (walking, standing still, going up stairs and going down stairs). The total number of segments (or cycles) captured is summarized in Table 1.

Table 1. Table of the number of segments per ADL.

ADL	Number of segments
Walking straight	**652**
Being still	**184**
Going up stairs	**170**
Going down stairs	**186**

3.2 Indicator Selection

The main goal of the segmentation process previously detailed is to reduce the dimension of the available data, and associate each *cycle* or segment to a particular ADL. For that purpose, a set of indicators associated to each segment have to be defined.

The selection of the indicator set is not a trivial task, and the effectiveness of the developed classifier can vary significantly depending on it. In this work, based on the preliminary analysis of the literature carried out in [13], 9 indicators are defined: Standard deviation of the gyroscope x axis; 25th, 50th and 75th percentile of the gyroscope x axis; Interquartile range of the gyroscope x axis; Interquartile range of the accelerometer z axis; Acceleration mean in z axis; Percentage of the stance phase in each cycle; and Cycle time.

While the first 7 indicators are of statistical nature, and have been derived from the most used indicators in the literature, the last two are defined specifically for assistive devices such as crutches or canes. The Percentage of the stance phase in each crutch cycle indicates the percentage of time the user of the crutch uses it to support himself. On the other hand, the cycle time is related with the time required to complete each crutch cycle (or window).

For each segment, these 9 indicators are calculated and stored in the database. These will be used to design the neural network classifier.

Ions is calculated. All these statistical data are taken from the windows resulting from the data obtained from the accelerometer and gyroscope. The selection of these statistical data and the selection of the most appropriate data for classification, based on these statistical indicators, has been carried out in a previous study [13].

4 ADL Classifier

Using the aforementioned indicators, a neural network based classifier has been developed to perform ADL classification. A Multi Layer Perceptron (MLP) architecture is selected, which will have the previously defined indicators as inputs and the identified ADL as output. In this section, the design procedure and its optimization is detailed.

4.1 Single Step ANN-Based Classifier

In a first step, all 9 indicators are calculated for each cycle, and will be fed into a single MLP ANN (see Fig. 3), which will provide the identified ADL (walking, going up stairs, going down stairs, standing still) based on a set of four binary outputs.

The previously defined database is adapted to generate the required training and validation sets, in which the ADL to be identified is codified by binary outputs (one for each ADL). As the database is based on the data provided by 13 individuals, only 9 are used to train the ANN, while the rest (4 individuals) is used as a Test Set to evaluate its validity.

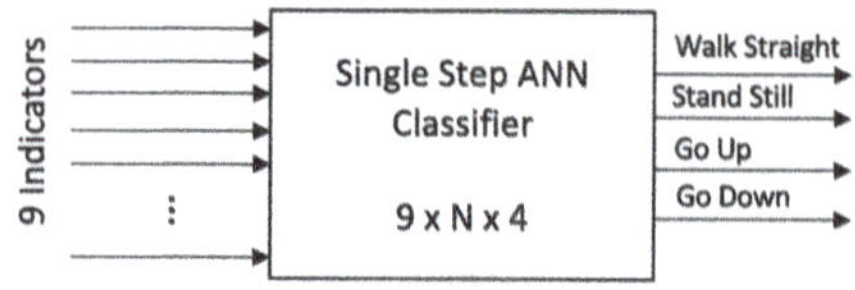

Fig. 3. Single step ANN-based classifier

A single layer MLP with 9 inputs and 4 outputs is defined as the best topology, and the number of hidden neurons (5, 10, 20, 30, 40, 50, 60, 70, 80, 90 and 100) is experimentally defined by testing network performance. For this purpose, each ANN topology is trained 50 times, selecting the one with the best performance as the representative for the topology.

Levenberg-Marquardt algorithm is used to train each ANN with the following parameters: maximum 500 iterations, 0 objective error, hyperbolic tangent sigmoid activation transfer function and $\mu = 1E - 5$. Early stopping to avoid overfitting is considered, and 70% of the data (of the 9 individual used for training) is used for training, while 30% is used for validation.

The evaluation of the trained networks is carried out considering the Test set, which comprises data from the 4 individuals not considered during training. The success rate metric is used to evaluate the classifier, this is, the percentage of times the classifier has properly classified an ADL.

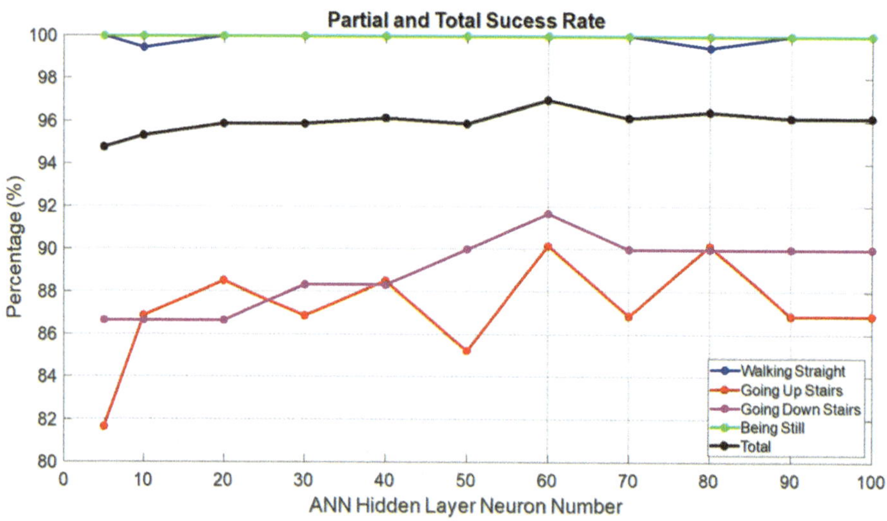

Fig. 4. Single step ANN ADL classifier. Partial and total success rate.

Results for the total and partial success rates are summarized in Fig. 4 for the 11 topologies analyzed. The total success rate indicates the percentage of times the MLP has correctly classified an ADL considering all four alternatives. As it can be seen the rate increases with the number of neurons, ranging from 94% to 97%, which is a quite high success rate. The best alternative seems the network with 60 hidden layer neurons.

However, if partial success rates are considered, i.e. the ones associated to each ADL classification, an uneven distribution is observed. For instance, *walking straight* and *standing still* cases have 100% of success rate, while the success rate of *going up stairs* and *going down stairs* is reduced to the 80–90% range.

These results indicate that some of the proposed indicators allow direct classification of some of the ADL. Hence, it would be possible to identify these ADL using a simpler approach, and then use an ANN to perform the rest of the work. This will be detailed next.

4.2 Two Step Based Classifier

In order to determine the relevance of each indicator to the classification system, a simple but effective tool is to analyze the statistical distribution for each analyzed ADL. From this analysis it has been detected that the *Percentage of the stance phase in each cycle* indicator provides significant information to the classifier.

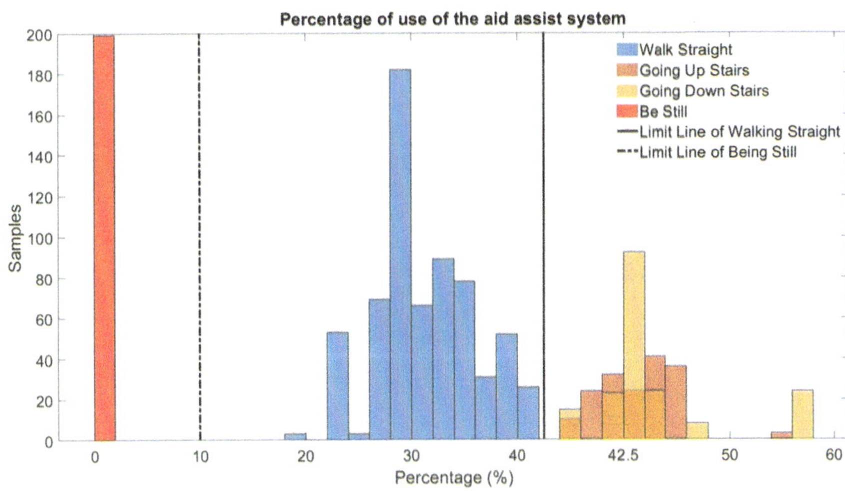

Fig. 5. Percentage of use of the walking aid system: distribution histogram

Figure 5 shows the statistical distribution of the *Percentage of the stance phase in each cycle* indicator for each ADL. As it can be seen, the *standing still* case can be directly identified using this indicator (value lower than 10%), while the *walking straight* case can be differentiated from the *going up/down stairs* case using also a simple range rule (10%– 42% range).

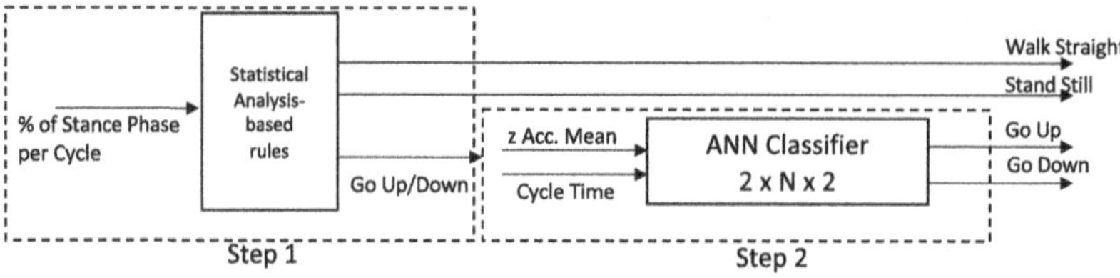

Fig. 6. Two step ANN-based classifier

Hence, a two step approach can be defined. First, using simple rules on the *Percentage of the stance phase in each cycle* indicator, the cases of *walking straight*, *standing still* and *going up/down stairs* are classified. Second, if the ADL falls into the *going up/down stairs* category, an ANN will be defined to classify between *going up stairs* and *going down stairs* (see Fig. 6).

In order to define the classifier ANN, the indicator set will also be modified. As the ANN only has to determine if the user is going up or down, the indicators related with the vertical motion (the *acceleration mean in z axis*), and cycle characteristics (*cycle time*) will be selected. A single output ANN topology will be defined, in which a binary value of 1 will be associated to *going up* and a binary value of 0 will be associated to *going down.*

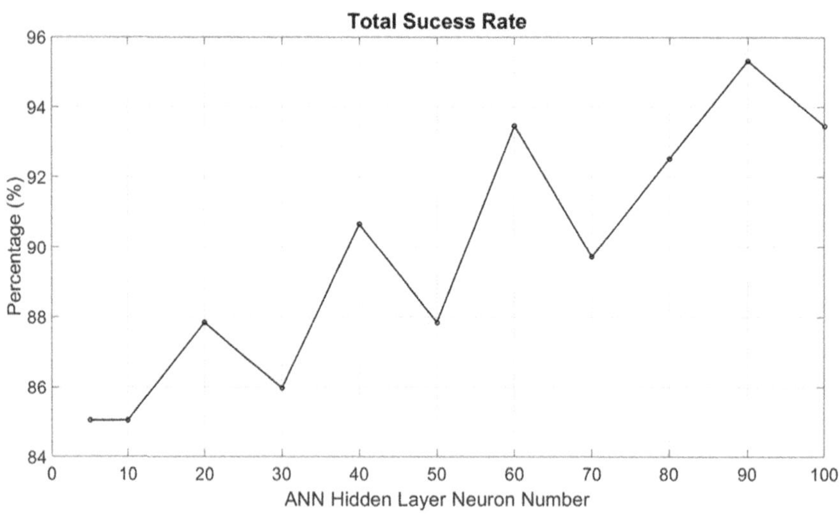

Fig. 7. Two step approach. Success rate of the going up/down classification ANN.

The same training methodology defined in the previous section is used to train this ANN. Results for the optimized ANN are shown in Fig. 7, where the success rate is shown.

As can be seen, the success rate associated to the *going up stairs* and *going down stairs* has increased in comparison with the previous classifier approach, with an average value in all networks of approximately 90%. Moreover, if the two step ADL classifier is considered as a whole (rules+ ANN), a 97% success rate is achieved, as the *walking straight* and *standing still* ADLs have a 100% success rate.

Hence, an optimal classification can be achieved with proper indicator selection, while reducing the computational cost by using simple rules and ANNs.

5 Conclusions

Individualization of the therapy in patients with lower-limb impairment is mandatory to increase their impact. Monitoring the Activities of the Daily Living (ADL) provides objective data of the routines that may help in defining proper therapies.

This work presents a first step to achieve this goal, proposing to develop an ADL recognition algorithm based on an ANN and the data provided by a smart crutch tip. The classifier is designed to be able to detect 4 ADLs (standing still, walking, going up and down stairs).

Two different approaches are analyzed. The first one is based on a single classifier ANN, which has a success rate of over 96%. The second approach tries to reduce the computational cost by classifying with ANN only those cases that cannot be easily differentiated using a single indicator, achieving a total success rate of 97%.

However, it is to be noted that this development has been carried out with healthy people. Hence, further research includes creating a database with patients that can allow to corroborate the proposed approaches, and analyze possible drawbacks in real scenarios.

Moreover, the performed analysis corroborates that there exist high dependency of the classifier performance depending on the selected indicators. Hence, future work will require to expand the database with new ADL, and define a methodology to define the most appropriate indicators to maximize the performance of ADL classifiers.

References

1. Sale, P., Russo, E.F., Russo, M., Masiero, S., Piccione, F., Calabrò, R.S., Filoni, S.: Effects on mobility training and de-adaptations in subjects with Spinal Cord Injury due to a Wearable Robot: a preliminary report. BMC Neurol. **16**(1), 12 (2016)
2. Lerner, Z.F., Damiano, D.L., Bulea, T.C.: The effects of exoskeleton assisted knee extension on lower-extremity gait kinematics, kinetics, and muscle activity in children with cerebral palsy. Sci. Rep. **7**(1), 1–12 (2017)

3. Latimer-Cheung, A.E., Pilutti, L.A., Hicks, A.L., Martin Ginis, K.A., Fenuta, A.M., Ann MacKibbon, K., Motl, R.W.: Effects of exercise training on fitness, mobility, fatigue, and health-related quality of life among adults with multiple sclerosis: a systematic review to inform guideline development. Arch. Phys. Med. Rehabil. **94**(9), 1800–1828.e3 (2013)
4. Cattaneo, D., Regola, A., Meotti, M.: Validity of six balance disorders scales in persons with multiple sclerosis. Disabil. Rehabil. **28**(12), 789–795 (2006)
5. Bethoux, F., Bennett, S.: Evaluating walking in patients with multiple sclerosis. Int. J. MS Care **13**(1), 4–14 (2011)
6. Shull, P.B., Jirattigalachote, W., Hunt, M.A., Cutkosky, M.R., Delp, S.L.: Quantified self and human movement: a review on the clinical impact of wearable sensing and feedback for gait analysis and intervention. Gait Posture **40**(1), 11–19 (2014)
7. Spain, R.I., St. George, R.J., Salarian, A., Mancini, M., Wagner, J.M., Horak, F.B., Bourdette, D.: Body-worn motion sensors detect balance and gait deficits in people with multiple sclerosis who have normal walking speed. Gait Posture **35**(4), 573–578 (2012)
8. Sardini, E., Serpelloni, M., Lancini, M., Pasinetti, S.: Wireless instrumented crutches for force and tilt monitoring in lower limb rehabilitation. Procedia Eng. **87**, 348–351 (2014)
9. Chamorro-Moriana, G., Sevillano, J., Ridao-Fernández, C.: A compact forearm crutch based on force sensors for aided gait: reliability and validity. Sensors **16**(6), 925 (2016)
10. Gadaleta, M., Merelli, L., Rossi, M.: Human authentication from ankle motion data using convolutional neural networks. In: 2016 IEEE Statistical Signal Processing Workshop (SSP). IEEE, June 2016
11. Watanabe, T., Yamagishi, S., Murakami, H., Furuse, N., Hoshimiya, N., Handa, Y.: Recognition of lower limb movements by artificial neural network for restoring gait of hemiplegic patients by functional electrical stimulation. In: 2001 Conference Proceedings of the 23rd Annual International Conference of the IEEE Engineering in Medicine and Biology Society. IEEE (2011)
12. Gyllensten, I.C., Bonomi, A.G.: Identifying types of physical activity with a single accelerometer: evaluating laboratory-trained algorithms in daily life. IEEE Trans. Biomed. Eng. **58**(9), 2656–2663 (2011)
13. Brull, A., Gorrotxategi, A., Zubizarreta, A., Cabanes, I., Rodriguez-Larrad, A.: Classification of daily activities using an intelligent tip for crutches. In: Robot 2019: Fourth Iberian Robotics Conference. Advances in Intelligent Systems and Computing, vol. 1093 (2020)
14. Zeng, W., Wang, C.: Classification of neurodegenerative diseases using gait dynamics via deterministic learning. Inf. Sci. **317**, 246–258 (2015)
15. Lei, L., Peng, Y., Zuojun, L., Yanli, G., Jun, Z.: Leg amputees motion pattern recognition based on principal component analysis and BP network. In: 2013 25th Chinese Control and Decision Conference (CCDC). IEEE, May 2013
16. Sesar, I., Zubizarreta, A., Cabanes, I., Portillo, E., Torres-Unda, J., Rodriguez-Larrad, A.: Instrumented crutch tip for monitoring force and crutch pitch angle. Sensors (Switzerland) **19**(13), 2944 (2019)

Hourly Air Quality Index (AQI) Forecasting Using Machine Learning Methods

Jose Antonio Moscoso-López[1(✉)], Daniel Urda[2], Javier González-Enrique[1], Juan Jesus Ruiz-Aguilar[1], and Ignacio J. Turias[1]

[1] Intelligent Modelling of Systems Research Group, Polytechnic School of Engineering (Algeciras), University of Cadiz, Avda. Ramon Puyol s/n, 11202 Algeciras, Spain
joseantonio.moscoso@uca.es

[2] Grupo de Inteligencia Computacional Aplicada (GICAP), Departamento de Ingeniería Informática, Escuela Politécnica Superior, Universidad de Burgos, Av. Cantabria s/n, 09006 Burgos, Spain

Abstract. Air Quality Index (AQI) is an index to inform the daily air quality. AQI is a dimensionless quantity to show the state of air pollution simplifying the information of concentrations in $\mu g/m^3$. Air quality indexes have been established for each of the five pollutants located in an interesting area to study in as Algeciras (Spain). Hourly data of air pollutants, available during 2010–2015, were analysed for the development of the proposed AQI. This work proposes a two-step forecasting approach to obtain future values, eight hours ahead, of AQI using Machine Learning methods. ANN, SVR and LSTM are capable of modelling non-linear time series and can be trained to accurately generalize when a new database is presented.

Keywords: Air Quality Index (AQI) · Air pollution forecasting · Machine learning · Deep learning · LSTM · ANN · SVR

1 Introduction

Nowadays, air pollution is a major and relevant concern in our societies as it has a huge and negative impact on human health and well being, as well as on ecosystems [4]. Based on the World Health Organization (WHO) reports, outdoor air pollution, also referred to as ambient air, was estimated to cause 4.2 million premature deaths worldwide in 2016. In addition, household air pollution could also cause around 3.8 million people a year to die prematurely. In this sense, there is a need for developing and carrying out a strategic plan and policies to control the level of pollutants and take actions on them. Many national environmental agencies and relevant authorities have focused and put their efforts on obtaining several air quality-related measures through monitoring networks in

Supported by MICINN (Ministerio de Ciencia e Innovación-Spain).

Á. Herrero et al. (Eds.): SOCO 2020, AISC 1268, pp. 123–132, 2021.
https://doi.org/10.1007/978-3-030-57802-2_12

order to allow a farther analysis of the concentration of atmospheric pollutants. Moreover, they have also established limit values over pollutants concentration in order to control emissions.

Air pollution data is composed of a wide variety of heterogeneous measures and understanding all the low-level details might be a complex task. To overcome this difficulty, pollutants concentration measurements are transformed onto an Air Quality Index (AQI) [22] which can be easily interpreted. Since policies vary from one to another country [18], one can find several air pollution indexes. The United States Environmental Protection Agency (EPA) started to use a daily air pollution index in 1976, first named Pollution Standards Index (PSI) and later on improved and renamed as AQI, which is based on five pollutants (O_3, NO_2, CO, SO_2 and PM_{10}). The Canadian government, through the Meteorological Service of Environment Canada, adopted an Air Quality Health Index (AQHI) which is based on three pollutants (O_3, PM and NO_2). Although the "Air quality guidelines for Europe" described by the WHO is a fundamental reference in order to set thresholds on pollutants levels, EU countries standard will vary according to their own goals. However, EU directives have to be complied by all countries (Directive 2008/50/EC) and through the European Environment Agency, which consists of 32 member countries, data is gathered and assessments are periodically produced on a wide range of topics related to the environment. The majority of AQIs do not take into account the combined effects of various pollutants [13], where AQI is usually defined as the value of the pollutant with the highest concentration relative to its standard on a given day.

Apart from institutional AQI, other approaches have been used in other studies. The European project [3] proposed a Common Air Quality Index (CAQI) which is defined in both hourly and daily values, and computed separately as "roadside" or "traffic" index or a "background" index . The Research and Advisory Institute for Hazardous Substance developed and tested an impact-related air quality index (DAQx) [16,17] which depend on emission and air mass exchange conditions . Bruno and Cocchi [2] used a synthetic AQI that is computed by aggregating values for the three dimensions that characterize pollution in a given area: time, space and type of pollutant. Kyrkilis et al. [13] used an aggregated AQI for an urban Mediterranean agglomeration while Murena [19] developed an specific air pollution index for the urban area of Naples. Güçlü et al. [7] used an AQI which is calculated from the geometric average of considered pollutants by assessing the air quality rating (AQR) of each pollutant.

1.1 State of the Art

Air quality forecasting has become the main goal for many governments and environmental agencies which require timely and very accurate future information to efficiently manage air pollution issues in advance [14,25,28]. To this end, the development of such an air quality forecasting tool in order to aid decision-making could also help these entities to collect useful information about environmental quality, air pollution variation or trends.

This paper aims at applying machine learning models to predict several environmental-related. In this sense, Jiang et al. [12] developed an artificial neural network (ANN) model for the air pollution index forecasting. Azid et al. [1] applied Principal Component Analysis (PCA) and ANNs to predict the air pollutant index in Malaysian. Lal Benjamin et al. [20] developed an air quality prediction model also using ANNs. Gonzalez-Enrique [5,6] applied Machine Learning techniques in air quality forecasting by using hourly time series of NO_2 and basic meteorological variables. In recent times, Long Short-Term Memory (LSTM) models are used to carry out air quality predictions with superior performance in comparison with traditional forecasting approaches [15,29].

The main contribution of this paper is to forecast the AQI considering the case study of Algeciras (Spain). Two neural network-based models, a classic ANN, a Support Vector Machine for Regression (SVR) and a Long Short-Term Memory (LSTM) network, will be applied to assist the index forecasting. The rest of the paper is organized as follows. Section 2 gives a brief description of the database and the environment of the case study. Section 3 discusses the methodology and the experimental procedure. Section 4 discusses the results. Finally, conclusions are drawn in Sect. 5.

2 Datasets

This study case is located in Algeciras in the southern part of Spain Fig. 1. This city belong to Algeciras Bay Metropolitan Area, being the most populated one with around 121,000 inhabitants. The port of Algeciras is one of the most important port in Europe, where approximately a number of 4,7M TEUs (twenty-foot equivalent unit) was handled and more than 28,000 vessels docked in 2018. Here, there are two predominant winds: from east to west and vice versa and it enjoys a Mediterranean climate. Furthermore, this region is a complex area as is one of the most significant industrial zones in Spain, where industries from different sectors are established such as an oil-refinery, an stain-less-steel factory, some power plants or different petrochemical factories. All the above described are a source of particulate and gaseous air pollution. In addition, the industrial and port activities generate a considerable vehicles traffic, which is other source of pollution. The dataset for the five pollutants used have been provided by the Environmental Agency of the Andalusian Government (research project RTI2018-098160-B-I00 supported by MICINN (Ministerio de Ciencia e Innovación-Spain)). The Regional Government has an air pollution monitoring station located in Algeciras which collects the database summarised in Table 1 and taking into account the European standards to obtain the AQI.

Table 1. Database features

Pollutants	Measure	Period	Register type	Location
PM_{10}, $PM_{2.5}$ CO, NO_2, O_3, SO_2	Concentration values ($\mu g/m^3$)	2010–2015	Hourly	Algeciras, Spain

Fig. 1. Location of the study region

3 Methods and Experimental Design

This section goes into details concerning the methods and experimental design used in this paper. Section 3.1 presents the requirements for obtaining AQI. Then, the methods proposed and evaluated in this work are described in Sect. 3.2 and, finally, the experimental design is described in Sect. 3.3.

3.1 Air Pollution Index

Several models have been used to develop an air quality index (AQI). However, the EPA model is one of the most widespread worldwide [24].

AQI is defined with respect to the five main common pollutants: carbon monoxide (CO), nitrogen dioxide (NO_2), ozone (O_3), particulate matter (PM_{10}) and sulphur dioxide (SO_2). The index levels is divided in six categories according to different level of health concerns, Table 2. The AQI scale is ranking of 0 to 500, related to daily concentrations of each of these five pollutants, which breakpoint concentrations have been defined by EPA [24]. This report shows the pollutants concentrations in different way of measure such us *ppm* or $\mu g/m^3$ in periods of 1 h concentration or 8/24 h moving average depending of the pollutant. These concentrations are converted into a numerical index by using linear interpolation from the Eq. (1):

$$I = \frac{I_{high} - I_{low}}{C_{high} - C_{low}}(C - C_{low} + I_{low}) \quad (1)$$

where: I = The (Air Quality) Index,
C = The pollutant concentration,
C_{low} = The concentration breakpoint that is $\leq C$,
C_{high} = The concentration breakpoint that is $\geq C$,
I_{low} = The index breakpoint corresponding C_{low},
I_{high} = The index breakpoint corresponding to C_{high}.

The overall index indicates the short-term air quality situation and is given by the maximum value of the individual pollutant AQIs.

Table 2. Air quality index standard values and related information

AQI levels	AQI values	Meaning
Good	0 to 50	Air quality is considered satisfactory, and air pollution poses little or no risk
Moderate	51 to 100	Air quality is acceptable; however, for some pollutants there may be a moderate health concern for a very small number of people who are unusually sensitive to air pollution
Unhealthy for sensitive groups	101 to 150	Members of sensitive groups may experience health effects. The general public is not likely to be affected
Very unhealthy	201 to 300	Health alert: everyone may experience more serious health effects
Hazardous	301 to 500	Health warnings of emergency conditions The entire population is more likely to be affected

In this paper, all the sub indexes and indexes here reviewed follow a data-driven approach, air pollution concentration data measured of the five key pollutants describes the current air quality situation at the monitoring station in Algeciras in the years 2010–2015.

3.2 Methods

In this work, Long Short-Term Memory Neural Networks and Artificial Neural Networks have been applied in order to predict the AQI at the Algeciras monitoring station, which is located in the southern region of Spain. Our goal in this work is to obtain predictions of each pollutant concentrations with 8-h ahead prediction horizons. Once these forecasted concentrations are obtained, the future AQI values are calculated. In this case, a combination of the past values of concentrations for each pollutant have been used as inputs. Additionally, auto regressive windows of different sizes have been used in a re sampling procedure in order to obtain the best future values.

Long Short-Term Memory Neural Network (LSTM). Recurrent Neural Network (RNNs) [27] are a type of Neural Network which excels at processing non-linear sequential data. However, RNNs show some important caveats (i.e. the vanishing gradient problem [21]) when they are applied to problems such as time series forecasting. In contrast, Long Short-Term Memory (LSTM) [10] allow us to learn from long input sequences by using memory cells. The main advantage produced by this fact is that LSTMs are able to learn long-term dependencies.

LSTM networks included in this paper used the following network topology: a sequence input layer, a LSTM layer, a dropout layer, a fully connected layer and a regression output layer. Regarding the hyper-parameters of the model, the number of neurons present in the hidden layer, the initial learning rate and the dropout rate were learned by performing Bayesian optimization. Additionally, Adam was used as the optimizer and the number of epochs employed was 200 (with early stopping set to 25 values). Default values were used for the remaining hyperparameters.

Artificial Neural Network (ANN). Artificial Neural Networks are computational models vaguely inspired by the biological neural networks. ANNs are universal approximators [11] and have the ability to learn complex nonlinear relations of information and generalize it for the prediction, classification and clustering activities [9].

In this study, feedforward neural network topology have been used with multi-layer perceptron (MLP). Backpropagation (BP) algorithm is applied as supervised learning procedure [23]. BPNNs use a single hidden layer and a different number of hidden neurons, this layer is interconnected between the input and output layers. The Levenberg–Marquardt algorithm has been used as the optimization algorithm, along with early stopping as regularization approach in order to provide robustness and velocity [8].

Support Vector Machine for Regression (SVR): SVR is based on statistical learning theory [26] and approximates an unknown function in order to map the input data into a high-dimensional feature space through a nonlinear mapping function, and then a linear regression problem is constructed in this new feature space. In the high-dimensional feature space, it theoretically exists a linear function, f, to formulate the non-linear relationship between input and output data. The performance of an SVR is determined by the type of kernel function and the setting of kernel parameters. In this work, an resampling procedure fit the best kernel function.

3.3 Experimental Procedure

As an initial step, the original database was normalized and divided into two disjoint groups. The first one included hourly pollutants (CO, NO_2, O_3, PM_{10}, SO_2) records going from 2010 to 2014 and was used as the training set. The second one included records belonging to 2015 and acted as the test set. In the forecasting approach, each pollutant was treated independently as a time series. The forecasting approach was divided into two steps, in the first step the concentration of each pollutant was obtained. The inputs of the forecasting approaches is a set of lagged values of the time series in the past. Different sizes of autoregressive windows are used in this study as 24, 48, 72, 96 and 120 h to obtain the best 8-h ahead forecasting values (output) for each pollutant. Once obtained the forecasting concentrations of the pollutants, the future AQI was obtained in a second step. All dataset was splited into three parts: training, validation and testing. A random resampling procedure using cross-validation was utilized in order to avoid the overfitting. Training-validation data is used to design the model. The test set, being unseen data, was used to assess the final performance.

The root mean squared error (*RMSE*) and the mean absolute error (*MAE*) were computed as performance indexes and are defined in the following equations:

$$RMSE = \sqrt{\frac{\sum_{i=1}^{n}(F_i - O_i)^2}{n}} \tag{2}$$

$$MAE = \frac{\sum_{i=1}^{n}|F_i - O_i|}{n} \tag{3}$$

The Eqs. 2–3 describe how the performances indexes are calculated given the observed (O) and forecasted (F) outcome and n is the times compared lower values of $RMSE$ and MAE imply more precise predictions.

4 Results and Discussion

In this work, an hourly database for full six years has used. This database contains the hourly concentration of five pollutants from the first of January 2010 until the end of December 2015 in the Algeciras monitoring station. The results represent the performance of the two steps forecasting approach. In Table 3 is shown the best performance for LSTM, SVR and ANN models for each pollutant. Furthermore, the best autoregressive windows (AW) is indicated for each forecasting method and each pollutant. In all contaminants, the better prediction performance is achieved by the ANN although with similar results obtained by LSTM and SVR.

AQI values are shown in Fig. 2, in red is represented the observed values of AQI while in blue, green and cyan are represented the forecasted AQI with LSTM, ANN and SVR respectively. The performance of the forecasted AQI (step 2) has been improved in comparison to the pollutants concentrations values (step 1). As is shown in Fig. 2 the forecasting AQI is well fitted in all cases.

Table 3. Pollutant forecasting performance indexes.

Pollutant	Prediction method	AW	RMSE	MAE
O_3	SVR	120	382.726	15.069
	LSTM	96	371.631	15.041
	ANN	120	370.292	14.962
PM_{10}	SVR	96	257.475	12.406
	LSTM	96	256.943	12.300
	ANN	120	245.131	12.015
CO	SVR	48	246.586	11.875
	LSTM	48	242.077	11.850
	ANN	120	222.015	11.459
SO_2	SVR	72	208.237	11.181
	LSTM	120	208.776	11.391
	ANN	120	197.297	10.964
NO_2	SVR	96	344.876	13.899
	LSTM	120	336.466	13.999
	ANN	120	321.565	13.465

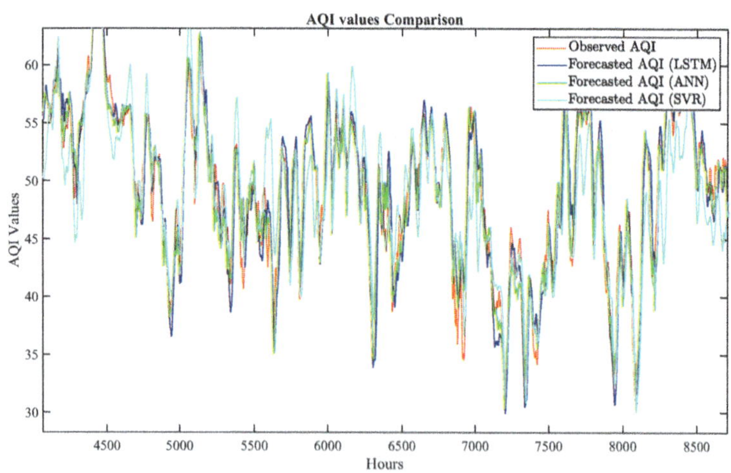

Fig. 2. AQI forecasted

5 Conclusions

Development AQI prediction model in metropolitan areas is a priority for environmental health research. The environmental management requires making decision tool to anticipate the negative impacts of air pollution.

In this work, a two-step procedure is proposed to forecast AQI values with eight hours in advance. LSTM, SVR and ANN have been used to obtain, in a first step, the concentration values of five pollutants and AQI in a second step. The experimental results indicate that ANN performance is slightly better than LSTM and SVR performance. In the same way, the values of AQI obtained with ANN is better than SVR and LSTM. The contribution of this study is to detect the values of AQI in enough time to activate defence protocols against air pollutants. This work could be further extended using a new methodology by bidirectional-LSTM and assess the influence of meteorological and traffic variables in Air Quality Index forecasting.

Acknowledgments. This work is part of the research project RTI2018-098160-B-I00 supported by MICINN (Ministerio de Ciencia e Innovación-Spain). The database has been kindly provided by the Environmental Agency of the Andalusian.

References

1. Azid, A., Juahir, H., Latif, M.T., Zain, S.M., Osman, M.R.: Feed-forward artificial neural network model for air pollutant index prediction in the southern region of Peninsular Malaysia. J. Environ. Prot. **04**(12), 1–10 (2013)
2. Bruno, F., Cocchi, D.: Recovering information from synthetic air quality indices. Environmetrics **18**(3), 345–359 (2007)
3. van den Elshout, S.: CiteairII. CAQI Air quality index. Comparing urban air quality across borders-2012 (October 2008), pp. 1–38 (2012)
4. European Environment Agency: Air quality in Europe — 2018 Report. Technical Report European Environment Agency, Copenhagen, Denmark (2018)
5. González-Enrique, J., Turias, I.J., Ruiz-Aguilar, J.J., Moscoso-López, J.A., Franco, L.: Spatial and meteorological relevance in NO_2 estimations: a case study in the Bay of Algeciras (Spain). Stoch. Environ. Res. Risk Assess. **33**(3), 801–815 (2019)
6. Gonzalez-Enrique, J., Turias, I.J., Ruiz-Aguilar, J.J., Moscoso-Lopez, J.A., Jerez-Aragones, J., Franco, L.: Estimation of NO2 concentration values in a monitoring sensor network using a fusion approach. Fresenius Environ. Bull. **28**(2), 681–686 (2019)
7. Güçlü, Y.S., Dabanlı, Şişman, E., Şen, Z.: Air quality (AQ) identification by innovative trend diagram and AQ index combinations in Istanbul megacity. Atmos. Pollut. Res. **10**(1), 88–96 (2019)
8. Hagan, M.T., Demuth, H.B., Beale, M.H.: Neural Network Design. Thomson Learning Stamford, CT (1996)
9. Hakimpoor, H., Arshad, K.A.B., Tat, H.H., Khani, N., Rahmandoust, M.: Artificial neural networks' applications in management. World Appl. Sci. J. **14**(7), 1008–1019 (2011)
10. Hochreiter, S., Schmidhuber, J.: Long short-term memory. Neural Comput. **9**(8), 1735–1780 (1997)
11. Hornik, K., Stinchcombe, M., White, H.: Multilayer feedforward networks are universal approximators. Neural Netw. **2**(5), 359–366 (1989)
12. Jiang, D., Zhang, Y., Hu, X., Zeng, Y., Tan, J., Shao, D.: Progress in developing an ANN model for air pollution index forecast. Atmos. Environ. **38**(40 SPEC.ISS.), 7055–7064 (2004)

13. Kyrkilis, G., Chaloulakou, A., Kassomenos, P.A.: Development of an aggregate air quality index for an urban Mediterranean agglomeration: relation to potential health effects. Environ. Int. **33**(5), 670–676 (2007)
14. Lauret, P., Heymes, F., Aprin, L., Johannet, A.: Atmospheric dispersion modeling using artificial neural network based cellular automata. Environ. Modell. Softw. **85**, 56–69 (2016)
15. Li, X., Peng, L., Hu, Y., Shao, J., Chi, T.: Deep learning architecture for air quality predictions. Environ. Sci. Pollut. Res. **23**(22), 22408–22417 (2016)
16. Mayer, H., Kalberlah, F., Ahrens, D., Reuter, U.: Analysis of indices for the assessment of the air. Gefahrstoffe Reinhaltung der Luft **62**, 177–183 (2002)
17. Mayer, H., Makra, L., Kalberlah, F., Ahrens, D., Reuter, U.: Air stress and air quality indices. Meteorol. Z. **13**(5), 395–403 (2004)
18. Mihăiţă, A.S., Dupont, L., Chery, O., Camargo, M., Cai, C.: Evaluating air quality by combining stationary, smart mobile pollution monitoring and data-driven modelling. J. Clean. Prod. **221**, 398–418 (2019)
19. Murena, F.: Measuring air quality over large urban areas: development and application of an air pollution index at the urban area of Naples. Atmos. Environ. **38**(36), 6195–6202 (2004)
20. lal Benjamin, N., et al.: Air quality prediction using artificial neural network. Int. Chem. Stud. **2**(4), 7–9 (2014)
21. Palangi, H., Ward, R., Deng, L.: Distributed compressive sensing: a deep learning approach. IEEE Trans. Signal Process. **64**(17), 4504–4518 (2016)
22. Plaia, A., Ruggieri, M.: Air quality indices: a review. Rev. Environ. Sci. Biotechnol. **10**(2), 165–179 (2011)
23. Rumelhart, D.E., Hinton, G.E., Williams, R.J.: Learning internal representations by error propagation. Parallel distributed processing: Exploration in the Microstructure of Cognition, pp. 318–362 (1986)
24. U.S. Environmental Protection Agency: Guidelines for the Reporting of Daily Air Quality – the Air Quality Index (AQI). Technical Report, Office of Air Quality Planning and Standards, Research Triangle Park, North Carolina (2006)
25. Van Fan, Y., Perry, S., Klemeš, J.J., Lee, C.T.: A review on air emissions assessment: transportation. J. Clean. Prod. **194**, 673–684 (2018)
26. Vapnik, V.: Statistical Learning Theory. Wiley, New York (1998)
27. Williams, R.J., Zipser, D.: A learning algorithm for continually running fully recurrent neural networks. Neural Comput. **1**(2), 270–280 (1989)
28. Yang, G., Huang, J., Li, X.: Mining sequential patterns of PM2.5 pollution in three zones in china. J. Clean. Prod. **170**, 388 – 398 (2018)
29. Zhou, Y., Chang, F.J., Chang, L.C., Kao, I.F., Wang, Y.S.: Explore a deep learning multi-output neural network for regional multi-step-ahead air quality forecasts. J. Clean. Prod. **209**, 134–145 (2019)

Interpretable Deep Learning with Hybrid Autoencoders to Predict Electric Energy Consumption

Jin-Young Kim and Sung-Bae Cho(✉)

Department of Computer Science, Yonsei University, Seoul, South Korea
{seago0828,sbcho}@yonsei.ac.kr

Abstract. As energy demand continues to increase, smart grid systems that perform efficient energy management become increasingly important due to environmental and cost reasons. It requires faster prediction of electric energy consumption and valid explanation of the predicted results. Recently, several demand predictors based on deep learning that can deal with complex features of data are actively investigated, but most of them suffer from lack of explanation due to the black-box characteristics. In this paper, we propose a hybrid autoencoder-based deep learning model that predicts power demand in minutes and also provides the explanation for the predicted results. It consists of an information projector that uses auxiliary information to extract features for the current situation and a model that predicts future power demand. This model exploits the latent space composed of the two different modalities to account for the prediction. Experiments with household electric power demand data collected over five years show that the proposed model is the best with a mean squared error of 0.3764. In addition, by analyzing the latent variables extracted by the information projector, the correlation with various conditions including the power demand is confirmed to provide the reason of the coming power demand predicted.

1 Introduction

As industrialization has progressed globally, world's electricity consumption is increased every year, reflecting the growth in the number of electric devices. A report published in 2018 [1] about energy consumption in the U.K. provides that electricity consumption in the U.K. has increased by 33%. Among the demanders of various energy sources, Streimikiene estimated that residential energy consumption would account for a large proportion by 2030 [2]. It is a reason that an energy management system (EMS) has been proposed to control the demand for soaring energy consumption. The smart grid, one of technologies for EMS, consists of a set of computers, controllers, automation and standard communication protocols, which are connected on the Internet, all of which are used in order to manage the generation and distribution of electricity to consumers through these digital technologies [3].

Smart grid emerged as a smart power grid that has recently achieved lot of popularity [4, 5]. Smart grid is usually performed on a Plan-Do-Check-Act cycle [6]. Formulating

Á. Herrero et al. (Eds.): SOCO 2020, AISC 1268, pp. 133–143, 2021.
https://doi.org/10.1007/978-3-030-57802-2_13

an energy plan is the first thing to do. This is the decision of the initial energy baseline, the energy performance indicators, the strategic and operative energy objectives and the action plans. Among the four stages, the "plan" phase is very important because it is the stage of establishing an energy use strategy and it includes an energy demand forecasting step. In addition, energy demand forecasting is well known as an important step for both companies and consumers in the smart grid, and energy storage systems based on individual demand forecasting results can build an effective smart city infrastructure [7–9]. Therefore, it is indispensable to study the electric energy demand prediction model to design an efficient EMS.

Besides, analyzing the cause for the predicted power demand value helps in efficient power demand planning. Kim and Cho proposed a method to predict future energy demand and interpret the results of predicted values through analyzing latent space [6]. However, they projected multiple information about energy consumption into only one latent space, resulting in entangled representation. In this paper, we propose a hybrid autoencoder-based model that combines a deep learning-based model that predicts minutely power demand with very complex features, and an information projector to help infer the result by inputting auxiliary information for interpretability. A projector which is one of components in predictive model receives past demands. A predictor that predicts future demand is placed by receiving the output of the projector and the pattern extracted through the information projector.

The rest of this paper is organized as follows. In Sect. 2, we discuss the related work on electric energy consumption prediction. Section 3 details the proposed hybrid autoencoder model. Section 4 presents experimental results and Sect. 5 concludes the paper.

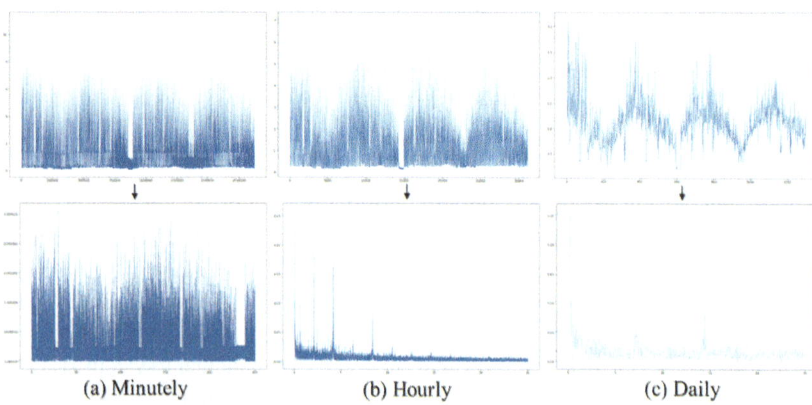

Fig. 1. (Top) The electric energy demand for each (a) minute, (b) hour and (c) date. The graphs in bottom row show the results of Fourier transform. The shorter the time unit, the greater the irregularity.

2 Backgrounds

2.1 Difficulties in Predicting Energy Consumption

There are two major problems in building a power demand forecasting model: irregular pattern of electric energy consumption and difficulty in providing evidence for predicted demands [6, 8]. As shown in Fig. 1, individual energy demand patterns are more complex for the shorter unit of time for collecting demand. Even if the periodicity is analyzed through the Fourier transform, it can be seen that there is no distinct pattern in the energy demand record collected by the minute unit. As a result of statistical analysis of the relationship between time and power demand, as shown in Fig. 2, the correlation is very low, indicating that the demand pattern is very complicated. Besides, as shown in Table 1, the statistical analysis conducted by Kim and Cho shows that the monthly, daily, and hourly time have low correlation with the power demand. Recently, many studies have been conducted to solve the problems by deep learning models that effectively extract complex features [6, 10–12]. However, since deep learning models are black-box, it is difficult to specify the reason for predicted results.

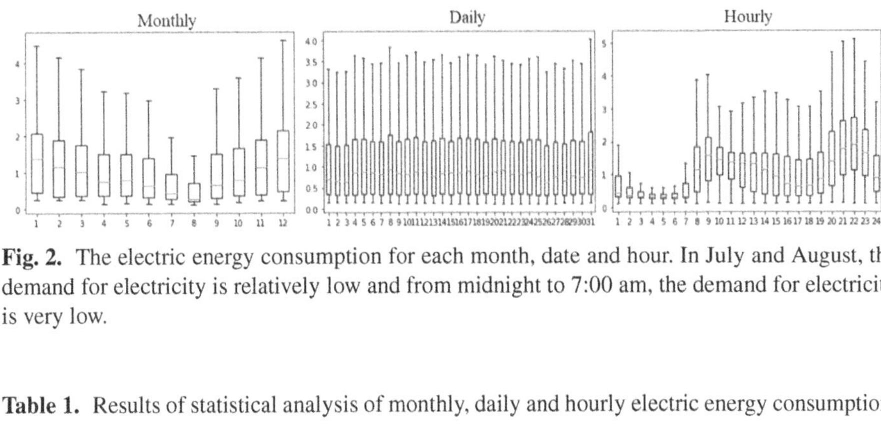

Fig. 2. The electric energy consumption for each month, date and hour. In July and August, the demand for electricity is relatively low and from midnight to 7:00 am, the demand for electricity is very low.

Table 1. Results of statistical analysis of monthly, daily and hourly electric energy consumption.

	Monthly	Daily	Hourly
p-value (t-test)	0.005	0.182	0.011
p-value (ANOVA)	0.000	6.986e−56	0.000

2.2 Related Works

Many researchers have conducted studies to predict energy demand. Table 2 shows the summary of the related works. In the past, statistical techniques were used mainly, but had limited performance due to the irregular patterns of energy consumption [13–15]. Dong et al. and Li et al. used a support vector machine (SVM) with weather condition or building's energy transfer coefficient respectively to predict the consumption of building energy [16, 17]. Xuemei et al. set the state for forecasting energy consumption through fuzzy c-means clustering and predicted demand with fuzzy SVM [18]. Ma forecasted energy consumption with specific population activities or unexpected events, as well as weather condition as inputs of the MLR model [19]. Although the above studies set the state and forecasted future consumption based on it, they lacked the mechanism to identify the state accurately.

In order to predict the energy consumption more accurately, many predictors based on deep learning model have been proposed. Ahmad et al. used a deep neural network (DNN) with the information of weather and building usage rate [20]. For more accurate time series modeling, Lee et al. predicted environmental consumption with recurrent neural network (RNN) [21]. Li et al. predicted energy demand with an autoencoder model consisting of fully connected layers [22]. Kim and Cho and Le et al. proposed more complex models including convolutional neural network, long short-term memory (LSTM) and Bi-LSTM [10, 11]. However, as mentioned previously, deep learning model is black-box so as to be difficult to provide the evidence of predicted results. To solve this problem, Kim and Cho proposed a state-explainable autoencoder that defines the state with the past consumption and predicts future demands based on it [6].

Table 2. Related works for predicting energy consumption.

Category	References	Method	Description
Statistical model	[13]	K-means clustering	Calculate the center value of the cluster and classify the time series into regular and irregular trend
	[14]	ARIMA	Present different forecasting methods to predict electricity consumption
	[15]	Statistical model	Apply statistical method of multiple linear regression to real-world trip and energy consumption data
Machine learning model	[16, 17]	SVM	Predict building energy demand or annual energy demand using SVM
	[18]	Fuzzy SVM	Present a novel short-term cooling load forecasting with conjunctive use of fuzzy C-mean clustering
	[19]	Linear regression	Propose model based on linear regression that predicts large-scale public building energy demand

(continued)

Table 2. (*continued*)

Category	References	Method	Description
Deep learning model	[20]	DNN	Predict energy consumption by using climate, date and building usage rate information
	[21]	RNN	Propose a pro-environmental consumption index using big data queries to measure the environmental consumption level for each country
	[22]	Autoencoder	Extract the building energy consumption and predict future energy consumption
	[10, 11]	CNN and LSTM	Predict household energy consumption with CNN-LSTM
	[6]	Autoencoder	Define the state with the past electric energy consumption and predict future demand based on it

3 The Proposed Method

The overall structure of the proposed model is shown in Fig. 3. It consists of three main components: general projector f, information projector h, and predictor g. The information projector h, which takes auxiliary information as input and adds explanatory power, outputs a latent variable and passes it to the predictor g. The predictor g receives the features extracted from the projector and the information projector as inputs and predicts future energy demand. There are many ways to deal with time series data, but f and g are based on LSTM, one of the RNN's, to handle time series data [23]. The information projector h consists of LSTM and fully connected layers. Kim and Cho predicted the future energy demand only with f and g so that the latent space (denoted as state in [6]) would be entangled with various factors such as patterns of energy consumption and auxiliary information. However, in this paper, we separate the latent space into two, and they are for power demand and auxiliary information, respectively. Each latent space is constructed with the general projector f and the information projector h.

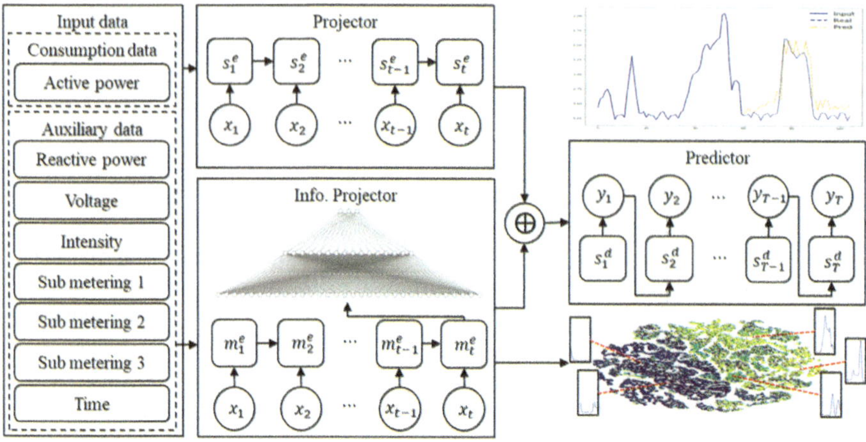

Fig. 3. The overall architecture of the proposed model.

3.1 Projector and Information Projector

We continuously update the latent variable with the auxiliary information during the time interval t as shown in Eq. (1). The extracted value m_i for the ith time is defined as follows:

$$m_i^e = f\left(x_i, m_{i-1}^e\right), \tag{1}$$

where x_i is the ith input of projector and $s_0 = 0$. $f(\cdot,\cdot)$ is a LSTM including input gate i_t, forget gate f_t, output gate i_t and memory cell c_t. Each value is computed as follows:

$$i_t = \sigma\left(W_{xi} * x_t + W_{mi} * m_{t-1}^e\right) \tag{2}$$

$$f_t = \sigma\left(W_{xf} * x_t + W_{mf} * m_{t-1}^e\right) \tag{3}$$

$$o_t = \sigma\left(W_{xo} * x_t + W_{mo} * m_{t-1}^e\right) \tag{4}$$

$$c_t = f_t * c_{t-1} + i_t * \sigma\left(W_{xc} * x_t + W_{mc} * m_{t-1}^e\right) \tag{5}$$

$$m_t^e = o_t * \sigma(c_t) \tag{6}$$

where W is weight matrix of the layer, σ is sigmoid function ($\sigma(x) = \left(1 + e^{-x}\right)^{-1}$) and $*$ means matrix multiplication. Outputs of input, forget and output gates are computed with activation function, weighted sum of input x_t and latent variable m_{t-1}.

In order to extract the patterns of previous energy consumption, we use a general projector as well as information projector. The extracted value s_t proceed in a similar manner to Eqs. (2)–(6) but we use s_t instead of m_t. The computed latent variable m_t and s_t is concatenated and transferred to the consumption predictor as shown in Fig. 3.

The extracted latent variables m_t and s_t by general and information projectors are concatenated to be used as an input of predictor. The dimension of m is set differently depending on the capacity of the power demand pattern to be expressed, but the dimension of s is set to two dimensions to facilitate analysis.

3.2 Consumption Predictor

In this section, we present how to forecast future demand by using computed variables. Unlike the projector, m is used instead of x, and y is used instead of m to calculate y in Eqs. (2)–(6).

$$y_t = W_{my} * m_{t-1}^d + W_{yy} * y_{t-1} \tag{7}$$

Here, y_t is the future power demand value and is calculated without the activation function. As shown in the predictor part in Fig. 3, we can see that the predicted demand y_i of the ith time-step is used as an input to compute y_{t+1}.

L2 norm-based loss function is used to train the proposed model as shown in Eq. (8) by sampling the data with time interval of t_x and t_y in energy consumption $\mathcal{X}$ and predicted values $\mathcal{Y}$, respectively.

$$\mathcal{L} = \sum_i \left[\left(y_{1:t_y} - g\left(f\left(x_{1:t_x}\right), h\left(m_{1:t_x}\right)\right)^2 \right] \tag{8}$$

3.3 Latent Space of Auxiliary Information for Interpretability

The cause for the predicted power demand is explained by the analysis of the latent space constructed from auxiliary information. We extract the latent variable m by entering auxiliary information into the information proejctor introduced in Sect. 3.1. By analyzing the electric energy consumption and auxiliary information values according to the location of m and the relationship between the two, it is possible to indirectly determine the cause of the high (or low) predicted demand value. Besides, since the number of dimensions of latent space is set to two, it is possible to confirm the relationship between the predicted value and the auxiliary information by visualizing the latent variables.

4 Experiments

4.1 Dataset and Experimental Settings

To verify the proposed model, we use a dataset on household electric power consumption [24]. There are about two million minutes of electric energy demand data from 2006 to 2010, and we use about first four years of data as training dataset and the rest as test dataset. It consists of eight attributes including date, global active power (GAP), global reactive power (GRP), global intensity (GI), voltage, sub metering 1, 2, and 3 (S1, 2, and 3), and the model predicts the GAP. S1 corresponds to the kitchen, containing manly a microwave, an oven, and a dishwasher. S2 corresponds to the laundry room, containing a refrigerator, a tumble-drier, a light, and a washing-machine. S3 corresponds to an air-conditioner and an electric water-heater.

To examine the performance of the prediction model, we use three evaluation metrics—the mean squared error (MSE) which can be calculated as follows.

$$MSE = \frac{1}{N}\Sigma_{i=1}^{N}\left(\overset{\wedge}{y_m} - y_m\right)^2 \quad (9)$$

4.2 Results of Demand Prediction

To verify the performance of the proposed model, we show the energy demand forecasting result using our model and compared with other conventional methods. Figure 4 is the result showing real and predicted energy demand values at the same time. The model predicts energy demand for 15, 30, 45, and 60 min with actual energy demand for 60 min. Although the model could not predict the energy demand perfectly, we confirm that the energy demand pattern predicted well.

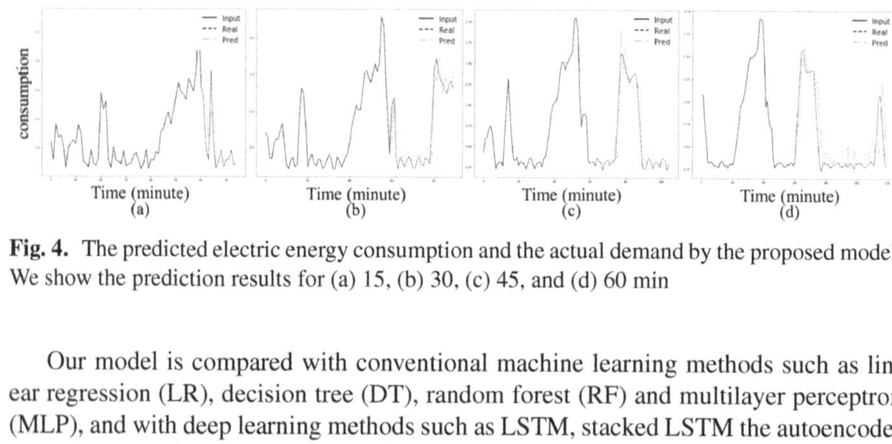

Fig. 4. The predicted electric energy consumption and the actual demand by the proposed model. We show the prediction results for (a) 15, (b) 30, (c) 45, and (d) 60 min

Our model is compared with conventional machine learning methods such as linear regression (LR), decision tree (DT), random forest (RF) and multilayer perceptron (MLP), and with deep learning methods such as LSTM, stacked LSTM the autoencoder model proposed by Li and state-explainable autoencoder proposed by Kim and Cho. The MSE measure of the experimental results for each model is shown in Fig. 5 as box plot. The results of the comparison with other models show that the proposed model outperforms other models. Some of deep learning methods are worse than machine learning methods, but our model yields the best performance.

4.3 Analysis on Evidence of Predicted Results

While training the latent space in an unsupervised way, the relationship between power demand and auxiliary information is automatically embedded on it. Therefore, the explainable reason can be added by visualizing each information on the latent space and finding the relationship. To analyze the latent space learned through the information projector, we visualize the latent variable using the t-SNE algorithm as shown in Fig. 6 [25]. In the graph at the upper left of Fig. 6, the case of using a lot of power is distributed at the top left. As introduced in Sect. 4.1, when we analyze the graphs of S1, S2, and S3 collected for each home appliance, we find that S3 is closely related to the power demand. In the case of S2, it is analyzed that there was no significant effect.

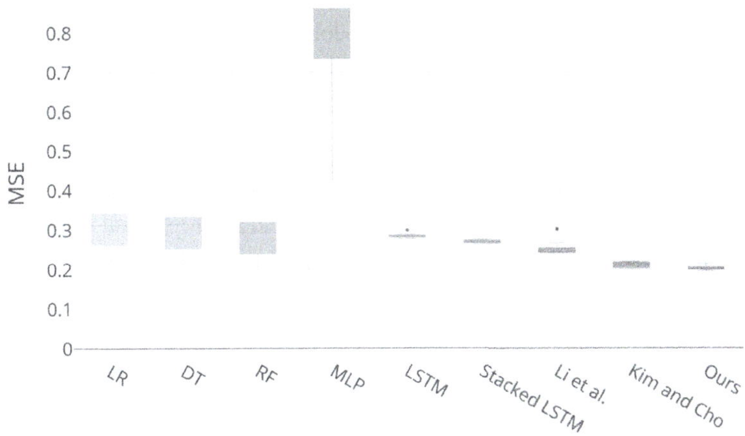

Fig. 5. The results of MSE of models. We show the MSE results for 15 min.

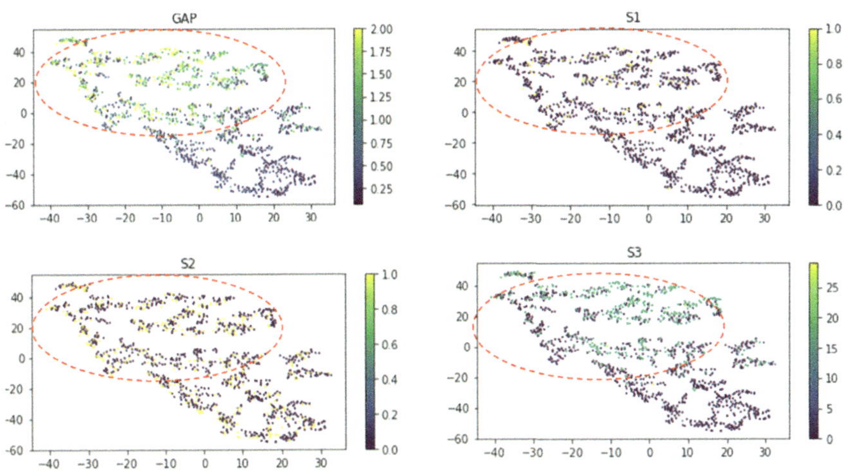

Fig. 6. Visualization of latent variables extracted by information projector. It can be seen that the higher the power demand value is predicted if the latent variable is located in the upper left. In addition, the predicted power demand value is most affected by S3, and S2 did not have a significant effect.

5 Conclusion

We have addressed the necessity and difficulty of predicting the future energy consumption. There are two main problems: irregular pattern of electric energy consumption and difficulty in providing evidence for predicted demands. To solve these problems, we have proposed a hybrid autoencoder-based model consisting of projector, predictor and information projector. Our proposed model has the best performance compared with the conventional models. Besides, by analyzing the latent space, we can confirm the correlation between energy demand and several specified consumption information.

Since the behavior of consumer is irregular, it is important to predict the future consumption based on various assumed environments. Therefore, we will forecast the energy demand by controlling the latent space. In addition, several experiments on different dataset with larger scale will be conducted. Finally, we will construct an efficient energy management system including the proposed prediction model.

Acknowledgement. This research was supported by Korea Electric Power Corporation (Grant number: R18XA05). J. Y. Kim has been supported by NRF (National Research Foundation of Korea) grant funded by the Korean government (NRF-2019-Fostering Core Leaders of the Future Basic Science Program/Global Ph.D. Fellowship Program).

References

1. Energy consumption in the U.K (2020). https://www.gov.uk/government/statistics/energyconsumption-in-the-uk. Accessed 27 Jan 2020
2. Streimikiene, D.: Residential energy consumption trends, main drivers and policies in Lithuania. Renew. Sustain. Energy Rev. **35**, 285–293 (2014)
3. Ferrag, M.A., Maglaras, L.A., Janicke, H., Jiang, J., Shu, L.: A systematic review of data protection and privacy preservation schemes for smart grid communications. Sustain. Cities Soc. **38**, 806–835 (2018)
4. Javaid, N., Hafeez, G., Iqbal, S., Alrajeh, N., Alabed, M.S., Guizani, M.: Energy efficient integration of renewable energy sources in the smart grid for demand side management. IEEE Access **6**, 77077–77096 (2018)
5. Hafeez, G., Javaid, N., Iqbal, S., Khan, F.: Optimal residential load scheduling under utility and rooftop photovoltaic units. Energies **11**(3), 611–637 (2018)
6. Kim, J.Y., Cho, S.B.: Electric energy consumption prediction by deep learning with state explainable autoencoder. Energies **12**, 739–752 (2019)
7. Metaxiotis, K., Kagiannas, A., Askounis, D., Psarras, J.: Artificial intelligence in short term electric load forecasting: a state-of-the-art survey for the researcher. Energy Convers. Manage. **44**(9), 1525–1534 (2003)
8. Hafeez, G., Javaid, N., Riaz, M., Ali, A., Umar, K., Iqbal, Z.: Day ahead electric load forecasting by an intelligent hybrid model based on deep learning for smart grid. In: Conference on Complex, Intelligent and Software Intensive Systems, pp. 36–49 (2019)
9. Ferrag, M.A., Maglaras, L.: DeepCoin: a novel deep learning and blockchain-based energy exchange framework for smart grids. IEEE Trans. Eng. Manage. **12**, 1–13 (2019)
10. Kim, T.Y., Cho, S.B.: Predicting residential energy consumption using CNN-LSTM neural networks. Energy **172**, 72–81 (2019)

11. Le, T., Vo, M.T., Vo, B., Hwang, E., Rho, S., Baik, S.W.: Improving electric energy consumption prediction using CNN and Bi-LSTM. Appl. Sci. **9**, 4237–4248 (2019)
12. LeCun, Y., Bengio, Y., Hinton, G.: Deep learning. Nature **521**(7553), 436–444 (2015)
13. Munz, G., Li, S., Carle, G.: Traffic anomaly detection using k-means clustering. In: GI/ITG Workshop MMBnet, pp. 13–14 (2007)
14. Kandananond, K.: Forecasting electricity demand in Thailand with an artificial neural network approach. Energies **4**, 1246–1257 (2011)
15. De Cauwer, C., Van Mierlo, J., Coosemans, T.: Energy consumption prediction for electric vehicles based on real-world data. Energies **8**, 8573–8593 (2015)
16. Dong, B., Cao, C., Lee, S.E.: Applying support vector machines to predict building energy consumption in tropical region. Energy Build. **37**, 545–553 (2005)
17. Li, Q., Ren, P., Meng, Q.: Prediction model of annual energy consumption of residential buildings. In: International Conference on Advances in Energy Engineering, pp. 223–226 (2010)
18. Xuemei, L., Yuyan, D., Lixing, D., Liangzhong, J.: Building cooling load forecasting using fuzzy support vector machine and fuzzy C-mean clustering. In: International Conference on Computer and Communication Technologies in Agriculture Engineering, pp. 438–411 (2010)
19. Ma, Y., Yu, J.Q., Yang, C.Y., Wang, L.: Study on power energy consumption model for large-scale public building. In: International Workshop on IEEE Intelligent Systems and Applications, pp. 1–4 (2010)
20. Ahmad, M.W., Mourshed, M., Rezgui, Y.: Trees vs neurons: comparison between random forest and ANN for high-resolution prediction of building energy consumption. Energy Build. **147**, 77–89 (2017)
21. Lee, D., Kang, S., Shin, J.: Using deep learning techniques to forecast environmental consumption level. Sustainability **9**, 1894–1910 (2017)
22. Li, C., Ding, Z., Zhao, D., Yi, J., Zhang, G.: Building energy consumption prediction: an extreme deep learning approach. Energies **10**, 1525–1544 (2017)
23. Hochreiter, S., Schmidhuber, J.: Long short-term memory. Neural Comput. **9**(8), 1735–1780 (1997)
24. Dua, D., Karra, T.E.: UCI machine learning repository Irvine, CA: University of California, School of Information and Computer Science (2007). http://archive.ics.uci.edu/ml
25. Maaten, L.V.D., Hinton, G.: Visualizing data using t-SNE. J. Mach. Learn. Res. **9**(11), 2579–2605 (2008)

On the Performance of Deep Learning Models for Time Series Classification in Streaming

Pedro Lara-Benítez[1(✉)], Manuel Carranza-García[1], Francisco Martínez-Álvarez[2], and José C. Riquelme[1]

[1] Division of Computer Science, University of Sevilla, 41012 Seville, Spain
plbenitez@us.es
[2] Data Science & Big Data Lab, Pablo de Olavide University, 41013 Seville, Spain

Abstract. Processing data streams arriving at high speed requires the development of models that can provide fast and accurate predictions. Although deep neural networks are the state-of-the-art for many machine learning tasks, their performance in real-time data streaming scenarios is a research area that has not yet been fully addressed. Nevertheless, there have been recent efforts to adapt complex deep learning models for streaming tasks by reducing their processing rate. The design of the asynchronous dual-pipeline deep learning framework allows to predict over incoming instances and update the model simultaneously using two separate layers. The aim of this work is to assess the performance of different types of deep architectures for data streaming classification using this framework. We evaluate models such as multi-layer perceptrons, recurrent, convolutional and temporal convolutional neural networks over several time-series datasets that are simulated as streams. The obtained results indicate that convolutional architectures achieve a higher performance in terms of accuracy and efficiency.

Keywords: Classification · Data streaming · Deep learning · Time series

1 Introduction

Learning from data arriving at high speed is one of the main challenges in machine learning. Over the last decades, there have been several efforts to develop models that deal with the specific requirements of data streaming. Traditional batch-learning models are not suitable for this purpose given the high rate of arrival of instances. In data streaming, incoming data has to be rapidly classified and discarded after using it for updating the model. Predicting and training have to be done as fast as possible in order to maintain a processing rate close to real-time. Furthermore, the models have to be able to detect possible changes in the incoming data distribution, which is known as concept drift.

Á. Herrero et al. (Eds.): SOCO 2020, AISC 1268, pp. 144–154, 2021.
https://doi.org/10.1007/978-3-030-57802-2_14

Despite the incremental learning nature of neural networks, there is little research involving deep learning (DL) models in the data streaming literature. Neural networks can adapt to changes in data by updating their weights with incoming instances. However, the high training time of deep networks presents challenges to adapt them to a streaming scenario. Very recently, our research group developed a deep learning framework for data streaming classification that uses an asynchronous dual-pipeline architecture (ADLStream) [14]. In this framework, training and classification can be done simultaneously in two different processes. This separation allows to use DL networks for data arriving at high speed while maintaining a high predictive performance.

The aim of this study is to evaluate how different DL architectures perform on the data streaming classification task using the ADLStream framework. Despite the promising results presented in [14], the experiments only considered convolutional neural networks, hence the suitability and efficiency of other types of deep networks is an area that has yet to be studied. In this work, we focus the experimental study on time-series data obtained from the UCR repository that have been simulated as streams. For this reason, we have designed DL models that are suitable for data having an inner temporal structure. The basic Multi-Layer Perceptron (MLP) is set as the baseline model and compared with other three architectures: Long-Short Term Memory network (LSTM), Convolutional Neural Network (CNN), and Temporal Convolutional Network (TCN). These models are evaluated in terms of accuracy and computational efficiency.

The rest of the paper is organised as follows: Sect. 2 presents a review on related work; Sect. 3 describes the materials used and the methodology; in Sect. 4 the experimental results obtained are reported; Sect. 5 presents the conclusions and future work.

2 Related Work

Over the last decades, there have been several efforts to develop models that deal with the specific requirements of data streaming. Traditional batch-learning models are not suitable for this purpose given the high rate of arrival of instances. In data streaming, incoming data has to be rapidly classified and then discarded after using it for updating the learning model. Predicting and training have to be done as fast as possible in order to maintain a processing rate close to real-time. Furthermore, the models have to be able to detect possible changes in the incoming data distribution, which is known as concept drift [1].

One of the most popular approaches has been to develop incremental or online algorithms based on decision trees, for instance, the Hoeffding Adaptive Trees (HAT) [3]. These models build trees incrementally based on the Hoeffding principle, that splits a node only when there is statistical significance between the current best attribute and the others. Later, ensemble techniques have been successfully applied to data stream classification, enhancing the predictive performance of single classifiers. ADWIN bagging used adaptive windows to control the adaptation of ensemble members to the evolution of the stream [3]. More recently, researchers have focused on building ensemble models that can deal effectively with concept drifts. The Adaptive Random Forest (ARF) algorithm proposes better resampling methods for updating classifiers over drifting data streams [11]. In [5], the authors proposed the Kappa Updated Ensemble (KUE) that uses weighted voting from a pool of classifiers with possible abstentions.

Despite the incremental learning nature of neural networks, there is little research involving DL models in the data streaming literature. Neural networks can adapt to changes in data by updating their weights with incoming instances. However, the high training time of deep networks presents challenges to adapt them to a streaming scenario in real-time. There have been proposals using simple networks such as the Multi-Layer Perceptron [10,16]. A deep learning framework for data streaming that uses a dual-pipeline architecture was developed in [14]. A more detailed description of the framework, which was the first using complex DL networks for data streaming, is provided in the next section.

3 Materials and Methods

3.1 ADLStream Framework

In this study, we use the asynchronous dual-pipeline deep learning framework (ADLStream) for data streaming presented in [14]. As can be seen in Fig. 1, the proposed system has two separated layers for training and predicting. This improves the processing rate of incoming data since instances are classified as soon as they arrive using a recently trained model. In the other layer, the weights of the network are constantly being updated in order to adjust to the evolution of the stream. This framework allows to use complex DL model, such as recurrent or convolutional, that would not be possible to use in a data streaming scenario if they are trained sequentially. The source code of ADLStream framework can be found at [13].

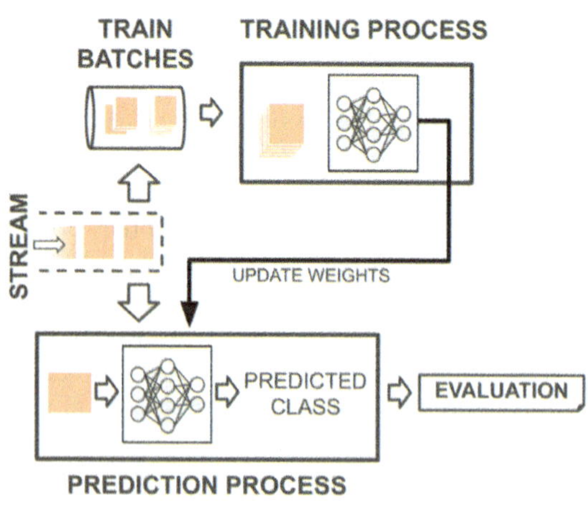

Fig. 1. Asynchronous dual-pipeline deep learning framework

3.2 Datasets

For the experimental study, 29 one-dimensional time series datasets from the UCR repository have been simulated as streams [6]. The selected datasets have different characteristics and are categorized into six different domains. Table 1 presents a detailed description of the number of instances, length of the time series instances, and the number of classes of each dataset.

Table 1. Datasets used for the study.

#	Dataset	Instances	Length	Classes	Type
1	TwoPatterns	5000	128	4	SIMULATED
2	CinCECGtorso	1420	1639	4	ECG
3	TwoLeadECG	1162	82	2	ECG
4	Wafer	7164	152	2	SENSOR
5	Pendigits	10992	16	10	MOTION
6	FacesUCR	2250	131	14	IMAGE
7	Mallat	2400	1024	8	SIMULATED
8	FaceAll	2250	131	14	IMAGE
9	Symbols	1020	398	6	IMAGE
10	ItalyPowerDemand	1096	24	2	SENSOR
11	ECG5000	5000	140	5	ECG
12	MoteStrain	1272	84	2	SENSOR
13	NonInvasiveFetalECGThorax1	3765	750	42	ECG
14	NonInvasiveFetalECGThorax2	3765	750	42	ECG
15	SwedishLeaf	1125	128	15	IMAGE
16	FordA	4921	500	2	SENSOR
17	Yoga	3300	426	2	IMAGE
18	UWaveGestureLibraryX	4478	315	8	MOTION
19	FordB	4446	500	2	SENSOR
20	ElectricDevices	16637	96	7	DEVICE
21	UWaveGestureLibraryY	4478	315	8	MOTION
22	UWaveGestureLibraryZ	4478	315	8	MOTION
23	HandOutlines	1370	2709	2	IMAGE
24	InsectWingbeatSound	2200	256	11	SENSOR
25	ShapesAll	1200	512	60	IMAGE
26	MedicalImages	1141	99	10	IMAGE
27	PhalangesOutlinesCorrect	2658	80	2	IMAGE
28	ChlorineConcentration	4307	166	3	SIMULATED
29	Phoneme	2110	1024	39	SENSOR

3.3 Experimental Study

In this section, we present the design of the different types of DL models selected for the experimental study. Furthermore, we also describe the details of the evaluation method used for the data streaming classification task.

3.3.1 Deep Learning Models

Our aim in this study is to evaluate the performance of different DL architectures within the ADLStream framework. Four different families of architectures are considered in the experiments: the Multi-layer Perceptron (MLP) which will serve as the baseline, recurrent networks using Long Short-Term Memory cells (LSTM), Convolutional Neural Networks (CNN), and Temporal Convolutional Networks (TCN). While the MLPs is unable to model the time relationships within the input data, the last three architectures are particularly indicated for dealing with data that has a temporal or spatial grid-like structure, such as the selected datasets. LSTM networks are one of the most popular types of recurrent neural networks. They connect each time step with the previous ones in order to model the long temporal dependencies of the data without forgetting the short-term patterns using special gates [9]. On the other hand, CNNs are networks based on the convolution operation, which creates features maps using sliding filters. They are also suitable for one-dimensional time series data since they are able to automatically capture repeated patterns at different scales [12]. Moreover, they have far less trainable parameters than recurrent networks due to their weight sharing scheme [4]. More recently, TCNs have emerged as a specialised architecture that can capture long-term dependencies more effectively by using dilated causal convolutions. With this operation, the receptive field of neurons is increased without the need for pooling operations, hence there is no loss of resolution [15]. Tables 2, 3, 4 and 5 provide a detailed description of the layers composing the four DL models considered. In these tables, the values of f and c are the number of features of the instances and the number of classes respectively. The baseline MLP model (Table 2) is composed of three dense layers with an increasing number of neurons. As can be seen, the other three models have a similar architecture since the convolutional or recurrent layers have the same number of maps or units and are followed by fully-connected layers with the same number of neurons. In the CNN (Table 3), two convolutional blocks with decreasing kernel size and max-pooling of stride 2 are applied before the dense layers. In the LSTM and TCN layers, the complete sequences are returned and connected to the next layers in order to use the information of all patterns extracted at different scales. In the TCN (Table 5), only one stack of residual blocks is used, and the dilated convolution is used with kernel ($k = 5$) and dilations ($d = \{1, 2, 4, 8, 16, 32, 64\}$). Another important element to consider is the use of a dropout with rate 0.2 on all dense layers in all models, with the aim of reducing over-fitting issues. The number of trainable parameters illustrates the computational cost of each model. The TCN has the highest number, which can be 37 times greater than the MLP model.

Table 2. Multi-Layer Perceptron architecture

MLP		
Layer	Type	Neurons
0	Input	f
1	Dense	32
2	Dense	64
3	Dense	128
4	Softmax	c
Params	$f \times 32 + 10240 + c \times 128$	

Table 3. Convolutional Neural Network architecture. k indicates the kernel size

CNN		
Layer	Type	Neurons
0	Input	f
1	Conv. ($k = 7$)	$f \times 64$ maps
2	Max-Pool ($k = 2$)	$f/2 \times 64$ maps
3	Conv. ($k = 5$)	$f/2 \times 128$ maps
4	Max-Pool ($k = 2$)	$f/4 \times 128$ maps
5	Dense	64
6	Dense	32
7	Softmax	c
Params	$f \times 2048 + 43648 + c \times 32$	

Table 4. Long Short-Term Memory Network architecture

LSTM		
Layer	Type	Neurons
0	Input	f
1	LSTM	$f \times 64$ units
2	LSTM	$f \times 128$ units
3	Dense	64
4	Dense	32
5	Softmax	c
Params	$f \times 8192 + 117760 + c \times 32$	

Table 5. Temporal Convolutional Network architecture. k indicates the kernel size

TCN		
Layer	Type	Neurons
0	Input	f
1	TCN ($k = 5$)	$f \times 64$ maps
2	Dense	64
3	Dense	32
4	Softmax	c
Params	$f \times 4096 + 372096 + c \times 32$	

3.3.2 Evaluation

For evaluating the results we use the prequential method with decaying factors, that incrementally updates the accuracy by testing the model with unseen examples [8]. The decaying factors are used as a forgetting mechanism to give more importance to recent instances for estimating the error, given the evolving nature of the stream. In our study, we use a decaying factor of $\alpha = 0.99$. The process of calculating the prequential accuracy can be formulated as follows, where L is the loss function and o and y are the real and expected output respectively.

$$P_\alpha(i) = \frac{\sum_{k=1}^{i} \alpha^{i-k} L(y_k, o_k)}{\sum_{k=1}^{i} \alpha^{i-k}} == L(y_i, o_i) + \frac{1}{\alpha} P_\alpha(i-1) \tag{1}$$

The metric selected is the Kappa statistic, that is more suitable than standard accuracy in data streaming due to the frequent changes in the class distribution of incoming instances [2]. The Kappa value can be computed as shown in the following equation, where p_0 is the prequential accuracy and p_c is the hypothetical probability of chance agreement.

$$k = \frac{p_0 - p_c}{1 - p_c} \tag{2}$$

4 Experimental Results

This section presents the Kappa accuracy results and the statistical analysis. The experiments have been carried out with an Intel Core i7-770K and two NVIDIA GeForce GTX 1080 8 GB GPU. The Apache Kafka server is used to reproduce the streaming scenario since it is the most efficient tool available [7].

4.1 Prequential Kappa

Table 6 presents the prequential kappa accuracy results obtained with the different models for each dataset. As can be seen, the CNN achieves the best performance for almost all the datasets considered, obtaining the highest average kappa accuracy value. The second model on average is the TCN, but closely followed by the LSTM that shows a similar performance. In general, the results prove that the ADLStream framework is able to achieve reliable results regardless of the deep learning architecture chosen.

4.2 Computation Time Analysis

In a data streaming environment, it is fundamental to analyse the efficiency of the architectures considered. The average processing rate of each model (average time to process each incoming instance) is provided at the end of Table 6. Logically, the MLP is the fastest model given its simple architecture. The second fastest model is the CNN, which has a significantly smaller number of parameters than the other two DL architectures. Thanks to the properties of parameter sharing, the CNN is able to process instances three times faster than the LSTM. The TCN is a more complex model with more convolutions which results in a processing rate of almost 8 times slower than the CNN.

4.3 Statistical Analysis

The ranking of the accuracy of the models obtained with the Friedman test is presented in Table 7. The CNN model leads the ranking, with a high difference in score with respect to the rest of the models. The TCN and LSTM obtain a similar score, while the MLP offers the worst performance. The null hypothesis is rejected since the p-value obtained (<0.001) is below the significance level ($\alpha = 0.05$).

In Bergmann-Hommel's post-hoc analysis, we perform pair-wise comparisons between all models. Table 8 reports the p-values and conclusions obtained. As can be seen, for the CNN all null hypothesis can be rejected since the p-values are always below the significance level. Therefore, it can be concluded that there is a statistical significance in the differences between the performance of the CNN and the other architectures considered. Nevertheless, there are no significant differences between the accuracy of LSTM and the TCN.

Table 6. Prequential kappa accuracy results

#	Dataset	MLP	LSTM	CNN	TCN
1	TwoPatterns	0.818	0.999	**1.000**	0.999
2	CinCECGtorso	0.348	**0.990**	0.994	0.933
3	TwoLeadECG	0.947	0.941	**0.991**	0.987
4	Wafer	0.581	0.995	**0.996**	0.710
5	Pendigits	0.728	0.987	**0.992**	0.953
6	FacesUCR	0.834	0.952	**0.974**	0.952
7	Mallat	0.963	0.920	**0.986**	0.978
8	FaceAll	0.841	0.953	**0.962**	0.948
9	Symbols	0.877	0.900	**0.949**	0.914
10	ItalyPowerDemand	**0.942**	0.921	0.935	0.934
11	ECG5000	0.881	**0.891**	0.888	0.890
12	MoteStrain	0.778	0.843	**0.878**	0.851
13	NonInvasiveFetalECGThorax1	0.851	0.862	**0.881**	0.873
14	NonInvasiveFetalECGThorax2	0.894	0.893	**0.901**	0.900
15	SwedishLeaf	0.679	0.775	**0.874**	0.844
16	FordA	−0.021	**0.691**	0.644	0.632
17	Yoga	0.213	0.659	**0.737**	0.689
18	UWaveGestureLibraryX	0.560	0.748	**0.761**	0.732
19	FordB	0.009	**0.654**	0.626	0.219
20	ElectricDevices	0.353	**0.803**	0.801	0.755
21	UWaveGestureLibraryY	0.506	0.641	**0.648**	0.608
22	UWaveGestureLibraryZ	0.496	0.646	**0.658**	0.613
23	HandOutlines	0.643	0.674	**0.721**	0.714
24	InsectWingbeatSound	0.605	0.602	**0.613**	0.598
25	ShapesAll	0.542	0.598	**0.606**	0.596
26	MedicalImages	0.304	0.566	**0.580**	0.503
27	PhalangesOutlinesCorrect	0.153	0.156	**0.474**	0.451
28	ChlorineConcentration	0.242	0.157	**0.900**	0.891
29	Phoneme	0.032	0.117	**0.182**	0.138
	Average kappa	**0.572**	**0.743**	**0.798**	**0.752**
	Average time per instance (ms)	**4.993**	**22.090**	**7.347**	**47.340**

Table 7. Friedman test ranking

Friedman test ranking	
CNN	1.200
TCN	2.533
LSTM	2.566
MLP	3.700

Table 8. Bergmann-Hommel's analysis

PostHoc analysis			
Comparison	p	z	Conclusion
MLP-CNN	<0.001	7.50	!=
LSTM-CNN	<0.001	4.10	!=
TCN-CNN	<0.001	4.00	!=
MLP-TCN	0.001	3.49	!=
MLP-LSTM	0.001	3.39	!=
LSTM-TCN	0.920	0.09	==

5 Conclusions

In this paper, the performance of several deep learning architectures for data streaming classification is compared using thc ADLStream framework. An extensive study over a large number of time-series dataset was conducted using multilayer perceptron, recurrent, and convolutional neural networks.

The research carried out for this study provided evidence that convolutional neural networks are currently the most suitable model for time series classification in streaming. Convolutional neural networks obtained the best results in terms of accuracy, with a very high processing rate. These characteristics present convolutional networks as the best alternative for processing data arriving at high speed. The other deep models, such as Long Short-Term Memory or Temporal Convolutional networks were not able to achieve such performance and their processing rate was slower.

Future work should study the behaviour of different deep learning models over concept drifts and their capacity to adapt to changes in the data distribution. Furthermore, a parameter optimization process could provide more specific architectures for the models and improve the performance. Future studies should also consider other less known models such as Echo State Networks, Stochastic Temporal Convolutional Networks or Gated Recurrent Units Networks.

Acknowledgements. This research has been funded by the Spanish Ministry of Economy and Competitiveness under the project TIN2017-88209-C2 and by the Andalusian Regional Government under the projects: BIDASGRI: Big Data technologies for Smart Grids (US-1263341), Adaptive hybrid models to predict solar and wind renewable energy production (P18-RT-2778). We are grateful to NVIDIA for their GPU Grant Program that has provided us high quality GPU devices for carrying out the study.

References

1. Anderson, R., Koh, Y., Dobbie, G., Bifet, A.: Recurring concept meta-learning for evolving data streams. Expert Syst. Appl. **138** (2019). https://doi.org/10.1016/j.eswa.2019.112832
2. Bifet, A., de Francisci Morales, G., Read, J., Holmes, G., Pfahringer, B.: Efficient online evaluation of big data stream classifiers. In: Proceedings of the 21th ACM SIGKDD International Conference on Knowledge Discovery and Data Mining, KDD 215, pp. 59–68. ACM, New York (2015). https://doi.org/10.1145/2783258.2783372
3. Bifet, A., Gavaldà, R.: Adaptive learning from evolving data streams. In: Adams, N.M., Robardet, C., Siebes, A., Boulicaut, J.F. (eds.) Advances in Intelligent Data Analysis VIII, pp. 249–260. Springer, Berlin (2009)
4. Borovykh, A., Bohte, S., Oosterlee, C.: Dilated convolutional neural networks for time series forecasting. J. Comput. Finance **22**(4), 73–101 (2019). https://doi.org/10.21314/JCF.2018.358
5. Cano, A., Krawczyk, B.: Kappa updated ensemble for drifting data stream mining. Mach. Learn. (2019). https://doi.org/10.1007/s10994-019-05840-z
6. Dau, H.A., Bagnall, A.J., Kamgar, K., Yeh, C.M., Zhu, Y., Gharghabi, S., Ratanamahatana, C.A., Keogh, E.J.: The UCR time series archive. CoRR abs/1810.07758 (2018)
7. Fernández-Rodríguez, J.Y., Álvarez García, J.A., Fisteus, J.A., Luaces, M.R., Magaña, V.C.: Benchmarking real-time vehicle data streaming models for a smart city. Inf. Syst. **72**, 62–76 (2017). https://doi.org/10.1016/j.is.2017.09.002
8. Gama, J., Sebastião, R., Rodrigues, P.P.: On evaluating stream learning algorithms. Mach. Learn. **90**(3), 317–346 (2013). https://doi.org/10.1007/s10994-012-5320-9
9. Gers, F.A., Schmidhuber, J., Cummins, F.: Learning to forget: continual prediction with LSTM. Neural Comput. **12**(10), 2451–2471 (2000). https://doi.org/10.1162/089976600300015015
10. Ghazikhani, A., Monsefi, R., Sadoghi Yazdi, H.: Online neural network model for non-stationary and imbalanced data stream classification. Int. J. Mach. Learn. Cybernet. **5**(1), 51–62 (2014). https://doi.org/10.1007/s13042-013-0180-6
11. Gomes, H.M., Bifet, A., Read, J., Barddal, J.P., Enembreck, F., Pfharinger, B., Holmes, G., Abdessalem, T.: Adaptive random forests for evolving data stream classification. Mach. Learn. **106**(9), 1469–1495 (2017). https://doi.org/10.1007/s10994-017-5642-8
12. Krizhevsky, A., Sutskever, I., Hinton, G.E.: ImageNet classification with deep convolutional neural networks. In: Proceedings of the 25th International Conference on Neural Information Processing Systems - Volume 1, NIPS 2012, pp. 1097–1105. Curran Associates Inc., Red Hook (2012)
13. Lara-Benítez, P., Carranza-García, M.: ADLStream: asynchronous dual-pipeline deep learning framework for online data stream mining. https://github.com/pedrolarben/ADLStream. Accessed 01 Apr 2020
14. Lara-Benítez, P., Carranza-García, M., García-Gutiérrez, J., Riquelme, J.: Asynchronous dual-pipeline deep learning framework for online data stream classification. Integr. Comput. Aided Eng., 1–19 (2020). https://doi.org/10.3233/ICA-200617

15. Yu, F., Koltun, V.: Multi-scale context aggregation by dilated convolutions. In: 4th International Conference on Learning Representations, ICLR 2016, Conference Track Proceedings, San Juan, Puerto Rico, 2–4 May 2016 (2016). http://arxiv.org/abs/1511.07122
16. Zhang, Y., Yu, J., Liu, W., Ota, K.: Ensemble classification for skewed data streams based on neural network. Int. J. Uncertain. Fuzz. Knowl. Based Syst. **26**(05), 839–853 (2018). https://doi.org/10.1142/S021848851850037X

An Approach to Forecasting and Filtering Noise in Dynamic Systems Using LSTM Architectures

Juan Pedro Llerena(✉), Jesús García, and José Manuel Molina

GIAA Group, University Carlos III of Madrid, Madrid, Spain
{jllerena,jgherrer,molina}@inf.uc3m.es

Abstract. Some of the limitations of state-space models are given by the difficulty of modelling certain systems, the filters convergence time or the impossibility of modelling dependencies in the long term. Having agile and alternative methodologies that allow the modelling of complex problems but still provide solutions to the classic challenges of estimation or filtering, such as the position estimation of a mobile with noisy measurements of the same variable, are of high interest. In this work, we address the problem of position estimation of 1-D dynamic systems from a deep learning paradigm, using Long-Short Term Memory (LSTM) architectures designed to solve problems with long term temporal dependencies, in combination with other recurrent networks. A deep neuronal architecture inspired by the Encoder-Decoder language systems is implemented, remarking its limits and finding a solution capable of making position estimations of a moving object. The results are finally compared with the optimal values from the Kalman filter, obtaining comparable results in error terms.

Keywords: Deep learning · LSTM · Parameter estimation · Filtering

1 Introduction

A wide variety of physical and scientific problems are based on the estimation of the state variables of a system that evolves with time, using for this purpose sensors that provide measurements with a certain level of uncertainty, so-called noisy observations.

To a large extent, these problems are formulated with state-space approximations. These approaches model the system behavior through a mathematical approximation mainly centered on a state vector, which is intended to contain all relevant and necessary information to describe it and make predictions. The sensors provide measurement or observation vectors that are related to the state vector of the analyzed system.

To analyze and infer a dynamic system, it is mainly required a model that describes the evolution of the states with time, and a second one that relates the observations with the states. These two large groups can be denominated from the state-space formulation as equations for state dynamics, and equations for observations (or likelihood), respectively.

In this context, many problems are tackled from the probabilistic formulation of the state space with Bayesian approximations, which provide a general solution for

Á. Herrero et al. (Eds.): SOCO 2020, AISC 1268, pp. 155–165, 2021.
https://doi.org/10.1007/978-3-030-57802-2_15

dynamic states estimation problems. Knowing the governing equations for dynamic systems allows forecasting, estimations, or control studies by structural stability analysis and bifurcations. However, when systems are very complex and/or when measurements are corrupted by not modelled errors [1], many complications may appear. In the work of H. H. Afshari et al. [2], a summary of different state estimation techniques from classical and Bayesian perspectives can be found.

It has been addressed that State Space Models (SSM), such as Hidden Markov Models (HMM) and Linear Dynamic Systems (LDS), have been and continue to be powerful tools for series modelling. However, these approaches are based on Markov assumption, while complex systems can actually have long-term dependencies that cannot be captured by these models, so their use is restricted.

Distinguishing the aforementioned cases by their probabilistic inference model, using artificial intelligence (AI) paradigms we can add intelligent inference methods. In [3] software sensors are treated as an alternative way to obtain estimators by means of classical methods. These AI-based estimators are computational algorithms designed to predict unmeasured parameters that are relevant for developing control laws or other applications.

LSTM neural architectures are not new [4], having been used in many applications where related with natural language processing [5] or attention [6] problems. Additionally, LSTM have shown good results in other scenarios, such as classification systems [7, 8], signal filtering after measurement [9], time estimations (e.g., oil production estimation [10]), traffic forecasting [11], stock index prediction [12] and system modeling [13], among others.

In Rassi et al. [14], they model highly non-linear systems that are restricted in the state space or centered around equilibrium points with ideal synthetic data. Rudy et al. work [1], models highly non-linear systems with noisy measurement information. The systems used are restricted in the state space by their equilibrium points or attractors.

In Zheng et al. work [15] present a new combined algorithm between LSTM and Monte Carlo for tracking, testing a continuous increasing function with noise (line) but limited to a specific time sequence.

In this paper we tackle the estimation/filtering problem with the position in a 1D moving object with an RNN inspired by the language encoder-decoder systems among others, comparing with the optimal solution of the Kalman filter. This work brings to neuro-estimators area, a new neural architecture, and we obtained comparable results in terms of error with respect to Kalman and opening new alternatives to problems not addressed by classical systems. Our work, in contrast to the majority of studies or problems in the literature such as [1, 14, 15] delocalizes the problem from a specific or limited estimation region and generalizes it, transforming in a recursive standardization-inference-unstandardization problem. For this purpose, the system is trained in a wide range of initial conditions.

This paper has been organized as follows. Section 2 presents the mathematical formulation of the problem. Section 3 introduces the used database. In Sect. 4 explained standardization process. Section 5 shows the LSTM neuro-estimator model and training parameters. The description and results of the numerical experiments are gathered in Sect. 6, concluding with the analysis, discussion and future works in Sect. 7.

2 Problem Formulation

A dynamic system is defined by the time dependency of its state variables, which may be observable or not:

$$dx(t)/\mathrm{dt} = f(x(t)) \tag{1}$$

The sensors used to monitor the observable variables of the system are dynamic model themselves. These sensors provide information regarding certain state variables, such as position, speed, among others. These observation systems can be defined as:

$$z(t) = \mathrm{H}x(t) + v(t) \tag{2}$$

where $x(t) \in \mathbb{R}^n$ is a state vector to be identified from a set of observations $z(t)$ with smaller or equal dimension than $x(t)$ and corrupted by an error parameter $v(t)$.

This paper proposes the estimation of state variables of an a-priori unknown noisy dynamic system, which may not depend only on a previous state but may present long term temporal dependencies for which information on temporary noisy measures of certain state variables is available in a supervised database that associates these measures with ideal values. For this purpose, we simulate the behavior of an ideal one-dimensional uniform rectilinear motion (URM), $W_k = [0\ ,\ 0]^T$ in which all parameters are controlled and distorted under constant Gaussian noise V_k, simulating measures z_k of the position state variable $H = [1\ 0]$:

$$\underbrace{\begin{bmatrix} p \\ v \end{bmatrix}_k}_{x_k} = \underbrace{\begin{bmatrix} 1 & \Delta T \\ 0 & 1 \end{bmatrix}}_{A} \overbrace{\begin{bmatrix} p \\ v \end{bmatrix}_{k-1}}^{x_{k-1}} + W_k \tag{3}$$

$$z_k = Hx_k + V_k \tag{4}$$

For this simple, short-term model, when all parameters are known, classical Markovian estimation techniques can be used. However, in the general case, LSTM architectures could generate better prediction of series using long term dependencies.

In this line, we propose to approach the problem from a deep learning (DL) perspective as a "*sequence-to-sequence*" problem, widely used in natural language processing problems. Where given a Z series composed by n features and k length time data, belonging to $\boldsymbol{\Phi}$ database, learn that another X series composed by n' features and k' length time data, associated to Z series but without noise, can belong to different dimensional spaces Z and X. Finally Z and X can be represented as: $Z = \{z1, z2, z3, \ldots zk\}$; $X = \left\{x1, x2, x3, \ldots, xk'\right\}$ for the ideal case of the same number of features $n = n'$ and length time data $k = k'$, $\boldsymbol{z} \cup \boldsymbol{x} = \Phi_i | z = \{\varphi_1, \varphi_2, \varphi_3, \ldots, \varphi_{k-1}\} and \boldsymbol{x} = \{\varphi_2, \varphi_3, \varphi_4, \ldots \varphi_k\}$ Where Φ_i is a full subset of the database $\boldsymbol{\Phi}$ and φ_k the characteristics vector at time position k. Finally, given a previous noisy sequence, our methodology allows to generate a filtered output.

Most dynamic systems are not restricted in their state space domain, while neuronal architectures are restricted systems defined by the functions that constitute each layer.

These layers are composed by the functions that define each of their units and, in greater depth, the activation functions of each of the artificial neurons. In this way, the regression problems will be limited to the training space unless a generalization is proposed to cover all the domain.

To address this issue, we propose the use of a recursive method of standardization based on the movement a time window through the data, maintaining a small overlap region with the previous window for network activation at each window shift. This overlap of network activations retains the long-term dependencies.

The activation process consists to introducing a small section of measured sequence into the network, so that the internal network architecture can adjust its internal weights to link them to the training data. These corrections are made by transitions, and the transitions happen when measurements are inserted.

3 Database

The position trajectories with respect to time of an URM are linear and they are generated with the model (3) to obtain ideal values. To simulate the measurement behavior, is added gaussian noise to the ideal values.

We consider positive and negative positions and speed. With previous descriptions it has been generated a database $\boldsymbol{\Phi}$, composed by a set of N = 1000 paths measured *Zi* and it is corresponding ideal *Xi* as a synthetic data set corresponding to URM according to the parameters of Table 1.

$$\Phi = \bigcup_{i=1}^{N} \Phi_i | \Phi_i = Z_i \cup X_i \quad (5)$$

Table 1. Synthetic data generation parameters

Parameter	Data generation range	
	Minimum	**Maximum**
Initial position [m] | Speed [m/s]	-25 | -55	25 | 55
Simulation end times [s] | Sampling time [s]	8 | 0.05	
Number of window data | Overlap [Nº data]	80 | 15	
Gaussian noise measurement $V_k \sim \mathcal{N}(0, \sigma_Z)$	0.9	

The speed range decision has been taken considering that the maximum speed of a vehicle for this problem is 198 km/h. The rest of the values have been considered in a heuristic way.

3.1 Setting up Data for Training

From the training point of view, we will have that the ideal data *X* which can also be called objective or "target" that corresponds to the goal to which the *Z* measurements must be adjusted and must be prepared by an expert system.

The input data Z and network target X are the same time size, are standardized and truncated as follows. The last value is removed from the measured Z set, while the target set corresponds to the X set, the first value is removed, so input and target keep the same temporal length, but a displaced temporal unit. In other words, given a Z' series learn that you get X'.

In this way, data are structured for a sequence-to-sequence architecture of the same input-output dimension, but shifted one unit of time, allowing the long term estimation of how long ("window size") $X = [x_2\, x_k]$ target from measured values $Z = \left[z_1, z_{k-1}\right]$ under a certain Gaussian noise. The training and validation subsets are obtained through two consecutive time windows of 80 samples of each signal. The first window is associated to the training set and the second to the validation set, obtaining two subsets with the same number of data.

4 Data Standardization

Considering the networks' sensitivity to data scaling, a data standardization is performed as in [10] but under a geometrical interpretation of them. The behavior of an URM, in general, shows an increasing tendency in absolute value, so this interpretation is essential for training and the model inference

So, the activation process can have certain previous information, a small region of overlap (Table 1) is used between the adjacent windows, defined by a set of data from the previous window that is used for the activation of the network at each window movement (overlap).

A translation is performed to transform for the second time window into the first, by subtracting the minimum (m) value of the signal from all its measurements. Then knowing the maximum (M) value of the window and the minimum, the normalization is done by dividing the set of data from which the minimum value has been subtracted by the amplitude of the signal in the window, which we can obtain as the difference between the maximum value minus the minimum (s), this normalization represents a scaling in geometric terms (Fig. 1).

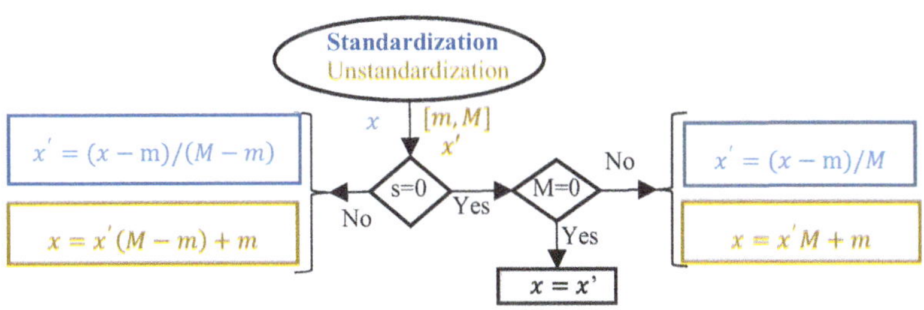

Fig. 1. Standardization/unstandardization algorithm

The m and M parameters required for unstandardization are essential for a good fitting between the results of the standardized space and the real space with which to obtain comparative metrics, so they will be specified in each of the experimental sections.

5 LSTM Neuro Position Estimator

In this section, we describe the artificial neural network architecture employed and training parameters used (Table 2).

Based on the good results in linear regression models with multilayer models of perception [16] and under the stability studies in recurrent neural networks of [17], the deepest part of the network is composed by a fully connected layer with ReLU activation functions and a dropout layer of 20%. According to reminding long term tendencies and under the good results in estimation problems with LSTM architectures like [18, 19] among others, and around encoder-decoder architectures concept for non Markovian models like [11, 14, 20] together with the good results in filtering problems [9] and for the system identification with noise [1, 14] the final architecture is composed by 8 set layers. Output network function is restricted because it is made up of each layer functions and they are restricted. This is because each layer's functions are functions composed by each hidden cell and they are restricted functions as the sigmoid function, among others.

Table 2. Listing of neural network layer: s = 80 is the number of samples per input trajectory.

Nr	Name and type	Activation/ prop.	Learnable	States
1	Sequence Input: 1x80	1	-	-
2	lstm_1: LSTM Hidden units: 400	State activation function: tanh Gate activation function: sigm	Input Weights: 1600x1 Recurrent Weights: 1600x400 Bias:1600x1	Hidden States: 400x1 CellState:400x1
3	fc_1: Fully connected	16	Weights: 16x400 Bias:16x1	-
4	relu_1: ReLU	16	-	-
5	Do: Dropout 20%	16	-	-
6	lstm_2: LSTM Hidden units: 200	State activation function: tanh Gate activation function: sigm	Input Weights: 800x16 RecurrentWeights:800x200 Bias:800x1	Hidden States: 400x1 CellState:400x1
7	fc_2: Fully connected	1	Weights: 1x200 Bias: 1x1	-
8	Regression output	Loss function: MSE	-	-

5.1 Setup Training Options

For training this model, we used Adam optimizer for the excellent result shown in multi-layer recurrent network training [20]. We train with 20 batches during 80 epochs starting from an initial learning rate of 0.005 and with a drop of the learning factor of 0.5 after the first 8 epochs. The training updates the individual weights using the Adam algorithm but with an $\mathcal{L}_2$ adjustment of the target function under the regularization factor of 10^{-4} with the intention of reducing overload in training. Loss training function:

$$\mathrm{RMSE} = \sqrt{\sum \frac{\left(x_{pred} - x_{ref}\right)^2}{N}} \tag{6}$$

6 Experiments

The following section present 3 different experiment. First, chapter 5 LSTM model validation whit Sect. 4 data set, loss position measurements simulation in filtering system and ending whit filtering system simulation which new measurements in feedback.

All experiment is comparing with Kalman filter. This Kalman filter consider zero process noise $W_k = [0\ ,0]^T$ and position measurement with gaussian noise $\mathcal{N}(0, \sigma_Z)$ (4) Table 1. The system model corresponds with Eq. (3). Kalman filter is initialized after two consecutive measurements to determine the unmeasured state (speed) as $v_2 = (p_2 - p_1)/\Delta T$ and covariance matrix start like this: $P_2 = \sigma_Z \begin{pmatrix} 1 & 100 \\ 100 & 2 \end{pmatrix}$.

6.1 LSTM Validation

To validate the model, we use a time window just like the training but from the validation set.

We use the RMSE (6) metric from the last filtered value from the validation set Fig. 2(a) and the first one after the overlap window Fig. 2(b) always in real space.

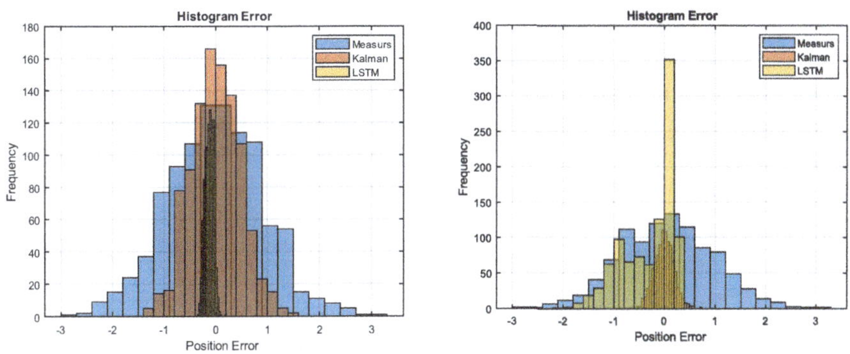

(a) (b)

Fig. 2. Histogram error: (a) first estimation after overlap measure area in 1st time window of 80 measures, (b) last estimate in 1st time window of 80 measurements.

The justification for using this metric in two different cases was given by the behavior of the Kalman filter, which improves its estimation when it acquires more measurements, so its best estimate in a time window will be the last filtered value, while the worst estimate will be made at the beginning of the estimates. Considering a temporal region of data to activate the network, this value is calculated just after that region for both systems, Kalman and LSTM.

The process of unstandardization of the network given data by the different validation series, the maximum (M) and minimum (m) values of each measured Z_i series are used and being saved in previous standardization phase.

The following figures illustrate the histograms obtained in prediction for a first-time window of data from the validation series.

6.2 Loss Position Measurements Effect Simulation

Table 3. LSTM and Kalman validation results

Model	Histogram RMSE	
	First predicted value	**Last predicted value**
Measures | Kalman | **LSTM**	0.9090 | 0.4750 | **0.1490**	0.9281 | 0.1969 | **0.5912**

This section shows the system evolution in the first and second time window when only one set of measures (overlap/activation) is used to make an estimate and then it is feedbacked with the previous estimate, both in the Kalman model and in the LSTM model. To do this, a series with the following initial conditions for the simulation of the URM is generated; $x0 = [-23.4897, -5.3815]$ and with the noise parameters indicated in Table 1. In the 2nd time window we use all the data from the 1st window to feedback in Kalman filter and only the overlap region to activate the neural architecture, later in both cases, we make an estimation without measurements.

The first window graph in Fig. 3(a), shows how the Kalman filter has not enough measures to reduce its error and it decouples when it does not receive new measures, increasing its error during the estimates, while LSTM architecture with few measures manages to make good estimations and gets in that window an RMSE lower order of magnitude than Kalman (Table 4). In Fig. 3(b), we see how Kalman with first window data has managed to improve its behavior, but will continue to increase its error with the estimates passage, while the LSTM architecture keeps its error limited, remembering that has been activated only with overlapping window data.

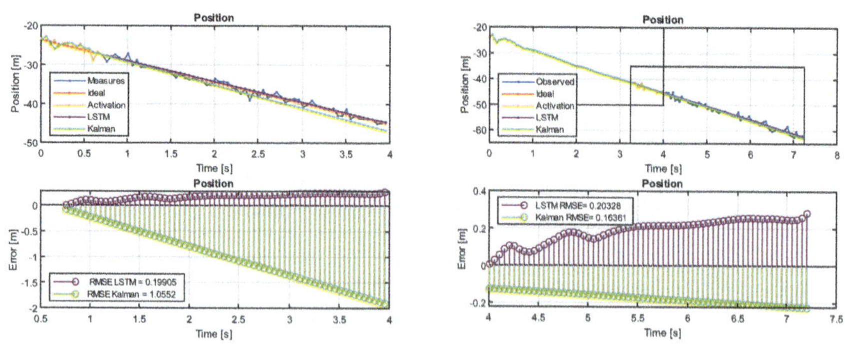

(a) (b)

Fig. 3. LSTM and Kalman without feedback measures. (a) first, (b) second, time-windows.

6.3 Filtering System Simulation with New Measurements in Feedback

In this following case, the systems are continuously in feedback with measures. In the case of the LSTM model, the network determines internally if the measurement is

Table 4. RMSE without feedback measures after overlap window

Model	RMSE	
	1St Window	**2Nd Window**
Kalman | **LSTM**	1.0552 | **0.1990**	0.1636 | **0.2033**

relevant or not to forecast next time step state to be forecast, while in Kalman's case this is used to reduce the filtering error (Table 5). In follow figure (Fig. 4), Kalman's filter tends to minimize his error when he receives new measures, but LSTM model too, getting in this first phase, an improved error regarding the Kalman filter.

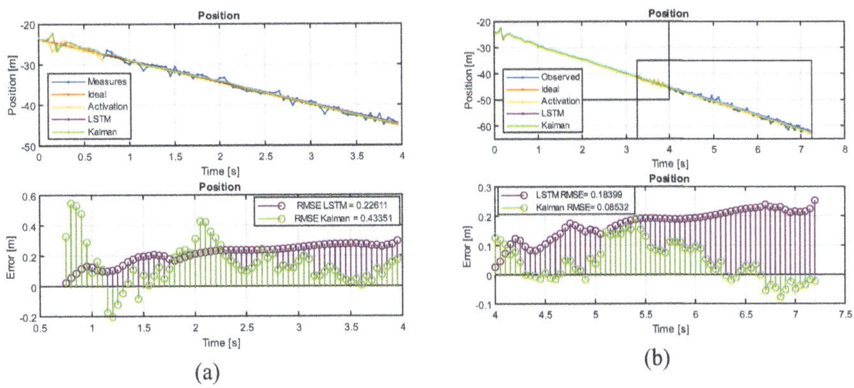

Fig. 4. LSTM and Kalman with feedback measures. (a) first, (b) second, time-windows.

In the first graph of Fig. 4(b) shows the evolution time in the second time window of the LSTM model and the Kalman filter. While the second shows the error evolution in that time window.

Table 5. RMSE with feedback measures after overlap window

Model	RMSE	
	1St Window	**2Nd Window**
Kalman | **LSTM**	0.4335 | **0.2261**	0.0853 | **0.1840**

7 Conclusions and Future Works

In this paper we implemented a neuro-estimator/filter architecture with recurrent LSTM layers and inspired by encoder-decoder systems for sequence-to-sequence problems able to estimate and filter position measurements of a URM. This model was compared with a Kalman filter along two time-windows, showing at first times the LSTM model improves the results in filtering and estimation than Kalman, also showing evidence of a limited error in the estimation/filtering process being able to interpret internally the measurement noise.

We have verified that with few initial measurements the LSTM system manages to extract the general trend of the trajectory, while the Kalman filter with few measurements may not be able to reduce their estimation error and the system is susceptible to decouple in the absence of measurements, Fig. 3(a). The magnitude orders of errors and RMSE are equivalent throughout this study between Kalman and LSTM, but it is noticeable how the LSTM model shows a minor magnitude in the RMSE at the first estimates, Fig. 2(a), Table 3.

It's important to mention the fact that in the processes of unstandardization for the neuronal architecture data for all experiments, we used (m) and (M) parameters obtained from the standardization ideal signal X associated with the series of measures Z, with the aim of making a first approximation with the lowest possible error level of these neural systems. So, in certain degree, the LSTM neural system is endowed with some additional information as compared to Kalman model.

In conclusion, the LSTM model shown may be a good proposal for an alternative or hybridization with a Kalman filter, but Kalman continues to be more robust method in long time ranges and continuous measurements for an URM.

Immediate future work includes application to non-linear or non-Gaussian problems, as well as multi-dimensional position estimation problems.

Acknowledgments. This work was supported by Ministry of Science, Innovation and Universities from Spain under grant agreement No. PRE-C-2018-0079.

References

1. Rudy, S.H., Kutz, J.N., Brunton, S.L.: Deep learning of dynamics and signal-noise decomposition with time-stepping constraints. J. Comput. Phys. **396**, 483–506 (2019)
2. Afshari, H.H., Gadsden, S.A., Habibi, S.: Gaussian filters for parameter and state estimation: a general review of theory and recent trends. Sig. Process. **135**, 218–238 (2017)
3. Ali, J.M., Hussain, M.A., Tade, M.O., Zhang, J.: Artificial intelligence techniques applied as estimator in chemical process systems - a literature survey. Expert Syst. Appl. **42**(14), 5915–5931 (2015)
4. Hochreiter, S., Schmidhuber, U.J.: Long short-term memory. Neural Comput. **9**(8), 1735–1780 (1997)
5. Bahdanau, D., Cho, K.H., Bengio, Y.: Neural machine translation by jointly learning to align and translate. In: 3rd International Conference on Learning Representation. ICLR 2015 - Conference Track Proceedings, pp. 1–15 (2015)
6. Gan, C., Wang, L., Zhang, Z., Wang, Z.: Sparse attention based separable dilated convolutional neural network for targeted sentiment analysis. Knowl.-Based Syst. **188**, 104827 (2019)
7. Wang, Y., Huang, M., Zhao, L., Zhu, X.: Attention-based LSTM for aspect-level sentiment classification. In: Proceeding Conference on Empirical Methods in Natural Language Processing. EMNLP 2016, pp. 606–615 (2016)
8. Arriaga, O., Plöger, P., Valdenegro-Toro, M.: Image captioning and classification of dangerous situations no. 1 (2017)

9. Arsene, C.T.C., Hankins, R., Yin, H.: Deep learning models for denoising ECG signals. In: 2019 European Signal Processing Conference, 2–6 September, vol. 2017, no. Iaa 220, pp. 1–5 (2019)
10. Song, X., et al.: Time-series well performance prediction based on long short-term memory (LSTM) neural network model. J. Pet. Sci. Eng. **186**, 106682 (2019)
11. Zhao, Z., Chen, W., Wu, X., Chen, P.C.V., Liu, J.: LSTM network: a deep learning approach for short-term traffic forecast. IET Image Process. **11**(1), 68–75 (2017)
12. Orimoloye, L.O., Sung, M.C., Ma, T., Johnson, J.E.V.: Comparing the effectiveness of deep feedforward neural networks and shallow architectures for predicting stock price indices. Expert Syst. Appl. **139**, 112828 (2020)
13. Zaheer, M., Ahmed, A., Smola, A.J.: Latent LSTM allocation joint clustering and non-linear dynamic modeling of sequential data. In: 34th International Conference on Machine Learning. ICML 2017, vol. 8, pp. 6040–6049 (2017)
14. Raissi, M., Perdikaris, P., Karniadakis, G.E.: Multistep neural networks for data-driven discovery of nonlinear dynamical systems, pp. 1–19 (2018)
15. Zheng, X., Zaheer, M., Ahmed, A., Wang, Y., Xing, E.P., Smola, A.J.L: State space LSTM models with particle MCMC inference, pp. 1–12 (2017)
16. Shapsough, S., Dhaouadi, R., Zualkernan, I.: Using linear regression and back propagation neural networks to predict performance of soiled PV modules. Procedia Comput. Sci. **155**(2018), 463–470 (2019)
17. Barabanov, N.E., Prokhorov, D.V.: Stability analysis of discrete-time recurrent neural networks. IEEE Trans. Neural Netw. **13**(2), 292–303 (2002)
18. Deng, L., Hajiesmaili, M.H., Chen, M., Zeng, H.: Energy-efficient timely transportation of long-haul heavy-duty trucks. IEEE Trans. Intell. Transp. Syst. **19**(7), 2099–2113 (2018)
19. Wu, Q., Lin, H.: A novel optimal-hybrid model for daily air quality index prediction considering air pollutant factors. Sci. Total Environ. **683**, 808–821 (2019)
20. Kingma, D.P., Ba, J.L.: Adam: a method for stochastic optimization. In: 3rd International Conference on Learning Representation - Conference Track Proceeding. ICLR 2015, pp. 1–15 (2015)

Novel Approach for Person Detection Based on Image Segmentation Neural Network

Dominik Stursa[1(✉)], Bruno Baruque Zanon[2], and Petr Dolezel[1]

[1] University of Pardubice, Studentska 95, 53210 Pardubice, Czech Republic
dominik.stursa@upce.cz

[2] University of Burgos, Calle Don Juan de Austria 1, 09001 Burgos, Spain

Abstract. With the rise of the modern possibilities in computer science and device engineering, as well as with growing population in big cities among the world, a lot of new approaches for person detection have become a very interesting topic. In this paper, two different approaches for person detection are tested and compared. As the first and standard approach, the YOLO architectures, which are very effective for image classification, are adapted to the detection problem. The second and novel approach is based on the encoder-decoder scheme causing the image segmentations, in combination with the locator. The locator part is supposed to find local maxima in segmented image and should return the specific coordinates representing the head centers in the original image. Results clearly report this approach with U-Net used as encoder-decoder scheme with the locator based on local peaks as the more accurately performing detection technique, in comparison to YOLO architectures.

Keywords: Person detection · Convolutional neural network · YOLO

1 Introduction and Related Work

With the rise of modern possibilities in computer science and device engineering as well as with growing population in big cities among the world, a lot of new approaches for person detection have become a very interesting topic. Person detection has an indispensable importance and is still increasingly needed for purposes of surveillance systems, safety in public transport, optimization in transport planning etc. As an initial part of person tracking, more accurate detection methods need to be developed.

These days, problems about person localization, detection and tracing are more in focus from academic and also corporate experts. Various researched

Supported from ERDF/ESF. "Cooperation in Applied Research between the University of Pardubice and companies, in the Field of Positioning, Detection and Simulation Technology for Transport Systems (PosiTrans)" (No. CZ.02.1.01/0.0/0.0/17_049/0008394).

Á. Herrero et al. (Eds.): SOCO 2020, AISC 1268, pp. 166–175, 2021.
https://doi.org/10.1007/978-3-030-57802-2_16

approaches to person detection are based on radar sensors [4], 3D scanners [2] or infra-red sensors [1]. However, these approaches often fail to detect every human passing through and are not able to track people precisely. For these difficulties, person tracking systems are still more often implemented using video processing algorithms and computer vision techniques [8]. Besides, the trend in image classification and object detection in visual data is clearly heading towards convolutional neural networks [15].

For person detection, not only technologies and methods interfere with implementation possibilities, but also laws of the country where the detection system will be applied. As such, methods, where identification of the person is not possible, are more attractive for the corporate environment. Thus, a monitoring system placed above passing humans should naturally solve the mentioned difficulty as shown in Fig. 1.

Fig. 1. Image captured from above heads (high angle - people cannot be identified).

As the view at each person is significantly limited, the main features to detect are heads and shoulders. Only a few approaches for person detection, tracking and counting with the video acquisition system placed above heads of people, have been proposed. Gao et al. [7] provide a technique combining convolutional neural networks and cascade Adaboost methods. The method based on combination of classical RGB and depth camera have been used in [6]. Both mentioned articles do not consider a strict vertical downward frame acquisition.

Round object detection based on image feature extraction using a histogram of oriented gradients in combination with pattern recognition network or SVM as a classifier, is shown in previous authors publications [5,16].

Sun et al. [18] proposed a method that utilizes the depth video stream and computes a normalized height image of the scene after removing the background. The height image is a projection of the scene depth below the camera, which helps for better segmentation of the scene. Therefore, based on the results [18], the scene segmentation seems to be a possible approach for object detection.

The paper is structured as follows. Firstly, the problem is properly formulated. In the following section, the used methods are described and the dataset acquisition is illustrated. Then, the experiments along with the results are presented and discussed. The article is finished with the conclusions.

2 Problem Formulation and Methodology

Vertical downward frames captured with a monocular camera create a special scene. If the camera sensor is placed indoors and statically at one place, the scene is composed by several different types of objects, which could be divided into a few categories like person, bags, floors, railings etc.

Hence, the idea of this work is to propose a novel method for human detection and positioning based on image segmentation neural networks and compare it to a well-established approach in order to develop an efficient person detector in real-life RGB images. As such, the provided images are supposed to be derived from a video, captured from above the person heads. Inputs of the detector, in the case considered in this contribution, are size normalized RGB images cropped from a video.

The process of person detection is realized and tested by two different methods based on neural networks. The comparative method uses the neural network architectures for object detection called the YOLO, where the object and its position are represented by a class and its bounding box defined by the specific coordinates. The proposed method is based on the encoder-decoder scheme, providing the image segmentation, supported with the locator. The locator part finds local maxima in segmented image and returns the specific coordinates representing the head centers.

2.1 Novel Approach for Person Detection

As mentioned above, our approach is composed of three parts: encoder, decoder and locator. The encoder part is based on a convolutional neural network (CNN) for image classification. Several topologies have been tested for selecting the one with best results to compare with YOLO architectures. The decoder part is created as the combination of feedforward neural network (FFNN) and CNN. The main reason for this structure is to create a segmented image where only heads are highlighted as radial gradients. Finally, the function returning coordinates of local peaks in image is used as a locator. For finding the local maxima, a maximum filter is used. This dilates the original image and merges neighboring local maxima closer than the size of the dilatation. Coordinates, where the dilated image is equal to the original image, are returned as a local maxima. The complete system is shown in Fig. 2. In this paper, the encoder-decoder part is examined in detail.

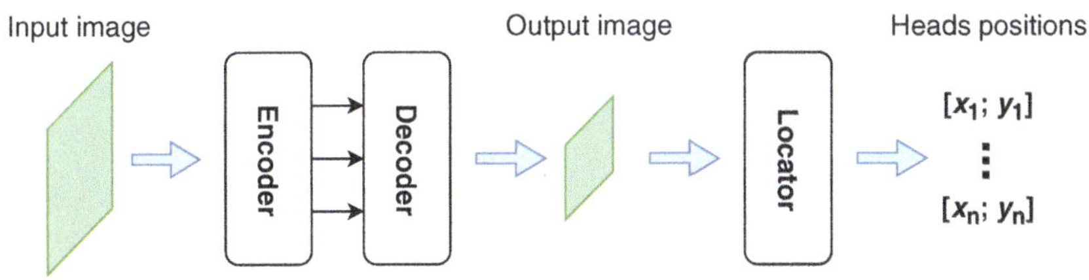

Fig. 2. Scheme of novel approach for person detection.

2.2 YOLO Architectures

In order to compare the proposed approach with a state-of-the-art model, we have selected the YOLO model [11]. The reasons for this decision are derived from the problem to tackle in the long term: the detection of person on a stream of video images. This task is on the one hand straightforward, as we only want to distinguish if there is a person present in the image or not, and on the other hand demanding in the sense that it must be completed in real time. YOLO is one of the fastest architectures for object recognition in images, while at the same time has a very similar correct recognition rates as more complex neural models, even on multi-class problems.

The main idea of the architecture is to frame the recognition problem as a regression one, instead of defining a convolutional window to swipe the image looking for recognizable shapes. This allows calculation of both the bounding box of the object to be recognized and the confidence percentage of that object which belongs to a certain class. The initial YOLO model is a deep neural network architecture which includes originally 24 convolutional layers that are used to extract the main features from the image and serve as inputs to 2 fully connected layers that are utilized to predict the final coordinates for objects and their probabilities of belonging to a class.

In our experiments we used the YOLOv2 model, which is a modification for faster performance of the original model [12]. The YOLOv2 includes many improvements to the initial architecture, in order to improve both its performance and its computational complexity. The main modification is the inclusion of pre-calculated anchor boxes, used to simplify calculations. Instead of predicting the coordinates of the bounding boxes of an object, the task is now to calculate the offset for one of the given boxes to match the detected object. This allows large simplification of the fully connected layers of the second phase of the model. In order to adapt the pre-calculated anchor boxes for a given recognition problem, a simple algorithm is used as a previous phase for the training of the model. This involves calculating a standard k-means on the sizes of the labeled objects in the images, to determine the most common sizes of the objects to detect.

In the original work, the base model for YOLOv2 includes 19 convolutional layers and 5 maxpooling layers alternated between them. This architecture uses batch normalization to stabilize training, speed up convergence, and regularize the model.

3 Dataset Creation

For the purposes of human detection with mentioned methods, specific datasets were created. As both methods are based on neural networks, each dataset was composed of an input-output pair series for use in supervised learning.

Two-dimensional matrices with three layers, representing RGB picture, were used as the input for both methods. The video sequence with person walking on the staircase was captured with a monocular camera. Then, the frames with significant shift between person head positions in two consecutive frames were selected. Selected frames were cropped and resized for the purposes of tested neural network architectures. Due to the difference between the tested methods, two types of outputs were prepared.

A labeled picture is supposed to be the output of the YOLO architectures. Therefore, the picture labeling was performed using the MATLAB tool called Image Labeler. The output from the Image Labeler was then modified to a proper structure necessary for the training of the YOLO.

For the proposed method, a special training set was prepared. In particular, output images were created, where every supposed center of a head was labeled by the value 1 and the surrounding values within the defined radius were gradually decreased to zero. The input and enlarged picture of the output for encoder-decoder training is shown in Fig. 3.

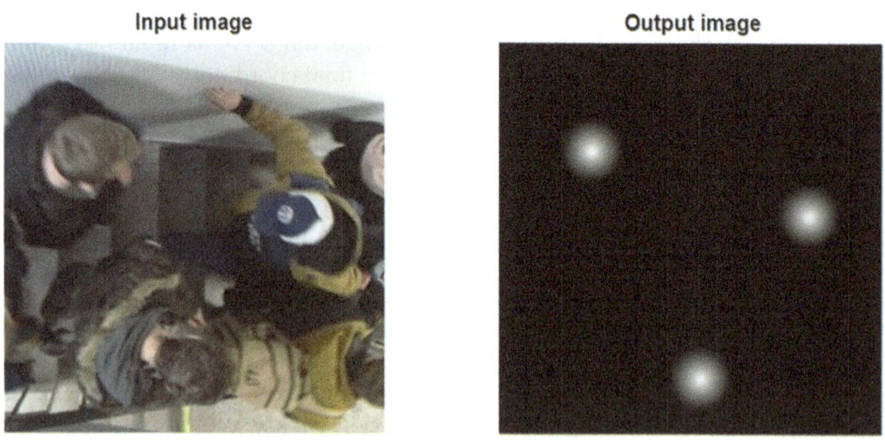

Fig. 3. Input-output pair for training of the encoder-decoder part of a novel approach.

4 Experiment Procedure

For both methods, specific datasets were created. Eventually, 1173 images from the captured video were selected. Images were size normalized, which made them ready as inputs for both methods. Then, for the every image, a corresponding expected output was created.

Datasets were split into 2 groups with the ratio of 3 to 1. The first group with a total of 881 input-output pairs was randomly selected from the dataset for the purposes of neural network training. The second group with a remaining 292 pairs was left for testing.

The YOLO architecture is well known and tested by its authors in [11,12]. Therefore, the training was realized for several of these structures with the specific data.

On the other hand, the topology considered in the case of the novel approach had to be tested first. Thus, totally 5 possible topologies were selected. Every topology was tested and evaluated 10 times. The total mean square error, defined as follows, was used as the metric.

$$E_{val} = \frac{1}{n \cdot N} \sum_{i=1}^{N} \sum_{j=1}^{n} [y_i(j) - \hat{y}_i(j)]^2, \quad (1)$$

where N is the number of the output samples in the testing set, n is the number of pixels in output, $y_i(j)$ is the desired value of pixel in the ith output, and $\hat{y}_i(j)$ is the predicted value of pixel in the ith actual output from the net.

All of the best tested topologies were then selected for comparison with the YOLO architecture.

4.1 Tested Encoder-Decoder Topologies

The encoder-decoder part of the approach was tested in two ways. In the first structure, the decoder part remained the same and the encoder part was progressively replaced by 4 tested topologies in total. The decoder part was composed of a flatten layer, two dense layers reshaped to a rectangle for possible connections to convolutional part, two convolution layers, max-pooling layer, and a convolution layer providing the output picture. Decoder was chosen on experiments with a few topologies, which were tested, but are not considered here.

Encoder topologies were selected based on authors previous experience. Net1 consists of two convolutional and one max-pooling layers. Net2 is similar, but it contains a more complex sequence of anterior layer. Both networks were adapted from [10]. The third is one of the pioneering architectures - LeNet-5 [3,9], while the fourth is probably the most cited topology - AlexNet [17].

The second structure was based on the U-Net, which is a symmetric fully convolutional network originally used for image segmentation in the medical sector and was defined in [13].

The best results of each tested structure with relative sizes (depth - number of layers, size - disk space) of the used networks are shown for comparison in Table 1.

Table 1. Resulting values of metric (1) for every tested structure

Network	E_{val}	Depth	Size	Parameters (Millions)
AlexNet	6.93E−03	12	227 MB	61.0
Net1	9.49E−03	8	285 MB	24.9
Net2	8.58E−03	10	535 MB	46.8
LeNet	9.39E−03	8	153 MB	13.4
U-Net	3.82E−03	24	355 MB	31.0

4.2 Tested YOLO Architectures

As discussed in Subsect. 2.2 the main architecture of the YOLOv2 approach includes several CNN initial layers for feature extraction and some final layers to perform the detection, as a particular type of regression task. With this architecture in mind, we have completed different tests using several pre-trained CNNs used for feature extraction present in literature for the first stage, while keeping the layers of the second stage unchanged. By choosing among the fastest performing CNNs and given that the recognition task is much simpler than those that initially designed for (multi-class detection), we expect to obtain a reasonable trade off between accuracy and low complexity for the recognition task. Relative sizes of the networks used are detailed for comparison in Table 2 including AlexNet, which was not used, but is included as a reference. All have been pre-trained on a subset of the ImageNet database [14] and then were performed a transfer learning operation using the same training set used in their encoder-decoder counterparts.

Table 2. Relative sizes of pre-trained models used as the part of the YOLO models tested.

Network	Depth	Size	Parameters (Millions)
squeezenet	18	4.6 MB	1.24
shufflenet	50	6.3 MB	1.4
mobilenetv2	53	13 MB	3.5
AlexNet	8	227 MB	61.0

5 Results and Discussion

The aim of this section is to evaluate both tested approaches represented by the YOLO and the novel approach.

5.1 Metrics Definition

At first, the overlap between two bounding boxes is defined and called the intersection over union (IOU). The ground truth bounding box and predicted bounded box are necessary to know in order to evaluate this metric. The intersection is given by dividing the overlapping area of these bounding boxes with the area of union between them, as shown in Fig. 4.

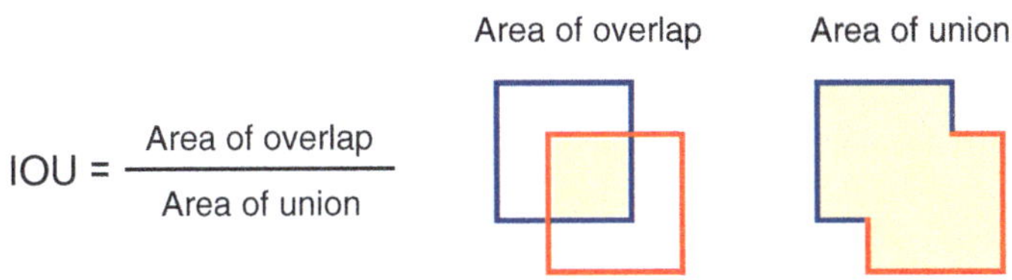

Fig. 4. Illustrative description of the IOU.

In addition, precision and recall are considered for further evaluation. The precision represents ability of a model to identify only the relevant objects. Hence, the percentage of correct positive predictions is given by following equation

$$Precision = \frac{TP}{TP + FP}. \tag{2}$$

The ability of a model to find all the relevant cases is called recall. It represents the percentage of true positive detection among all relevant ground truths given by following equation

$$Recall = \frac{TP}{TP + FN}. \tag{3}$$

In the equations above, TP means true positive, FP means false positive and FN means false negative.

5.2 Results

The best topology of every structure was tested over testing dataset. Then, the IOU (accuracy), precision and recall was calculated with defined threshold of 0.75. The resulting values of all the selected metrics, evaluated over the testing set, are summarized in Table 3.

5.3 Discussion

Results obtained in the previous section clearly report U-Net, with the locator based on local peaks, as the most accurately performing detection technique in terms of IOU, precision and recall. However, other architectures (LeNet,

Table 3. Resulting values of all the selected metrics

Metric	AlexNet	LeNet	Net1	Net2	U-Net	YOLOv2 (squeezenet)	YOLOv2 (shufflenet)	YOLOv2 (mobilenet.v2)
IOU	0.755	0.115	0.122	0.165	0.908	0.737	0.791	0.780
Precision	0.829	0.118	0.126	0.314	0.949	0.942	0.936	0.954
Recall	0.887	0.195	0.178	0.183	0.960	0.949	0.833	0.902

AlexNet, Net1, Net2) used as encoders, fail to over-perform the YOLOv2 architecture, which is a generally accepted standard for object detection using deep learning. Furthermore, Table 1 and 2 obviously indicate, that the number of parameters for learning, as well as the memory necessary to store the detector, is unnecessarily big in the case of U-Net. Hence, the detectors used by the YOLOv2 approach are simpler, and probably, more computationally efficient.

Therefore, future work needs to include several elements in order to provide satisfactory grounds for the introduced approach. Firstly, the U-net encoder-decoder architecture should be optimized to reduce the memory size and computational complexity. Then, the time consumption of the performance needs to be evaluated. And consequently, the approach has to be tested and analyzed under operating conditions with proprietary hardware.

6 Conclusion

A deep convolutional neural network based method for person detection is proposed in this paper. The proposed method is intended to be used for person a flow monitoring system in public transport. Contrary to other approaches, the proposed method uses a convolutional neural network for image segmentation. The segmented image is then processed using the local peaks approach in order to provide the positions of the people in the image. The experiments using a custom dataset provided a precision rate of more than 98% and recall rate of 96%. The YOLOv2 approach with various detectors was used as a competitive approach. When using the same dataset and considering all the metrics, the best performing of the new approach versions (the U-Net version) clearly outperforms an established model as the YOLOv2.

However, the work presented in this contribution is only one step in the development of the complex and robust person flow monitoring system. The future work includes neural network architecture optimizing, computational complexity testing and, obviously, testing under operational conditions.

References

1. Ahmed, A., Siddiqui, N.: Design and implementation of infra-red based computer controlled monitoring system (2005). https://doi.org/10.1109/SCONEST.2005.4382890

2. Akamatsu, S.I., Shimaji, N., Tomizawa, T.: Development of a person counting system using a 3D laser scanner, pp. 1983–1988 (2014). https://doi.org/10.1109/ROBIO.2014.7090627
3. Bottou, L., et al.: Comparison of classifier methods - a case study in handwritten digit recognition. In: Proceedings of the 12th IAPR International Conference on Pattern Recognition, vol. II - Conference B: Pattern Recognition and Neural Networks. The International Association for Pattern Recognition; IEEE Computer Society; The Information Processing Association of Israel (1994)
4. Choi, J.W., Quan, X., Cho, S.H.: Bi-directional passing people counting system based on IR-UWB radar sensors. IEEE Internet Things J. (2018). https://doi.org/10.1109/JIOT.2017.2714181
5. Dolezel, P., Stursa, D., Skrabanek, P.: On possibilities of human head detection for person flow monitoring system. In: Advances in Computational Intelligence, IWANN 2019, PT II. Lecture Notes in Computer Science (2019). https://doi.org/10.1007/978-3-030-20518-8_34
6. Fu, H., Ma, H., Xiao, H.: Real-time accurate crowd counting based on RGB-D information (2012). https://doi.org/10.1109/ICIP.2012.6467452
7. Gao, C., Li, P., Zhang, Y., Liu, J., Wang, L.: People counting based on head detection combining Adaboost and CNN in crowded surveillance environment. Neurocomputing **208**, 108–116 (2016)
8. He, M., Luo, H., Hui, B., Chang, Z.: Pedestrian flow tracking and statistics of monocular camera based on convolutional neural network and Kalman filter. Appl. Sci. Basel (2019). https://doi.org/10.3390/app9081624
9. LeCun, Y., Bottou, L., Bengio, Y., Haffner, P.: Gradient-based learning applied to document recognition. Proc. IEEE (1998). https://doi.org/10.1109/5.726791
10. Millstein, F.: Deep Learning with Keras. CreateSpace Independent Publishing Platform, Scotts Valley (2018)
11. Redmon, J., Divvala, S., Girshick, R., Farhadi, A.: You only look once: unified, real-time object detection. In: 2016 IEEE Conference on Computer Vision and Pattern Recognition (CVPR) (2016). https://doi.org/10.1109/CVPR.2016.91
12. Redmon, J., Farhadi, A.: YOLO9000: better, faster, stronger. Computing Research Repository (CoRR) (2016). http://arxiv.org/abs/1612.08242
13. Ronneberger, O., Fischer, P., Brox, T.: U-Net: convolutional networks for biomedical image segmentation. In: Lecture Notes in Computer Science, vol. 9351 (2015). https://doi.org/10.1007/978-3-319-24574-4_28
14. Russakovsky, O., et al.: ImageNet large scale visual recognition challenge. Int. J. Comput. Vis. (2015). https://doi.org/10.1007/s11263-015-0816-y
15. Sharma, P., Singh, A.: Era of deep neural networks: a review. In: 8th International Conference on Computing, Communications and Networking Technologies, ICCCNT 2017 (2017). https://doi.org/10.1109/ICCCNT.2017.8203938
16. Skrabanek, P., Dolezel, P.: Robust grape detector based on SVMs and HOG features. Comput. Intell. Neurosci. **2017** (2017). https://doi.org/10.1155/2017/3478602
17. Sun, S., Akhtar, N., Song, H., Zhang, C., Li, J., Mian, A.: Benchmark data and method for real-time people counting in cluttered scenes using depth sensors. IEEE Trans. Intell. Transp. Syst. **20**(10) (2019). https://doi.org/10.1109/TITS.2019.2911128
18. Sun, S., Akhtar, N., Song, H., Zhang, C., Li, J., Mian, A.: Benchmark data and method for real-time people counting in cluttered scenes using depth sensors. IEEE Trans. Intell. Transp. Syst. (2019). https://doi.org/10.1109/TITS.2019.2911128

An Adaptive Cognitive Model to Integrate Machine Learning and Visual Streaming Data

Esteban García-Cuesta(✉), Jose M. López-López, Daniel Gómez-Vergel, and Javier Huertas-Tato

Data Science Lab, School of Architecture, Engineering and Design, Universidad Europea de Madrid, 28670 Madrid, Spain
{esteban.garcia,josemanuel.lopez,daniel.vergel, javier.huertas}@universidadeuropea.es

Abstract. In this paper, we present our current work towards developing a context aware visual system with capabilities to generate knowledge using an adaptive cognitive model. Our goal is to assist people in their daily routines using the acquired knowledge in combination with a set of machine learning tools to provide prediction and individual routine understanding. This is useful in applications such as assistance to individuals with Alzheimer by helping them to maintain a daily routine based on historical data. The proposed cognitive model is based on simple exponential smoothing technique and provides real time detection of objects and basic relations in the scene. To fulfill these objectives we propose the integration of machine learning tools and memory based knowledge representation.

Keywords: Assistive technology · Robotics · Intelligent environment · Time series · Deep neural networks · Cognitive architectures

1 Introduction

Cognitive psychology and Artificial Intelligence (AI) have been intertwined to mimic human problem solving and to understand environment by computer systems. The understanding usually relies on characterizing the problem space as a combination of symbolic or sub-symbolic inferences, pattern matching, and machine learning methods. Robotics and its extension to any sensor with data acquisition capabilities has emerged as an AI domain where the knowledge representation is needed to provide more complex reasoning and prediction capabilities. As mentioned in [1] the research in AI and Robotics has concentrated on expanding existing theory (neural networks and its brute force counterpart deep neural networks) and dimensionality reduction (Principal Component Analysis or more sophisticated methods based on subspace learning such as SLMVP [2]) which are limited approaches and do not address underlying theoretical issues of adaptability and generalizability which are key of human cognition. An intrinsic characteristic of sensors is that the collection of data is permanent (streaming data) and thereof the creation of the knowledge that best represent it is expected to be also on real time in order to take full advantage of data.

Á. Herrero et al. (Eds.): SOCO 2020, AISC 1268, pp. 176–185, 2021.
https://doi.org/10.1007/978-3-030-57802-2_17

Data acquisition and integration information are commonly used with the purpose of enrich original data with other external sources. Usually, enriched data improves the results obtained by machine learning methods and consequently also the applications where are used as in biomedicine domain [5] or recommendation systems [6]. However, some applications need a more robust approach able to work under non-stable environments as may occur in robotics data acquisition scenes and real time integration of machine learning models in dynamic environments. These capacities to adapt the cognitive models and tools build upon it (such as machine learning models) were introduced by Newell as five of the twelve desirable criteria for artificial cognitive architectures: i) "flexible behavior", ii) "real-time performance", iii) "adaptive behavior", iv) "vast knowledge base", and v) "dynamic behavior" [7] and was renamed as "The Newell Test" [8] though passed into oblivion.

On the applicability side, intelligent assistive technologies (IATs) have a tremendous potential of offering innovative solution to mitigate dementia problems that is one of the most important causes of disability in the elderly [9]. Home automation sensors provide important monitor information and allows prediction of near future behavior that can be use by caregiver to prevent anomalous behaviors when the behavioral pattern is learned [3, 9]. As reviewed at [9] these works use different techniques to extract regular behaviors (e.g. echo state networks) but they use raw data directly obtained from sensors without any cognitive model.

A problem that appears in visual knowledge representation on real time environment applications (in our case with applications on dementia care [3]) is the unexpected events such as the opacity between two objects. An object N_1 may be in the scene but it can be hidden from the camera because an object N_2 appears just in from ahead of it. This does not imply that object N_1 is not in the scene anymore but our certainty should decrease gradually given the actual data collected by the camera. Thereof, real-time identification of the objects of interest is not sufficient for our purposes. It is necessary to build, keep and constantly update a robust model of the world. Such a model should tell apart whether a connection between two objects is just incidental or, even worse, due to a wrong identification of objects. Conversely, an object that is not detected in a single frame of the video should not lead to a well-established connection involving that object to vanish immediately. Therefore, providing robustness to the model is our primary objective to represent the visual context and solve these limitations.

In this work, we present a cognitive model based on simple exponential smoothing technique [4] to add adaptability and stability to the knowledge representation process overcoming some of the above-mentioned limitations. We tested this approach in a visual computer task for extracting and predict individuals' daily behavior in a "toy" simulated Alzheimer's domain and using Aldebaran NAO Robot. In following, at Sect. 2 we present the cognitive model, at Sect. 3 we explain how machine learning tools are integrated with the cognitive model, then Sect. 4 describes the design of the experiments and the obtained results, and finally at Sect. 5 we present the conclusions.

2 Proposal of an Adaptive Cognitive Model for Knowledge Representation

Our world domain consist of N objects of interest within the domestic environment, and up to N^2 possible relations among them (all these objects are detected using computer vision techniques as explained later in Sect. 3.1). Each relation is characterized by an evolving weight $0 \leq w_{ij}(t) \leq 1$ with $i, j = 1 \ldots N$, that can be interpreted as an estimation of the relation being true in a given moment. For instance, if objects with $i = 1$ and $j = 2$ represent "a table" and "a burger", respectively, then $w_{12}(t)$ would be an estimation of our confidence at time t for the statement "the burger is on the table" to be true. Due to the nature of video processing, we assume that the input of our model should consist of a set of discrete-time signals, with the same sampling period Δt. Hence, $t = k\Delta t$, where $k \in \mathbb{N}$ is the epoch number k (we use the term epoch here as an abstraction of time).

In a given epoch, the weights of every relation should be updated in order to incorporate new information about the world. In our model, the primary source of information is the overlapping between identified objects in that epoch. Let us assume that the detection algorithm detects two objects, namely i and j, with confidences (interpreted as probabilities) p_i and p_j, respectively. Then, a first estimation of the strength of the relation "object j is on object i" is given by:

$$s_{ij}(t) = \begin{cases} p_i\, p_j; & \text{if } i \text{ and } j \text{ are overlapping} \\ 0; & \text{otherwise} \end{cases} \tag{1}$$

Please notice that the set of sequences $s_{ij}(t)$ provide an instantaneous model of the world that does not take into account past events. Such a representation is inherently unstable and full of spurious relations that introduce a large amount of noise. This is due to, not only occasional misidentifications of objects, but also most importantly, the situation of two correctly identified objects with a casual overlapping that is not due to an actual connection (e.g. a user carrying a mug incidentally passes in front of a table, the visual algorithms may detect an overlapping between the objects "mug" and "table", but the model should not infer that "the mug is on the table"). On the other hand, when a connection between two objects is consistently found for consecutive epochs, our confidence on the event increases. Therefore, we define and updating rule for the weights of the relation "object j is on object i":

$$w_{ij}(t + \Delta t) = w_{ij}(t)\left(1 - \frac{1}{\lambda}\right) + \frac{s_{ij}(t)}{\lambda} \tag{2}$$

The persistency super-parameter λ measures the inertia of the model against any change. It takes a row of λ positive reinforcements to learn a new relation and conversely, it also takes λ consecutive negative reinforcements to forget an existing relation. This approach can also be understood, as shown in (2), as a lowpass filter that introduces a simple exponential smoothing of the signal with a smoothing factor equals to $1/\lambda$ [10].

3 Integration of Machine Learning Tools

In this section we present the integration of the computer vision and prediction machine learning tools into the proposed cognitive model and how they interact each other.

3.1 Computer Vision

For the purpose of this study, we need to achieve real-time identification and monitoring of elements of the environment that may be of interest to the patient related with the detection of regularities and recurring patterns in his interactions with them. To detect the different objects in a scene we relied on visual object detection algorithms and learning models based on deep neural networks (DNN). When deciding upon a choice of object detectors best suited for our model, we investigated the accuracy/speed trade-offs of some well-known convolutional neural networks such as Faster R-CNN [11], SSD [12], and YOLO [13]. A complete comparison between these architectures can be found at [14]. Due to speed limitations of the integration framework that we are working on (memory and CPU capacities mainly) and after testing all of these three we chose the SSD MobileNet architecture –a light-weight DNN that sacrifices a reasonable amount of accuracy in favor of lower latency and higher speed [15]– specially suitable for our real-time processing. Although MobileNet's performance may suffer with small objects' detection compared to other architectures, it showed near-optimal accuracy and speed at this stage of our model's development. Figure 1 shows the working framework of this work.

To integrate this computer vision model with the cognitive model we used the confidence value provided by the learned model for a given object that is in the scene. This value assigns the confidence (interpreted as probability) on the detection of given object i, and it is passed to the cognitive model as formulated in (1) as p_i. In this way, we can assure that, on real time, we are updating our knowledge of the world considering also the uncertainty provided by the automatic detection system.

a) b) c)

Fig. 1. a) Simulated scenario using small sized housing objects; b) Aldebaran NAO Robot vision recognition system; c) Aldebaran NAO Robot

3.2 Deep Neural Networks

To provide prediction capabilities to our robot we use a Long Short Term Memory Networks (LSTM) that is a type of recurrent neural network (RNN) [16] design for time series knowledge prediction. The prediction is done over the knowledge extracted from the computer vision system and modeled with the cognitive model. Knowledge extracted from visual representations is codified as a set of objects and relationships. The objects in the image are represented as a set of probabilities over the given objects, which represent the probability P(oi) of an object *o* appearing in a given image *i*. The relationships are represented by another set of probabilities P(ri) and a set of vertices representing their location loc(ri) also shown in Fig. 2.

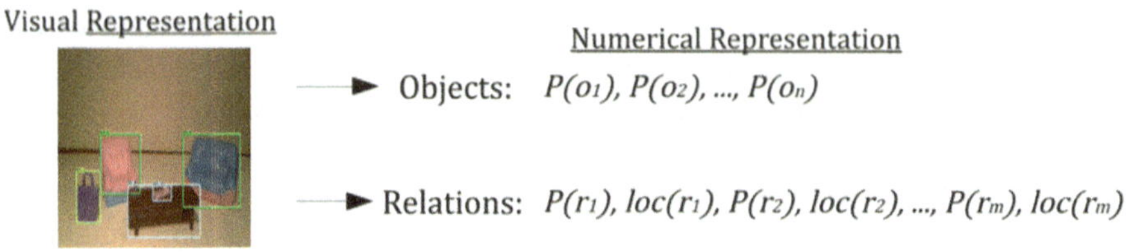

Fig. 2. Representation of objects and relations as numerical attributes for the machine learning model.

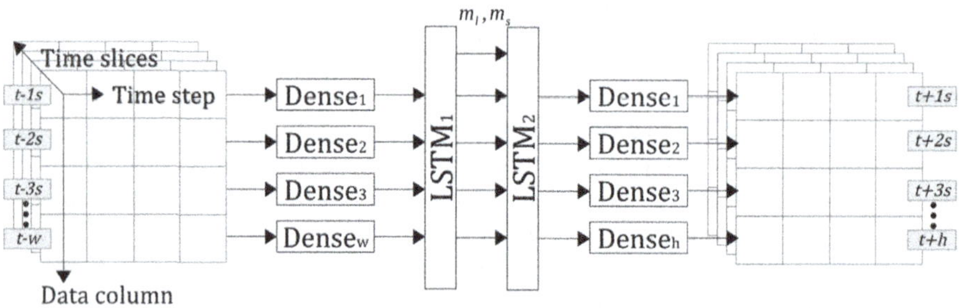

Fig. 3. Network architecture representation. Depth in the matrixes are each time slice, every row is an event (time step), and each column is a set of features (probability of object 1, for example). t represents the time instant, w the window, h the horizon. Note that LSTMs are connected via m_l and m_s, which represent the memory state of the previous LSTM.

Each codified representation is registered at a time instant t, and it is desired to find the future states up to time instant $t + h$ (z representing the lookup horizon). Information from the past is included by looking up to the past state at $t - w$ (w representing the past window). The number of time instants taken before time t are specified by a time step s, therefore the number of instants taken will be w/s for looking back and h/s looking into the future. The architecture proposed for this problem is described in Fig. 3. The network receives as inputs a time slice from t to $t - w$, where every codified state is processed through an individual densely connected neural layer. The outputs of the dense layer are processed in turn through a LSTM neural layer. This produces a sequence of outputs and two states c and h with the outputs of each time step and the memory of the LSTM. This information is again connected to another LSTM, which outputs the sequence of forecasts that are converted into the final outputs with another densely connected layer (for each time step). Between every layer there is a batch normalization process that regularizes the inputs to the following layer. In addition, each layer has a Rectified Linear Unit (ReLU) activation. The output dense layers have sigmoid activation that keep final values between 0 and 1. The output from this network is a sequence of time steps (up to $t + h$) predicting future configurations of objects and relations.

4 Design of the Experiments

We have simulated a set of experiments that include daily routine events for an individual (it is worth to recall that this work intention is to test a system that can be used to help individuals with Alzheimer to maintain its daily routine). In our recordings, one second of actual video footage amounts to 0.006 h of real time for the patient. This enables us to reduce an entire day of the patient's life to approximately 71 min of testing. We registered a total of 7 recordings, one for each day of the week (Fig. 4 shows the timeline for one day). Each recording contained the following events: i) we place a characteristic object on the table –a piece of food/burger,—to simulate the eating time and taking of medication time; ii) we place a second object (remote) that is used daily and meanwhile the individual is sleeping is located in the table, and the rest of the day is on the couch; iii) we place randomly three times the remote in other places, and also another unforeseen object.

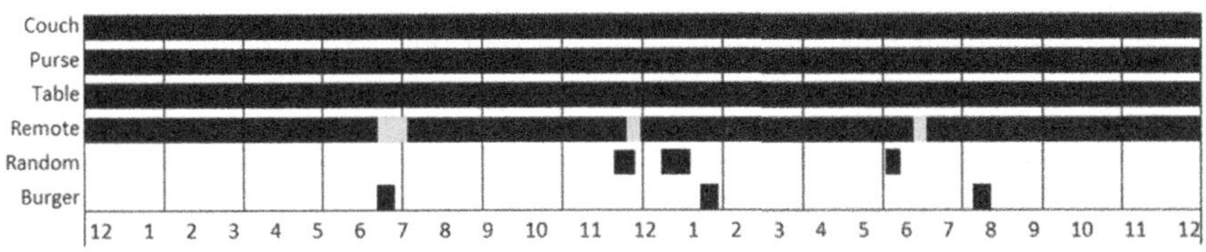

Fig. 4. Example of 1-Day (Wednesday) timeline for the different simulated events (for remote object grey color shows the random placements).

We have simulated a complete week with a total 27414 records (1 every second approx. or every 20 s in real time scale). Assuming all data is approximately evenly spaced (this time depends on the execution time of the computer vision algorithm for detecting objects and it has some small variations); the dataset has been imputed to fill the missing gaps in time with a period of 3 s at a minimum, leaving a training dataset with 25328 slices, each slice containing 72 events. Any event at a time step has 55 fields of data composed of object and relation probabilities and relation positions.

4.1 Results

In this section, we present the obtained results after the integration of machine learning models (object detection and behavior prediction) with the cognitive model. We also have included some insights into the cognitive model behavior to proof its robustness and adaptability.

Object Recognition Results

The SSD MobileNet model was pre-trained on a dataset created specifically for this project consisting of 5 object categories of approximately 100 images each. These images were labeled using the LabelImg[1] graphical annotation tool and minimum of 3 negative

[1] https://github.com/tzutalin/labelImg.

images has been establish in the deep neural network architecture in order to avoid false positive recognitions due to background. The final model was obtained using cross-validation with K = 10 to find the optimal parameters for 300 × 300 resized images. The detection accuracy results are shown in Table 1. It indicates the success or failure in real-time recognition of our five specific categories to distance to the camera from 40 to 140 cm and at different spatial rotations.

Table 1. Visual object detection results

Object	Couch	Fruit	Purse	Table	Remote control	Average
Accuracy (%)	99,91	94,58	99,65	99,92	99,04	98,61

The background image in our tests was carefully considered when calibrating the neural network to avoid recognizing it as a false positive. This notably improved the model's accuracy, achieving a success rate close to 99%. To populate our cognitive model we extracted the following semantic information in JSON format:


```
{  "Timestamp": "wed 29 jan 12:38:22 2020",
   "Objects": [{
         "Table": {
            "Detection Probability": "0.6196",
            "Cognitive Model Probability": "0.3027"  }
      },{
         "Fruit": {
            "Detection Probability": "0.2701",
            "Cognitive Model Probability": "0.4504"  }
   }],
   "Relations": [{
      "Fruit-Table": {
         "Cognitive Model Probability": "0.3032",
         "Vertex1": "390", "Vertex2": "478", "Vertex3": "219", "Vertex4": "367",
            "Vertex5": "327","Vertex6": "413","Vertex7": "291", "Vertex8": "362"}
   }]
}
```


Deep Neural Network Prediction Results

The inputs to the network were scaled featurewise between 0 and 1 and the network was optimized by the Adam [17] algorithm with a learning rate of $\alpha = 5 * 10^{-3}$. Each dense layer has 128 neurons and both LSTM layers have 128 neurons too. The model was trained across 1000 epochs. These parameters have been manually tuned until convergence on the test set was observed. The network has been developed with a window w = 6 h, a lookup horizon h = 6 h and a time step of s = 5 min. To measure the quality of this model, five quality metrics are extracted: MAE, RMSE, normalized MAE and RMSE and R2 score as shown in Table 2. The implementation of this model has been developed in Python 3.7, by using the Tensorflow 2.1 backend with a NVIDIA Tesla K80 GPU. Metrics have been measured on the standardized and non-standardized data to give a fair representation of performance for positions and probabilities (because they have different magnitudes).

Table 2. Summary of experimental results

Metric	Normalized data	Raw data
MAE	0.018	4.549
RMSE	0.044	11.904
nMAE	0.025%	0.02%
nRMSE	0.061%	0.052%
R2	98%	99.5%

Results on Standardized and Non-Standard data do not differ when normalizing metrics (nMAE and nRMSE) while MAE and RMSE differ by a large margin, this is due to the different magnitudes of the data, which was to be expected. The R2 score is very high (98% and 99.5%) in any case, while the normalized error are extremely low. This highlights the strengths of the designed architecture for the described problem. Normalized metrics are relevant due to the differing magnitudes of each event in the dataset (positions and probabilities).

Cognitive Model Parameters and Results

The effect of the persistency parameter defined in (2) can be appreciated in the situation shown in Fig. 5. There, three discrete-time signals are plotted: *i*) the probability of detection for a large stable object (a table) $p_{\text{table}}(t)$, as it is provided by the computer vision algorithm; *ii*) the strength of the relation "a burger is on the table" $s_{\text{table,burger}}(t)$; and *iii*) the weight of that same relation $w_{\text{table,burger}}(t)$ obtained for the persistency parameter $\lambda = 10$. The figure corresponds to a video track where a burger was placed on a table and, after some time, was removed. As expected, the detection of the large object is quite stable and reliable, with a probability consistently close to unity. Nevertheless, the performance fluctuates when the burger is removed during the simulation because the view of a relevant region of the table is blocked during manipulation. For the smaller object (the burger), random fluctuations are much more important because smaller objects are usually harder to identify in real environments. We can see that, at some point around frame number 115, $s_{\text{table,burger}}(t)$ abruptly goes to zero. Then, after a few additional frames, it partially recovers, also abruptly. This is an artefact because the operator blocked the view of the burger during the process of its removing. This large and meaningless fluctuations would introduce noise and worsens prediction models accuracy. As can be seen, random fluctuations, on the contrary, do not appear in the smooth output signal corresponding to our model with $\lambda = 10$.

We want to highlight there are other cases similar to the one presented here. For instance, a user could just walk in between the camera and the object, blocking the view for a few frames. This would produce a highly fluctuating $s_{ij}(t)$ signal, but would hardly affect the $w_{ij}(t)$ values. Only when the number of affected consecutive frames surpasses the persistency parameter, the model would reflect the changes.

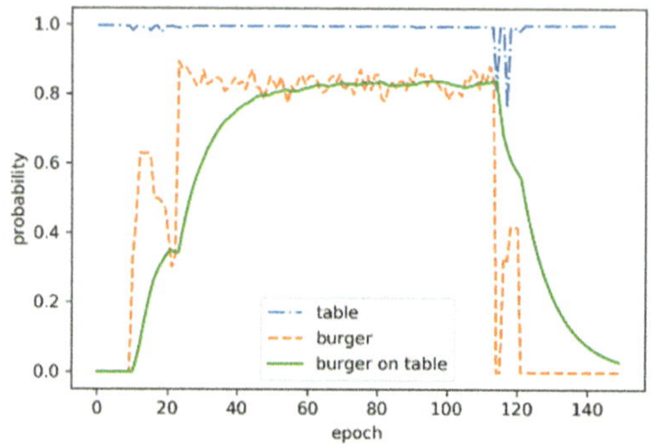

Fig. 5. Two input discrete-time signals obtained during the experiment, a table (blue dash-dot line) and an overlapping burger (orange dashed line), and the corresponding output discrete-time signal "the burger is on the table" (green line) with a persistency parameter $\lambda = 10$.

We set the persistency super-parameter to 10 because it provides a good balance between smoothing the input signal and reflecting the actual changes. The process was supervised, as humans can easy tell apart whether an event is meaningful. These parameters can also be adjusted quite straightforward in an unsupervised manner once a day, during "sleeping time" and parameters adjusted accordingly similar to that of the human brain [18].

5 Conclusions

The proposed adaptive cognitive model satisfies some of the Newell' desirable criteria for artificial cognitive architectures as i) flexible behavior, ii) real-time performance, iii) adaptive behavior, iv) vast knowledge base capabilities, and v) dynamic behavior. This model is based on the simple exponential smoothing technique and its integration with sensorial streaming data collected from a camera to create a dynamic knowledge representation of the world in a robotics context. We also have integrated higher-level machine learning models that make use of the cognitive models providing prediction capabilities to model regular behaviors of individuals. Our experiments show that the models are robust in evolving and dynamic data streams even when unexpected events occurs. In addition, the proposed model can be integrated easily with other input sensors using the confidence over the detected objects as a probability and making the whole model easily scalable.

References

1. Kelley, T., Lebiere, C.: From cognitive modeling to robotics: how research on human cognition and computational cognitive architectures can be applied to robotics problems. In: 9th AHFE Conference, pp. 273–279. Springer (2019). http://doi.org/10.1007/978-3-319-94223-0_26
2. García-Cuesta, E., Iglesias, J.A.: User modeling: through statistical analysis and subspace learning. Expert Syst. Appl. **39**(5), 5243–5250 (2012)
3. Ienca, M., Fabrice, J., Elger, B., Caon, M., Scoccia Pappagallo, A., Kressig, R.W., Wangmo, T.: Intelligent assistive technology for Alzheimer's disease and other dementias: a systematic review. J. Alzheimers Dis. **56**(4), 1301–1340 (2017)
4. Chatfield, C., Koehler, A., Ord, K., Snyder, R.: A new look at models for exponential smoothing. J. Roy. Stat. Soc. Ser. D (Stat.) **50**(Part 2), 147–159 (2001)
5. Aparicio, F., Morales-Botello, M.L., Rubio, M., Hernando, A., Muñoz, R., López-Fernández, H., Glez-Peña, D., Fdez-Riverola, F., de la Villa, M., Maña, M., Gachet, D., de Buenaga, M.: Perceptions of the use of intelligent information access systems in university level active learning activities among teachers of biomedical subjects. Int. J. Med. Inform. **112**, 21–33 (2018)
6. García Cuesta, E., Gómez Vergel, D., Gracia Exposito, L.M., Vela Pérez, M.: Prediction of user opinion for products: a bag-of-words and collaborative filtering based approach. In: Proceedings of the 6th ICPRAM, vol. 1, pp. 233–238 (2017). https://doi.org/10.5220/0006209602330238
7. Newell, A.: Physical symbol systems. Cogn. Sci. **4**, 135–183 (1980). https://doi.org/10.1207/s15516709cog0402_2
8. Anderson, J.R., Lebiere, C.: The Newell test for a theory of cognition. Behav. Brain Sci. **26**, 587–637 (2003)
9. Lotfi, A., Langensiepen, C., Mahmoud, S.M., et al.: J. Ambient Intell. Human Comput. **3**, 205 (2012). https://doi.org/10.1007/s12652-010-0043-x
10. Oppenheim, A.V., Schafer, R.W.: Discrete-Time Signal Processing, International edn. Prentice-Hall, Inc., Upper Saddle River (1989)
11. Ren, S., He, K., Girshick, R., Sun, J.: Faster R-CNN: towards real-time object detection with region proposal networks. arXiv preprint arXiv:1506.01497 (2015)
12. Liu, W., Anguelov, D., Erhan, D., Szegedy, C., Reed, S.: SSD: single shot multibox detector. arXiv:1512.02325 (2015)
13. Redmon, J., Divvala, S., Girshick, R., Farhadi, A.: You only look once: unified, real-time object detection. arXiv preprint arXiv:1506.02640 (2015)
14. Huang, J., Rathod, V., Sun, C., Zhu, M., Korattikara, A., Fathi, A., Fischer, I., Wojna, Z., Song, Y., Guadarrama, S., Murphy, K.: Speed/accuracy trade-offs for modern convolutional object detectors. arXiv:1704.04861 (2017)
15. Howard, A., Zhu, M., Chen, B., Kalenichenko, D., Wang, W., Weyand, T., Andreetto, M., Adam, H.: Mobilenets: efficient convolutional neural networks for mobile vision applications. arXiv preprint arXiv:1611.10012 (2016)
16. Hochreiter, S., Schmidhuber, J.: Long short-term memory. Neural Comput. **9**(8), 1735–1780 (1997). https://doi.org/10.1162/neco.1997.9.8.1735
17. Kingma, D.P., Ba, J.: Adam: a method for stochastic optimization. arXiv:1412.6980 (2014)
18. Todorova, R., Zugaro, M.: Isolated cortical computations during delta waves support memory consolidation. Science **366**(6463), 377–381 (2019)

Smart Song Equalization Based on the Classification of Musical Genres

Jesus Iriz[1], Miguel A. Patricio[2(✉)], Jose M. Molina[2], and Antonio Berlanga[2]

[1] BQ Engineering Team, Calle Sofía, 10. Európolis, 28232 Las Rozas - Madrid, Spain
jesus.gonzalez@MasMovil.com

[2] Applied Artificial Intelligence Group (GIAA), Universidad Carlos III de Madrid, Madrid, Spain
mpatrici@inf.uc3m.es, {molina,aberlan}@ia.uc3m.es

Abstract. Music equalization is the process of trimming or raising specific frequencies (or a range of frequencies, called "frequency bands") to increase the quality of a song. This paper presents an architecture that obtains a smart song equalization for a song based on a prediction of its musical genre. This prediction is based on a neural model designed in this work for the classification of musical genres in separated segments of a song. Each song needs its own equalizer setting that will not be the same throughout the song. For each segment, a distribution of probabilities of belonging to different genres is calculated, and then the song is equalized based on this distribution and different configuration options such as genres taken into account and using interpolation for the segments or not. These configurations are discussed in the paper. Both automatic music equalization and song genre classification are poorly researched matters, a gap we try to cover by investigating previous works in the area of study. Different proofs of concept are presented in the work to show the operation of the automatic equalizer. Adapting the solution for a smart equalizer can produce a system capable of automatically improving the music millions of people listen to daily in their mobile devices or applications such as Spotify and Youtube.

Keywords: Parametric equalization · Neural networks · Music classification

1 Introduction

This paper presents an smart song equalization system [12]. It is essential first to understand the concepts behind an audio track and what makes a song sound in one way or another. In this paper, a neural model will be introduced that are capable of determining the genre to which a song corresponds, in order to use equalization profiles for each segment of the song.

We can continue to delve in this way into the characteristics of music. For example, electric guitars and drums are common in different genres such as rock

Á. Herrero et al. (Eds.): SOCO 2020, AISC 1268, pp. 186–195, 2021.
https://doi.org/10.1007/978-3-030-57802-2_18

and metal, how can both be distinguished? The different musical genres are not limited to differing only based on their instruments, but there are different techniques and sonorities used in each.

Musical genres can also be understood as a tree diagram, in which there is a hierarchy depending on whether some genres are derived from others [4]. For example, jazz is a genre derived from the mix of rock and roll and African-American music.

There are different types of equalization [12]. In our specific case we will focus on parametric equalization. Therefore, we understand by equalization the process of altering the amplitude of each of the frequencies of an audio in order to make some frequency ranges more noticeable, and blur others.

Digital music players usually have features to establish the equalization pattern that you want to set. Likewise, there are pre-established patterns for different musical styles. Although these patterns allow an optimal equalization for a musical style, each song has an optimal equalization, which is different from the rest of the songs. These systems propose the most useful equalization for each genre, even if they are not perfect for each song. In addition, equalization patterns are set for the entire song, regardless of the different variations it may suffer. This work presents an architecture for smart equalization of songs, which identifies, for a certain segment of the song, the most likely musical styles. Once

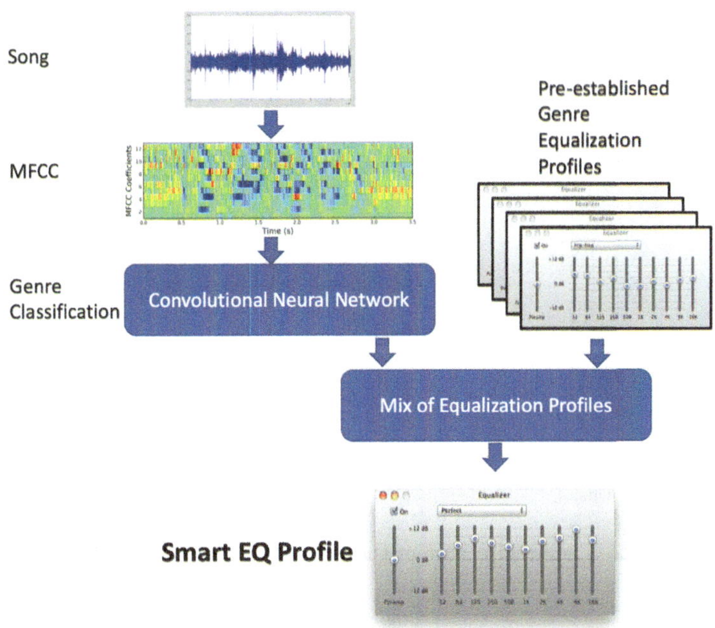

Fig. 1. Architecture of the smart equalizer proposed in this work.

identified, it performs a mix of equalization patterns. This mixture is weighted according to the probability of belonging to a particular musical style. In this way, we can have an smart song equalization.

2 Related Works

The classification of musical genres has been treated by various supervised classification methods (Gaussian Mixture Models [6], Hidden Markov Models [10], Support Vector Machines [1], Artificial Neural Networks [7], and Convolutional Neural Networks).

In [5], also with the aim of classifying music by genres, the authors use a dataset of 400 songs of two genres, Indian music and Classical music, with 200 songs each of them. The data extracted from each song is the MFCC (Mel Frequency Cepstral Coefficients) coefficients of the entire song.

The authors in [3] uses a convolutional neural network with the aim of analyzing whether it is possible to use a neural network to create playlists of recommendations on platforms such as Spotify based on the music they usually listen to. Again, use the MFCCs as input parameters, but this time by extracting them from just a fragment of each song. The output consists of a genre classification.

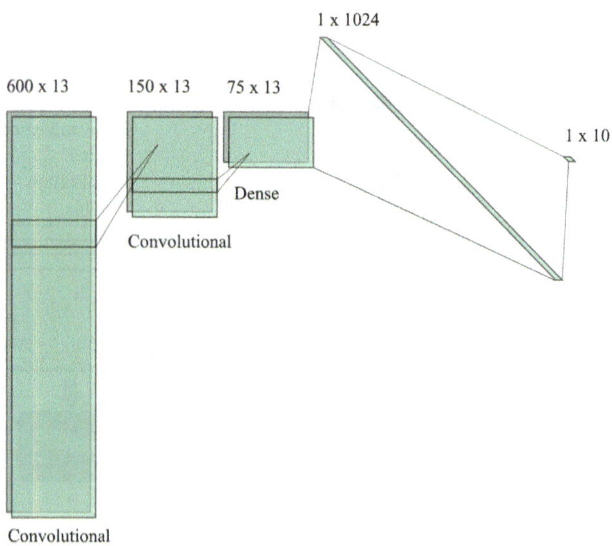

Fig. 2. The architecture of the neural network implemented.

3 Proposed Architecture

This section shows the architecture of the smart equalizer proposed in this work. In Fig. 1, this architecture is depicted.

The first step is obtain the Cepstral Coefficients of Mel Frequencies (MFCC) of a song (MFCCs) [8]. The MFCCs are coefficients representing a sound wave derived from the coefficients of the scale of Mel. The Mel scale is a transformation applied to frequencies, transforming it from a linear scale to a perceptual scale.

To obtain the MFCCs, the process is as follows: (1) Segment the sound into fixed length sections; (2) To each section, apply the discrete Fourier transform to separate it into frequencies and obtain the spectral power (or relative energy) of each frequency in the segment; (3) Apply the Mel scale to the spectra obtained in the previous point; (4) Take the Neperian logarithm of each Mel coefficient obtained; and (5) Apply the discrete cosine transform to each of these logarithms.

Thus, we obtain temporal information for each song with each segment, and frequency information with each coefficient obtained in each segment. In this way, although handling more data, the temporal information is blurred less than simply using Mel coefficients. As the result of calculating the MFCCs is a two-dimensional matrix, this makes them suitable for working with a convolutional neural networks.

According to studies such as [11] the optimal number of frequencies to be taken to calculate the MFCCs is between 10 and 20. In the case of our work, 13 will be used following the indications of studies such as [3]. The consensus on how many windows to take in each fragment is about one window for every 0.05 s of audio. In this way, if we use samples of 30 s duration, we will have 600 window shots per song. The neural network will take as input a total of 600 * 13 parameters.

As we will see later, our work is based on the GTzan dataset [9] that provides 1,000 fragments of songs of 30 s each, divided into 10 genres: Blues, Classical, Country, Disco, Hip-hop, Jazz, Metal, Pop, Reggae and rock.

In [3] a convolutional network with 3 hidden layers is used. In our proposal, we want to have a simpler neural model that can be embedded in platforms with fewer resources (for instance, smart speakers or smartphones). Although we are aware that precision is lost, as will be seen later, great precision is not needed when mixing equalizations of musical genres. Therefore, we proceed to make a simplification using only 2 hidden layers.

In this way, each of these layers will apply a max pooling and a dropout. The max pooling consists in applying a reduction of parameters, in order that as enters the network, fewer data and network begins to take shape in a given output (for example, rated 7800 parameters in just 10 obtained genres). On the other hand, the dropout consists in eliminating a certain percentage of connections between layers in each iteration, so that the network begins to adapt not only to its function of classifying, but also that each layer begins to learn to solve errors of previous layers if there would be. When handling both data you have to be careful, since a very extreme value of them could lead to poor results of the neural network.

Finally and just before the final output, a dense (or fully connected) layer will be used in which no max pooling will be applied, that is, it will maintain all connections with the previous layer.

Finally, this layer applies a ReLU (Rectified Linear Unit) function, which is equal to 0 when its input is less than or equal to 0, and is linear when the input is positive.

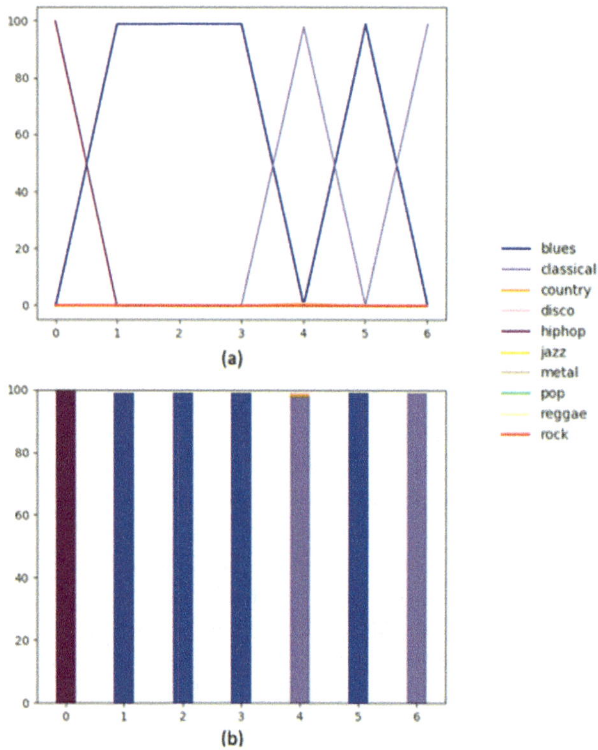

Fig. 3. Results of the song segment classification with the Softmax output: (a) in linear format, and (b) in bar format.

After these layers, the final output of the network is already obtained, a numerical value for each of the 10 genres to which a Softmax function can be applied as a smooth approximation to the most probable genre. In Fig. 2, the architecture of the model implemented is depicted. Finally, with the results obtained by the musical genre classifier, a mixture of the equalization profiles of the musical genres is carried out, weighted by their corresponding value of the output of the neural network. In this way, a smart equalization profile is obtained, adapted to a particular song. As we will see in the experimentation section, this process is not carried out with the entire song, but is applied to segments of the song, producing an equalization more adapted to the music that is heard in each moment.

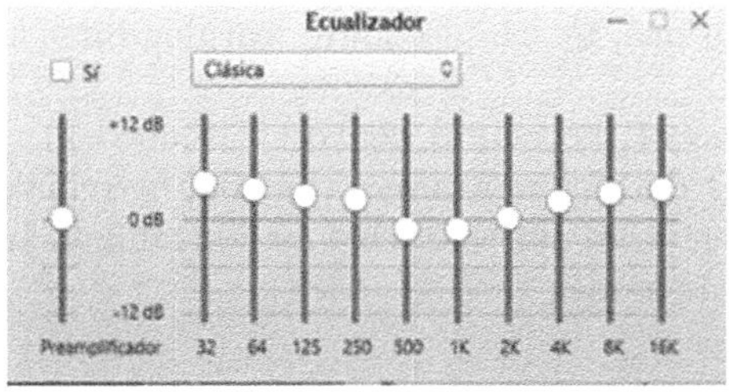

Fig. 4. Equalization profile for Classical music.

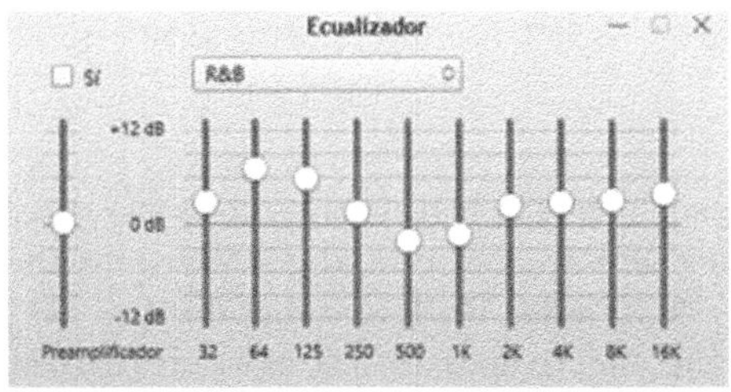

Fig. 5. Equalization profile for Blues music.

4 Experimentation

4.1 Datasets

In order to design the proposed neuronal model, the present work has been based on the GTzan dataset [9]. The dataset consists of 1,000 fragments of songs of 30 s each, divided into 10 genres: Blues, Classical, Country, Disco, Hip-hop, Jazz, Metal, Pop, Reggae and Rock.

In the other hand, MSD (Million Song Dataset) [2] is a very broad dataset that has a total of one million songs. This dataset has been used to validate the neural model designed and trained with the GTzan dataset. This dataset only contains the labels of each song with a series of features already extracted.

4.2 Training

For the training process, we have 1000 songs which are segmented in 30-s windows. The training process consists of 100,000 iterations. In each iteration we use 800 random songs of the 1,000 of which the dataset consists. At the end of the 100,000 iterations, the state of the network in which a better result of between 100,000 was achieved, which will not necessarily be the final state. We opted for such a high number of iterations since training the net is a very time-draining process. In that case, we considered better training once for a high iteration count, than training multiple times for lower iterations.

Each iteration divides the training set (800 songs) in batches of 64 songs and executes the training of the model for each of the batches. Since we are using a high number of iterations, we can use lower learning rates for more precision, in this case 0,001.

For the validation of the neural network obtained in the learning process, songs from the MSD dataset were used, that is, the remaining 200 songs from the original 1000 that were not selected for the training. In Table 1 we can see the success rate of the model.

Table 1. Success rate of the neural model designed.

Genre	Blues	Classical	Country	Hiphop	Jazz	Metal	Pop	Reggae	Rock
Rate	54.46%	93.20%	53.30%	50.16%	87.34%	34.90%	85.426%	73.88%	45.72%

For some genres, such as classical music songs, the success rate even reaches 93.2% of success, but there are genres and genre groups that greatly reduce the average rate of up to 40.31% for Rock and metal songs. As indicated above, the data used to verify the effectiveness of the network corresponds to a subset of the MSD dataset. Some genres are shared by both GTzan and MSD, but for different genres, songs obtained by searching in the most popular songs section of Spotify have been used. In MSD, Pop/Rock and Jazz/Blues had been merged into a single genre, so the remaining where obtained from Spotify by looking for "Rock" and "Jazz" playlist in the app. In total, 1.000 songs (100 for each genre) have been used for the evaluation, both from MSD and Spotify. As you can see later, these success rates will be sufficient to obtain the desired result, which is the automatic equalization of the songs. Most songs can't be labeled as just a single genre, most of them are a mixture of genres (Or a mixture of characteristics common to different genres) In the datasets, songs are labeled as the most likely genre, so the classification may not be the same as the previous classification. In these cases, missing the label means that the song may not belong 100% to a single genre, but to a mixture of them. In our strategy we will use genre mixes and we will consider equalization equally among those that the neural network determines to be the most likely.

4.3 Smart Song Equalization

As proof of concept the results of automatic equalization of a song will be displayed. In this test, the results are shown with the song "Gravity" of the musical group "Against the current". The song is divided into segments of 30 s and for each segment you get the output of the network that indicates the corresponding musical genre.

In Fig. 3, the results for the Softmax case can be observed. Both graphs represent the output of the neural model (y axis) for each time segment (x axis) of corresponding to one genre or another. Each color corresponds to a genre according to its color. The first graph includes a representation of the individual output of each genre at each instant, while the second represents the outputs in a cumulative bar chart, where each genre appears alongside the others in each bar. The duration of the song used is 3 min and 42 s, so 7 fragments of 30 s each are generated. Leaving the last 12 s unprocessed. In this proof of concept, the Softmax function is used to obtain the genres, and it is observed that the genres obtained for each segment each obtain almost 100% probability. For this song, there is clearly a structure in three parts, the first detected as Hip-hop, the second as Blues, and the end of the song between Classical and Blues. In the fifth segment (number 4) there is a small segment that corresponds to country.

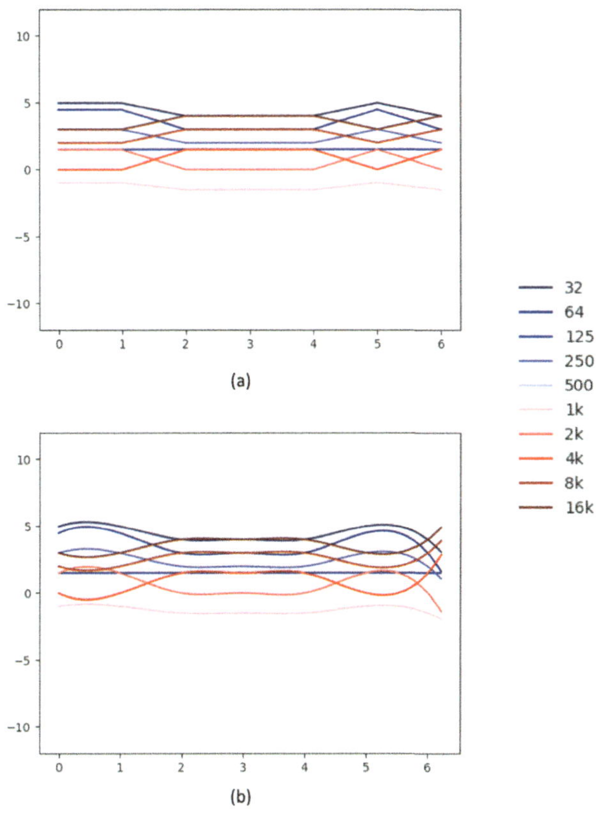

Fig. 6. Results of automatic equalization using a weighted mixture of the Softmax output: (a) without interpolation of genres; (b) with interpolation of genres.

If we look at the equalization profiles for Classical and Blues music (Figs. 4 and 5), it can be seen that, even if they are confused, they really respond to very similar equalization needs, so, if the objective is not to classify, but to equalize based on the classification, So we can say that this is a good result.

For the intermediate and final segments, the important confusion occurs between Blues and Classical. Even if we compare the profiles of Blues, Classical, Hip-hop and Disco, the four follow a similar structure in the form of 'V' that accentuates bass and treble, but maintains or attenuates intermediate frequencies.

Finally, we want to know the results of automatic equalization if what we do is a weighted interpolation of the genres detected in each segment, which was the work intended in this paper. In Fig. 6, graphs represent the gain of each frequency band (y axis) with respect to each time segment (x axis). The bands are represented with colored lines, garnets and red when it comes to high frequencies, and blue when it comes to low frequencies. With this proof of concept the suspicion of the structure of the song in three parts is confirmed. Comparing the interpolation system, in some moments it seems that precision is lost, for example, between segments 0 and 1 the frequency of 2k appears above that of 8k (higher gain) in the second graph, but appears below in the first. Even so, applying interpolation is considered beneficial because it provides smoothness to transitions between segments.

5 Conclusions

In this work a system has been presented that is able to equalize a song adaptively. Unlike the traditional equalization systems that establish a profile for each song based on the musical genre to which it belongs, our proposal divides the song into segments. For each segment, calculate the probability of belonging to a musical genre through a neural model designed and trained for the identification of musical genres. Starting from a pre-established profile of each musical genre, a mixture of equalization profiles weighted by the probability of belonging to a particular musical genre is carried out. It is also verified how the proposed interpolation method softens transitions and is beneficial for the final equalization result. Music equalization, just as image enhancement is a subjective matter. For some people, the equalized version of a song may be better and for some may be worse. That's why we use equalization profiles, those profiles are crafted and designed to satisfy most of their users, so the result that we get should be close to what a final user may want from an equalizer. As future works, the authors are working on different research lines. On the one hand, we are working on a system that allows us to include the feedback of users who listen to a song in order to adapt to their preferences. Another research line is in the detection of significant elements. Sometimes, a specific instrument or even an individual note needs to be more highlighted in a song than the rest of the sounds, regardless of the genre. In this line it is intended to train a neuronal model that detects these situations with which to enter this information in the mix of equalization profiles according to gender.

Acknowledgements. This work was funded by the private research project of Company BQ and the public research projects of the Spanish Ministry of Economy and Competitiveness (MINECO), references TEC2017-88048-C2-2-R, RTC-2016-5595-2, RTC-2016-5191-8 and RTC-2016-5059-8.

References

1. Elbir, A., İlhan, H.O., Serbes, G., Aydın, N.: Short time Fourier transform based music genre classification. In: 2018 Electric Electronics, Computer Science, Biomedical Engineerings' Meeting, Istanbul, pp. 1–4 (2018)
2. Bertin-Mahieux, T., Ellis, D.P.W., Whitman, B., Lamere, P.: The million song dataset. In: Proceedings of the 12th International Society for Music Information Retrieval Conference, ISMIR 2011 (2011)
3. Dieleman, S.: Recommending music on Spotify with deep learning (2014). https://benanne.github.io/2014/08/05/spotify-cnns.html
4. George, J., Shamir, L.: Unsupervised analysis of similarities between musicians and musical genres using spectrograms. Artif. Intell. Res. (2015). https://doi.org/10.5430/air.v4n2p61
5. Goel, A., Sheezan, M., Masood, S., Saleem, A.: Genre classification of songs using neural network. In: Proceedings - 5th IEEE International Conference on Computer and Communication Technology, ICCCT 2014 (2015). https://doi.org/10.1109/ICCCT.2014.7001506
6. Kaur, C., Kumar, R.: Study and analysis of feature based automatic music genre classification using Gaussian mixture model. In: 2017 International Conference on Inventive Computing and Informatics (ICICI), pp. 465–468 (2017)
7. Mandal, P., Nath, I., Gupta, N., Jha Madhav, K., Ganguly Dev, G., Pal, S.: Automatic music genre detection using artificial neural networks. In: Intelligent Computing in Engineering, pp. 17–24. Springer, Singapore (2020)
8. Sahidullah, M., Saha, G.: Design, analysis and experimental evaluation of block based transformation in MFCC computation for speaker recognition. Speech Commun. (2012). https://doi.org/10.1016/j.specom.2011.11.004
9. Sturm, B.L.: An analysis of the GTZAN music genre dataset. In: MIRUM 2012 - Proceedings of the 2nd International ACM Workshop on Music Information Retrieval with User-Centered and Multimodal Strategies, Co-located with ACM Multimedia 2012 (2012). https://doi.org/10.1145/2390848.2390851
10. Li, T., Choi, M., Fu, K., Lin, L.: Music sequence prediction with mixture hidden Markov models. In: IEEE International Conference on Big Data (Big Data), Los Angeles, CA, USA, pp. 6128–6132 (2019)
11. Tjoa, S.: Mel Frequency Cepstral Coefficients (MFCCs) (2018). https://musicinformationretrieval.com/mfcc.html
12. Välimäki, V., Reiss, J.D.: All about audio equalization: solutions and frontiers (2016). https://doi.org/10.3390/app6050129

Special Session: Contributions of Soft Computing to Precision Agriculture

Machine Learning in Classification of the Wax Structure of Breathing Openings on Leaves Affected by Air Pollution

Aleš Procházka[1,2(✉)], Martina Mudrová[1], Pavel Cejnar[1], and Jan Mareš[1]

[1] University of Chemistry and Technology, 166 28 Prague 6, Czech Republic
A.Prochazka@ieee.org
[2] Czech Technical University, 160 00 Prague 6, Czech Republic

Abstract. Texture analysis and classification of image components belong to common problems of the interdisciplinary area of digital signal and image processing. The paper is devoted to the pattern matrix construction using features evaluated by the discrete Fourier transform (DFT) or the discrete wavelet transform (DWT) using the relative power in selected frequency bands or scale levels, respectively. Image features are then used to recognize groups of similar pattern vectors by self-organizing neural networks forming a mathematical tool for cluster analysis. Further classification methods including the decision tree, support vector machine, nearest neighbour method and neural networks are then applied for construction of specific models and evaluation of their accuracy and cross validation errors. The proposed algorithm is applied for analysis of given microscopic images representing wax structures covering breathing openings on leaves affected by environmental pollution in different locations. The classification accuracy depends upon the method used and it is higher than 92% for all experiments.

Keywords: Image processing · Machine learning · Classification · Vegetation health monitoring · Air pollution

1 Introduction

Classification of image components [2] forms a fundamental problem in many areas of information engineering, natural sciences, biomedicine, and robotics. Datasets recorded by different sensor systems including RGB or thermal cameras [1,11] are mostly preprocessed at first to reduce the noise and artifacts [8] that decrease the information content of observed signals.

The work has been supported by the research grant No. LTAIN19007 Development of Advanced Computational Algorithms for Evaluating Post-surgery Rehabilitation.

Á. Herrero et al. (Eds.): SOCO 2020, AISC 1268, pp. 199–206, 2021.
https://doi.org/10.1007/978-3-030-57802-2_19

Specific image components are then associated with their features evaluated by different methods in time, frequency or scale domains in many cases. These feature vectors can be then organized in the pattern matrix used for their classification.

The present paper is devoted to extraction of image features evaluated by the discrete Fourier transform (DFT) or wavelet transform (DWT) using the relative power in selected frequency bands or scale levels, respectively. The classification of these features is then performed by self-organizing and self-creating neural networks allowing clustering with no preliminary information about the number of classes. Then further classification methods including the decision tree, support vector machine (SVM), k-nearest neighbour method (k-NN) and neural networks (NN) are applied for construction of specific models and evaluation of their accuracy and cross validation errors.

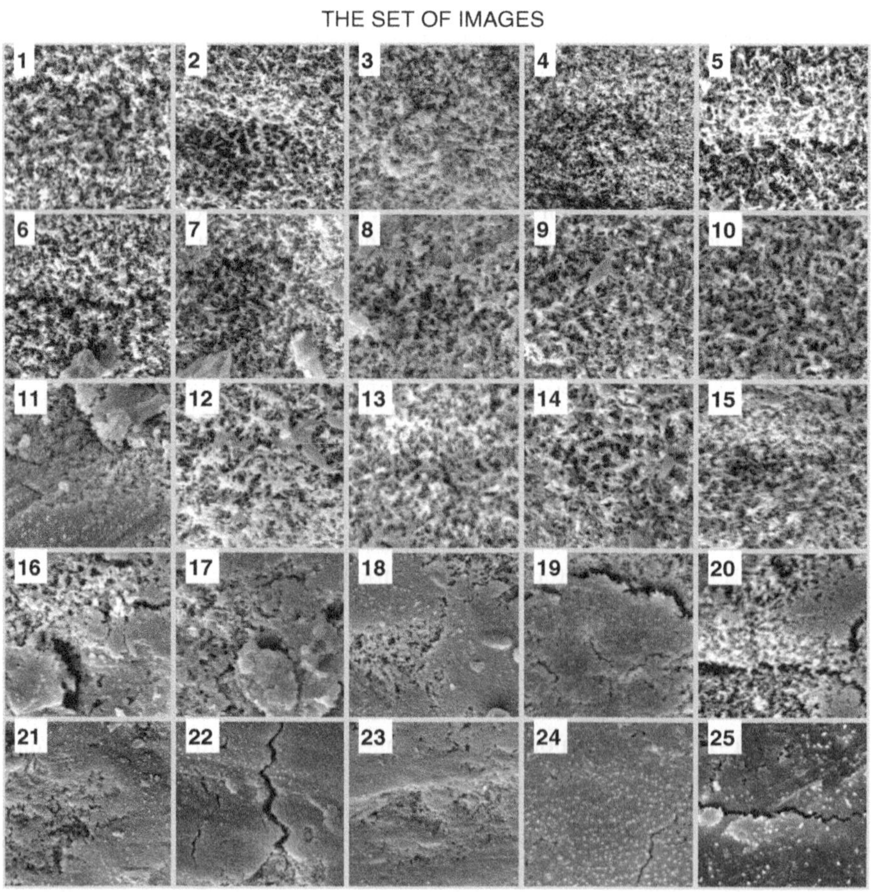

Fig. 1. The set of 25 images used for feature extraction and classification.

The paper presents the application of proposed methods to the investigation of stomata images of the picea abies leaves. The leaves are taken from trees observed on a long-term basis in selected regions distinguished by a low or high pollution level, respectively. Figure 1 presents acquired microscopic images belonging to different classes with organic and inorganic particles covering parts of their surface.

Research of stomata of spruce leaves, or to be more accurate, of structural signs of the stomata wax, represents a manner of evaluation of the given locality load by pollutants. Stomata images, obtained upon the previous modification of the leaves using an electron microscope are classified into the selected number of classes according to structural changes of the stomata wax. The current goal of interest is to find an algorithm of automatized texture classification of images corrupted by randomly placed artifacts into selected number of classes.

2 Methods

2.1 Real Data Description

The set of images of the size 1024 by 768 pixels representing microscopic wax structures covering leaves use pixel size 0.063 μm by 0.116 μm. Quality of the images varies significantly and their processing is hindered by the changing sharpness level and angle of the stomata view. Both of these factors introduce errors into the images evaluation. A significant problem is posed by presence of various inorganic and organic particles of different sizes and shapes on the stomata surface. These very small particles cannot be removed from the leaves before making an image, and their effect on the image processing can be suppressed by digital filtering methods only. In a number of images, these impurities cover a small area only. It can be thus assumed that in spite of their presence, correct classification can be made using the procedure chosen. However, in a non-negligible portion of images, the impurities cover a large part of the stomata surface, making their classification more complex.

In order to verify the processing procedures chosen, a testing set of images was created, which includes images with the minimum amount of impurities and disturbing components. The set of 25 images presented in Fig. 1 were selected from locations both with high and low concentration of air pollution in the Czech Republic.

2.2 Feature Extraction

Image classification assumes a proper image preprocessing, de-noising, and enhancement at first using appropriate methods including digital filtering, numerical interpolation, and gradient algorithms [4–7,9]. Further functional transforms are then often applied to estimate image features.

The DFT method is based upon power estimation in selected frequency bands while DWT ensembles more sophisticated image analysis allowing to distinguish finer and coarser wax structures in the given application. Features can then be associated with the power at each decomposition scale.

The set of wavelet functions [3] is usually derived from the initial (mother, basis) wavelet $h(t)$ which is dilated by value $a = 2^m$, translated by constant $b = k\ 2^m$ and normalized so that

$$h_{m,k}(t) = \frac{1}{\sqrt{a}}\, h(\frac{t-b}{a}) = \frac{1}{\sqrt{2^m}}\, h(2^{-m}\ t - k) \tag{1}$$

for integer values of m and k. Using the signal processing point of view the discrete wavelet transform is defined by a bank of bandpass filters and the complementary low pass filter (scaling function) for the lowest frequencies.

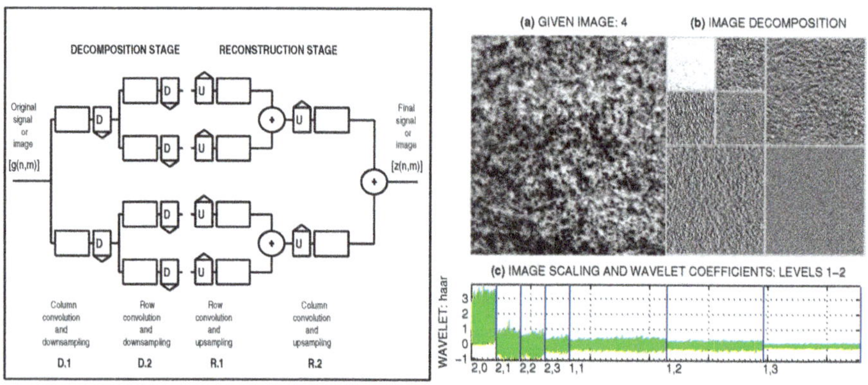

Fig. 2. Wavelet image decomposition and reconstruction principle presenting **(a)** the given image, **(b)** resulting subimages after the decomposition into the second level and **(c)** coefficients of the second level decomposition in the row vector using Haar wavelet functions.

Application of Haar wavelet decomposition into the second level of a selected image is presented in Fig. 2. The high-pass filter and complementary low-pass filters are applied to image columns and then to its rows followed by downsampling after each processing unit at first. This step of decomposition using dilation index $m = 1$ results in four subimages representing both approximation and horizontal, vertical and diagonal detail coefficients. In the second step for dilation index $m = 2$ the same algorithm is applied to the subimage of approximation coefficients after the previous level of decomposition with its result presented in Fig. 2(b). The same process can then be repeated again. Resulting coefficients can be used for image analysis or for its reconstruction again. Image downsampling by the scaling factor two in each decomposition level resulted in the total size of decomposition coefficients equal to the size of the original image without any redundancy.

Summed squared coefficients or their variance at selected levels can then be used as image features for their classification. This process allows the use of a wide range of wavelet functions including complex wavelets [8] and selection of decomposition levels.

2.3 Image Classification

Classification of Q images using R features organized in the pattern matrix $\mathbf{P}_{R,Q}$ was realized by self-organizing neural networks to estimate image clusters at first. Then different classification methods including the decision tree, 3-nearest neighbour method, support vector machine, and the two layer $R - S1 - S2$ neural network were used. The best results were achieved by neural networks [10] with $S1$ elements (and the sigmoidal transfer function $f1$) in the first layer and $S2$ elements (and softmax function $f2$ based on the Bayes theorem) in the second layer, respectively. Layer outputs were evaluated by the following relations:

$$\mathbf{A1}_{S1,Q} = f1(\mathbf{W1}_{S1,R}\, \mathbf{P}_{R,Q}, \mathbf{b1}_{S1,1}) \quad (2)$$

$$\mathbf{A2}_{S2,Q} = f2(\mathbf{W2}_{S2,S1}\, \mathbf{A1}_{S1,Q}, \mathbf{b2}_{S2,1}) \quad (3)$$

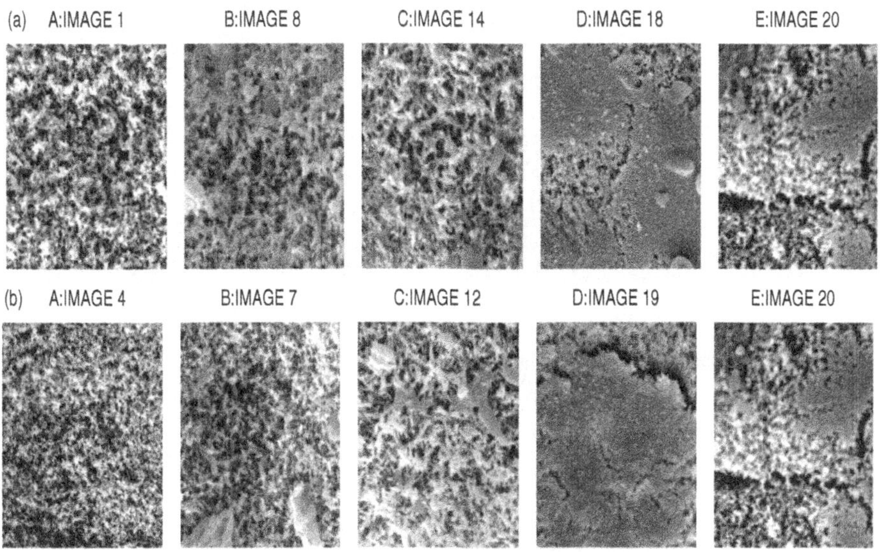

Fig. 3. Typical representatives of separate classes using **(a)** the discrete Fourier transform and **(b)** the discrete wavelet transform for features associated with selected decomposition coefficients.

The network coefficients of the two layer system included the elements of the matrices $\mathbf{W1}_{S1,R}$, $\mathbf{W2}_{S2,S1}$ and associated vectors $\mathbf{b1}_{S1,1}$, $\mathbf{b2}_{S2,1}$. For each column vector in the pattern matrix, the corresponding target vector has one unit element in the row pointing to the correct target value.

3 Results

The method presented above has been applied for classification of real images of wax structure presented in Fig. 1. In the initial stage both DFT and DWT image features were evaluated end self-creating neural networks used to deter-

Table 1. Comparison of image segments classification into five classes using two features evaluated by DFT and DWT using Haar wavelet function and horizontal (H), vertical (V) or diagonal (D) decomposition up to the second level.

Method	Class members (typical class images in bold)				
	A	B	C	D	E
DFT	**1**, 2, 4, 5, 6	**2**, 8, 10, 11	13, **14**	17, **18**, 19, 21 22, 23, 24, 25	3, 7, 9, 12, 15, 16, **20**
DWT	1, **2**, 4, 5, 6	**7**, 9	3, 10 **12**	17, 18, **19**, 21 22, 23, 24, 25	8, 11, 13, 14, 15, 16, **20**

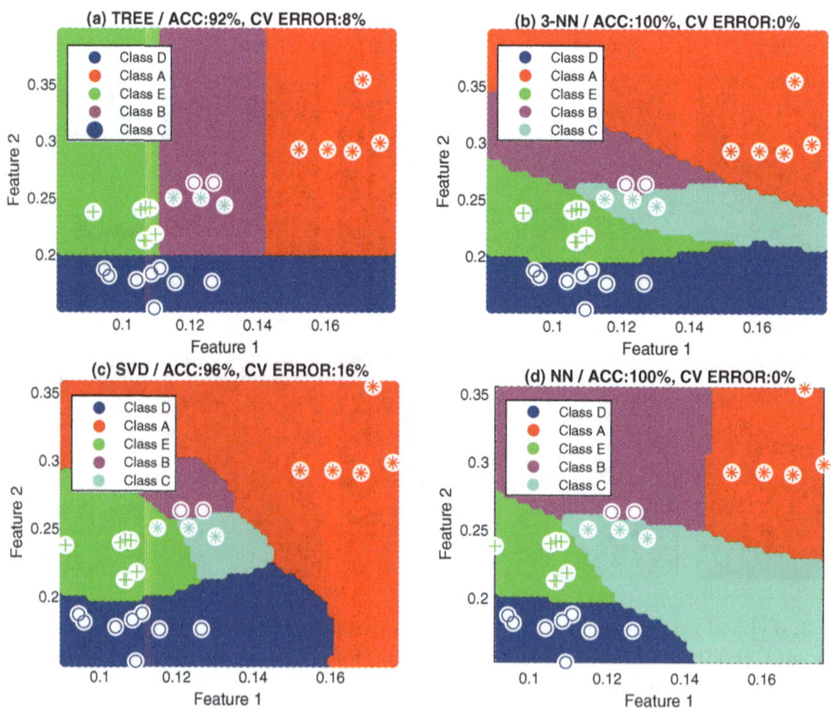

Fig. 4. Results of classification of 25 images into five classes with class boundaries using features resulting from the discrete Haar wavelet transform in the second decomposition level using diagonal and vertical decomposition coefficients by (a) the decision tree, (b) 3-nearest neighbour method, (c) support vector machine, and (d) 2-10-5 neural network method.

mine image clusters specified in Table 1. Typical class representatives are presented in Fig. 3 with the lowest distances of their features from the image cluster centres.

Further studies were devoted to analysis of features obtained by the DWT with different wavelet functions and different number of decomposition levels. The flexibility of the DWT allowed the selection of image features on different scales and the construction of more compact and better separated clusters in comparison with the DFT use.

Table 2. Accuracy [%] and cross-validation errors [%] of the classification of image features evaluated by the DFT and DWT methods by selected classification methods.

Method	DFT		DWT	
Dec. tree	92	8	92	8
3-NN	96	16	100	0
SVM	92	16	96	16
NN	100	0	100	0

Table 2 presents the classification accuracy and cross-validation errors of the subsequent classification into five classes using DFT and DWT features and different classification methods. Results show that the two layer neural network provides the best classification results in both cases. Figure 4 presents distribution of individual DWT features and classification results with boundaries of individual classes. Owing to the limited size of the dataset the learning process was very fast and it was completed after 40 epochs with the sufficient accuracy in case of the neural network use.

4 Conclusion

The novelty of the contribution is in the use of wavelet transform for image classification and comparison of results with that obtained by the discrete Fourier transform. Mathematical basis of the discrete wavelet transform and the following numerical experiments proved that image features based on wavelet transform coefficients can be used very efficiently for image classification and artifacts rejection.

The initial self-organizing clustering methods enabled the construction of classification models with their accuracy higher than 92%. Best results were obtained by the neural network classifier.

It is assumed that further research will be devoted to further methods of image features acquisition using special methods for image de-noising and artifacts rejection. Deep learning methods will be used for image segments classification as well.

References

1. Charvátová, H., Procházka, A., Vaseghi, S., Vyšata, O., Vališ, M.: GPS-based analysis of physical activities using positioning and heart rate cycling data. Signal Image Video Process. **11**(6), 251–258 (2017)
2. Choi, D.I., Park, S.H.: Self-creating and organizing neural networks. IEEE Trans. Neural Netw. **5**(4), 561–575 (1994)
3. Daubechies, I.: The wavelet transform, time-frequency localization and signal analysis. IEEE Trans. Inf. Theory **36**, 961–1005 (1990)
4. Dong, J., Han, Z., Zhao, Y., Wang, W., Procházka, A., Chambers, J.: Sparse analysis model based multiplicative noise removal with enhanced regularization. Signal Process. **137**(8), 160–176 (2017)
5. Hošťálková, E., Vyšata, O., Procházka, A.: Multi-dimensional biomedical image denoising using Haar transform. In: Proceedings of the 15th International Conference on Digital Signal Processing, Cardiff, UK, pp. 175–179. IEEE (2007)
6. Jerhotová, E., Švihlík, J., Procházka, A.: Biomedical image volumes denoising via the wavelet transform, pp. 435–458. INTECH (2011)
7. Kavalcová, L., Škába, R., Kyncl, M., Rousková, B., Procházka, A.: The diagnostic value of MRI fistulogram and MRI distal colostogram in patients with anorectal malformations. J. Pediatr. Surg. **48**(8), 1806–1809 (2013)
8. Kingsbury, N.G.: Complex wavelets for shift invariant analysis and filtering of signals. J. Appl. Comput. Harmonic Anal. **10**(3), 234–253 (2001)
9. Langari, B., Vaseghi, S., Procházka, A., Vaziri, B., Aria, F.: Edge-guided image gap interpolation using multi-scale transformation. IEEE Trans. Image Process. **25**(9), 4394–4405 (2016)
10. Procházka, A., Charvátová, H., Vaseghi, S., Vyšata, O.: Machine learning in rehabilitation assessment for thermal and heart rate data processing. IEEE Trans. Neural Syst. Rehabil. Eng. **26**(6), 1209–12141 (2018)
11. Procházka, A., Charvátová, H., Vyšata, O., Kopal, J., Chambers, J.: Breathing analysis using thermal and depth imaging camera video records. MDPI Sensors **17**, 1408:1–1408:10 (2017)

Software Sensors for the Monitoring of Bioprocesses

Pavel Hrnčiřík(✉)

University of Chemistry and Technology, Technická 5, 16628 Prague, Czech Republic
Pavel.Hrncirik@vscht.cz

Abstract. This paper presents various software based approaches suitable for the design of knowledge-based monitoring of biotechnological production processes. These processes require special treatment with respect to the complexity in biochemical reactions which make the design and construction of reasonably complex and practically usable mathematical models rather difficult. Additional complexity arises from the lack of industrially viable sensors for on-line measurement of key process variables. Software sensors which often use tools from the field of artificial intelligence represent one of the suitable approaches for the overcoming of the above mentioned limitations for its ability to utilize effectively both quantitative and qualitative knowledge about the monitored bioprocess. This approach is shown in practice using two different case studies of knowledge-based software sensors.

Keywords: Software sensors · Process monitoring · Bioprocesses

1 Introduction

The term "software sensor" or "soft sensor" already represents an established term in the field of monitoring of production processes. The attribute "software" expresses the fact that the output signal is largely the result of more or less complex calculations performed in the program module. The term "sensor" then means that the entire software sensor ultimately provides information about the monitored process, similar to traditional hardware sensors [1].

The basic principle of software sensors is to use a set of relatively easily online measurable process variables to estimate other variables or process indicators that are difficult to measure in the on-line mode or can only be measured with very long sampling periods (see Fig. 1).

The interest in the application of software sensors in the monitoring of production bioprocesses is increasing in proportion to the increasing demands on the quality of the production process and the resulting products. Compared to costly and relatively complex analytical technologies, application of software sensors is often a more advantageous solution for monitoring especially those bioprocesses that are operated as fed

Á. Herrero et al. (Eds.): SOCO 2020, AISC 1268, pp. 207–215, 2021.
https://doi.org/10.1007/978-3-030-57802-2_20

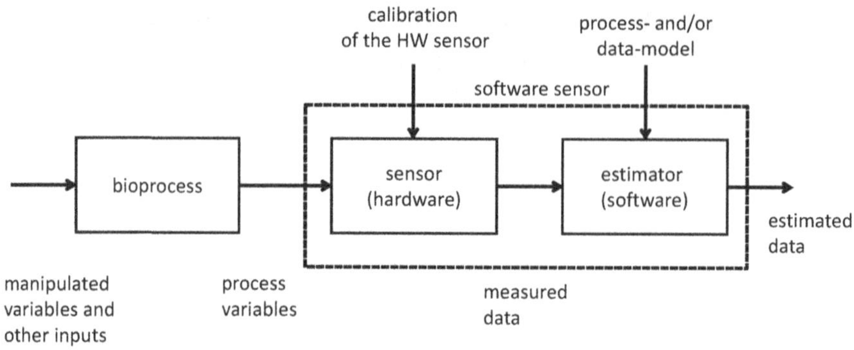

Fig. 1. General scheme of software sensor for monitoring of biotechnological processes.

batch cultures that are characterized by complex process dynamics, considerable variability due to variable feedstock composition and frequent changes in the production bioprocesses as a result of the production of various products. In these cases, software sensors are successfully used not only for monitoring of the production cultures, but also for evaluating the quality of feedstocks and the seed microbial cultures at the very beginning of the production process [2–4].

Fermentation processes are also important in modern sustainable agriculture, especially in the form of fermentations for the processing of agricultural bio-waste as a source of renewable energy within the framework of distributed decentralized on-site energy generation at the waste source (farms, dairies etc.). Advanced monitoring of these fermentation processes by software sensors has a considerable potential to contribute substantially to the improvement of their operation.

In principle, it is possible to distinguish between two basic types of software sensors [5, 6]:

- "gray box" type sensor, also referred to as "model-driven" - based on a mathematical model of a process based on physical, chemical or biological relationships with experimental identification of unknown parameters from historical process data;
- "black box" type sensor, also referred to as a "data-driven" – a priori, there is no known mathematical model of the relationship between inputs and outputs of the software sensor, the mathematical description of this relationship must therefore be obtained from historical process data using suitable computational tools, *e.g.* regression analysis, neural networks, etc.

Software sensors based on mathematical models that are used in the field of chemical and biotechnological processes are typically based on mass or energy balances often supplemented by kinetic relationships, all in combination with estimating algorithms such as Kalman filter or extended Kalman filter. The main problem in using this type of sensor in the field of bioprocesses is the difficulty of deriving sufficiently accurate models of cultivation processes. For this reason, this type of sensor is not very widespread in the field of bioprocess monitoring [1]. In addition to the above-mentioned complications associated with modeling, other typical properties of bioprocesses also complicate the design [6]:

- sensitivity of the production microbial culture to changes in bioreactor conditions;
- during cultivation, the microbial culture passes through various physiological states, which usually manifest themselves in a different type of culture behavior;
- main parameters of bioprocess models usually change during cultivation, while on-line measurement or at least estimation of these changes is very complicated.

An example of a software sensor based on a mathematical model of the process is the on-line estimator for the fed batch culture process of the microorganism *Saccharomyces cerevisiae*, which was designed by the author of this paper. This estimator uses a modified extended Kalman filter (the modification consists in supporting alternative models in the form of a tree structure) to estimate 6 state variables (concentration of main reactants and batch volume) based on 4 on-line measured process variables. A major obstacle to the successful deployment of this system in practice was the considerable number of model parameters that had to be identified experimentally for each variant of the bioprocess and whose consistency during the cultivation process could not be guaranteed [7].

For the above-mentioned reasons, software sensors of the second type ("black box" or "data-driven") are more suitable for the monitoring of bioprocesses. Neural networks, fuzzy systems [8], PLS and PCR multivariate statistical methods [9], or other variants of regression analysis methods, recently *e.g.* support vector regression (SVR), are used successfully for their design [6].

In addition to selecting a suitable calculation method, a key factor for the functionality of a given software sensor is the selection of appropriate input on-line measured quantities. Software sensors that use on-line measurement of off-gas composition from a bioreactor are very common [10]. These on-line measurements are successfully used for on-line calculation of so-called derived quantities such as oxygen uptake rate (OUR), CO_2 production rate (CPR), respiratory quotient (RQ), or oxygen transfer coefficient from gaseous to liquid phase in a bioreactor (k_La). More sophisticated software sensors are able to estimate key bioprocess indicators such as biomass concentration, biomass growth rate, or concentrations and rates of production of the main products [11–15].

The author of this paper participated in the design of software "data-driven" sensors for a number of bioprocesses from the category of fed batch cultures:

- estimation of biomass concentration (*Saccharomyces cerevisiae* cultivation, provitamin D2 – ergosterol production process) [16]
- estimation of biomass concentration and microbial culture state in terms of feeding sufficiency (*Streptomyces noursei* cultivation, Nystatin antibiotic production process) [17]
- calculation of the indicator of biopolymer content in cells (*Pseudomonas putida* cultivation, PHA bioplastic production process) [18]

In the next part of the paper, the first two examples of the use of software sensors to estimate the biomass concentration and culture state in a bioreactor will be presented in more detail.

2 Case 1: ANN-Based Software Sensor for Biomass Concentration Estimation in a Yeast Cultivation Process

Sterols are among the materials forming the cell wall of a eukaryotic cell. In the case of yeast, the dominant sterol is ergosterol, also known as provitamin D2. Ergosterol is an important substance for the pharmaceutical industry. Its importance is increasing in connection with the development of anti-cancer drugs. *Saccharomyces cerevisiae* yeast cultures are generally used to produce ergosterol on an industrial scale. The efficiency of its production depends mainly on the final amount of yeast biomass and the ergosterol content in the cells. An equally important parameter is the purity of ergosterol, which has an important effect on the processing of yeast biomass after the cultivation process [19]. On-line monitoring of yeast biomass concentration development during production cultures is therefore one of the key tasks of advanced control of this production bioprocess.

The design of knowledge-based control of the cultivation process of the yeast *Saccharomyces cerevisiae* for the production of provitamin D2 (ergosterol) hence also involved solving the role of on-line estimation of the values of important cultivation indicators - especially the concentration of yeast biomass - using software sensors [16]. Following the results of preliminary studies, it was decided to address the question of on-line estimation of yeast biomass concentration by a software sensor based on artificial neural network. The main innovation of the proposed solution compared to standard sensors of this type was the inclusion of information on the current metabolic state of the microbial culture (inferred on-line by a rule-based system) as one of the data inputs into the software sensor.

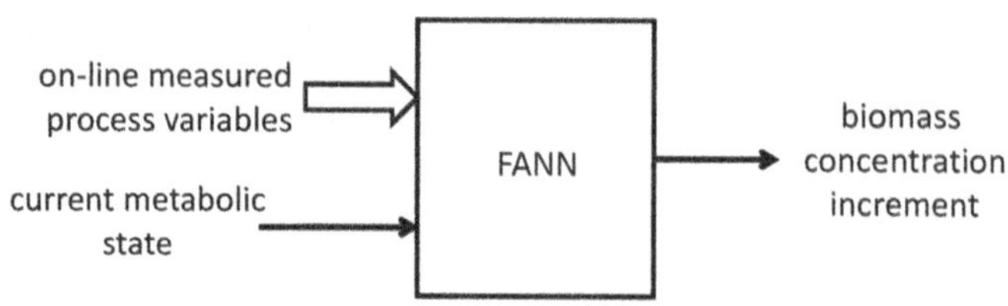

Fig. 2. Block diagram of feedforward artificial neural network based sensor software for yeast biomass concentration estimation using metabolic state data as extra data input.

In the first variant the software sensor consisted of one neural network and the metabolic state value was one of the inputs into this network (see Fig. 2). In the second variant, the software sensor consisted of a set of neural networks for individual metabolic states and input information on the current metabolic state served to switch between individual neural networks. The artificial neural networks used in the design of the software sensor were of two types - a classical multilayer feedforward neural network trained using the Levenberg-Marquardt algorithm and a cascade correlation artificial neural network taking advantage of the automatic design of its topology running parallel to the learning process. Specifically, the multilayer feedforward neural network used in the first variant of the software sensor took the form of a 3-layer network with one hidden layer (number of neurons in each layer: 4-5-1). The input to the software sensor consisted of 4 on-line measured variables (O_2 and CO_2 concentration in the off-gases, ethanol concentration and volume flow of nutrients at the entrance to the bioreactor) and metabolic state of the microbial culture.

The only output from the software sensor was the biomass concentration increment per sampling period (1 min). The choice of increment in contrast to the absolute biomass concentration proved to be more appropriate during the design to ensure independence from the initial cultivation conditions. Testing of the resulting software sensors confirmed that the inclusion of metabolic state information in the sensor inputs significantly contributed to improving the quality of biomass concentration estimation. On average, the error of estimation decreased by 54% compared to a sensor of the same type without using the metabolic state as input. Both considered variants of the software sensor with metabolic state (one network vs. several networks for individual states) provided comparable results. However, the single-network variant has proved to be more appropriate not only due to its simpler structure, but also because it provided a smoother output signal (see Fig. 3) [16].

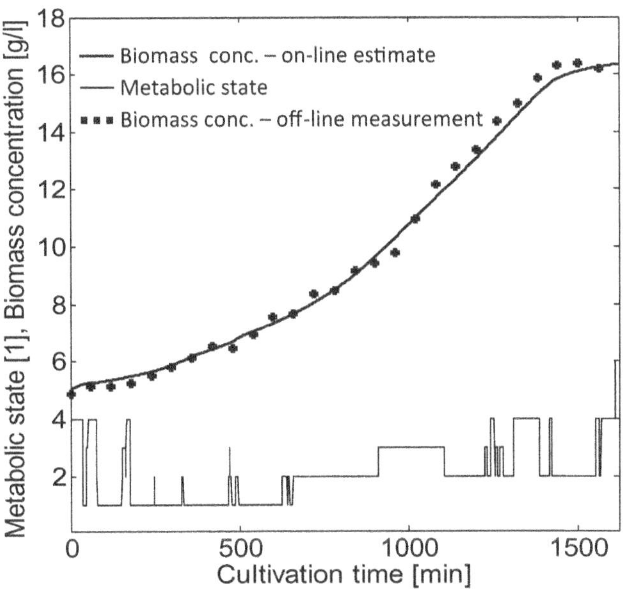

Fig. 3. Yeast biomass concentration estimation using software sensor based on single FANN using metabolic state data as extra data input.

3 Case 2: Set of Two Software Sensors for the Monitoring of Biomass Growth in a Filamentous Bacterial Cultivation Process

The process used in the second study is the industrial production of antibiotics Nystatin, which is a polyene antifungal medication that is produced as a secondary product from a filamentous bacterium *Streptomyces noursei*. The production process can be divided into two distinct phases. In the initial phase the main objective is the maximization of the cellular growth. In the subsequent second phase the filamentous microbial culture produces the secondary product Nystatin. Particularly in the initial phase timely information related to cell growth are important for efficient process monitoring and hence the application of software sensors for on-line estimation of biomass concentration presents a cost-efficient solution that can substantially improve the quality of the Nystatin production process because the biomass concentration is not usually measured on-line.

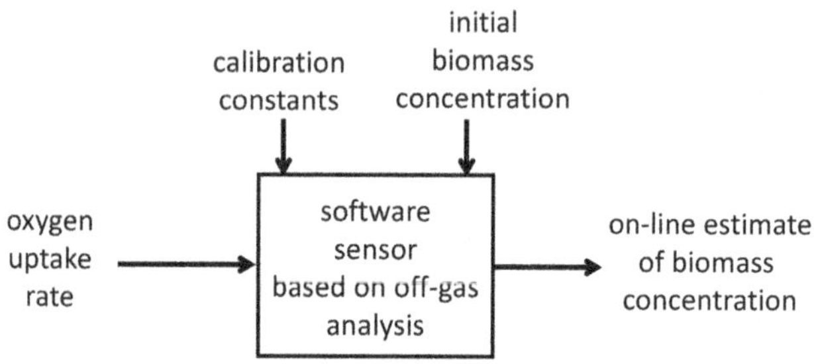

Fig. 4. Block diagram of sensor software for filamentous bacteria biomass concentration estimation based on off-gas analysis.

The solution proposed in this case consists of a combination of two software sensors for on-line estimation of yeast biomass concentration in the bioreactor. The first of the software sensors is based on the estimation of biomass concentration from the on-line measured composition of off-gases from the bioreactor (see Fig. 4). From this composition it is possible to continuously calculate the oxygen uptake rate (OUR) and then integrate this rate into the form of cumulative oxygen consumption by biomass since the beginning of cultivation (COC). The linear dependence between the square root of the COC and the biomass concentration can then be used to calculate the biomass concentration estimate (see Eqs. 1 and 2).

$$\Delta c_{BIO\_s1}(t) = k1 \cdot \sqrt{COC(t)} + k2 \tag{1}$$

$$c_{BIO\_s1}(t) = \Delta c_{BIO\_s1}(t) + c_{BIO\_s1}(0) \tag{2}$$

where $k1$, $k2$ are calibration constants, t is cultivation time and c_{BIO_s1}, Δc_{BIO_s1} are the biomass concentration estimate and the increment of the biomass concentration estimate, respectively. This relatively computationally simple software sensor is able to estimate the biomass concentration value in an online mode with an error less than 10% of the concentration measurement range (see Fig. 5). Due to preserving manufacturing secrets, unit-scale representations are used for all data sets in the charts related to this process published in this paper [17].

The second sensor is based on biocalorimetry, *i.e.* the on-line calculation of the heat generated by the biomass from general energy balance of the bioreactor. Based on the

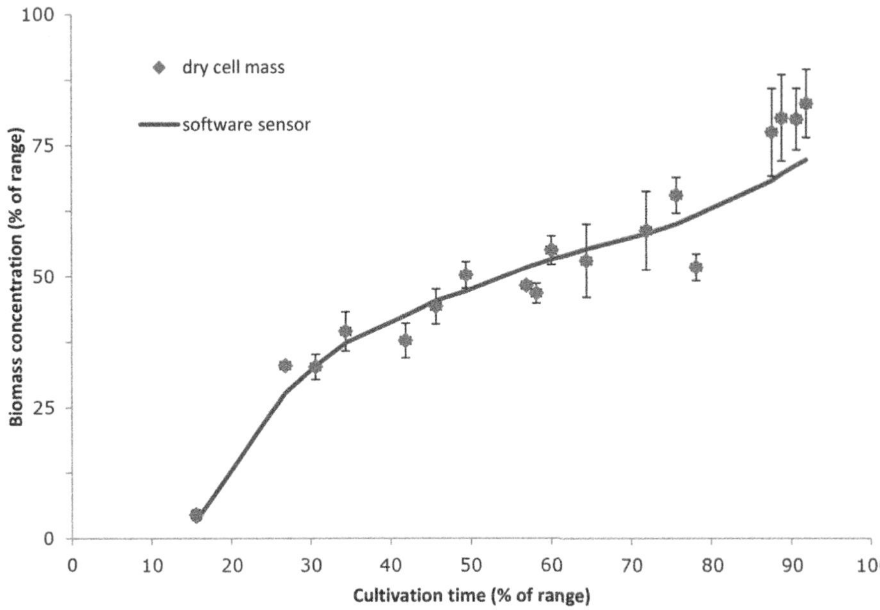

Fig. 5. Filamentous bacteria biomass concentration estimation using software sensor based on off-gas analysis.

knowledge of the specific biomass heat production as a calibration parameter obtained by laboratory assays, an estimate of the biomass concentration in the bioreactor can then be calculated on-line (see Fig. 6 and Eq. 3).

$$c_{BIO\_s2}(t) = \frac{1}{V_L} \cdot \frac{Q_{BIO}(t)}{q_{sbio}} \tag{3}$$

where t is cultivation time, c_{BIO_s2} is the biomass concentration estimate, Q_{BIO} is the bioprocess heat production in the bioreactor, V_L is volume of the fermentation broth in the bioreactor and q_{sbio} is the specific biomass heat production.

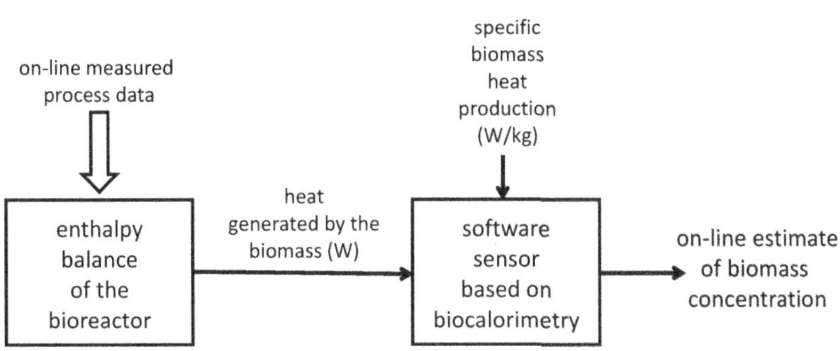

Fig. 6. Block diagram of sensor software for filamentous bacteria biomass concentration estimation based on biocalorimetry.

However, the biocalorimetry-based biomass concentration software sensor is very sensitive to changes in microbial biomass state. In fact, if the biomass is not in an optimal state, the specific heat value decreases compared to the normal values. Since this parameter is of key importance in estimating the biomass concentration (see Eq. 3), this decrease leads to an underestimation of the estimated concentration. This phenomenon occurs especially when the nutrient supply is insufficient. For these reasons, the biocalorimetric software sensor is not suitable for biomass concentration estimation, but it can be used in combination with the software sensor based on off-gas analysis to monitor the biomass condition. In particular, the ratio of both estimates can serve as an indicator of underfeeding (see Fig. 7), and its decrease below a suitably chosen threshold can serve as a timely warning of the risk of onset of this undesired condition. This indicator can serve as a suitable supplement to the standard operation procedure for the monitoring of the nutritional state of the filamentous culture, which consists of laboratory off-line determination of carbohydrate substrates concentration in the bioreactor, measured as reducing substances [17].

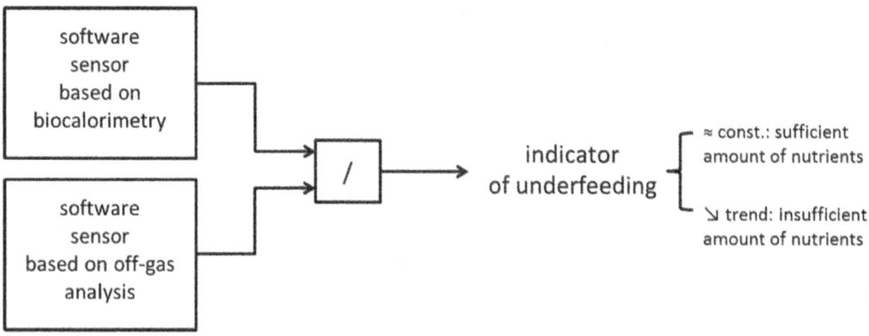

Fig. 7. Block diagram of a combination of two software sensors (off-gas analysis, biocalorimetry) for on-line filamentous bacteria biomass monitoring.

4 Conclusion

The aim of this paper was to introduce the possibilities and to show the potential of software sensors for advanced bioprocess monitoring. Their main application is in the field of monitoring of the growth of microbial biomass and related phenomena such as the state of biomass in terms of nutrient sufficiency. In this context, two specific applications of software sensors were presented, both using unique approaches. The first case was a software sensor based on a feedforward artificial neural network, using as an extra input data on the current metabolic state of yeast culture, which is continuously inferred by a knowledge-based system. In the second case, on the contrary, the solution had the form of two connected software sensors, on the basis of which it is possible to monitor the biomass state from the perspective of sufficient nutrients. One of these software sensors can also be used to directly estimate biomass concentration based on on-line measurement of bioreactor off-gas composition.

References

1. Kadlec, P., Gabrys, B., Strandt, S.: Data-driven soft sensors in the process industry. Comput. Chem. Eng. **33**(4), 795–814 (2009)
2. Faergestad, E.M., Oyaas, J., Kohler, A., Berg, T., Naes, T.: The use of spectroscopic measurements from full scale industrial production to achieve stable end product quality. J. Food Sci. Technol. **44**(10), 2266–2272 (2011)
3. Gao, Y., Yuan, Y.J.: Comprehensive quality evaluation of corn steep liquor in 2-keto-L-gulonic acid fermentation. J. Agric. Food Chem. **59**(18), 9845–9853 (2011)
4. Cunha, C.C.F., Glassey, J., Montague, G.A., Albert, S., Mohan, P.: An assessment of seed quality and its influence on productivity estimation in an industrial antibiotic fermentation. Biotechnol. Bioeng. **78**(6), 658–669 (2002)
5. Luttmann, R., Bracewell, D.G., Cornelissen, G., Gernaey, K.V., Glassey, J., Hass, V.C., Kaiser, C., Preusse, C., Striedner, G., Mandenius, C.F.: Soft sensors in bioprocessing: a status report and recommendations. Biotechnol. J. **7**, 1040–1048 (2012)
6. Sharma, S., Tambe, S.S.: Softsensor development for biochemical systems using genetic programming. Biochem. Eng. J. **85**, 89–100 (2014)
7. Hrnčiřík, P., Náhlík, J., Havlena, V.: State estimation of baker's yeast fed-batch cultivation by extended Kalman filter using alternative models. In: Georgakis, C. (ed.) Dynamics & Control of Process Systems 1998 (DYCOPS 5), IFAC, pp. 601–606. Pergamon Press, Oxford (1999)
8. Glassey, J., Montague, G.A., Ward, A.C., Kara, B.: Enhanced supervision of recombinant *E.coli* fermentations via artificial neural networks. Proc. Biochem. **29**, 387–398 (1994)
9. Ödman, P., Lindavald Johansen, C., Olsson, L., Gernaey, K.V., Eliasson Lantz, A.: On-line estimation of biomass, glucose and ethanol in *S. cer.* cultivations using in-situ multi-wavelength fluorescence and software sensors. J. Biotechnol. **144**(2), 102–112 (2009)
10. Aehle, M., Kuprijanov, A., Schaepe, S., Simutis, R., Luebbert, A.: Simplified off-gas analyses in animal cell cultures for process monitoring and control purposes. Biotechnol. Lett. **33**(11), 2103–2110 (2011)
11. Chéruy, A.: Software sensors in bioprocess engineering. J. Biotechnol. **52**, 193–199 (1997)
12. Montague, G.A., Morris, A.J., Tham, M.T.: Enhancing bioprocess operability with generic software sensors. J. Biotechnol. **25**, 183–201 (1992)
13. Ignova, M., Glassey, J., Ward, A.C., Montague, G.A.: Multivariate statistical methods in bioprocess fault detection and performance forecasting. Trans. Inst. MC **19**(5), 271–279 (1997)
14. Albiol, J., Robustr, J., Casas, C., Poch, M.: Biomass estimation in plant cell cultures using an extended Kalman filter. Biotechnol. Prog. **9**(2), 174–178 (1993)
15. Arnold, S.A., Crowley, J., Woods, N., Harvey, M.L.: In-situ near infrared spectroscopy to monitor key analytes in mammalian cell cultivation. Biotechnol. Bioeng. **84**(1), 13–19 (2003)
16. Vaněk, M., Hrnčiřík, P., Vovsík, J., Náhlík, J.: On-line estimation of biomass concentration using a neural network and information about metabolic state. Bioprocess Biosyst. Eng. **27**(1), 9–15 (2004)
17. Hrnčiřík, P., Moucha, T., Mareš, J., Náhlík, J., Janáčová, D.: Software sensors for biomass concentration estimation in filamentous microorganism cultivation process. Chem. Biochem. Eng. Q. **33**(1), 141–151 (2019)
18. Hrnčiřík, P., Vovsík, J., Náhlik, J.: A new on-line indicator of biopolymer content in bacterial cultures. IFAC Proc. Vol. **43**(6), 192–196 (2010)
19. Náhlík, J., Hrnčiřík, P., Mareš, J., Rychtera, M., Kent, C.A.: Towards the design of an optimal strategy for the production of ergosterol from *Saccharomyces cerevisiae* yeasts. Biotechnol. Prog. **33**(3), 838–848 (2017)

RGB Images Driven Recognition of Grapevine Varieties

Pavel Škrabánek[1(✉)], Petr Doležel[2], Radomil Matoušek[1], and Petr Junek[1]

[1] Brno University of Technology, Brno, Czech Republic
pavel.skrabanek@vut.cz
[2] University of Pardubice, Pardubice, Czech Republic

Abstract. We present a grapevine variety recognition system based on a densely connected convolutional network. The proposed solution is aimed as a data processing part of an affordable sensor for selective harvesters. The system classifies size normalized RGB images according to varieties of grapes captured in the images. We train and evaluate the system on in-field images of ripe grapes captured without any artificial lighting, in a direction of sunshine likewise in the opposite direction. A dataset created for this purpose consists of 7200 images classified into 8 categories. The system distinguishes among seven grapevine varieties and background, where four and three varieties have red and green grapes, respectively. Its average per-class classification accuracy is at 98.10% and 97.47% for red and green grapes, respectively. The system also well differentiates grapes from background. Its overall average per-class accuracy is over 98%. The evaluation results show that conventional cameras in combination with the proposed system allow construction of affordable automatic selective harvesters.

Keywords: Recognition of grapevine varieties · Densely connected convolutional network · In-field images · Agriculture mechanization

1 Introduction

In last years, we can observe unprecedented progress of agriculture mechanization towards its full automation. The rapid development in areas such as computer vision and machine learning likewise affordability of powerful hardware and precise manipulators allowed construction of autonomous robotic systems, e.g. for weed control [18], precise spraying [2,21] and harvesting. Robots capable to crop greenhouse vegetables, apples, grapes [2], sweet peppers [1] and even strawberries [24] have been presented. One of the directions of their further development is selective harvesting. A good example is harvesting of grapevines according to their varieties. The basic prerequisite for such a selective harvester is correct recognition of grapevine varieties.

Á. Herrero et al. (Eds.): SOCO 2020, AISC 1268, pp. 216–225, 2021.
https://doi.org/10.1007/978-3-030-57802-2_21

Recognition of grapevine varieties can be carried out different ways. A traditional recognition method is ampelometry [4]. As the method is visual, it is non-destructive. However, it requires involvement of an expert with extensive training, even when using a specialized software [20]. Accuracy of this method is strongly dependent on skills and experience of the expert. More objective is for example DNA analysis [15]; however, this method, as well as other wet chemistry techniques, is destructive, time-consuming, labour-demanding and requires involvement of an expert. Automation of grapevine variety identification enabled the current development of computer vision methods and availability of advance image sensors. Methods processing data provided by a spectrometer [3,5] or a hyperspectral camera [6] are automatic, non-destructive and fast. Measurements of an interaction of electromagnetic radiation with matter at many different spectral bands allow accurate recognition of grapevine varieties. The main disadvantage of this approach is a high purchase price of a spectrometer or a hyperspectral camera, which is several orders of magnitude higher compared to a conventional camera. Implementation of such sensors into a selective harvester would significantly increase its price.

Traditional methods aimed at recognition of grapevine varieties are limited by human senses. For example, ampelometry uses eyesight by the grapevine variety recognition. Despite that the human perceives only visible light in mostly three bands, experts are capable to recognize tens of varieties. Conventional cameras provide images of comparable attributes. We expect that the images keep information that allows an accurate classification of grapes according to their varieties.

An image-based classification of grapevines according to their varieties is a complex task which requires extraction of many discriminative features. An extensive diversity of an outdoor environment further increases the complexity of the feature extraction. The overall complexity of this task requires employment of a state-of-the-art image categorization system.

The state-of-the-art image categorization systems are based on deep convolutional networks (deep ConvNets) [12]. Deep ConvNets allow creation of self-contained image categorization systems which ensure both feature extraction and classification of object images. Key factors influencing performance of such a system are a learning capacity of a deep ConvNet and the quality of a training set. The capacity of the network is given by its topology. Modern topologies control the capacity by varying width or depth of networks [11]. Enlarging a deep ConvNet capacity through increasing its width is used e.g. in GoogLeNet [23]. The second approach is to increase number of network's layers (the network depth), while retaining the data processing linearity. Topologies, such as Highway Networks [22], Residual Networks [8], Deep Pyramidal Residual Networks [7], Densely Connected Convolutional Networks (DenseNets) [9] and Cross-Layer Neurons Networks [26] can have tens to hundreds of layers.

Factors such as selection of training samples, their correct categorization, proportional representation of samples with respect to their categories, as well as the total number of samples in a training set predetermine the quality of the training set [13]. To increase the number of samples, data augmentation techniques, such as image translations, horizontal reflections [25], and rotations [17] are used.

Herein, we present a variety recognition system. We based the system on a DenseNet topology. A dense connectivity pattern used in DenseNets alleviates a vanishing-gradient problem and it allows creation of very deep networks with high learning capacity [9]. For a training and evaluation of the system, we form a dataset based on in-field photos captured under various lighting conditions.

2 Materials and Methods

2.1 Data Collection

We captured colour photos (individual images) of common grapevine during a harvest. The data collection was carried out within two days in the morning and in the afternoon in August 2015. We selected various locations in vineyards in Čejkovice, Czech Republic. We used no artificial lighting and we captured the photos in a direction of sunshine likewise in the opposite direction (both days was partly sunny). The resulting collection of photos includes Welschriesling, Saint Laurent, Gewürztraminer, Pinot noir, Riesling Weiss, Pinot gris, and Veltliner Grün varieties (names of varieties according to Vitis International Variety Catalogue [14]).

We used camera bodies CANON EOS 1000D and CANON EOS 1100D with CANON ZOOM lenses EF-S 18–55 mm f/3.5-5.6 II and IS II, respectively. Resolutions of the photos are 1936×1288 pixel (px) and 4272×2848 px, respectively. The photos use RGB colour model with 24 bits bit depth. We placed the cameras perpendicular to vineyard rows (in terms of an axis of a lens), in a distance about 1.4 m from the rows, at an altitude of 1.25 m from the ground. A focal length varied between 18 mm and 24 mm.

2.2 Dataset

For training and evaluation of the variety recognition system, we form a dataset of RGB images of resolution 120×120 px. For each variety, we create 900 images. Grape clusters in the images cover at least 70% of their surface. Further, we create 900 images capturing a background, i.e. the final dataset consists of 7200 images classified into 8 categories (Fig. 1).

The images in the dataset are cut-outs of grapevine photos acquired within the data collection. For this purpose, we randomly select between 12 and 14 photos (depending on a density of grape clusters in photos) of each variety. In Table 1, we provide information about the number of selected photos with respect to grapevine varieties (first column), camera bodies (first row) and focal lengths (second row).

(a) Veltliner Grün (b) Riesling Weiss (c) Welschriesling (d) Gewürztraminer

(e) Pinot gris (f) Pinot noir (g) Saint Laurent (h) Background

Fig. 1. Example images for the categories in the dataset.

Table 1. Number of images selected for forming of the dataset. For each variety (first column), number of used images is stated with respect to the focal length (second row) and the camera body (first row).

Camera body	EOS 1000 D			EOS 1100 D	
Focal length	18	21	24	18	23
Gewürztraminer	8	–	4	–	–
Veltliner Grün	8	4	–	–	–
Pinot gris	8	2	–	–	4
Pinot noir	8	–	4	–	–
Riesling Weiss	6	–	4	4	–
Saint Laurent	12	–	0	–	–
Welschriesling	10	–	4	–	–

2.3 Densely Connected Convolutional Networks

As in other deep ConvNets [12], convolutional, pooling and fully connected layers are arranged in a feed-forward manner to form a DenseNet. Regular patterns occurring in DenseNets allow us to simplify description of their topologies. Let us define two composite building elements which will be used to describe a topology of the presented variety recognition system: a *dense blocks* (DB) and a *transition layer* (TL).

Dense Block. Let us consider a n-th DB of d_n layers that is built in a DenseNet of L layers. The input and the output of the n-th DB are placed at i_n-th and o_n-th levels of the network, respectively, i.e. $d_n = o_n - i_n + 1$. Feature maps produced at the ℓ-th level of the network, where $\ell \in [i_n, o_n]$, are given as

$$\mathbf{x}_\ell = H_\ell([\mathbf{x}_{i_n-1}, \ldots, \mathbf{x}_{\ell-1}]), \tag{1}$$

where $H_\ell(\cdot)$ is a non-linear transformation performed at the ℓ-th level, $\mathbf{x}_{i_n-1}$ are feature maps at the input of the n-th DB, $\mathbf{x}_i$ for $i \in [i_n, \ell - 1]$ are feature maps produced at preceding levels of the n-th DB, and $[\mathbf{x}_{i_n-1}, \ldots, \mathbf{x}_{\ell-1}]$ denotes their concatenation.

Two variants of the non-linear transformation $H(\cdot)$ can be used in DBs: a basic and a bottleneck version [9]. The basic version is a composite function which consists of a batch normalization (BN) [10], a rectified linear unit (ReLU), and a convolution (Conv) [12], respectively. Using a short notation, the basic version of $H(\cdot)$ can be written as BN-ReLU-Conv($h \times w, f, s$), where s is stride of convolutional filters, f is number of the filters, and h and w are their height and width, respectively. The bottleneck version of $H(\cdot)$ is defined as BN-ReLU-Conv($1 \times 1, 4f, 1$)-BN-ReLU-Conv($h \times w, f, s$). If necessary, convolutions are zero-padded to keep the feature-map size fixed. For both versions of the composite function $H(\cdot)$, the parameters h, w, s, f are identical for all layers within a DB. We use abbreviations DBa and DBb for DBs with the basic and the bottleneck version of $H(\cdot)$, respectively.

Transition Layer. Let us consider a TL connected at the output of the n-th DB (i.e. the TL is placed at the $(o_n + 1)$-th level of the network). The $(o_n + 1)$-th TL produces feature maps

$$\mathbf{x}_{o_n+1} = H_{o_n+1}([\mathbf{x}_{i_n-1}, \mathbf{x}_{i_n}, \ldots, \mathbf{x}_{o_n}]), \tag{2}$$

where $[\mathbf{x}_{i_n-1}, \mathbf{x}_{i_n}, \ldots, \mathbf{x}_{o_n}]$ denotes the concatenation of all feature maps that appear in the n-th DB. H_{o_n+1} is a composite function BN-ReLU-Conv($1 \times 1, f, 1$)-AP($2 \times 2, 2$), where AP($2 \times 2, 2$) denotes an average pooling with pools 2×2 and stride 2 [9].

Compactness of the network is controlled by the number of the 1×1 convolutional filters f incorporated in TLs. The number of feature maps produced by the $(o_n + 1)$-th TL is given as $f_{o_n+1} = \lfloor \theta m_n \rfloor$, where θ is a compression factor, $\theta \in [m_n^{-1}, 1]$ and m_n is the number of feature maps produced by the n-th DB.

2.4 Variety Recognition System

The presented variety recognition system is a DenseNet. The network classifies RGB images of dimensions 120×120 px according to varieties of grapes captured in the images. We control number of filters f in DBs by a variable k, where $k = 20$. The network is opened by one DBa which consists of one layer ($d = 1$) with $2k$ convolutional filters ($f = 2k$) with kernels of size 7×7 px ($h = w = 7$), stride by 2 px ($s = 2$). The following layer is a max pooling layer (MPL) with pools 3×3 px ($h = w = 3$) stride by 2 ($s = 2$). The inner parts of the network consist of two DBbs with 6 and 9 layers, respectively. At each layer of a DBb, k filters with kernels of size 3×3 px stride by 1 px ensure the feature extraction. Each DBb in the network is followed by one TL. The network is closed by a global average pooling (GAP) and a classifier, respectively. The classifier consists of one fully connected layer of eight neurons followed by a softmax function. We setup the compression factor θ at 0.5. The topology of the network is summarized in Table 2.

Table 2. Topology of the variety recognition system. Building elements which form the system are listed with respect to their placement in the network in the first row of the table (the first block is the leftmost one), where DBa and DBb denote the basic and the bottleneck versions of the dense block; MPL is a max pooling layer; TL is a transition layer, GAP denotes a global average pooling, and C is used for a classifier that consists of one fully connected layer followed by a softmax function. The parameters h and w are a height and a weight of a filter kernel or of a pool; s is stride of the kernel or of the pool; f is the number of filters at one convolution in a dense block; and d is the number of layers in the dense block.

	DBa	MPL	DBb	TL	DBb	TL	GAP	C
h	7	3	3	–	3	–	5	–
w	7	3	3	–	3	–	5	–
s	2	2	1	–	1	–	5	–
f	$2k$	–	k	–	k	–	–	–
d	1	–	6	–	9	–	–	–

We use MATLAB R2018b and Deep Learning Toolbox to train and evaluate the system. We split randomly the dataset into a training and an evaluation set, where the training set consists of 750 samples of each category. The rest of images (150 samples of each category) form the evaluation set. We train the system using ADAM optimizer [16] for 500 epochs with mini batches of 400 samples, minimizing a cross entropy function. We setup a learning rate, and an exponential decay rate for first and second moment estimates at 10^{-3}, 0.95 and 0.999, respectively. We shuffle images in the training set every epoch.

We use data augmentation techniques to bring more variability into the training set. We utilize a function `imageDataAugmenter`, where we use a random rotation (range of a rotation angle: $\pm 20^\circ$), a random reflection in the left-right direction, a random horizontal and vertical translation (range of a translation distance: ± 3 px), and a random horizontal and vertical shear (range of a shear angle: $\pm 20^\circ$).

3 Results and Discussion

We summarize the evaluation results in a confusion matrix (Table 3), where rows and columns represent instances in actual and predicted classes, respectively. We extend the matrix about average per-class accuracies of all classes (last column of Table 3). The average per-class accuracy of the i-th class is given as $\mathrm{acc}_i = l^{-1}\sum_{i=1}^{l}\frac{|\mathrm{TP}_i|+|\mathrm{TN}_i|}{n}$, where l is the number of classes, $|\mathrm{TP}_i|$ is the number of correctly classified samples of the i-th class, $|\mathrm{TN}_i|$ is the total number of correctly classified samples of complementary classes to the class i, and n is the total number of samples in the evaluation set [19]. Calculating arithmetic means of the accuracies for the red and the green grapes, we get accuracies at 98.10% and 97.47%, respectively. The overall average accuracy (arithmetic mean of all classes) is at 98.02%.

Table 3. Confusion matrix. Rows and columns represent instances in actual and predicted classes, respectively. Average per-class accuracies of the classes are summarized in the last column. Distinguished classes are Gewürztraminer (GT), Veltliner Grün (VG), Pinot gris (PG), Pinot noir (PN), background (BG), Riesling Weiss (RW), Saint Laurent (SL), and Welschriesling (WR).

	GT	VG	PG	PN	BG	RW	SL	WR	acc
GT	148	0	2	0	0	0	0	0	0.9950
VG	0	140	0	0	1	6	0	3	0.9717
PG	1	0	141	3	0	0	5	0	0.9875
PN	0	0	0	139	0	0	11	0	0.9742
BG	2	0	1	0	146	1	0	0	0.9933
RW	0	8	0	0	0	140	0	2	0.9783
SL	1	0	3	17	2	0	127	0	0.9675
WR	0	16	0	0	1	9	0	124	0.9742

A detail analysis of the performance using the confusion matrix (Table 3) shows that the system well distinguishes between grapes and background (4 from 150 images of background miss classified as grapes, and 4 from 1050 images of grapes miss classified as background). Confusions occur only among grapes of the same colour (no green grape was classified as red and vice versa).

The system best recognizes Gewürztraminer (only 2 from 150 images of Gewürztraminer miss classified as Pinot gris, and only 4 images of another class miss classified as Gewürztraminer). The second-best recognized variety is Pinot gris (141 from 150 images of Pinot gris correctly classified, and only 6 images of another class miss classified as Pinot gris). For the varieties Pinot noir, Veltliner Grün, and Riesling Weiss, we observe similar performance (around 140 from 150 images per class classified correctly, and 20, 24 and 16 images of another class miss classified as Pinot noir, Veltliner Grün, and Riesling Weiss, respectively).

We observe a confusion between the varieties Veltliner Grün and Riesling Weiss (6 images of Veltliner Grün miss classified as Riesling Weiss, and 8 images of Riesling Weiss miss classified as Veltliner Grün).

The most difficult variety is Saint Laurent (127 from 150 images of Saint Laurent correctly classified). The system mostly confuses this variety with Pinot noir (11 images of Pinot noir miss classified as Saint Laurent, and 17 images of Saint Laurent miss classified as Pinot noir). Also, Pinot gris is mostly miss classified as Saint Laurent (5 from 9 miss classifications). The second problematic variety is Welschriesling (124 from 150 images of Welschriesling correctly classified). The system has difficulty to distinguish Welschriesling from Veltliner Grün (16 miss classifications) and from Riesling Weiss (9 miss classifications). The positive thing is that only 5 images of another class are miss classified as Welschriesling.

4 Conclusion

We show that in-field colour images of ripe grapes acquired by a conventional camera can be used for classification of grapevines according to their varieties. The presented variety recognition system is capable to distinguish among seven grapevines varieties, where four and three varieties have red and green grapes, respectively. The system also well differentiates grapes from background. Its overall average per-class accuracy is over 98% on images captured without any artificial lighting, in a direction of sunshine likewise in the opposite direction. Considering all these facts, we conclude that the proposed solution allows construction of affordable automatic selective harvesters.

Acknowledgments. The work was supported from ERDF/ESF "Cooperation in Applied Research between the University of Pardubice and companies, in the Field of Positioning, Detection and Simulation Technology for Transport Systems (PosiTrans)" (No. CZ.02.1.01/0.0/0.0/17_049/0008394).

References

1. Bac, C.W., Hemming, J., van Tuijl, B., Barth, R., Wais, E., van Henten, E.J.: Performance evaluation of a harvesting robot for sweet pepper. J. Field Robot. **34**(6), 1123–1139 (2017). https://doi.org/10.1002/rob.21709. https://onlinelibrary.wiley.com/doi/abs/10.1002/rob.21709
2. Bontsema, J., Hemming, J., Pekkeriet, E., Saeys, W., Edan, Y., Shapiro, A., Hočevar, M., Oberti, R., Armada, M., Ulbrich, H., Baur, J., Debilde, B., Best, S., Evain, S., Gauchel, W., Hellström, T., Ringdahl, O.: CROPS: clever robots for crops. Eng. Technol. Ref. **1**(1) (2015). https://doi.org/10.1049/etr.2015.0015
3. Fernandes, A., Utkin, A., Eiras-Dias, J., Silvestre, J., Cunha, J., Melo-Pinto, P.: Assessment of grapevine variety discrimination using stem hyperspectral data and adaboost of random weight neural networks. Appl. Soft Comput. **72**, 140–155 (2018). https://doi.org/10.1016/j.asoc.2018.07.059

4. Galet, P.: A Practical Ampelography: Grapevine Identification, 1st edn. Comstock Pub. Associates, Ithaca (1979)
5. Gutiérrez, S., Tardaguila, J., Fernández-Novales, J., Diago, M.: Data mining and NIR spectroscopy in viticulture: applications for plant phenotyping under field conditions. Sensors (Switzerland) **16**(2) (2016). https://doi.org/10.3390/s16020236
6. Gutiérrez, S., Fernández-Novales, J., Diago, M.P., Tardaguila, J.: On-the-go hyperspectral imaging under field conditions and machine learning for the classification of grapevine varieties. Front. Plant Sci. **9**, 1102 (2018). https://doi.org/10.3389/fpls.2018.01102
7. Han, D., Kim, J., Kim, J.: Deep pyramidal residual networks. In: 2017 IEEE Conference on Computer Vision and Pattern Recognition (CVPR), pp. 6307–6315, July 2017. https://doi.org/10.1109/CVPR.2017.668
8. He, K., Zhang, X., Ren, S., Sun, J.: Deep residual learning for image recognition. In: 2016 IEEE Conference on Computer Vision and Pattern Recognition (CVPR), pp. 770–778, June 2016. https://doi.org/10.1109/CVPR.2016.90
9. Huang, G., Liu, Z., van der Maaten, L., Weinberger, K.Q.: Densely connected convolutional networks. In: 2017 IEEE Conference on Computer Vision and Pattern Recognition (CVPR), pp. 2261–2269, July 2017. https://doi.org/10.1109/CVPR.2017.243
10. Ioffe, S., Szegedy, C.: Batch normalization: accelerating deep network training by reducing internal covariate shift. In: Proceedings of the 32nd International Conference on Machine Learning, ICML 2015, Proceedings of Machine Learning Research, vol. 37, pp. 448–456. PMLR (2015)
11. Krizhevsky, A., Sutskever, I., Hinton, G.E.: ImageNet classification with deep convolutional neural networks. Commun. ACM **60**(6), 84–90 (2017)
12. LeCun, Y., Bengio, Y., Hinton, G.: Deep learning. Nature **521**, 436–444 (2015). https://doi.org/10.1038/nature14539
13. Lemnaru, C., Potolea, R.: Imbalanced classification problems: systematic study, issues and best practices. In: Enterprise Information Systems, pp. 35–50. Springer, Heidelberg (2012)
14. Maul, et al.: Vitis international variety catalogue (2020). www.vivc.de
15. Pelsy, F., Hocquigny, S., Moncada, X., Barbeau, G., Forget, D., Hinrichsen, P., Merdinoglu, D.: An extensive study of the genetic diversity within seven French wine grape variety collections. Theor. Appl. Genet. **120**(6), 1219–1231 (2010). https://doi.org/10.1007/s00122-009-1250-8
16. Ruder, S.: An overview of gradient descent optimization algorithms. CoRR abs/1609.04747 (2016)
17. Shorten, C., Khoshgoftaar, T.M.: A survey on image data augmentation for deep learning. J. Big Data **6**(1), 60 (2019). https://doi.org/10.1186/s40537-019-0197-0
18. Slaughter, D., Giles, D., Downey, D.: Autonomous robotic weed control systems: a review. Comput. Electron. Agric. **61**(1), 63–78 (2008). https://doi.org/10.1016/j.compag.2007.05.008
19. Sokolova, M., Lapalme, G.: A systematic analysis of performance measures for classification tasks. Inf. Process. Manag. **45**(4), 427–437 (2009). https://doi.org/10.1016/j.ipm.2009.03.002
20. Soldavini, C., Schneider, A., Stefanini, M., Dallaserra, M., Policarpo, M.: Super ampelo, a software for ampelometric and ampelographic descriptions in vitis. Acta Horticulturae **827**, 253–258 (2009). https://doi.org/10.17660/ActaHortic.2009.827.43

21. de Soto, M.G., Emmi, L., Perez-Ruiz, M., Aguera, J., de Santos, P.G.: Autonomous systems for precise spraying - evaluation of a robotised patch sprayer. Biosyst. Eng. **146**, 165–182 (2016). https://doi.org/10.1016/j.biosystemseng.2015.12.018
22. Srivastava, R.K., Greff, K., Schmidhuber, J.: Training very deep networks. In: Proceedings of the 28th International Conference on Neural Information Processing Systems - Volume 2, NIPS 2015, pp. 2377–2385. MIT Press, Cambridge (2015)
23. Szegedy, C., Vanhoucke, V., Ioffe, S., Shlens, J., Wojna, Z.: Rethinking the inception architecture for computer vision. In: 2016 IEEE Conference on Computer Vision and Pattern Recognition (CVPR), pp. 2818–2826, June 2016. https://doi.org/10.1109/CVPR.2016.308
24. Xiong, Y., Peng, C., Grimstad, L., From, P.J., Isler, V.: Development and field evaluation of a strawberry harvesting robot with a cable-driven gripper. Comput. Electron. Agr. **157**, 392–402 (2019). https://doi.org/10.1016/j.compag.2019.01.009. http://www.sciencedirect.com/science/article/pii/S0168169918312456
25. Xu, Y., Jia, Z., Ai, Y., Zhang, F., Lai, M., Chang, E.I.: Deep convolutional activation features for large scale brain tumor histopathology image classification and segmentation. In: 2015 IEEE International Conference on Acoustics, Speech and Signal Processing (ICASSP), pp. 947–951, April 2015. https://doi.org/10.1109/ICASSP.2015.7178109
26. Yu, Z., Li, T., Luo, G., Fujita, H., Yu, N., Pan, Y.: Convolutional networks with cross-layer neurons for image recognition. Inf. Sci. **433–434**, 241–254 (2018). https://doi.org/10.1016/j.ins.2017.12.045

Discovering Spatio-Temporal Patterns in Precision Agriculture Based on Triclustering

Laura Melgar-García[1](✉), Maria Teresa Godinho[2,3], Rita Espada[4], David Gutiérrez-Avilés[1], Isabel Sofia Brito[5,6], Francisco Martínez-Álvarez[1], Alicia Troncoso[1], and Cristina Rubio-Escudero[7]

[1] Data Science & Big Data Lab, Pablo de Olavide University, 41013 Seville, Spain
{lmelgar,dgutavi,fmaralv,atrolor}@upo.es

[2] Department of Mathematical and Physical Sciences, Polytechnic Institute of Beja, Beja, Portugal
mtgodinho@ipbeja.pt

[3] Center for Mathematics, Fundamental Applications and Operations Research, University of Lisboa, Lisbon, Portugal

[4] Associação dos Agricultores do Baixo Alentejo, Beja, Portugal
rita.espada.25@gmail.com

[5] Department of Engineering, Polytechnic Institute of Beja, Beja, Portugal
isabel.sofia@ipbeja.pt

[6] Instituto de Desenvolvimiento de Novas Tecnologias - Centre of Technology and Systems, Lisbon, Portugal

[7] Department of Computer Languages and Systems, University of Seville, Seville, Spain
crubioescudero@us.es

Abstract. Agriculture has undergone some very important changes over the last few decades. The emergence and evolution of precision agriculture has allowed to move from the uniform site management to the site-specific management, with both economic and environmental advantages. However, to be implemented effectively, site-specific management requires within-field spatial variability to be well-known and characterized. In this paper, an algorithm that delineates within-field management zones in a maize plantation is introduced. The algorithm, based on triclustering, mines clusters from temporal remote sensing data. Data from maize crops in Alentejo, Portugal, have been used to assess the suitability of applying triclustering to discover patterns over time, that may eventually help farmers to improve their harvests.

Keywords: Triclustering · Spatio-temporal patterns · Precision agriculture · Remote sensing

Á. Herrero et al. (Eds.): SOCO 2020, AISC 1268, pp. 226–236, 2021.
https://doi.org/10.1007/978-3-030-57802-2_22

1 Introduction

It is a well-established fact that shortage of natural resources endangers our future. Public awareness of these problems urges local authorities to intervene and impose tight regulations on human activity. In this environment, reconciling economic and environmental objectives in our society it is mandatory.

Precision agriculture (PA) has an important role in the pursuit of such aspiration, as the techniques used in PA permit to adjust resource application to the needs of soil and crop as they vary in the field. In this way, specific-site management (that is the management of agricultural crops at a spatial scale smaller than the whole field) is a tool to control and reduce the amount of fertilizers, phytopharmaceuticals and water used on site, with both ecological and economic advantages. Indeed, being able to characterize how crops behave over time, extracting patterns and predicting changes is a requirement of utmost importance for understanding agro-ecosystems dynamics [1].

One of the major concerns associated to the shortage of natural resources is the enormous consumption of water associated to farming activities. Water is a scarce resource worldwide and this problem is particularly acute in the South of Europe, where the Alentejo (Portugal) and Andalusia (Spain) regions are located. Both regions are mainly agriculture-dependent and thus farmers and local authorities are apprehensive about the future.

In this paper, an algorithm is proposed to delineate management zones by measuring the variability of crop conditions within the field through the analysis of time series of geo-referenced vegetation indices, obtained from satellite imagery. In particular, the well-known normalized difference vegetation index (NDVI), indicator for vegetation health and biomass, is used to analyze how the crop varies over time in order to find patterns that may help to improve its production. There are more vegetation indices as GNDVI, SAVI, EVI or EVI2 [2,3] which should be used in extended works.

A triclustering method, based on an evolutionary strategy called TriGen [4] has been applied to a set of satellite images indexed over time from a particular maize crop in Alentejo, Portugal. Although the method was originally designed to discover gene behaviors over time [5], it has also been applied to other research fields such as seismology [6]. The TriGen is a genetic algorithm, and therefore the fitness function is a key aspect since it leads to the discovery of triclusters of different shapes and aspects. The multi-slope measure (MSL) [7], the three-dimensional mean square residue (MSR3D) [8] and the least squared lines (LSL) [9] are the available fitness functions to mine triclusters in TriGen. Furthermore, the TRIclustering quality (TRIQ) index [10] was proposed to validate the results obtained from the aforementioned fitness functions.

The rest of the paper is structured as follows. In Sect. 2, the recent and related works are reviewed and the process of data acquisition and preprocessing is described. In Sect. 3 the proposed algorithm and its adaption to this particular problem are described. In Sect. 4 the results are presented and discussed. Finally, in Sect. 5, the conclusions of this work and point directions for future work are presented.

2 Related Works

This section reviews the most recent and relevant works published in the field of spatio-temporal patterns in precision agriculture.

The spatio-temporal pattern discovery issues for satellite time series images are discussed in [11]. The authors introduced how to perform an automatic analysis of these patterns and the problem of determining its optimal number. Unfortunately, these questions are still open issues in the literature and it is unlikely that a general consensus can be reached in the near future.

The estimation of spatio-temporal patterns of agricultural productivity in fragmented landscapes using AVHRR NDVI time series was analyzed in [12]. Four different approaches were applied to eight years of Australian crops, including calculation of temporal mean and standard deviation layers, spatio-temporal key NDVI patterns, different climatic variables and relationships between productivity and production.

In Fung et al. [13], the authors proposed a novel spatio-temporal data fusion model for satellite images using Hopfield Neural Networks. Synthetic and real datasets from both Hong Kong and Australia, respectively, were used to assess the method performance, showing remarkable results and outperforming some of other existing methods.

The use of convolutional neural networks (CNN) is being currently applied in a wide range of spatio-temporal patterns discovery applications [14]. Hence, Tan et al. [15] enhanced an existing CNN model for image fusion by proposing a new network architecture and a novel loss function. Results showed superior performance in terms of accuracy and robustness. Ji et al. [16] proposed a 3D CNN dealing with multi-temporal satellite images. In this case, the method was designed for crop classification. After discussing the results achieved, outperforming existing well-established methods, the authors claimed that it is especially suitable for characterizing crop growth dynamics.

An ensemble model for making spatial predictions of tropical forest fire susceptibility using multi-source geospatial data can be found in [17]. The authors evaluated the Lao Cai region, Vietnam, through several indices including NDVI.

Bui et al. [18] proposed an approach based on deep learning for predicting flash flood susceptibility. Real data from a high frequency tropical storm area were used to assess its performance.

Clustering-based approaches with application to precision agriculture can also be found in the literature. Thus, clustering tools for integration of satellite imagery and proximal soil sensing data are described in [19]. In particular, a novel method was introduced with the aim of determining areas with homogeneous parts in agricultural fields.

The application of triclustering to georeferenced satellite images time series can be also found in [20]. However, the authors addressed a different problem: the patterns analysis of intra-annual variability in temperature, using daily average temperature retrieved from Dutch stations spread over the country.

3 Methodology

This section introduces the *TriGen* algorithm, the methodology used to extract behavior patterns from satellite images along with the time points when they were taken. This methodology is applied to a 3D dataset (composed of rows, columns, and depths) that represents the X-axis coordinates (rows) and the Y-axis coordinates (columns) of each satellite image taken at a particular instant (depth). *TriGen* is a genetic algorithm that minimizes a fitness function to mine subsets of X-axis coordinates, Y-axis coordinates, and time points, called triclusters, from 3D input datasets. The *NDVI* values in the yielded subsets of $[X, Y]$ coordinates along with the subset of time points, share similar behavior patterns.

In general terms, *TriGen* is explained from two main concepts, presented in the following sections: the triclustering model applied to the case study (Sect. 3.1) and the inputs, output and algorithm workflow of *TriGen* (Sect. 3.2).

3.1 Triclustering

The case study presented has been modeled as a triclustering problem, in which 3-dimensional patterns are extracted from an original dataset. Prior to explaining this development, it is necessary to distinguish between two types of dataset:

- D_{2D} (2-dimensional dataset): a matrix with a set of instances (rows) and a set of features (columns).
- D_{3D} (3-dimensional dataset): a 3D matrix with a set of instances (rows) and features (columns), taken at a particular time points (depths).

Clustering algorithms are applied to D_{2D} datasets performing a complete partition it; for each yielded clusters, the values of the grouped instances share a behavior pattern through all features. In contrast, the triclustering algorithms work with D_{3D} datasets and group not only subsets of instances, but also subsets of features and time points. In this case, for each yielded tricluster, the values of grouped instances for the particular grouped features share a behavior pattern through a group of time points.

Thus, for this case study, the application of the *TriGen* algorithm to a D_{3D} dataset of satellite images where the instances are the Y coordinates of the space, the features are the X coordinates of the area and, the time points are the moment at the images where taken, will yield a set of triclusters representing, each of them, a behavior pattern of NDVI, for a particular subspace (subset of Y and X coordinates) through a specific set of times (subset of time points).

3.2 The *TriGen* Algorithm

In order to mine the triclusters from the D_{3D} dataset of satellite images, the *TriGen* algorithm is applied. *TriGen* is based on the genetic algorithm paradigm;

therefore, it evolves a population of individuals employing genetic operators during a specific number of generations to optimize an evaluation function.

The inputs of *TriGen* are two: the D_{3D} dataset of satellite images and the initial configuration of the genetic process. The parameters that can be set are the number of triclusters to mine (N), the number of generations of the genetic process (G), the size of the initial population (I), the fraction of population that promoted to the next generation (Sel) and, the probability of mutation (Mut). A complete analysis of the influence of these parameters in the performance of the algorithm can be consulted in [4,7,8].

Each individual in the genetic process is represented as a tricluster and composed of a subset of instances of D_{3D}, a subset of features of D_{3D} and, a subset of time points of D_{3D}; the individuals (triclusters) with the best fitness function value are the output of the algorithm.

The genetic operators allow for searching among the individuals to obtain better solutions for each generation. For the *TriGen* algorithm, the description of them is the following:

- Initial population. The individuals are generated with three methods. The first method consists in a random selection of the elements of the individuals. The second one, considering the rows and columns of D_{3D} as a geographical area, performs a random selection of a rectangular sub-area and time points. The last one selects the elements of the individuals taking into account the rows, columns, and time points of D_{3D} visited in already extracted solutions in order to explore the most number of elements of D_{3D}.
- Evaluation. This operator applies the fitness function to the population in order to asses the quality of each individual. The fitness function used in the present case study is *MSL*.
- Selection. A tournament selection algorithm is applied to promote the individuals with the best evaluation to the next generation. The rest of individuals in the next population are generated by crossing and mutations.
- Crossover. Two individuals are combined to generate another two ones. The crossover used is the one point crossing. Each of the three elements of the two involved individuals (parents), are split in two and the four parts are combined two new individuals (offspring).
- Mutation. This operator modifies an individual to obtain variability in the next generation. Three actions have been used: insertion of a new coordinate $[X, Y]$ or time point, deletion of an existing coordinate $[X, Y]$ or time point and change of an existing coordinate $[X, Y]$ or time point.

4 Results

This section reports and discusses the results achieved after the application of the proposed methodology to a particular dataset. Thus, Sect. 4.1 describes the high resolution remote sensing imagery used in this study and Sect. 4.2 introduces the validation function used to evaluate the quality of the triclusters obtained. Finally, Sect. 4.3 reports the spatio-temporal patterns obtained and discusses its physical meaning.

4.1 Dataset Description

Located in the Baixo Alentejo region of Portugal, the site under study is a 63.82 ha maize plantation, with center at coordinates $(38°08'12''N, 7°53'42''W)$, as shown in Fig. 1. The site was monitored between sowing (April of 2018) and harvesting (September of the same year) and it is characterized by a set of nineteen images retrieved at time intervals of five, ten and fifteen days, from the Sentinel 2 Mission. The research site was irrigated using a central pivot irrigation system.

Fig. 1. Location of the research site.

Vegetation indices are, by definition, algebraic combinations of the measured canopy reflectance of different wavelength bands [21]. The use of Vegetation Indices in this context is based on the fact that healthy and unhealthy plants reflect light differently. Due to this difference, crop canopy multispectral reflectance, which is detectable remotely through aerial or satellite imagery, can be used to monitor the state of the crop [22]. For these reasons, one of the most widely used indices is applied to the images: the Normalized Differential Vegetation Index (NDVI). The NDVI can be calculated as follows:

$$NDVI = \frac{NIR - Red}{NIR + Red}, \tag{1}$$

where Red and NIR stand for the spectral reflectance measurements acquired in the red (visible) and near-infrared regions, respectively, and $NDVI \in [-1, 1]$.

As pointed out in [23], the NDVI index has proven to be quite useful in monitoring variables such as crop nutrient deficiency, final yield in small grains, and long-term water stress. All these variables are very important to the case study presented here. Figure 2 illustrates how the NDVI of the target area varies over time, including images at six different chronologically ordered time stamps.

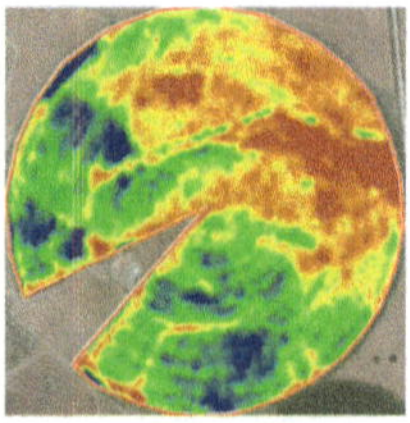

(a) June 19th, 2018

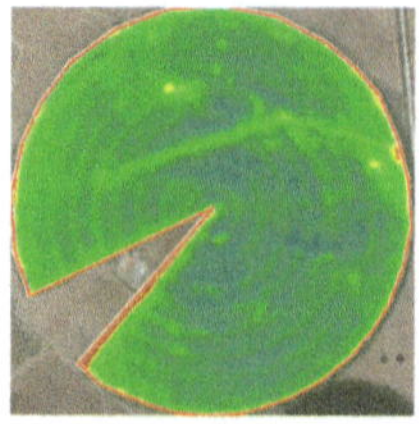

(b) August 3rd, 2018

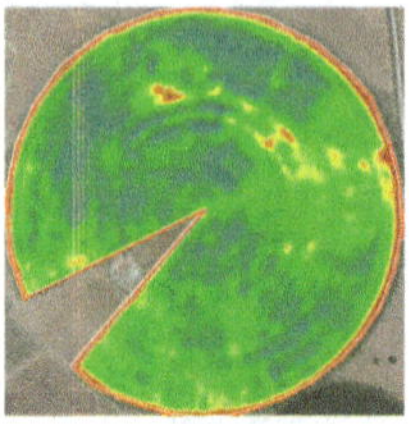

(c) September 22nd, 2018

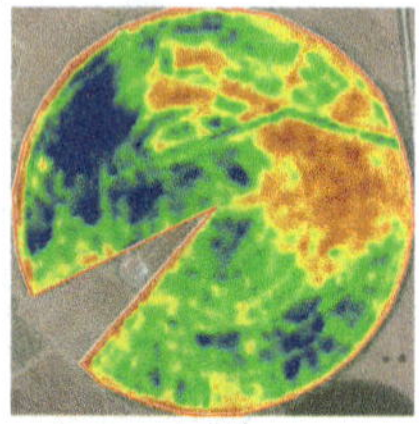

(d) October 17th, 2018

Fig. 2. Sample NDVI values for the research site, chronologically ordered.

4.2 Behaviour Patterns Quality, the *TRIQ* Measure

The *TRIQ* index has been used in order to measure the quality of the yielded triclusters in this case study, that is, the quality of the behavior pattern that a tricluster depicts. *TRIQ* measures the quality of a tricluster based on three elements: the similarity of the behavior patterns of the grouped $[X, Y]$ points along with the grouped time points and the Pearson's and Spearman's correlation indexes between all the $[X, Y]$ time series of the tricluster. $TRIQ$ values rank in the $[0, 1]$ interval; $TRIQ$ is a measure to maximize. A full description, definition, development, and performance of *TRIQ* can be consulted in [10].

4.3 Discovery of Spatio-Temporal Patterns in Maize Crops

TriGen analyzes the evolution of NDVI indices in each specific area and discovers triclusters of similar behavior patterns. Thus, the dataset with the NDVI indices of the satellite images over time is the first input of the algorithm.

TriGen has some configuration parameters, above-mentioned in Sect. 3.2. The algorithm has been run several times with different settings for each parameter. The configuration parameters that fit the best to these images are: G = 10, I = 200, Sel = 0.8 and Mut = 0.1. The number of triclusters to find is 4 and the fitness function used is MSL. Therefore, these values are the second input of the algorithm.

Each of the 4 discovered triclusters has a $TRIQ$ measure. The first one has a $TRIQ$ of 0.803, the second has 0.753, the third has 0.827 and the fourth has 0.742. These high values lead to confirm the good quality of all the triclusters. However, this measure itself does not guarantee the meaningfulness of the triclusters discovered. In order to interpret the evolution of the triclusters in an accurate way, field's farmers provided additional information about the plantations site-specific conditions, such as irrigation or fungicide, for the same period. This information confirmed that triclusters were meaningful also in geophysical terms.

The triclusters discovered are represented in Figs. 3a, 3b, 3c and 3d. Each graph represents the evolution of the NDVI of the selected $[X, Y]$ components over time. The black dashed line added in each graph represents the mean value of all components. Triclusters components share a similar behavior. The first tricluster corresponds to areas with high NDVI values that remain almost constant over time. The components of the second tricluster are fields that start with a high NDVI and experiment a sudden decrease for the rest of the dates studied. The beginning of the third tricluster is similar to the previous one but with a recovery of the initial values after mid September. The last tricluster is formed by areas with constant low NDVI over time.

The changes of the NDVI values identified by triclusters 1, 2 and 3 during the first samples seem to be related with the use of fertilizers and the increase of the amount of water for the irrigation process. The third tricluster and some components of the first one show a change in their behaviour at mid September. It could be related to the application of fungicide by the farmers during August.

The proposed algorithm contributes in finding areas of similar crop conditions over the NDVI vegetation index using satellite images in different times. In addition, as TriGen includes the time dimension, the evolution over time of

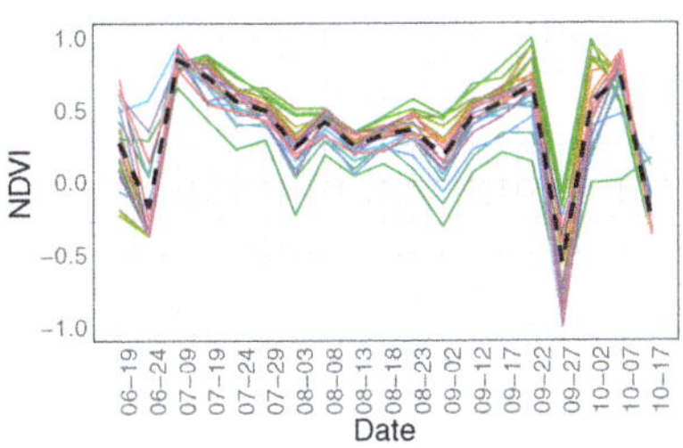

(a) Tricluster 1.

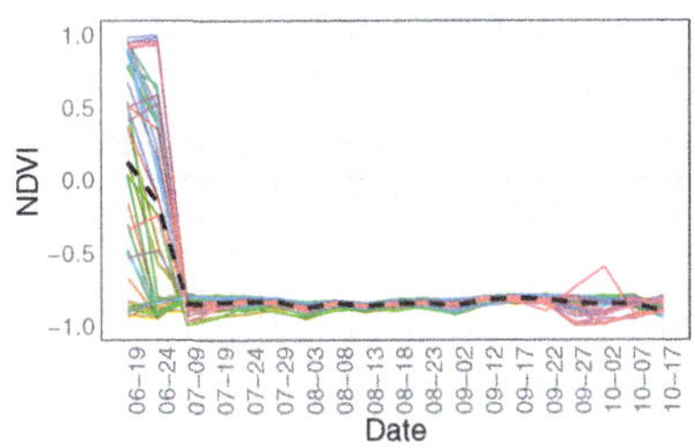

(b) Tricluster 2.

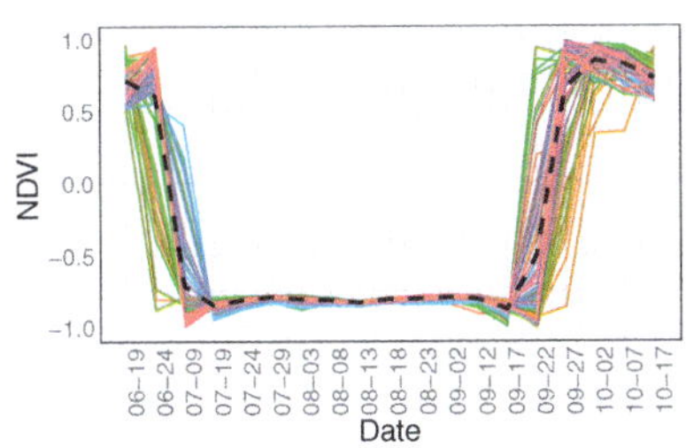

(c) Tricluster 3.

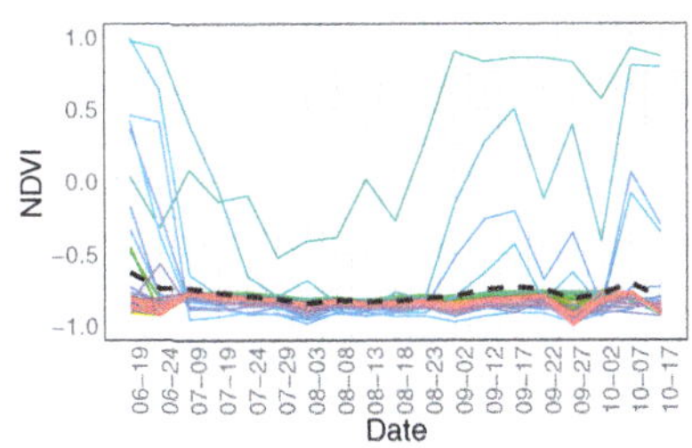

(d) Tricluster 4.

Fig. 3. Triclusters found by *TriGen* in 2018.

each tricluster's features can be analyzed. Nevertheless, the interpretation of the results needs the validation of a specialist as the TRIQ measure does not consider neither geographical nor environmental features.

5 Conclusions

The suitability of applying triclutstering methods to discover spatio-temporal patterns in precision agriculture has been explored in this work. In particular, a set of satellite images from maize crops in Alentejo, Portugal, has been analyzed in terms of its NVDI temporal evolution. Several patterns have been found, identifying zones with tendency to obtain greater production and others in which human interventions are required to improve the soil properties. Several issues remain unsolved and are suggested to be addressed in future works. First, these patterns may help to identify the most suitable moments to apply fertilizers or pesticides. Second, the forecasting of maize production could be done based on such patterns. Third, additional crop production features such as amounts and characteristics of the fertilizers, phytopharmaceuticals and water used throughout the season (moister probes placed 30 cm underground were used to access the soil need for water before irrigation, when needed), would help to discover more robust patterns. Fourth, more images records during more years and a specific measure to assess the quality and meaning of precision agriculture triclusters would improve the application of the proposed algorithm to agricultural production. Fifth, more vegetation indices should be used.

Acknowledgements. The authors would like to thank the Spanish Ministry of Economy and Competitiveness for the support under project TIN2017-88209 and Fundação para a Ciência e a Tecnologia (FCT), under the project UIDB/04561/2020. The authors would also like to thank António Vieira Lima for giving access to data and Francisco Palma for his support to the whole project.

References

1. Tan, J., Yang, P., Liu, Z., Wu, W., Zhang, L., Li, Z., You, L., Tang, H., Li, Z.: Spatio-temporal dynamics of maize cropping system in Northeast China between 1980 and 2010 by using spatial production allocation model. J. Geog. Sci. **24**(3), 397–410 (2014)
2. Jurecka, F., Lukas, V., Hlavinka, P., Semeradova, D., Zalud, Z., Trnka, M.: Estimating crop yields at the field level using landsat and modis products. Acta Universitatis Agriculturae et Silviculturae Mendelianae Brunensis **66**, 1141–1150 (2018)
3. Jiang, Z., Huete, A., Didan, K., Miura, T.: Development of a two-band enhanced vegetation index without a blue band. Remote Sens. Environ. **112**, 3833–3845 (2008)
4. Gutiérrez-Avés, D., Rubio-Escudero, C., Martínez-Álvarez, F., Riquelme, J.C.: Trigen: A genetic algorithm to mine triclusters in temporal gene expression data. Neurocomputing **132**, 42–53 (2014)

5. Melgar, L., Gutiérrez-Avilés, D., Rubio-Escudero, C., Troncoso, A.: High-content screening images streaming analysis using the STriGen methodology. In: Proceedings of the 35th Annual ACM Symposium on Applied Computing, pp. 537–539 (2020)
6. Martínez-Álvarez, F., Gutiérrez-Avilés, D., Morales-Esteban, A., Reyes, J., Amaro-Mellado, J.L., Rubio-Escudero, C.: A novel method for seismogenic zoning based on triclustering: application to the Iberian peninsula. Entropy **17**(7), 5000–5021 (2015)
7. Gutiérrez-Avilés, D., Rubio-Escudero, C.: MSL: a measure to evaluate three-dimensional patterns in gene expression data. Evol. Bioinform. **11**, 121–135 (2015)
8. Gutiérrez-Avilés, D., Rubio-Escudero, C.: Mining 3D patterns from gene expression temporal data: a new tricluster evaluation measure. Sci. World J. **2014**, 1–16 (2014)
9. Gutiérrez-Avilés, D., Rubio-Escudero, C.: LSL: a new measure to evaluate triclusters. In: Proceedings of the IEEE International Conference on Bioinformatics and Biomedicine, pp. 30–37 (2014)
10. Gutiérrez-Avilés, D., Giráldez, R., Gil-Cumbreras, F.J., Rubio-Escudero, C.: TRIQ: a new method to evaluate triclusters. BioData Min. **11**(1), 15 (2018)
11. Radoi, A., Datcu, M.: Spatio-temporal characterization in satellite image time series. In: Proceedings of the International Workshop on the Analysis of Multitemporal Remote Sensing Images, pp. 1–4 (2015)
12. Hill, M.J., Donald, G.E.: Estimating spatio-temporal patterns of agricultural productivity in fragmented landscapes using AVHRR NDVI time series. Remote Sens. Environ. **84**(3), 367–384 (2003)
13. Fung, C.H., Wong, M.S., Chan, P.W.: Spatio-temporal data fusion for satellite images using Hopfield neural network. Remote Sens. **11**(18), 2077 (2019)
14. Kamilaris, A., Prenafeta-Boldú, F.: A review of the use of convolutional neural networks in agriculture. J. Agric. Sci. **156**(3), 312–322 (2018)
15. Tan, Z., Di, L., Zhang, M., Guo, L., Gao, M.: An enhanced deep convolutional model for spatiotemporal image fusion. Remote Sens. **11**(18), 2898 (2019)
16. Ji, S., Zhang, C., Xu, A., Shi, Y., Duan, Y.: 3D convolutional neural networks for crop classification with multi-temporal remote sensing images. Remote Sens. **10**(1), 75 (2018)
17. Tehrany, M.S., Jones, S., Shabani, F., Martínez-Álvarez, F., Bui, D.T.: A novel ensemble modeling approach for the spatial prediction of tropical forest fire susceptibility using logitboost machine learning classifier and multi-source geospatial data. Theoret. Appl. Climatol. **137**, 637–653 (2019)
18. Bui, D.T., Hoang, N.-D., Martínez-Ávarez, F., Ngo, P.-T.T., Hoa, P.V., Pham, T.D., Samui, P., Costache, R.: A novel deep learning neural network approach for predicting flash flood susceptibility: a case study at a high frequency tropical storm area. Sci. Total Environ. **701**, 134413 (2020)
19. Saifuzzaman, M., Adamchuk, V., Buelvas, R., Biswas, A., Prasher, S., Rabe, N., Aspinall, D., Ji, W.: Clustering tools for integration of satellite remote sensing imagery and proximal soil sensing data. Remote Sens. **11**(9), 1036 (2019)
20. Wu, X., Zurita-Milla, R., Izquierdo-Verdiguier, E., Kraak, M.-J.: Triclustering georeferenced time series for analyzing patterns of intra-annual variability in temperature. Ann. Am. Assoc. Geogr. **108**, 71–87 (2018)

21. Schueller, J.: A review and integrating analysis of spatially-variable control of crop production. Fertil. Res. **33**, 1–34 (1992)
22. Xue, J., Su, B.: Significant remote sensing vegetation indices: a review of developments and applications. J. Sens. **17**, 1353691 (2017)
23. Govaerts, B., Verhulst, N.: The normalized difference vegetation index (NDVI) GreenSeekerTM handheld sensor: toward the integrated evaluation of crop management. CIMMYT (2010)

Counting Livestock with Image Segmentation Neural Network

Petr Dolezel(✉), Dominik Stursa, Daniel Honc, Jan Merta, Veronika Rozsivalova, Ladislav Beran, and Ivo Hora

University of Pardubice, Studentska 95, 53210 Pardubice, Czech Republic
petr.dolezel@upce.cz
http://www.upce.cz/fei

Abstract. Livestock farming industries, as well as almost any industry, want more and more data about the operation of their business and activities in order to make the right decisions. However, especially when considering very large animal farms, the precise and up-to-date information about the position and numbers of the animals is rather difficult to obtain. In this contribution, a novel engineering approach to livestock positioning and counting, based on image processing, is proposed. The approach is composed of two parts. Namely, a fully convolutional neural network for input image transformation, and a locator for animal positioning. The transformation process is designed in order to transform the original RGB image into a gray-scale image, where animal positions are highlighted as gradient circles. The locator then detects the positions of the circles in order to provide the positions of animals. The presented approach provides a precision rate of 0.9842 and a recall rate of 0.9911 with the testing set, which is, in combination with a rather suitable computational complexity, a good premise for the future implementation under real conditions.

Keywords: Livestock counting · Fully convolutional neural network · U-Net · Precision agriculture

1 Introduction

Livestock farming industries, as well as almost any industry, want more and more data about the operation of their business and activities, in order to make the right decisions, at the right location, at the right time, and at the right intensity. These days, with the development of precision agriculture [3], farmers can acquire more data then ever before, including soil moisture and acidity, ground and air temperature, individual stock or crop increments, etc.

The work has been supported by SGS grant at Faculty of Electrical Engineering and Informatics, University of Pardubice, Czech Republic. This support is very gratefully acknowledged.

Á. Herrero et al. (Eds.): SOCO 2020, AISC 1268, pp. 237–244, 2021.
https://doi.org/10.1007/978-3-030-57802-2_23

Nevertheless, especially when considering very large farms, the precise and up-to-date information about the position and numbers of the animals is still difficult to obtain. Counting livestock is often performed once in a time period, and animals have to be led through a drafting race or a narrow choke point, while being counted manually or using some sort of smart collar [5]. This approach does not provide the information continuously and it can also be uncomfortable for the animals. Hence, continuous, or at least more frequent, livestock counting directly in the pastureland is a desired task, which can be then applied for monitoring of animal numbers, animal growth, animal distress, distribution of the herds, etc.

In order to solve the task, several challenges, such as species characteristic, diversity of background, variable light conditions, overlapping of animals, animal reaction to monitoring and the mosaicking process appear [1,12]. Hence, various approaches are proposed to deal with those challenges. Some of them are based on classical statistical techniques [11]. Others use more recent methods, such as K-means clustering [10], histogram of oriented gradients and local binary pattern [6], power spectral based methods [13], support vector machines with various sound processing approaches [4], etc.

Considering image or video as the input signal for livestock counting, deep learning techniques become one of the major approaches for implementation. Very good results were provided especially by the methods based on convolutional neural networks [7,8,16]. However, the cited approaches were tested on images, where animals occupy a substantial part of the image and each animal is depicted in high resolution. Contrary to these approaches, Farah Sarwar and Anthony Griffin published an approach to deal with images having hundreds of small animal silhouettes per image [15]. Their testing experiments provided a precision rate of 95.6% and recall rate of 99.5%. However, the dataset used in the cited work did not contain spacial clusters of animals and the diversity of the background was rather low.

Therefore, we focus on counting livestock animals from an Unmanned Aerial Vehicle (UAV) video, considering high angle take from an altitude bigger than 50 m. We consider various types of background, different size of animals, big variance of animal numbers including crowded animals stocks in the source signal.

The rest of the contribution is structured as follows. In the next chapter, the aim of the paper is formulated and the proposed solution is described. Then, the implementation part follows and the results are presented. The paper is finished with some conclusions.

2 Methodology

In this section, the aim of this paper is defined and a method, which is based on image segmentation neural network, is presented to solve the problem.

2.1 Problem Formulation

The aim of the paper is to design a monitoring system for livestock positioning and counting, using images acquired by the UAV. The monitoring system should be robust enough to handle various light conditions and background types, size of animals and both crowded and blank situations. The examples of these variants are shown in Fig. 1.

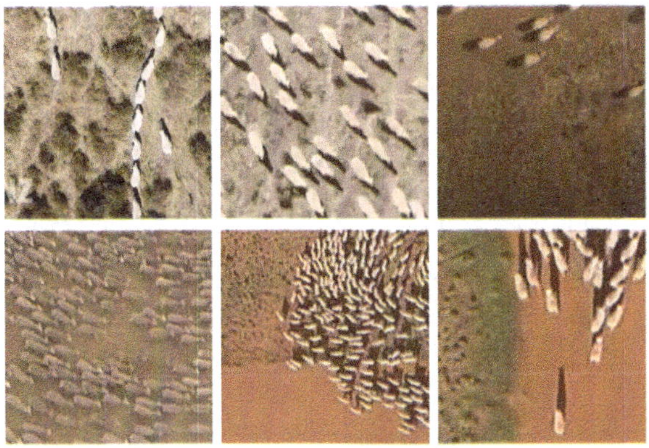

Fig. 1. Examples of livestock considered to be processed by the proposed system.

The proposed monitoring system is supposed to work as shown in Fig. 2. The UAV continuously takes the images of the area. The images are then properly combined together through mosaicking. During mosaicking, the images are also geometrically distorted to fit to the detailed map of the scanned area. Subsequently, the resulting complex image is processed to provide positions of each individual animal. Such information can be then used for various analyses required by farmers.

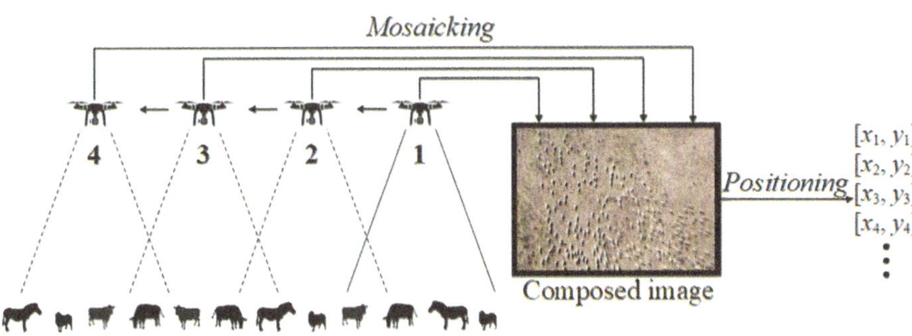

Fig. 2. Proposed functionality of the monitoring system.

In this particular paper, the process of mosaicking is not discussed in detail. It is clearly a non-trivial process and has the potential to produce various kinds of errors [1]. Hence, it will be comprehensively dealt with in authors' future work. Here, the positioning of the animals in the combined image is dealt with. In other words, we propose a novel approach to detect, to locate and to count livestock animals in visual data.

2.2 Proposed Solution

Recent advance in convolutional neural networks (CNNs) has resulted in them outperforming traditional computer vision techniques in object detection as well as image classification tasks across various benchmark datasets [17,18]. The proposed approach uses CNN for the original image transformation into a gray-scale image. The reason for this transformation is to create a segmented image, where animals are highlighted as radial gradients and the rest of the image remains black. The approach is depicted in Fig. 3.

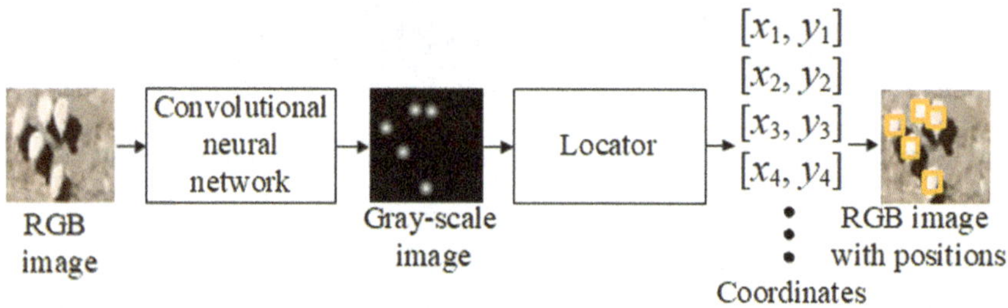

Fig. 3. Proposed positioning approach.

It is obvious that the function of the Locator is a routine process of finding local maxima in a gray-scale image. The transformation of the RGB image is, on the other hand, a tricky issue. One of the possible solutions is the implementation of a representative from the family of image segmentation neural networks. These neural networks are often based on deep encoders and decoders. It involves encoding the input image into low dimensions, and then recovering it with orientation invariance capabilities in the decoder. If the training set is prepared correctly, the approach should generate a segmented image at the decoder end. Although the image segmentation process is mostly supposed to produce images covered with monotonous areas of colors, the authors believe, that a similar approach is able to code gradient circles, as required in the presented method. Therefore, the image segmentation neural network, which transforms an RGB image into a gray-scale image where animals are highlighted as radial gradients, is defined in the next section.

3 Image Segmentation Neural Network

In recent years, many types of deep neural networks, including CNNs, fully convolutional networks, ensemble learning systems, and SegNet neural networks, were introduced [2]. Based on the authors' previous experience, from the wide set of possibilities, the U-Net is selected for the initial testing. It is robust, reasonably fast, and it is supposed to need only a small training set to be sufficiently trained.

The U-Net is a fully convolutional neural network developed originally for biomedical image segmentation. It consists of a contracting path (left side) and an expansive path (right side). The contracting path follows the typical architecture of a convolutional network. It consists of the repeated application of two 3×3 convolutions (unpadded convolutions), each followed by a rectified linear unit (ReLU) and a 2×2 max pooling operation with stride 2 for downsampling. Every step in the expansive path consists of an upsampling of the feature map followed by a 2×2 convolution that halves the number of feature channels, a concatenation with the correspondingly cropped feature map from the contracting path, and two 3×3 convolutions, each followed by a ReLU. The cropping is necessary due to the loss of border pixels in every convolution. At the final layer, a 1×1 convolution is used to map the output. In total, the network has 23 convolutional layers [14].

3.1 Dataset for Training and Validation

The dataset for training and validation is prepared in order to fulfill the conditions described in Sect. 1. Therefore, several videos taken by the UAV were processed, and images, which satisfied the conditions, were extracted. Altogether, 440 images $[288 \times 288]$ px, which cover livestock (sheep and cows), taken from the height 50–100 m, were selected. These images were then divided into a training and testing set. The overall information is summarized in Table 1.

Table 1. Parameters of dataset

–	Training set	Testing set	Together
Images	330	110	440
Animals	10621	3707	14328

Then, the target images for the training and validation needed to be prepared. This process was performed manually by a custom tagging application. For each input image, a gray-scale target image, where animal positions were highlighted by a gradient circle, was prepared. The examples of input-target pairs are shown in Fig. 4.

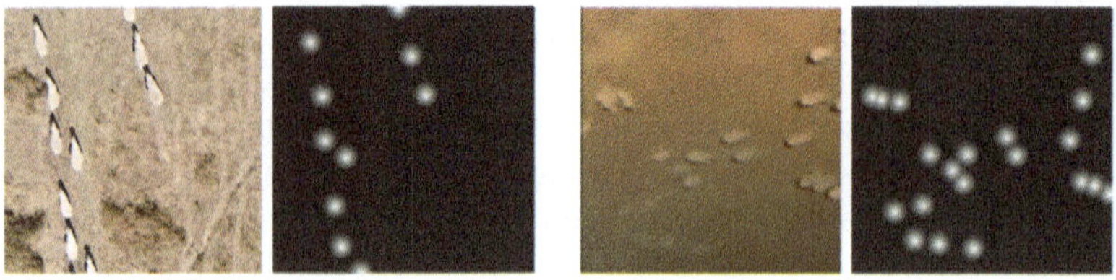

Fig. 4. Examples of input-target pairs for training set. The dimensions are 288 × 288 px.

3.2 U-Net Training

Consequently, the training of the U-Net architecture was performed. The ADAM algorithm is implemented as an optimizer based on its generally acceptable performance [9]. Initial weights were set randomly with normal distribution (mean = 0, standard deviation = 0.05). The experiments are performed twenty times due to a stochastic character of training. All the parameters are shown in Table 2.

Table 2. Parameters of the training

Input shape	288 × 288 × 3
Training algorithm	ADAM algorithm
Initialization	Normal distribution (mean = 0, std = 0.05)
Maximum epochs	50
Stopping criterion	Maximum epochs reached
Learning rate α	0.001
Exponential decay rate 1 β_1	0.9
Exponential decay rate 2 β_2	0.999

3.3 Results

In this section, the performance of the best U-Net, trained according to the previous paragraph, is introduced. A good practice for the evaluation is to determine the accuracy over the testing set. However, two additional metrics, precision and recall, are added. The metrics are described by the following equations.

$$\text{Accuracy} = \frac{\text{TP}}{\text{TP} + \text{FP} + \text{FN}}, \tag{1}$$

$$\text{Precision} = \frac{\text{TP}}{\text{TP} + \text{FP}}, \tag{2}$$

$$\text{Recall} = \frac{\text{TP}}{\text{TP} + \text{FN}}, \tag{3}$$

Table 3. Results

Dataset	Accuracy	Precision	Recall
Testing	97.56%	98.42%	99.11%
Training	100.00%	100.00%	100.00%
Testing (50–66 m)	97.55%	98.58%	98.93%
Testing (66–83 m)	96.37%	97.22%	99.10%
Testing (83–100 m)	98.34%	99.18%	99.13%

where TP (true positive) is the number of correctly classified animals, FN (false negative) is the number of misclassified animals, FP (false positive) is the number of incorrectly detected animals. All these quality indicators are expressed as percentage values in the rest of the article.

The results obtained by the best U-Net for image transformation, and the local maxima detector for livestock positioning, are summarized in Table 3, where the testing set is divided into three subsets according the height of the camera.

According to the results, the proposed method obviously works perfectly on the training images, and shows only slight degradation on the test images. In addition, there is no obvious relevance between the accuracy (and other metrics) and the height of the UAV with the camera.

4 Conclusion

A novel engineering approach to livestock positioning and counting is proposed in this contribution. The approach is composed of two parts. Firstly, a fully convolutional neural network for input image transformation, and secondly, a locator for animal positioning. The transformation process is designed in order to transform the original RGB image into a gray-scale image, where animal positions are highlighted as gradient circles. After a set of experiments, the U-Net was selected for the transformation. In combination with the local maxima function for positioning, the U-Net provides a precision rate of 0.9842 and a recall rate of 0.9911 with the testing set.

The presented contribution should be understood as a first step in the development of a robust livestock counting device. Work in the near future will include convolutional neural network architecture optimizing, and computational complexity testing in order to prepare the approach for implementation.

References

1. Arnal Barbedo, J.G., Koenigkan, L.V.: Perspectives on the use of unmanned aerial systems to monitor cattle. Outlook Agr. **47**(3), 214–222 (2018). https://doi.org/10.1177/0030727018781876
2. Badrinarayanan, V., Kendall, A., Cipolla, R.: SegNet: a deep convolutional encoder-decoder architecture for image segmentation. IEEE Trans. Pattern Anal. Mach. Intell. **39**(12), 2481–2495 (2017). https://doi.org/10.1109/TPAMI.2016.2644615

3. Bhakta, I., Phadikar, S., Majumder, K.: State-of-the-art technologies in precision agriculture: a systematic review. J. Sci. Food Agric. **99**(11), 4878–4888 (2019). https://doi.org/10.1002/jsfa.9693
4. Bishop, J.C., Falzon, G., Trotter, M., Kwan, P., Meek, P.D.: Livestock vocalisation classification in farm soundscapes. Comput. Electron. Agric. **162**, 531–542 (2019). https://doi.org/10.1016/j.compag.2019.04.020
5. Cowlar: Streamline your dairy business! (2020). https://www.cowlar.com/
6. Dhulekar, P.A., Gandhe, S.T., Bagad, G.R., Dwivedi, S.S.: Vision based technique for animal detection. In: 2018 International Conference on Advances in Communication and Computing Technology (ICACCT), pp. 344–348, February 2018. https://doi.org/10.1109/ICACCT.2018.8529660
7. Ismail, Z.H., Chun, A.K.K., Shapiai Razak, M.I.: Efficient herd – outlier detection in livestock monitoring system based on density – based spatial clustering. IEEE Access **7**, 175062–175070 (2019). https://doi.org/10.1109/ACCESS.2019.2952912
8. Kellenberger, B., Marcos, D., Courty, N., Tuia, D.: Detecting animals in repeated UAV image acquisitions by matching CNN activations with optimal transport. In: IGARSS 2018 - 2018 IEEE International Geoscience and Remote Sensing Symposium, pp. 3643–3646, July 2018. https://doi.org/10.1109/IGARSS.2018.8519012
9. Kingma, D.P., Ba, J.: Adam: a method for stochastic optimization. CoRR abs/1412.6980 (2014). http://arxiv.org/abs/1412.6980
10. Li, X., Xing, L.: Use of unmanned aerial vehicles for livestock monitoring based on streaming K-means clustering. IFAC PapersOnLine **52**(30), 324–329 (2019). https://doi.org/10.1016/j.ifacol.2019.12.560
11. McKinlay, J., Southwell, C., Trebilco, R.: Integrating count effort by seasonally correcting animal population estimates (ICESCAPE): a method for estimating abundance and its uncertainty from count data using Adelie penguins as a case study. CCAMLR Sci. **17**, 213–227 (2010)
12. Nyamuryekung'e, S., Cibils, A.F., Estell, R.E., Gonzalez, A.L.: Use of an unmanned aerial vehicle-mounted video camera to assess feeding behavior of Raramuri Criollo cows. Rangeland Ecol. Manag. **69**(5), 386–389 (2016). https://doi.org/10.1016/j.rama.2016.04.005
13. Parikh, M., Patel, M., Bhatt, D.: Animal detection using template matching algorithm. Int. J. Res. Mod. Eng. Emerg. Technol. **1**(3), 26–32 (2013)
14. Ronneberger, O., Fischer, P., Brox, T.: U-Net: convolutional networks for biomedical image segmentation. CoRR abs/1505.04597 (2015). http://arxiv.org/abs/1505.04597
15. Sarwar, F., Griffin, A., Periasamy, P., Portas, K., Law, J.: Detecting and counting sheep with a convolutional neural network. In: 2018 15th IEEE International Conference on Advanced Video and Signal Based Surveillance (AVSS), pp. 1–6, November 2018. https://doi.org/10.1109/AVSS.2018.8639306
16. Seo, J., Sa, J., Choi, Y., Chung, Y., Park, D., Kim, H.: A yolo-based separation of touching-pigs for smart pig farm applications. In: 2019 21st International Conference on Advanced Communication Technology (ICACT), pp. 395–401, February 2019. https://doi.org/10.23919/ICACT.2019.8701968
17. Sharma, P., Singh, A.: Era of deep neural networks: a review. In: 8th International Conference on Computing, Communications and Networking Technologies, ICCCNT 2017 (2017). https://doi.org/10.1109/ICCCNT.2017.8203938
18. Xu, Y., Zhou, X., Chen, S., Li, F.: Deep learning for multiple object tracking: a survey. IET Comput. Vision **13**(4), 355–368 (2019). https://doi.org/10.1049/iet-cvi.2018.5598

Smart, Precision or Digital Agriculture and Farming - Current State of Technology

Daniel Honc(✉) and Jan Merta

Faculty of Electrical Engineering and Informatics,
University of Pardubice, Pardubice, Czech Republic
daniel.honc@upce.cz

Abstract. Together with smart homes, cities and factories, energy hubs and self-driving cars the smart agriculture or farming could be a way how to increase yields and efficiency as well as improve the welfare of farm animals, grow high quality crops and preserve the natural resources. Smart, precision and digital agriculture and farming current state survey with the technical challenges, interesting applications and future prospects is the aim of the paper. Worldwide and UE view is presented and compared with the situation in the Czech Republic. Authors are seeking for used or at least potential agriculture and farming applications of soft computing methods like fuzzy logic, machine learning and evolutionary computation.

Keywords: Smart · Precision · Digital · Agriculture · Farming · Soft computing

1 Introduction

Climate changes are becoming a real problem and they can cause a decrease in the agriculture production. Growing population increases demand for production while arable landscapes are shrinking due to the urbanization. Fresh water supplies are going to be vital. Farmers compete against each other, trying to reduce costs and differentiate. Weather predictions and monitoring, crop monitoring, insect detection, soil analysis and much more - all connected to IoT network informing the farmers or answering them questions like when should the farmers seed or harvest, what pesticides should they deploy or how to prepare the soil. Use of mobile laboratory or drone for land surveys or crop monitoring is another source of information for their right decisions called as high-tech or precision farming. Through the livestock monitoring ranchers can gather data regarding the health, well-being, and location of their cattle. This can help them to identify sick animals and it lowers the labour costs connected with cattle localization. Monitoring plant and soil conditions is another use case - sensing for soil moisture and nutrients, controlling water usage for optimal plant growth, determining custom fertilizer profiles based on the soil chemistry, determining the optimal time to plant and harvest, reporting the weather conditions.

Á. Herrero et al. (Eds.): SOCO 2020, AISC 1268, pp. 245–254, 2021.
https://doi.org/10.1007/978-3-030-57802-2_24

In the article authors bring survey about smart, precision and digital agriculture and farming current state and perspectives worldwide, find out and describe the situation in the Czech Republic. Authors do not want to concentrate to the economical, ethical, societal, political or other conditions or impacts. They are searching for offered or used smart, precision or digital agriculture or farming solutions and look for the potential areas for soft computing techniques as fuzzy logic, machine learning and evolutionary computation. Author do not want to give review of published papers, but they are aiming to the commercial applications. Paper is structured as follows: chapter one is a short introduction, worldwide, EU and Czech Republic view is given in chapter two, technology and typical applications are given in chapter four, interesting solutions are described in chapter five and chapter six gives conclusions.

2 Worldwide, EU and Czech Republic View

One of Precision Agriculture (PA) analyses was made by Hexa Report company - Precision Agriculture Market Analysis By Component (Hardware, Software & Services), By Technology (Variable Rate Technology, Remote sensing, Guidance Systems), By Application, By Region, And Segment Forecasts, 2014–2025 [1]. The global precision agriculture market is expected to reach 43.4 billion USD by 2025. American management consulting firm McKinsey & Company brings different analytics to broad area of the industries. They see efficiency opportunities for emerging economies of PA in the food chain. They define PA or farming in report on How big data will revolutionize the global food chain [2] as a technology-enabled approach to farming management that observes, measures, and analyses the needs of the individual fields and crops. According to them PA development is being shaped by two technological trends: big-data and advanced-analytics capabilities on the one hand, and robotics - aerial imagery, sensors and sophisticated local weather forecasts - on the other.

In the publication Precision agriculture in Europe, Legal, social and ethical considerations – Study [3] authors analyse different ways in which the current EU legislative framework may be affected by the digitisation and automation of the farming activities and the respective technological trends. According to EU publication Precision agriculture and the future of farming in Europe, Scientific foresight study - Study [4] the PA is defined as a modern farming management concept using digital techniques to monitor and optimise agricultural production processes. The following four main future opportunities and concerns regarding PA, or precision farming, in the EU are stated: 1. PA can actively contribute to food security and safety; 2. PA supports sustainable farming; 3. PA will trigger societal changes along with its uptake; 4. PA requires new skills to be learned. The wide diversity of agriculture throughout the EU, regarding particularly farm size, types of farming, farming practices, output and employment, presents a challenge for European policy-makers. European policy measures therefore should differentiate between Member States, taking into account that the opportunities and concerns vary highly from one country to another. Two annexes are complementary to this study: Annex 1: Technical Horizon Scan [5] and Annex 2: exploratory scenarios [6]. The aim of those publications is study, analyse, inform and guide the discussion to identify and explore policy actions in the European Parliament.

Only two small references to PA can be found in the official publication of Ministry of Agriculture of the Czech Republic in Concept of research, development and Innovation of the Ministry of Agriculture for 2016–2022 [7]. One in the section V. 2. Sustainable production of healthy and quality food and feed of plant origin, paragraph (h) the application of precision agriculture elements in the technological systems of cultivation in order to optimize benefits nutrients in mineral fertilizers and optimizing the dosage, timing and application of pesticides. Second and last PA reference is in the section VI. 3. Technology for livestock production, welfare, systems, ethics and economics of livestock breeding, paragraph (f) innovation and development of technological processes for new types of livestock farming, including aquaculture, the use of automated livestock management systems ('precision livestock farming'), focusing on the level and quality of production; the health and satisfaction of the physiological needs of animals. Czech ministry of agriculture organizes meetings and conferences about PA, allocates and manages European Union subsidy programs. One of them gave arise to a smart and precision agriculture demonstration farm.

3 Terminology and Technology

There are no strong borders and definitions between smart, precision and digital agriculture and farming. At the end it is nothing else than the way how to use hardware and software to give the farmers information needed to make their decisions based on the information rather than on the intuition and help them to control and automate their business. Smart is very popular word and usually we think about acting differently from the normal or standard way with use of the modern technology. In agriculture and farming smart can be anything connected with the sensors, actuators, images from drones, planes, satellites, vehicles, advanced algorithms, weather forecasts with the output to the mobile phones, web servers or clouds. Precision farming is a special part aimed to the precision manipulation with the irrigation, fertilization, pesticides for the crop production and the feeding and pharmaceuticals for the animal production. Practically every variable is processed in digital systems nowadays. The digitalization is standard also for the transmission and the storage. Interconnected systems covering whole farms or even bigger geographical areas can be created. Higher number of information requires different methods for the data processing and visualization (Big Data challenge). The technological changes in the agriculture can be compared to the concept of Industry 4.0. But the processes are different - agriculture is determined by the natural and biological factors. The idea and technology are the same with the same goals - lower costs, higher returns, ecology aspects, product quality, etc. Automotive industry is mostly located in the technologically advanced countries and the application areas are limited within the factory. Agriculture has huge diversity from the cultural, social, geographical or other aspects. Agriculture can be very conservative with high ratio of the labour and being operated on the brink of the economic profitability very frequently. Therefore, consideration and planning of the investments to the new technology is not easy task and the process will be gradual. Successful applications or good practice will help in the decision-making of the farmers. Technologies with the potential for smart or precision agriculture or farming are listed below.

Smart farming is an application of Information and Communication Technologies (ICT) solutions such as precision equipment, Internet of Things (IoT), sensors and actuators, geo-positioning systems, Big Data, Unmanned Aerial Vehicles (UAVs, drones), robotics, etc. Radio Frequency Identification (RFID) or Near Field Communication (NFC) is used for the shortest distance communication. Bluetooth or Bluetooth Low Energy (BLE) is used for the objects 10 m far or closer. For applications sending data over hundreds or even thousands of meters Low Power Wide-Area Network (LPWAN) options are needed - such as Narrowband IoT (NB-IoT), Long Range LoRa, Long Term Evolution (4G) Machine Type Communication (LTE-M) and SigFox. Wide range of sensors are installed in the fields, on smart vehicles and weather stations measuring e.g. humidity, moisture, light, air temperature, CO_2. Information from sensors, weather forecast and satellites allow to build models and predictions how the crop will grow, build harvesting strategy, etc.

Instead of applying equal amount of the fertilisers over an entire field, precision agriculture involves measuring the within-field soil variations and adapting the fertiliser strategy accordingly. This leads to the optimised fertiliser usage, saving costs and reducing the environmental impact.

Smart sprinklers activate the automated irrigation, connected coolers and heaters in the storage and transportation facilities enable sustainable conditions for the product and help reduce the waste. Intelligent LED lighting automatically adjusts to the changing conditions and ensures every part of the greenhouse or storage space gets the right amount of light.

IoT-driven smart greenhouses can intelligently monitor as well as control the climate, eliminating the need for the manual intervention. They become a self-regulating, microclimate-controlled environment for the optimal plant growth.

In most cases, drones work as an IoT-based monitoring system in the smart agriculture, as the tools for the farm mapping and on-demand irrigation and pesticide treatment.

Sensors have been developed for the various parts of the cattle, including the neck, hooves, stomach and tail in an effort to reduce the mortality rate associate with birth, prevent cattle rustling and illness, monitor the dietary health, and monitoring the amount of the nitrogen in a cow's urine. Additionally, the implementation of IoT combined with Artificial Intelligence (AI) can help improve breeding and identify which breeds are "top producers" in order to increase both the quality and the volume of the milk.

Large farm owners can use wireless IoT applications to monitor the location, well-being and health of their cattle. With this information, they can identify sick animals, so that they can be separated from the herd to prevent the spread of disease. Farm owners are keep tabs on their herd through GPS sensors. These devices allow farm owners to track their cattle over long distances, even in remote areas. Using different scenarios/options they can provide this information in different ways:

- Geo-fencing – when the device leaves a pre-defined area
- Chrono-fencing – the device sends a status update after staying at a location for a certain pre-defined time
- On-demand location request – request to device to update its status and position.

Vehicles are equipped with the precise GPS and machine learning algorithms to enable self-driving, sensors, computer vision for precision farming capabilities.

Categories of vehicles control level are:

- Guidance – operator is needed - maintain row-to-row vehicle accuracy in the field in order to reduce overlaps and skips
- Coordination & Optimization - vehicle and environmental data are used to develop an information hub for operator
- Operator Assisted Autonomy - operator is monitoring automated functions and only providing backup and support when needed
- Supervised Autonomy - in-field supervision is needed while unmanned vehicles perform designated tasks
- Full Autonomy - remote supervision - such as from the farm office - or via AI.

Companies like GEOSYS, planet or ASTRO DIGITAL deliver daily imagery from the satellites for precision agriculture purposes. For example Sentinel satellites from ESA allow the creation of maps of the spatial variability of many measurements and variables - crop yield, terrain features/topography, organic matter content, moisture levels, nitrogen levels etc.

4 Soft Computing Techniques Application Areas

Areas for use of the soft computing techniques as fuzzy logic, machine learning and evolutionary computation are practically unlimited for the smart, digital or precision agriculture or farming. Fuzzy logic and neural networks help to create models of the farmers' behaviour and expert knowledge to build the decision and analytical systems, to predict the future trends, consider case studies or carry out on-line optimisations. Evolutionary algorithms can be used for the optimisation of operational, investment or logistic decisions and operations. Data, methods and hardware for the artificial intelligence are available and can be applied for the image processing or other tasks connected with the agriculture and farming.

5 Commercial PA and Smart Agriculture Solutions

PA solution are stated so the reader can get the ideas and information about the application possibilities of the new technologies in the agriculture and farming especially connected with the soft computing techniques.

CropX [8] – CropX app can help figure out exactly how much to irrigate the field, by providing an irrigation prescription that is constantly adapting to the changing conditions of the field. By analysing crop growth against crop models, prediction the crop's needs and expected growth, detect any deviation and identify early-stage field variability and non-uniformity of the crop growth. Integration of crop models, satellite imagery and weather forecast data alongside the soil data creation map nutrient distribution across the field and zone-specific nitrogen application recommendations (Fig. 1).

Fig. 1. CropX platform [8]

Arable [9] – Arable Mark 2 is all-in-one weather and crop monitor - precipitation, evapotranspiration, radiation, plant health, weather, harvest/event timing with cellular connectivity.

Gamaya [10] – patented ultracompact hyperspectral imaging camera with machine learning engine for precision farming and global crop intelligence based on agronomic insights (Fig. 2).

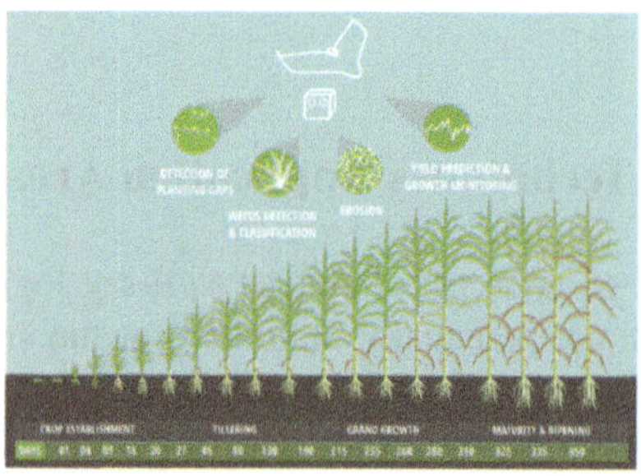

Fig. 2. Canefit – smart solution for sugarcane cultivation [10]

Ceres Imaging [11] – irrigation management, nutrient management, pest and disease management, labour management with high-resolution multispectral imagery of chlorophyll, colour infrared, NDVI, thermal and water stress.

Mothive [12] – devices installed next to the plants collect environmental and soil data. Bespoke Machine Learning models predict crop growth conditions, diseases and crop harvest. Recommendations and alerts delivered via dashboard, SMS, email. Live & historical data, intelligent automation (irrigation & ventilation) and specific tasks delivered to robots in the future (Fig. 3).

Fig. 3. Mothive Ladybird [12]

PrecisionHawk [13] – autonomous UAV that collects high-quality data through a series of sensors that are used for the surveying, mapping, and imaging of agricultural land. Equipped with sensors – lidar, thermal, multispectral, hyperspectral.

AgEagle [14] – aerial imagery-based data collection and analytics solutions - "tractor tough," precision drones capable of capturing thousands of ultra-high resolution images and producing actionable intelligence (Fig. 4).

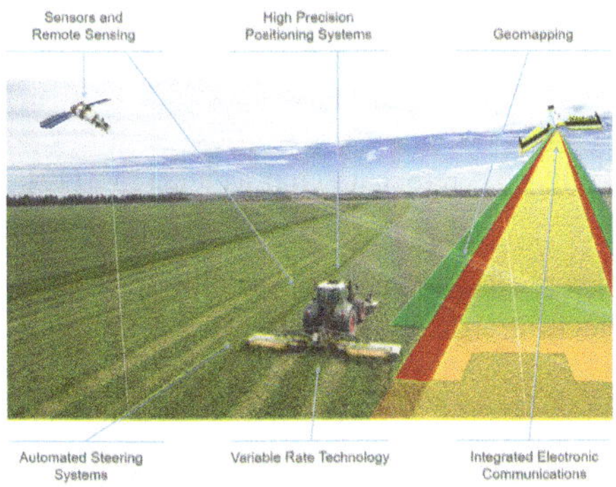

Fig. 4. AgEagle – precision farming [14]

Phytech [15] – plant-based application for the optimized irritation for corn, almonds, citrus, cotton and apple and other crops.

WaterBit [16] – automated irrigation solution - one field, many microblocks, remote irritation control, planning and scheduling (Fig. 5).

Fig. 5. WaterBit – precision farming [16]

Aker Technologies Inc. [17] – accurate crop monitoring of disease, insects, and other stresses. AkerScout - crop scouting to help document and prioritize in-season crop damage, imagery and analytics (Fig. 6).

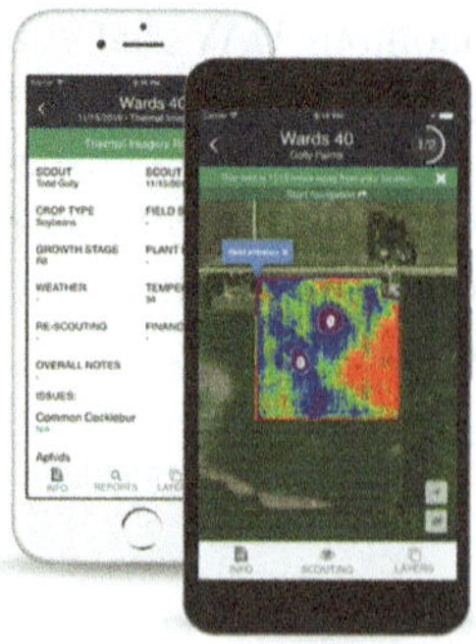

Fig. 6. AkerScout - crop scouting tool [17]

JMB North America [18] – cow-monitoring solutions for American beef and dairy producers – calving detection, heat detection, nutritional monitoring, health monitoring (Fig. 7).

Fig. 7. JMB North America [18]

Case IH [19] – producer of agriculture machinery with advanced farming systems, displays, autoguidance, section & rate control.

AgroCares [20] – Lab-in-the-box - direct on-site access to testing services, Scanner - instant, on-the-spot monitoring of nutrients in soil, feed and leaf, Scoutbox - digitally determination, counting and location of harmful insects (Fig. 8).

Fig. 8. AgroCares - Lab-in-the-box, Scanner, Scoutbox [20]

CleverFarm [21] – Czech company dealing with PA, online records of agronomic activities, sensors, satellite imagery and land registry.

Digital Garden Lab [22] – open-source community exploring new forms of digital augmentation to facilitate urban community gardening and urban landscapes.

6 Conclusions

Digitalization trends and smart technologies are everywhere around us. Agriculture or farming belongs historically to one of the most rigid human activities. Together with the climate changes, population grow, water resources decrease the smart or precise technologies give chance to cope with the coming challenges with respect to the production quality, operate environmentally friendly and humanely to the animals. Another factor is also to help the farmers to do informed decisions even without extensive experience or education. Images from the satellites, planes or drones in broader frequency range than only within visible light bring new source of the information. Similarly sensors on the fields or carrying on the animals have not been used in the past. Accurate and localized weather forecast are also improved much in the recent years. But the information alone is not enough. Data analysis techniques, models, optimization methods are theoretically known and waiting for their application. The potential of the soft computing methods is indisputable. Image and other data analysis are made by the machine learning techniques. Fuzzy logic and neural networks build the models and experts systems used for the analysis, optimization and predictions. Evolutionary algorithms are used for the optimization in economical, logistical and technological areas. Big Data algorithms process information from huge amounts of sensors, imagery and weather forecast. The data and results are stored in cloud services, displayed on the mobile phones or tablets and sent to the workers, machinery, robots, drones and planes to close the feedback actuate and control. Security of the system must be also a priority for the future. State or funding support for the introduction of the new technologies will be needed together with the new legislation. The solution must be in balance between traditional approaches and possibilities of the new technologies. For example, is it worth to use advance technology and do not carry out deep plowing because of the fuel savings and let the water and fertilizer run off the fields? The new technology will work if we see the bigger picture and all activities play together.

Acknowledgment. This research was supported by SGS grant at Faculty of Electrical Engineering and Informatics, University of Pardubice.

References

1. Hexareports: Precision Agriculture Market Analysis By Component. http://www.hexareports.com/report/precision-agriculture-market. Accessed 18 Apr 2020
2. McKinsey: How big data will revolutionize the global food chain. https://www.mckinsey.com/business-functions/mckinsey-digital/our-insights/how-big-data-will-revolutionize-the-global-food-chain. Accessed 18 Apr 2020
3. Publication office of the EU: Precision agriculture in Europe. https://op.europa.eu/en/publication-detail/-/publication/1d338444-1783-11e8-9253-01aa75ed71a1/language-en/format-PDF/source-search. Accessed 1 Feb 2020
4. Publication office of the EU: Precision agriculture and the future of farming in Europe. https://op.europa.eu/en/publication-detail/-/publication/40fe549e-cb49-11e7-a5d5-01aa75ed71a1/language-en. Accessed 1 Feb 2020

5. Publication office of the EU: Precision agriculture and the future of farming in Europe. https://op.europa.eu/en/publication-detail/-/publication/6a75e0ac-90ae-11e9-9369-01aa75ed71a1/language-en/format-PDF/source-search. Accessed 1 Feb 2020
6. Publication office of the EU: Precision agriculture and the future of farming in Europe. https://op.europa.eu/en/publication-detail/-/publication/77b851b0-90b1-11e9-9369-01aa75ed71a1/language-en/format-PDF/source-search. Accessed 1 Feb 2020
7. eAGRI: Koncepce výzkumu, vývoje a inovací Ministerstva zemědělství na léta 2016–2022. http://eagri.cz/public/web/file/461417/Koncepce_vyzkumu__vyvoje_a_inovaci_Ministerstva_zemedelstvi_na_leta_2016_2022.pdf. Accessed 1 Feb 2020
8. CropX. https://www.cropx.com/. Accessed 1 Feb 2020
9. Arable. http://www.arable.com/. Accessed 1 Feb 2020
10. Gamaya. https://gamaya.com/. Accessed 1 Feb 2020
11. Ceres Imaging. https://www.ceresimaging.net/. Accessed 1 Feb 2020
12. Mothive. https://www.mothive.com/. Accessed 1 Feb 2020
13. PrecisionHawk, agriculture. https://www.precisionhawk.com/agriculture. Accessed 1 Feb 2020
14. AgEagle. https://www.ageagle.com/. Accessed 1 Feb 2020
15. Phytech. https://www.phytech.com/. Accessed 1 Feb 2020
16. WaterBit. https://www.waterbit.com/. Accessed 1 Feb 2020
17. Aker. https://aker.ag/. Accessed 1 Feb 2020
18. JMB North America, technology. http://cowmonitor.com/technology/. Accessed 1 Feb 2020
19. Case IH, Advanced Farming Systems. https://www.caseih.com/northamerica/en-us/innovations/advanced-farming-systems. Accessed 1 Feb 2020
20. AgroCares. https://www.agrocares.com/en. Accessed 1 Feb 2020
21. CleverFarm. https://www.cleverfarm.org/. Accessed 1 Feb 2020
22. Digital Garden Lab. https://digitalgardenlab.cz/. Accessed 18 Apr 2020

An Automated Platform for Microrobot Manipulation

Jan Vrba[1(✉)], Charlie Maslen[1], Ivan Rehor[1,2], and Jan Mares[1]

[1] Faculty of Chemical Engineering, University of Chemistry and Technology Prague, Technická 5, 16628 Prague 6, Czech Republic
{jan.vrba,maslenc,ivan.rehor,jan.mares}@vscht.cz
[2] Institute of Organic Chemistry and Biochemistry, Flamingovo Namesti 2, 16000 Prague 6, Czech Republic
https://www.vscht.cz

Abstract. This paper presents hydrogel microrobots (100 µm) that are directed to specific locations in their environment by an automated platform. The microrobots are actuated by focused laser light and crawl in aqueous environments. The platform consists of a stage, manipulated by stepper drivers and controlled by a Raspberry PI 4. This positions the laser light in the desired locations to move microrobots towards a goal location. The microrobot localisation is done via a microscope camera and repetitive usage of a template matching algorithm. Instead of a path planning algorithm, the optimal position for the laser is chosen before every step so that the disk reaches the goal as fast as possible.

Keywords: Robotics · Image processing · Pattern recognition · Motion control · Microrobots

1 Introduction

Population growth, climate change, and water and resource management impose novel problems that will require new technologies to solve. The development of microrobotic systems that can perform useful work promises to advance technologies in a wide variety of fields just as conventional, human-scale robotics has transformed all industries in modern society. Microrobots have the potential to remotely access areas at length-scales that are currently only reachable by invasive methods or in lab controlled environments [1]. Untethered locomoting robots, hold a great promise to revolutionise healthcare as they may operate autonomously inside the human body, serving for diagnostic and therapeutic purposes [2–4]. The in-field applications of locomoting microrobots have been already proposed for environmental applications, such as pollutant degradation and removal, bacteria killing or dynamic environmental monitoring [5]. Heavy metals are toxic and their removal from water is one key application for which

Á. Herrero et al. (Eds.): SOCO 2020, AISC 1268, pp. 255–265, 2021.
https://doi.org/10.1007/978-3-030-57802-2_25

the deployment of a large group of microrobots provides a technological leap [6]. An actively moving robot can, for example, overcome diffusive mass transport in aqueous environments, which inspired the use of microswimmers for water remediation applications [7,8]. Moreover, it can gather and report information of the chemical composition of its surrounding and serve as an autonomous mobile microsensor [9]. These early-stage applications of microrobots in environmental care appear transferable to the tightly connected applications in agriculture.

To produce functional actuating robots on the micro-scale requires a shift in the paradigm of what technologies are used. Conventional robots are constructed from multiple components such as actuators, sensors and controllers. However, these components have a limit to how small they can be miniaturised and assembled. As such, current research is tending towards the use of alternative approaches, exploiting soft materials that can actuate by deformation in response to a stimulus such as light [11], or chemical changes in their environment [12]. These microrobots can locomote through an environment by cyclic periodic deformation of their bodies. Their navigation through the environment has been so far provided by some external measurement system such as microscope imaging and control of the stimulus. However, implementing mechanisms for autonomous navigation, similar to chemo- or photo-taxis, exploited by microorganisms, or self-propelling microswimmers is currently at the centre of broad research interest [13].

In this research, we present crawling, disk-shaped hydrogel microrobots that are powered by absorption of focused laser light. The microrobots are driven by localised shrinking and expanding cycles coupled to a friction hysteresis between their bodies and surface. The resulting motion can thus be likened to the crawling mechanism of inchworms and the microrobots have full translational freedom owing to their radial symmetry. We developed a driving mechanism for this robot relying on image analysis coupled to the microposition system and pulsed focused irradiation. Crawling robots are promising candidates for agricultural applications, such as managing biofilms [14], delivering anti-pesticides or antibiotics or sensing local changes in chemicals within a crop or slurry [15]. However, our current findings are too early-stage to be directly applied at the moment.

2 Microrobots Specification

2.1 Composition

The microrobots are composed of a thermo-responsive polymer - poly-n-isopropylacrylamide (PNIPAM) - cross-linked by poly(ethylene glycol) diacrylate (PEGDA). PNIPAM is a widely researched polymer owing to its thermoresponsive properties - at temperatures greater than 32 °C it transitions from a hydrophilic to a hydrophobic state. The hydrogel network formed by crosslinking PNIPAM thus de-swells reversibly when heated above 32 °C as water is expelled from the network. This is observed as isotropic contraction of the microgel to approximately 50% of the original volume. The stimulus inducing shrinking response can be transduced from heat, to visible light by incorporating gold nanospheres (15 nm) [16]. The nanoparticles absorb laser light at a resonant wavelength (532 nm) and photothermally heat the gel by plasmonic absorption. The shrinking response occurs rapidly and locally in the focal point of a focussed 532 nm laser. Removal of the laser and subsequent cooling of the network causes the gel to re-swell to its original volume.

2.2 Production

Microrobots can be produced using a microgel synthesis technique, stop-flow lithography [17,18], in which a photo-polymerizable pre-cursor solution is pumped into a microfluidic channel. The pressure is then removed, stopping the flow of pre-cursor. With the flow stationary, UV light is focussed by a microscope objective on the channel through a photo-mask which polymerizes the pre-cursor - forming a hydrogel with shape given by the photomask. The thickness is given by the thickness of the channel (30 μm in our case). The pre-cursor used in this synthesis is a 100 μL water solution of *n*-isopropylacrylamide (37 mg), PEGDA (20 mg) and a water soluble photoinitiatior - lithium phenyl-2,4,6-trimethylbenzoylphosphinate (LAP) (1.5 mg). The water contains a 1.5% wt/wt% suspension of 15 nm gold nanospheres. Microfluidic channels are produced using poly-dimethylsiloxane (PDMS) moulded from SU-8 on silicon photoresists. After production, the microrobots are stored in polystyrene wells with flat bottoms in a 1.25% Tween-20 surfactant solution. A glass slide is placed over the top of the well to remove any meniscus which would distort the laser beam.

2.3 Microrobots Response to Laser Beam Pulse

The shrinking-expanding actuation of the hydrogel microrobots is coupled to a hysteresis in the friction between the hydrogel and the surface during the cycles. Whilst shrinking, the friction is reduced and the shrinking area slides easily to the centre of mass. When re-expanding, the friction is enhanced and so the actuating side becomes an anchor point. The expansion thus drives the non-actuating side away from the centre of mass which results in a net displacement with each actuation cycle, away from the irradiation centre 1. The origin of this hysteresis will be explored in greater detail in a separate publication. Because the displacement occurs in a direction away from the irradiation, we can produce circular disk-shaped microrobots (diameter = 100 μm) which have full translational freedom. We propose 8 regions where the laser pulse should be aimed to displace the microrobot in a desired direction. The proposed regions are depicted in Fig. 2. The regions can be divided into two groups. The first group (positions 11, 13, 31, 33) cause microrobot to move in both axes. The second group (positions 12, 21, 23, 32) cause the microrobot to move in single axis direction. When moving the microrobot in X-axis direction the average displacement in X-axis is approximately 4.5 px with standard deviation 2.5 px and average displacement in Y-axis is approximately 0.13 px with standard deviation 2.5 px. When moving the microrobot in Y-axis direction the average displacement in X-axis is approximately 0.36 px with standard deviation 1.39 px and average displacement in the Y-axis is approximately 3.6 px with standard deviation 1.69 px (Fig. 1).

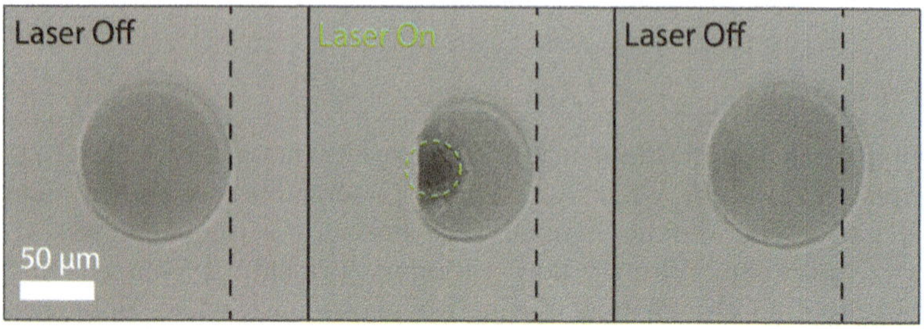

Fig. 1. The sequence of photos depicting the microrobots response to the laser pulse.

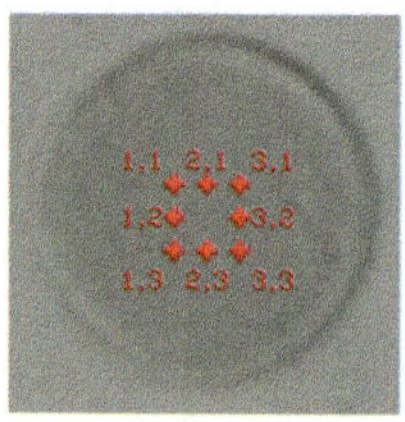

Fig. 2. Detailed photo of the disk-shaped microrobot with the proposed illumination regions marked by red crosses. The annotation of the illumination regions is included.

3 Automated Platform Overview

The platform consists of microscope with positioning stage, camera, step motors and laser controlled by Raspberry PI 4 and personal computer. The scheme of the system is depicted in the following Fig. 3 and the photo of the actual workplace is in the Fig. 4. The microscopes camera (DMK 23UX174) is connected to the personal computer (PC) via USB. It supports the adjustment of the brightness, gain, exposure and frames-per-second parameters. The acquired image has 0,48 megapixels with the resolution 800 × 600 pixels. A 200 mW laser is inserted into the condenser lens of the microscope such that it can be manually focused. Additionally, the focal point is fixed in relation to the camera which allows the position of the laser, with respect to the microrobots, to be controlled by the step motors positioning the stage. The x, y coordinates of the laser in the image and the goal are inputted manually and stored. The image is processed on the

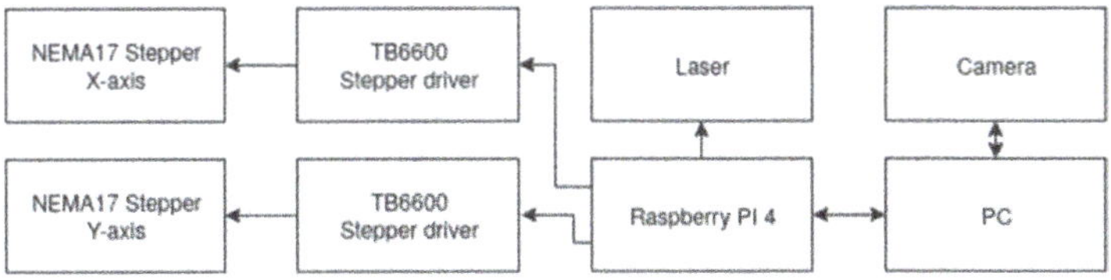

Fig. 3. The block diagram of the automated platform for the microrobots manipulation.

PC to obtain the positions of disks and calculate the desired number of steps to be done by the step motors in order to move the microrobot in the direction of the goal. The desired number of steps is sent to the Raspberry PI 4 via Ethernet, using network socket. The stage is moved into the desired position for illumination, with the position of laser at the illumination region which will provide the greatest displacement towards the goal. The Raspberry PI 4 controls the TB6600 stepper drivers, that are connected to NEMA17 step motors. Those steppers are connected via micrometric bolts to the stage, so precise positioning in two independent axis (X-axis, Y-axis) is achieved. When the movement is done, the laser pulse is performed (0.8 s), the microrobot displaces and an acknowledgement is sent to the PC. Then, the new image can be acquired and the result of the illumination is evaluated.

Fig. 4. The photo of the workplace with automated platform for microrobot manipulation. A - laser, B - camera, C - movable stage with wells containing microrobots, D - microscope, E - NEMA17 steppers, F - personal computer with running software, G - TB6600 drivers, H - Raspberry PI 4, I - ATX power supply used for powering TB6600 drivers.

3.1 Software Overview

In order to control the disk crawling and camera settings, a Python 3.7 [19] based software with graphical user interface was developed. A screenshot of the software is found in Fig. 5. The main functions of the software are:

- camera settings
- image processing
- laser-camera calibration
- setting the goal and disk to move
- template for pattern matching acquisition
- saving the images and creating video sequences

All the image processing is done with NumPy and OpenCV [20] libraries. The graphical user interface was created using PyQt5 [21] framework that enables the portability between Windows/Linux machines.

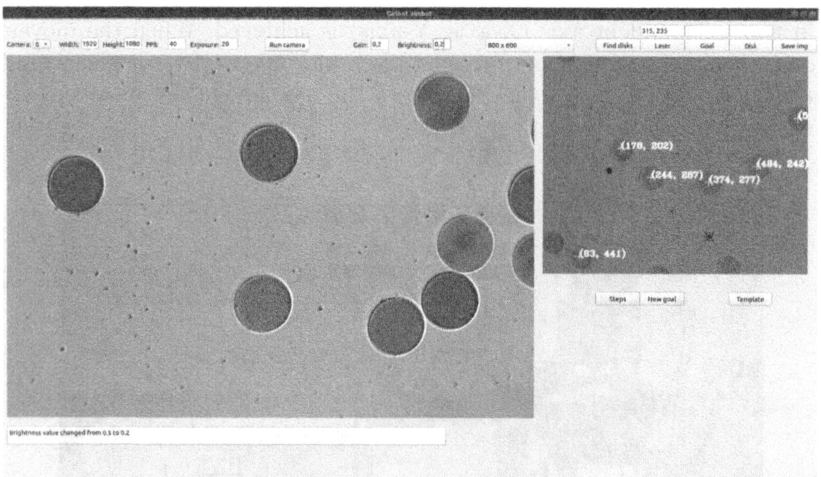

Fig. 5. The screenshot from the developed software for the automated platform for microrobot manipulation.

3.2 Precision of the Positioning

The NEMA17 step motors are controlled with a TB6600 driver in 1/4 micro-stepping mode. The 1/4 micro-stepping mode combined with the micrometric bolts allows precise positioning. Approximately 6.667 steps are required to change the location in the image by 1 pixel in the desired axis. This resolution is the same for both axes.

3.3 Camera Setting

The camera is placed into the optical axis of the microscope. Due to high homogeneity of the disks and liquid we use the same settings of the DMK 23UX174 camera for all experiments. The exposure time is set to 40 ms, brightness to 0.2, the gain is set to 0.2 dB and frames-per-second is 20. A 4× objective is used in the microscope and the image size is 800×600 pixels. With these settings we need only limited number of disks templates, so we can successfully perform the disks localisation without need of acquiring new templates with every new experiment.

3.4 Disk Localisation Algorithm

The acquired colour images from the camera are transformed into grayscale images to avoid working with multiple channels and because the colour information is redundant. The disk matching is done via normalised correlation coefficient, that is given as follows

$$R(x,y) = \frac{\sum_{x',y'}(T(x',y') \cdot I(x+x',y+y'))}{\sqrt{\sum_{x',y'} T(x',y')^2 \cdot \sum_{x',y'} I(x+x',y+y')^2)}} \tag{1}$$

where T is the template, I is the image and $x' = 0...w-1$, $y' = 0...h-1$ where h and w are image height and width. An object is recognised as a disk, when $R(x,y) \geq 0.7$. When a disk is found, the area in the original image I around the centre is replaced by the background colour, that has typically intensity value 178 (Fig. 7), and the search for disks continues. To obtain highly accurate disk detection, we used 20 different disk templates. The algorithm of disk matching is described in the following pythonic pseudocode.

Algorithm 1: Disk localisation algorithm

```
diskLocations = []
for temple in templates do
    while max(R(x,y)) ≥ 0.7 do
        compute R(x,y) for all x,y
        if max(R(x,y) ≥ 0.7 then
            if [x_max,y_max] not in diskLocations then
                diskLocations.append([x_max,y_max])
                Image[x_max-30:x_max+30,y_max-30:y_max+30] = 178
            end
        end
    end
end
```

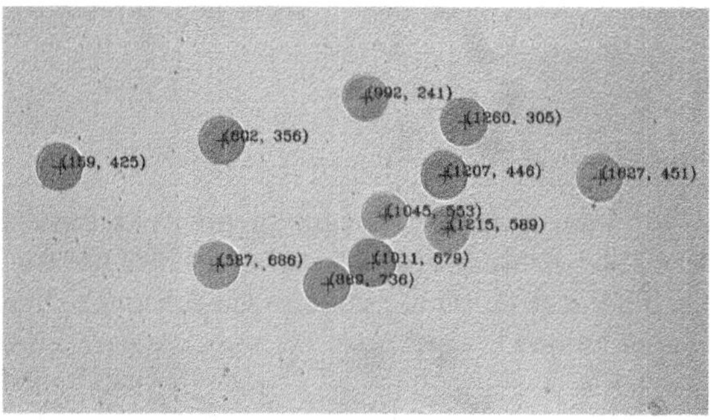

Fig. 6. Result of the disk localisation algorithm. The black cross is placed in the centre of the disk. The centres coordinates are displayed to the left of it.

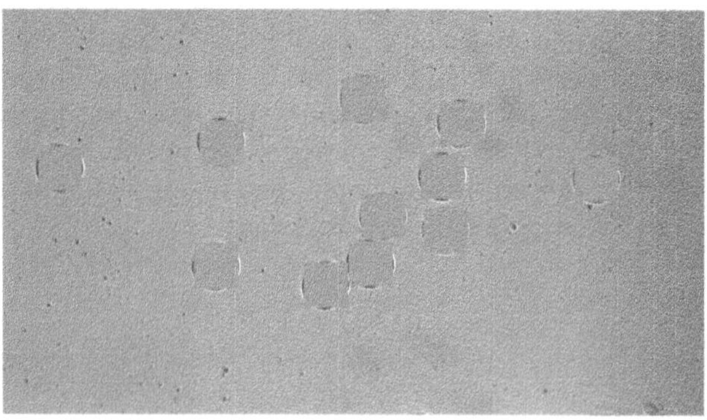

Fig. 7. Partial result of the disk localisation algorithm. It shows the effect of the mask after the disk are found.

The result of the proposed disk localisation algorithm is depicted in following Figs. 6, 7. The successful detection of the disk was performed in approximately 95% of occasions.

3.5 Automated Control of Microrobots Movement

The overall function of the automated platform for microrobot manipulation can be described by the following algorithm.

Algorithm 2: workflow of the automated platform

```
1  user sets laser coordinates
2  user sets goal coordinates
3  find all microrobots in the image
4  user selects the microrobot that should crawl to the goal
5  while goal not reached do
6      select the optimal region of target microrobot for laser illumination
       (equations 2,3,4,5)
7      set the laser to desired position by moving steppers
8      illuminate with the laser
9      evaluate new position of target microrobot
10     recompute goal coordinates
11 end
12 go to 2
```

The user has to calibrate laser coordinates in the image, choose the goal and select the microrobot that is going to reach the goal. All other steps are performed automatically. The coordinates region i.e. where the microrobot is illuminated by the laser, is estimated as follows:

$$dx = x(goal) - x(disk) \tag{2}$$

$$x(laser) = x(disk) - 5 \cdot sgn(dx) \tag{3}$$

$$dy = y(goal) - y(disk) \tag{4}$$

$$y(laser) = y(disk) - 5 \cdot sgn(dy) \tag{5}$$

The $x(goal)$ is x coordinate of the goal, $x(disk)$ is the x-axis coordinate of target disk, $x(laser)$ is desired x-axis coordinate of the laser, $y(goal)$ is y-axis coordinate of the goal, $y(disk)$ is the y-axis coordinate of target disk and $y(laser)$ is desired y-axis coordinate of the laser.

4 Conclusions and Further Work

The automated platform for microrobots manipulation has been built and tested and software for its control has been developed. The microrobots crawling is achieved via laser pulses that illuminate 8 different regions of the microrobot. These regions are found experimentally. As it is hard to predict the direction, the path planning algorithm was not implemented and, instead of it, the optimal position where to direct the microrobot is chosen according to relative position of microrobot to its goal. The position is chosen such that the microbot is approaching the goal as fast as possible. The video of crawling microbot can be viewed at https://www.youtube.com/watch?v=BNxAqCNGisc.

4.1 Further Works

In the future we would like to implement the convolution neural network to improve the successful disk recognition. We currently obtain 95% success of the disk detection. However, for robust and highly efficient manipulation, a success rate of >99% is desirable. Another problem that arises with the contamination of the liquid during disk preparation and manipulation is the presence of obstacles that consist of dirt and damaged disks. So far we did not concern about the possible obstacles in the liquid. In the future we would like to develop a robust control algorithm that takes into account stochastic behaviour of the microbots and existence of obstacles in the environment. Additionally, the porosity of hydrogels makes them ideal candidates for heavy metal removal. We aim to use our robots to absorb and remove heavy metal ions by including chemical moieties suited for this purpose.

Acknowledgements. Financial support from specific university research (MSMT No 21-SVV/2020). Financial support from MEYS INTER-EXCELLENCE project LTAIN19007. Financial support from Junior GACR project nr. 18-19170Y. I. Rehor acknowledges his J.E. Purkyne fellowship.

References

1. Sitti, M., Ceylan, H., Hu, W., Giltinan, J., Turan, M., Yim, S., Diller, E.: Biomedical applications of untethered mobile milli/microrobots. Proc. IEEE **103**, 205–224 (2015)
2. Cianchetti, M., Laschi, C., Menciassi, A., Dario, P.: Biomedical applications of soft robotics. Nat. Rev. Mater. **3**, 143–153 (2018)
3. Palagi, S., Fischer, P.: Bioinspired microrobots. Nat. Rev. Mater. **3**, 113–124 (2018)
4. Sitti, M.: Miniature soft robots - road to the clinic. Nat. Rev. Mater. **3**, 74–75 (2018)
5. Jurado-Sánchez, B., Wang, J.: Micromotors for environmental applications: a review. Environ. Sci. Nano **5**, 1530–1544 (2018)
6. Vilela, D., Parmar, J., Zeng, Y., Sanchez, S.: Graphene-based microrobots for toxic heavy metal removal and recovery from water. Nano Lett. **16**(4), 2860–2866 (2016)
7. Liu, W., Ge, H., Chen, X., Lu, X., Gu, Z., Li, J., Wang, J.: Fish-scale-like intercalated metal oxide-based micromotors as efficient water remediation agents. ACS Appl. Mater. Interfaces **11**, 16164–16173 (2019)
8. Soto, F., Lopez–Ramirez, M., Jeerapan, I., Esteban–Fernandez de Avila, B., Mishra, R., Lu, X., Chai, I., Chen, C., Kupor, D., Nourhani, A., Wang, J.: Rotibot: use of rotifers as self–propelling biohybrid microcleaners. Adv. Funct. Mater. **29**, 1900658 (2019)

9. Li, J., Esteban-Fernández de Ávila, B., Gao, W., Zhang, L., Wang, J.: Micro/nanorobots for biomedicine: delivery, surgery, sensing, and detoxification. Sci. Rob. **2**, eaam6431 (2017)
10. Chen, X., Jang, B., Ahmed, D., Hu, C., De Marco, C., Hoop, M., Mushtaq, F., Nelson, B., Pané, S.: Small-scale machines driven by external power sources. Adv. Mater. **30**, 1705061 (2018)
11. Zeng, H., Wasylczyk, P., Wiersma, D., Priimagi, A.: Light robots: bridging the gap between microrobotics and photomechanics in soft materials. Adv. Mater. **30**(24), 1703554 (2018)
12. Maeda, S., Hara, Y., Sakai, T., Yoshida, R., Hashimoto, S.: Self-walking gel. Adv. Mater. **19**, 3480–3484 (2007)
13. Plutnar, J., Pumera, M.: Chemotactic micro- and nanodevices. Angewandte Chemie Int. Ed. **58**, 2190–2196 (2018)
14. Velmourougane, K., Prasanna, R., Saxena, A.: Agriculturally important microbial biofilms: present status and future prospects. J. Basic Microbiol. **57**, 548–573 (2017)
15. Felekis, D., Muntwyler, S., Vogler, H., Beyeler, F., Grossniklaus, U., Nelson, B.: Quantifying growth mechanics of living, growing plant cells in situ using microrobotics. Micro Nano Lett. **6**, 311 (2011)
16. Sershen, S., Mensing, G., Ng, M., Halas, N., Beebe, D., West, J.: Independent optical control of microfluidic valves formed from optomechanically responsive nanocomposite hydrogels. Adv. Mater. **17**, 1366–1368 (2005)
17. Dendukuri, D., Gu, S., Pregibon, D., Hatton, T., Doyle, P.: Stop-flow lithography in a microfluidic device. Lab Chip **7**, 818 (2007)
18. Rehor, I., van Vreeswijk, S., Vermonden, T., Hennink, W., Kegel, W., Eral, H.: Biodegradable sensors: biodegradable microparticles for simultaneous detection of counterfeit and deteriorated edible products. Small **13**(39), 1701804 (2017)
19. Rossum, G., Drake, F.: Python 3. SohoBooks, United States (2009)
20. Bradski, G., Kaehler, A.: Learning OpenCV. O'Reilly Media Incorporated, Cambridge (2016)
21. Summerfield, M.: Rapid GUI programming with Python and Qt. Prentice Hall, Upper Saddle River (2012)

Growth Models of Female Dairy Cattle

Jaroslav Marek(✉), Alena Pozdílková, and Libor Kupka

University of Pardubice, Studentská 95, Pardubice, Czech Republic
jaroslav.marek@upce.cz

Abstract. Different methods of representing animal growth are possible and are defined for different animal categories. In this paper, weight measuring of female dairy cattle will be modelled by several nonlinear models. The most commonly used methods for describing the growth of animals are: Gompertz function, logistic function, Schmalhausen function, Brody function, Weibull function, Wood function and Von Bertalanffy function. Measured weight values and estimated parameters of growth curves will be analyzed using regression analysis methods. We will work with the weight measurements of 10 calves under 25 months of age from cowsheds in village Záluží in the Czech Republic. A comparison of several growth curves will be done. The suitability of individual models will be evaluated not only by the index of determination, but also by the intrinsic curvature according to Bates and Watts. This curvature affects the size of the linearization areas in which initial solution will ensure convergence of nonlinear regression.

Keywords: Growth models · Animal growth · Nonlinear regression · Bates and Watts curvature

1 Introduction

1.1 Problem of Growth Modelling

This contribution deals with problem of growth modelling, when the need arises to choose a suitable nonlinear function for approximation of the growth curve. The booming importance of the fitting of a growth curve is observed in a large number of studies on this topic. Our paper serves as a survey of growth models designed since the 19th century. That is why we also work with original old sources. We would like to highlight an article that covers all models. It's an article [5]. An overview of the functions is also in the paper [16].

If the approximation function is nonlinear in parameters, then linearization is used so that the problem can be posed as a linear one, and a well-known apparatus of linear statistical models is used. However, past papers are not devoted to providing a view of examining the dependence of quality approximation and the curvature of regression function. Various functions with a known analytical form for fitting of the growth curve were formed, but research on the issue of the Bates and Watts curvature was not carried out yet in any article. In parameter estimation for nonlinear regression models we need to

Á. Herrero et al. (Eds.): SOCO 2020, AISC 1268, pp. 266–275, 2021.
https://doi.org/10.1007/978-3-030-57802-2_26

know the initial values of unknown parameters. Thus, we must know whether uncertainty in the initial solution is essential in estimation, or whether it can be neglected. If a nonlinear regression model is linearized in a nonsufficient small neighborhood of the true parameter, then all statistical inferences may be deteriorated.

The subject of our research will be to study the suitability of application of growth models. So construction of a linearization domain for all models will be the main subject of investigation in this paper. If the linearization region is large, there are no problems in the calculation with the initial solution and in the quality of estimation of unknown parameters of the regression function.

2 Growth Models

Of course, the growth curves vary from crop to crop or from animal to animal. To describe the growth of a particular individual, it is necessary to firstly test a suitable model, unless there has been a general consensus in the choice of the model. Our goal will be to find the function best suited to describe the growth of cattle.

2.1 Functions Overview

For growth description, various models have been developed and tested in the past, since the beginning of the 19th century. A detailed analysis of the used functions is performed in [5]. Some models have been rediscovered and published by several scientists under different names. Other models (e.g. Feller, Weiss and Kavanau, Fitzhugh, Laird, Parks) are listed in the book [9].

1) Gompertz Function (1825)

It is first published in [4] as a mortality curve. It is used as a growth curve in [15]

$$y = \beta_1 e^{-\beta_2 e^{-\beta_3 x}}. \tag{1}$$

2) Logistic Function (1845)

Published by Verhulst in 1845 [14] later many times "rediscovered" by other authors.

$$y = \beta_1/(1 + \beta_2 e^{-\beta_3 x}). \tag{2}$$

3) Schmalhausen Function (1927)

Probably the third oldest growth model, published in 1927 [11]. Cf. [8, 12]

$$y = \beta_1 x^{\beta_2}. \tag{3}$$

4) Brody Function (1945)

This function was first published in 1945 [3]. It is also sometimes referred to as [13, 16] Monomolecular (and elsewhere, e.g. [5]) and also as the Mitscherlich function.

$$y = \beta_1\left(1 - e^{-\beta_2(x-x_0)}\right), \tag{4}$$

and in equivalent form

$$y = \beta_1\left(1 - \beta_2 e^{-\beta_3 x}\right). \tag{5}$$

5) Weibull Function (1951)

It is published in the form

$$y = \beta_1(1 - e^{-\beta_2 x^{\beta_3}}). \tag{6}$$

6) Wood Function (1967)

Very popular model with the equation

$$y = \beta_1 x^{\beta_2} e^{-\beta_3 x}. \tag{7}$$

7) Von Bertalanffy Function (1957)

In the original article from 1957 [2] the function is presented in the form

$$y = \beta_1\left(1 - e^{-\beta_2(x-x_0)}\right). \tag{8}$$

In later publications [13] it is presented in the form of another formula

$$y = \beta_1\left(1 - \beta_2 e^{-\beta_3 x}\right), \tag{9}$$

or [16] or [5],

$$y = \beta_1\left(1 - \beta_2 e^{-\beta_3 x}\right)^3. \tag{10}$$

Remark: Both Brody (m = 1) and von Bertalanffy (m = 3) functions are referred to as a special case of the Richards function [10]

$$y = \beta_1\left(1 - \beta_2 e^{-3x}\right)^m.$$

3 Linearization of Nonlinear Model and Curvature of Nonlinear Model

3.1 Nonlinear Model, Linearization and Estimates

All nonlinear models f try to explain the dependence of variable Y (grow characteristics) on variable x (time from the time origin):

$$Y = f(\boldsymbol{\beta}, x) + \varepsilon. \tag{11}$$

where $\boldsymbol{\beta} = (\beta_1, \beta_2, \ldots, \beta_k)'$ is unknown vector parameter.

Measurements of grow variable Y from dataset of monthly measurement are at our disposal. Our problem actually lies in estimating the values of the vector parameter $\boldsymbol{\beta}$ based on nonlinear regression. Cf. [6].

Estimates of unknown parameters can be computed by the method of nonlinear regression. We can estimate the values of unknown parameters occurring in studied nonlinear models by linearization of the model and by the ordinary least squares method.

The criterion for estimation is minimization of functional

$$S_e = \sum_{i=1}^{n} \left(Y_i - \hat{Y}_i\right)^2, \hat{Y}_i = f\left(\hat{\boldsymbol{\beta}}, x_i\right). \tag{12}$$

If we carry out a Taylor expansion of $f(\boldsymbol{\beta}, x)$ about point $\boldsymbol{\beta}_0$, where $\boldsymbol{\beta}_0$ is a vector of suitable initial values, we can transform the nonlinear model (11) to a linear model

$$\boldsymbol{Y} - \boldsymbol{Y}_0 \sim N[\boldsymbol{F\beta}, \boldsymbol{\Sigma}], \tag{13}$$

where $\boldsymbol{Y}_0 = f\left(\boldsymbol{\beta}_0, \boldsymbol{x}\right)$ and $\boldsymbol{F} = \frac{\partial f(\boldsymbol{\beta}, \boldsymbol{x})}{\partial \boldsymbol{\beta}'}$.

Notation $\boldsymbol{Y} \sim N[\boldsymbol{F\beta}, \boldsymbol{\Sigma}]$ means that observation vector $\boldsymbol{Y}$ (with elements $Y_1, \ldots, Y_n$) has (symbol $\sim$) multinomial normal distribution with a mean value $\boldsymbol{F\beta}$ and covariance matrix $\boldsymbol{\Sigma}$.

For example, in model (3) the i-th row of matrix $\boldsymbol{F}$ takes the form of

$$\boldsymbol{F}_i = \frac{\partial f(\boldsymbol{\beta}_0)}{\partial \boldsymbol{\beta}'} = \left(\frac{\partial f}{\partial \beta_1}, \frac{\partial f}{\partial \beta_2}\right)\Bigg|_{\beta_0} = \left(x_i^{\beta_2^0}, \beta_1^0 x_i^{\beta_2^0} \ln x_i\right).$$

The elements of this matrix can be easily computed through deriving functions f described in (1–10).

In our linearized model $\boldsymbol{Y} - \boldsymbol{Y}_0 = \boldsymbol{F}\left(\boldsymbol{\beta} - \boldsymbol{\beta}_0\right)$ is a correction $\delta\hat{\boldsymbol{\beta}}$ of initial vector $\boldsymbol{\beta}_0$ in a form

$$\delta\hat{\boldsymbol{\beta}} = \left(\boldsymbol{F}'\boldsymbol{\Sigma}^{-1}\boldsymbol{F}'\right)^{-1}\boldsymbol{F}'\boldsymbol{\Sigma}^{-1}(\boldsymbol{Y} - \boldsymbol{Y}_0), \hat{\boldsymbol{\beta}} = \boldsymbol{\beta}_0 + \delta\hat{\boldsymbol{\beta}}. \tag{14}$$

The covariance matrix of the $\hat{\boldsymbol{\beta}}$ estimator is given by

$$var\left(\hat{\boldsymbol{\beta}}\right) = \left(\boldsymbol{F}'\boldsymbol{\Sigma}^{-1}\boldsymbol{F}\right)^{-1}. \tag{15}$$

We can now place the estimate as a new initial vector. The iterative process is continued until fulfillment of the stopping criterion.

The special issue in our calculations is the choice of initial estimate. Initial values may be gained by information in fitting a similar growth curve, or a using values suggested as "about right" by the experimenter, based on past experience and knowledge.

The linearization method has possible drawbacks: the sum of squares may not converge for all cows. So, the sum of squares may oscillate or increase without bound. It is known, that if the model contains strong nonlinearity, this will cause impossibility of linearization and bad statistical properties of estimates. In this context, linearization regions are constructed, cf. [6].

The measure of nonlinearity is described by several characteristics. The intrinsic curvature is a key tool in nonlinear regression analysis [1].

Given a real-valued function $f(\boldsymbol{\beta}, x)$, Bates and Watts intrinsic curvature at point $\boldsymbol{\beta}_0$ is

$$C^{(int)}\left(\boldsymbol{\beta}_0\right) = \sup\left\{\frac{\sqrt{\kappa'(\delta\boldsymbol{\beta})\Sigma^{-1}\boldsymbol{M}_F^{\Sigma^{-1}}\kappa(\delta\boldsymbol{\beta})}}{\delta\boldsymbol{\beta}'C\delta\boldsymbol{\beta}} : \delta\boldsymbol{\beta} \in \boldsymbol{R}^k\right\}, \boldsymbol{C} = \boldsymbol{F}'\boldsymbol{\Sigma}^{-1}\boldsymbol{F}. \tag{16}$$

The projection matrices are given by formulas $\boldsymbol{P}_F = \left(\boldsymbol{F}'\boldsymbol{\Sigma}^{-1}\boldsymbol{F}\right)^{-1}\boldsymbol{\Sigma}^{-1}\boldsymbol{F}$ and $\boldsymbol{M}_F^{\Sigma^{-1}} = \boldsymbol{I} - \boldsymbol{P}_F$.

Functional $\kappa(\delta\boldsymbol{\beta})$ is intended by

$$\kappa(\delta\boldsymbol{\beta}) = \left(\boldsymbol{H}_1'\boldsymbol{\Sigma}^{-1}\boldsymbol{H}_1, \ldots, \boldsymbol{H}_n'\boldsymbol{\Sigma}^{-1}\boldsymbol{H}_n\right). \tag{17}$$

So it is necessary to prepare the matrices of second partial derivatives $\boldsymbol{H}_1, \ldots, \boldsymbol{H}_n$, to which we will gradually substitute individual observations.

For example, in model (3) matrix $\boldsymbol{H}_i$ take the form of

$$\boldsymbol{H}_i = \frac{\partial^2 f(\boldsymbol{\beta}_0)}{\partial\boldsymbol{\beta}\partial\boldsymbol{\beta}'} = \left(\frac{\partial^2 f}{\partial\beta_1\partial\beta_2}\right)\Bigg|_{\beta_0} = \begin{pmatrix} 0, & x_i^{\beta_2^0}\ln x_i \\ x_i^{\beta_2^0}\ln x_i, & \beta_1^0 x_i^{\beta_2^0}\ln^2 x_i \end{pmatrix}.$$

In [6] a test of intrinsic linearity of model $H_0 : K^{(int)} = 0$ is considered. This test can be used to derive the following linearization criterion. Set O_b, observes bias of the linear estimator. If $\delta\boldsymbol{\beta}$ in O_b, where

$$O_b = \left\{\delta\boldsymbol{\beta} : \delta\boldsymbol{\beta}'\boldsymbol{F}'\Sigma^{-1}\boldsymbol{F}\delta\boldsymbol{\beta} < \frac{2\sqrt{\delta_{max}}}{C^{(int)}(\boldsymbol{\beta}_0)}\right\}, \tag{18}$$

then the bias of special function $\boldsymbol{h}'\beta, \boldsymbol{h} \in |\boldsymbol{R}^k$, is smaller then ($\varepsilon$ is chosen by a user).

$$\varepsilon \cdot \sqrt{\boldsymbol{h}'\boldsymbol{C}^{-1}\boldsymbol{h}}.$$

If the intrinsic curvature of the nonlinear regression model is too big, then the situation may arise that the model cannot be linearized. To assess the possibility of linearization, the confidence domain is rendered; that is, it is compared with the confidence domain.

An algorithm published by Kubáček can be used for calculation of $C^{(int)}$, cf. [6], Remark 5.1. In the first step, we choose an arbitrary vector $\delta\boldsymbol{u}_1 \in \boldsymbol{R}^k$, such that $\delta\boldsymbol{u}_1'\delta\boldsymbol{u}_1 = 1$. After that, we determine the vector δs defined as

$$\delta s = \left(\boldsymbol{F}'\Sigma^{-1}\boldsymbol{F}\right)^{-1}(\boldsymbol{H}_1\delta\boldsymbol{u}_1, \boldsymbol{H}_2\delta\boldsymbol{u}_1, \ldots, \boldsymbol{H}_n\delta\boldsymbol{u}_1)\Sigma^{-1}\boldsymbol{M}_F^{\Sigma^{-1}}\kappa(\delta\boldsymbol{u}_1). \tag{19}$$

Then, we identify the vector $\delta\boldsymbol{u}_2 = \frac{\delta s}{\sqrt{\delta s'\delta s}}$. In the last step, we verify the inequality given as $\delta\boldsymbol{u}_2'\delta\boldsymbol{u}_2 \geq 1 - tol$, where tol is a sufficiently small positive number. If the stopping criterion is satisfied, we terminate the iterative process and intrinsic curvature is given after substitution $\delta\boldsymbol{\beta} = \delta\boldsymbol{u}_2$ into (16). If the inequality is not satisfied, we return to the first step of the algorithm where we update the vector $\delta\boldsymbol{u}$ by $\delta\boldsymbol{u}_2$.

If the true value of parameter $\boldsymbol{\beta}$ lies in the linearization set, the nonlinear model can be replaced by a linear model. Often it is contemplated that linearization can be used, if the confidence domain is covered with a linearization domain.

The confidence domain (see [7]) for the parameter $\boldsymbol{\beta}$ is a set in parametric space of $\boldsymbol{\beta}$, which covers the true value of $\boldsymbol{\beta}$ with a given probability $1 - \alpha$.

The formula for $(1-\alpha)\%-$ confidence domain is given by

$$\varepsilon_{1-\alpha}(\boldsymbol{\beta}) = \left\{\boldsymbol{u} : \boldsymbol{u} \in \boldsymbol{\Theta}_{\boldsymbol{\beta}}, \left(\boldsymbol{u} - \hat{\boldsymbol{\beta}}\right)\left[var\left(\hat{\boldsymbol{\beta}}\right)\right]^{-1}\left(\boldsymbol{u} - \hat{\boldsymbol{\beta}}\right)' \leq \chi_k^2(1-\alpha)\right\}. \quad (20)$$

The symbol $\chi_k^2(1-\alpha)$ denotes the $(1-\alpha)-$ quantile of an χ^2 distribution with k degrees of freedom.

4 Numerical Study

4.1 Data Processing

In today's cowshed, processes of growth measurement are performed only once a month. Therefore, we need an approximation of the growth curve. Based on this approximation. The estimate of total growth in a given day may be obtained. The study will be conducted for 10 selected cows. The corresponding pairs of observations of three cows are given in Table 1. Notice that the numbers of daily measurements are different.

Table 1. Data.

	Number 01/9623, day of birth 1.10.2017								
Days from birth	0	120	211	365	393	421	449	477	729
Weight	40	210	333	506	538	592	628	648	702
	Number 02/9624, day of birth 1.10.2017								
Days from birth	0	120	211	365	393	421	449	477	729
Weight	37	184	304	487	520	580	630	651	712
	Number 03/9625, day of birth 4.10.2017								
Days from birth	0	117	208	362	390	418	446	474	726
Weight	40	193	317	516	548	598	644	662	716

On these data we will present the numerical and graphical results of estimation and we will analyze linearization features of all models.

In the next figures, approximations of the growth curves of 1st cow are presented. In Fig. 1 are functions whose graph does not pass the origin of the coordinate system, i.e. on day 0 the value of these functions is generally non-zero. The functions shown in Fig. 2 have a graph that always passes through the origin of the coordinate system, i.e. on day 0 the value of these functions is zero. The indexes of determination varied approximately by 0.98 in all models. Figure 3 analyzes the residue behavior for all cows and functions using boxplots. It can be seen that the smallest quartile range has residues in the Schmalhausen function.

Figures 3, 4, 5, 6 and 7 show a domain of linearization and confidence for all used functions: Gompertz, Logistic, Schmalhausen, Brody, Weibull, Wood and Von Bertalanffy function. It can be seen, that the graphs have different scales on the x and y axes for different models, some graphs are on a much larger scale than others.

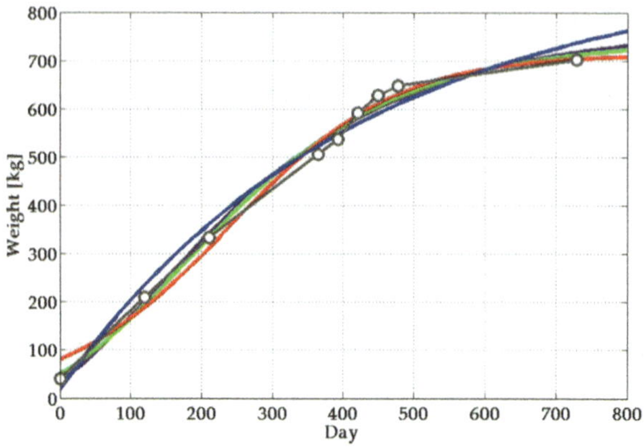

Fig. 1. Gompertz (green), Logistic (red), Brody (blue), Bertalanffy (purple) function: Cow No.1.

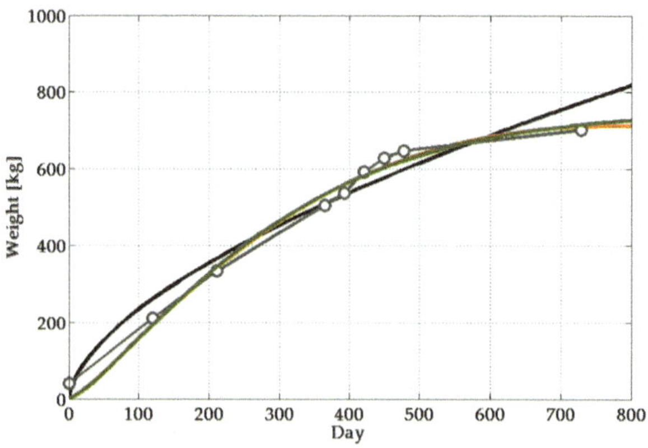

Fig. 2. Schmalhausen (black), Weibull (olive), and Wood (orange) function: Cow No.1.

The linearization is possible even in the case that we can provide an initial solution lying in this domain. The linearization region of all models is large in comparison with the confidence ellipse. The nonlinear model can be linearized in all situations, where we can choose an initial solution from the linearization domain. In practice, a small linearization domain brings biased estimates.

To compare the models, we selected 4 criteria (c_1: determination index, c_2: linearization area size, c_3: quartile range, c_4: intersection with origin). We ranked the models in descending order from 7 to 1 for c_1, c_2, and c_3. For criterion 4, we gave one point if the function did not pass the origin.

The largest indexes of determination reached Gompertz function and Von Bertalanffy function. From the boxplot can be seen that zero mean values were achieved for functions Logistic, Schmalhausen, and Weibull function, Weibull function and Wood function go through the beginning. If we sort the functions by the size of the linearization area, we get the following order: Logistic function, Brody function, Schmalhausen function, Von Bertalanffy function, Wood function, Gompertz function and Weibull function.

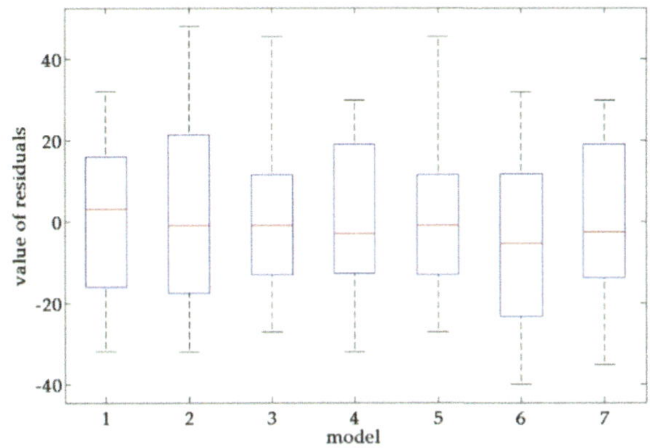

Fig. 3. Graph of residuals for all 10 cows and 7 models

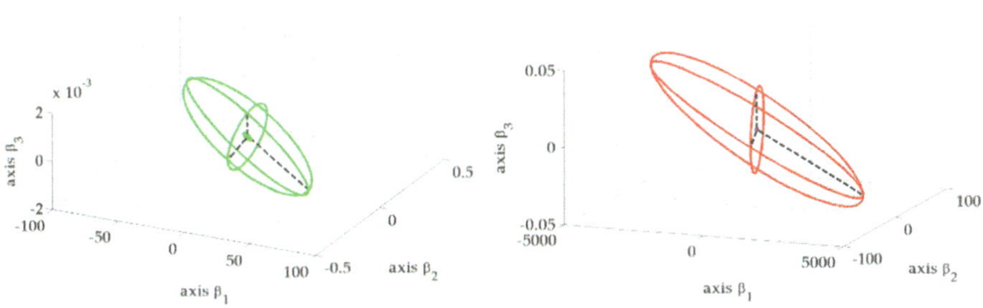

Fig. 4. Linearization and confidence domain: Gompertz function and Logistic function

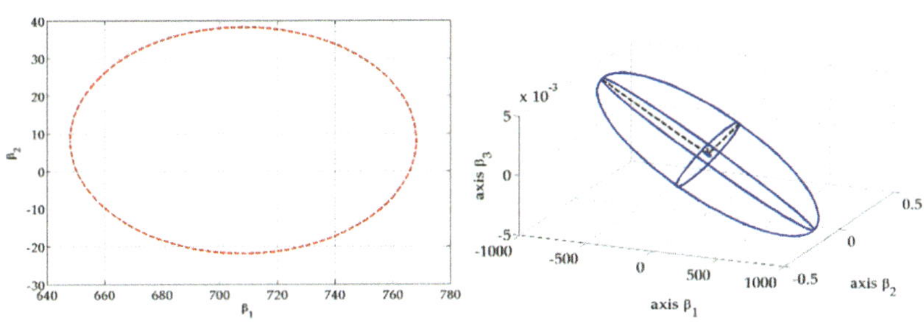

Fig. 5. Linearization and confidence domain: Schmalhausen function and Brody function

Considering Table 2, we conclude that logistic function and Von Bertalanffy (both with the sum of 16 points) are the most appropriate.

5 Conclusion Remark

A linearization of nonlinear functions causes an uncertainty in an estimation of unknown parameters of the regression model. Various growth models are differently sensitive to

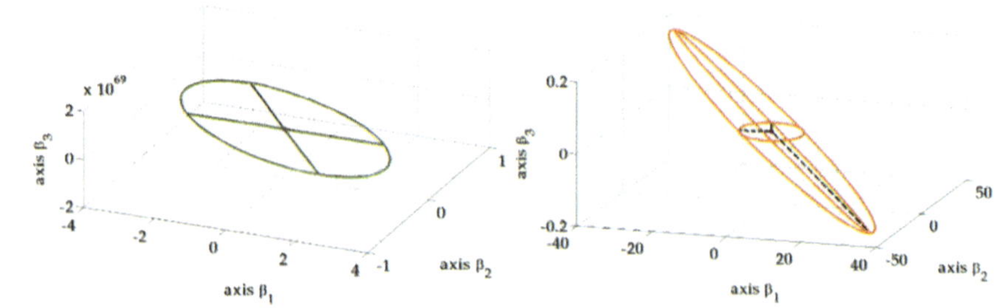

Fig. 6. Linearization and confidence domain: Weibull function and Wood function

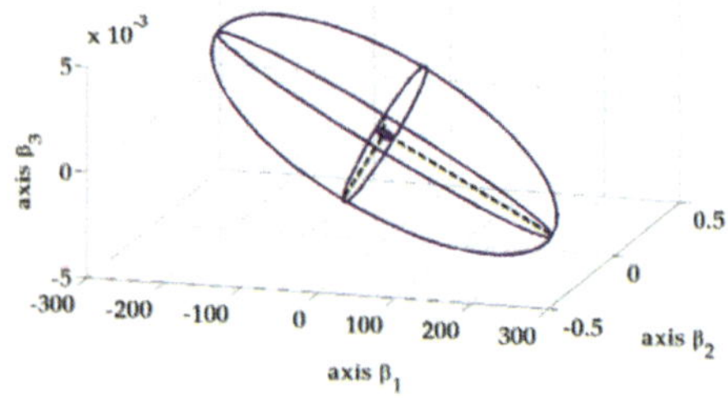

Fig. 7. Linearization and confidence domain: Von Bertalanffy function

Table 2. Quality of fitting.

Model	Estimates of parameters	Average I	c_1	c_2	c_3	c_4	$\sum_{i=1}^{4} c_i$
Gompertz	(736.7991, 2.7475, 0.0058)	0.9963	7	2	3	1	13
Logistic	(708.0358, 8.2764, 0.0088)	0.9952	5	3	7	1	16
Schmalhausen	(14.3594, 0.6071)	0.9764	1	7	4	0	12
Brody	(889.3802, 0.9824, 0.0024)	0.9898	2	5	6	1	14
Weibull	(738.5050, 0.0005, 1.3427)	0.9937	3	6	1	0	10
Wood	(0.4315, 1.3376, 0.0018)	0.9945	4	1	2	0	7
Von Bertalanffy	(758.0516, 0.6472, 0.0048)	0.9957	6	4	5	1	16

the quality of the initial solution. On the basis of the Bates Watts curvature, the best models for approximation of the growth curve are logistic, Brody and Von Bertalanffy models. Great care is also necessary for their use. If the initial solution does not lie in the (very small!) linearization domain, then uncertainty in the initial solution is essential in estimation, and it leads to a completely wrong estimate of the growth curve. This fact causes a large proportion of not fitted growth curves in previous studies.

Acknowledgement. This contribution has been supported by institutional support of the University of Pardubice, Czech Republic.

References

1. Bates, D.M., Watts, D.G.: Relative curvature measures of nonlinearity. J. Roy. Stat. Soc. B **42**, 1–25 (1980)
2. Von Bertalanffy, L.: Quantitative laws in metabolism and growth. Q. Rev. Biol. **32**(3), 217–231 (1957)
3. Brody, S.: Bioenergetics and growth: with special reference to the efficiency complex. In: Domestic Animals. Reinhold Publishing Corp., New York (1945)
4. Gompertz, B.: On the nature of the function expressive of the law of human mortality, and on a new mode of determining the value of life contingencies. Phil. Trans. R. Soc. Lond. **115**, 513–583 (1825)
5. Koya, P., Goshu, A.: Generalized mathematical model for biological growths. Open J. Model. Simul. **1**, 42–53 (2013)
6. Kubáček, L.: On a linearization of regression models. Appl. Math. **40**(1), 61–78 (1995)
7. Kubáčková, L.: Joint confidence and threshold ellipsoids in regression models. Tatra Mt. **7**, 157–160 (1996)
8. Martyushev, L.M., Terentiev, P.S.: Universal Model of Ontogenetic Growth: Substantiation and Development of Schmalhausen's Model (2014). https://arxiv.org/abs/1404.4318
9. Parks, J.R.: A theory of feeding and growth of animals. In: Advanced Series in Agricultural Sciences. Series, vol. 11. Springer, Heidelberg (1982)
10. Richards, F.J.: A flexible growth function for empirical use. J. Exp. Bot. **10**, 290–300 (1959)
11. Schmalhausen, I.: Beiträge zur quantitativen Analyse der Formbildung. II. Das Problem des proportionalen Wachstums. Roux' Archive für Entwicklungsmechanik der Organismen **110**(1), 33–62 (1927)
12. Schmalhausen, I.: Das Wachstumsgesetz und die Methode der Bestimmung der Wachstumskonstante. W. Roux' Archiv f. Entwicklungsmechanik **113**(3), 462–519 (1928)
13. Ünal, D., Yeldan, H., Gül, E., Ergüç, N.D., Adiyan, M.: Gompertz, logistic and brody functions to model the growth of fish species Siganus rivulatus. Acta Biologica Turcica **30**(4), 140–145 (2017)
14. Verhulst, P.-F.: Recherches mathématiques sur la loi d'accroissement de la population" [Mathematical Researches into the Law of Population Growth Increase]. Nouveaux Mémoires de l'Académie Royale des Sciences et Belles-Lettres de Bruxelles (1845)
15. Winsor, C.: The Gompertz curve as a growth curve. Proc. Natl. Acad. Sci. U.S.A. **18**(1), 1–8 (1932)
16. Zeide, B.: Analysis of growth equations. Forest Sci. **39**, 594–616 (1993)

A Preliminary Study on Crop Classification with Unsupervised Algorithms for Time Series on Images with Olive Trees and Cereal Crops

Antonio Jesús Rivera[1(✉)], María Dolores Pérez-Godoy[1], David Elizondo[2], Lipika Deka[2], and María José del Jesus[1]

[1] University of Jaén, Jaén, Spain
{arivera,lperez,mjjesus}@ujaen.es
[2] De Montfort University, Leicester, England
{elizondo,lipika.deka}@dmu.ac.uk

Abstract. Satellite imagery has been consolidated as an accurate option to monitor or classify crops. This is due to the continuous increase in spatial-temporal resolution and the availability of free access to this kind of services. In order to generate crop type maps (a valuable preprocessing step to most remote agriculture monitoring application), time series are built from remote sensing images, and supervised techniques are widely used to classify them. However, one of the main drawbacks of these methods is the lack of labelled data sets to carry out the training process. Unsupervised classification has been less frequently used in this research field. The paper presents an experimental study comparing traditional clustering algorithms (with different dissimilarity measures) for the classification of olive trees and cereal crops from time series remote sensing data. The results obtained provide crucial information for developing novel and more accurate crop mapping algorithms.

Keywords: Unsupervised learning · Clustering · Crop mappings · Time series classification · Satellite imagery

1 Introduction

The increase in the world's population and the effects of global warming have attracted interest in new trends that can transform agricultural practices. These trends often involve [10] a close monitoring of crop lands with the aim of testing agricultural parameters, evaluating the impact of changing policies, predicting how climate change influences the harvest or forecasting crop yields.

Remote sensing satellite data [5] is one of the main sources used in the agricultural data science field thanks to the continuous increase in spatial-temporal resolution or the availability of free access to this kind of service. Satellite con-

Á. Herrero et al. (Eds.): SOCO 2020, AISC 1268, pp. 276–285, 2021.
https://doi.org/10.1007/978-3-030-57802-2_27

stellations such as MODIS, LandSat or Sentinel are typically employed in the knowledge extraction process. A key characteristic of these missions is that they obtain multi-spectral data/images with bands in the visible, near infrared, and short wave infrared part of the spectrum. These bands can be directly used to monitor crops or can be combined in order to obtain new indices that are commonly known as Vegetation Indices (VIs) [11]. The aim of these indices is to comprise the electromagnetic wave reflectance information captured from satellite sensors. This reflectance value of light spectral changes according to the plant type, the chlorophyll level and the water content within tissues, among other factors. This reflectance also changes during the growth cycle of the plant. Therefore, a time series per pixel can be obtained and analyzed in order to apply the corresponding method. Examples of these VIs include [11]: Ratio Vegetation Index (RVI), Difference Vegetation Index (DVI), Perpendicular Vegetation Index (PVI), Green Vegetation Index (GVI) or Normalized Difference Vegetation Index (NDVI). The latter is one of the most widely used indices and therefore has been chosen for the experimentation in this study.

Machine learning methods [3,6,9,10] have been successfully applied to remote sense imagery, allowing for large-scale automated analysis in both monitoring and crop identification. The methodology of these techniques consists of building time series from satellite images. Although supervised methods have been extensively used to classify these series, it can be difficult to find correctly labelled data sets (or crop type maps) to train models with. On the other hand, unsupervised methods remain relatively unexplored [10]. These methods present disadvantages such as their vulnerability to some traits of the data (outliers, high dimensionality, data inaccuracy, etc.) or may require initialization data that is not available. They have, however, a highlighted advantage in the research field: unsupervised methods can create crop type maps without the need for previously labelled data.

The present study tests clustering (i.e. unsupervised) methods for time series in order to create crop type maps on areas cultivated with olive trees and cereals. More specifically, the results of the well-known K-means and Hierarchical clustering algorithm are analyzed. In addition to the typical Euclidean distance, a more time series specialized distance, the Dynamic Time Warping, is studied.

This paper is organized as follows: Sect. 2 describes the clustering algorithms and dissimilarity measures tested. Section 3 shows the methodology, experimentation and results obtained, and, an analysis of these results is carried out. Finally, some conclusions are outlined.

2 Clustering Algorithms

Clustering is an unsupervised technique that allows to group data. Clusters are established on the basis of a similarity or distance measure. The elements within a cluster are more similar to each other than to the elements in another cluster. In this paper two clustering algorithms, K-means and Hierarchical, are tested to create crop type maps. These algorithms are described as follows.

The K-means algorithm is one of the most well-known clustering algorithms [4]. Each cluster is identified by means of a central point, called a centroid. The algorithm groups the elements so that the sum of distances between each element and the centroid of its group is minimized. The process can be summarized in the following steps:

1. The algorithm starts with a number k of centroids that are selected randomly.
2. Each data sample is assigned to the cluster with the closest centroid.
3. The centroids of each cluster are recalculated using the average of the data samples in the cluster.
4. Steps 2 and 3 are repeated until the changes to the centroids are stabilized.

The Hierarchical algorithm [2] creates a cluster hierarchy that can be represented using a tree structure, known as a dendogram. The root of the tree is a single cluster containing all the elements and the leaves represent single-element clusters. These algorithms can be classified as either agglomerative (bottom-up) or divisive (top-down). The agglomerative algorithm takes the leaves of the tree and then, in each iteration, the two nearest clusters are grouped together. In the divisive algorithm, the tree root is divided into two clusters. This division is continued in each iteration until the leaves are reached. The hierarchical agglomerative method [2] is the most widely used and therefore has been chosen for the experimentation of this paper. The main steps are as follows:

1. Each data sample is considered a cluster.
2. In each iteration, the two closest clusters are combined into one.
3. Step 2 is repeated until all data samples are contained with in a single cluster.

As mentioned above, in any case the construction of the clusters is based on a dissimilarity or distance measure [1] between the data samples. In this study, clustering algorithms have been tested using the following two distances: the typical Euclidean distance and a more specific time series distance such as Dynamic Time Warping [7].

The Euclidean distance is the most widely used distance measure used. The Euclidean distance of two vectors, X and Y in $\mathbb{R}^M$ is defined in the Eq. 1.

$$d(X,Y) = \| X - Y \| = \sqrt{\sum_{i=1}^{M}(X_i - Y_i)^2} \tag{1}$$

The Dynamic Time Warping (DTW) distance allows the dissimilarity between two sequences that can vary in time or space to be measured [7]. A distance matrix is constructed from the two series $X \in \mathbb{R}^M$ and $Y \in \mathbb{R}^N$. This matrix, DM, calculates the Euclidean distance between each pair of elements in the series.

When considering DM matrix, a *warping path* (wp) can be defined as follows (Eq. 2):

$$\begin{aligned} wp = (p_1, p_2, p_L) \text{ with } p_r = DM(i,j) \in [1, M]\text{x}[1, N], \\ \forall r \in [1, L] \text{ with } \max(M, N) \leq L \leq M + N - 1 \end{aligned} \tag{2}$$

The DTW distance, showed in Eq. 3, is calculated as the minimum warping path of contiguous elements in DM.

$$DTW(X,Y) = min\left(\sqrt{\sum_{r=1}^{L} wp_r}\right) \quad (3)$$

The definition of this warping path is completed with constrains such as continuity or monotonicity.

3 Methodology, Experimentation and Results

For this preliminary analysis, the study area is located in the north of the province of Jaén (Spain). It is, more specifically, a rectangle where the top, left coordinates are (long,lat) = (–3.12739983, 37.99600858) and the bottom, right are (long,lat) = (–3.12104593, 37.99122638). As can be seen from the RGB image obtained from Google Maps in Fig. 1 (left), it is an area composed of olive trees, with different densities, as well as of cereal crops.

The images/data used to carry out unsupervised classification were downloaded from Copernicus Open Access Hub (https://scihub.copernicus.eu/), which provides completely free and open access to different Sentinel missions. Sentinel 2 mission was chosen for the experimentation in this paper. Sentinel 2 is a constellation of two polar-orbiting satellites with the aims of monitoring the land surface. For this mission the revisit time is 10 days at the equator with one satellite, and 5 days with 2 satellites under cloud-free conditions. The spatial resolutions available are 10 m, 20 m and 60 m per pixel over land and coastal waters. A key characteristic of Sentinel 2 is that it returns multi-spectral data with 13 bands in the visible, near infrared, and short wave infrared part of the spectrum.

In this study, the downloaded images belong to the time period from April 2018 to March 2019. These images have a spatial resolution of 20 m that implies 783 pixels per image. From this set of images we have discarded those that show blurring due to cloud coverage within the area of interest. Thus a total of 22 images, distributed in the different months of the year, were considered. With the NIR (band 8) and RED (band 2) spectral bands, the NDVI [11] vegetation index, see Eq. 4, is calculated for each pixel of each image. This way, an NDVI value for each pixel and for each one of our 22 images was obtained. A time series, that contains chlorophyll changes, was built for each (x,y) NDVI pixel from the entire ordered set of images. Therefore, a total of 783 times series with a length of 22 values were obtained.

$$NDVI = \frac{(NIR - RED)}{(NIR + RED)} \quad (4)$$

In order to classify the crops in an area in an unsupervised way, the time series obtained will be classified using clustering algorithms. Specifically K-means and Hierarchical clustering algorithms are tested along with two distance measures

Table 1. Classification rate of clustering algorithms

Algorithm	DTW distance	Euclidean distance
Kmeans	0.9655	0.9553
Hierarchical	0.8212	0.8519

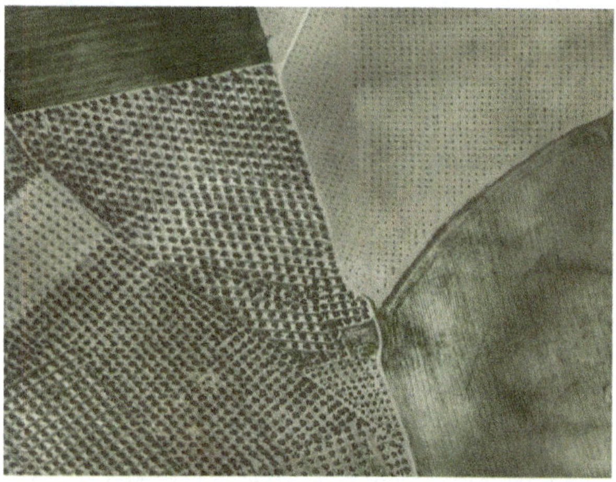

RGB image

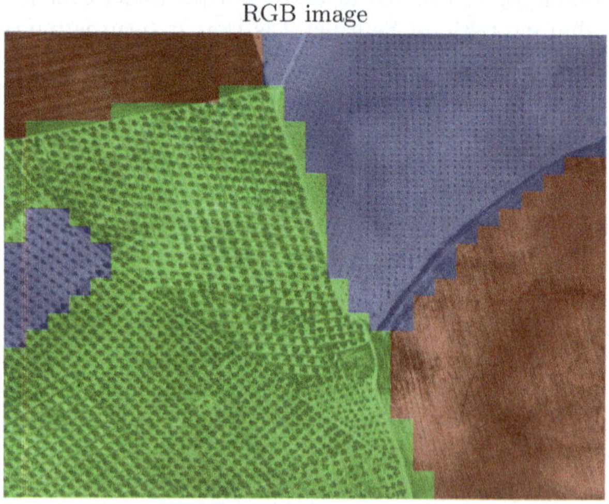

RGB image with color labels

Fig. 1. RGB image base and color labels. Green represents dense olive trees, blue sparse olive trees and red cereal crops.

for each algorithm: the traditional Euclidean distance and the DTW distance, a distance that is more specifically used for time series. The number of clusters was set to three, considering the characteristics of the study zone and the use of the Elbow method [8]. The packages tslearn (Python) and TSclust (CRAN) were respectively used for running the K-means and the Hierarchical clustering algorithms.

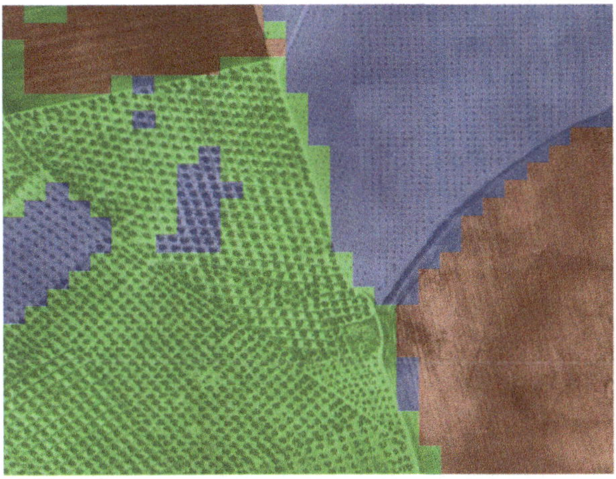

Predicted labels obtained using DTW distance

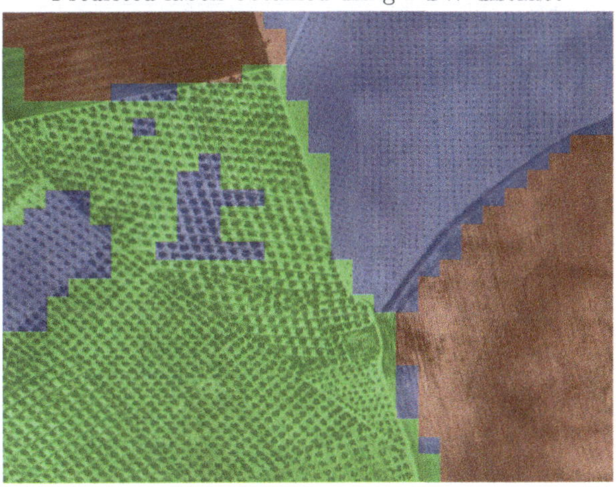

Predicted labels obtained using Euclidean distance

Fig. 2. Colored image with K-means clustering algorithm. Green represents dense olive trees, blue sparse olive trees and red cereal crops.

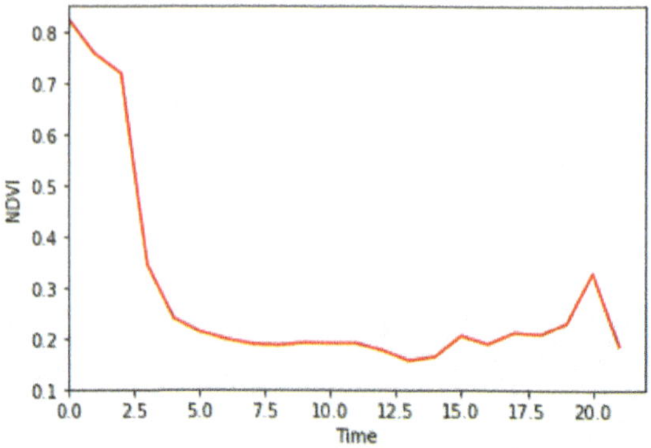

Cluster prototype for cereal crop

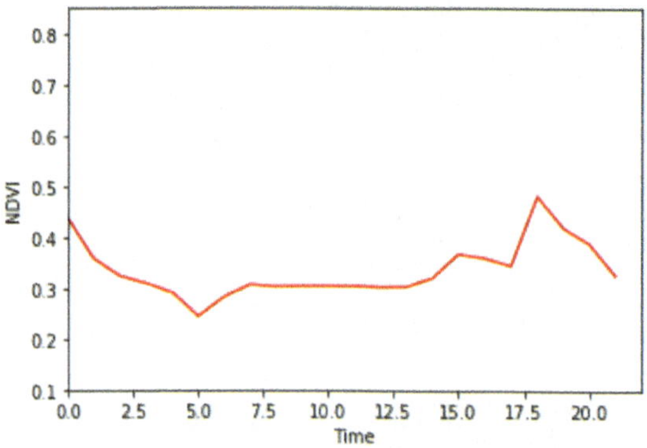

Cluster prototype for dense olive tree

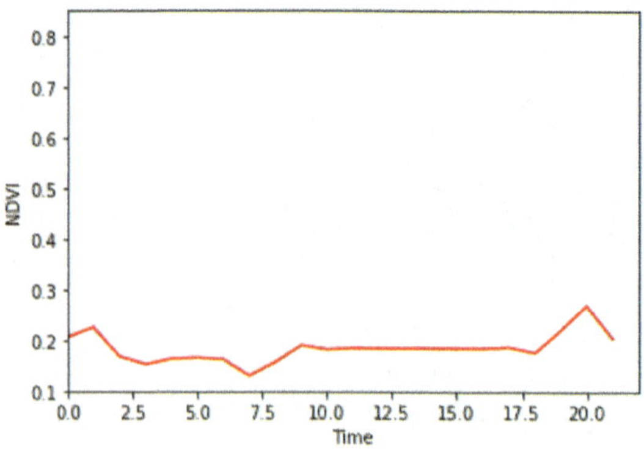

Cluster prototype for sparse olive tree

Fig. 3. Prototypes obtained by K-means algorithm with DTW distance.

To measure the efficiency of these algorithms, the classification rate (the ratio between correct predictions and total number of examples) quality measure was

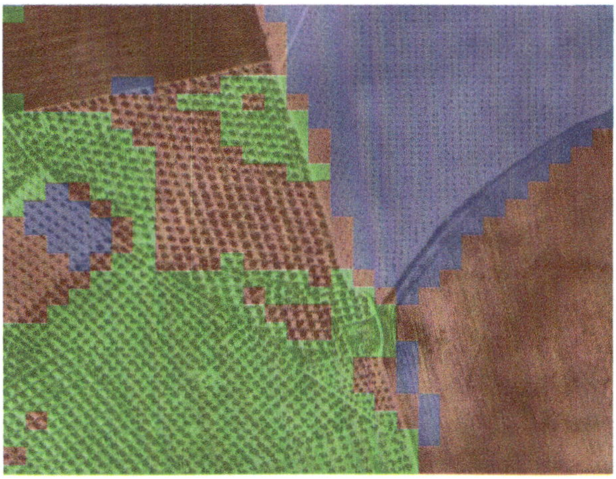

Predicted labels obtained using DTW distance

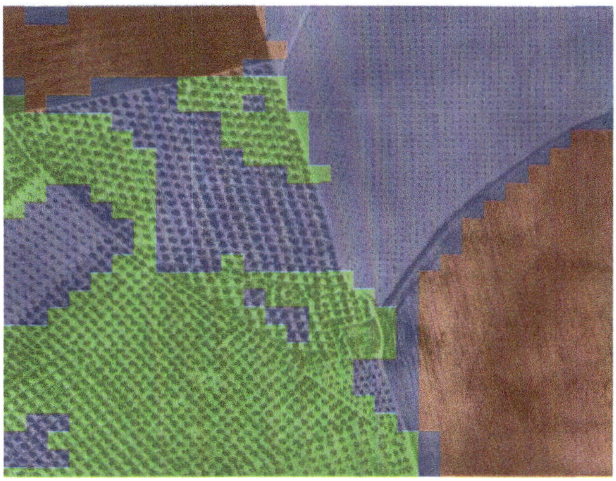

Predicted labels obtained using Euclidean distance

Fig. 4. Colored image with Hierarchical clustering algorithm. Green represents dense olive trees, blue sparse olive trees and red cereal crops.

used. The correct classification of the test area was labelled manually. These labels can be seen in Fig. 1 where green depicts dense olive trees, blue sparse olive trees and red cereal crops.

The results obtained for the four tested combinations are shown in Table 1. As can be seen, if the focus is in the clustering algorithm, the K-means clustering algorithm, which is the most classical clustering proposal, outperforms the Hierarchical clustering algorithm for any distance measure. Thus, the K-means operation, which consists of applying a whole procedure that iteratively adjusts the cluster prototype, works better than the Hierarchical operation, which consists in aggregating instances to clusters depending on the distances. Regarding the distances measures, the more specialized DTW time series distance outperforms the typical Euclidean distance only when it is used by the K-means clustering algorithm.

Next, the graphic results of the clustering algorithms are analyzed. The label predictions obtained for K-means are shown graphically in Fig. 2. As can be observed, both versions of the K-means carry out an accurate prediction of the true labels. The most problematic zone is located in the centre of the image where dense olive trees are classified as sparse olive trees. In any case this is an understandable mistake because, as the RGB image base shows, dense olive trees are a bit sparse in this zone. In order to explain the operating mode of the K-means algorithm with the DTW distance, its cluster prototypes are shown in Fig. 3. The cluster prototype for the cereal crop is on the right and shows high NDVI values in Spring. The cluster prototype for dense olive trees can be found in the centre of the image and for sparse olive trees is on the right. As can be seen, the NDVI shape is similar for both types of olive trees but higher NDVI values are obtained for dense olive trees than for sparse olive trees.

Finally, label predictions obtained for the Hierarchical algorithm are shown graphically in Fig. 4. In this case, both versions of these algorithms show more errors in their predicted labels, specifically in the central zone of the image. For the Hierarchical-DTW combination the aggregation operating mode works the worst and obtains confusing and mixed labels for the central zone of the image or for the cereal crop cultivated areas.

4 Conclusions

In this paper two clustering algorithms, K-means and a Hierarchical agglomerative version, are tested in order to generate crop type maps. Also, for each clustering algorithm two dissimilarity measures, Euclidean and DTW, are used. From the analysis of the results it can be concluded that K-means, with any distance measure, obtain better results than the Hierarchical clustering algorithm. Regarding the dissimilarity measures, the DTW measure only outperforms the Euclidean distances when the K-means algorithm is used. The analysis presented in this paper provides a valuable contribution as a pre-processing step for numerous agriculture monitoring applications. In the future, these techniques can be applied as a basis of different methodologies to monitor the growth of crops or to forecast their harvest.

Acknowledgments. This work is partially supported by the Spanish Ministry of Economy, Industry and Competitiveness under project PID2019-107793GB-I00.

References

1. Fakhrazari, A., Vakilzadian, H.: A survey on time series data mining, pp. 476–481 (2017)
2. Ferstl, F., Kanzler, M., Rautenhaus, M., Westermann, R.: Time-hierarchical clustering and visualization of weather forecast ensembles. IEEE Trans. Vis. Comput. Graph. **23**(1), 831–840 (2017)
3. Gonçalves, R.R.V., Zullo, J., Amaral, B.F., Coltri, P.P., Sousa, E.P.M., Romani, L.A.S.: Land use temporal analysis through clustering techniques on satellite image time series. In: 2014 IEEE Geoscience and Remote Sensing Symposium, pp. 2173–2176 (2014)
4. Hartigan, J.A., Wong, M.A.: Algorithm as 136: a k-means clustering algorithm. J. Roy. Stat. Soc. Ser. C (Appl. Stat.) **28**(1), 100–108 (1979)
5. Huang, Y., Chen, Z.X., Yu, T., Huang, X.Z., Gu, X.F.: Agricultural remote sensing big data: management and applications. J. Integr. Agric. **17**(9), 1915–1931 (2018)
6. Kamir, E., Waldner, F., Hochman, Z.: Estimating wheat yields in Australia using climate records, satellite image time series and machine learning methods. ISPRS J. Photogram. Rem. Sens. **160**, 124–135 (2020)
7. Keogh, E., Ratanamahatana, C.: Exact indexing of dynamic time warping. Knowl. Inf. Syst. **7**(3), 358–386 (2005)
8. Ketchen Jr., D., Shook, C.: The application of cluster analysis in strategic management research: an analysis and critique. Strateg. Manag. J. **17**(6), 441–458 (1996)
9. Singh, S., Ambegaokar, S., Champawat, K.S., Gupta, A., Sharma, S.: Time series analysis of clustering high dimensional data in precision agriculture. In: 2015 International Conference on Innovations in Information, Embedded and Communication Systems (ICIIECS), pp. 1–8 (2015)
10. Wang, S., Azzari, G., Lobell, D.B.: Crop type mapping without field-level labels: random forest transfer and unsupervised clustering techniques. Rem. Sens. Environ. **222**, 303–317 (2019)
11. Xue, J., Su, B.: Significant remote sensing vegetation indices: a review of developments and applications. J. Sens. **2017**, 1–17 (2017)

Special Session: Soft Computing Methods in Manufacturing and Management Systems

Blocks of Jobs for Solving Two-Machine Flow Shop Problem with Normal Distributed Processing Times

Wojciech Bożejko[1(✉)], Paweł Rajba[2], and Mieczysław Wodecki[3]

[1] Department of Control Systems and Mechatronics, Wrocław University of Science and Technology, Janiszewskiego 11/17, 50-372 Wrocław, Poland
wojciech.bozejko@pwr.edu.pl

[2] Institute of Computer Science, University of Wrocław, Joliot-Curie 15, 50-383 Wrocław, Poland
pawel@cs.uni.wroc.pl

[3] Department of Telecommunications and Teleinformatics, Wrocław University of Science and Technology, Janiszewskiego 11/17, 50-372 Wrocław, Poland
mieczyslaw.wodecki@pwr.edu.pl

Abstract. We consider strongly NP-hard problem of two-machine task scheduling with due dates and minimizing of the total weighted tardiness. Task execution times are random variables. We propose methods of intermediate review of solutions, the so-called 'block properties', which we use in the tabu search algorithm. From computational experiments carried out it follows that the use of blocks significantly speeds up calculations.

Keywords: Scheduling · Metaheuristics · Uncertain parameters

1 Introduction

In a two-machine flow shop problem with minimizing the sum of lateness costs (total tardiness, in short, F2T problem), each of the n tasks must be completed on the first machine and then on the second machine. The time of completing the tasks and the due dates (on the second machine) are given. Exceeding this due date will result in a penalty, which depends on the size of the delay (so called tardiness) and a fixed penalty factor (weight). The problem consists on determining the order of tasks (the same on both machines) which minimizes the sum of penalties. In the literature this problem is denoted by $F2||\sum w_i T_i$. It is a generalization of the NP-hard single-machine problem with the minimalization of sum of penalties for tardiness $1||\sum w_i T_i$ – a detailed description and algorithm of its solution is provided in the work of Bożejko et al. [4].

Á. Herrero et al. (Eds.): SOCO 2020, AISC 1268, pp. 289–298, 2021.
https://doi.org/10.1007/978-3-030-57802-2_28

There are relatively few papers devoted solely to the F2T problem and methods of solving it. Some theoretical results, as well as approximation algorithms are presented in the papers: Gupta and Harari [8], Lin [11], Bulfin and Hallah [7]. Various variants of this problem were also considered by Ahmadi et al. [10], Al-Salem et al. [1], Ardakan et al. [2] and Bank et al. [3]. Two-machine flow problem with $C_{\max}$ criterion (minimizing the end of execution time of all tasks, $F2||C_{\max}$) is a problem with polynomial computational complexity (Johnson's algorithm [9]).

The research of discrete optimization problems conducted for many years concerns the vast majority of deterministic models, in which the basic assumption is the uniqueness of all parameters. To solve these types of problems, which mostly belong to the class of strongly NP-hard problems, a number of effective approximate algorithms in which specific properties of problems are applied. However, in many areas of the economy we are dealing with random processes, e.g. transport, agriculture, trade, construction, etc. Effective management of such processes often leads to optimization models with random parameters. Already for the deterministic case solving these problems is very difficult, because they usually belong to the NP-hard class. The inclusion of parameter uncertainty in the model causes additional complications. Hence, the problems with random parameters are much less frequently studied. In this work we are considering a random problem with times of tasks execution. We present some properties of the problem (the so-called block elimination properties) accelerating the search of neighborhoods. Due to their implementation, it is possible to eliminate inferior solutions without having to calculate the value of the criterion function (intermediate review method). First, we will describe the case of a problem with deterministic task execution times, and then with the durations represented by random variables.

2 Two-Machine Problem with Due Dates

Two-machine flow shop scheduling problem with minimalization of total weighted tardiness (denoted by $F2||\sum w_i T_i$) can be formulated as follows:

F2T Problem. A set of tasks is given $\mathcal{J} = \{1, 2, \ldots, n\}$, and a set of machines $\mathcal{M} = \{1, 2\}$. A task $i \in \mathcal{J}$ consists of two operations O_{i1} and O_{i2}. An operation O_{ik} corresponds to performing a task i on a machine k. For a task $i \in \mathcal{J}$ we define:
p_{ik} – *execution time* (duration) of the operation O_{ik},
d_i – *requested completion time* (due date),
w_i – *weight of penalty function for exceeding the due date* (being tardy).

Each task should be executed on both machines and there must be fulfilled the following constraints:

(a) each task must be completed on the first and then on the second machine,
(b) the task cannot be interrupted,

(c) the request can be performed simultaneously on only one machine,
(d) the machine cannot perform more than one task at the same time,
(e) the task order, on both machines, must be the same.

For the fixed execution order on machines, let S_{ij} be the time of beginning of the operation O_{ij} $(i \in \mathcal{J},\ j = 1, 2)$. From constraints (b) and (c) it follows, that $C_{ij} = S_{ij} + p_{ij}$ is the finishing time of the operation O_{ij}. These moments can be determined from the following recursive relationships:

$$C_{i,j} = \max\{C_{(i-1),j},\ C_{i,(j-1)}\} + p_{i,j}, \quad i = 1, 2, \ ...\ .n, \quad j = 1, 2, \tag{1}$$

with starting conditions $C_{0,j} = 0,\ j = 1, 2$ and $C_{i,0} = 0,\ i = 1, 2, \ldots, n$. By $\mathcal{C}_i = C_{i2}$ we denote finishing time of execution of the task i (operation O_{i2}). Therefore

$$T_i = \max\{0, \mathcal{C}_i - d_i\} \tag{2}$$

is *tardiness* of execution of the task i, $f_i = w_i \cdot T_i$ *penalty* for being tardy (*cost of a task execution*). If $T_i = 0$ than this task is called *early*, otherwise – *tardy*.

Any solution, i.e. the order in which the tasks are to be carried out (the same on both machines) can be represented by the permutation of tasks from the set $\mathcal{J}$. Let Π be a set of all such permutations. For any permutation $\pi = (\pi(1), \ldots, \pi(n)),\ \ \pi \in \Pi$, penalty for tardiness of tasks execution (cost)

$$\mathcal{T}(\pi) = \sum_{i=1}^{n} w_{\pi(i)} \cdot T_{\pi(i)}. \tag{3}$$

In the F2T problem under consideration, the order of tasks execution should be determined, which minimizes the sum of penalties for tardy tasks, i.e. optimal permutation $\pi^* \in \Pi$, for which

$$\mathcal{T}(\pi^*) = \min\{\mathcal{T}(\pi) :\ \pi \in \Pi\}. \tag{4}$$

In the introduction we wrote that the two-machine flow problem with the $C_{\ max}$ criterion belongs to the P class. Johnson's algorithm [9] is used for solving this problem.

Any sequence of immediately following elements in we will call *sub-permutation*. If

$$\eta = (\pi(u), \pi(u+1), \ldots, \pi(v)),\ 1 \le u \le v \le n, \tag{5}$$

is a sub-permutation of a permutation π, then the cost of tasks execution from η

$$\mathcal{T}_\pi(\eta) = \sum_{i=u}^{v} (w_{\eta(i)} \cdot (\mathcal{C}_{\eta(i)} - d_{\eta(i)})), \tag{6}$$

where $\mathcal{C}_{\eta(i)}$ is a *finishing time* of execution of the task $\eta(i)$ in the permutation π. By $\mathcal{Y}(\eta)$ we denote a set of elements of sub-permutation η, i.e.

$$\mathcal{Y}(\eta) = \{\pi(u), \pi(u+1), \ldots, \pi(v)\}.$$

Let

$$\alpha = (1, 2, \ldots, a-1),\ \beta = (a, a+1, \ldots, b-1, b),\ \gamma = (b+1, \ldots, n), \tag{7}$$

where $1 \le a \le b \le n$ be sub-permutations in π.

Therefore permutation $\pi = (\alpha, \beta, \gamma)$ is a concatenation of three sub-permutations, and its cos

$$\mathcal{T}(\pi) = \mathcal{T}_\pi(\alpha) + \mathcal{T}_\pi(\beta) + \mathcal{T}_\pi(\gamma). \tag{8}$$

3 Random Task Execution Times

In this section we consider the probabilistic version of F2T problem – scheduling tasks on two machines with due dates. We assume that task execution times are independent random variables. Extensive review of methods and algorithms for solving optimization combinatorial problems with random parameters was presented by Vondrák in a monograph [12] and newer [13]. Some practical problems are also considered in the works of Bożejko et al. [6] and [5]. We will now introduce the necessary definitions and notions.

If X is a continuous random variable, we will use the following symbols later in this work:

F_X – cumulative distribution function of a random variable X,
$E(X)$ – expected value of a random variable X.

We consider, described in the previous chapter probabilistic version of the F2T problem, in which the task execution times $\tilde{p}_{ij}$ are independent random variables, and the remaining task parameters w_i and d_i $(i = 1, 2, \ldots, n)$ are deterministic. This problem we will briefly refer to as PF2T.

If tasks durations $\tilde{p}_{ij}$ are independent random variables, than for any tasks execution order $\pi \in \Pi$, a time of a task $\pi(k)$ finishing $\tilde{C}_{\pi(k)}$, tardiness $\tilde{T}_{\pi(k)} = \max\{0, \tilde{C}_{\pi(k)} - d_{\pi(k)}\}$ and cost function

$$\tilde{\mathcal{T}}(\pi) = \sum_{i=1}^{n} w_{\pi(i)} \cdot \tilde{T}_{\pi(i)}. \tag{9}$$

are also random variables.

There is a necessity in algorithms for solving optimization problems comparing the value of the criterion function for various acceptable solutions (e.g. permutation). In case this function is a random variable (9) we will be they used her expected value. Therefore, the following function will be used as comparative criteria for solutions:

$$\mathcal{L}(\pi) = E(\tilde{\mathcal{T}}(\pi)) = \sum_{i=1}^{n} w_{\pi(i)} \cdot E(\tilde{T}_{\pi(i)}). \tag{10}$$

In the rest of the work we present the methods for calculating the value of criterion function (10).

If $\beta = (\pi(a), \pi(a+1), \ldots, \pi(b))$, where $1 \leq a \leq b \leq n$ is a sub-permutation of a permutation $\pi \in \Pi$,

$$\mathcal{L}(\beta) = \sum_{i=a}^{b} w_{\pi(i)} \cdot E(\tilde{T}_{\pi(i)}). \tag{11}$$

is a cost of tasks execution of a sub-permutation β.

4 Blocks of Tasks

We consider a permutation $\pi \in \Pi$ – a solution of PF2T problem. If an expected value of the execution finishing time of a task $\pi(i)$, $E(\tilde{C}_{\pi(i)}) \leq d_{\pi(i)}$ than this task $\pi(i)$ we call *early*, otherwise, if $E(\tilde{C}_{\pi(i)}) > d_{\pi(i)}$, *late* (tardy).

Later in this chapter we present a method of constructing sub-permutations (called blocks) containing only early or tardy tasks.

4.1 Blocks of Early Tasks

Let a permutation $\pi \in \Pi$ define a sequence (7) of three sub-permutations, i.e. $\pi = (\alpha, \beta, \gamma)$. For tasks from sub-permutation $\beta = (a, a+1, \ldots, b-1, b)$ we assume the duration of the task on the machine

$$p_{ij} = E(\tilde{p_{ij}}),\ i \in \mathcal{J},\ j \in \mathcal{M}. \tag{12}$$

Then, we use the described in Sect. 2 Johnson's algorithm. In this way we set a new order of tasks from the set $\mathcal{Y}(\beta)$, i.e. sub-permutation

$$\beta' = (a', a'+1, \ldots, b'-1, b'). \tag{13}$$

We will call it *Johnson optimal*, in short *J-opt*. One can easily prove that this is the optimal order, due to the minimization of the expected date value of completion of all tasks in the set $\mathcal{Y}(\beta)$.

We consider permutations $\pi = (\alpha, \beta, \gamma)$ and $\pi' = (\alpha, \beta', \gamma)$. It's easy to show that if sub-permutation β' is *J-opt*, then the finishing time of the last task in β' is not greater than the finishing time of the last task in β.

Theorem 1. *Let β be J-opt sub-permutation in the permutation $\pi = (\alpha, \beta, \gamma)$, $\pi \in \Pi$. If permutation $\sigma = (\alpha, \delta, \gamma)$ was generated from π by changing the order of tasks in sub-permutation β, then expected value of the execution finishing time for any task from γ in the permutation σ is not less than the time of finishing of this task in the permutation π.*

The proof should use the assumption: sub-permutation β is $J - opt$.

Definition 1. *If all the tasks from J-opt sub-permutation β after insertion into the first position in β are on-time, then we call β* ***block of early tasks*** *(in short* ***T-block****).*

Theorem 2. *(Elimination T-block property) If a permutation σ was generated from $\pi \in \Pi$ by changing the order of tasks in a T-block, then*

$$\mathcal{L}(\sigma) \geq \mathcal{L}(\pi). \tag{14}$$

One should take advantage of *T-block* definition in proof.

Remark 1. *While generating new permutations from π one can omit these of them, which were generated by changing the order in any T-block. They do not give an improvement in the cost function value.*

4.2 Blocks of Tardy Tasks

Let a permutation of tasks order $\pi = (1, 2, \ldots, a-1, a, a+1, \ldots, b-1, b, b+1, \ldots, n) = (\alpha, \beta, \gamma)$, where $\alpha = (1, 2, \ldots, a-1)$, $\beta = (a, a+1, \ldots, b-1, b)$, $\gamma = (b+1, \ldots, n)$. We assume then that a sub-permutation β of all the tasks is tardy, and additionally

$$\forall i \in \beta,\ d_i < E(\tilde{C}_{a-1} + \tilde{p}_{i,2}). \tag{15}$$

If $a = 1$ we assume $E(\tilde{C}_{a-1}) = 0$. It follows from the above inequality that any task from the sub-permutation β inserted into the first position of β, i.e. position a is late (tardy). Let us assume, that the tasks from β fulfill inequalities (15). We generate two new permutations from π: $\pi' = (\alpha, \beta', \gamma)$ and $\pi'' = (\alpha, \beta'', \gamma)$, where $\mathcal{Y}(\beta) = \mathcal{Y}(\beta') = \mathcal{Y}(\beta''))$. We define sub-permutations β' and β'' occuring in both permutations as follows:

(a) in β' we set the order of the elements using the Johnson algorithm (it is therefore subpermutation *J-opt*),
(b) we construct the sub-permutation β'' by setting the tasks from the set $\mathcal{Y}(\beta)$, according to non-growing values $w_v/E((\tilde{p}_{v,1} + \tilde{p}_{v,2}))$, $v \in \mathcal{Y}(\beta)$.

Definition 2. *Sub-permutation β'' defined in (b) we call a* ***block of tardy tasks****, in short* ***D-block****, if*

$$(E(\tilde{C}_{b''}) - E(\tilde{C}_{b'}))/E(\tilde{C}_{b''}) \leq \varphi,$$

where φ is a parameter which we assign experimentally.

While, $E(\tilde{C}_{b''})$ and $E(\tilde{C}_{b'})$ are expected values of finishing time of the last task in sub-permutation β' and β'', respectively, which were defined in (a) and (b).

A parameter φ (whose value is determined experimentally) is a measure enabling estimation of the difference between expected values of finishing times of tasks from $\mathcal{Y}(\beta)$ in order β'' and β'. For a small value of φ (e.g. 0.1) they differ only 'a little'. Then, β'' sub-permutation is quasi optimal for tasks from the set $\mathcal{Y}(\beta)$, both due to the expected value of the finishing time and the cost.

A *D-block* does not meet the elimination block property: '*reordering elements in block does not generate solutions with a smaller value of criterion function*'.

Despite this fact we will use them to eliminate certain solutions from neighborhood, due to its empirical advantage.

Any permutation can be partitioned into sub-permutation such that each is a *T-block* or a *D-block*. The algorithm for determining blocks is similar to the one presented in the paper [4] and has computational complexity $O(n^2)$.

5 Tabu Search Algorithm

Standard version of the tabu search method was used to solve the considered T2FS problem, with the neighborhood generated by insert type moves. The procedure of generating the environment uses blocks so that some elements are omitted without the necessity to calculate their criterion function value. In order to diversify the search process, a 'backtrack jump' mechanism was applied that resumes the process searches from remembered promising solutions. It is implemented through the introduction of the so-called long-term memory, abbreviated to LTM). A return jump (to the last saved LTM element) is executed in a case where through a certain number of iterations there is no improvement of the best solution. The algorithm terminates after a fixed number of iterations.

6 Random Tasks Execution Times

Let $\delta = (\tilde{p}_{ij}, w_i, d_i)$, $i = 1, 2, \ldots, n$, $j = 1, 2$ be a data instance for the PF2T problem, where tasks execution times $\tilde{p}_{ij}$ are independent random variables with normal distribution, i.e. $\tilde{p}_{ij} \sim N(p_{ij}, \lambda p_{ij})$, and λ a fixed parameter. For the simplification, let's assume that the order of performing tasks (the same on both machines)

$$\beta = (1, 2, \ldots, n). \tag{16}$$

To calculate the value of the cost function

$$\mathcal{L}(\beta) = E(\tilde{T}(\beta)) = \sum_{i=1}^{n} w_{\beta(i)} \cdot E(\tilde{T}_{\beta(i)}), \tag{17}$$

where $\tilde{T}_{\beta(k)} = \max\{0, \tilde{C}_{\beta(k)} - d_{\beta(k)}\}$, it is necessary to determine tasks execution finishing times $\tilde{C}_{\beta(k)}$. We introduce additional values:

$$C'_{i,j} = \begin{cases} \sum_{k=1}^{i} p_{i,j}, & \text{for } j = 1, \\ C'_{i,j-1} + p_{i,j}, & \text{for } i = 1, j = 2, \\ \max\{C'_{i,j-1}, C'_{i-1,j}\} + p_{i,j}, & \text{for } i > 1, j = 2. \end{cases}$$

Finally random variable representing the time of finishing of the i-th task

$$\tilde{C}'_i \sim N(C'_{i2}, \lambda\sqrt{C'_{i2}}).$$

Let

$$\mu_i = p_{1,1} + p_{2,1} + \ldots, +p_{i,1} + p_{i,2} \text{ and } \sigma_i = \lambda\sqrt{p_{1,1}^2 + p_{2,1}^2 + \ldots, +p_{i,1}^2 + p_{i,2}^2}.$$

When calculating the expected value $E(\tilde{T}_i)$ appearing in the definition of the criterion function (17), we will use the following theorem.

Theorem 3. *If tasks duration are independent random variables with normal distribution* $\tilde{p}_{ik} \sim N(p_{ik}, \lambda \cdot p_{ik})$ $(i = 1, 2, \ldots, n,\ k = 1, 2)$ *then an expected value of tardiness of the task* $i \in \mathcal{J}$

$$E(\tilde{T}_i) = (1 - F_{\tilde{C}'_i(d_i)}) \left(\frac{\sigma_i}{\sqrt{2\pi}} e^{\frac{-(d_i - \mu_i)^2}{2\sigma_i^2}} + (\mu_i - d_i) \left(1 - F_{N(0,1)}(\frac{d_i - \mu_i}{\sigma_i}) \right) \right).$$

The proof of this theorem is similar to the one given in Bożejko et al. [6].

7 Computational Experiments

Computational experiments were carried out on two versions of the tabu search algorithm for solving the PF2T probabilistic problem:

1. PTS – an algorithm with neighborhood generated by swap moves,
2. PTR+b – an algorithm with elimination block properties application.

The main goal of the experiments was to examine the stability of algorithms, i.e. the resistance of solutions to random data disturbances (times of operations). The exact algorithm stability was described in the paper of Bożejko et al. [6]. The starting solution for both algorithms was the natural permutation $\pi = (1, 2, \ldots, n)$, and in addition: the length of the list of tabu moves: n, number of algorithm iterations: $2n$. The algorithms have been implemented in C++ and run on a PC with a 2.8 GHz clock.

Because in the literature there are no examples of tests for the problem under consideration, for the needs of execution computational experiments were generated randomly. Times for completing tasks on individual machines were randomly designated, in accordance with uniform distribution from the set $\{1, 2, \ldots, 99\}$, and weights the penalty function w_i, from the set $\{1, 2, \ldots, 9\}$. The values requested deadlines for completing tasks have been set based on two parameters: T – *latency factor* and R – *timeliness range.* These terms, according to the uniform distribution, were randomly selected from the range $[P(1 - T - R/2), (1 - T + R/2)]$, where $P = \sum_{i=1}^{n} \sum_{j=1}^{2} p_{i,j}$. Test examples were generated for each pair of parameter values $T = 0.2, 0.4$ and $R = 0.2, 0.4$ (the larger the coefficients, the more difficult the generated examples). There are four such pairs in total. The examples were generated for the number of tasks $n = 100$ and 500. For each pair of T and R values 25 examples were generated. Ultimately, 200 examples were used for the performed computational experiments whose collection (the so-called *deterministic data*) is denoted by Ω. Then, for each example (p_{ij}, w_i, d_i), $i = 1, 2, \ldots, n$, $j = 1, 2$ of the deterministic problem two examples of probabilistic data were determined $(\tilde{p}_{ij}, w_i, d_i)$, where the operation duration was a random variable with a normal distribution $\tilde{p}_{ij} \sim N(d_i, \lambda \cdot p_{ij})$ for $\lambda = 0.02, 0.05$. The collection of this data (the so-called *probabilistic data*) is denoted by $\widetilde{\Omega}$. To examine the stability of probabilistic algorithms, a disturbed data set $\Omega^{\approx}$ was generated. For each example of deterministic data (p_{ij}, w_i, d_i) 50 examples of disturbed data were generated. The disorder consists in changing the duration of operation p_{ij} into randomly designated value, according to the distribution of $N(p_{ij}, \lambda \cdot p_{ij})$. In total, 10,000 examples were given. They were then solved by the tabu search algorithm. The results obtained were the basis for determination of the stability coefficient of probability algorithms PTS and PTS+b. Aggregate results are included in the Table 1.

Table 1. Stability coefficient of probabilistic algorithms.

Instance	n	$\lambda = 0.02$		$\lambda = 0.05$	
		PTS	PTS+b	PTS	PTS+b
$T = 0.2, R = 0.2$	100	3.41	3.37	2.24	2.27
$T = 0.2, R = 0.4$	100	3.82	3.87	3.06	2.99
$T = 0.4, R = 0.2$	100	4.78	4.77	4.02	3.99
$T = 0.4, R = 0.4$	100	5.12	5.01	3.52	3.35
$T = 0.2, R = 0.2$	500	4.68	4.55	4.79	4.66
$T = 0.2, R = 0.4$	500	5.34	5.93	6.13	6.31
$T = 0.4, R = 0.2$	500	7.11	6.58	8.09	8.01
$T = 0.4, R = 0.4$	500	10.28	9.54	12.64	10.07
Average		**5.57**	**4.80**	**5.56**	**4.32**

The main purpose of the carried out experiments was to examine individual stability of algorithms, i.e. the robustness of solutions determined by these algorithms for random changes (disturbances) of parameters. Among the probabilistic algorithms tested, he proved to be more stable the 'with blocks' algorithm PTS+b. Its stability factor is 4.56. That is, the data disorder (according to the described random procedure) causes average relative deterioration criterion (in relation to the best solution of this example) of 5.56%. The PTS+b algorithm stability factor is 5.57. In addition, it turned out that the use of blocks in a probabilistic algorithm resulted in a shortening of the average calculation time by about 30%. The results obtained prove the high efficiency of the blocks.

8 Conclusions

The paper examines the problem of scheduling tasks on two machines, in which the times of task execution are random variables. Blocks of tasks were introduced to eliminate the use of solutions with the environment generated by swap movements that require in the taboo algorithm search. Computational experiments were carried out in order to study the impact of blocks on computation and analize the times of designated solutions. The results obtained are clearly available, the use of blocks significantly reduces calculation time and improves the stability of solutions. Application of elements of probability in the adaptation of tabu search methods allows one to solve uncertain data problems. These are very difficult optimization problems, much better describing reality than deterministic models.

Acknowledgments. This work was partially funded by the National Science Centre of Poland, grant OPUS no. 2017/25/B/ST7/02181 and a statutory subsidy 049U/0032/19.

References

1. Al-Salem, M., Valencia, L., Rabadi, G.: Heuristic and exact algorithms for the two-machine just in time job shop scheduling problem. Math. Prob. Eng. **5**, 1–11 (2016)
2. Ardakan, M., Beheshti, A., Hamid Mirmohammadi, S., Ardakani, H.D.: A hybrid meta-heuristic algorithm to minimize the number of tardy jobs in a dynamic two-machine flow shop problem. Numer. Algebra Control Optim. **7**(4), 465–480 (2017)
3. Bank, M., Fatemi, S., Ghomi, M.T., Jolai, F., Behnamian, J.: Two-machine flow shop total tardiness scheduling problem with deteriorating jobs. Appl. Math. Model. **36**(11), 5418–5426 (2012)
4. Bożejko, W., Grabowski, J., Wodecki, M.: Block approach tabu search algorithm for single machine total weighted tardiness problem. Comp. Ind. Eng. **50**(1–2), 1–14 (2006)
5. Bożejko, W., Hejducki, Z., Wodecki, M.: Flowshop scheduling of construction processes with uncertain parameters. Arch. Civil Mech. Eng. **19**, 194–204 (2019)
6. Bożejko, W., Rajba, P., Wodecki, M.: Stable scheduling of single machine with probabilistic parameters. Bull. Pol. Acad. Sci. Tech. Sci. **65**(2), 219–231 (2017)
7. Bulfin, R.L., M'Hallah, R.: Minimizing the weighted number of tardy jobs on two-machineflow shop. Comput. Oper. Res. **30**, 1887–1900 (2003)
8. Gupta, J.N.D., Hariri, A.M.A.: Two-machine flowshop scheduling to minimize the number of tardy jobs. J. Oper. Res. Soc. **48**, 212–220 (1997)
9. Johnson, S.M.: Optimal two- and three-stage production schedules with setup times included. Naval Res. Logist. Q. **1**, 61–68 (1954)
10. Ahmadi-Darani, M., Moslehi, G., Reisi-Nafchi, M.: A two-agent scheduling problem in a two-machine flowshop. Int. J. Ind. Eng. Comput. **9**(3), 289–306 (2018)
11. Lin, B.M.T.: Scheduling in the two-machine flowshop with due date constraints. Int. J. Prod. Econ. **70**, 117–123 (2001)
12. Vondrák, J.: Probabilistic methods in combinatorial and stochastic optimization. PhD, MIT (2005)
13. Cai, X., Wu, X., Zhou, X.: Optimal Stochastic Scheduling. Springer, New York (2014)

Soft Computing Analysis of Pressure Decay Leak Test Detection

Ander Garcia[1](✉), Juan Luis Ferrando[1], Ander Arbelaiz[1], Xabier Oregui[1], Andoni Bilbao[2], and Zelmar Etxegoien[2]

[1] Vicomtech Foundation, Basque Research and Technology Alliance (BRTA), Mikeletegi 57, 20009 Donostia-San Sebastián, Spain
agarcia@vicomtech.org

[2] Gaindu. Inzu Group, Elgoibar, Spain

Abstract. Leak detection is a common and relevant step of manufacturing processes, which takes place throughout the production line. While several leak test methods have been proposed, pressure decay testing is still widespread due to its lower cost, simplicity and sensitivity in relatively small volumes. However, pressure decay testing is very sensitive to external parameters, mainly temperature. While high-end leak test machines can compensate temperature variations integrating specialized hardware, this paper analyses the viability of applying soft computing models on a regular leak test machine to perform the same compensation. Gaindu, an automation company which sells leak test stations, has customized a leak test station to measure and publish key test data. This data has been stored on a database to be analyzed. Moreover, a model compensating temperature variations has been developed and validated. Results encourage to further vary parameters that may affect the leak test, such as part temperature or humidity, to extend the model and integrate it on commercial leak test machines.

Keywords: Leak test · Process modelling · Industry 4.0

1 Introduction

Leak detection is a common manufacturing quality measurement method applied in several industries. A product leak is material flow from or into a product (a control volume) during a given time, in excess of allowable limits. Product leaks are caused by open flow paths, such as pinholes, broken seals or material porosity. In most cases, a product leak is a very small flow. The process of quantifying a product leak is called leak testing [4].

Due to its relevance, several methods have been designed to measure leaks, such as bubble immersion, helium sniffing, ultrasonic, and differential pressure decay. Differential pressure decay testing (DPDT) is widely used in the plumbing, aerospace and automation industries due to its lower cost, simplicity and sensitivity in relatively small volumes.

Á. Herrero et al. (Eds.): SOCO 2020, AISC 1268, pp. 299–308, 2021.
https://doi.org/10.1007/978-3-030-57802-2_29

However, DPDT measurements has been traditionally affected by external environmental conditions, mainly temperature. While regular DPDT leak test machines are affected by these conditions, some machines are able to compensate them with advanced hardware. This functionality positions the second group as high-end components with a competitive advantage and higher price range.

This paper focuses on DPTD, the main leak test method used by the leak test machines sold by Gaindu Automation, an industrial automation company. Software based temperature compensation would give Gaindu a competitive advantage without a cost increase for each leak test machine. Thus, this paper focuses on analyzing the viability of integrating this compensation on regular DPTD leak test machines with soft computing models.

First, key data from leak test equipment and the environment has been identified. Then, data from a customized real leak test machine has been captured by a PLC and published through an OPC UA server. Finally, this data has been stored on a database to be and analyzed.

After an initial analysis, a soft computing model compensating changes in test temperature has been designed and validated, encouraging further data acquisition and the extension of the model to be integrated on commercial leak test machines.

This paper is organized as follows: after presenting related work, key leak test data and data acquisition flow are shown. Then, the data acquisition setup, the initial data analysis and the model generation and validation step are described. The paper finalises presenting some future work and conclusions.

2 Literature Review

The pressure decay method is sensitive to the volume of the test part and the pressure decay rate. Any correlation between the leak flow rate and pressure decay must be performed with the same volume that was used during product testing. In addition, engineers must allow enough time for a steady decay to develop. The pressure decay rate is temperature sensitive, because gas density depends on pressure and temperature [4].

In a leak detection based on differential pressure measurement, leakage is detected by measuring pressure difference between a reference and a tested component using a differential pressure sensor. As compared with only measurement of pressure inside the tested component chamber, measurement of differential pressure between the dual chambers, in which a leak-tight master is used as a reference, has several advantages [3]. A regular DPDT cycle can be divided in four periods: charge, balance, measure and vent (Fig. 1).

The relation between the temperature recovery time and its relevance on the accuracy and repeatability of air leak detection has been studied [3]. The waiting time for detection can be shortened, thermal instability can be reduced and influence of external environment change can be counterbalanced.

Due to its industrial relevance, leak test research has lead both to registered patents and scientific publications. For example, [2] proposed a method to

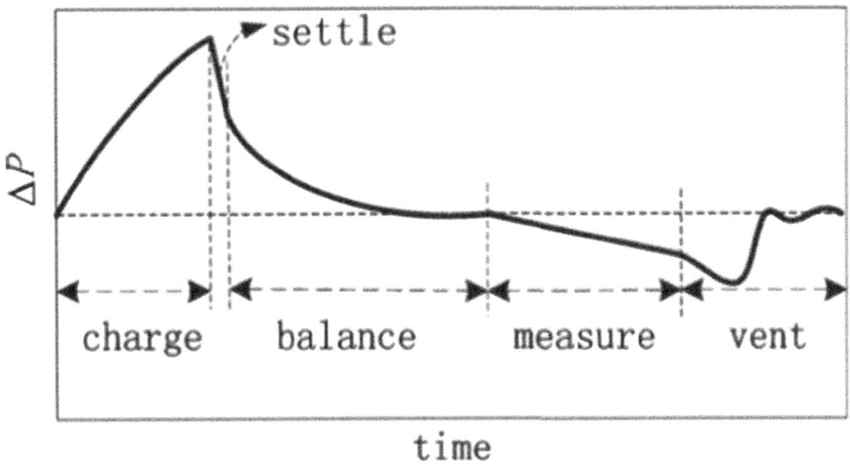

Fig. 1. DPDT periods [6].

accurately predict the minimum required temperature recovery for various test volumes and applied pressures.

A method for temperature effect compensation to improve testing efficiency has been proposed [6]. They achieved an accuracy less than 0.25 cc min^{-1} for various volumes, reducing the minimum measurement time with temperature compensation to four times the theoretical thermal-line constant.

Recently, researchers have proposed data-driven approaches to improve leak detection. However, most of these novel approaches are focused on pipeline leaks detection, an application area with a completely different requirements from automation industry. For example, [1] present a novel data-driven algorithm for pipeline leak detection and localization.

Finally, although [7] also focus on pressure decay tests for automotive batteries, they target IIoT data dashboards and integration of the test results into production line workflow, they do not apply any modelling approach.

3 Leak Test Data

Automation of leak test data handling has required a thorough extensive design, development and integration work. First, key data from the leak test machine must be identified, such as configuration parameters (times, part volume. . .) or relevant commands. Then, this data must be read and written in order to control the test from a PLC integrated on a manufacturing station. This process is dependent on the provider of the leak test machine and its data communication protocol. Moreover, the PLC also must capture external data influencing the leak test detection, such as air or part temperatures. Finally, the PLC must publish this data in order to be stored and analyzed.

Key data from the leak test machine has been divided in categories. The first category is related to input data of the leak test machine: configuration parameters of the leak test program. The second one is composed by summary results of the leak test: mainly data that is shown on the display of leak test machine while the test is executing. The third category references the evolution of the test values varying in real time, such as the current pressure. This data has been enriched with sensors measuring the ambient temperature, part temperature, and temperature of the air injected on the leak test machine.

A Siemens 1500 PLC has been programmed to read all this data and automate the leak test. The TIA Portal program of the PLC has defined several UDTs (User-defined Data Types) according to previous categories. Finally, this data has been published by the OPC UA Server build-in inside the Siemens PLC, being available to any OPC Client connected to the PLC.

Next, an OPC UA client has been designed and developed. This client subscribes to the key data of machine leak tests, reading and storing it on a database. Leak test data of each test has been stored inside a NoSQL database containing both the summary data of the test and the real time data. The latter has been stored with a frequency of ten values per second.

4 Data Analysis

4.1 Test Setup

Gaindu has customized a leak test station (Fig. 2) to capture the real data as described on previous section. First, in order to avoid any physical risk, the pneumatic component of a regular machine has been disabled. As the objective of this component is to generate a vacuum space on the part that is going to be measured, a calibrated part has been connected instead. This calibrated part has been the one measured during the data acquisition step of the paper.

An air heating equipment has also been installed to control the temperature of the air injected to the leak test measurement equipment. Moreover, temperature sensors have been deployed to measure the temperature of the controlled part and the ambient temperature.

An automation expert from Gaindu has programmed the PLC to control the customized leak test station through an HMI, including automatic working modes to continuously perform leak tests on the controlled part. An industrial PC has also been integrated on the station, being connected to the PLC to store data about tests. The leak test equipment integrated on the customized machine (Fig. 3) is an Ateq model F620, which has been calibrated and configured with the volume of the cavity of the controlled part.

The customized leak test station has been working for several days running tests continuously. Each day, the temperature of the air injected into the Ateq machine has been set to a different value and the ambient and part temperatures have been measured.

(a) General view (b) HMI

Fig. 2. Leak test station

4.2 Model Generation

After previous analysis, tests have been run continuously capturing part temperature, ambient temperature, temperature of injected air, pressure decay and minimum pressure of the stabilization time (just before the testing time starts). Test duration has been configured to 60 s, with 2 s measurement time and 20 s stabilization time. 1720 tests have been performed to acquire enough data to train the machine learning models. The model has been integrated on the following workflow (Fig. 4) to analyze its viability to compensate the temperature.

During the development of the model, first, input data has been analyzed to detect outliers. These outliers have been considered as errors on the data acquisition system and have been discarded. Several outlier removal methods have been tested: one-class support vector machine (SVM) [5], local outlier factor, isolation forest and elliptic envelope. The best accuracy detecting outliers (erroneous measurements) in the dataset has been obtained applying one-class SVM method.

To clean the captured data and reduce the noise during the training process of the machine learning models one class SVM outlier detector was applied to the captured dataset. Figure 5 shows in blue the outliers detected and orange the normal observations. The plot also shows the correlation between the different variables. For example, the air temperature and part temperature have a very high correlation because the part temperature was under environmental conditions. In addition, grid plot also shows the distribution for the different variables for the outlier and normal observations.

Fig. 3. Leak test measurement equipment.

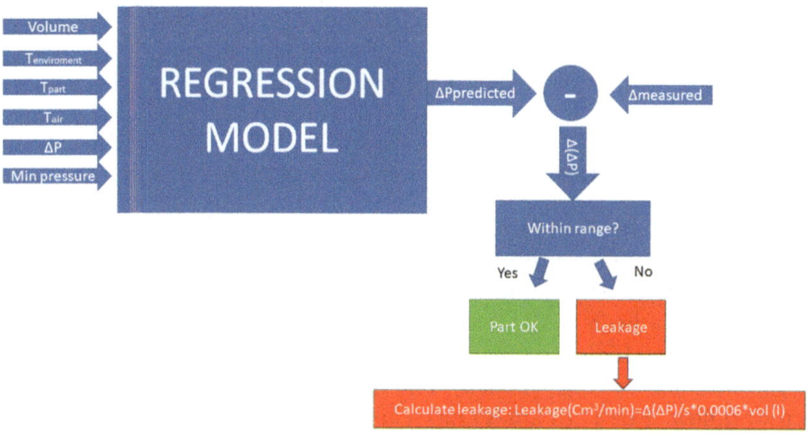

Fig. 4. Temperature compensation workflow.

With the clean data, several regressive predictive methods have been tested: decision trees, artificial neural networks (multilayer perceptron), SVM, polynomic regression and k-neighbours. K-fold technique was applied to increase the accuracy of the results, the dataset was split 80% for training purposes and 20% for validation. The best results were obtained by decision trees, configured with a maximum depth of 11 levels and 'Best' as splitter as shown in Fig. 6. The RMS error obtained was just below 0.10 Pa that converted to mbar*l/sec was 0.0003, taking into account that the volume of the controlled test part was 0.31. This result validates the model under the temperature variation range of the test setup.

It is worth noting the relatively low R2 parameter, just above 0.45. This is attributed to the high uncertainty of measurement due to the low stabiliza-

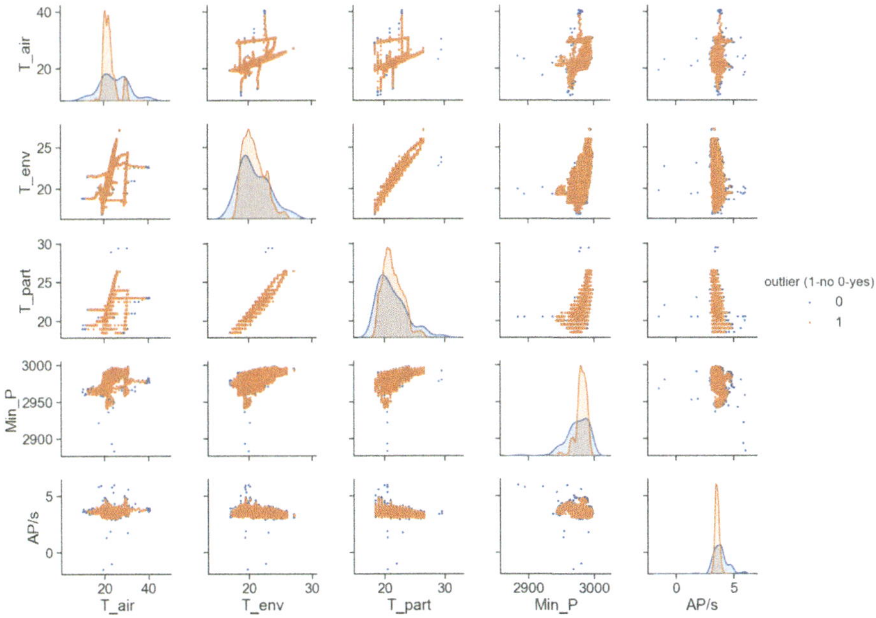

Fig. 5. Predicted and real outputs.

tion time selected. As previously mentioned, if the model accuracy needs to be improved, the stabilization time should be increased. Figure 7 shows the error distribution with the developed compensation model.

4.3 Impact of Measure Time

After the model has been developed, the impact of the measurement time to the accuracy of the test has been evaluated. As leak test duration has an impact on the cycle time of the production line where it is integrated, if overall test time is shortened so it is the cycle time. However, reducing the measure time affects the accuracy and repeatability of the test. Thus, a compromise between accuracy and testing time has to be found to select a measure time value.

To compare the impact of the measure time of the test, the same test with varying measure times ranging from 60 ms to 15 s have been run. Figure 8 shows the result of the test. It clearly shows and asymptotic relationship between the testing time and error of the model. After 2 s, the impact of the measurement time is close to 0.

5 Future Work

Test results of the model have validated the viability of the approach of the paper. However, in order to integrate it on commercial leak test machines further work is required.

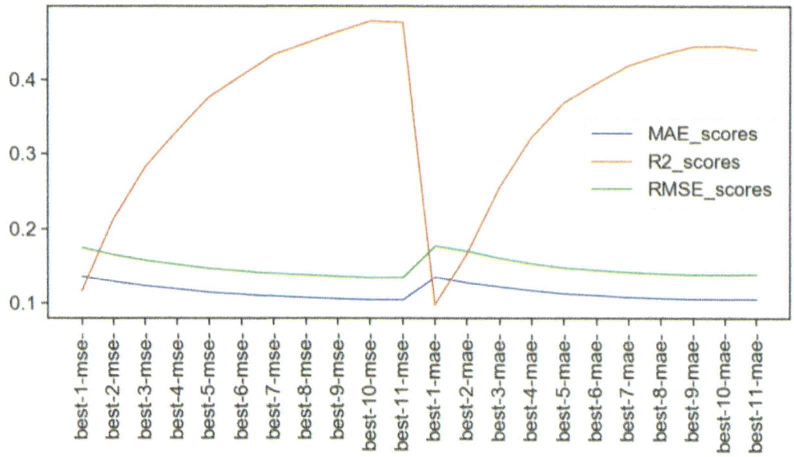

Fig. 6. Decision tree configuration.

First, input temperature parameters (ambient and part) have varied in a restricted range, temperatures of the factory from Gaindu. However, Gaindu sells leak test machine all over the world, even to places where temperature can change almost 40° within the same work shift. Moreover, at some manufacturing lines, the station preceding the leak test heats the part to high temperatures. Thus, the customized test setup has to be extended to be able to acquire data simulating these real conditions.

Second, the pneumatic system of the leak station must be enabled to test parts under real operation conditions. Finally, parts with different materials and

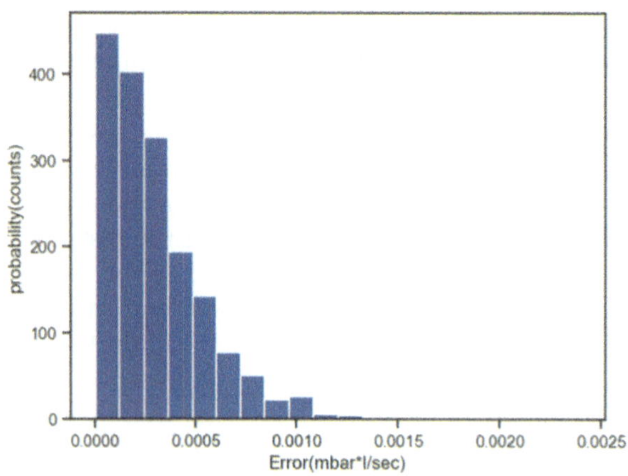

Fig. 7. Error of the generated model.

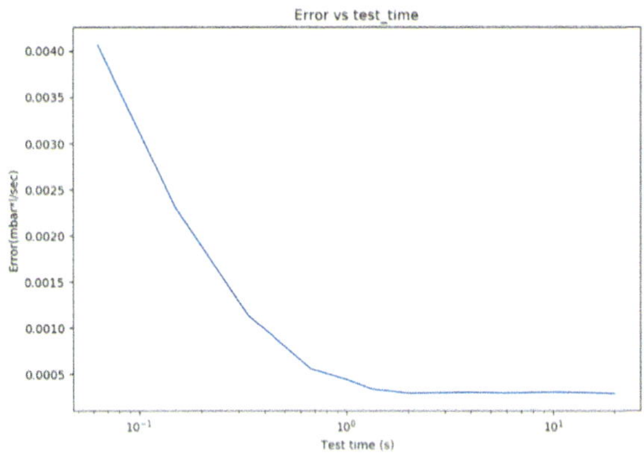

Fig. 8. Impact on error of varying measure time.

cavities with different volumes should be included on a further validation before integrating the model in a commercial leak test station.

6 Conclusion

Leak detection is a common and relevant step of manufacturing processes. Although there are several leak detection methods, differential pressure decay testing (DPDT) is widespread due to its advantages and cost. However, DPDT measures are affected by temperature changes.

This paper analyses the viability of generating a soft computing model to compensate the impact of temperature changes by software. First, a leak test detection station has been customized to capture and store key test data.

Then, automatic tests have been configured to continuously run and capture data, varying the temperature of the injected air once per day and measuring the ambient and part temperature.

Finally, this data has been analyzed to remove input data outliers and to generate a model based on decision tress to compensate temperature changes. The model has been validated with a k-fold approach, obtaining an error below 0.0010 mbar*l/sec.

Test results of the model have validated the viability of the approach of the paper, encouraging to further develop the customized leak test machine. However, in order to integrate it on commercial leak test machines further work is required.

First, input temperature parameters (ambient and part) have varied in a restricted range, temperatures of the factory from Gaindu. However, Gaindu sells leak test machine all over the world, even to places where temperature can change almost 40° within the same work shift. Moreover, at some manufacturing lines, the station preceding the leak test heats the part to high temperatures. Thus, the customized test setup has to be extended to be able to acquire data simulating these real conditions.

Second, the pneumatic system of the leak station must be enabled to test parts under real operation conditions. Finally, parts with different materials and cavities with different volumes should be included on a further validation before integrating the model in a commercial leak test station.

Acknowledgment. This research was partially supported by the Centre for the Development of Industrial Technology (CDTI) and the Spanish Ministry of Economy and Competitiveness (IDI-20150643).

References

1. Arifin, B.M.S., et al.: A novel data-driven leak detection and localization algorithm using the Kantorovich distance. Comput. Chem. Eng. **108**, 300–313 (2018)
2. Harus, L.G., et al.: Determination of temperature recovery time in diffierential-pressure- based air leak detector. Meas. Sci. Technol. **17**(2), 411–418 (2006)
3. Harus, L.G., et al.: Characteristics of leak detection based on diferential pressure measurement. In: Proceedings of the JFPS International Symposium on Fluid Power, vol. 2005, pp. 316–321 (2005)
4. Sagi, H.: Advanced Leak Test (2001). https://www.assemblymag.com/articles/83578-advanced-leak-test-methods. Accessed 02 Dec 2020
5. Schölkopf, B., et al.: Support vector method for novelty detection. In: Advances in Neural Information Processing Systems, May 2014, pp. 582–588 (2000)
6. Shi, Y., Tong, X., Cai, M.: Temperature effect compensation for fast differential pressure decay testing. In: Measurement Science and Technology, vol. 25, no. 6 (2014)
7. Titmarsh, R., Harrison, R.: Automated leak testing for cylindrical cell automotive battery modules: enabling data visibility using industry 4.0. In: 2019 23rd International Conference on Mechatronics Technology, ICMT 2019, pp. 1–4 (2019)

Fuzzy FMEA Application to Risk Assessment of Quality Control Process

Dagmara Łapczyńska and Anna Burduk(✉)

Wroclaw University of Science and Technology, Wybrzeze Wyspianskiego 27, 50-370 Wroclaw, Poland
{dagmara.lapczynska,anna.burduk}@pwr.edu.pl

Abstract. All of processes that are being performed are connected with the risk. Thus, manufacturing companies need to evaluate and react to these risks, as well as it is possible. One of the method recommended for risk assessment in production companies is Failure Mode and Effects Analysis (FMEA), which allows to calculate the risk and prioritize it. However, the FMEA is expert-knowledge based method, which makes it susceptible for the human-factor mistakes. The solution that allow to avoid uncertainty of FMEA is using the fuzzy sets, which is called fuzzy FMEA (fFMEA). The discussed case study is about the company that produces components being used in delivery vans – the production of these components need to end by the overall Final Quality Control (FQC), which means that 100% of components need to be controlled. This FQC process, like every else, is connected with the risk of mistakes. In the paper, the example of performing fuzzy FMEA in industry was described. In involves the analysis of FQC, which is very important, especially in automotive industry, where some of the possible risks or defects can result in danger for humans health or even a life. The aim of the research was to perform the risk evaluation of Final Quality Control (FQC) process, basing on the experts knowledge. The aim was reached by implementing the fuzzy FMEA method.

Keywords: Risk assessment · Quality management system · Fuzzy FMEA · Quality measurements · Quality control

1 Introduction

Nowadays, customers expect to be able to use the products they purchased safely, without taking the risk [1–4]. That is understandable, especially in the automotive industry, where the quality of products is closely related to safety of the customer (for example braking systems, airbags etc.). In the case of an accident (especially causing users death of health damage), which was caused by defective components of vehicle, the producer need to face not only legal consequences, but also the lowering of the overall brand value. Thus, quality measurement system in automotive industry need to be efficient enough to prevent releasing defective components to the customers, as much as it is possible to be done.

Á. Herrero et al. (Eds.): SOCO 2020, AISC 1268, pp. 309–319, 2021.
https://doi.org/10.1007/978-3-030-57802-2_30

The reliability of production process is one of the most important factor in automotive industry – due to detailed requirements of both, customers and legal standards (for example IATF 16949, which includes the standards of performing production process especially for the automotive area). Thus, quality assurance in the area of manufacturing vehicle components is very important task, that need to be done in every company [5–8]. In order to provide the suitable components quality, the companies perform quality measurements, that allow to verify if the characteristics of the manufactured product meet the requirements of legal standards or customers. However, each process is associated with the risk. It also applies to the process of measuring that is being performed at the end of the production process, which is Final Quality Control (FQC).

Quality management system (QMS) includes many different elements and all of them are to ensure the proper quality of the product. These may be formal documents (for example procedures, instructions, standards etc.), implementing the continuous improvement methods, performing trainings for operators, measuring tools etc. The details about QMS are also included in the ISO 9001 standards series. Thus, many of companies in automotive industry decide to certify their processes, to be able to ensure the customers about the proper quality of their products. One of the QMS parts is FQC process, that is usually carried in the companies from automotive industry [9, 10]. It usually include measurements of characteristics of final products. In order to consider the product as correct and able to send to the client, the results of these measurements need to be compatible to requirements, within the assumed limits.

The aim of the research was to perform the risk evaluation of Final Quality Control (FQC) process, basing on the experts knowledge. The aim was reached by implementing the fuzzy FMEA method.

2 Classical and Fuzzy Failure Mode and Effects Analysis (FMEA and Fuzzy FMEA)

The Failure Mode and Effects Analysis is one of the most popular methods that is being used in risk assessment. It is also one of the methods of risk assessment that is being recommended to be performed in manufacturing companies by the ISO/IEC 31010 [11]. It is based on experts knowledge, and is being done by valuing the three factors, which are detectability (D), severity (S) and occurance (O). After, the Risk Priority Number (RPN) is calculated. It is the result of multiplying all three values:

$$RPN = S \times O \times D \tag{1}$$

The value of RPN is within the 1 and 1000. The higher the RPN of potential defect is, the sooner the risk should be analysed.

The FMEA process is being done by the group of experts (process engineers, quality controllers etc.), which means that it can be affected by the subjective character of opinions. It is especially important while analysing the non-specific phrases like "very low" or "low" – because one risk can be assigned to both of the groups by different experts. This kind of data are known in fuzzy logic as linguistic variables.

The legitimacy of using the fuzzy sets in FMEA is confirmed in various types of processes and a lot of the authors emphasize the advantages of this solution in risk assessment process. The application of fuzzy FMEA includes for example in paper mill industry [12], sterilization units [13], maintenance of technical systems in mining [14], ships systems [15], water mounting process [16] and many more. The fuzzy FMEA, analogically to the classical FMEA, is the universal method that can be used actually in every area, after the proper base rules preparing.

The fuzzy rule base is actually the clue of using the fuzzy sets in risk assessment. Despite of the fact that a lot of researchers are performing the studies where they are trying to decrease the needed rules amount, it is often being said that the accuracy of the whole fuzzy analysis depends proportionally on the rules number. In the FMEA example, the number of rules is closely related to number of classes that are being used: with the 3 class evaluation (i.e. *low, moderate, high*) there need to be 27 of rules and with the 5 class evaluation (i.e. *very low, low, moderate, high, very high*) – 125 of rules. The most popular operator that is being used in rule base creating is Mamdani implication [12–16], because of its simplicity and well-fitting to the experts kind of problems. The used implication operator is called *minimum*, which is being based on the assumption that the truth of conclusion (output value) cannot be higher than the lowest membership of the premise (one of the input values) – it can be shown as:

$$\mu_{A \to B}(x, y) = MIN(\mu_A(x), \mu_B(y)) \quad (2)$$

where:

μ – membership function
x_i – input data
A_i – fuzzy set of A (premise)
y – output data
B_i – fuzzy set of B (conclusion)

This type of fuzzy FMEA performing was implicated in the case study in this paper. The fuzzy FMEA was chosen because of its advantages claimed by researches, which are [12, 16]:

- risk evaluation and prioritization basing on experts knowledge,
- ability to customize the risk assessment process to the character of process,
- high accuracy of risk evaluation (with well-defined fuzzy rule base).

3 Research of FQC in Automotive Industry

The research was performed in the automotive industry company, which is a producer of components to delivery vans in Poland. In this case, the customer is the other company that use these components in vans that they sell to the final user (see Fig. 1). Thus, the company, in addition to the requirements of the legal standards, must meet the requirements of vehicle producer.

The bolded part of the scheme (Fig. 1) is the company, which FQC process was examined.

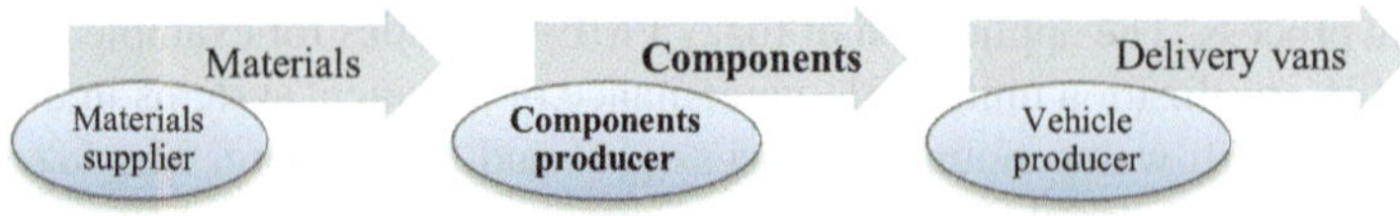

Fig. 1. Supply chain scheme

3.1 Final Quality Control (FQC)

The process of final quality control (FQC) includes measurements of final products, which can concern different amount of samples – in the discussed case, this process need to be done to every single product that was manufactured. That means, that the quality control process is being performed overall and include 100% of produced components. The process flow of FQC is shown in Fig. 2.

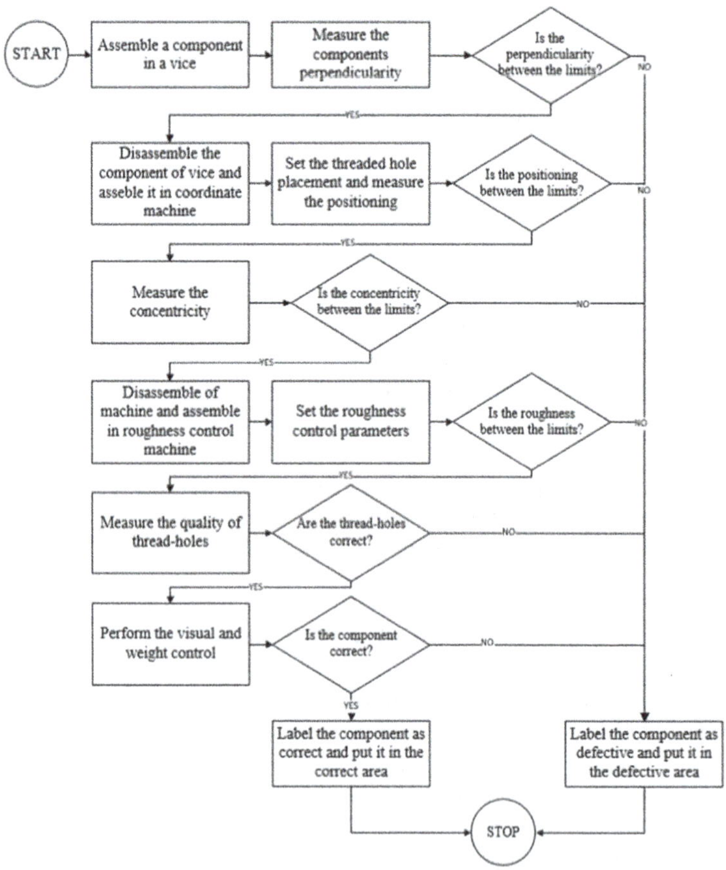

Fig. 2. Final quality control process flow

In order to evaluate the risk of performing the Final Quality Control process in the wrong way, the failure mode and effects analysis (FMEA) was performed.

3.2 FMEA of FQC Process

The FMEA method of risk assessment was chosen because of its verified ability to identify the risks in manufacturing companies [17]. There are three variables (severity, occurance, detectability) being considered while performing FMEA, which need to be specified by using linguistic variables (i.e. very low, low, moderate, high and very high) and numerical value assigned, within the 1–10 scale (see Table 1).

Table 1. Severity, detectability and occurance values defining

Variable	Meaning	Linguistic variable	Numerical value
Detectability (D)	What is the possibility of defect/risk detection?	Very low	1–3
		Low	3–5
		Moderate	5–7
		High	7–8
		Very high	9–10
Severity (S)	How severe is the defect/risk?	Very low	1–3
		Low	3–5
		Moderate	5–7
		High	7–8
		Very high	9–10
Occurance (O)	What is the possibility of defect/risk occurance?	Very low	1–3
		Low	3–5
		Moderate	5–7
		High	7–8
		Very high	9–10

The classic FMEA results of potential risks in performing FQC process are shown in Table 2.

In this case, the most important are two or three situations:

- **Mistake in reading the results of perpendicularity** (RPN = 360) – it is especially dangerous because of its very low detectability and very high occurance. It is a human mistake, which is not currently being under the control by other people than the measurement operator.
- **Inaccurate component positioning in machine in positioning measurements** (RPN = 360) – it is dangerous because of its occurance and severity. The wrong positioning of threaded hole can lead to loses in the vehicle can be the reason of accidents (severity)

Table 2. FMEA of Final Quality Control (FQC)

Process	Possible defect	Detectability (D)	Occurance (O)	Severity (S)	Risk Priority Number (RPN)
Perpendicularity measurements (R1)	Wrong assembly in a vice (R1.1)	1	9	9	81
	The use of damaged template (R1.2)	8	4	9	288
	Mistake in reading the results (R1.3)	8	5	9	360
Positioning measurements (R2)	Wrong assembly in a coordinate machine (R2.1)	3	8	7	168
	Measuring tip breaking (R2.2)	1	3	10	30
	Inaccurate component positioning in machine (R2.3)	5	8	9	360
Concentricity measurements (R3)	Wrong assembly in a coordinate machine (R3.1)	1	5	9	45
	Measure tip breaking (R3.2)	1	3	10	30
Surface roughness measurements (R4)	Wrong assembly in a roughness control machine (R4.1)	1	2	5	10
	The use of wrong surface control parameters (R4.2)	3	1	7	21

(*continued*)

Table 2. *(continued)*

Process	Possible defect	Detectability (D)	Occurance (O)	Severity (S)	Risk Priority Number (RPN)
Thread quality control (R5)	The use of wrong templates (R5.1)	2	1	8	16
	Lack of the grease in thread holes (R5.2)	2	1	2	4
	Wrong amount of thread holes (R5.3)	3	3	9	81
	Omission of thread hole (R5.4)	1	5	6	30
Visual and weight control (R6)	The use of wrong variant template (R6.1)	6	1	2	12
	The use of damaged weight (R6.2)	6	2	2	24

and the machine will not detect minor mistakes in positioning of the component (it can occur often). However, if the mistake is high enough, the result of measurement will show that the component is defective – even if it is not. It is less dangerous situation (the component need to be unassembled and assembled again), but it results in costs of re-manufacturing the correct product.

– **The use of damaged template in perpendicularity measurements** (RPN = 288) – this situation is dangerous because of its low detectability (analogical to the first one) and high severity (defects in perpendicularity results in not being able to assembly the component in the van, without the possibility to repair the component – it needs to be wasted). However, this situation occurs rather rarely, because the damage on the template is usually visual and the operator sees it.

The low RPN (especially with detectability 1) are the situations, where there is almost no possibility for the situation to be. These are for example the breaking of the measuring tip (the machine will not start the measurement process with broken tips) or wrong assembly in a vice (it is almost impossible to overlook, because in this situation the component is unstable which can be easily seen).

3.3 Fuzzy FMEA of Final Quality Control (FQC)

To perform the fuzzy FMEA in the components producer case, the scale of the linguistic variables were re-considered (Fig. 3). In this paper, the 5 class evaluation was used, including linguistic variables: *very low, low, moderate, high* and *very high.*

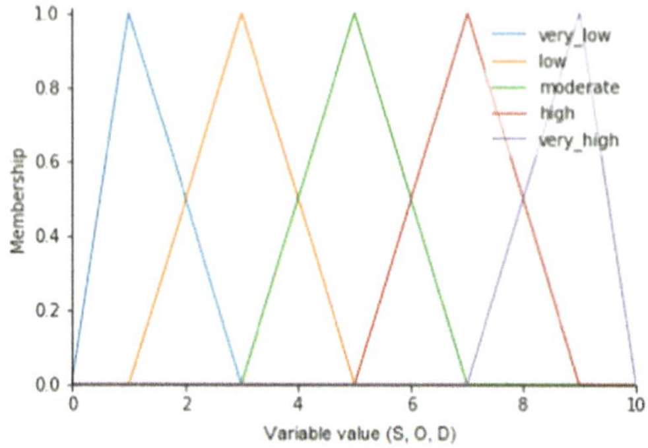

Fig. 3. Variable values (S, O, D)

Because of using the 5 class evaluation of risks/potential defects, the rule base had to include 125 rules in total. A few rules were shown in Fig. 4 as examples.

```
rule0=ctrl.Rule(occurance['very_low'] & detectability['very_low'] & severity['very_low'], rpn['very_low'])
rule1=ctrl.Rule(occurance['very_low'] & detectability['very_low'] & severity['low'], rpn['very_low'])
rule2=ctrl.Rule(occurance['very_low'] & detectability['very_low'] & severity['moderate'], rpn['low'])
rule3=ctrl.Rule(occurance['very_low'] & detectability['very_low'] & severity['high'], rpn['low'])
rule4=ctrl.Rule(occurance['very_low'] & detectability['very_low'] & severity['very_high'], rpn['low'])
rule5=ctrl.Rule(occurance['very_low'] & detectability['low'] & severity['very_low'], rpn['very_low'])
rule6=ctrl.Rule(occurance['very_low'] & detectability['low'] & severity['low'], rpn['low'])
```

Fig. 4. Example of implemented rules

The defuzzification was done by using the centroid (also known as center of area, COA) method, which characterizes with the finding of the center of the area of output data. The way of performing the COA method can be shown as:

$$y^* = y_c = \frac{\int y\mu_{output}(y)dy}{\int \mu(y)dy} \tag{3}$$

where:

y^* – crisp output

y_c – center of area (COA)

μ – membership function

The result of performing fuzzy FMEA was named the fuzzy RPN (FRPN) and listed in the Table 3.

The FRPN are calculated basing on the minimum of inputs membership functions, which was described in detail before. The visual example of this membership function was shown for the risk R2.3 (Fig. 5).

It can be seen, that the R2.3 risk is both – very high and high, with the stronger membership to very high value.

Table 3. FMEA of FQC process

Defect/risk type		Value			FMEA results	Fuzzy FMEA results	
		(D)	(O)	(S)	RPN	FRPN	Linguistic variable
R1	R1.1	1	9	9	81	750,00	High
	R1.2	8	4	9	288	779,76	High/very high
	R1.3	8	5	9	360	902,78	Very high
R2	R2.1	3	8	7	168	750,00	High
	R2.2	1	3	10	30	499,99	Moderate
	R2.3	5	8	9	360	902,78	Very high
R3	R3.1	1	5	9	45	499,99	Moderate
	R3.2	1	3	10	30	499,99	Moderate
R4	R4.1	1	2	5	10	249,99	Low
	R4.2	3	1	7	21	250	Low
R5	R5.1	2	1	8	16	375	Moderate/high
	R5.2	2	1	2	4	220,24	Very low/low
	R5.3	3	3	9	81	500	Moderate
	R5.4	1	5	6	30	500	Moderate
R6	R6.1	6	1	2	12	250	Low
	R6.2	6	2	2	24	375	Low/moderate

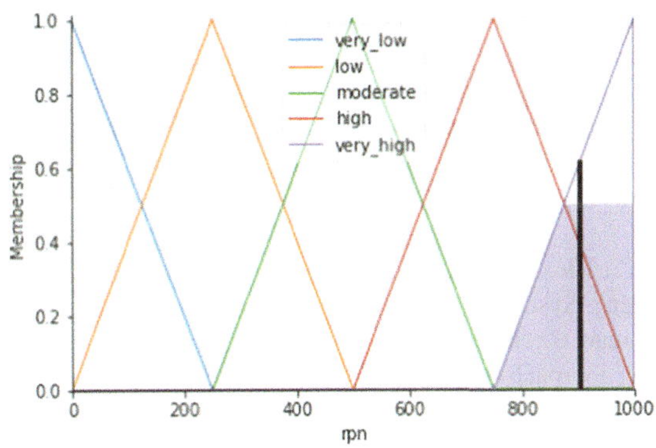

Fig. 5. The membership of R2.3 example

4 Conclusion

The fuzzy FMEA allowed to perform the risk assessment process. The main advantages of using the fuzzy rules in classical FMEA are:

- **the method is less susceptible for human factor mistakes**: the experts can assign one risk to two scales: i.e. the risk can be *low* and *very low* at the same time, and the membership function support to find out which one is more accurate,
- **the method should be more accurate in risk evaluation**: but only if the rule base is prepared well enough. In classical FMEA, the experts need to evaluate the RPN as low or high etc. basing on their knowledge and experience, while the fuzzy FMEA allow to mathematically count the closest value,
- **the method allows to assign different weights to the input values**: it was not performed in this study, but there is a possibility to choose which one of the input values (S, O, D) is most important – this can be very helpful in analysing specific processes, for example in medicine and hospitals FMEA the highest severity usually results in death, which possibly makes this value more important than detectability and occurance.

The implementation of fuzzy FMEA usually requires to use the software that includes the tool designed to perform the fuzzification and deffuzification processes on data. The main difficulty is the fact that it takes time and require the programming skills (in case of using own software) or money and time to learn how to use the tools (in case of using the available software, i.e. Matlab). The next important task is to build the rule base, which can take different amount of time (depending on used tool) and require the knowledge about risks that are being analysed. However, once these tasks are done, it the tool can be used all over again to different processes (with updating the rule base in every case, depending on the character of process that is being analysed).

References

1. Mrugalska B., Tytyk E.: Quality control methods for product reliability and safety. In: 6th International Conference on Applied Human Factors and Ergonomics (AHFE 2015) and the Affiliated Conferences, AHFE 2015. Procedia Manuf. **3**, 2730–2737 (2015)
2. Myers, A.: Complex System Reliability: Multichannel Systems with Imperfect Fault Coverage. Springer-Verlag, London (2010)
3. Górny, A.: Minimum safety requirements for the use of work equipment (for example of control devices). In: Occupational Safety and Hygiene – Sho 2013, pp. 227–229 (2013)
4. Nakagawa, T.: Advanced Reliability Models and Maintenance Policies. Springer-Verlag, London (2008)
5. Xu, K., Tang, L.C., Xie, M., Ho, S.L., Zhu, M.L.: Fuzzy assessment of FMEA for engine systems. Reliab. Eng. Syst. Saf. **75**, 17–29 (2002)
6. Stylidis, K., Wickman, C., Söderberg, R.: Defining perceived quality in the automotive industry: an engineering approach. In: CIRP 25th Design Conference Innovative Product Creation. Procedia CIRP, vol. 36, pp. 165–170 (2015)
7. Schmitt, R., Quattelbaum, B., Falk, B.: Distribution of customer perception information within the supply chain. Oper. Supply Chain Manage. **3**(2), 94–104 (2010)
8. Burduk, A., Kochańska, J., Górnicka, D.: Calculation of labour input in multivariant production with use of simulation. In: Information Systems Architecture and Technology Proceedings. Advances in Intelligent Systems and Computing, vol. 1051, pp. 31–40 (2020)
9. Reis, D., Vanxo, F., Reis, J., Duarte, M.: Discriminant analysis and optimization applied to vibration signals for the quality control of rotary compressors in the production line. Arch. Acoust. **44**(1), 79–87 (2019)

10. Nahmias, S., Olsen, T.L.: Production and Operations Analysis: Strategy, Quality, Analytics. Application. Waveland Press, Long Grove (2015)
11. ISO/IEC 31010:2009 Risk management—Risk assessment techniques. The International Organization for Standardization and The International Electrotechnical Commission (2009)
12. Sharma, R., Kumar, D., Kumar, P.: Systematic failure mode effect analysis (FMEA) using fuzzy linguistic modelling. Int. J. Qual. Reliab. Manage. **22**, 986–1004 (2005)
13. Dagsuyu, C., Gocmen, E., Narli, M., Kokangul, A.: Classical and fuzzy FMEA risk analysis in a sterilization unit. Comput. Ind. Eng. **111**, 286–294 (2016)
14. Petrovic, D.V., Tanasijevic, M., Milic, V., Lilic, N., Stojadinovic, S., Svrkota, I.: Risk assessment model of mining equipment failure based on fuzzy logic. Expert Syst. Appl. **41**, 8157–8164 (2014)
15. Nguyen, H.: Fuzzy methods in risk estimation of the ship system failures based on the expert judgements. J. KONBiN **43**, 393–403 (2017)
16. Tay, K.M., Lim, C.P.: Fuzzy FMEA with a guided rules reduction system for prioritization of failures. Int. J. Qual. Reliab. Manage. **23**(8), 1047–1066 (2006)
17. Almannai, B., Greenough, R., Kay, J.: A decision support tool based on QFD and FMEA for the selection of manufacturing automation technologies. Robot. Comput. Integr. Manuf. **24**, 501–507 (2008)

Similarity of Parts Determined by Semantic Networks as the Basis for Manufacturing Cost Estimation

Grzegorz Ćwikła[1(✉)] and Krzysztof Bańczyk[2]

[1] Faculty of Mechanical Engineering, Silesian University of Technology, Konarskiego 18A, 44-100 Gliwice, Poland
grzegorz.cwikla@polsl.pl

[2] Research and Development Centre for Mechanical Appliances OBRUM Ltd., Toszecka 102, 44-117 Gliwice, Poland

Abstract. The method of estimating the production costs proposed in the article is based on the hypothesis that the cost of producing a newly introduced element is similar to the production cost of a previously manufactured element, provided that the elements are similar in terms of design, structure and manufacturing technology. The semantic web method was used to determine the similarity of the elements. In proposed method, the shape as well as structural and technological features of the element are recorded in the form of a graph. The element is divided into functional surfaces to which quantitative and qualitative parameters and technological features can be assigned. Networks describing specific elements can be compared by pairs, resulting in obtaining a factor of structural and technological similarity (s&t similarity). The ability to set the weights of semantic network's branches allows to fine-tune the method to the requirements of different users, according to specific technical and organizational conditions in the company. In order to verify the thesis, the estimated costs of a selected group of gear-housing-type elements were compared with the costs calculated by another method.

Keywords: Cost estimation · Semantic networks · Functional surfaces · Structural and technological similarity

1 Introduction

The need for quick response to market demand means that companies devote considerable attention to developing tools that can shorten the time of production preparation (understood as the acceptance of an inquiry for the production of specific elements, preparation of a price offer, waiting for the ordering party's response, and if it is positive, the order is carried out) and accelerate its individual stages [2]. Small and medium-sized enterprises (SMEs) from the machining industry can be in a more difficult situation, because they frequently do not have a fixed production program and the main part

Á. Herrero et al. (Eds.): SOCO 2020, AISC 1268, pp. 320–330, 2021.
https://doi.org/10.1007/978-3-030-57802-2_31

of their activity is the realization of small, various orders, which are not necessarily repeated [4, 5]. One of the most important parts of this process is the preparation of offers as answers to requests for manufacturing specific products. The answer to the inquiry should be given as soon as possible and at the same time must meet several conditions [6]. The manufacturer must know the actual manufacturing cost of the specific part. If the part has already been produced, it is necessary to just check and update the cost of its production in the past, calculated by precise methods [7]. If the element has not been manufactured so far, it is necessary to use either exact (calculation based on machining times) or approximation methods.

Cost calculation based on machining times of mechanical elements requires time-consuming machining process planning and a lot of calculations of the demand for machine time and labour [6]. Literature sources indicate that only 20–30% of inquiries result in ordering proposed product, so conducting a full process planning procedure leads to unnecessary waste of resources needed for this process [9]. In that situation it is usually preferred to use one of the approximation methods. Classic estimation methods are in many cases based on the experience and intuition, so their accuracy can be problematic [3]. This paper presents approach based on assumption that production costs of previously manufactured elements are known. Proposed hypothesis states, that it production costs of new parts can be estimated, if it is possible to find an element similar to a new one in the database, containing previously produced parts. If similarity factor of these elements is higher than threshold, their production costs are similar too in the degree allowing for fast and safe pre-offer preparation.

The theoretical foundations of the proposed method and the first practical example of its application were presented in [1], in this article only some information will be repeated for greater clarity of the text, while new elements are the extension of the method to other, more complex types of machine parts, whose structural and technological similarity, and then the similarity of costs will be estimated. In the previous text, similarities of *shaft-type* machine parts were examined, now the proposed algorithms will be used for *gear-housing-type* parts.

1.1 Method for Determining the Similarity of Parts

This study proposes a solution based on the assumption that the documentation and accurate final calculations of the manufacturing cost of parts that have already been produced are available. Among these objects, those that are similar to the new one are searched, taking into account the features that have the greatest impact on manufacturing costs. If the database contains description of previously produced parts, whose degree of structural and technological similarity (s&t similarity) to the new object is high enough, we can assume that the cost of producing the new part will be similar to the cost of producing the part retrieved from the database (Fig. 1). This assumption finds additional support in the principles of group technology, often used in companies producing similar mechanical parts [9]. According to the assumptions of group technology, a group of similar elements is sought, from which the most complex representative is selected, and then the reference technological process is developed for him. The technological documentation of other elements is prepared by modifying and simplifying the reference process.

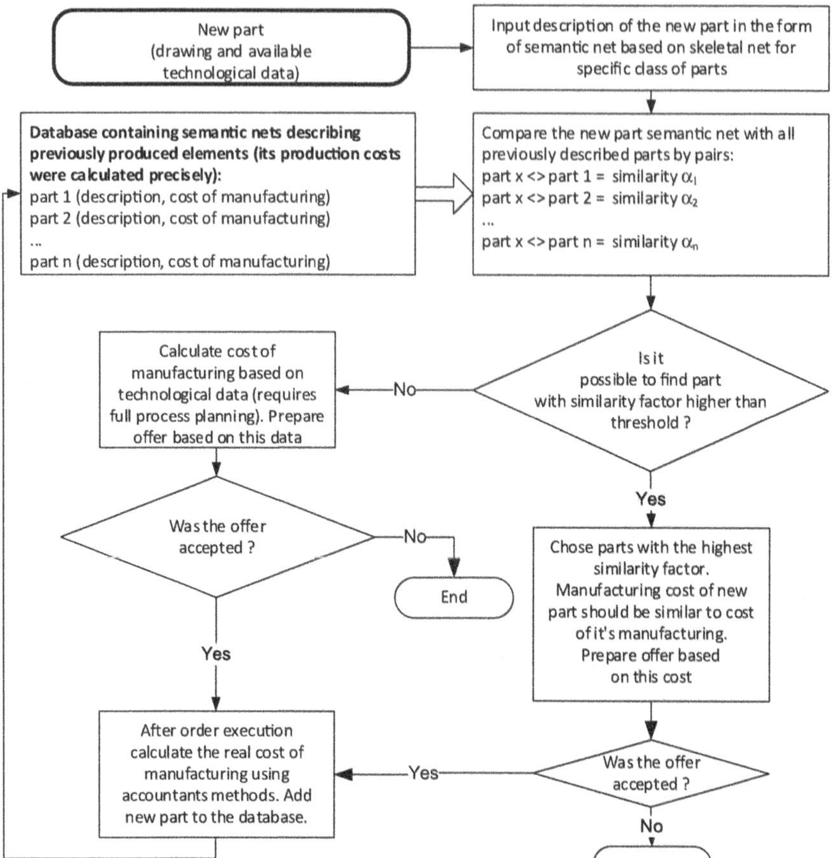

Fig. 1. The cost estimation algorithm [1].

In the case of practical application of the described method, to obtain correct results, it is necessary to take into account the fact that a period of time has already passed since the production of the reference elements, which means that due to inflation and other external and internal factors, data on the costs of their production are no longer valid. Therefore, their direct use is inadvisable, these data should be somehow updated to reflect current manufacturing costs.

2 Description of Gear-Casing-Type Parts Using Semantic Networks and Estimating Similarity of Semantic Networks

The part description method based on semantic networks [8] and justification for the selection of semantic networks is presented in [1]. The previous text presented a method of describing and determining the similarity of shaft-type elements, which are usually relatively simple and unambiguous. Housing-type parts (e.g. gear-housings) are much more complicated in terms of technology, because they can have many different elementary surfaces and other features, usually it is necessary to machine them in many positions, from different directions, on many machines. In addition, when gear-casing is composed of two parts, some technological operations must be performed simultaneously on both parts constituting a single gear-housing.

Due to the high variety of elements that can be classified as housing-type, it was decided to take into account the group of gear-housings produced by the selected manufacturer, similar to each other in terms of materials, degree of complexity, shape and dimensions. All the proposed gear-housings can be produced using machines owned by the said manufacturer. As with shaft-type parts, the description of the features of the gear-housing parts was divided into two sections, the description of gear-housing's technological and structural features and the description of gear-housing's elementary functional surfaces.

2.1 Description of Gear-Housing's Technological and Structural Features

Features describing form and technological features of a part are presented below:

- type of form of gear-housing (*single-body*, *divided*),
- dimensions of a part (*length*, *width*, *height* – max values od dimensions),
- weight of a part,
- type of a preform/stock (*cast*, *welded*, *other*),
- expected machining except mandatory milling (*grinding of flat surfaces*, *lapping of flat surfaces*, *grinding of holes*, *reaming of holes*, *threading*),
- heat treatment (*necessary* or *unnecessary*),
- type of material (*cast iron*, *cast steel*, *light alloy*, *other*),
- volume of production (*single-piece*, *series production*, *mass production*).

2.2 Description of Gear-Housing's Elementary Functional Surfaces

To define elementary surfaces of the gear-housing the user has to describe every entity belonging to one of three planes: XY, YZ and XZ (Fig. 2). The entities are as follows: *base surface*, *division surface* (only for divided gear-housings), *secondary positioning holes in the base surface*, *secondary fastening holes in the division surface* (only for divided gear-housings), *secondary holes fastening cover* and *main hole*. Each of entities can be defined by the following relative values: length, width, thickness and diameter (calculated relatively to the total width of the gear-housing), as well as technological features (e.g. dimensional accuracy, roughness, method of machining). The *main hole* feature has large influence on production cost and have to be described more precisely. It is described by following features: *form of the main hole* (divided, undivided, port, blind hole), *relative position*, *diameter* and *length*, *coaxial elementary holes*, *dimensional accuracy*, *roughness*, *machining of the face* (interior and exterior faces are described more precisely) and *associated surfaces* (surfaces that can be associated to a main hole are as follows: *bored groove*, *groove for spring ring*, *sloped edge*, *perpendicular oiling hole*.

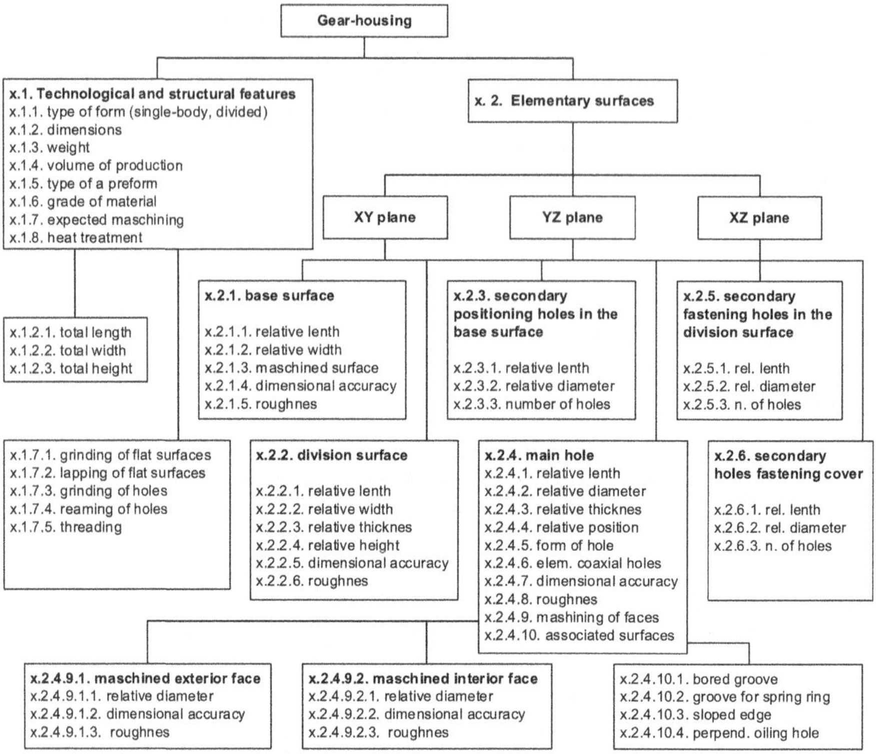

Fig. 2. The skeletal network describing a gear-housing [source: own].

Because we took into consideration only the selected group of gear-housings it is possible to pass over entities like an *inspection opening cover*, *deaerator*, *oil level gauge* and *drain plug hole.* Each gear-housing manufactured by chosen producer has to have these features and its influence on cost of machining is similar. Due to the fact that depending on the owned machine park and the degree of employee training, the impact of individual features on the final cost of the product may be different, the possibility of weighting for individual network branches, and individual structural and technological features, has been introduced. The introduction of weights allows tuning the system to the needs of a specific producer. The application of the described system in a specific enterprise requires a numerous tests to be carried out in order to select the appropriate weights for individual features and network branches.

2.3 Assessment of the Similarity of Semantic Networks

The process of comparing set of two nets describing elements starts from lowest level and steps up to the top of the network, where the overall similarity factor (s&t similarity) of the compared networks is obtained. If one of the corresponding nodes does not exist, the similarity of these nodes equals zero. Total similarity factor of networks is based on partial ones, calculated as a weighted average of node similarities [8]. The result of the comparison is a number from [0, 1] range, where 1 means that elements are identical. Calculation method of the corresponding node similarities depends on the type of node. Equations allowing to calculate similarities of different corresponding types of nodes were presented in our article [1].

3 An Application of Proposed Method

The input data were drawings of 16 gear-housing parts, obtained from a medium-sized company specializing in machining of such parts. The sketch of an example element is shown in Fig. 3, Table 1 presents its main parameters and features.

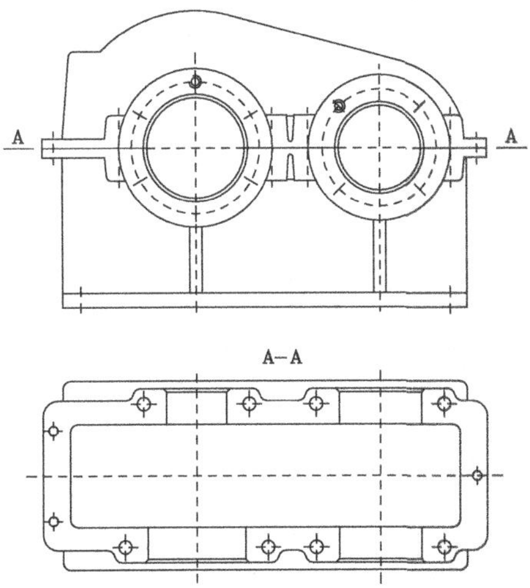

Fig. 3. An example of gear-housing-type part [source: own].

Table 1. Selected parameters and features of an example gear-housing

Max dimensions [mm]			Weight [kg]	Volume of production	Type of preform	Material	Additional machining
Height	Length	Width					
300	416	180	33	series	cast	cast iron	threading
Plane	*Elementary surface*			*Main hole feature*	*Dimensional accuracy*	*Roughness*	*Associated features*
XZ	A plane with the main hole			Hole	High	2,5	Chamfer
				Hole	High	2,5	Chamfer
XZ	A plane with the main hole			Hole	High	1,25	Chamfer
				Hole	High	1,25	Chamfer
XY	Base surface				Medium	20	
XY	Division surface				Medium	2,5	

The developed algorithm has been implemented in the Prolog language, an interactive program has been developed that allows to enter a gear-housing description, as well as calculate s&t similarity of a specific pair. Not all surfaces of a typical gear-housing have to be machined, which simplifies the description, because only machined surfaces need to be included in the description.

3.1 Cost Similarity

The *real production cost similarity factor* was again defined to compare *s&t similarity* results with real production costs similarities [1]. In addition to the drawings of 16 parts, the company provided data on the cost of their manufacturing, calculated by classical methods. Real manufacturing costs of provided gear-housings were calculated based on the following partial costs: material cost (casting or welding cost), labour cost (calculated from workers earnings), overheads (87.5% of labour cost, e.g. cost of social security), department cost (680% of overheads) and plant cost (82% of department cost). Total cost of element in this company is the sum of these components. Table 2 shows real cost calculation of provided gear-casing-type parts.

Table 2. Example parts real cost calculation

Part name	Material cost	Labour cost	Over-heads	Department cost	Plant cost	Total cost
Gear-casing1	138.60	58.15	50.88	345.97	283.69	**877.28**
Gear-casing2	172.20	66.68	58.35	396.77	325.35	**1019.36**
Gear-casing3	180.60	62.21	54.44	370.16	303.53	**970.94**
Gear-casing4	159.60	56.93	49.81	338.71	277.74	**882.79**
Gear-casing5	325.50	68.55	59.99	407.90	334.48	**1196.42**
Gear-casing6	987.00	101.57	88.88	604.35	495.57	**2277.37**
Gear-casing7	722.40	91.08	79.70	541.93	444.39	**1879.50**
Gear-casing8	180.60	68.31	59.77	406.45	333.29	**1048.42**
Gear-casing9	168.00	76.44	66.89	454.84	372.97	**1139.14**
Gear-casing10	588.00	97.51	85.32	580.16	475.73	**1826.71**
Gear-casing11	155.40	75.22	65.82	447.58	367.01	**1111.04**
Gear-casing12	630.00	80.27	70.23	477.58	391.61	**1649.69**
Gear-casing13	168.00	64.57	56.50	384.19	315.04	**988.30**
Gear-casing14	252.00	66.68	58.35	396.77	325.35	**1099.16**
Gear-casing15	424.20	80.51	70.45	479.03	392.81	**1446.99**
Gear-casing16	504.00	85.39	74.72	508.06	416.61	**1588.78**

3.2 S&T Similarity and Cost Similarity Comparison Results

S&t similarities between all elements and cost similarities were calculated. Resultant s&t and cost similarities were compared. High similarity of both results means that system of cost estimation works properly. In the first series of calculations all weights were set to 1. Results are presented on the chart (Fig. 4). First curve marked with rhombus shows *s&t similarity*, second curve marked with squares shows *cost similarity* and third one, marked with circles shows *difference between these similarities*. Results acquired in 1-st set of calculations were not satisfying. The difference between similarities is high only if c&t similarity is higher than 0.96. This means that with this weights setting it is difficult to get useful results.

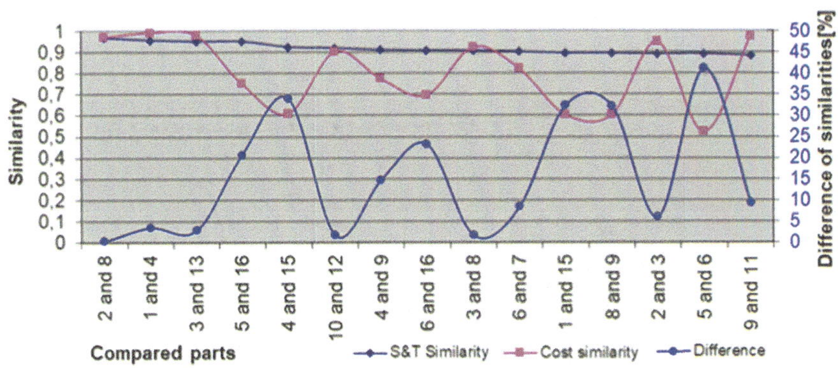

Fig. 4. S&T and cost similarity in the 1-st series of calculations

In the second calculation series weights of following features were increased: "dimensions" and "weight" in node "technological and structural features", "base surface", "division surface", "main hole", feature "elementary main hole" in node "main hole". Results of 2-nd series (Fig. 5) were better. Number of pairs of elements having high s&t similarity and low cost similarity goes down. Threshold similarity factor guarantying proper results of cost estimation can be set to 0,92.

It was decided to introduce further changes in the weights, the importance of such features as dimensional accuracy, roughness, and many others was increased. These changes caused worsening of results (Fig. 6), comparing to the 2-nd series. Difference between similarities has increased. As a result, some changes were withdrawn while experimenting with other settings. A total of 5 series of calculations were carried out, the best results were obtained in the 5-th series (Fig. 7). The curve representing the difference between c&t similarity and cost similarity is smooth. There are no sudden changes of the similarities difference. Cases of pairs of elements having high s&t similarity factor and low cost similarity were reduced. If the threshold similarity factor is equal 0,9, the cost estimation accuracy is about ±7%. This is accuracy sufficient for early calculation of offers.

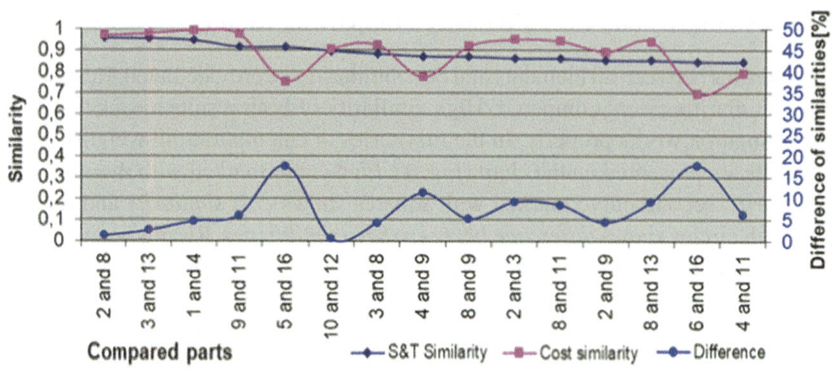

Fig. 5. S&T and cost similarity in the 2-nd series of calculations

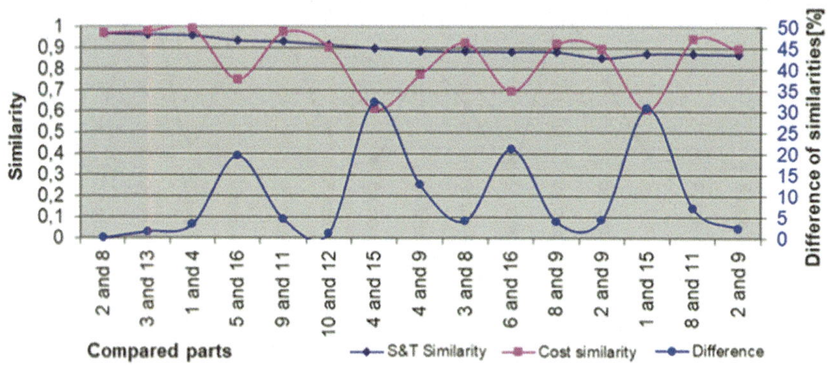

Fig. 6. S&T and cost similarity in the 3-nd series of calculations

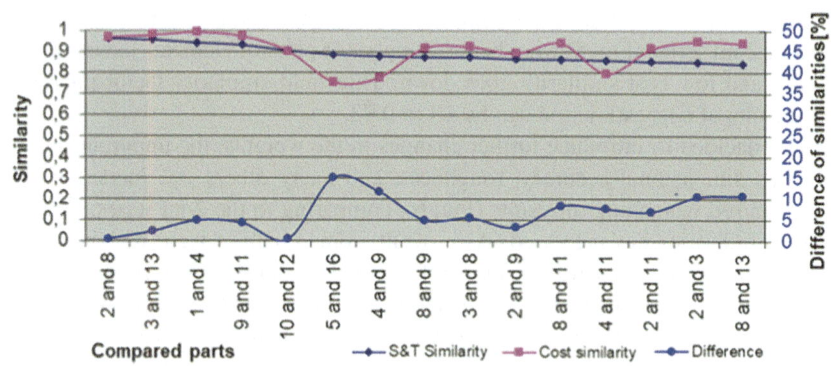

Fig. 7. S&T and cost similarity in the 5-th series of calculations

4 Summary

Presented method enables for relatively quick assessment of pair of elements structural and technological similarity, and in consequence - production costs estimation. It takes the quantitative and qualitative features of elements into account. It is possible to compare elements belonging to the same class using the semantic net for specific class or group of parts, as long as a skeletal semantic net for this class were created and programmed. Proposed method can be easily applied for any group of axial-symmetry elements. Description of housing-type elements is more difficult because of a large variety of parts of this type and its higher degree of complication. Development of skeletal semantic network describing any housing-type element is very difficult, because housings are manufactured using many methods and have wide range of shape and dimensions. In this situation skeletal network describing the selected group of gear-housings has been created and tested.

Possibility of changing the semantic network weights factors allows to tune the system for specific manufacturer conditions. Extensive database containing descriptions of elements is needed for proper system operation. If weights of semantic network branches are set correctly, accuracy of cost estimation is proportional to the similarity factor. The procedure of setting proper nodes weights of semantic network consists of calculating s&t similarities of elements and comparing them with the cost similarity. The cost similarity is based on real production costs. Weights of nodes having high influence on production cost have to be increased, the problem is to find the most important features. The time consumption of the algorithm has not been measured, because currently there are too few described items in the database. The need to compare a new element with each in the database means that time consumption will increase proportionally to the number of elements in the database.

Płatne ze środków Ministerstwa Nauki i Szkolnictwa Wyższego na podstawie umowy nr 12/DW/2017/01/1 z dnia 07.11.2017 r. Paid from the funds of the Ministry of Science and Higher Education, contract No. 12/DW/2017/01/1 of 07.11.2017.

References

1. Ćwikła, G., Grabowik, C., Bańczyk, K., Wiecha, Ł.: Assessment of similarity of elements as a basis for production costs estimation. In: Martínez Álvarez, F., Troncoso Lora, A., Sáez Muñoz, J.A., Quintián, H., Corchado, E. (eds.) SOCO 2019. AISC, vol. 950, pp. 386–395. Springer, Cham (2020)
2. Davidrajuh, R., Skolud, B., Krenczyk, D.: Performance evaluation of discrete event systems with GPenSIM. Computers **7**(1), 8 (2018). https://doi.org/10.3390/computers7010008
3. Kempa, W.M., Paprocka, I., Kalinowski, K., Grabowik, C., Krenczyk, D.: Study on transient queueing delay in a single-channel queueing model with setup and closedown times. In: Dregvaite, G., Damasevicius, R. (eds.) ICIST 2016. CCIS, vol. 639, pp. 464–475. Springer, Cham (2016). https://doi.org/10.1007/978-3-319-46254-7_37
4. Krenczyk, D., Skolud, B., Herok, A.: A heuristic and simulation hybrid approach for mixed and multi model assembly line balancing. In: Advances in Intelligent Systems and Computing, vol. 637, pp. 99–108 (2018). https://doi.org/10.1007/978-3-319-64465-3_10
5. Paprocka, I.: The model of maintenance planning and production scheduling for maximizing robustness. Int. J. Prod. Res. (2018). https://doi.org/10.1080/00207543.2018.1492752

6. Roy, R., Souchoroukov, P., Shehab, E.: Detailed cost estimating in the automotive industry: Data and information requirements. Int. J. Prod. Econ. **133**, 694–707 (2011)
7. Salmi, A., David, P., Blanco, E., Summers, J.D.: A review of cost estimation models for determining assembly automation level. Comput. Ind. Eng. **98**, 246–259 (2016)
8. Song, S., Lin, Y., Guo, B., Di, Q., Lv, R.: Scalable distributed semantic network for knowledge management in cyber physical system. J. Parallel Distrib. Comput. **118**, 22–33 (2018)
9. Więcek, D., Więcek, D.: Production costs of machine elements estimated in the design phase. In: Intelligent Systems in Production Engineering and Maintenance – ISPEM 2017. Advances in Intelligent Systems and Computing, vol. 637. Springer, Cham (2018)

A Simulated Annealing Based Method for Sequencing Problem in Mixed Model Assembly Lines

Damian Krenczyk and Karol Dziki(✉)

Faculty of Mechanical Engineering, Silesian University of Technology, Gliwice, Poland
{damian.krenczyk,karol.dziki}@polsl.pl

Abstract. The paper proposes a method to solve the mixed-model assembly line sequencing problem based on the Simulated Annealing Optimization algorithm. Achieving full line synchronization, by creating the appropriate model version sequence, becomes increasingly difficult at current levels of product complexity. The method of generating the candidate sequence by repeatedly swapping two random positions depending on the current temperature value was used. The search area is relatively large in the early phase of the algorithm. In addition, the conditions for resetting the temperature indicator if the local point candidate solutions are not improved have been added. It was also necessary to create a search objective function, taking into account specific aspects related to the mix-model sequencing problem. The proposed approach is based on binary coding of the input sequence and a suitably modified method of determining the boundaries of the search area. This increases the chance to avoid local optima trapping.

Keywords: Mixed-model · Simulated annealing · Sequencing · Assembly line

1 Introduction

With the growing need to adapt products to customer requirements in today's market and the growing demand for diversified goods, production systems must reach an increasingly high level of complexity. For this reason, among other things, the most dynamically developing concepts of production systems are Mixed-Model Assembly Lines (MMAL). In such systems, it becomes necessary to solve problems in the areas of technology and organization related to the production of many models on the same line. MMAL is based on the concept of product flow, during which individual features of the product version are processed in subsequent stages of production. New methods of acquiring process data required in planning and control systems are also being sought [1]. These products must, therefore, be designed in such a way as to achieve the maximum level of line flexibility. Most often they contain a common basic part, to which additional components are mounted, and additional functions are assigned in accordance with the requirements of a given variant or model.

Á. Herrero et al. (Eds.): SOCO 2020, AISC 1268, pp. 331–341, 2021.
https://doi.org/10.1007/978-3-030-57802-2_32

The MMAL design process requires that two basic problems be solved first [2]. The first concerns line balancing (Mixed Model Assembly Line Balancing Problem MMALBP), i.e. determining the allocation of operations to assembly stations. The allocation should be made in such a way as to perform variable production in accordance with the fixed cycle time or the fixed number of assembly workstations. It is usually solved by searching for the minimum number of assembly stations and their operation when a new system is designed and external demand is well estimated [3, 4]. This situation occurs in modern systems in the automotive industry, in which the system cycle (which is also often a derivative of market demand) must be synchronized with other departments. However, when a given assembly line already exists and needs to be optimized - solving the balancing problem usually requires minimizing the sum of operation times [5].

In the literature, the most common approaches to solve this type of problem consider the time of assembly operations as average values for the entire population of variants [6, 7]. Depending on the type and complexity of the product, when at certain stages of the assembly task times may differ significantly or be omitted, the use of average values may cause problems with the so-called smoothness of the solution. The second problem is related to determining the order in which each version of the product will be assembled, which boils down to creating a production sequence that meets the demand requirements at a fixed settlement time. Therefore, it is required to set intervals determining the amount of time between the given versions, while taking into account critical parameters, such as, for example, inventory levels, internal logistics restrictions or additional elements limiting smooth production. The specified sequence should be verified by determining deviations from the nominal tact time of line between subsequent operations on the workstation and analyzing their impact on the execution date of the order. Both of the above problems, in general, are NP-hard. The solution to these problems is, therefore, especially in the automotive industry, where dozens and sometimes hundreds of operations can be carried out in different variants, depending on the model currently being performed, in many cases impossible using accurate methods (branch and bound, dynamic programming, etc.) [8, 9].

Published results of research in this area are based on artificial intelligence methods of searching for near-optimal solutions, for example, heuristic methods [10], machine learning algorithms [11], genetic algorithms [12], tabu search [13] or based on simulated annealing (SA) [14]. The approach based on a modified SA algorithm was proposed in this study. The basic difference from the previously published research is related to the proposed method of generating the candidate solution and resetting the temperature in the process of algorithm execution. Methods of generating candidate solutions found in the literature are generally based on a simple or complex (from 2–3 steps) process of transforming the current sequence, e.g. exchanging two successive units, pairwise swapping, three-way swapping, inversion or insertion in the entire area of feasible sequences [14–17]. In this study, the candidate sequence is generated by repeatedly swapping two random positions, which depends on the current temperature value. This allows for a relatively large range of search area in the early phase of the algorithm. In addition, conditions have been added to allow the temperature indicator to be reset if the candidate solutions for the local point is not improved. The paper is organized as follows: In

Sect. 2, the MMALSP is defined. In the following subsections, the simulated annealing algorithm with its modifications for MMALSP is presented. The proposed objective function is also defined. Section 3 presents a computational example, which is solved by the proposed algorithm. Finally, Sect. 4 concludes the research with directions for future work.

2 Mixed-Model Assembly Line Sequencing Problem

The stage of solving the problem of assembly line optimization connected with determining the sequence of assembled versions (mix-model), which is performed after carrying out the line balance is considered. The basic assumption is that the number of workstations and the nominal cycle time of the line are specified. Transportation times of parts between workstations are ignored as insignificant compared to the cycle time of the line. Only one product can occupy a given workstation and the station is occupied until the next station is ready to receive it (the next operation will not start sooner). Once determined, the product version position in the sequence is fixed and matching takes place in a specific order.

2.1 The Simulated Annealing Algorithm

The fundamental descriptions of the Simulated Annealing algorithm were provided by S. Kirkpatrick's et al. [18], Eglese [19] and Goldberg [14, 20]. The concept of this method was inspired by the simulation of the physical metallurgy annealing process. The generated solutions reflect the slow cooling of the heated material in order to achieve equilibrium at the smallest grain sizes. The feature of annealing is that achieving a better effect (smaller grain) is possible by reheating the material to the initial temperature and cooling it again in other conditions, e.g. by changing parameters (cooling rate). Table 1 illustrates the characteristic steps in implementing Simulated Annealing based algorithms.

Simulated Annealing Optimization (SAO) gives good results in process optimization problems as well as operational research, as it has a built-in mechanism to avoid local optima. Unlike genetic algorithms or motion-based swarms of points or insects, it does not require a large amount of memory to record previous iteration parameters in order to return to stages where there are doubts that they have directed the algorithm toward the local optimum.

Table 1. Characteristic stages in implementing SA-based algorithms.

Description of the stage	Stage illustration
Step 1. Generate a space of feasible solutions, initial and final values of the control parameter temperature (*T*), the cooling rate (cr) and the iteration counter (*it*).	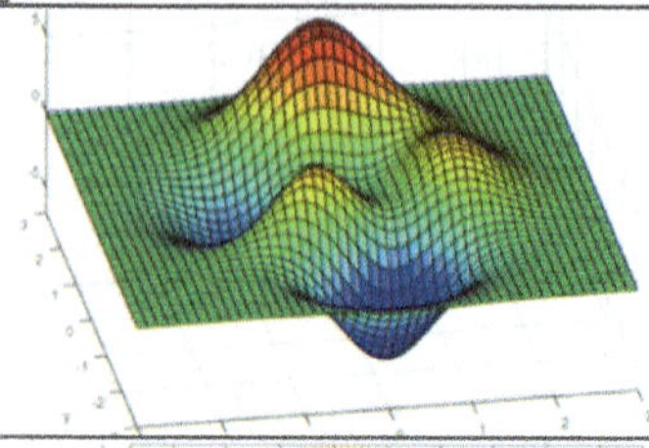
Step 2. The initial solution is selected (f_0). (The middle point of the area or a random point).	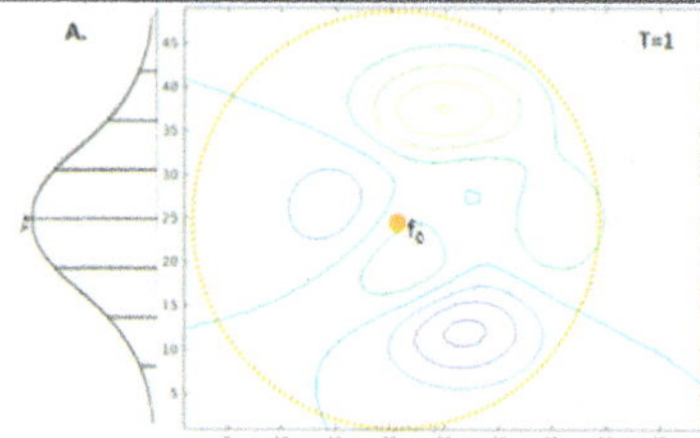
Step 3. Generate a feasible candidate neighboring solution (f_1).	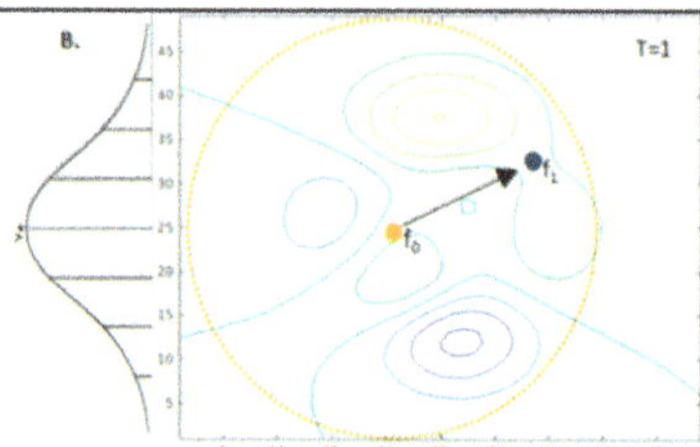
Step 4. For current and neighboring solution (f_c, f_1) the objective function value are calculated $(g(f))$. If $g(f_1) < g(f_c)$, then the point becomes a new starting point.	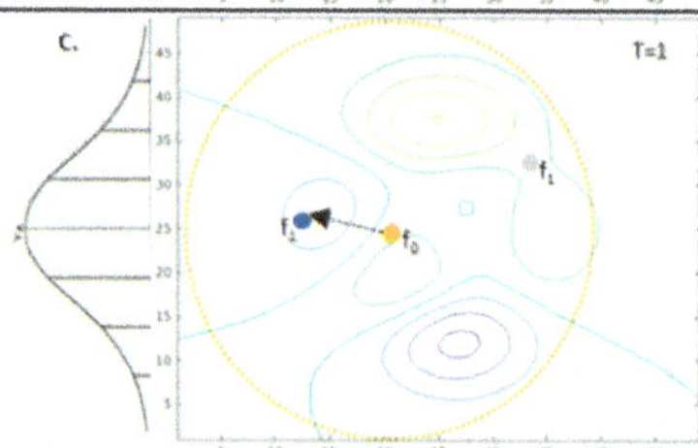
Step 5. In case the value $g(f_n)$ is not less than the best value, in order to avoid the "trap" of the local optimum, the algorithm attempts to determine the probability *P(A)* at which the new solution is accepted.	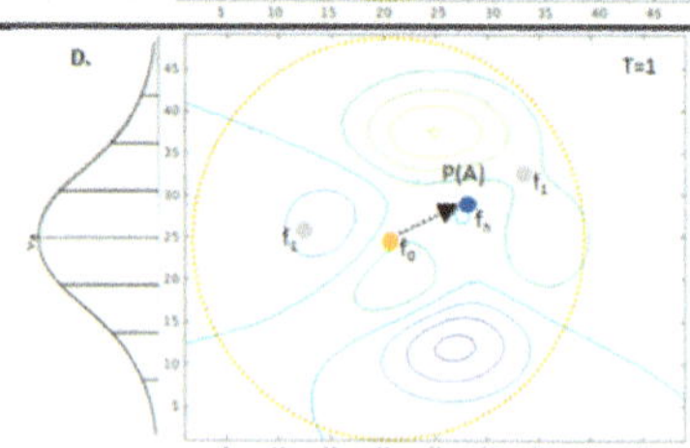

(*continued*)

Table 1. *(continued)*

Step 6. After a specified amount of iteration scans (*it*) inside each temperature level (*T*), the area is modified by a new temperature value that changes the range of the search area.	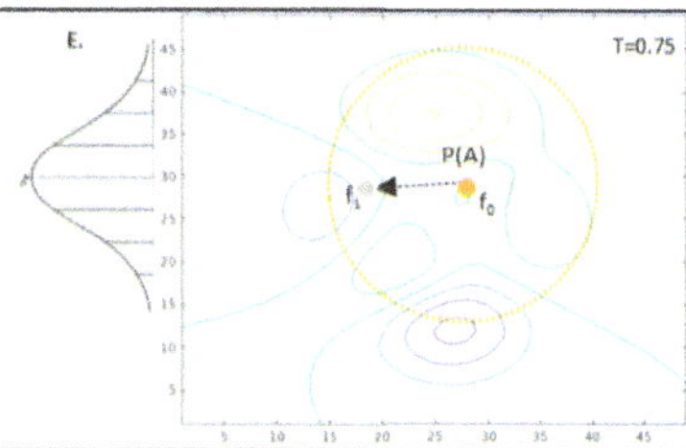
Step 7. The algorithm does not search the entire space for feasible solutions. In this situation, the algorithm may get stuck in the local optimum.	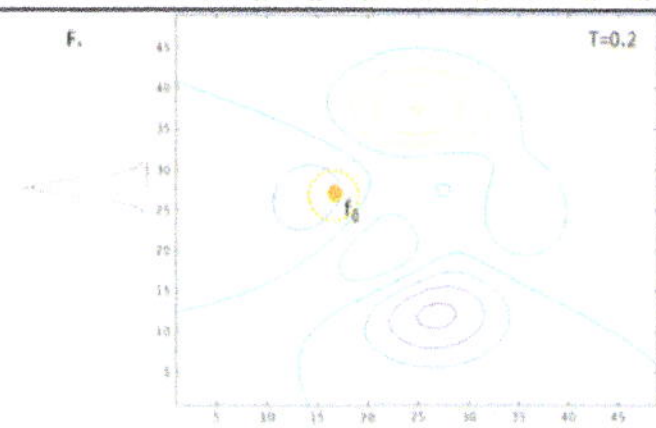
Step 8. Depending on the initial parameters set, the algorithm triggers a mechanism that resets the T value and enlarges the search area. In this case, the algorithm has a chance to find a global minimum, which can now be found in the search area.	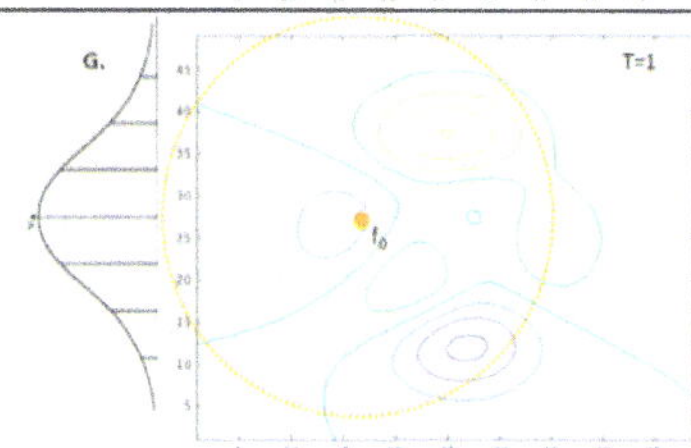
Step 9. The graph generated from the best results in each iteration shows whether the algorithm tried to leave the local maximum. With more complex problems related to, among others with the organization of production, this is often associated with a similar chart to that shown.	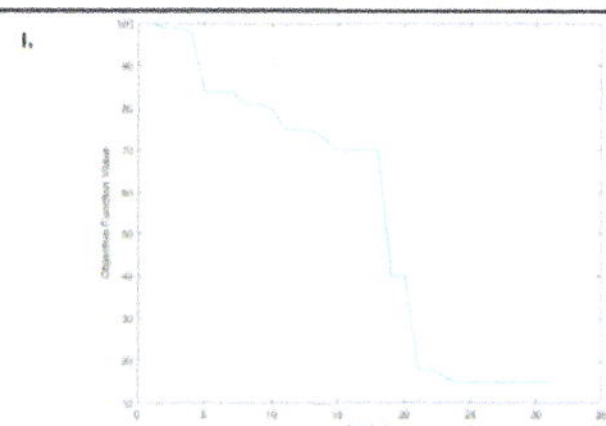

2.2 Modified SA Algorithm for MMALSP

Standard SAO is used to look for near-optimal solutions for complex objective functions. Typically, the dimensions of the space of feasible solutions represent the parameters affecting the result of the objective function for the problem under consideration. In the case of MMALSP, their solutions are sequences of model versions in which the assembly on the line is carried out. It could be assumed that each possible sequence is a separate value in space. For example, for two models (A and B) where the demand is 2 and 3 pieces, respectively, the possible sequences are ABABB, BABBA, AABBB, BBAAB, etc. In the space created from such solutions, it is necessary to look for a near-optimal sequence by means of SAO due to the purpose function. Just for a simple example, where four models are produced with a demand 6, 4, 3 and 2, respectively, the number of possible sequences is over 6.3 million.

The proposed modification of the standard SAO algorithm involves randomizing the sequence as a solution for a given set of model versions and checking whether it meets the conditions imposed by the need to maintain the appropriate smoothness of the solution. A randomly generated sequence replaces, in the proposed solution, the standard feasible neighbouring solution determined for the distance from the current point. Based on the algorithm presented below, the matrices are generated that represent the sequence and the value of the objective function is checked for them. The sequence is represented by a matrix with the dimensions of the number of model versions and the total demand for products:

$$S_n = \begin{bmatrix} s_1^1 & \cdots & s_{Z^c}^1 \\ \vdots & \ddots & \vdots \\ s_1^W & \cdots & s_{Z^c}^W \end{bmatrix}, \tag{1}$$

where:

S_n – the sequence generated for the nth iteration,
s_i^w – binary value representing the occurrence (1) or not (0) of the wth version at the ith position in the sequence, $i = 1, \ldots Z^c$,
Z^c – total demand during the production period, $Z^c = \sum_{w=1}^{W} z_w^c$,
z_w^c – total demand for the wth version,
W – number of model versions

An example S_n matrix with the encoded random sequence is shown below:

$$S_1 = \begin{bmatrix} 1\,0\,0\,1\,0\,0 \\ 0\,1\,1\,0\,0\,1 \\ 0\,0\,0\,0\,1\,0 \end{bmatrix}, \tag{2}$$

1st sequence consists of 6 elements: two elements of model No. 1 in the 1st and 4th positions, three elements of model No. 2 in the 2nd, 3rd and 6th positions and one piece of model No. 3 in the 5th position.

Randomly generated initial sequences for subsequent iterations should meet additional conditions related to the nature of MMALS problem under consideration:

– the number of non-zero elements in each row of the S_n matrix must be equal to total demand for the w-th version z_w^c,

$$z_w^c == \sum_{i=1}^{Z^c} s_i^w,$$

– in each S_n matrix column (corresponding to the position in the assembly version sequence) there should be exactly one non-zero element,

$$\sum_{w=1}^{W} s_i^w == 1.$$

The binary numbers in the rows of the matrix corresponding to each models' version represent their order of execution in sequence. Generating the candidate solution is carried out by repeatedly swapping two randomly selected columns in the S matrix. This repeatability varies during the execution of the algorithm and depends on the current value of temperature T:

```
for i = 1:floor(T)+1
    swap_col1 = ceil(rand()*nrcols);
    while 1
         swap_col2 = ceil(rand()*nrcols);
         if swap_col2~= swap_col1
             break
        end
    end
     temp = sequence(:,swap_col1);
     sequence(:,swap_col1) = sequence(:,swap_col2);
     sequence(:,swap_col2) = temp;
end
```

In case the objective function of the candidate sequence solution $f(S_{cnd})$ is worse than the current point $f(S_{crn})$, the algorithm attempts to determine the probability $P(A)$ to accepts a worse sequence as the new current one (see Table 1, Step 5):

$$P(A) = \frac{1}{1 + \exp\left(\left(100\frac{f(S_{cnd}) - f(S_{cur})}{f(S_{cur})}\right)/T\right)} \tag{3}$$

The algorithm, in addition to determining the probability P(A), has also been enriched with additional protection against choosing the local optimum as the final value. For the set number of iterations required for each temperature reset (*ti*), if later tested solutions are not better than the current ones, the current temperature is reset to the initial value.

Determined as a result of the algorithm, the S_n matrix is a near-optimal sequence that minimizes waiting times between assembly line workstations. For very complex systems, an additional algorithm step can be added that changes the order of model versions when generating initial sequences to provide additional search capabilities. In its basic version, the models' versions are assigned to the rows of the Sn matrix in a constant order resulting from the assigned markings. This method was used in the calculation example shown in the last section of this paper.

2.3 Objective Function

The application of the SA algorithm to solve a specific optimization problem, which is the sequencing of a mixed-model assembly line in the analysed case, requires the creation of an adequate objective function. The proposed objective function takes into account deviations of assembly operations times for individual versions and refers to the sum of times at individual workstations. It includes, for each operation, at each workstation, exceeding the tact time and the required waiting time for the intermediate from the previous station. To simplify the calculations, it is assumed that the workstation is idle when waiting for a part or when it waits for the next workstation to be available (possibility of transferring the intermediate to the next station). The value of delays is normalized in relation to the total average execution (assembly) times of all versions on workstations in accordance with the formulas:

$$\text{Minimize} f(S_n) = \frac{\sum_{i=1}^{Z^c} \sum_{j=1}^{s} (\varphi_{i,j} + \omega_{i,j})}{Tr}, \tag{4}$$

where:

$\varphi_{i,j}$ – workstation idle time related to waiting for the product from the previous workstation:

$$\varphi_{i,j} = \begin{cases} Tp_{\sum_{w=1}^{W}(s_{i+1}^{W} w),j-1} - Tp_{\sum_{W=1}^{W}(s_i^{W} w),j} & \text{if } i < Z^c \text{ and } j > 1 \text{ and } Tp_{\sum_{w=1}^{W}(s_{i+1}^{W} w),j-1} > Tp_{\sum_{w=1}^{W}(s_i^{W} w),j} \\ 0 & \text{if } i = Z^c \text{ or } j = 1 \end{cases}, \tag{5}$$

$Tp_{w,j}$ – duration of the assembly in the wth version at the jth station, $j = 1, \ldots, s$,
s – number of workstations,
$\omega_{i,j}$ – workstation idle time related to waiting for the possibility of transferring the product to the next workstation:

$$\omega_{i,j} = \begin{cases} Tp_{\sum_{w=1}^{W}(s_{i-1}^{W} w),j+1} - Tp_{\sum_{w=1}^{W}(s_i^{W} w),j} & \text{if } i > 1 \text{ and } j < s \text{ and } Tp_{\sum_{w=1}^{W}(s_{i-1}^{W} w),j+1} > Tp_{\sum_{w=1}^{W}(s_i^{W} w),j} \\ 0 & \text{if } i = 1 \text{ or } j = s \end{cases}, \tag{6}$$

Tr – total average execution (assembly) times:

$$Tr = \max_j \left(\frac{\sum_{w=1}^{W} Tp_{w,j}}{W} \right) Z^c s.$$

3 An Illustrative Example

The proposed simulated annealing-based algorithm for solving mixed-model assembly line sequencing optimization problem is illustrated below. The duration of assembly operations on workstations, determined depending on the model version, are shown in Table 2 (all values expressed in seconds). It has been assumed that the transfer times between workstations are negligibly small, and the product can leave the workstation only when the next workstation has completed the assembly operations of the previous product.

Table 2. Duration of assembly operations.

Workstation	Model 1	Model 2	Model 3	Model 4
W-1	489	363	861	470
W-2	792	324	665	458
W-3	376	799	761	443
W-4	365	787	449	452
W-5	420	816	894	605
W-6	499	424	937	387

The part of the assembly line under consideration consists of 6 workstations. Within the available production time, a total of 15 products in 4 variants should be produced. Demand for particular versions is as follows: $z_1^c = 6$, $z_2^c = 4$, $z_3^c = 3$ and $z_4^c = 2$. The initial temperature is 100 the maximum number of iterations is set to 500. Cooling rate is 0,95. Figure 1 shows a graph illustrating the results obtained in each individual iteration of the algorithm carried out with the assumed parameters. The analysis of the obtained results clearly confirms the assumed large area of the variability of generated sequences for high temperature.

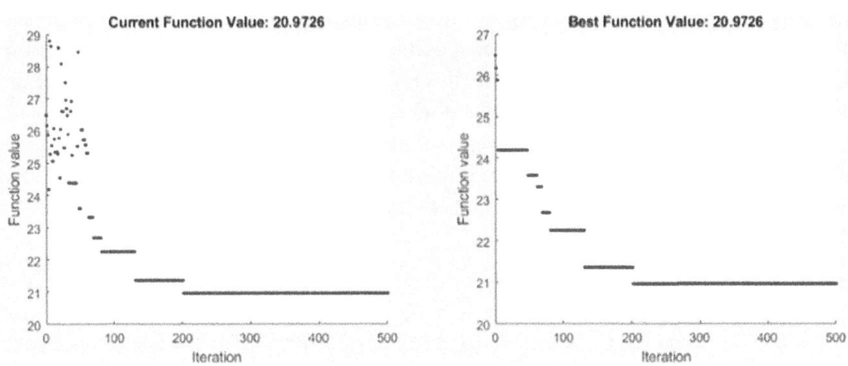

Fig. 1. The results obtained in each individual iteration of the algorithm.

The decrease in the probability of choosing the tested worse sequence as the starting point for further iterations is noticeable. This leads to a very fast decline in the value of the objective function, but at the same time causes the algorithm to "stick" to the local

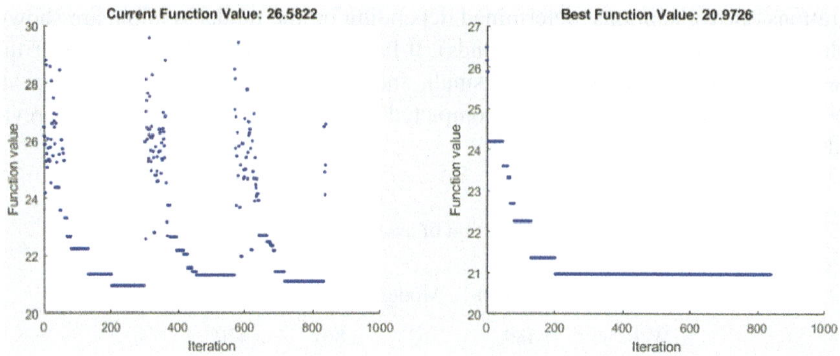

Fig. 2. The SA results - temperature reset mechanism.

minimum. In this case, the temperature reset mechanism ($ti = 70$, max. iterations: 900) works well (Fig. 2), however, this does not guarantee protection against returning to the same local area (even for an increased number of iterations). The best sequence $S_{best} = [1, 4, 4, 1, 1, 1, 3, 3, 3, 2, 2, 2, 2, 1, 1]$, with the result 20,97, was adopted as near-optimal result for the shortest idle time on a workstation.

4 Summary

The paper presents a method of solving one of the basic problems related to production planning in mix-model assembly lines, which is determining the sequence of model versions. Achieving full line synchronization, by creating the appropriate model sequence, becomes increasingly difficult at current levels of product complexity. The presented algorithm is based on SAO, a widely used approach provided near-optimal solutions to combinatorial optimization problems. However, to be able to apply this approach, it is necessary to develop dedicated algorithms of sequence generation and create a search objective function, taking into account specific aspects related to the MMALS problem. The main distinguishing feature of the presented solution is the method of generating a feasible neighboring solution, the range of which depends on the temperature factor. This increases the chance to avoid local optima trapping. The ability to reset the temperature value for a larger search range is also important In this area, further studies are needed to determine the additional stopping conditions and evaluate generated sequences.

References

1. Ćwikła, G., Grabowik, C., Kalinowski, K., Paprocka, I., Banaś, W.: The initial considerations and tests on the use of real time locating system in manufacturing processes improvement. IOP Conf. Ser. Mater. Sci. Eng. **400**, 1757–8981 (2018)
2. Golz, J.: Part feeding at high-variant mixed-model assembly lines. Flex. Serv. Manuf. J. **24**, 119–141 (2011)
3. Akpinar, S., Bayhan, G.M.: A hybrid genetic algorithm for mixed model assembly line balancing problem with parallel workstations and zoning constraints. Eng. Appl. Artif. Intell. **24**, 449–457 (2011)

4. Scholl, A., Voß, S.: Simple assembly line balancing—heuristic approaches. J. Heuristics **2**(3), 217–244 (1996)
5. Simaria, A.S., Vilarinho, P.M.: A genetic algorithm based approach to the mixed-model assembly line balancing problem of type II. Comput. Ind. Eng. **47**, 391–407 (2004)
6. Şeker, Ş., Özgürler, M., Tanyaş, M.A.: Weighted multiobjective optimization method for mixed-model assembly line problem. J. Appl. Math. **2013**, 1–10 (2013). Article ID 531056
7. Krenczyk, D., Skolud, B., Herok, A.: A heuristic and simulation hybrid approach for mixed and multi model assembly line balancing. In: Advances in Intelligent Systems and Computing, vol. 637, pp. 99–108 (2018). https://doi.org/10.1007/978-3-319-64465-3_10
8. Hamzadayi, A., Yildiz, G.: A simulated annealing algorithm based approach for balancing and sequencing of mixed-model U-lines. Comput. Ind. Eng. **66**, 1070–1084 (2013)
9. Krenczyk, D., Dziki, K.: A hybrid heuristic algorithm for multi-manned assembly line balancing problem with location constraints. In: Advances in Intelligent Systems and Computing, vol. 950, pp. 333–343 (2020)
10. Kundua, K.: A study of a Kanban based assembly line feeding system through integration of simulation and particle swarm optimization. Int. J. Ind. Eng. Comput. **10**, 421–442 (2019)
11. Cohen, Y., Naseraldin, H., Chaudhuri, A., Pilati, F.: Assembly systems in Industry 4.0 era: a road map to understand Assembly 4.0. Int. J. Adv. Manuf. Technol. **105**, 4037–4054 (2019)
12. Hyun, C.J., Kim, Y., Kim, Y.K.: A genetic algorithm for multiple objective sequencing problems in mixed model assembly lines. Comput. Oper. Res. **25**, 675–689 (1998)
13. Zhang, X., Gao, L., Wen, L., Huang, Z.: A hybrid algorithm based on tabu search and large neighbourhood search for car sequencing problem. J. Cent. South Univ. **25**, 315–330 (2018)
14. McMullen, P.R., Frazier, G.V.: A simulated annealing approach to mixed-model sequencing with multiple objectives on a just-in-time line. IIE Trans. **32**(8), 679–686 (2000)
15. Liu, Z., Wang, C., Sun, T.: Production sequencing of mixed-model assembly lines based on simulated annealing algorithm. In: International Conference of Logistics Engineering and Management, ICLEM 2010, vol. 387, pp. 1803–1808 (2010)
16. Xiaobo, Z., Ohno, K.: Algorithms for sequencing mixed models on an assembly line in a JIT production system. Comput. Ind. Eng. **32**, 47–56 (1997)
17. Dong, J., Zhang, L., Xiao, T., Mao, H.: Balancing and sequencing of stochastic mixed-model assembly U-lines to minimise the expectation of work overload time. Int. J. Prod. Res. **52**(24), 7529–7548 (2014)
18. Kirkpatrick, S., Gelatt, C.D., Vecchi, M.P.: Optimization by simulated annealing. Science **220**(4598), 671–680 (1983)
19. Eglese, R.W.: Simulated annealing: a tool for operational research. Eur. J. Oper. Res. **46**, 271–281 (1990)
20. Goldkrg, D.E.: Genetic Algorithms in Search, Optimization and Machine Learning. Addison-Wesley, Reading (1989)

The Concept of Genetic Algorithm Application for Scheduling Operations with Multi-resource Requirements

Iwona Paprocka, Krzysztof Kalinowski(✉), and Barbara Balon

Department of Engineering Processes Automation and Integrated Manufacturing Systems, Faculty of Mechanical Engineering, Silesian University of Technology, Konarskiego 18A, 44-100 Gliwice, Poland
{iwona.paprocka,krzysztof.kalinowski,barbara.balon}@polsl.pl

Abstract. The paper presents the concept of a genetic algorithm for solving the problem of scheduling production processes, in which there are operations requiring the interaction of resources from at least two, different groups of competences. The considered system is based on flexible flow shop and the objective function is associated with minimizing the flow time of tasks. The general schedule generation procedure using the genetic algorithm is presented. Three sub-chromosomes are proposed for describing an individual. First of them represents a precedence feasible order of production tasks. Numbers of parallel machines are coded by the second sub-chromosome of the individual. Numbers of production employees able to execute operation on the set of parallel machines are coded by the third sub-chromosome. The order crossover and shift mutation procedures are described for the proposed chromosome differentiation and selection. Implementation of the developed concept enables parallel planning of positions and human resources (or any groups of resources) and improve practical usability in relation to hierarchical methods of resource planning.

Keywords: Production scheduling · Genetic algorithm · Multi-resources planning

1 Introduction

The high level of complexity of scheduling tasks in complex manufacturing systems motivates the search for methods that allow obtaining acceptable but not necessarily optimal solutions. Soft computing methods, using elements of fuzzy logic, neural and evolutionary calculations, etc. play an invaluable role among the methods used in such cases. Evolutionary algorithms (EA), as an adaptive heuristic search algorithms based on the evolutionary ideas of natural selection, are especially widely used to solve scheduling problems. The basic concept of EA is designed to simulate processes in a natural system, necessary to preserve evolution processes and adhere to the principles of Darwinian evolution.

Á. Herrero et al. (Eds.): SOCO 2020, AISC 1268, pp. 342–351, 2021.
https://doi.org/10.1007/978-3-030-57802-2_33

Works on improving optimization algorithms and practical development of supported models are constantly developing. Over the past few decades, many methodologies have been introduced to solve problems directly related to production scheduling. The considered area of production planning relates mainly to the optimal allocation of resources, especially those that limit the operation of processes over time [1]. These are, for example, bottleneck heuristics, local search methods or metaheuristics [2]. Currently, metaheuristics are considered to be the most powerful scheduling techniques [3], mainly those that use evolutionary genetic algorithms [4–6]. Known examples that have been successfully carried out include: simulated annealing [6, 7], problem search [8–10] and optimization of ant colonies [11–14]. Soft computing has also found wide application in planning problems in supply chain systems and in considerations on scheduling theory. In 15 authors propose optimization algorithms based on hormone modulation mechanism. Based on presented research they conclude that elaborated neuroendocrine-inspired optimization algorithms possess the merits of global exploration, fast convergence, and robustness to solve the job shop scheduling problems. In 16 the Google DeepMind's Deep Q Network agent algorithm for Reinforcement Learning was implemented to production scheduling. In their solution agents, operating in reinforcement learning environment use deep neural networks to optimize scheduling with selected, user-defined objectives. Available methods and modern solutions giving the possibility to apply selected concepts from industry 4.0, like mass customization, Cyber-Physics Systems, Digital Twin, and SMAC (Social, Mobile, Analytics, Cloud are also analyzed [17]. Human factors and interactions between the scheduling and workforce development are considered in [18]. This study integrates human factors in the form of worker competence into production scheduling. They discuss the factors of worker learning rate, professional skills, experience in each type of operation, and operation specifications in production process and proposes a two-hierarchy scheduling optimization model to synchronize the production scheduling with workforce and to optimize the scheduling from both the short-term (time-oriented) and long-term (competence-oriented) views.

The problem discussed at this work concerns concurrent planning of machines and human resources. Such a situation arises in the real production systems where both the machines and human resources are limited and staff are not assigned to specific workplace per shift.

2 Machines and Crews Scheduling Problem

A flow shop scheduling problem is taken into account where tasks are allocated to sets of resources: machines and employees. Production systems are described by: (a) tasks, (b) machines, (c) employees (d) operation sequence resulting from technological route, (e) operation times, (f) deadlines for performing the processes. The production tasks are executed in the exclude like mode, and operations are non-preemptive.

Let i denotes task index where $i \in [1, \ldots, I]$ and j presents index of operation where $j \in [1, \ldots, J]$. A set of employees assigned to operation j can consist of K employees where $k \in [1, \ldots, K]$. A set of parallel machines assigned to operation j can consist of N employees where $n \in [1, \ldots, N]$. The production model assumptions state as follows:

1. Production crew of employees (k) with the given competences is assigned for a certain production task i during parallel machines (n) working times.

$$Max\sum_{i=1}^{I}\sum_{j=1}^{J}\sum_{k=1}^{K}\sum_{n=1}^{N}\left(p_{i,j,n,m}\times\beta_{i,j,k,m}\times t_{i,j}\right). \tag{1}$$

Let the binary decision variable, $p_{i,j,n}$ denotes that parallel machine n is busy due to the execution of operation j of production task i; where $p_{i,j,n,m}$ equals one if operation j of production task i is assigned to machine n at time m, and zero otherwise. Let the binary variable $\beta_{i,j,k}$ denotes the assignment of operation j of task i to production crew k where $\beta_{i,j,k,m}$ equals one if production employee k is assigned to operation j of task i at time m, and zero otherwise. The production practice requires a maximum number of tasks to allocate to restricted set of parallel machines and production crew in the shortest possible time.

2. The objective function is presented by the Eq. (2). For parallel machines n, for production crew k, the total of production task's duration must be minimized:

$$F=\sum_{i=1}^{I}\sum_{j=1}^{J}\sum_{k=1}^{K}\sum_{n=1}^{N}\left(p_{i,j,n,m}\times\beta_{i,j,k,m}\times t_{i,j}\right)\rightarrow \text{min.}\,\forall i,\ j,\ k,\ n \tag{2}$$

Let $t_{i,j}$ denotes duration time of operation (j) of task (i) executed on machine set (n) by production crew (k). The objective function is subject to the following constraints:

- For the set of parallel machines the execution time of task i must not exceed the predefined deadline of production task i, D_i.

$$\sum_{i=1}^{I}\sum_{j=1}^{J}\sum_{n=1}^{N}\left(p_{i,j,n,m}\times t_{i,j}\right)\leq D_i. \tag{3}$$

- The completion time of scheduled task i for production crew k should not exceed the deadline of the production task.

$$\sum_{i=1}^{I}\sum_{j}^{J}\sum_{k=1}^{K}\left(\beta_{i,j,k,m}\times t_{i,j}\right)\leq D_i.\,\forall i,\ j,\ k \tag{4}$$

- To ensure that each production task is executed once, condition (5) should be met.

$$\sum_{i=1}^{I}\sum_{j=1}^{J}\sum_{k=1}^{K}\sum_{n=1}^{N}\left(p_{i,j,n,m}\times\beta_{i,j,k,m}\times t_{i,j}\right)=\sum_{i=1}^{I}\sum_{j=1}^{J}t_{i,j}.\,\forall i,\ j,\ k,\ n \tag{5}$$

- To ensure that each production task can be executed on set of machines (n) at time m under the condition that production crew k is also available at the time.

$$\sum_{i=1}^{I}\sum_{j=1}^{J}\sum_{n=1}^{N}\left(p_{i,j,n,m} \times t_{i,j}\right) - \sum_{i=1}^{I}\sum_{j=1}^{J}\sum_{k=1}^{K}\left(\beta_{i,j,k,m} \times t_{i,j}\right) = 0.\forall m \tag{6}$$

- The assignment of operation j of production task i to machine n takes a value of zero or one, thus $p_{i,j,n}$ is a binary variable: $p_{i,j,n} = \{0,1\}$
- The assignment of operation j of production task i to production employee k takes a value of zero or one, thus $\beta_{i,j,k}$ is a binary variable: $\beta_{i,j,k} = \{0,1\}$

The genetic algorithm is proposed to solve the described problem.

3 A Basic Schedule Generation Using the Genetic Algorithm

The GA consists of the following modules: data interface, individuals coding, genetic optimization and selection and individuals decoding.

In GA, an individual represents a solution of the flow shop scheduling problem – schedules for machines and employees, while a fitness function is a measure used to estimate the quality of the schedules. The fitness function of an individual is flow time minimization of tasks (2). The pseudo code of the GA is presented in Fig. 1. The steps of the algorithm are explained in following subsections.

Step 1: Genes coding in the DNA Library

Step 2: Generation of initial population η

While number of iterations Γ (τ = 0, …, Γ) is higher than 0 in the genetic selection process do

Step 3: Evaluate individuals ρ_η from η using the fitness function F (2),

Step 4: Clone initial population η in order to create mating population λ

Step 5: Select matting individual cλ from λ for each ρ_η

Step 6: Crossover in order to create offspring population λ^*

Step 7: $\rho_\eta \leftarrow c_\lambda^*$ if $F(c_\lambda^*) < F(\rho_\eta)$ in the selection process

Step 8: Shift Mutation in order to create offspring population λ^{**},

Step 9: $\rho_\eta \leftarrow c_\lambda^{**}$ if $F(c_\lambda^{**}) < F(\rho_\eta)$ in the selection process

Step 10: The number of the best individuals ϑ survive in η for criterion F. Remaining individuals are randomly selected in η

Fig. 1. Pseudo code of the GA algorithm

3.1 Encoding and Decoding

Usually, in flow shop scheduling problems, job-based representation is a common scheme to encode a solution. The complexity of the problem presented requires more sophisticated coding practice for an individual. Each individual ρ_η is represented by three sub-chromosomes. First sub-chromosome of the individual represents a precedence feasible order of production tasks. The order of a production task is randomly generated between 1 and a total number of production tasks accepted for execution. By scanning the permutation code from left to right, the occurrence of production task number (gene) indicates the priority of the task (Fig. 2).

$[1\ \ 3\ \ 2]$ ← random permutation of tasks
$[1\ \ 2\ \ 3]$ ← priority indicators of tasks

Fig. 2. First sub-chromosome of the individual for the problem of scheduling three tasks

To transform the individual to a feasible solution a production task (according to the permutation) is scheduled at the earliest feasible time according to the precedence and resource (machines and employees) constraints.

Numbers of parallel machines that constitute a set of parallel machines are coded by the second sub-chromosome of the individual. To transform the individual to a feasible solution, operation j of production task i is scheduled on a set of parallel machines at the earliest feasible time according to the precedence and resource (machines and employees) constraints. A set of parallel machines can consist of n machines where $n \in [1, \ldots, N]$ randomly selected. Consider the first row of the matrix $p_{i,j,n}$ where first operation $j = 1$ of first task ($i = 1$) must be executed by the first set of machines, and exactly by the first, second and third parallel machine, m_1, m_2 and m_3 (Fig. 3).

$p_{i,j,n}$ $\quad [m_1\ \ m_2\ \ m_3]$

$$i=1 \quad \begin{bmatrix} j=1 \\ j=2 \\ j=3 \end{bmatrix} \quad \begin{bmatrix} 1 & 1 & 1 \\ 1 & 0 & 1 \\ 0 & 1 & 1 \end{bmatrix}$$

← first set of machines consists of first, second and third machine
← second set of machines consists of first and third machine
← third set of machines consists of second and third machine

Fig. 3. The second sub-chromosome of the individual codes sets of parallel machines

Numbers of production employees able to execute operation on set of parallel machines are coded by the third sub-chromosome of the individual. To transform the individual to a feasible solution, operation j of production task i mast be executed by set of employees at the earliest feasible time according to the precedence and set of parallel machines constraints. Each set of production employees can consists of k employees, where $k \in [1, \ldots, K]$ randomly selected. Consider the first row of the matrix $\beta_{i,j,k}$ where first operation $j = 1$ of first task ($i = 1$) must be executed by the first set of employees, and exactly by the first and third employee, e_1 and e_2 (Fig. 4).

$$\beta_{i,j,k} \qquad \begin{bmatrix} e_1 & e_2 & e_3 \end{bmatrix}$$

$$i=1 \quad \begin{bmatrix} j=1 \\ j=2 \\ j=3 \end{bmatrix} \quad \begin{bmatrix} 1 & 0 & 1 \\ 1 & 0 & 0 \\ 0 & 1 & 1 \end{bmatrix}$$

← first set of employees consists of first and third employee
← second set of employees consists of first employee
← third set of employees consists of second and third employee

Fig. 4. The third sub-chromosome of the individual codes sets of production employees

Consider the production system which consists of three machines M = 3, three employees K = 3 and two tasks accepted for realization, each production task consists of three operations, thus the chromosome can be represented by the decimal and binary string as presented in Fig. 5.

I	II		III	
[2 1]	[010 101 111]	[010 111 111]	[110 101 111]	[010 011 100]
{1 2}	{2,1 2,2 2,3}	{1,1 1,2 1,3}	{2,1 2,2 2,3}	{1,1 1,2 1,3}

Fig. 5. The chromosome of the first individual from population η

The decoding process is started for the second task which is first assigned to machines $[\,010\ 101\ 111\,]$ and employees $[\,110\ 101\ 111\,]$. First operation of task $i = 2$ is executed on the second machine (010) only and two employees need to be available at the same time: first and second (110). The second operation ($j = 2$) of task $i = 2$ is executed on the first and third machine at parallel (101) by two employees (101), e_2 and e_3, and so on. Let us assume that each operation of production tasks takes two minutes, thus the Gantt charts for machines and employees are presented in Fig. 6 and 7.

3.2 Initialization

Genes, stored in the DNA Library, represents tasks, machines and employees used in the production system. A set of randomly generated solutions serves as the initial population: permutation representations of tasks for the first sub-chromosome and binary selection for the second and third chromosome.

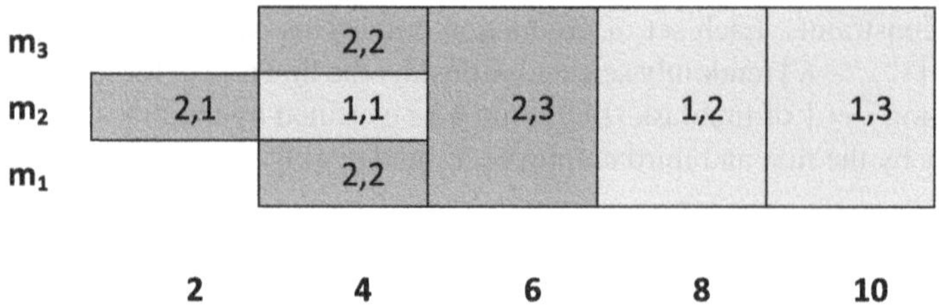

Fig. 6. The Gantt's chart of production tasks assigned to machines after decoding individual ρ_1 (Fig. 5)

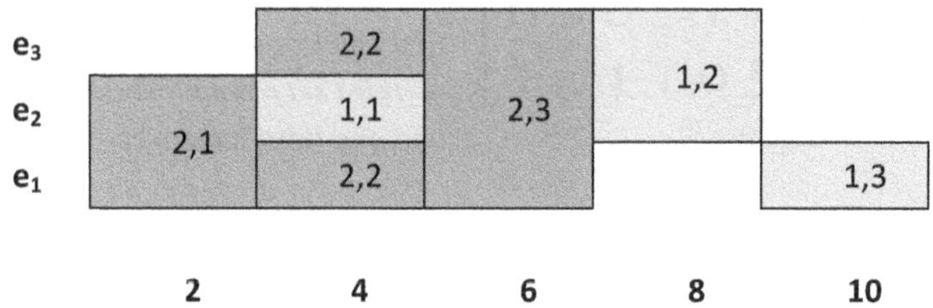

Fig. 7. The Gantt's chart of production tasks assigned to employees after decoding individual ρ_1 (Fig. 5)

3.3 Chromosome Differentiation and Selection

The set of chromosomes for reproduction is created by copying from the initial population and pairing the best-matched individuals. The individual selection procedure used at this stage is described in detail in [14]. The fitness function is transformed into a new fitness equation:

$$new\_fit(c_\lambda) = 1 - \frac{F(c_\lambda)}{\sum_{\lambda=1}^{\Lambda} F(c_\lambda)} \tag{7}$$

The new fitness parameter is converted into frequency of selection $fr(c_\lambda)$:

$$fr(c_\lambda) = \frac{new\_fit(c_\lambda)}{\sum_{\lambda=1}^{\Lambda} new\_fit(c_\lambda)} \tag{8}$$

The probability that individual c_λ survives and evolves depends on accumulation:

$$a(c_\lambda) = \begin{cases} fr(c_\lambda), \; if \; \lambda = 1 \\ a(c_{\lambda-1}) + fr(c_\lambda), \; else \; if \; \lambda > 1 \end{cases} \tag{9}$$

A number (r) between 0 and 1 is randomly selected for each individual. Individual c_λ is the second parent if the following condition is met:

$$a(c_\lambda) \leq r \geq a(c_{\lambda+1}) \tag{10}$$

The above procedure guarantees selection of the best matched individual. The most-matched individuals are used in the reproduction, and their descendants inherit the best features, encoded in the genes. The most matched chromosomes have many copies, the worst ones 'die'.

The Order Crossover (OX) procedure is adopted to create a new solution in the differentiation of chromosomes. The OX procedure starts from the selection of a gene sub-sequence in the chromosome of the first parent. Offspring is produced by copying the selected gene up at the appropriate positions on its chromosome. Selected genes are removed from the second parent's chromosome. As a result, the genes required to complete the offspring are obtained. Moving from left to right, the genes are copied according to the sequence resulting from the chromosome of the second parent [15, 16]. Genes represents numbers of tasks for the first sub-chromosome, and binary numbers of machines and employees for the second and third sub-chromosomes respectively.

Assume that the number of genes that undergo OX procedure is two for the second sub-chromosome. The selected genes (of the chromosome of the first parent) are removed from the chromosome of the second parent and copied in the corresponding positions of the offspring's chromosome, as is presented in Fig. 8. The remaining bit of the offspring is copied from the chromosome of the second parent. OX procedure is performed within the range of the selected operation and set of machines or employees for the second and third sub-chromosome respectively.

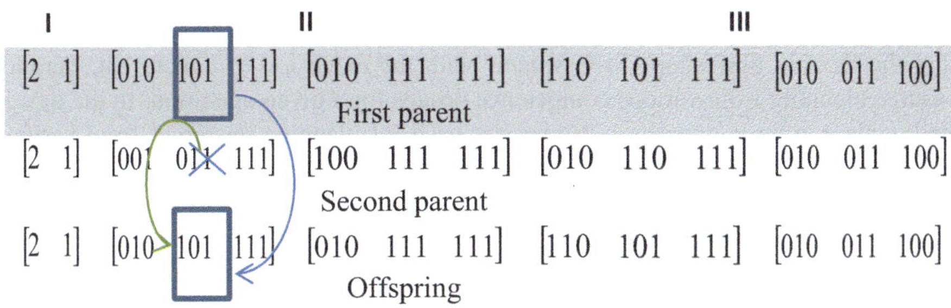

Fig. 8. The chromosome of the first individual from population η

Next, the fitness function F is calculated for each individual. In the elite selection procedure, the best individual does not change from the pair: parent and child.

The parents undergo Shift mutation (SM) procedure. In the SM procedure, a task, machine or employee (gene) is randomly selected, then is swapped with the preceding gene. By using the SM, the emphasis of losing the genetic material is low. Also, the elite selection is repeated. The best individuals are unchanged and survive to the next generation.

Assume that the number of genes which undergo mutation procedure is one for each sub-chromosome, thus the chromosome of the first individual (Fig. 5) after Shift mutation procedure is presented in Fig. 9.

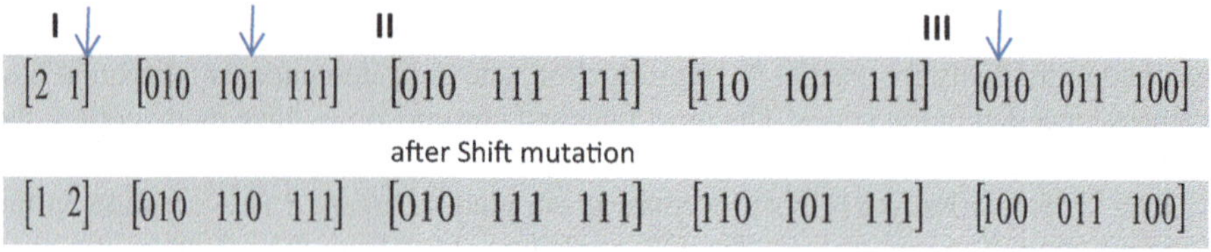

Fig. 9. The chromosome of the first individual after Shift mutation procedure

3.4 Ordering Selection Procedure and Terminal Condition

In the procedure of ordering selection, a fixed number of the best individuals ϑ for each criterion F create a new initial population. The remaining individuals are randomly selected from a feasible solution space. High selection pressure is balanced with random generation of chromosomes in order to escape from a local optima.

Executing a given number of iterations Γ meets a termination condition. The best of the designated solutions, closest to the optimal, is in the last generation.

4 Summary

The article describes the problem of the simultaneous allocation of different types of resources according to the resource requirements of the scheduled processes and a genetic algorithm was proposed to solve it. The configuration of the adopted system is derived from the flexible flow shop class systems with the extension of concurrent, parallel resource planning from various competence groups for a given operation. In the developed method, a three-part chromosome was used for planning resources from various competence groups. The different parts of the chromosome describe order precedence, a set of machines and set of employees respectively.

The solution used significantly expands the planning possibilities in production systems to support various resource groups and the use of a genetic algorithm allows the determination of good quality solutions. The most important directions of further research are focused on the development of the algorithm in order to enable planning in systems derived from the flexible job shop class. It is also expedient to extend the objective function and to take into account additional criteria related to e.g. deadlines for tasks and costs parameters.

References

1. Zweben, M., Fox, M.S.: Intelligent Scheduling. Morgan Kaufman Publishers, Burlington (1994)
2. Pirlot, M.: General local search methods. Eur. J. Oper. Res. **92**, 493–511 (1996)
3. Jain, A.S., Meeran, S.: Deterministic job-shop scheduling: past, present and future. Eur. J. Oper. Res. **113**, 390–434 (1999)
4. Holland, J.H.: Adaptation in Natural and Artificial Systems. The University of Michigan Press, Ann Arbor (1975)

5. Catrysse, D., Van Wassenhove, L.N.: A survey of algorithms for the generalized assignment problem. Eur. J. Oper. Res. **60**, 260–272 (1992)
6. Kirkpatrick, S., Gelatt Jr., C.D., Vecchi, M.P.: Optimization by simulated annealing. Science **220**(4598), 671–680 (1983)
7. Van Laarhooven, P.J.M., Aarts, E.H.L., Lenstra, J.K.: Job-shop scheduling by simulated annealing. Oper. Res. **40**(1), 113–125 (1992)
8. Glover, F., Laguna, M.: Tabu Search. Kluwer Academic Publishers, Boston (1997)
9. Laguna, M., Glover, F.: Integration target analysis and tabu search for improved scheduling systems. Exp. Syst. Appl. **6**, 287–297 (1993)
10. Nowicki, E., Smutnicki, C.: A fast taboo search algorithm for the job-shop problem. Manag. Sci. **42**(2), 797–813 (1996)
11. Dorigo, M., Maniezzo, V., Colorni, A.: The ant system: optimization by a colony of cooperating agents. IEEE Trans. Syst. Man Cybern. B Cybern. **26**(1), 29–41 (1996)
12. Blum, C.: Beam-ACO-hybridyzing ant colony optimization with beam search: an application to open shop scheduling. Comput. Oper. Res. **32**(6), 1565–1591 (2005)
13. Merkle, D., Middendorf, M., Schmeck, H.: Ant colony optimization for resource-constraint project scheduling. IEEE Trans. Evol. Comput. **6**(4), 333–346 (2002)
14. Shang, J., Tian, Y., Liu, Y., Liu, R.: Production scheduling optimization method based on hybrid particle swarm optimization algorithm. J. Intell. Fuzzy Syst. **34**(2), 955–964 (2018)
15. Tang, D., Zheng, K., Gu, W.: Hormone regulation based algorithms for production scheduling optimization. In: Adaptive Control of Bio-Inspired Manufacturing Systems, pp. 19–45. Springer, Singapore (2020)
16. Waschneck, B., Reichstaller, A., Belzner, L., Altenmüller, T., Bauernhansl, T., Knapp, A., Kyek, A.: Optimization of global production scheduling with deep reinforcement learning. Procedia CIRP **72**(1), 1264–1269 (2018)
17. Zhang, J., Ding, G., Zou, Y., Qin, S., Fu, J.. Review of job shop scheduling research and its new perspectives under Industry 4.0. J. Intell. Manuf. **30**(4), 1809–1830 ((2019))
18. Wang, Z., Hu, H., Gong, J.: Modeling worker competence to advance precast production scheduling optimization. J. Constr. Eng. Manag. **144**(11), 04018098 (2018)

Special Session: Soft Computing Applications for the Management of Industrial and Environmental Enterprises

Comparative Analysis of Clustering Techniques for a Hybrid Model Implementation

María Teresa García-Ordás[1], Héctor Alaiz-Moretón[1], José-Luis Casteleiro-Roca[2(✉)], Esteban Jove[2], José Alberto Benítez-Andrades[1], Isaías García-Rodríguez[1], Héctor Quintián[2], and José Luis Calvo-Rolle[2]

[1] Department of Electrical and Systems Engineering, University of León, Escuela de Ingenierías, Campus de Vegazana, 24071 León, Spain
[2] CTC, Department of Industrial Engineering, CITIC, University of A Coruña, Avda. 19 de febrero s/n, 15405 Ferrol, A Coruña, Spain
jose.luis.casteleiro@udc.es

Abstract. This research is oriented to compare the performance of two clustering methods in order to know what is the best one for archiving high quality hybrid models. For testing purposes, a real dataset has been obtained of a bio-climate house located in Sotavento Experimental Wind Farm, in Xermade (Lugo) in Galicia (Spain). Between several systems installed in the house, the thermal solar generation system has been the chosen one for studying its behaviour and experimenting with the clustering techniques.

Two approaches have been utilized for establishing the quality of each clustering algorithm. The first one is based on typical error measurements implied in a regression procedure such as Multi Layer Perceptron. Whereas, the second one, is oriented to the utilization of three unsupervised learning metrics (Silhouette, Calinski-Harabasz and Davies-Bouldin).

Keywords: Clustering · Prediction · Learning metrics · Spectral Clustering · Gaussian Mixture Clustering

1 Introduction

Some are the different hot topics in general terms and for all the possible applications, and regardless of the field of application. Examples of them are: sustainability, ecological, zero impact, environment safety, and so on [4,13]. Commonly, these issues goes in opposition with other terms like comfort, benefits, luxury, etc. [15,16]. Then, it is a challenge the compromise between the two trends; for instance, people like comfort homes, and therefore, it is desirable this achievement comes from renewable energies.

For an optimal performance of the renewable energy systems, due to some different reasons, commonly it is necessary make predictions of the used variables for the facility right management [14]. There are many techniques to make

Á. Herrero et al. (Eds.): SOCO 2020, AISC 1268, pp. 355–365, 2021.
https://doi.org/10.1007/978-3-030-57802-2_34

predictions, from the traditional ones to the most advanced through the middle ones between both [3]. When a specific system to be modeled has a performance with a very non-linear component for instance, the modeling based on hybrid systems frequently gives very satisfactory results [5,7–9,17,21].

When hybrid systems are used for modeling tasks, during the clustering stage frequently is used K-means method as an standard [24]. However, there are many clustering techniques with a satisfactory performance, and in a lot o cases with a better performance versus k-means technique [24]. The present research accomplish a performance study of two clustering technique, Gaussian Mixture and Spectral Clustering. For comparing their behaviour, two approaches have been implemented. Firstly a set of error non-supervised measurements and following a MLP regressor for establishing the quality when a hybrid model is developed. The work has been accomplished over a real system based on a solar thermal panel, installed in a bioclimatic house.

The rest of the document is structured as follows. Section 2 describes briefly the case of study. After Sect. 2, the techniques applied to achieve the fault detection one-class classification are explained. Section 4 details the experiments and achieved results and finally, the conclusions and future works are exposed in Sect. 5.

2 Case of Study

This research is based on a dataset from the boiclimatic house built by Sotavento Galicia Foundation. This foundation studies different types of renewable energies and, with this aim, this mentioned house has several installation for researching propose.

2.1 Sotavento Bioclimatic House

The house is used to show new systems as a part of demonstrative project; it tries to use a few energy as possible to reduce the energy comsumption. The real house is shown in Fig. 1, and it is in the *Sotavento Experimental Wind Farm*, in Xermade (Lugo) in Galicia (Spain).

Fig. 1. Sotavento bioclimatic house

The thermal energy system of the bioclimatic house can be divided in three different parts: Generation, Accumulation and Consumption. The house has also electrical, geothermal and biomass energy, but as this research is focused in the thermal solar generation, the explanation will be only for this part. In Fig. 2 it is shown this whole thermal part. This paper uses only the sensors S1, S2, S3, S4 and the Flow-meter. Moreover, it is necessary to used the radiation sensor outside the bioclimatic house.

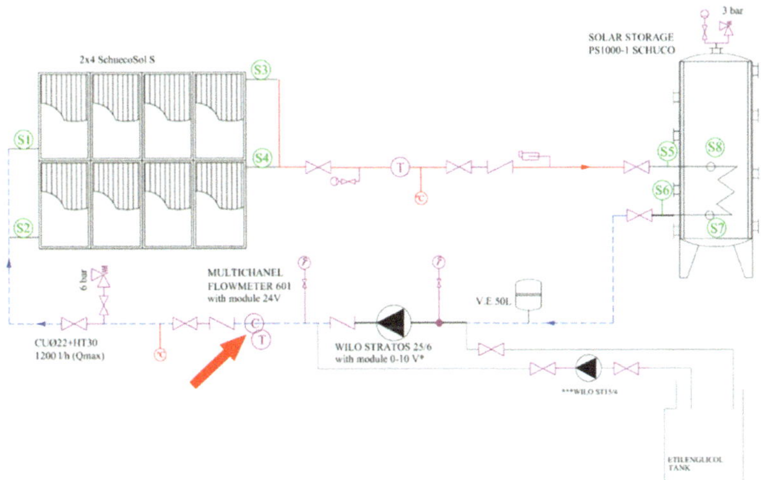

Fig. 2. Solar thermal energy layout

2.2 Model Approach

The research takes into account the input temperatures of solar panels, the radiation and the flow. Despite the dataset has also the output temperatures of the panels, the clustering techniques will not be applied to this measurements. The clustering performance of the different techniques, is measured with three error metrics, but this paper uses one more specific metric. This metric is based on regression, applying a hybrid model to predict the output temperature of the panel, and calculate the regression error with the different hybrids models.

3 Used Techniques

In the data processing phase, a pre-processing step has been performed before clustering by applying MinMax normalization. Subsequently, two different clustering algorithms have been applied, evaluating each one of the through three metrics and typical error measurements as results of regression based on a MLP. To improve the visualization of results, a LDA technique has been implemented. The methods mentioned above will be briefly explained in this section.

3.1 Preprocessing

The MinMax normalization method modifies the data so that it is in the [0,1] range based on its maximum and minimum values. This process is done by following the Eq. 1.

$$\hat{x_i} = \frac{x_i - x_{min}}{x_{max} - x_{min}} \tag{1}$$

When applying clustering [11] and Multi Layer Perceptron techniques with regression purposes, it is advisable to apply this standardization to obtain better results [2].

3.2 LDA Projection

Sometimes, applying a logistic regression on the model, the classes are well separated, but the estimation parameters for the model are unstable. In these cases, Linear Discriminant Analysis (LDA) does not suffer from this problem and it is a more recommendable technique to use. Thanks to the application of LDA, the separability of classes is maximized.

On the other hand, LDA allows transforming the data, achieving that the classes are as separated as possible, being thus a good projection technique. This technique has been frequently used as a two-dimensional projection method, an example is the one that can be read in [19].

3.3 Clustering

Spectral Clustering: Spectral Clustering algorithm [20] tries to divide a dataset based on a similarity graph of its samples. With this graph, we obtain the adjacency matrix and the degree matrix, which indicate the relation between samples and the number of relationships respectively. After that, we calculate the associated Laplacian matrix extracting the adjacency matrix from our degree matrix. Last step consists of running K-means over the eigenvectors of the Laplacian matrix in order to arrange the samples into clusters. As it uses K-Means, it is also necessary to determine the number of centroids previously.

Gaussian Mixture Clustering: Gaussian Mixture Clustering [18] define each cluster by its centroid, covariance and the weight of the cluster. We can define Gaussian Mixture models as a mixture of K gaussian distributions. In order to find those distributions, an Expectation-Maximization algorithm [10] is carried out to determine the correct values for the mean, covariance and weight of each of them. Whereas K-Means algorithm just take into account the mean information, the gaussian mixture clustering also considers the variance of the data.

3.4 Error Metrics

For evaluating the clustering methods three unsupervised metrics: Silhouette coefficient, Calinsky-Harabasz and Davies-Bouldin, have been studied.

Silhouette: The Silhouette coefficient is a metric to evaluate the quality of the clustering obtained with clustering algorithms. The objective of Silhouette is to identify what is the optimal number of clusters.

In unsupervised learning algorithms, the number of clusters may be an input parameter of the algorithm or may be determined automatically by the algorithm. In the first case, as with the K-Means algorithm, the determination of the optimal number of clusters has to be done by some measure external to the algorithm. The Silhouette coefficient is an indicator of the ideal number of clusters. A higher value of this index indicates a more desirable case of the number of clusters.

The Silhouette coefficient for an observation i is denoted as $s(i)$ and defined as:

$$s(i) = \frac{b - a}{max(a, b)} \tag{2}$$

where:

- **a** is the average of the dissimilarities (or distances) of observation i with the other observations of the cluster to which i belongs.
- **b** is the minimum distance to another cluster that is not the same as observation i. That cluster is the second best choice for i and is called the neighbourhood of i.

The Silhouette coefficient is a value between -1 and 1.

A value of $s(i)$ that is close to zero will indicate that the i observation is on the border of two clusters.

And if the value of $s(i)$ is negative, then the i observation should be assigned to the nearest cluster.

In short:

- $s(i) \approx 1$, the i observation is well assigned to its cluster.
- $s(i) \approx 0$, the i observation is between two clusters.
- $s(i) \approx -1$, the i observation is bad assigned to its cluster.

Calinski-Harabasz: The Calinsky-Harabasz index is calculated according to the following formula (3):

$$C = \frac{\frac{BGSS}{K-1}}{\frac{WGSS}{N-K}} = \frac{N-K}{K-1}\frac{BGSS}{WGSS} \tag{3}$$

where N is the number of observations and K is the number of clusters and where

$$BGSS = \sum_{k=1}^{K} n_k ||G^k - G||^2 \tag{4}$$

(being G^k represents the dispersion of the barycenters of each cluster, G the barycenter of the whole set of data and n_k the number of samples in cluster C_k)

$$WGSS = \sum_{k=0}^{K} WGSS^k \tag{5}$$

$$WGSS^k = \sum_{i \in I_k} ||M_i^k - G^k||^2 \tag{6}$$

(being M_i^k the coefficients of the i-th row of the data matrix for a cluster C_k and I_k, the set of the indices of the observations belonging to the cluster C_k)

Davies-Bouldin: Davies-Bouldin is a metric for the evaluation of clustering algorithms. It is an internal evaluation scheme, where the validation of how well the clustering has been done is done using quantities and characteristics inherent in the data set. This metric is defined as the mean value, among all the clusters, of the samples M_k (See 7).

$$DB = \frac{1}{K} \sum_{k=1}^{K} M_k \tag{7}$$

This expression can also be defined as (8):

$$DB = \frac{1}{K} \sum_{k=1}^{K} max_{k' \neq k} \left(\frac{\delta_k + \delta_{k'}}{\triangle_{kk'}} \right) \tag{8}$$

where δ_k is the mean distance of the points which belong to cluster C_k to their barycenter G_k and $\triangle_{kk'}$, the distance between barycenters G^k and $G^{k'}$ (See Eq. 9).

$$\triangle_{kk'} = d(G^k, G^{k'}) = ||G^k - G^{k'}|| \tag{9}$$

Small values for the DB index indicate compact clusters, whose centers are well separated from each other. Consequently the number of clusters minimized by the DB index is taken as the optimum.

3.5 Regression Method

Multi-Layer Perceptron: Next, a Multi-Layer Perceptron (MLP) has been implemented, with the main objective of obtaining a metric to evaluate the clustering algorithms explained above.

Among the various supervised learning techniques available, MLP is one of the most widely used. This algorithm learns thanks to the function: $Fun(\cdot) : X^N \rightarrow X0$. To implement this technique, the Python Scikit-Learn library has been used.

In order to obtain the optimal number of neurons in the hidden layer, and the best activation function associated with each one, a cross validation procedure has been used. Thanks to this procedure, it has been possible to train the MLP with different parameters (number of neurons in the hidden layer and the activation function), selecting the best combination of parameters to obtain the best regression model [1,6,12,23].

After the application of cross-validation procedure, the optimal number of neurons is selected in the range of 3 to 30. The best options in terms of activation function are 'Rectified Linear Unit' and 'Tanh function'.

4 Experiments and Results

4.1 Cluster

For determining, possible groupings of the unsupervised data, two different clustering techniques have been evaluated: Spectral Clustering and Gaussian Mixture Clustering. After this, the assigned group of each sample is used as the class for a supervised regression. We carried out a hyperparameters study varying the number of clusters. Three different unsupervised metrics were taken into account to determine the best configuration: Silhouette, Calinski-Harabasz and Davies-Bouldin scores. In Table 1 we can see the results achieved with the selected hyperparameter.

Table 1. Best hyperparameter scoring using Gaussian Mixture and Spectral Clustering

Custering	Best number os clusters	Silhouette	Calinski-Harabasz	Davies-Bouldin
Gaussian Mixture	4	0.4450	32735.4139	0.7654
Spectral Clustering	4	0.4936	40391.5038	0.6354

Showing the results, it can conclude that the optimum value for both cases is four clusters. Although both methods obtains similar scores, Spectral Clustering slightly outperforms Gaussian mixture in all the evaluated techniques.

In order to get a projected visualization of the data, a 2D mapping was done by training a LDA model using the cluster assigned to each sample as its class. In Fig. 3, we can see the 2D projection for both clustering techniques.

4.2 Regression

The main objective of this work is to know which is the best clustering algorithm, being the regression procedure, complementary to unsupervised clustering metrics for knowing what clustering technique is better. For regression purposes, an MLP architecture has been chosen coupled of a cross-validation, oriented to look for a series of neurons in the hidden layer, as well as their activation function.

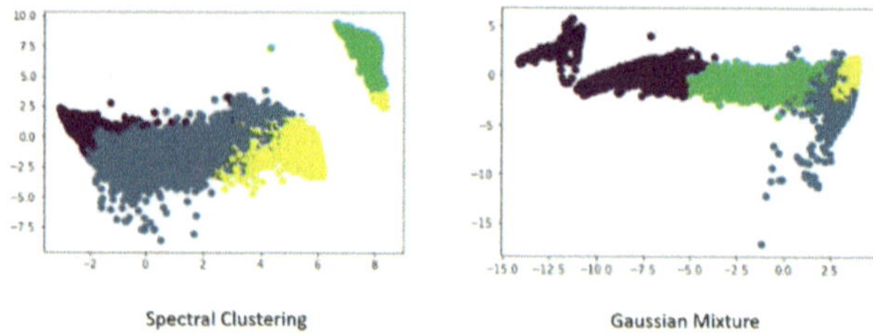

Fig. 3. 2D representation of the dataset for each clustering technique.

Error measurement chosen for Grid Search implantation in the cross validation has been the *Mean Squared Error* [22].

Two different approaches have been used, which can be seen in the final results. The first is a hybrid model, based on the Gaussian Mixture clustering method, while the second is also a hybrid model, but based on the Spectral Clustering method. Error measurements per each cluster are shown in Tables 2 and 3. A weighted average proportional to the size of each grouping has also been included in these tables.

Table 2. MLP error measurements for Gaussian Mixture clustering with 4 clusters

Cluster	1	2	3	4	Weighted average
MSE	24,4448	0,57695	60,7936	15,5296	**28,8981**
MAE	3,074782	0,56373	6,01444	2,79303	**3,3216**

Table 3. MLP error measurements for Spectral clustering with 3 clusters

Cluster	1	2	3	Weighted average
MSE	27,356834	2,35818	57,28238	**35,6353**
MAE	3,283114	0,89186	5,87033	**3,9488**

Figures 4 and 5 display the graphical representation for each clustering technique, with the real output represented in blue and the predicted output represented in red. For visualization purposes the "X" axis represents only 100 elements from each data sample, from the final test data set, formed by 20% of the cases in each cluster. The validation division is made up of 26665 elements, divided, in turn, into 4 groupings for Gaussian Mixture and 3 for Spectral. This fact can be very tedious to observe the quality of the regression if all the elements of the validation division are plotted. In addition to this, the "Y" axis represents the output value, which refers to the output temperature of the upper solar panel.

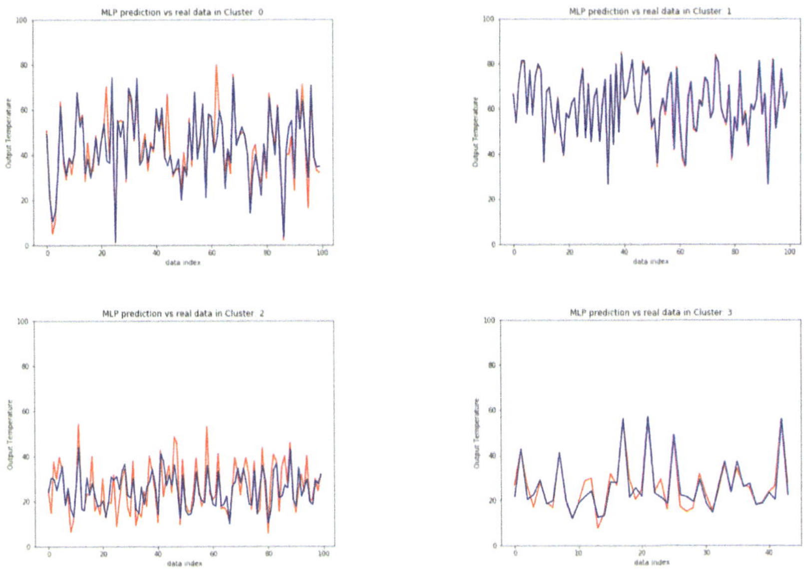

Fig. 4. Real data vs. MLP predictions for Gaussian Mixture clustering

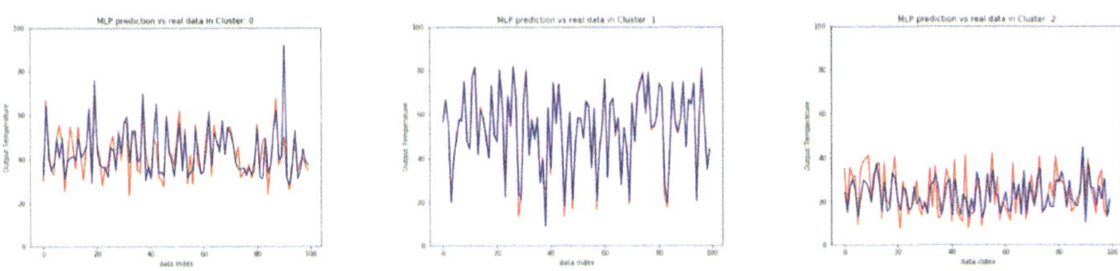

Fig. 5. Real data vs. MLP predictions for Spectral clustering

5 Conclusions and Future Works

In this work, two different clustering algorithms have been evaluated for predicting the temperature in thermal solar panels: Gaussian Mixture Clustering and Spectral Clustering. Although the selected clustering methods are based on different aggrupation techniques, an hyperparameter evaluation determined that the best performance is achieved with four clusters in both cases. This evaluation was carried out taking into account three different scoring metrics: Silhouette, Calinski-Harabasz and Davies-Bouldin. Comparing both algorithms, Spectral Clustering achieved a better grouping of the data for all the three scores. After that, an MLP neural network have been implemented in order to predict the temperature merging all the features with the cluster assigned by the unsupervised algorithm. With this information, the best results for regression were obtained with the Gaussian Mixture clustering addition outperforming the Spectral Clustering in a 18.91% taking into account the MSE error and in a 15.88%

with respect to MAE. With all this information, we can conclude that although the best clustering technique was the Spectral Clustering, the Gaussian Mixture approach provides more information for the temperature prediction purpose.

Future works will be based on apply new clustering methods to new datasets. Besides, authors will explore new ways for implementing robust hybrid models with the application of new clustering techniques.

As a main limitation of this work, it can be highlighted that the dataset is not large enough when it is separated in four cluster, in order to use the latest deep learning techniques.

References

1. Alaiz-Moreton, H., Fernández-Robles, L., Alfonso-Cendón, J., Castejón-Limas, M., Sánchez-González, L., Pérez, H.: Data mining techniques for the estimation of variables in health-related noisy data. In: International Joint Conference SOCO'17-CISIS'17-ICEUTE'17, Proceeding, León, Spain, 6–8 September 2017, pp. 482–491. Springer, Heidelberg (2017)
2. Bacong, J.R., Juanico, D.E.: Performance analysis of multi-layer perceptron regression model with mixed-rate sensor data inputs. In: Proceedings of the Samahang Pisika ng Pilipinas (2018)
3. Bishop, C.M.: Pattern Recognition and Machine Learning. Springer, Heidleberg (2006)
4. Blackburn, W.R.: The Sustainability Handbook: The Complete Management Guide to Achieving Social, Economic and Environmental Responsibility. Routledge, Abingdon (2012)
5. Calvo-Rolle, J.L., Casteleiro-Roca, J.L., Quintián, H., del Carmen Meizoso-Lopez, M.: A hybrid intelligent system for PID controller using in a steel rolling process. Exp. Syst. Appl. **40**(13), 5188–5196 (2013)
6. Castejón-Limas, M., Alaiz-Moreton, H., Fernández-Robles, L., Alfonso-Cendón, J., Fernández-Llamas, C., Sánchez-González, L., Pérez, H.: Coupling the paella algorithm to predictive models. In: International Joint Conference SOCO'17-CISIS'17-ICEUTE'17, Proceeding, León, Spain, 6–8 September 2017, pp. 505–512. Springer, Heidelberg (2017)
7. Casteleiro-Roca, J.L., Calvo-Rolle, J.L., Méndez Pérez, J.A., Roqueñí Gutiérrez, N., de Cos Juez, F.J.: Hybrid intelligent system to perform fault detection on BIS sensor during surgeries. Sensors **17**(1), 179 (2017)
8. Casteleiro-Roca, J.L., Jove, E., Gonzalez-Cava, J.M., Pérez, J.A.M., Calvo-Rolle, J.L., Alvarez, F.B.: Hybrid model for the ANI index prediction using remifentanil drug and EMG signal. Neural Comput. Appl. **32**(5), 1–10 (2018)
9. Cecilia, A., Costa-Castelló, R.: High gain observer with dynamic deadzone to estimate liquid water saturation in PEM fuel cell. Revista Iberoamericana de Automática e Informática Ind. **17**(2), 169–180 (2020)
10. Dempster, A.P., Laird, N.M., Rubin, D.B.: Maximum likelihood from incomplete data via the EM algorithm. J. Roy. Stat. Soc. Ser. B (Methodological) **39**(1), 1–22 (1977)
11. Ding, C.H., He, X., Zha, H., Gu, M., Simon, H.D.: A min-max cut algorithm for graph partitioning and data clustering. In: Proceedings 2001 IEEE International Conference on Data Mining, pp. 107–114. IEEE (2001)

12. Duan, K., Keerthi, S.S., Poo, A.N.: Evaluation of simple performance measures for tuning SVM hyperparameters. Neurocomputing **51**, 41–59 (2003)
13. Epstein, M.J.: Making Sustainability Work: Best Practices in Managing and Measuring Corporate Social, Environmental and Economic Impacts. Routledge, Abingdon (2018)
14. Kalogirou, S.A.: Artificial neural networks in renewable energy systems applications: a review. Renew. Sustain. Energ. Rev. **5**(4), 373–401 (2001)
15. Kapferer, J.N., Michaut-Denizeau, A.: Are millennials really more sensitive to sustainable luxury? a cross-generational international comparison of sustainability consciousness when buying luxury. J. Brand Manag. **27**(1), 35–47 (2020)
16. Karaosman, H., Perry, P., Brun, A., Morales-Alonso, G.: Behind the runway: extending sustainability in luxury fashion supply chains. J. Bus. Res. **117**, 652–663 (2018)
17. Marrero, A., Méndez, J., Reboso, J., Martín, I., Calvo, J.: Adaptive fuzzy modeling of the hypnotic process in anesthesia. J. Clin. Monit. Comput. **31**(2), 319–330 (2017)
18. McLachlan, G., Peel, D.: Finite Mixture Models. Wiley Series in Probability and Statistics, John Wiley & Sons, Inc., Hoboken (2000), https://doi.org/10.1002/0471721182
19. Mika, S., Ratsch, G., Weston, J., Scholkopf, B., Mullers, K.R.: Fisher discriminant analysis with kernels. In: Neural networks for signal processing IX: Proceedings of the 1999 IEEE signal processing society workshop (cat. no. 98th8468). pp. 41–48. Ieee (1999)
20. Ng, A.Y., Ng, A.Y., Jordan, M.I., Weiss, Y.: On spectral clustering: analysis and an algorithm. In: Advances in Neural Information Processing Systems, vol. 14, pp. 849—-856 (2001). http://citeseerx.ist.psu.edu/viewdoc/summary?doi=10.1.1.19.8100
21. Quintián, H., Calvo-Rolle, J.L., Corchado, E.: A hybrid regression system based on local models for solar energy prediction. Informatica **25**(2), 265–282 (2014)
22. Tuchler, M., Singer, A.C., Koetter, R.: Minimum mean squared error equalization using a priori information. IEEE Trans. Signal Process. **50**(3), 673–683 (2002)
23. Grid search cross validation (2019). http://scikit-learn.org/stable/modules/generated/sklearn.model_selection.GridSearchCV.html. Accessed 22 Apr 2019
24. Wagstaff, K., Cardie, C., Rogers, S., Schrödl, S., et al.: Constrained k-means clustering with background knowledge. In: Icml, vol. 1, pp. 577–584 (2001)

Data Balancing to Improve Prediction of Project Success in the Telecom Sector

Nuño Basurto[1], Alfredo Jiménez[2], Secil Bayraktar[3], and Álvaro Herrero[1](✉)

[1] Grupo de Inteligencia Computacional Aplicada (GICAP), Departamento de Ingeniería Informática, Escuela Politécnica Superior, Universidad de Burgos, Av. Cantabria s/n, 09006 Burgos, Spain
{nbasurto,ahcosio}@ubu.es

[2] Department of Management, KEDGE Business School, Bordeaux, France
alfredo.jimenez@kedgebs.com

[3] Department of Human Resources Management and Business Law, TBS Business School, 1 Place Alphonse Jourdain, 31068 Toulouse, France
s.bayraktar@tbs-education.fr

Abstract. Investments in the telecom industry are often conducted through private participation projects, allowing a group of investors to build and/or operate large infrastructure projects in the host country. As governments progressively removed the barriers to foreign ownership in this sector, these investment consortia have become increasingly international. Obviously, an accurate and early prediction of the success of such projects is very useful. Softcomputing can certainly contribute to address such challenge. However, the error rate obtained by classifiers when trying to forecast the project success is high due to the class imbalance (success vs. fail). To overcome such problem, present paper proposes the application of classifiers (Support Vector Machines and Random Forest) to data improved by means of data balancing techniques (both oversampling and undersampling). Results have been obtained on a real-life and publicly-available dataset from the World Bank.

Keywords: Classification · Data balancing · Oversampling · Undersampling · Private-Participation projects · Internationalization

1 Introduction and Previous Work

Large infrastructure projects typically require not only specific expertise to manage the project, but also a significant amount of resources to invest. For this reason, governments all over the world have frequently employed privatizations to attract investors who can provide such expertise and resources. Given the large requirements of this type of projects, investors typically are unable to conduct them on their own, and instead they create a private participation project, a consortium of investors together with other interested companies [1, 2]. One of the sectors in which this phenomenon has occurred the most is the telecommunication sector.

Á. Herrero et al. (Eds.): SOCO 2020, AISC 1268, pp. 366–373, 2021.
https://doi.org/10.1007/978-3-030-57802-2_35

Given the type of sector and service provided, telecommunications have been frequently subject to strong regulation and under high scrutiny of governments, who many times did not allow foreign investors to participate. However, despite still under a great regulation, foreign companies have been allowed to enter multiple countries, both developed and developing, in order to access state of the art technology and managerial expertise [1, 2]. As a result of this more favorable environment to foreign investors, the composition of private participation project in the telecommunication sector has greatly diversified and internationalized. This, in turn, has increased the scholarly attention towards the specific factors that affect the performance and success of this specific instrument of investment. Despite the majority of projects are successful, because obviously investors bid only in privatization projects in which they expect to obtain a positive return, and also because governments tend to choose the most experienced and capable consortia of investors, previous literature has investigated how multiple factors can have an important impact on private participation projects (see [3] for a review of the literature on relevant institutional factors). This high interest is justified not only because these investments are themselves very large in economic terms, but also because predicting the success of the projects is crucial because this sector provides a service that is key for the rest of the economy and therefore failures should be prevented or minimized as much as possible. Thus, in this study we analyze 9176 private participation projects in the telecommunications sector in 32 host countries from 2004 to 2013 and investigate techniques to improve the prediction of success of these projects.

A wide range of soft-computing techniques have been applied to enterprise management [4–7] so far. Conversely, very few supervised machine-learning models have been applied to problems similar to the one above described. That is the case of [8], where corporate credit rating analysis is conducted based on Support Vector Machine (SVM) and Artificial Neural Networks (ANN). These classifiers are applied to data from for the United States and Taiwan markets trying not only to forecast but also to get a model with better explanatory power. More recently, [9] combined SVM together with fuzzy logic as a real case study in construction management. This hybrid system tried to predict project dispute resolution outcomes (i.e., mediation, arbitration, litigation, negotiation, and administrative appeals) when the dispute category and phase in which a dispute occurs are known during project execution.

In [10] *k*-Nearest Neighbor (k-NN) is compared to ANN, Discriminant Analysis, Quadratic Discriminant Analysis, and Multinomial Logistic Regression Analysis to provide input to managers who make business decisions. These models were applied to retail department store data, showing that they are most useful when uncertainty is high and a priori classification cannot be made with a high degree of reliability. Additionally, [11] proposed the application of k-NN to multi-criteria inventory classification in order to manage inventory more efficiently. k-NN are compared to SVM, ANN, and Multiple Discriminant Analysis when applied to 4 benchmark datasets. SVM was identified as the most accurate among all of them due to its high generalization capability, as well as its use of kernel functions to increase the learning efficiency.

As a seminal work of present research, [12] proposed different classifiers (SVM, k-NN, and Random Forest) to check their ability to predict the final success of Private Participation Projects (PPP) involving infrastructures. Going one step further, present paper proposes the application of data balancing techniques to improve the classifier performance when applied to such imbalanced PPP datasets. This proposal is validated trough a dataset by the World Bank.

The rest of this paper is organized as follows: the applied techniques are described in Sect. 2, the dataset, setup of experiments and obtained results are described in Sect. 3. Finally, the conclusions of present study and future work are stated in Sect. 4.

2 Soft-Computing Techniques

As stated, data balancing techniques (described in Subsect. 2.1) are proposed in order to improve the performance of some popular classifiers (described in Subsect. 2.2).

2.1 Data-Balancing

There are different methods designed to pre-process data, prior to a subsequent supervised learning stage. They can be classified in three categories that are described in the following paragraphs: undersampling, oversampling and hybrid.

Undersampling methods obtain a balanced number of instances per class by creating a new subset of data in which some instances (from the majority class) are removed. The most popular of such methods is known as Random Under Sampling (RUS), that gets the target subset by randomly selecting those instances to be deleted.

On the contrary, oversampling methods get a balanced number of instances per class by artificially generating new data instances (from the minority classes) that were not in the original dataset. In this case, the most popular method is known as Random Over Sampling (ROS), that randomly selects the data instances to be duplicated. Based on this idea, there is a more complex and widely-used oversampling method called Synthetic Minority Oversampling TEchnique (SMOTE) [13]. This method introduces new data samples artificially created by interpolating values taken from pre-existing instances of the minority class. The base instances used to generate the new ones are selected by k-NN.

Finally, hybrid methods combine both oversampling and undersampling techniques in order to reduce the impact in only one of the classes that the single methods have. In present paper, the combination or ROS and RUS (ROS + RUS) has been applied. Additionally, RUS is also combined with the SMOTE oversampling method (SMOTE + RUS).

2.2 Classifiers

Based on previous results [12], both Support Vector Machines (SVMs) [14, 15], and Random Forests (RFs) [16] are applied in present work. The used class information is the success of the projects (true or false), as defined in Subsect. 3.1.

SVM show good generalization performance so they have been applied to wide range of real-life problems [17], including multi-class classification. They try to find the optimal hyperplane that not only separates the classes with no error but also maximizes the distance to closest point (for either class).

SVMs can be seen as classifiers where the loss function is the Hinge function, defined as:

$$L[y, f(x)] = max[0, 1 - yf(x)] \tag{1}$$

Being x an observation from input features, y the class x belongs to, and $f(x)$ the output of the classifier. Additionally, there is the gamma parameter that states the influence of a single training example; i.e. a low value means a far influence while a high value means close. It can also be seen as the inverse of the radius of influence of samples selected by the model as support vectors.

On the other hand, classification trees [18] are well-known inductive learning methods. They contain two types of nodes:

- Inner nodes: they are associated to differentiate responses (branches) for a given question regarding the values of a feature from the original training dataset. All of them have at least two child nodes.
- Leaf nodes: they are designed for taking the final decision (prediction).

Labels are assigned to the archs connecting a node to one of its child nodes (their content is related to the responses to the node question) and leaf nodes (their content is one of the classes in the training dataset).

A RF can be seen as an aggregation of a number of classification trees such that each one of them depends on the values of a random vector. This vector is sampled independently and with the same distribution for all trees in the forest. One of the main advantages, when compared to a single classification tree schema, is the reduction of variance. In the case of RF, the prediction is obtained for a new data by aggregating (through majority voting) the predictions made by all the single trees. That is, the new data is assigned to the class that was most often predicted by the individual trees.

3 Experiments and Results

The analyzed dataset is presented in Subsect. 3.1, while the results generated through the different experiments are described in Subsect. 3.2.

3.1 Dataset

We obtained our sample from the World Bank's Private Participation in Infrastructure (PPI) dataset. Projects from all over the world in the Telecommunication sector are analyzed. This sector has been chosen because of its critical impact on the economy and also because it is one of the most imbalanced sectors in the PPI dataset. There are 9176 projects from this sector in 32 host countries, from 2004 to 2013, and most of them (9043–98,55%) succeeded.

Drawing on prior literature on the private participation projects field, we conceptualize project success as the completion of the bidding process, fulfillment of binding agreements, and access to the required capital. Empirically, we employ a dichotomous variable based on the project status as reported in the data source, as previously done in earlier studies [3, 19, 20]. We consider successful projects those whose project status is reported as either "operational", "merged" or "concluded" and, conversely, we consider failed projects those reported as "cancelled" or "distressed" (i.e. when the investors or the government have requested the termination of the project respectively).

We collected information on a number of explanatory variables, both at the country and at the project levels. Specifically, we follow prior empirical studies and replicate the measures used by [3]. Thus, we include in the analyses macro variables such as GDP (log), the rate of GDP growth, unemployment (log), political stability (Polconv index), and corruption (World Bank Worldwide Governance Indicators). As it is standard practice in the literature, we reversed this latter variable to simplify the interpretation of results, so more corruption is associated to higher figures of the variable. Furthermore, we also account for the size of the project as measured by the total investment (log), its age since the year it was started, the time lag difference between the project commitment and project closure, and whether it is a project started from scratch (greenfield) or already pre-existing (brownfield). We also include a number of features regarding the composition of the consortia of investor, such as whether a foreign investor is the leading one, at least one investor in the consortium is from the host country, the host government is included in the consortium as an investor, and whether it is a publicly traded project or not. All in all, 13 features are compressed in each one of the datasets for all the project instances.

3.2 Results by SVM

Results obtained when applying SVM after balancing the dataset with the techniques described in Sect. 2.1 are shown in Table 1. For comparison purposes, results without applying any data-balancing technique are also shown (referred as "None").

Table 1. AUC results by SVM per data-balancing algorithm and different values of gamma (0.01, 0.05, and 0.1).

	0.01	***0.05***	***0.1***
None	0.9193	0.8916	0.4977
SMOTE	**0.9754**	0.8844	**0.8850**
SMOTE + RUS	0.9655	0.8926	0.8746
ROS	0.9744	**0.9002**	0.8513
RUS	0.8463	0.8189	0.8678
ROS + RUS	0.9704	0.8811	0.8363

From these results, it can be said that classification results are improved thanks to the balancing of the data; for all the gamma values, AUC results are improved by any of the data balancing algorithms. More precisely, SMOTE (alone in the case of the gamma values 0.01 and 0.1), and ROS (in the case of the 0.05 gamma value) have proved to be the best techniques for the analyzed dataset when using SVM.

On the other hand, it is worth mentioning that results are worsen when applying some of the balancing algorithms. In the case of gamma value 0.01, the RUS algorithm has led to a worse AUC result than the one obtained without balancing (None). In the case of gamma value 0.05, only the techniques ROS and SMOTE + RUS have obtained better results than None.

3.3 Results by RF

Results obtained when applying RF after balancing the dataset with the techniques described in Sect. 2.1 are shown in Table 2. Similar to previous SVM results, scores obtained without applying any data-balancing technique are also shown (referred as "None").

Table 2. AUC results by RF per data-balancing algorithm and different numbers of trees (100, 200, 500, and 1000).

	100	*250*	*500*	*1000*
None	0.9300	0.9168	0.9078	0.9002
SMOTE	0.9245	0.9385	0.9315	0.9349
SMOTE + RUS	0.9382	0.9553	0.9260	0.9612
ROS	0.9184	0.9060	0.9208	0.9324
RUS	**0.9670**	**0.9749**	**0.9754**	**0.9696**
ROS + RUS	0.9186	0.9346	0.9653	0.9631

As it has been discussed in the case of SVM results, the classification results are improved thanks to the balancing of the data; for all the numbers of trees under analysis, AUC results are improved by any of the data balancing algorithms. Differentiating from the results obtained by SVN, it is now the RUS technique the one obtaining the best results in all cases. SMOTE combined with RUS in the case of the smallest numbers of trees (100 and 250) and ROS combined with RUS in the case of the highest numbers of trees (500 and 1000) are those that have obtained second best results. It can be concluded that undersampling are the best methods, as they outperforms the other ones when classifying by RF.

Logically, oversampling methods have obtained the worst results. Once again, one of the techniques (ROS) has led to worse results when compared to the raw data (None) for 100 and 250 trees.

4 Conclusions and Future Work

By analyzing obtained results, it can be stated that thanks to the data-balancing techniques, prediction results may be improved. When selecting the appropriate technique, the prediction of success in PPP can be greatly improved. Furthermore, acceptable results can be obtained even from a highly imbalanced dataset as the one analyzed in present study. Oversampling in the case of SVM and undersampling in the case of RF outperform all the other techniques when balancing the dataset for subsequent classification.

By identifying techniques that allow a more accurate prediction of project success, our paper makes an important contribution with repercussions for investors and governments. On the one hand, investors participating in private participation projects need to raise significant amount of funds, and the higher predictability of success can reduce the cost of borrowing from financial institutions. On the other, the higher predictability of the projects can allow governments to attract more suitable firms interested in the privatization, allowing the government to receive better bids and choose the one that allows a better functioning of the telecommunication sector in the country and ensuring positive multiplier effects and synergies in other sectors of the host economy.

Future work will focus on considering some other sectors where private participation projects are also imbalanced and comparing some additional classifiers.

References

1. Ramamurti, R., Doh, J.P.: Rethinking foreign infrastructure investment in developing countries. J. World Bus. **39**, 151–167 (2004)
2. García-Canal, E., Guillén, M.F.: Risk and the strategy of foreign location choice in regulated industries. Strateg. Manag. J. **29**, 1097–1115 (2008)
3. Jiménez, A., Russo, M., Kraak, J.M., Jiang, G.F.: Corruption and private participation projects in Central and Eastern Europe. Manag. Int. Rev. **57**, 775–792 (2017)
4. Herrero, Á., Jiménez, A.: Improving the management of industrial and environmental enterprises by means of soft computing. Cybern. Syst. **50**, 1–2 (2019)
5. Jiménez, A., Herrero, Á.: Selecting features that drive internationalization of Spanish firms. Cybern. Syst. **50**, 25–39 (2019)
6. Simić, D., Svirčević, V., Ilin, V., Simić, S.D., Simić, S.: Particle swarm optimization and pure adaptive search in finish goods' inventory management. Cybern. Syst. **50**, 58–77 (2019)
7. Herrero, Á., Jiménez, A., Bayraktar, S.: Hybrid unsupervised exploratory plots: a case study of analysing foreign direct investment. Complexity **2019**, 6271017 (2019)
8. Huang, Z., Chen, H., Hsu, C.-J., Chen, W.-H., Wu, S.: Credit rating analysis with support vector machines and neural networks: a market comparative study. Decis. Support Syst. **37**, 543–558 (2004)
9. Chou, J.-S., Cheng, M.-Y., Wu, Y.-W.: Improving classification accuracy of project dispute resolution using hybrid artificial intelligence and support vector machine models. Exp. Syst. Appl. **40**, 2263–2274 (2013)
10. Malhotra, M.K., Sharma, S., Nair, S.S.: Decision making using multiple models. Eur. J. Oper. Res. **114**, 1–14 (1999)
11. Yu, M.C.: Multi-criteria ABC analysis using artificial-intelligence-based classification techniques. Exp. Syst. Appl. **38**, 3416–3421 (2011)
12. Herrero, Á., Jiménez, A.: One-class classification to predict the success of private-participation infrastructure projects in Europe, pp. 443–451. Springer, Heidelberg (2020)
13. Chawla, N.V., Bowyer, K.W., Hall, L.O., Kegelmeyer, W.P.: SMOTE: synthetic minority over-sampling technique. J. Artif. Intell. Res. **16**, 321–357 (2002)
14. Boser, B.E., Guyon, I.M., Vapnik, V.N.: A training algorithm for optimal margin classifiers. In: 5th Annual Workshop on Computational Learning Theory, pp. 144–152. ACM (1992)
15. Cortes, C., Vapnik, V.: Support-Vector networks. Mach. Learn. **20**, 273–297 (1995)

16. Breiman, L.: Random forests. Mach. Learn. **45**, 5–32 (2001)
17. Byun, H., Lee, S.-W.: Applications of support vector machines for pattern recognition: a survey, pp. 213–236. Springer, Heidelberg (2002)
18. Safavian, S.R., Landgrebe, D.: A survey of decision tree classifier methodology. IEEE Trans. Syst. Man Cybern. **21**, 660–674 (1991)
19. Jiang, Y., Peng, M.W., Yang, X., Mutlu, C.C.: Privatization, governance, and survival: MNE investments in private participation projects in emerging economies. J. World Bus. **50**, 294–301 (2015)
20. Jiménez, A., Jiang, G.F., Petersen, B., Gammelgaard, J.: Within-country religious diversity and the performance of private participation infrastructure projects. J. Bus. Res. **95**, 13–25 (2019)

Demand Control Ventilation Strategy by Tracing the Radon Concentration in Smart Buildings

Roberto Casado-Vara[1(✉)], David García-Retuerta[1], Alvaro Bartolomé[1], Esteban Jove[2], Jose Luis Calvo-Rolle[2], Angel Martin-del Rey[3,4], and Juan M. Corchado[1,5]

[1] BISITE Research Group, University of Salamanca, 37008 Salamanca, Spain
{rober,alvarob96,dvid,corchado}@usal.es

[2] Department of Industrial Engineering, University of A Coruña, Ferrol, A Coruña, Spain
{esteban.jove,jose.rolle}@udc.es

[3] Department of Applied Mathematics, University of Salamanca, Calle del Parque 2, 37008 Salamanca, Spain
delrey@usal.es

[4] Department of Applied Mathematics, Institute of Fundamental Physics and Mathematics, University of Salamanca, Calle del Parque 2, 37008 Salamanca, Spain

[5] IoT Digital Innovation Hub (Spain), Air Institute, Calle Segunda 4, 37188 Salamanca, Spain

Abstract. Ensuring air quality should be a mandatory premise in every building, since if not, its occupants are on high risk. In fact, Radon pollutants are stated to be the second main cause among all lung cancer patients in the United States. Radon is a noble gas which seeps up through the ground and accumulates there, making it hard to be identified. A proper ventilation system needs to be installed on industrial plants so that the Radon exhaled from building materials is properly dispelled, ensuring fresh, quality air. In order to keep a proper air quality level in smart buildings, a control ventilation strategy should be defined so that the exhaled Radon is ensured to be dispelled keeping the indoor air quality high. In the proposed paper, the diffusion-advecntion method has been studied in order to propose a solution on Radon concentration tracing on smart buildings ventilation system. Diffusion-advecntion is a mathematical method that will determine whether Radon will propagate or not, based on the concentration of Radon, the diffusion constant and the advecntion velocity of the indoor air, which can lead to a recommendation for the smart building ventilation system to be activated or not, respectively. In this paper a new ventilation strategy for smart buildings based on the Diffusion-advecntion equation has been proposed to improve air quality. The results of this new ventilation strategy have been tested in a real case study in a smart building in the city of Salamanca. The

Á. Herrero et al. (Eds.): SOCO 2020, AISC 1268, pp. 374–382, 2021.
https://doi.org/10.1007/978-3-030-57802-2_36

main outcome of this new strategy is the improvement in response times of the current systems.

Keywords: Indoor air quality · Demand control ventilation · Radon

1 Introduction

Radon (^{222}Rn) is a noble gas, which means that it does not react combined with other substances as it seeps up through the ground and accumulates there [14]. In industrial plants, Radon pollutants exhaled from building materials will be accumulated on the building, so that the safety of anyone close to the building may be on risk due to the accumulated Radon pollutants [9] as it is stated that Radon is the second leading cause of lung cancer in the United States [13]. Air ventilation in buildings is required so that the air quality is ensured to be proper for building occupants, which depends on the quality of the brought-in air if it comes from a non-polluted ambient source which ensures a better indoor quality [8]. Ventilation is one method to maintain good indoor air quality. The more fresh air is brought into the indoor environment, the better the indoor air quality can be achieved if the fresh air comes from non-polluted ambient source. IoT [6,7], mathematical models [5] and automatic control [3,4] technologies are success case studies in other fields such as temperature control in smart buildings, therefore, this research will use these technologies to improve indoor air quality in smart buildings.

The proposed study aims to explain how do pollutants spread in industrial plants based on the convection-diffusion equation, which will lead to define the best location of pollution filters leading to the improvement of air quality in benefit of the employees, making the industrial plant cleaner while saving energy at the same time. According to the National Association of Clean Air Agencies (NACAA) air pollution is caused by many types of sources of every size, where industrial plants are the dominant emitters of mercury (50%), sulfur dioxide (60%), acid gases (over 75%) and arsenic (63%) in the United States. The importance of controlling ventilation system in order to dispel pollutants from buildings lies in the need of also caring about people's health.

The importance of this work field lies in its potential applications to polluted industrial complexes, garages and any environment rich in noxious fumes. Several studies have discovered interesting relations between poor indoor air quality and productivity, health and welfare. For example, it can reduce the performance of office work by 6–9% and some of the caused negative effects include headaches and concentration problems [17]. Currently, the main proposals for dealing with this problem are based on choosing the optimal materials and improving the efficiency of the filters. However, they lack a mathematical model for understanding the propagation of certain gases in indoors situations which can become the cornerstone of any future optimisation method. On top of the previous reasons, the importance in this particular case lies in the dangers associated with radon gas.

The event of an expansion of radon gas has been modelled using a new convection–diffusion based algorithm. It describes the physical phenomena of particles' transfer in a certain location. Its solution provides an approximation of the particles distribution on the considered space, which produces great amounts of useful information for later optimization processes. The results of this research work show that the problem of modeling the distribution of polluting gases can be satisfactory solved using differential equations. Numerical methods provide effective tools for solving the proposed equations and are implemented for obtaining a fast performance. It results in a great energy optimisation which is now available for environments were a lot of exhume fumes are generated, and subsequently need to be aired out. Improved energy performance and better air quality is therefore within reach.

The paper is organised as follows: Sect. 1 provides an introduction to the topic and states the key facts of this research, Sect. 2 shown the detailed description of the algorithm and Sect. 3 presents the results of this research. Finally, Sect. 4 provides the conclusions.

2 Indoor Air Quality Algorithm

This section describes the hybrid algorithm for automatic Radon gas ventilation in smart buildings. The algorithm will receive the concentration of the Radon gas (c), the diffusion constant (d) and the advecntion velocity (v) of the air inside the smart buildings. A new mathematical Diffusion-advecntion model is solved in the first part of the algorithm, both the concentrations and whether the gas will propagate or not are passed to the algorithm as an input. Finally, the recommendation algorithm then sends the recommendations to the smart building as to whether or not ventilation is necessary. This algorithm determines the propagation of radon gas through the smart building. Thus a new strategy for automatic smart building ventilation can be designed so that the concentration of radon gas inside the building is within the range that the World Health Organization (WHS) considers good for humans. An IoT network is used as input of the algorithm and as output you have a concentration vector depending on the mesh that has been used to discretize the problem. A new demand control ventilation strategy designed in this paper will be used to control the automatic ventilation of smart buildings from this concentration vector. The operation of this algorithm is shown in Fig. 1. The main contribution of this paper is the new ventilation strategy we have designed. This strategy uses the Diffusion-advecntion model to calculate the concentrations of radon gas in its propagation through the smart building. In this way the algorithm that controls the building ventilation can anticipate the measurements of the IoT devices and offer a better operation.

2.1 Mathematical Model

The mathematical model of this algorithm uses diffusion-advecntion equation to determine the concentration of the gas in one of the smart building plants and

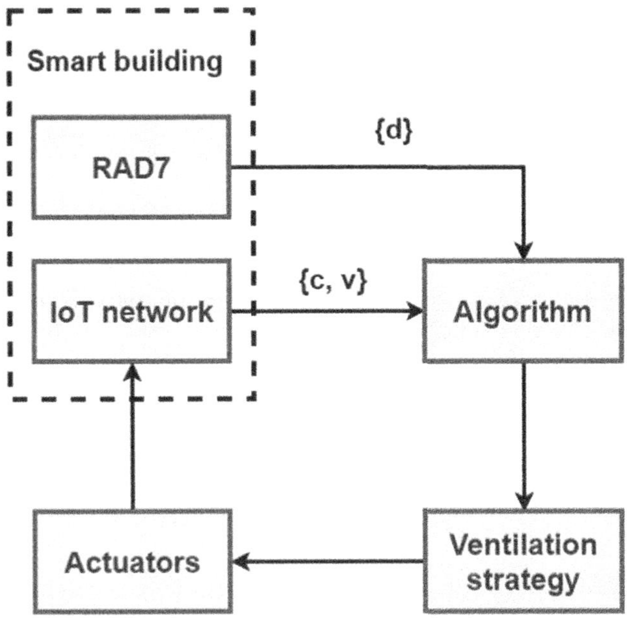

Fig. 1. Flowchart of the hybrid algorithm proposed in this paper The algorithm has as input the radon concentration (c), the radon diffusion constant (d) and the wind speed inside the smart building (v). It processes these data and sends a concentration vector to the ventilation strategy that decides which is the best option for each moment and sends the instructions to the actuators.

whether the radon will continue to propagate or dissipate [1,11,15]. This model, is a predictive model, therefore, will allow increasing energy savings.

$$\frac{\partial c}{\partial t} = -v\frac{\partial c}{\partial x} + d\frac{\partial^2 c}{\partial x^2} \\ \text{with} \quad c(x,0) = c_0(x), \quad 0 \leq x \leq L, \quad c(0,t) = c(L,t) = 0 \tag{1}$$

where $c(x,t)$ is the concentration of radon gas at the point x at the instant t, L the length of the smart building, v the wind speed and d the diffusion constant. To solve this equation we will integrate it as Carnahan *et al.* does [2].

Since it is economically unfeasible to put several radon gas concentration measurement sensors in smart buildings, in this work we are going to study the radon gas propagation from the porous soils through which the radon gas filters. In order to solve this equation, it is necessary to discretize it. Assuming the spatial step Δx and the temporal step Δt, and using three point operators for the second spatial derivatives and two point backwards operators for the first derivatives. The advance in time is solved with the explicit Euler method. Furthermore, it is assumed that the concentration must be zero in the windows and doors - since it dissipates - the radon will be dispersed until it disappears.

Equation 1 takes the form

$$\frac{\partial c}{\partial t} = f(c, x, t) \tag{2}$$

discretizing the second part of the equation in space and time

$$f(c, x, t) = -v\frac{c}{\Delta x} + d\frac{\partial^2 c}{\partial x^2} \approx -\frac{v}{\Delta x}(c_j^n - c_{j-1}^n) + \frac{d}{\Delta x^2}(c_{j+1}^n - c_j^n + c_{j-1}^n) \tag{3}$$

where j is the index of space and n is the index of time. Thus, we solve the equation using Euler's explicit scheme.

$$c_j^{n+1} = c_j^n - \Delta t f(c^{(n)}, x, t^{(n)}) \tag{4}$$

in other words

$$c_j^{n+1} = c_j^n - v\frac{\Delta t}{\Delta x}(c_j^n - c_{j-1}^n) + d\frac{\Delta t}{\Delta x^2}(c_{j+1}^n - 2c_j^n + c_{j-1}^n) \tag{5}$$

replacing $v\frac{\Delta t}{\Delta x} = \alpha$ and $d\frac{\Delta t}{\Delta x^2} = \beta$, one has

$$c_j^{n+1} = (1 - \alpha - 2\beta)c_j^n + (\alpha + \beta)c_{j-1}^n + \beta c_{j+1}^n \tag{6}$$

Therefore, by placing the Eq. 6 in matrix form

$$\begin{pmatrix} c_1 \\ c_2 \\ \vdots \\ c_n \end{pmatrix}^{(n+1)} = \begin{pmatrix} (1-\alpha-2\beta) & \beta & 0 & 0 \\ (\alpha+\beta) & (1-\alpha-2\beta) & \beta & 0 \\ 0 & (\alpha+\beta) & (1-\alpha-2\beta) & \beta \\ 0 & 0 & (\alpha+\beta) & (1-\alpha-2\beta) \end{pmatrix} \begin{pmatrix} c_1 \\ c_2 \\ \vdots \\ c_n \end{pmatrix}^{(n)} \tag{7}$$

2.2 Details of the Demand Control Ventilation Strategy

Based on the findings in the site measurements, a new type of demand control ventilation strategy was developed. Great amounts of fresh air are required to dilute radon to an acceptable level -among other contaminants such as CO_2- is selected as the control signal in our control strategy. From our former site measurements, radon was found to be the dominant indoor pollutant in the smart building. In this demand control ventilation strategy, the radon level is monitored all the times to determine whether there is a health hazard or not. In that case, the control strategy will switched to a ventilation strategy.

There is one set point in this demand control strategy: randon level. An indoor radon level of 200 Bq m^{-3} is recommended by the Hong Kong Environmental Protection Department as the Level 2 criterion in the newly established guidance notes [12] for public health, and in thus chosen as the set point in our control algorithm. Regarding the location of the sensors, as thorough mixing can

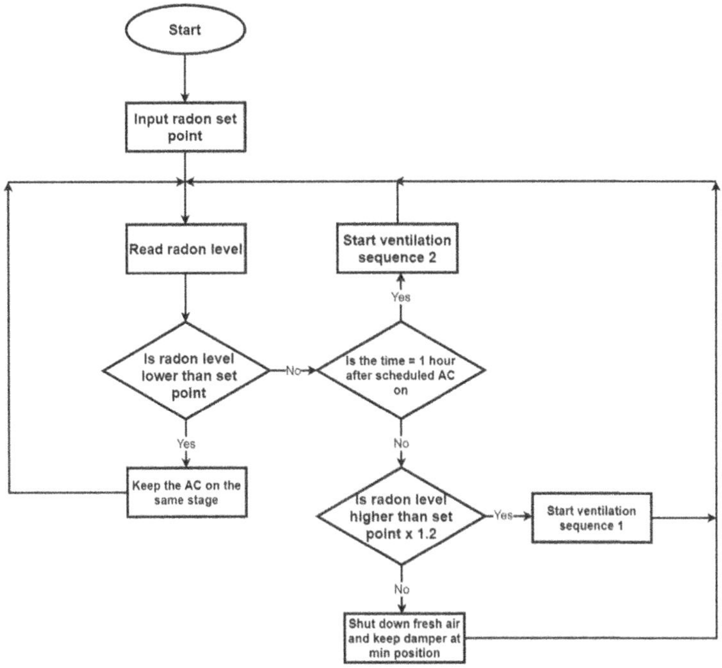

Fig. 2. Flowchart of radon based demand control ventilation strategy.

never be fully achieved in building, the most important point is to achieve an acceptable indoor air quality in the occupant's breathing zone. In this control strategy radon levels will forecast by the algorithm at the breathing zones. The strategy flowchart are shown in Fig. 2. The two kind of ventilation sequence are sequence 1 and sequence 2. The first ventilation sequence is one-hour ventilation mode in order to ensure that the occupants will not be exposed to undesirable levels of radon, the AC is set in full power. The ventilation sequence 2 is designed to cater to the situation in which the premises are not in hazzard for publich health, so the ventilation sequence are set for relatively long period of time. Also, a pre-ventilation sequence 1 is set 1 h before the occupant come to work.

3 Results and Discussion for the Operation of the Demand Control Ventilation Strategy

In this section we introduce the case study and the results achieved during the entire experiments. This ventilation strategy uses data collected from the IoT network, radon sensors and radon concentration results obtained from the indoor air quality algorithm. The case study has been carried out in a smart building in the city of Salamanca (Spain) which is a city where radon gas levels are high. The ventilation strategy keeps radon levels in the building under control, which improves the quality of life of the people working there.

3.1 Experimental Setup

The aim of this experiment is to test the efficiency of the new ventilation strategy proposed in this paper. We also want to prove that the energy consumption of the building does not vary much since its implementation. In order to address these goals this experiment has been developed in a smart building. The lecture's room of the smart building is on the ground floor of the R&D building of the University of Salamanca. It has a total area of about 150 m^2, a volume of about 500 m^3 and maximum capacity of 110 occupants. As there is variable occupancy, the venue is very suitable for the study of demand strategy ventilation. A HVAC system was used to serve only this lecture's room of the smart building. It is a single-zone, variable-air-volume system. A direct digital controller is used in this system to control the chilled water valve and the supply air inlet guide vane actuator to maintain the desired supply air temperature and static pressure. A fresh air damper is used to control the fresh air intake.

3.2 Indoor Radon Level

During the experiment, major indoor air pollutants such as radon and CO_2 were measured. Figure 3 shows the average time variations of CO_2 and radon levels, which indicate the demand of fresh air to dilute the occupant-related indoor air pollutants. From Fig. 3, it can be seen that the radon level continuously increased during the non-occupied hours. As the radon emanation rate in the lecture in the smart building was not so high, the case when more than one hour is requited to bring the radon level down to 200 Bq m^{-3} was not found. However, this situation may occur in other buildings where the radon emanations rates are relatively

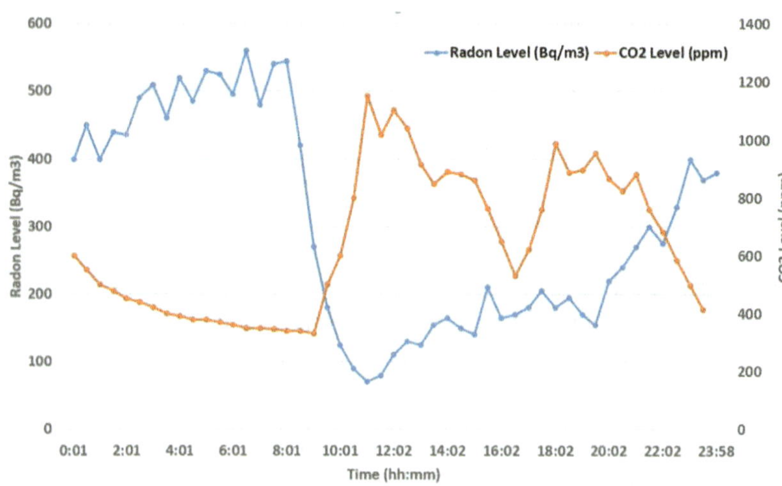

Fig. 3. Average radon and CO_2 level profiles in the smart building while using the demand control ventilation strategy.

higher and the ventilation rates are relatively lower. In the case study smart building, the scheduled pre-ventilation, ventilation 1, started one hour before the occupied hours. In Fig. 3 we can see that after starting ventilation sequence the indoor radon level decreased rapidly. At the beginning of the occupied hours it had decreased to about 200 Bq m^{-3}.

During the lecture hours, by the radon PID controller, the radon level was never found to exceed 200 Bq m^{-3}, which was acceptable as the Hong Kong Environmental Protection Department had set the 200 Bq m^{-3} as the upper limit of Level 2 in the newly established guidance notes [16] and Dai *et al.* research [10]. The experimental results show that the pre-ventilation plus the real-time modulation can efficiently prevent the occupants from exposing to an undesirable radon level. In most of the case study hours, the CO_2 level was found to be below or around 1000 ppm. The highest CO_2 level was about 1180 ppm. By controlling the CO_2 at such levels is just enough to dilute the occupant-related indoor air pollutants to acceptable level.

4 Conclusion

In this paper, the results of average measurements made in a smart building in the city of Salamanca are reported. The experimental results show that using only radon based demand control ventilation is sufficient to guarantee acceptable levels of pollutants in the smart building. In fact, the ventilation itself reduces CO_2 levels. Occupants can be exposed to undesirable pollution levels for a relatively long period. The reason is that in the unoccupied hours the ventilation system is inactive and pollutants accumulate. Based on the measurements of the radon level in the smart building a new strategy for ventilation on demand is designed in this paper using an algorithm that calculates the propagation of radon gas allowing the ventilation strategy to be more efficient. To verify the performance of our ventilation strategy a case study is carried out in a smart building. The results show that an acceptable level of air is obtained inside the buildings with the new ventilation strategy. In future work the ventilation strategy will be optimised taking into account more gases such us CO_2 and considering the periods of occupancy of the building. Future work will investigate the implementation of soft-computing techniques to enhance the efficiency of the control strategy. One of the limitations of the current strategy is that it only accounts for the current situation, and AI techniques could be used to allow the strategy to predict concentrations rather than reacting to high concentrations detected by IoT devices.

Acknowledgements. This work was developed as part of "Virtual-Ledgers-Tecnologías DLT/Blockchain y Cripto-IOT sobre organizaciones virtuales de agentes ligeros y su aplicación en la eficiencia en el transporte de última milla", ID SA267P18, project cofinanced by Junta Castilla y León, Consejería de Educación, and FEDER funds.

References

1. Baetens, K., Ho, Q., Nuyttens, D., De Schampheleire, M., Endalew, A.M., Hertog, M., Nicolaï, B., Ramon, H., Verboven, P.: A validated 2-D diffusion-advection model for prediction of drift from ground boom sprayers. Atmos. Environ. **43**(9), 1674–1682 (2009)
2. Carnahan, B., Luther, H., Wilkes, J.O., Maynar, M.M., de Miguel Anasagasti, E.: Cálculo numérico: métodos, aplicaciones. Rueda (1979)
3. Casado-Vara, R., Chamoso, P., De la Prieta, F., Prieto, J., Corchado, J.M.: Non-linear adaptive closed-loop control system for improved efficiency in IoT-blockchain management. Inf. Fusion **49**, 227–239 (2019)
4. Casado-Vara, R., Novais, P., Gil, A.B., Prieto, J., Corchado, J.M.: Distributed continuous-time fault estimation control for multiple devices in IoT networks. IEEE Access **7**, 11972–11984 (2019)
5. Casado-Vara, R., Prieto-Castrillo, F., Corchado, J.M.: A game theory approach for cooperative control to improve data quality and false data detection in WSN. Int. J. Robust Nonlinear Control **28**(16), 5087–5102 (2018)
6. Casado-Vara, R., Martin-del Rey, A., Affes, S., Prieto, J., Corchado, J.M.: IoT network slicing on virtual layers of homogeneous data for improved algorithm operation in smart buildings. Future Gener. Comput. Syst. **102**, 965–977 (2020)
7. Casado-Vara, R., Vale, Z., Prieto, J., Corchado, J.M.: Fault-tolerant temperature control algorithm for IoT networks in smart buildings. Energies **11**(12), 3430 (2018)
8. Chao, C.Y.H., Hu, J.: Development of a dual-mode demand control ventilation strategy for indoor air quality control and energy saving. Build. Environ. **39**(4), 385–397 (2004)
9. Chen, J., Schroth, E., MacKinlay, E., Fife, I., Sorimachi, A., Tokonami, S.: Simultaneous ^{222}Rn and ^{220}Rn measurements in Winnipeg, Canada. Radiat. Protect. Dosim. **134**(2), 75–78 (2009)
10. Dai, D., Neal, F.B., Diem, J., Deocampo, D.M., Stauber, C., Dignam, T.: Confluent impact of housing and geology on indoor radon concentrations in Atlanta, Georgia, United States. Sci. Total Environ. **668**, 500–511 (2019)
11. El-Zein, A.: Exponential finite elements for diffusion-advection problems. Int. J. Numer. Methods Eng. **62**(15), 2086–2103 (2005)
12. Mui, K., Wong, L., Hui, P., Law, K.: Epistemic evaluation of policy influence on workplace indoor air quality of Hong Kong in 1996–2005. Build. Serv. Eng. Res. Technol. **29**(2), 157–164 (2008)
13. Pawel, D., Puskin, J.: The US environmental protection agency's assessment of risks from indoor radon. Health Phys. **87**(1), 68–74 (2004)
14. Stidworthy, A.G., Davis, K.J., Leavey, J.: Radon emissions from natural gas power plants at the Pennsylvania State University. J. Air Waste Manag. Assoc. **66**(11), 1141–1150 (2016)
15. Taigbenu, A., Liggett, J.A.: An integral solution for the diffusion-advection equation. Water Resour. Res. **22**(8), 1237–1246 (1986)
16. Thomson, S.: Governance and digital transformation in Hong Kong. In: Redesigning Organizations, pp. 229–238. Springer (2020)
17. Wyon, D.P.: The effects of indoor air quality on performance and productivity. Indoor Air **14**, 92–101 (2004)

Implementation of a Statistical Dialogue Manager for Commercial Conversational Systems

Pablo Cañas[1] and David Griol[2(✉)]

[1] École Polytechnique Fédérale de Lausanne (EPFL), Route Cantonale, 1015 Lausanne, Switzerland
pablo.canascastellanos@epfl.ch

[2] Department of Languages and Computer Systems, University of Granada, Periodista Daniel Saucedo Aranda sn, Granada, Spain
dgriol@ugr.es

Abstract. Conversational interfaces have recently become an ubiquitous element in both the personal sphere by improving individual's quality of life, and industrial environments by the automation of services and its corresponding costs savings. However, designing the dialogue model used by these interfaces to decide the next response is a hard-to-accomplish task for complex conversational interactions. In this paper, we propose a statistical-based dialogue manager architecture, which provides flexibility to develop and maintain this module. Our proposal has been integrated with DialogFlow, a natural language understanding platform provided by Google to design conversational user interfaces. The proposed architecture has been assessed with a real use case for a train scheduling domain, proving that the user experience is of a high value and it can be integrated for commercial setups.

Keywords: Conversational systems · Dialogue management · Machine learning · DialogFlow

1 Introduction

Conversational interfaces are systems that emulate interactive conversations with humans [2,9]. These systems use natural language to provide dialogue capabilities with different purposes, such as performing transactions, responding to questions, or simply to chat.

These interfaces have become a key research subject for many organizations that have understood the potential revenue of introducing it in society's mainstream. Virtual personal assistants, such as Google Now, Apple's Siri, Amazon's

The research leading to these results has received funding from the European Union's Horizon 2020 research and innovation programme under grant agreement No 823907 (MENHIR project: https://menhir-project.eu).

Á. Herrero et al. (Eds.): SOCO 2020, AISC 1268, pp. 383–393, 2021.
https://doi.org/10.1007/978-3-030-57802-2_37

Alexa or Microsoft's Cortana, allow users to perform a wide variety of tasks, from setting an alarm to updating the calendar, finding nearest restaurants, preparing a recipe or reporting news [7].

In addition, a growing amount of entities are using conversational interfaces to automate their services while increasing customer experience and satisfaction. Among others, these agents are being used for making appointments, providing legal advice, self-help therapy, and answer FAQ about the COVID-19 pandemic [1]. Such systems are making these range of services more efficient without the need for human resources, hence generating a potential billion-dollar industry around them [2].

Spoken conversational interfaces are usually made up of four different components: an automatic speech recognizer (ASR), which records the sequence of words uttered by the speaker; a natural language understanding module (NLU), which obtains the semantics from the recognized sentence by performing morphosyntactic analysis; a dialogue manager (DM), which decides the next response of the system, interpreting the semantic representation of the input in the context of the dialogue; and a text-to-speech synthesizer (TTS), which transforms the response in natural language into synthesized speech.

The dialogue manager is the main object of study of this paper. This module considers the information provided by the user during the history of the dialogue, the dialogue context and the results of the queries to the data repositories of the system to return a meaningful response.

Early models for dialogue strategies were implemented using rule-based methods and predefined dialogue trees [8]. This methodology consists of manually determining the response that the agent should retrieve to each of the user inputs. Such approach, which is still broadly used nowadays, can be appropriate for very simple use cases; for instance, systems answering a reduced set of isolated frequently asked questions. However, more complex dialogue systems usually require several user-system interactions for a successful interaction, thus making a rule-based approach unfeasible both in terms of maintainability and scalability.

As a solution to this problem, new methodologies for statistically dialogue modeling have been proposed during the last years [4]. Recent literature includes proposals based on Partially Observable Markov Decision Processes [11] and Reinforcement Deep Learning [3], which generate user-system interaction simulations to learn the appropriate response for every input. Supervised-learning-based solutions have also been proposed, including the use of Neural Networks [5], Stochastic Finite-State Transducers [6], and Bayesian Networks [10].

There currently exist several frameworks that ease the task of building industrial conversational agents, being Google's DialogFlow[1] one of the most popular ones. Most of these toolkits allow specifying tree-based implementations for the dialogue manager, in which the system will respond to the specified user utterances [7]. However, some toolkits, like DialogFlow, also allow developers

[1] https://dialogflow.com/.

to integrate their own statistical model of the dialogue manager for the agent implementation. This brings a huge potential to develop and maintain such module for commercial and industrial setups.

To achieve this objective, in this paper we propose a practical framework to develop statistical-based dialogue managers that can be easily integrated in toolkits like Dialogflow. As a proof of concept, we have implemented a practical conversational system for a train scheduling domain, in which we use the functionalities provided by DialogFlow for natural language understanding and a statistical dialogue manager developed using our proposal with a dialogue corpus acquired for the task.

The remainder of the paper is as follows. Section 2 describes the main features of the DialogFlow platform to create conversational interfaces. Section 3 presents our proposal to integrate statistical dialogue management functionalities in a conversational system designed using this platform. In Sects. 4 and 5 we describe the application of this proposal to develop a conversational system providing railway information and the results of its integration and preliminary evaluation. Finally, Sect. 6 presents the conclusions and future research lines.

2 The DialogFlow Platform

DialogFlow allows the development of conversational interfaces by automatically implementing the natural language understanding module by means of providing training phrases for each intent (end-user's intention for a conversation turn) and defining the dialogue manager by means of the use of context conditions for each intent and the responses to return to the end-users for each one of them.

DialogFlow currently supports 32 languages and dialects[2]. Conversational interfaces developed using this toolkit can be integrated into wearable devices, cars, intelligent speakers, web plugins, and other mobile applications.

Regarding the NLU module development, DialogFlow has three basic primitives:

1. Intents: An intent is a specific action that users can invoke by using sentences that match their NLU model. Developers must provide a set of training phrases for each intent. As a result, depending on the user input, the agent maps each user response to a specific intent in order to provide a system response. Therefore, each intent represents a dialogue turn within the conversation.
2. Entities: An entity represents a term or object that is relevant for the intents and that provides a specific context for them. The entities are usually keywords used for identifying and extracting useful data from user inputs. DialogFlow provides a wide variety of predefined system entities, such as dates, times, cities, colors, or units of measure, but developers can also define their own domain-dependent entities. An entity consists of an entity type (e.g. fruit) and entity values (e.g. banana, strawberry, orange).

[2] https://cloud.google.com/dialogflow/docs/reference/language.

3. Contexts: They represent the current state of the interaction and allow agents to carry information from one intent to another. They can be combined to control the conversational path in order to define conditions required to access an intent (input contexts) or defined after accessing them (output contexts).

Developing a dialogue manager in DialogFlow implies defining a set of possible responses for each intent. However, this set of responses is static and hence limits the flexibility of the chatbot. As an example, let's imagine that a user requests information to buy a train ticket. The user could start the interaction querying for different pieces of information: origin and destination cities, departure and arrival dates, price range, duration, services, train type, etc. A possible option is to define a single intent for these requests, as all of them are related to the same user's intention to book a train ticket. However, the number of combinations of parameters to consider becomes exponentially large for practical domains. Moreover, if the user does not provide all the information pieces, the agent needs to ask for the remaining data, and the range of context possibilities to take into account becomes huge. Another option is to define unique intents for each of the pieces of information, but again taking into account all the different combinations makes the dialogue management definition unfeasible.

By means of the fulfillment functionality provided by DialogFlow it is possible to connect natural language understanding and processing for each intent to any kind of business logic, such as querying databases, access third party APIs, or using a Machine-Learning-based model to predict the most adequate response given the dialogue context. In our proposal, we use this functionality to integrate more scalable statistical dialogue models for the dialogue manager.

3 Proposed Framework for Statistical Dialogue Management

Figure 1 shows the architecture for the proposed statistical DM framework. As it can be observed, the framework integrates Firebase applications[3] to provide cloud functions, realtime databases, and hosting. Nonetheless, other internal or third party services can be used to facilitate these services.

DialogFlow's NLP module is used to select the user intents and the entities in their utterances. Instead of defining a tree-based model, intents are not used to retrieve a predefined response, but to extract the context information to appropriately feed the statistical DM model with the dialogue history. The context is sent to the cloud function, which will first obtain the dialogue state from the previous interaction with such user.

The statistical dialogue model selects the next system response according to the dialogue state defined by the context (for instance, confirm a particular piece of information, request for additional information pieces, or inform about the results of a query). Depending on the type of response, the framework could

[3] https://firebase.google.com/.

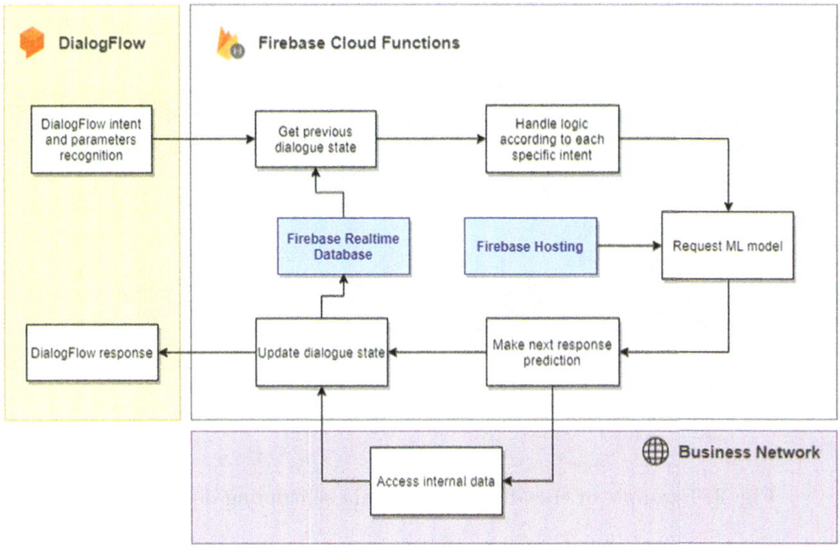

Fig. 1. Architecture of the proposed framework for the statistical dialogue manager implementation

require accessing a third party or internal database for completing the request (e.g., to inform about the ticket price for a specific train).

The dialogue state is updated with the data gathered and crafted during the interaction, in order to be ready for the next user input. The response selected by the statistical model is sent to DialogFlow so that the TTS module concludes the dialogue cycle.

The architecture of the framework provides modularity, scalability, speed, domain-independence, ability to handle complex and long interactions, and easiness for assembling with the rest of the modules required by complex conversational systems.

4 Use Case: Train Scheduling Domain

As a proof of concept, the proposed framework has been used to develop a conversational system for a train scheduling domain, in which users can ask the system about information to plan a rail trip around Spain. The corpus that has been used to train the statistical dialogue manager is described in [5]. It consists of 4,006 training instances, which are labelled with one of the 23 possible system responses.

The number of input features used to define the context for the dialogue states is 19. They are encoded as Fig. 2 shows: the previous action taken by the system; five attributes encoded as 0 or 1 depending on the type of request the user is asking for (timetables, price, train type, order number or service list); ten

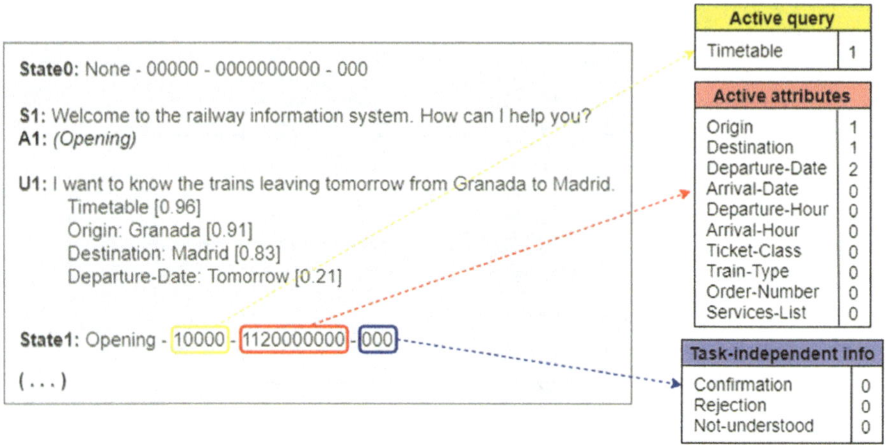

Fig. 2. Example of encoding for the train scheduling domain corpus

attributes denoting the type of information mentioned in the dialogue, encoded with 0 if the concept is unknown, 1 if the concept is known with a high confidence score, or 2 if the concept is known with a low confidence score; and three task-independent attributes denoting acceptance, rejection or not understanding.

The statistical model used for training was a standard Multilayer Perceptron architecture, suitable for this task as found in previous work [5], implemented in TensorFlow[4]. It consisted of 1 densely-connected layer of 256 neurons, trained using a step size of $\alpha = 0.0005$, ADAM optimizer, and early stopping to avoid overfitting. The trained model was stored as a JSON object in a Firebase Hosting instance.

DialogFlow's NLP module was created by defining the set of intents, parameters and entities required for the use case. A total of 13 intents were defined, each of them related to a specific request or piece of information that the user could ask. Table 1 shows an example of some of the training phrases that were defined for two of those intents, the one providing information about the departure schedule, and the one asking for the duration of the trip.

The set of entities corresponds to the 10 attributes that Table 2 shows. Seven of these entities were already predefined in DialogFlow.

A specific handler for each of the different DialogFlow intents was defined for the cloud function Following the previously described architecture, the first step is to access the Firebase Realtime Database instance to obtain the previous

[4] https://www.tensorflow.org/.

Table 1. Intent examples for the train scheduling domain.

Intent name	Training phrases	Parameters
Say-Departure-Date	Para mañana *(For tomorrow)* Me gustaría salir el 2 de abril *(I would like to depart April the 2nd)* Para mañana a las 3 *(For tomorrow at 3)* Salgo el 4 de marzo a las 8 de la tarde *(I depart March the 4th at 8 pm)* Me gustaría coger el tren a las 5 y cuarto de hoy *(I would like to take the train today at quarter past 5)* Me gustaría salir el 2 de abril a las 16:00 *(I would like to depart April the 2nd at 16:00)*	departureDate departureHour
Ask-Route-Length	Sí, ¿cuál es la duración del trayecto? *(Yes, what is the route length?)* ¿Cuál es el tiempo de recorrido? *(What is the route length?)* Sí, me gustaría saber el tiempo que se tarda *(Yes, I would like to know how long does it take)* ¿Cuánto se tarda? *(How long does it take?)* ¿Cuánto tarda el tren en llegar? *(How long does the train take to arrive?)*	

Table 2. Parameters and entities defined for the train scheduling domain

Parameter name	Entity type	Entity values
origin		
destination	city (system)	Madrid, Barcelona, Vigo...
departureDate		
arrivalDate	date (system)	2020-05-04, tomorrow...
departureHour		
arrivalHour	time (system)	09:30, 4pm, noon...
ticketClass	ticketClass (crafted)	tourist, preference
trainType	trainType (crafted)	AVE, Alvia, Cercanías...
services	services (crafted)	cafeteria, wifi, newspaper...

system response, as well as all the information that it was already stored for the interaction. After this, depending on each specific intent, new information is added to the state (e.g. for the *Say-Departure-Date* shown in Table 1, departure schedule data).

The dialogue state is then encoded and sent to the statistical dialogue model, that uses this information as input to predict the next system response. Depending on the type of response (e.g., to provide the schedule for a train route), a new request to a third party or internal business layer can be required to inform about the trains fulfilling the conditions required by the user.

After this, the updated state is inserted in the Firebase Realtime Database, together with the system response, so that this information is available for the next interaction. Finally, the new system response is sent to DialogFlow.

5 Integration and Evaluation

One of the main reasons that makes DialogFlow ideal for industrial applications is its straightforward integration with a wide collection of popular chat environments that allows to deploy real conversational user interfaces instantly. Integrations include Facebook Messenger, Twitter, Slack, Skype, Telegram, Google Assistant, and Amazon Alexa, among others, as well as embeddings for web applications.

A preliminary evaluation process was followed to validate the architecture implemented and the quality of the conversational agent. As a result, a total of 20 people from diverse ages, gender, educational and technological background were selected and interviewed to evaluate the train scheduling chatbot.

First, an objective evaluation of the system was completed, analyzing seven different metrics extracted from the interactions between the user and the system. Table 3 shows the metrics and results of the evaluation. The results show that the success rate is very acceptable and the turn coherence is high. The percentage of unique dialogs also points that the system is quite versatile.

Table 3. Results of the objective evaluation

Metric	Evaluation value
Dialogue success rate	80%
Turn coherence rate	78%
Average #turns	7
Average #requests	2.89
#Turns shortest dialogue	10
#Turns longest dialogue	5
% Different dialogs	55%

Users were also asked to provide their subjective opinion on the system's performance with seven questions, scoring from 1 (lowest) to 5 (highest). Results, presented in Table 4, show a positive perception of the application. While the weakest point is the error recovery capability, users believe that the interaction with the system is clear and fast. The overall satisfaction is also high, with a large percentage of returning customers. This information validates the viability of the proposed solution for industrial purposes.

Figure 3 shows an example of a successful dialogue extracted from one of the tests. Although the user speaks with a colloquial wording, providing unnecessary extra information and other tags such as *'more or less'* or *'everything has become clear to me'*, the system is able to retrieve very accurate responses, and successfully complete the interaction.

Table 4. Results of the subjective evaluation

Question	Avg. score	Std Dev.
How well did you understand the system messages?	5	0
How well did the system understand you?	3.80	0.83
Was it easy for you to get the requested information?	4	1.12
Was the interaction with the system quick enough?	4.60	0.71
If there were system errors, was it easy for you to correct them?	3.30	1.41
In general, are you satisfied with the performance of the system?	4.40	0.73
Would you use this system to schedule your future train rides?	4.20	0.68

U1: Muy buenas tardes. *(Good afternoon.)*
S1: Bienvenido, ¿en qué puedo ayudarte? *(Welcome, how can I help you?)*

U2: ¿Podría decirme los horarios de trenes de Madrid a Zaragoza, por favor? *(Could you please tell me the train schedule from Madrid to Zaragoza?)*
S2: ¿En qué fecha desearías viajar? *(In which date would you like to travel?)*

U3: Para mañana, si es posible. *(For tomorrow, if possible.)*
S3: Buscando los trenes disponibles, espere un segundo... Hay un tren disponible de Madrid a Zaragoza el día 26/5/2019 a las 16:00. ¿Puedo ayudarte en algo más? *(Searching for available trains, please wait... There is an available train from Madrid to Zaragoza on the 26/5/2019 at 16:00. Can I help you in anything else?)*

U4: ¿Cuánto me costaría ese billete? *(How much would that ticket cost me?)*
S4: El trayecto seleccionado tiene un coste de 31 euros en clase turista. ¿Necesitas algo más? *(The selected route has a cost of 31 euros in tourist class. Do you need anything else?)*

U5: ¿Cuánto tiempo dura el viaje, más o menos? *(How long does the trip take, more or less?)*
S5: El tiempo de duración del tren seleccionado es de 90 minutos. ¿Te puedo ayudar en algo más? *(The selected train duration is 90 minutes. Can I help you in anything else?)*

U6: ¿Y qué tipo de trenes hay disponibles en ese billete? *(And what type of trains are available with that ticket?)*
S6: El tipo de tren del vehículo seleccionado es AVE. ¿Tienes alguna otra consulta? *(The train type of the selected vehicle is AVE. Do you have any other enquiry?)*

U7: No, muchas gracias, me ha quedado todo muy claro. Muy amable. *(No, thank you, everything has become clear to me. Very kind of you.)*
S7: Espero haber sido de ayuda. Que tenga un buen día. *(I hope I have been helpful. Have a nice day.)*

Fig. 3. Example of a successful dialogue extracted from the evaluation

6 Conclusions and Future Work

In this paper we have presented a framework to automatize the processes to develop commercial conversational interfaces by means of statistical dialogue models. We have integrated this framework within the DialogFlow platform, easily assembling the set of components in the interface, and deploying a ready-to-use industrial application which can be integrated into different environments and devices.

We have validated our proposal developing and evaluating a dialogue system based on a real use case, a train scheduling domain. This implementation has covered the end-to-end process of developing a personalized conversational interface learning a statistical dialogue manager for the task and integrating it with DialogFlow using our framework. The results of the evaluation show the viability and potential value of our proposal to develop commercial conversational systems.

This document contains some limitations that will be addressed in future research work. As more advanced algorithms for text processing have recently emerged, it would be interesting to address their suitability for dialogue management tasks. Moreover, it would be valuable to further explore the scalability of our proposal for larger and more complex datasets, as well as detailed comparisons to other similar work for such scenarios.

Future work also includes the completion of a more detailed evaluation of our framework integrating other statistical approaches for dialogue management, and the automation of the processes required for creating the structure of intents and entities in toolkits such as DialogFlow.

References

1. Androutsopoulou, A., Karacapilidis, N., Loukis, E., Charalabidis, Y.: Transforming the communication between citizens and government through AI-guided chatbots. Gov. Inf. Q. **36**(2), 358–367 (2019)
2. Bavaresco, R., Silveira, D., Reis, E., Barbosa, J., Righi, R., Costa, C., Antunes, R., Gomes, M., Gatti, C., Vanzin, M., Junior, S.C., Silva, E., Moreira, C.: Conversational agents in business: a systematic literature review and future research directions. Comput. Sci. Rev. **36**, 100239 (2020)
3. Cuayáhuitl, H., Keizer, S., Lemon, O.: Strategic Dialogue Management via Deep Reinforcement Learning. CoRR abs/1511.08099 (2015)
4. Gao, J., Galley, M., Li, L.: Neural Approaches to Conversational AI. Now Publishers, Boston (2019)
5. Griol, D., Callejas, Z., López-Cózar, R., Riccardi, G.: A domain-independent statistical methodology for dialog management in spoken dialog systems. Comput. Speech Lang. **28**(3), 743–768 (2014)
6. Hurtado, L., Planells, J., Segarra, E., Sanchis, E., Griol, D.: A stochastic finite-state transducer approach to spoken dialog management. In: Proceedings of the 11th Annual Conference of the International Speech Communication Association (InterSpeech 2010), pp. 3002–3005. Makuhari, Chiba, Japan (2010)
7. Janarthanam, S.: Hands-On Chatbots and Conversational UI Development: Build chatbots and Voice User Interfaces with Chatfuel, Dialogflow. Twilio, and Alexa Skills. Packt Publishing, Microsoft Bot Framework (2017)
8. Lopes, J., Eskenazi, M., Trancoso, I.: From rule-based to data-driven lexical entrainment models in spoken dialog systems. Comput. Speech Lang. **31**(1), 87–112 (2015)

9. McTear, M., Callejas, Z., Griol, D.: The Conversational Interface: Talking to Smart Devices. Springer, Heidelberg (2016)
10. Thomson, B., Yu, K., Keizer, S., Gasic, M., Jurcicek, F., Mairesse, F., Young, S.: Bayesian dialogue system for the Let's Go Spoken Dialogue Challenge. In: Proceedings of the IEEE Spoken Language Technology Workshop (SLT 2010), pp. 460–465, Berkeley, USA (2010)
11. Young, S., Gašić, M., Keizer, S., Mairesse, F., Schatzmann, J., Thomson, B., Yu, K.: The hidden information state model: a practical framework for POMDP-based spoken dialogue management. Comput. Speech Lang. **24**, 150–174 (2010)

Special Session: Optimization, Modeling and Control by Soft Computing Techniques (OMCS)

Wind Turbine Pitch Control with an RBF Neural Network

Jesus Enrique Sierra-García[1] and Matilde Santos[2(✉)]

[1] University of Burgos, Burgos, Spain
jesierra@ubu.es
[2] Institute of Knowledge Technology, University Complutense of Madrid, Madrid, Spain
msantos@ucm.es

Abstract. There are many control challenges in wind turbines: controlling the generator speed, blade angle adjustment (pitch control), and the rotation of the entire wind turbine (yaw control). In this work a neuro-control strategy is proposed to control the pitch angle of the wind turbine. The control architecture is based on an RBF neural network and an on-line learning algorithm. The neural network is not pre-trained but it learns from the system response (power output) in an unsupervised way. Simulation results on a small wind turbine show how the controller is able to stabilize the power output around the rated value for different wind ranges. The controller has been compared with a PID regulator with encouraging results.

Keywords: Neural networks · Control · Wind turbine · Pitch angle · RBF network · On-line learning

1 Introduction

Wind turbines (WT) harvest the natural wind resource to generate clean energy [1]. In the nacelle of a wind turbine, the rotor with the blades captures the wind energy and transforms it to rotational torque; the generator transforms this mechanical energy into electricity, and the gearbox couples the rotor speed to the required by the generator [2].

Wind electricity generation capacity depends on the wind speed and the size of the wind turbine. In general, there are three operating regions (Fig. 1). The cut-in speed is the minimum wind speed required to start rotating the wind turbine and thus when the turbine starts generating power (around 3–4 m/s). From that wind speed the turbine is run at the maximum efficiency to extract all power. With wind speed over around 10–17 m/s, the turbine gets the rated turbine power. The cut-out speed is the maximum operating limit of the turbine (around 25 m/s).

The control system of a wind turbine is designed to seek the highest efficiency of energy generation and to ensure safe operation under all wind conditions. In order to either optimize or limit power output of the wind turbine, there are different control methods. It is possible to control a turbine by controlling the generator speed, the blade

Á. Herrero et al. (Eds.): SOCO 2020, AISC 1268, pp. 397–406, 2021.
https://doi.org/10.1007/978-3-030-57802-2_38

angle adjustment, and the rotation of the entire wind turbine. Blade angle adjustment and turbine rotation are also known as pitch and yaw control, respectively. In this work we are going to focused on pitch control of the wind turbine.

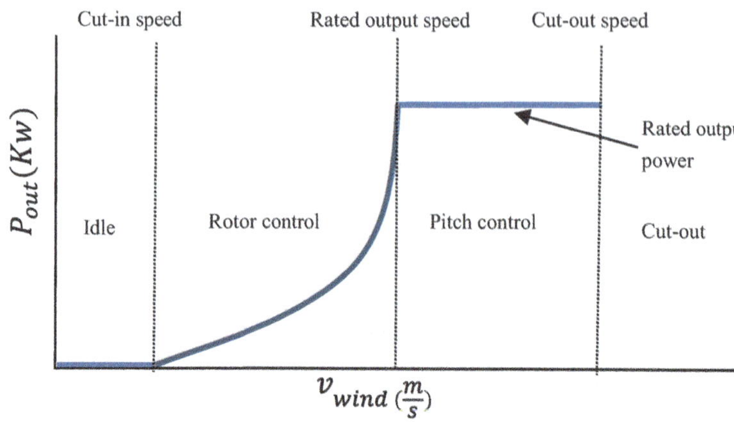

Fig. 1. Operating regions in a wind turbine

A pitch control strategy based on neural networks is proposed. The goal is to maintain the optimum blade angle to achieve certain rotor speeds or power output (rated power). Pitch angle adjustment is the most effective way to limit output power by changing aerodynamic force on the blade at high wind speeds.

However, the generation of the pitch control signal is not trivial, due to the is highly non-linear dynamics of the system, the coupling of the internal variables, unknown parameters and, above all, the random external conditions the wind turbines are subjected to [3]. These are critical issues specially for floating offshore wind turbines (FOWT) as the harsh environmental conditions produce vibration and fatigue [4]. That is why this control problem has been addressed in the literature using different artificial intelligence techniques, mostly Soft Computing ones [5–9].

Indeed, fuzzy control has been widely applied to this control problem. In [10], pitch angle fuzzy control is proposed and compared to a PI controller for real weather characteristics and load variations. The goal is to hold steady the output generator and accomplished aerodynamic braking effectively. Rocha et al. [11] apply a fuzzy controller to a variable speed wind turbine and compared the results with a classical proportional controller in terms of system response characteristics. In [12] a hybrid intelligent learning based adaptive neuro-fuzzy inference system (ANFIS) is proposed for online estimation of effective wind speed from instantaneous values of wind turbine tip speed ratio (TSR), rotor speed and mechanical power. Rubio et al. [13] presents the development of a fuzzy-logic based control system, that considers the effects of wave converters, for the control of a wind turbine installed on a semi-submersible platform. It is compared with a PI regulator. From a different point of view, in [14] authors propose an information management system for a wind power producer having an energy storage system and participating in a day-ahead electricity market. But the works that use neural network in WT control are quite scarce.

The RBF neural network used in this work is able to generate the pitch control signal without being previously trained. Simulation results show how the proposed neuro control strategy stabilizes the power output of the wind turbine to the rated power even with changing wind conditions.

The paper is organized as follows. In Sect. 2 the mathematical description of the system dynamics is presented. Section 3 describes the neuro control strategy implemented. Simulation results are shown and discussed in Sect. 4. The paper ends with the conclusions and future works.

2 Wind Turbine Mathematical Description

In this work the model of a small turbine of 7 kW is used. For the sake of simplicity, the ratio of the gearbox is fixed to 1, thus the torque in the rotor will be the same than the mechanical torque T_m (Nm) in the generator, given by [15] (1):

$$T_m = \frac{C_p(\lambda, \theta) \cdot \rho \cdot A \cdot v^3}{2 \cdot w}. \tag{1}$$

Where C_p is the power coefficient, i.e., the ratio of the electrical power produced by the wind turbine divided by the wind power into the turbine; ρ is the air density (Kg/m^3), A is the area swept by the turbine blades (m^2), v is wind speed (m/s), and w is the angular rotor speed. The blade swept area can be approximated by $A = \pi R^2$, where R is the radius or blade length.

The power coefficient is normally determined experimentally for each turbine. In the wind turbine literature there are different expressions to approximate C_p, in our case it has been calculated a function of the tip speed ratio λ and the blade pitch angle θ (rad), that is

$$C_p(\lambda, \theta) = c_1\left(\frac{c_2}{\lambda} - c_3\theta - c_4\theta^{c_5} - c_6\right)e^{-\frac{c_7}{\lambda}}, \tag{2}$$

where the values of the coefficients c_1 to c_7 depend on the characteristics of the wind turbine. The pitch angle θ is defined as the angle between the plane of rotation and the blade cross section chord, and the tip-speed ratio is defined by Eq. (3).

$$\lambda = \frac{w \cdot R}{v}. \tag{3}$$

From Eq. (3) it is possible to observe how C_p decreases with the pitch angle. Indeed, when $\theta = 0$ the blades are pitched so the blade is all out and producing at its full potential, but with $\theta = \frac{\pi}{2}$ (rad) the blades are out of the wind.

On the other hand, the relation between the rotor angular speed w and the mechanical torque T_m in a continuous current generator is given by the following expressions [15]:

$$J\frac{dw}{dt} = T_m - T_{em} - K_f w, \tag{4}$$

$$T_{em} = K_g \cdot K_\phi \cdot I_a, \tag{5}$$

where T_{em} is the electromagnetic torque (Nm), J is the rotational inertia (Kg.m^2), K_f is friction coefficient (N•m•s/rad), K_g is a dimensionless constant of the generator, K_ϕ is the magnetic flow coupling constant (V•s/rad), and I_a is the armature current (A).

The armature current of the generator is given by the Eqs. (6–7)

$$L_a \frac{dI_a}{dt} = E_a - V - R_a I_a, \tag{6}$$

$$E_a = K_g \cdot K_\phi \cdot w, \tag{7}$$

where L_a is the armature inductance (H), E_a is the induced electromotive force (V), V is the generator output voltage (V), and R_a is the armature resistance (Ω). For simplicity it is commonly assumed that the load is purely resistive, given by R_a. Thus, $V = R_L I_a$ and the output power (W) is $P_{out} = R_L I_a^2$.

The following expressions derived from the combination of Eqs. (1–7) summarizes the behavior of the system (8–10):

$$\dot{I}_a = \frac{1}{L_a}\left(K_g \cdot K_\phi \cdot w - (R_a + R_L) I_a\right), \tag{8}$$

$$\begin{aligned} \dot{w} =& \frac{c_1}{2 \cdot J \cdot w}\left(\left(\frac{v \cdot c_2}{w \cdot R} - c_3\theta - c_4\theta^{c_5} - c_6\right)e^{-\frac{v \cdot c_7}{w \cdot R}} \cdot \rho \cdot \pi R^2 \cdot v^3\right) \\ & - \frac{1}{J}\left(K_g \cdot K_\phi \cdot I_a + K_f w\right), \end{aligned} \tag{9}$$

$$\boldsymbol{P_{out}} = R_L \cdot I_a^2. \tag{10}$$

Regarding the control problem, I_a and w are considered the state variables, θ is the control input and P_{out} is the controlled variable.

The system wind turbine parameters used during the simulations are shown in Table 1 [16].

Table 1. Parameters of the wind turbine model

Parameter	Description	Value/Units
L_a	Inductance of the armature	13.5 mH
K_g	Constant of the generator	23.31
K_ϕ	Magnetic flow coupling constant	0.264 V/rad/s
R_a	Resistance of the armature	0.275 Ω
R_L	Resistance of the load	8 Ω
J	Inertia	6.53 kg m^2
R	Radio of the blade	3.2 m
ρ	Density of the air	1.223 kg/m^3
K_f	Friction coefficient	0.025 N m/rad/s
$[c_1, c_2, c_3]$	C_p constants	[0.73, 151, 0.58]
$[c_4, c_5, c_6, c_7]$	C_p constants	[0.002, 2.14, 13.2, 18.4]

3 Description of the Neuro-Controller

3.1 Control Architecture

The control architecture (Fig. 2) includes an RBF neural network that is used to generate the pitch control signal.

The input of the wind turbine (at the right of the figure) is the pitch angle θ and its output is the power P_{out}. The power reference P_{ref} is given by the rated power of the turbine. The error is then calculated as the difference between this reference and the real power output P_{out}. The error and its derivative, $\dot{P}_{err}$, are introduced in a saturator to limit their values. These saturated error signals, P_{err_S} and $\dot{P}_{err_S}$, feed the RBF neural network whose output is the pitch angle θc. In addition, a bias of $\frac{\pi}{4}$ (rad) has been included to the neural network output, that is, half of the maximum pitch control value. The input to the wind turbine is the pitch angle that results of subtracting the offset and the pitch calculated by the RBF neural network, that is, $\frac{\pi}{4} - \theta c$.

The RBF network is not pre-trained with real data, but it learns to generate the correct output while it is working with the on-learning algorithm. The learning algorithm updates the weights of the RBF, $\bar{W}$, based on the saturated error P_{err_S}.

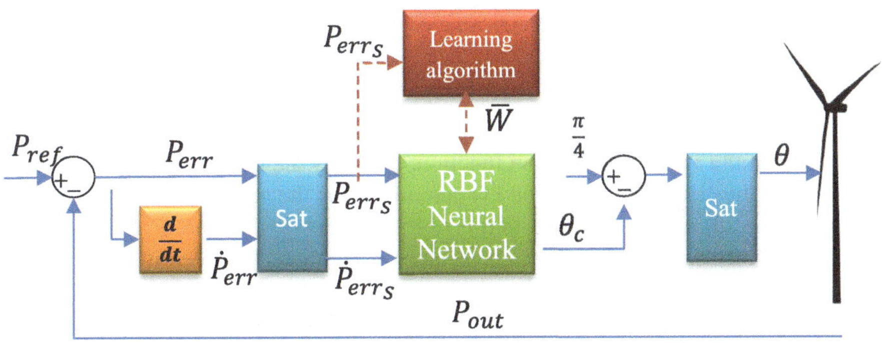

Fig. 2. Architecture of the neuro controller

The equations of this neuro control strategy are the following (11–17):

$$P_{err}(t_i) = P_{ref}(t_{i-1}) - P_{out}(t_{i-1}), \tag{11}$$

$$\dot{P}_{err}(t_i) = \frac{P_{err}(t_i) - P_{err}(t_i - T_c)}{T_c}, \tag{12}$$

$$P_{err_S}(t_i) = MIN\left(P_{err_{MAX}}, MAX\left(P_{err_{MIN}}, P_{err}(t_i)\right)\right), \tag{13}$$

$$\dot{P}_{err_S}(t_i) = MIN\left(\dot{P}_{err_{MAX}}, MAX\left(\dot{P}_{err_{MIN}}, \dot{P}_{err}(t_i)\right)\right), \tag{14}$$

$$\theta_c(t_i) = f_{RBF}\left(P_{err_S}(t_i), \dot{P}_{err_S}(t_i), \bar{W}(t_{i-1})\right), \tag{15}$$

$$\bar{W}(t_i) = f_{learn}\left(P_{err_S}(t_i), \bar{W}(t_{i-1})\right), \tag{16}$$

$$\theta(t_i) = MIN\left(\frac{\pi}{2}, MAX\left(0, \frac{\pi}{4} - \theta_c(t_i)\right)\right), \tag{17}$$

where T_c is the control sample period (s), the set $[P_{err_{MIN}}, P_{err_{MAX}}, \dot{P}_{err_{MIN}}, \dot{P}_{err_{MAX}}] \in \mathcal{R}^4$ is a set of constants to adjust the range of the controller, with the constraints $P_{err_{MIN}} < P_{err_{MAX}}$ and $\dot{P}_{err_{MIN}} < \dot{P}_{err_{MAX}}$; f_{RBF} is the RBF function and f_{learn} denotes the function of the learning algorithm.

It is important to note that the variables $P_{err}, \dot{P}_{err}, P_{err_S}, \dot{P}_{err_S}, \theta_c, \bar{W}, \theta$ in Eqs. (11–17) are updated each T_c seconds, otherwise their values remain constant.

3.2 RBF Neural Network Calculation

The output of the RBF neural network (15) is given by the following expressions (variable t_i has been omitted for sake of clarity)

$$f_{RBF}\left(P_{err_S}, \dot{P}_{err_S}, \bar{W}\right) = \sum_{i=1}^{M} W_i \cdot e^{-\frac{dist\left(P_{err_S}, \dot{P}_{err_S}, c_{i1}, c_{i2}\right)}{\sigma_i}}, \tag{18}$$

$$dist\left(P_{err_S}, \dot{P}_{err_S}, c_{i1}, c_{i2}\right) = \sqrt{\frac{(c_{i1} - P_{err_S})^2}{P_{err_{MAX}}^2} + \frac{(c_{i2} - \dot{P}_{err_S})^2}{\dot{P}_{err_{MAX}}^2}}, \tag{19}$$

where M is number of neurons in the hidden layer, W_i is the weight of the i-neuron, σ_i is the width of the i-neuron activation function, normally the same for all neurons (here set to 1), and the center of the neuron i is determined by the pair (c_{i1}, c_{i2}).

There are different methods to initialize the centers of the neurons. Though it is usually randomly done, in this work the centers are uniformly spaced in the ranges of the error signals, $\left[P_{err_{MIN}}, P_{err_{MAX}}\right]$ and $\left[\dot{P}_{err_{MIN}}, \dot{P}_{err_{MAX}}\right]$. That is, the centers of the neurons are initialized to the values obtained by Eqs. (20) and (21).

$$c_{i1} = i \cdot \frac{P_{err_{MAX}} - P_{err_{MIN}}}{M - 1} + P_{err_{MIN}} \quad \forall i \in \mathcal{N} \cup 0 | i < M - 1 \tag{20}$$

$$c_{i2} = i \cdot \frac{\dot{P}_{err_{MAX}} - \dot{P}_{err_{MIN}}}{M - 1} + \dot{P}_{err_{MIN}} \quad \forall i \in \mathcal{N} \cup 0 | i < M - 1 \tag{21}$$

Once obtained, these centers are not updated by the learning algorithm. The parameter M has been set to 49. This value has been obtained by trial and error, after testing different squared numbers (16, 25, 36, 64…). This value gives a good balance between control performance and computational complexity.

3.3 Learning Algorithm

The weights W_i of the RBF neural network have been initialized to the same value in order to give all the neurons the same importance and to homogeneously distribute the input space. However, the weights are continuously updated by the learning algorithm.

The weights are updated following the learning rule given by Eq. (22), that corresponds to function f_{learn} of Eq. (16).

$$W_j(t_i) = W_j(t_{i-1}) + \mu \cdot P_{err_S}(t_i) \cdot e^{-\frac{dist\left(P_{err_S}(t_i), \dot{P}_{err_S}(t_i), c_{i1}, c_{i2}\right)}{\sigma_j}} \quad \forall j \in \mathcal{N} \cup 0 | i < M - 1 \tag{22}$$

where μ, the learning rate, has been set to 0.00015 by trial and error.

The learning rule that updates the weights in an RBF works with the error, usually defined in a supervised learning scheme as the difference between the current output of the network and the desired value. Nevertheless, in this case we do not know the desired output, i.e., the right control signal, thus instead of working with the error we use the saturated error signal, P_{err_S}, that estimates how good is the control performance. The network learns trying to reduce P_{err_S}, to zero.

As the exponential term of Eqs. (18) and (22) is the same, once calculated is saved to be used in both and thus save computational time.

4 Simulation Results

The simulation results have been obtained with Matlab/Simulink software. The duration of each simulation is 100 s. In order to reduce the discretization error the simulations have been carried out with a variable step size, being the maximum step size 10 ms. The control period T_c is 100 ms.

The performance of the proposed approach is compared with the application of a PID regulator. In order to make a fair comparison, the PID output has been scaled to adjust its range to $\left[0, \frac{\pi}{2}\right]$ and it has been also biased by the term $\frac{\pi}{4}$. The equation of the biased PID controller is (23).

$$\theta = \frac{\pi}{4} - \frac{\pi}{4P_{err_{MAX}}}\left[KP \cdot P_{err} + KD \cdot \frac{d}{dt}P_{err} + KI \cdot \int P_{err}dt\right]. \tag{23}$$

The tuning parameters $[KP, KD, KI]$ have been determined by trial and error and their values are [1, 0.2, 0.9], respectively. The power output interval $\left[P_{err_{MIN}}, P_{err_{MAX}}\right]$ is [−1000,1000] W and the limits of the derivative are $\left[\dot{P}_{err_{MIN}}, \dot{P}_{err_{MAX}}\right] =$ [−100,100] W/s. The wind turbine nominal power is 7 kW, thus the reference P_{ref} = 7000 W.

The wind is randomly generated between an average velocity range of $[v_{min}, v_{max}]$, several simulations have been carried out with different range values.

In Fig. 3 left, the power output obtained with different control strategies is shown. The rated power is represented in green, the output power when the pith angle is 0 (rad) in blue, and when the pitch is $\frac{\pi}{2}$ (rad), in red; the PID control response is shown in orange and the RBF control system response in purple. The wind velocity range was set to [11.3, 14.3] m/s during the simulation. As it is possible to see, both the classical controller and the neural one are able to get the desired rated power output.

Nevertheless, in Fig. 3 right the same system responses are shown but they have been zoomed and the first three seconds have been ruled out. It is possible to notice now that the PID overshoot is much larger than the NN one. In addition, as expected, with $\theta = \frac{\pi}{2}$ the power output is below the nominal power and with $\theta = 0$ the power output is over the rated one.

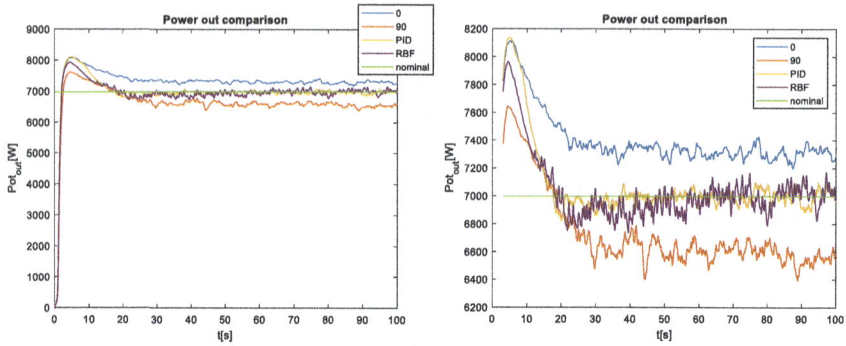

Fig. 3. Power output comparison for different control techniques

Figure 4 represents the comparison of the pitch control signal (degrees) generated by the two control techniques (blue line, PID and red line with the neuro-control). At the beginning of the simulation the power output is 0 W and both controllers generate low pitch angles to increase the power. Rapidly the power grows and overpass the rated power, then both controllers set the pitch angle around 90° to reduce the power. The neuro controller starts at 30° due to the initialization of the weights $\bar{W}$ (see 3.2). Once the pith reaches 90°, it starts to decrease until it stabilizes about 50°. In general, the pitch generated by the neuro controller is noisier than the one given by the PID. It would be possible to reduce this noise with a low pass filter at the output of the neural network or to vary the learning rate of the training algorithm to decrease it.

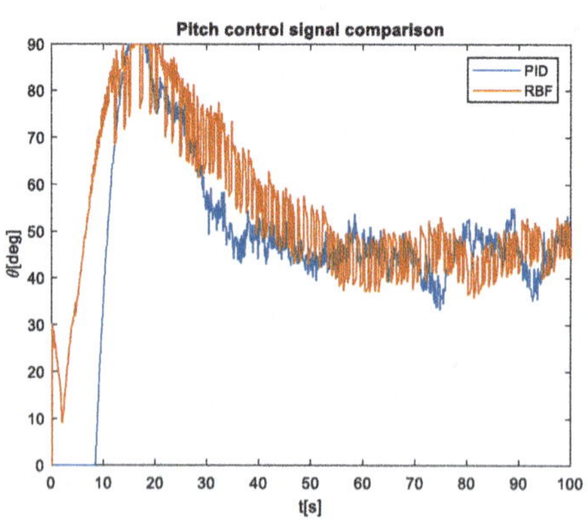

Fig. 4. Pitch control comparison

Numerical results have been also obtained at three different wind velocity ranges (m/s) to test the performance of the controllers (Table 2). The best results have been boldfaced. The MSE is calculated from the third second because before the MSE is too high as the initial value of the reference is 7 kW and the initial output power is 0 W, this way it is possible to better see the differences between the controllers. The neuro-control strategy gives smaller overshoot and thus smaller MSE values. The rise time does not depend on the controller but on the wind. As expected, higher wind speeds produce larger values of MSE, settling time and rise time.

Table 2. Performance of the controllers with different wind velocities

Wind $[v_{min}, v_{max}]$	Parameter	RBF	PID
[11, 14]	Rise time (s)	2.22	2.22
	Settling time (s)	**15.9**	19.5
	Overshoot (%)	**10.86**	12.71
	MSE (W) [3 < t < 100]	**199.8455**	242.6648
[11.3, 14.3]	Rise time (s)	1.79	1.79
	Settling time (s)	21.27	**19.19**
	Overshoot (%)	**13.46**	16.31
	MSE (W) [3 < t < 100]	**240.7804**	289.6240
[11.5, 14.5]	Rise time (s)	1.9	1.9
	Settling time (s)	**25.71**	28.01
	Overshoot (%)	**14.77**	17.87
	MSE (W) [3 < t < 100]	**266.6342**	316.7998

5 Conclusions and Future Works

In this work, a first approach to apply online neuro-control to the pitch angle of a wind turbine is presented. The main contribution of this paper is that the neural network is not pre-trained but it learns online from the system response.

The motivation to apply this artificial intelligent technique is given by the highly non-linearity behavior of the system, that has been mathematically modelled. The RBF neural network has been applied to obtain the pitch angle that stabilizes the power output around its rated value for different wind ranges.

Simulation results validate the effectiveness of this stabilization controller, even with changing wind conditions. In addition, the wind turbine response with the neural control strategy has been compared to the one obtained with a classical PID controller with encouraging results.

Among other possible future works it would be desirable to test this approach under real conditions and perform a deep analysis on its computational complexity, and to study the effects of the pitch control on the vibration of the wind turbine.

Acknowledgement. This work was partially supported by the MCI/AEI/FEDER Project number RTI2018-094902-B-C21.

References

1. Mikati, M., Santos, M., Armenta, C.: Electric grid dependence on the configuration of a small-scale wind and solar power hybrid system. Renew. Energy **57**, 587–593 (2013)
2. Burton, T., Jenkins, N., Sharpe, D., Bossanyi, E.: Wind Energy Handbook. Wiley, Hoboken (2011)
3. Li, Z., Adeli, H.: Control methodologies for vibration control of smart civil and mechanical structures. Exp. Syst. **35**(6), e12354 (2018)
4. Tomás-Rodríguez, M., Santos, M.: Modelado y control de turbinas eólicas marinas flotantes. Revista Iberoamericana de Automática e Informática Industrial **16**(4), 381–390 (2019)
5. Navarrete, E.C., Perea, M.T., Correa, J.J., Serrano, R.C., Moreno, G.R.: Expert control systems implemented in a pitch control of wind turbine: a review. IEEE Access **7**, 13241–13259 (2019)
6. Sierra, J.E., Santos, M.: Wind and payload disturbance rejection control based on adaptive neural estimators: application on quadrotors. Complexity **2019**, 1–20 (2019)
7. Menezes, E.J.N., Araújo, A.M., da Silva, N.S.B.: A review on wind turbine control and its associated methods. J. Clean. Prod. **174**, 945–953 (2018)
8. Saenz-Aguirre, A., Zulueta, E., Fernandez-Gamiz, U., Lozano, J., Lopez-Guede, J.M.: Artificial neural network based reinforcement learning for wind turbine yaw control. Energies **12**(3), 436 (2019)
9. Sierra, J.E., Santos, M.: Modelling engineering systems using analytical and neural techniques: hybridization. Neurocomputing **271**, 70–83 (2018)
10. Hassan, S.Z., Li, H., Kamal, T., Abbas, M.Q., Khan, M.A., Mufti, G.M.: An intelligent pitch angle control of wind turbine. In: 2017 International Symposium on Recent Advances in Electrical Engineering (RAEE), pp. 1–6. IEEE (2017)
11. Rocha, M.M., da Silva, J.P., De Sena, F.D.C.B.: Simulation of a fuzzy control applied to a variable speed wind system connected to the electrical network. IEEE Latin Am. Trans. **16**(2), 521–526 (2018)
12. Asghar, A.B., Liu, X.: Adaptive neuro-fuzzy algorithm to estimate effective wind speed and optimal rotor speed for variable-speed wind turbine. Neurocomputing **272**, 495–504 (2018)
13. Rubio, P.M., Quijano, J.F., López, P.Z., Lozano, J.J.F., Cerezo, A.G., Casanova, J.O.: Control inteligente para mejorar el rendimiento de una plataforma semisumergible híbrida: sistema de control borroso para la turbina. Revista Iberoamericana de Automática e Informática Industrial **16**(4), 480–491 (2019)
14. Gomes, I.L.R., Melicio, R., Mendes, V.M.F., PousInHo, H.M.I.: Wind power with energy storage arbitrage in day-ahead market by a stochastic MILP approach. Logic J. IGPL **28**(4), 570–582 (2019)
15. Ackermann, T.: Wind Power in Power Systems. Wiley, Hoboken (2005)
16. Mikati, M., Santos, M., Armenta, C.: Modelado y simulación de un sistema conjunto de energía solar y eólica para analizar su dependencia de la red eléctrica. Revista Iberoamericana de Automática e Informática Industrial **9**(3), 267–281 (2012)

MIMO Neural Models for a Twin-Rotor Platform: Comparison Between Mathematical Simulations and Real Experiments

Kerman Viana(✉), Mikel Larrea(✉), Eloy Irigoyen(✉), Mikel Diez(✉), and Asier Zubizarreta(✉)

University of the Basque Country (UPV/EHU), Leioa, Spain
{kviana,m.larrea,eloy.irigoyen,mikel.diez,asier.zubizarreta}@ehu.eus

Abstract. This work presents a neural model developed for a multivariable system with complex nonlinear dynamics, obtained through a tight methodology used both in simulation and in the real platform. In addition, this neural model has been studied and designed to meet the requirements of a predictive control strategy. A Twin-Rotor platform is used as an example of a Multi-Input Multi-Output (MIMO) system to study and analyse how a neural network is able to reproduce its non-linear coupled dynamics and accurately estimate future system outputs. An in-depth study of the neural structures and their performance in the prediction of future states has been developed. Results show with comparisons, the modelization inaccuracies that appears when a proposal made just on the basis of a mathematical simulation is used to conclude the good performance of these MIMO neural models.

Keywords: Non-linear modelling · ANN · MIMO systems · Twin-Rotor

1 Introduction

The study and design of new control strategies that offer efficient and accurate solutions ensuring that the complexity of the problem is correctly addressed still offers a considerable challenge within many industrial system and processes. These challenges derive mainly due to non-linearities, coupled dynamics, variable randomisation, disruptions and other inherent characteristics to real systems.

Pursuing a solution to all these factors, *Soft Computing* techniques have shown to be an appropriate approach to control strategies [1]. Along with these techniques, many traditional control strategies have been tested to obtain robust and efficient solutions for complex system management, but always with the cost of simplifications and working environment delimitation. These modifications have result on a strategy that lacks on flexibility and viability for variable range of operating points.

Á. Herrero et al. (Eds.): SOCO 2020, AISC 1268, pp. 407–417, 2021.
https://doi.org/10.1007/978-3-030-57802-2_39

In order to approach a strategy without any of the previous limitations, predictive control strategies have been implemented to many multiobjective systems. The main aim of these strategies are based on human-like "intelligence" implementation in order to anticipate control actions to variable actions in the system [4,12]. Furthermore, the use of *Soft Computing* techniques offers a main inaccuracy tolerance management, key aspect of intelligent control strategies.

At the same time, the use of predictive control is an example of strategies, along with many others, that needs the nonlinear modelling of the real system. This kind of modelling allows to include nonlinear and coupled effects when traditional models prove to be impractical or impossible to implement. Furthermore, newer non-linear modelling approaches have included non-parametric methods which don't need any previous information about the study system's relations; but depend on production data or experimental results over time. Therefore, the extensive study of this models leads to the importance of obtaining a predefined amount of prediction steps ahead with precision, known as the "prediction horizon", which enables a more accurate control action definition within the predictive control strategy optimisation.

Specifically for the Multi-Input Multi-Output (MIMO) systems, identification is based on the relation between the input and corresponding output data, from which a parameter estimation problem is derived. Therefore, traditional identification methods such as least mean square or instrumental variable are proved methods of solving some of these system modelling [2]. Besides that, in the last two decades interest on the use of feedforward multilayer Artificial Neural Networks (ANNs) for this same purpose has clearly increased, particularly in terms of recurrent networks (feedback) whose learning potential is improved by the information received as input coming from different time-steps and sources, including network's output past computations. Furthermore, long time research had already proved that ANNs can be used as universal function approximator [3]. Therefore, Multi-Layer Perceptrons (MLP) with feedback loops and a single or multiple hidden layers [5,6], Radial Basis Function neural networks [10], ADaptative LINear Element neural networks [13], or Evolving Neuro-Fuzzy networks [9] are examples of commonly used multilayer feedforward structures for various MIMO system modelling.

A system extensively used to study both non-linearities and coupled dynamics between variables is the Twin-Rotor MIMO system (TRMS) platform developed by Feedback Instruments Ltd. The TRMS recreates a simplified behaviour of a real helicopter with two degrees of freedom, both vertical and horizontal angles, while rotors angle of attack remains fixed, and hence the control is achieved by manipulating the rotor's speed. Many studies have reported TRMS system identification proposals using analytical techniques [8], and much more extensively by means of ANN with various structures [7,9,11,14]. Even then, almost all of the studies referenced above analyse the Twin-Rotor system not as a MIMO structure, but as a SISO system on which a single degree of freedom is identified; mostly the horizontal or elevation angle. Moreover, most of these works do not consider to map their studies out from simulation to the real platform and also they tend to avoid some of the system's known nonlinearities.

Therefore, this paper aims to present a methodology for nonlinear system's modelling with the objective of implementing the result in predictive control strategies; particularly, using ANN structures where both degrees of freedom from the Twin-Rotor system are taken into account, along with the effect of the coupled dynamics. With that purpose, Sect. 2 will briefly study the dynamics of this particular system, pointing out its non-linearities and the differences between real and mathematical dynamics. Section 3 settles the system's modelling using ANNs along with the proposed methodology. Finally, Sect. 4 exposes the results of the implementations of this methodology as well as the derived conclusions from this study; which will be condensed again in Sect. 5.

2 Twin-Rotor System Dynamics

2.1 Mathematical Model

The Twin-Rotor mechanical model consists of two rotors placed on a beam together with a counterbalance; while the whole unit is attached to a tower allowing safe control experiments, as shown in Fig. 1.

Apart from the mechanical units, the electrical unit consisting on signal measurement and control signal applied towards the activation of the DC motors is also needed. Therefore, both mechanical and electrical units interactions enable the correct functioning of the Twin-Rotor.

Thus, non-linear equations for both degrees of freedom (θ for the *pitch* or elevation angle, and ϕ for the *yaw* or azimuth angle) can be derived from the momentum balance in each movement. Being for the vertical movement or *pitch* angle as shown in (1).

$$I_1 \cdot \ddot{\theta} = M_1 - M_{FG} - M_{B\theta} - M_G. \tag{1}$$

Where I_1 reflects the vertical rotor moment of inertia, M_1 is the nonlinear static characteristic moment, M_{FG} the gravity momentum, $M_{B\theta}$ the friction forces momentum and M_G the gyroscopic momentum; all for the vertical movement. Besides, the horizontal plane motion or *yaw* angle is described in (2).

$$I_2 \cdot \ddot{\phi} = M_2 - M_{B\phi} - M_R. \tag{2}$$

Where I_2 reflects the horizontal rotor moment of inertia, M_2 is the nonlinear static characteristic moment and $M_{B\phi}$ the friction forces momentum; all for the horizontal movement.

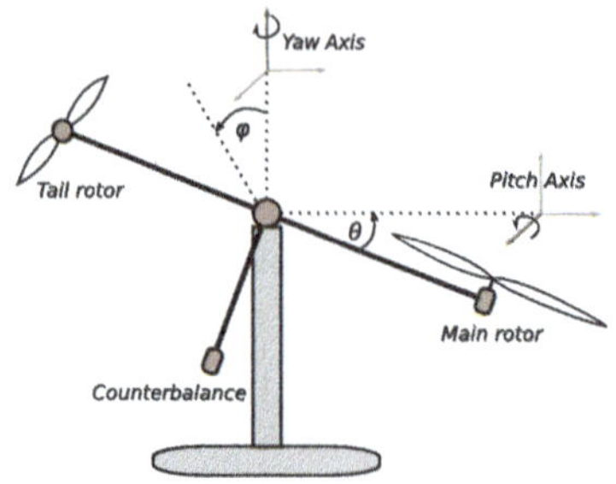

Fig. 1. Twin-Rotor system diagram

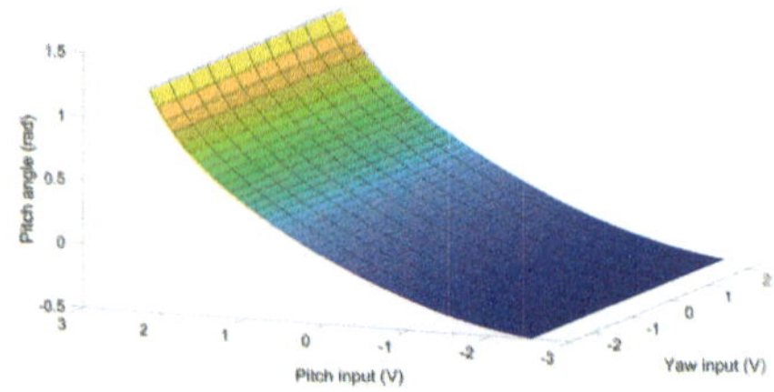

Fig. 2. 3D representations of the workspace for *pitch* angle.

According to the previous equations, both momentum balance equations are composed of various components that have been identified many times in works such as [2] as well as in the Feedback Instrument Ltd. manual; which exposes that clear non-linearities and interactions between both degrees of freedom are part of the Twin-Rotor dynamics. Therefore, working areas on which these clear non-linearity and interaction effects manifest above other dynamics must also be included in the range of interest and action for an appropriate system modelling.

2.2 Study of the Range of Non-linear Working Points

Non-linearity effects appear all along the range of action in the Twin-Rotor movement, but can only be exposed in a meaningful and clear way around certain working areas. These working areas have been proved to be close to the boundaries of the Twin-Rotor both mechanical and electrical units and thus not commonly noticed in regular movement behaviours.

Anyway, the study and delimitation of these areas is needed to form a complete model of the non-linear system, which would include not only the most linear behaviour zones. For the study of these non-linear zones, both simulation and real platform environments have been studied along a wide mesh of input points in order to plot the correspondent surface of working points.

The results show a significant non-linear behaviour along the *pitch* angle which grows exponentially when reaching the boundaries of the electrical unit input for his own angle; while the cross action of the *yaw* angle input exposes very limited variation along its range of action, all shown in Fig. 2. Therefore, it may be accepted that the main non-linearity source for the pitch angle are the equations corresponding to its own input values, when these reach zones next to the boundaries of the system. These zones next to these boundaries will be included in the modelling of the system aiming to include all non-linearities present within the physical range of operating points of the system.

On the other hand, from the *yaw* angle cross reaction and own equations it is clear that both its own and *pitch* angles inputs manifest a non-linear reaction on its working points. Furthermore, same identification experiments have shown that these effects are not significant in any of the range that the *pitch* angle boundaries delimitate; and thus this aspect will be considered while planning the experiments referring to the *pitch* non-linearities.

3 Neural Network Structures

The Twin-Rotor movement behaviour can be defined as an input-output relationship in which the output *pitch* and *yaw* angles are a result of the electrical inputs, u_{pitch} and u_{yaw}, to the DC motors. Aiming a proper modelling of the system along with sources of discrepancy within the real platform, ANNs are proposed as widely used function approximators. Many ANN structures have been proposed, as it has already been exposed previously, for these kind of environments although this work specifically experiments with two of the most common and best performing structures:

- **NARX-NOE**: Besides being a MLP structure, in NARX networks, along with the input values of the system, output values of the previous instants are inserted to the input layer with a specific *output delay* value in a feedback loop. Furthermore, NARX-NOE (hereafter NARX) structures also implement previous input values with a specific *input delay* value; as shown in Fig. 3, copying a parallel configuration, where past outputs are taken from the own model.
- **Elman Recurrent Network**: Given the previous layers with similar behaviour, Elman Recurrent Network implement the context layer consisting of nodes supplied with data from the hidden layer computational output. Therefore, the straight feedforward structure is broken by a feedback loop, as shown in Fig. 4 which increases the learning capability of the network.

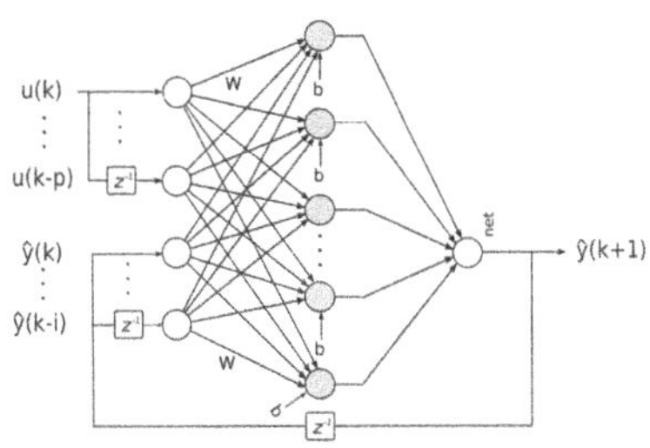

Fig. 3. NARX-NOE structure

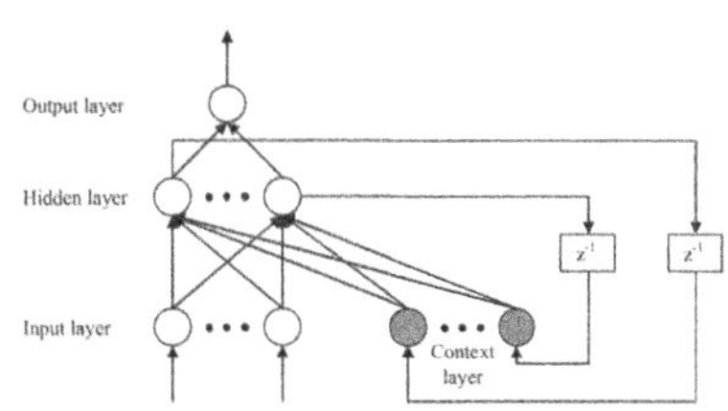

Fig. 4. Elman recurrent network structure.

Along with the networks above, regardless of the structure for the network, modelling strategy faces the dispute between studying both degrees of freedom from the Twin-Rotor independently with a single ANN, which would lead towards a MIMO system; or using two ANN, one for each degree of freedom including cross-interactions, consisting of two MISO systems. Both possibilities prove to be viable and thus commonly used for these kind of systems.

3.1 Neural Network Model Structure Comparison

According to preliminary studies, both in simulation and real platform environments, the network structure and configuration that has proved to obtain the best performance in prediction problems of a single sample is the NARX single network taking into account of both degrees of freedom; this is, a MIMO system. As Table 1 shows, all of 4 systems are initially studied, where the objective reference and prediction horizons are not too challenging. Anyway, choosing the structure that has obtained the best performance in these early stages directs this work to improve the performance of one single case, instead of four different possibilities. According to the results obtained in the preliminary experiments, the rest of the work will be performed with MIMO NARX.

Table 1. Comparison of structure and problem type performance; where prediction horizon is set to 5 steps (0.5 s) on an objective reference of a single transition curve between 2 differing operation points for all cases.

Network structure	Problem type	Performance (max. error)
NARX	MISO	6.667e-07
NARX	MIMO	8.082e-08
Elman	MISO	1e-03
Elman	MIMO	2.9796e-07

3.2 Experimental Methodology

Following the aim of achieving consistent results which could lead to a set of conclusions valid for any of the Twin-Rotor models, same guidelines have been set for all the experiments, both for the simulation and real platform environment, explained as follows:

- **Delimiting the region of interest**: Studying and modelling nonlinear behaviour is one of the main objectives of this work. Therefore, this phenomenon must be considered in each study region, establishing the limits and conditions for each experiment, both in simulation and real platform.
- **Data acquisition**: To avoid the overfitting problem during neural network training, two different data bases of examples have been selected to carry out the training and validation processes. For the training dataset a thorough sweep of operation points has been taken, whereas for the validation process has not been required so exhaustive.
- **Network structure comparison**: Aiming to obtain the optimal structure, in the iterative batch training process, the ANN performing based on the number of hidden layer neurons (5...50), along with the input and output delays (1...5), has been analysed.

- Reference profiles: In order to assure that main dynamics and operating points are all considered, smooth transition curves between differing operation points were established. In this way, fast and sharp shifts are avoided and similarities with human riding behaviour are taking into account.
- Number of predicted steps and sample time: For a well-trained neural model, valid for predictive control strategies, sample time measures must match the amount of predicted steps that the model is able to predict according to the system dynamics; and resulting on a real time predictive model always valid within a range of time step prediction during training or validation.
- Statistical approach: For structural optimisation on the studied neural networks, single training and validations sets prove to be inconsistent due to the strong statistical dependency. Therefore, 5 batches of successive training and validations will be performed with the same neural network structures for performance comparisons and consistency studies.

4 Experiments and Results

With a sample time on the data acquisition of 0.1 s, validation experiments have been set to a prediction of 40 steps in simulation as well as real platform studies; which satisfies both predictive control strategies models needs and differing operation point transition curves appropriate characterisation, as noted in Sect. 3.2 reference profiles explanation.

4.1 Simulation

Successive training and validation performed along all the studied structures on the simulation environment have lead to obtain the following 10 optimised neural models as well as their MSE error through 5 different experiment sets, as Table 2 summarises; where NARX topology's first numbers refers to the neuron quantity at the hidden layer, the second number refers to input delay amount and the last number refers to output delay amount.

Low MSE values shown in Table 2 during all validation sets are mainly due to the predominant stationary positions on which the Twin-Rotor lays along with the reference profiles. Furthermore, it has been proved that, excluding clearly non viable network structures, most of the studied neural models are able to perform a consistently small MSE error during prediction in all experiment sets. Anyway, most of these structures, although they perform correctly during stationary zones, expose a lack of consistency on the transitionally curves; making them unsustainable for proper predictive control strategies.

From the same Table 2 results, it is inferred that even if from the statistical analysis an optimal structure is obtained (with 50 neuron in the hidden layer, 3 input delays and 1 output delay), the following networks are also able to achieve almost the same performance. Therefore, it is possible that subsequent experiment sets may establish a new optimal structure.

Table 2. The best ten network structures found after the batch training, and the corresponding error

NARX topology	MSE 1	MSE 2	MSE 3	MSE 4	MSE 5	MSE average
N-50-3-1	0.4625	0.3734	0.3817	0.373	0.3858	0.39528
N-22-3-1	0.4291	0.5028	0.3961	0.3827	0.3915	0.42044
N-10-4-1	0.4677	0.4772	0.3939	0.3838	0.3888	0.42228
N-26-4-1	0.4	0.4698	0.3585	0.4641	0.453	0.42908
N-24-3-1	0.4982	0.4959	0.3372	0.4893	0.4818	0.46048
N-26-3-1	0.3733	0.4862	0.3727	0.37	0.9726	0.51496
N-16-5-1	0.9467	0.4697	0.4076	0.3781	0.3885	0.51812
N-46-4-1	0.3977	0.9429	0.371	0.4228	0.4809	0.52306
N-44-2-1	0.3535	0.2813	0.9757	0.4387	0.482	0.52624
N-46-2-1	0.4182	0.513	0.3395	0.3822	0.9833	0.52724

Figure 5 shows the prediction performance for a horizon of 40 steps (4 s) of the optimal neural model trained and validated using data from the real platform; along with it's analogous network using data from the simulation environment.

Anyway, all of the best performance structures fulfil with the requirements established previously for a prediction control strategies models. Therefore, all are valid Twin-Rotor non-linear models, even if they show different network structures; which leads into a further analysis on the behaviour of these models once the validation requirements become more challenging, as the following chapter will explain.

4.2 Structure Comparison and Prediction Degradation

Once the optimal or set of optimal networks are obtained from the training and validation experience, the boundaries of these same models are analysed; not only to reach the prediction limit of the models, but also to understand the reasons behind the optimal performance of the models.

Following this last idea, it is remarkable that all the best networks offer the same amount of output delays in their structures; as well as the rest of networks, with multiple delays for the output offer a much worse performance. Therefore, it is concluded that for NARX structures predicting Twin-Rotor behaviour, the models with minimum amount of output delays will always perform better; which is confirmed by the results at Table 2.

For the other structure variables, conversely, no strong dependency is proved to be shown, according to their performance. Therefore, the optimal structure, as said above, is only achieved after a series of statistical analysis on experiment data and even if structures with higher amount of input delays and neurons on the hidden layer seem to obtain better performance than their reduced competitors, these improvements are not clear enough to make a strong statement.

On the other hand, when the optimal structures are validated into longer prediction steps, it is proved that the models still offer a good performance with very little degradation; even if we double the prediction time. Therefore, all things considered, it seems that the training methodology has allowed us to obtain neural models that predict the behaviour of non-linear models, such as the Twin-rotor, with a much better performance and for further horizons that regular predictive control strategy models need in simulation environments.

4.3 Simulation vs Real Platform

As exposed above, Twin-Rotor real platforms differ from the simulation environments not only because of the external agent's effects, but also because of the real elements that take part on the simulation own limitations. This said, it is easily expected that same experiment sets based on data from the real-platform may translate on different optimal structures from the ones obtained from the simulation environments.

But even after considering this matters, the results obtained during the real platform experiments lead towards two main conclusions, that share the nature of the simulation environment:

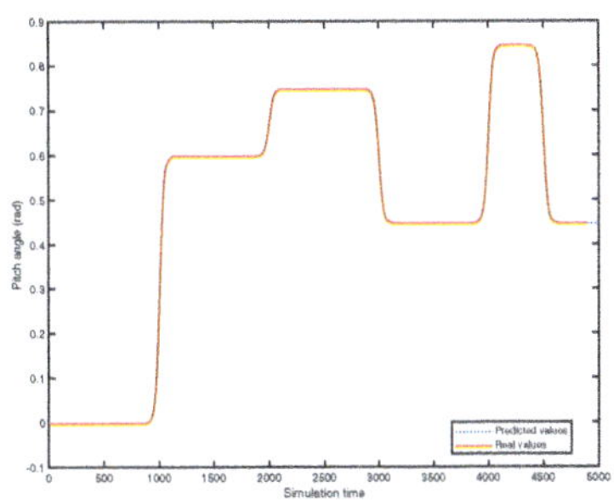

Pitch prediction on simulation

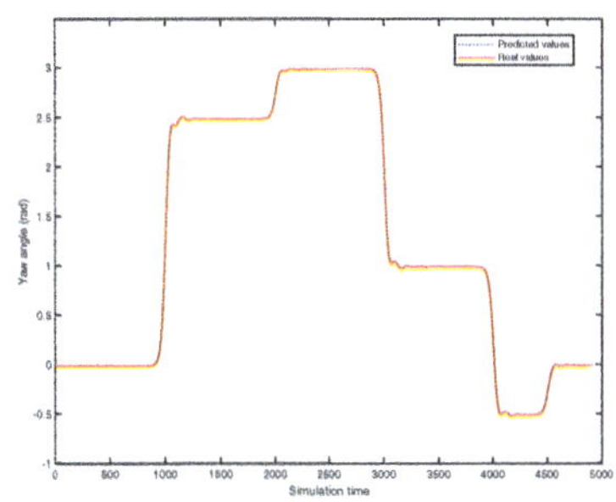

Yaw prediction on simulation

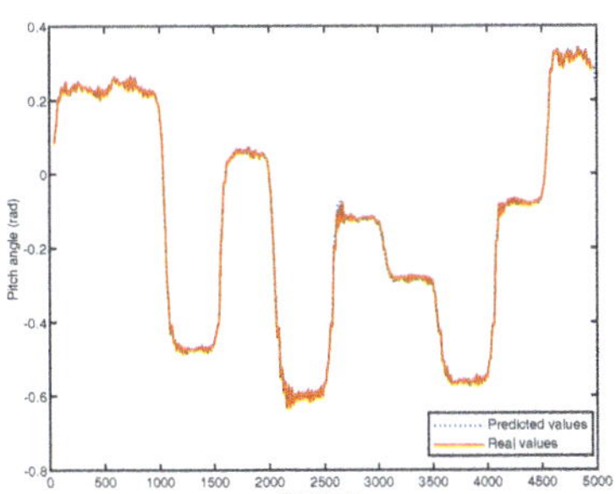

Pitch prediction on the real platform

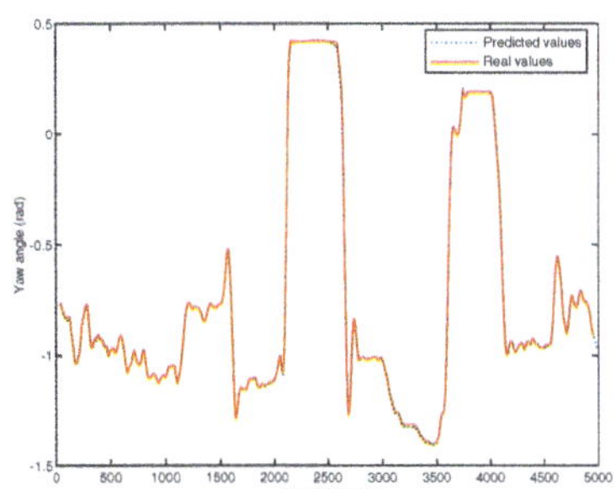

Yaw prediction on the real platform

Fig. 5. Prediction on a 4 s horizon (40 sample time-steps).

- **Statistical dependency**: along with the optimal results on simulation environments, it is clear that no exact structure expose a clear dominance over its competitors in terms of performance; due to a amount of them obtaining almost as good results as the optimal network.
- **Viable model formation**: even if the performance offered in real time optimal models seem worse than the ones on simulation environments, still perform as well as a predictive control strategy model needs and, therefore, fulfils the purpose of its formation.

5 Conclusions

This work presents the modelling of the non-linear system Twin-rotor based on artificial neural networks. A laborious analysis of the system has been performed, and as result the step by step procedure has been described in this work to obtain a non-linear model based on ANN; specifically, a MIMO NARX model. The obtained ANN optimised structures have proved to be viable for system modelling, but at the same time expose clear statistical dependencies due to various valid models. Anyway, the use of any of the best and optimised MIMO NARX models is viable for the objectives set in this paper.

Acknowledgements. This work comes under the framework of the project IT1284-19 granted by the Regional Government of the Basque Country to the Computational Intelligence Group (GIC) from the UPV/EHU and in the form of PIC 269-19 contract.

References

1. Bonissone, P.: Soft computing: the convergence of emerging reasoning technologies. Soft Comput. **1**, 6–18 (1997)
2. Gilson, M., den Hof, P.V.: Instrumental variable methods for closed-loop system identification. Automatica **41**, 241–249 (2005)
3. Hornik, J., Stinchcombe, M., White, H.: Neural Networks, pp. 359–366. Pergamon Press, Oxford (1989)
4. Irigoyen, E., Larzabal, E., Valera, J., Larrea, M.: Primeros resultados de un control genético predictivo sobre maqueta de helicóptero. Jornadas de Automática (2014)
5. Jagannathan, S., Lewis, F.L., Pastravanu, O.: Model reference adaptative control of nonlinear dynamical systems using multilayer neural networks. In: Proceedings of 1994 IEEE International Conference on Neural Networks (ICNN 1994) (1994)
6. Kim, H., Parker, J.K.: Hidden control neural network identification-based tracking control of a flexible joint robot. In: 1993 International Joint Conference on Neural Networks (1993)
7. Madhusanka, A., de Mel, R.: Artificial neuronal network-based nonlinear dynamic modelling of the twin-rotor MIMO system. J. Autom. Syst. Eng. (2011)
8. Rahideha, A., Shaheeda, M.H., Huijbertsa, H.J.C.: Dynamic modelling of a TRMS using analytical and empirical approaches. Control Eng. Pract. **16**, 241–259 (2008)

9. Silva, A., Caminhas, W., Lemos, A., Gomide, F.: Real-time nonlinear modeling of a twin rotor MIMO system using evolving neuro-fuzzy network. In: 2014 IEEE Symposium of Computational Intelligence in Control and Automation (CICA) (2014)
10. Slama, S., Errachdi, A., Benrejeb, M.: Model reference adaptive control for MIMO nonlinear systems using RBF neural networks. In: 2018 International Conference on Advanced Systems and Electric Technologies (IC ASET) (2018)
11. Subudhi, B., Jena, D.: Nonlinear system identification of a twin rotor MIMO system. In: TENCON 2009 - 2009 IEEE Region 10 Conference (2009)
12. Tayyebi, S., Alishiri, M.: A novel adaptive three stages model predictive control based on fuzzy systems: application in MIMO controlling of MED-TVC process. J. Franklin Inst. **356**, 9344–9363 (2019)
13. Wenle, Z.: MADALINE neural network for parameter estimation of LTI MIMO systems. In: Chinese Control Conferences (2010)
14. Yu, Z.R., Yang, T.C., Juang, J.G.: Application of CMAC and FPGA to a twin rotor MIMO system. In: 2010 5th IEEE Conference on Industrial Electronics and Applications (2010)

Fuzzy-Logic Based Identification of Conventional Two-Lane Roads

Felipe Barreno[1], Matilde Santos[1(✉)], and Manuel G. Romana[2]

[1] Computer Science Faculty, Complutense University of Madrid, Madrid, Spain
{febarren,msantos}@ucm.es
[2] Civil Engineering School, Technical University of Madrid, Madrid, Spain
manuel.romana@upm.es

Abstract. This paper presents a Soft Computing based system to identify and classify conventional two-lane roads according to their geometrical characteristics. The variability of input information and the uncertainty generated by the overlapping of this information make fuzzy logic a suitable technique to address this problem. A fuzzy rule-based Mamdani-type inference system and a neuro-fuzzy system are applied. The roads geometrical features are measured by vehicle sensors and are used to classify the roads according to their real conditions. The conventional two-lane roads used for this research are located in the Madrid Region, Spain. The good results obtained with the fuzzy system suggests this intelligent system can be used to update the road databases; the theoretical class of road assigned to each road should be updated according to their present characteristics, as this is key to estimate the recommended speed for a safety and comfortable driving.

Keywords: Soft Computing · Fuzzy logic · Neuro-fuzzy · Identification · Classification · Two-lane road · Road geometry

1 Introduction

User experience on roads, regarding both comfort and driving safety, depends largely on road conditions. Both pavement and road infrastructure may deteriorate due to multiple causes: vegetative wear and tear, rain, heavy vehicle traffic, etc. But the geometry of a road also have a significant impact on making the driving safer or on how much safe a road is considered.

The geometrical characteristics of a road are used to define different types of roads, according to which some traffic regulations are set. These geometric criteria are, among others, the number of lanes, width of the shoulders, camber, gradient, curve radii, etc.

The speed limit is usually set in the road design phase, according to the assigned road class and other local section features. But, over time, road conditions may change, either because it was never built as originally thought or due to degradation, erosion, aging, encroachment of vegetation, new roadside buildings, etc. This may do the initial road categorization incorrect, and in such case the current road section must be reclassified.

Á. Herrero et al. (Eds.): SOCO 2020, AISC 1268, pp. 418–428, 2021.
https://doi.org/10.1007/978-3-030-57802-2_40

Therefore, it is important the keep updated the correct assignment of the road type, as the speed regulation will depend on it.

In this paper Soft Computing techniques, particularly fuzzy and neuro-fuzzy systems, are applied to classify sections of two-lane roads based on their current geometrical characteristics. As far as we know, the identification problem of two-lane roads using fuzzy logic has not been addressed before and, therefore, the fuzzy perception of the classification of roads is novel. These techniques have been proved useful when facing similar tasks [1, 2]. Results obtained on conventional two-lane roads of the Madrid Region, Spain, are encouraging and allow the determination of a more appropriate speed for a comfortable and safer driving. Thus, these tools may be used to develop an intelligent driving speed recommender [3, 4].

Regarding the available literature on the subject, in [5] authors use fuzzy logic to identify roads using Ikonos satellite images. In [6], a road detection algorithm based on fuzzy techniques is described, using satellite images from GIS. In [7], the road geometry is analysed to establish an adaptive cruise control. To provide a vehicle with that functionality, geometrical characteristics (radius and slope) and GPS information about speed limit are used in [8]. A flexible logic-based approach is applied in [9], where this qualitative reasoning is applied to maintain the allowed speed as a function of some geometrical factors such as the road slope. A preliminary study of classification of roadway surface disruptions based on threshold is presented in [10].

These works support the interest of using the geometric characteristics of two-lane roads to adjust the traveling speed, but, unlike the one presented here, they do not use fuzzy geometrical variables, and they are mainly focused on vehicle cruise control.

This paper is structured as follows. Section 2 describes two-lane road classes and their geometrical characteristics, which constitute the basis for the identification. In Sect. 3 the fuzzy system applied is presented and applied to real conventional roads. Results are discussed in Sect. 4 where fuzzy and neuro-fuzzy identification systems are compared. Conclusions and future works end the paper.

2 Road Classes and Geometric Characteristics

2.1 Two-Lane Road Classes

Roads are usually considerably long, and are divided into sections that have similar characteristics. For example, a road section may have road barriers, or a median, and these characteristic do not hold along all the road length. Moreover, these sections may not correspond to the type originally assigned to the whole road. This is important because traffic regulations differ depending on the type of road considered.

According to the official regulation [11], the classification of the roads is as follows (Table 1). The road classes are named with a letter followed by a number. The "A" letter stands for highways and "C" for conventional two-lane roads. The number that follows the letter indicates the design speed (*Vp*), expressed in kilometers per hour (km/h), regardless of the maximum speed allowed by the regulation.

The design speed of a section (*Vp*) is linked to its geometrical characteristics to assure comfort and safety driving conditions. It is calculated as follows:

$$Vp^2 = 127R\left(f_t + \frac{\rho}{100}\right) \tag{1}$$

where R is the radius (m), f_t the friction coefficient, and ρ is the inclination of the cross slope.

In addition, two-lane roads are classified according to their use and location:

- Class I (intercity);
- Class II (accessibility);
- Class III (suburban area);
- Class IV (urban area)

Some two-lane roads may belong to more than one type (Table 1). For example, C-40 and C-50 roads may belong to classes II and IV.

Table 1. Road types and correspondence with conventional sections.

Road type	Two-lane road	Class [13]
Intercity	C-90	I
Suburban	C-80, C-70, C-60	III
Accessibility	C-50, C-40	II
Urban	C-50, C-40	IV

2.2 Geometric Characteristics of a Road

The geometric features of a road section include both the cross section and the plan and vertical geometry. Roads are defined by geometrical characteristics that determine whether a vehicle can travel at a certain speed with adequate degree of comfort and safety. These features are mainly the slope, the camber, horizontal radius and carriageway dimensions.

The grade of a road indicates the inclination of the road surface relative to the horizontal plane. The value of the angle between the road and the horizontal plane is the slope. In order to neutralize the centripetal force that appears in sections of curved alignments, the road is fitted with a cross slope or camber, which is the transversal inclination of the road, causing that one of the shoulders remains at a higher elevation than the other.

From the construction point of view, the travelled section of a road is composed of lanes and shoulders. The lane is part of the road intended for vehicular traffic; a road consists of a certain number of lanes. The exterior of the road is called shoulder (or sidewalk, if the road is in an urban environment). The shoulders do not belong to the road and vehicles cannot circulate on them in normal conditions. On roads with divided carriageway the median separates the vehicles by direction. The platform width is the total of the right and left shoulders and the lane widths.

In Fig. 1, some of these geometric characteristics of the M-519 two-lane road are represented. According to its official assignment, it is an intercity road (class I). In the upper image the cross slope (‰) is shown; in the middle the slope (‰) is presented (given per thousand to better discriminate its value); in the bottom image the radius of curvature (m) is shown.

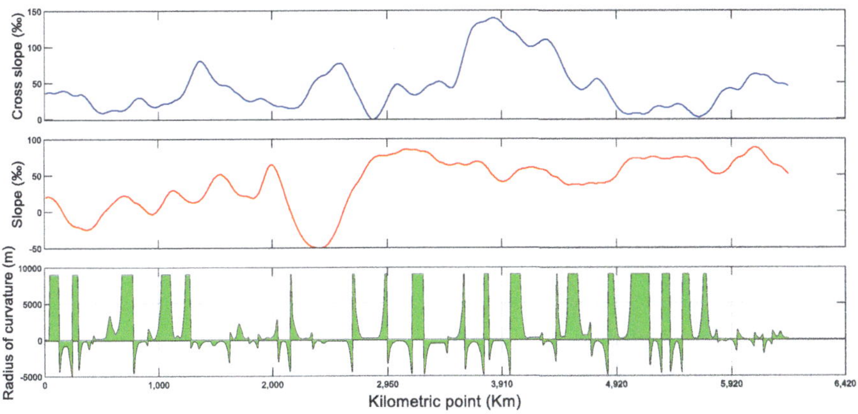

Fig. 1. Cross slope, slope and radius of curvature of M-519 two-lane road.

3 Identification of Two-Lane Roads by a Fuzzy Mamdani System

The available data used here belong to the road inventory records of the Madrid Region (Spain) [12]. The registers have the following information: road name, mileage post, number of lanes, additional lanes, width of the lane and shoulders, radius of curvature, camber and slope. Each data is collected every 10 m. The on-board diagnostic (OBD) of the vehicle measures the status of the various vehicle sub-systems through the sensors the car is equipped with, among others, lidar, laser, 3D cameras, accelerometer, GPS, etc.

This work has been carried out with the geometrical information of the following two-lane roads: M-607, M-519, M-852, M-618, M-305, M-509 and M-601, all of them in the Madrid Region. Roads are associated with a particular class. However, as discussed, they are composed of road sections which can be of different types along the same road.

Table 2 shows this fact. Roads M-607, M-519 and M-509 are assigned to class I, roads M-305 and M-601 to class III. Road M-618 is made up of road sections of at least two classes, I and III, and M-852 of classes II and III. These classes were assigned by experts. In this work we want to verify whether this information is still correct or if road class has changed due to road deterioration.

Table 2. Classes associated to roads under study.

Two-lane road	Road type	Class
M-607	Intercity*	I
M-519	Intercity*	I
M-852	Accessibility, suburban	II, III
M-618	Intercity, suburban	I, III
M-305	Suburban	III
M-509	Intercity*	I
M-601	Suburban	III

* In Intercity roads, users expect to travel at relatively high speeds.

3.1 Variables of the Fuzzy Inference System

The input variables of the fuzzy identification system that represent the road geometry are the following:

- Right and left shoulder width.
- Lane width.
- Width of platform (it is a function of lane and shoulders dimensions and thus it will not be considered as an independent variable).
- Radius of curvature.
- Cross slope (it is a function of the radius).
- Longitudinal slope.

Some of the variables are considered to be "restrictive", whereas others have been considered to be "informative". The former do not allow to discriminate the class of road, so they cannot be used for road identification. They may be the same for different classes of roads. These are radius, cross slope and slope.

Informative variables, however, indicate the variability of the road itself. These variables give the maximum and minimum dimensions that each "class" of road must have. Therefore, informative variables will be used as linguistic input variables of the fuzzy identification system, namely right and left shoulder widths and lane widths, being the platform width the sum of all of them.

3.2 Mamdani-Type Fuzzy Classification System

The first identification system proposed for the classification of two-lane roads is shown in Fig. 2. The inputs are real numerical values of the road dimensions taken by the sensors that are incorporated into the vehicle while travelling.

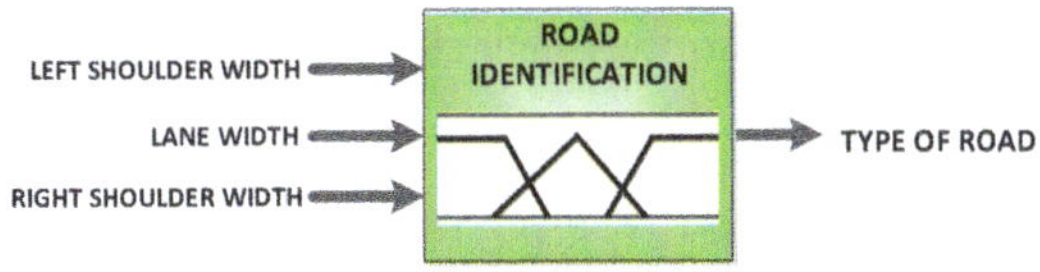

Fig. 2. Fuzzy rule-based system classifier of two-lane roads.

The linguistic terms of these variables are represented by trapezoidal membership functions. Each input variable -left shoulder, lane, right shoulder- is assigned three fuzzy sets: {N, narrow; M, medium; W, wide} (Fig. 3 and Fig. 4, left). The variable "lane width" is within the range [0–165 dm] and the shoulders in the range [0–65 dm].

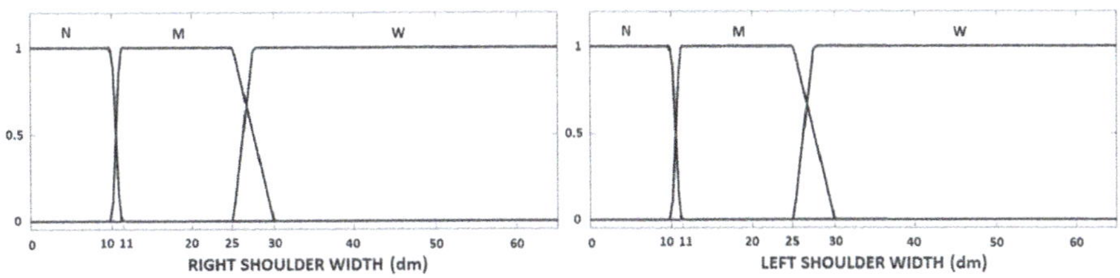

Fig. 3. Membership functions: right shoulder (left) and left shoulder (right) width.

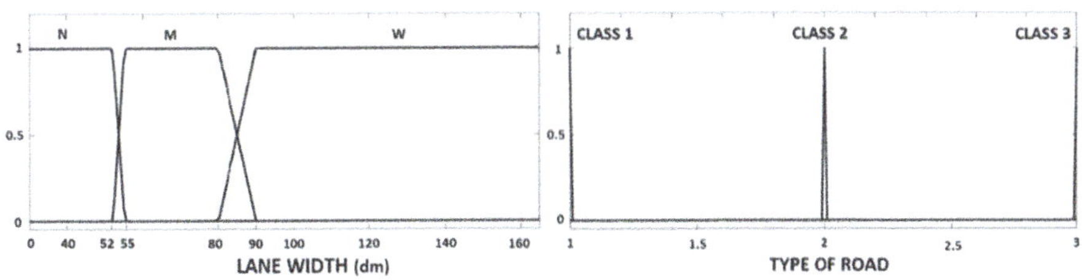

Fig. 4. Membership functions of lane width (left) and type of road output (right).

The output is the class of road, which can be I (intercity), II (accessibility) and III (suburban). The value obtained after the defuzzification process may not be an integer. In that case, a threshold is applied to determine the closest class according to the membership degree. For example, an output value of 1.2 would be rounded to the closest integer, and thus assigned to class I (intercity).

The knowledge is represented by if-then rules that have three antecedents and one consequent, where v_{ij} is the corresponding linguistic variable.

Ri: **IF** (*in_1 is* v_{1j}) **AND** (*in_2 is* v_{2j}) **AND** (*in_3 is* v_{3j}) **THEN** (*out is Class_x*)

The approximate reasoning implemented takes into account the following knowledge-based criteria:

- Class I: two or more lanes, wide shoulders, high speed.
- Class II: one or two lanes, lane and shoulders narrower than in class I and class III; there may not even be any shoulders, low speed.

- Class III: lane and shoulders narrower than class I but not than class II.

The rules combine road dimensions such as: if at least one of the shoulders and lane are medium or wide, output is Class I; if the lane and shoulders are narrow, output is Class II; if the lane is medium and the shoulders are narrow, output is Class III.

3.3 Discussion of the Results with the Mamdani Fuzzy System

The results are given in terms of the value of P and Ps, defined as the ratio between the samples of road correctly classified over the total number of samples (accuracy) (2), and the same ratio for samples of the road sections (accuracy per section) (3):

$$P = \left(\frac{correct\ samples}{\sum total\ samples}\right) \bullet 100 \tag{2}$$

$$Ps = \left(\frac{correct\ samples\ per\ section}{\sum total\ samples\ per\ section}\right) \bullet 100 \tag{3}$$

The results of applying the Mamdani-type fuzzy classifier are shown in Table 3. For each road there are a larger number of samples correctly classified (bolded). When this percentage is greater than 80% it is considered that the type of these roads has been well identified. This happens with roads M-607, M-618, M-305, M-509 and M-601. That is, in 5 out of the 7 cases studied.

However, for roads M-519 and M-852 the larger number of samples classified in a class is smaller than 80%. This may mean that there are road sections misclassified, or, alternatively, that the road has sections of different classes.

Comparing Table 2 and Table 3, roads M-607, M-601, M-509 and M-852 are identified correctly. In contrast, roads M-519, M-618 and M-305 are misclassified. In some cases, a specific road is only partially wrongly classified, as M-618 road, which has sections that are class I and class III. As the percentage of sections classified as class I is so small, it is only assigned to class III.

Table 3. Classification results with the fuzzy system.

Road section	% correct class			% correct detection per section		
	Class I	Class II	Class III	Class I	Class II	Class III
M-607	**95.58**	4.19	0.23	**95.58**	0	0
M-519	28.30	2.78	**68.92**	**28.30**	0	0
M-852	6.21	**77.40**	16.39	0	**67.85**	**90.60**
M-618	6.86	5.50	**87.64**	**6.95**	0	**94.74**
M-305	**91.19**	0	8.81	0	0	**8.81**
M-509	**82.83**	6.64	10.53	**82.83**	0	0
M-601	12.69	2.94	**84.37**	0	0	**83.76**

Figure 5 shows the types of road found along the M-607 road, which is mainly class I. The system also identifies some sections as belonging to class II (accessibility, 4.19%). These could be considered outliers and this would not change the class of the road, which is uniform along its entire length regarding the geometric characteristics. Most of the road section has a medium or wide lane and medium or wide shoulders along most of its length, thus, the fuzzy system correctly identifies it as class I.

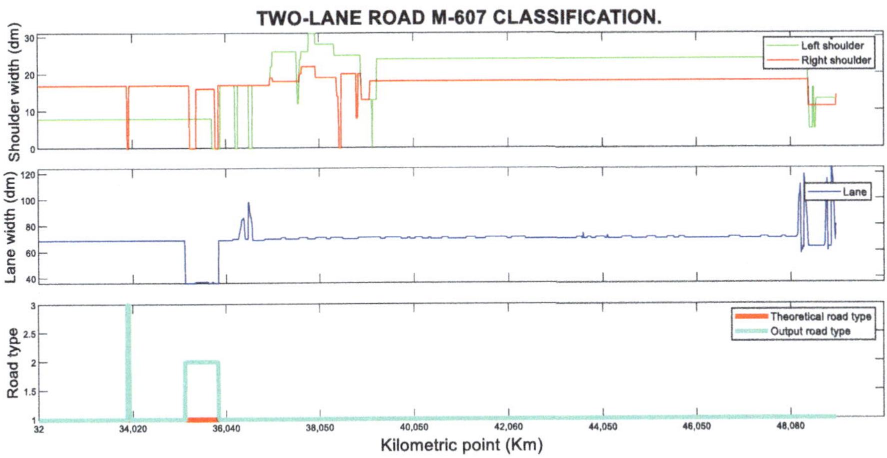

Fig. 5. Two-lane road M-607 classification.

Road M-852 (Fig. 6) is more complex. According to Table 1 it belongs to classes II and III, so sections of this road belong to different classes. Most of the road has narrow lane and shoulders that the fuzzy system identifies as class II (77,40%). There are also some road sections that have a medium lane with narrow shoulders, identified as class

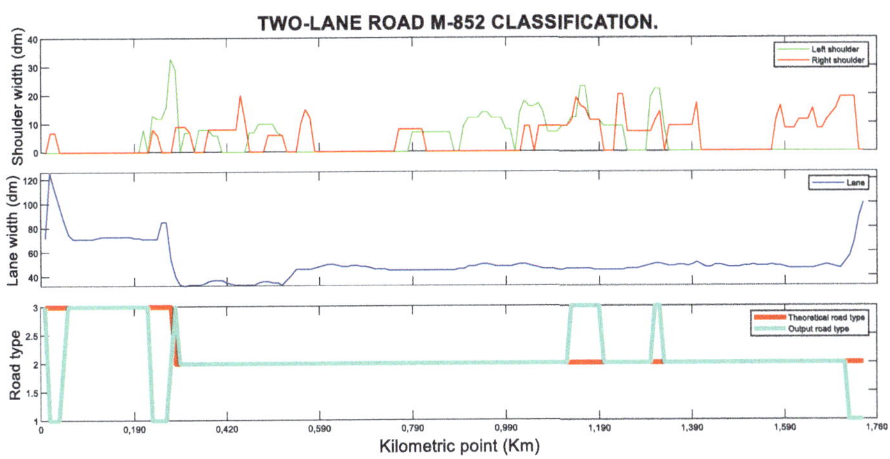

Fig. 6. Two-lane road M-852 classification.

III (16,39%). But taking into account only the sections, the classification is much better (see Table 3, last three columns). The hits are 67,83% and 90,60%.

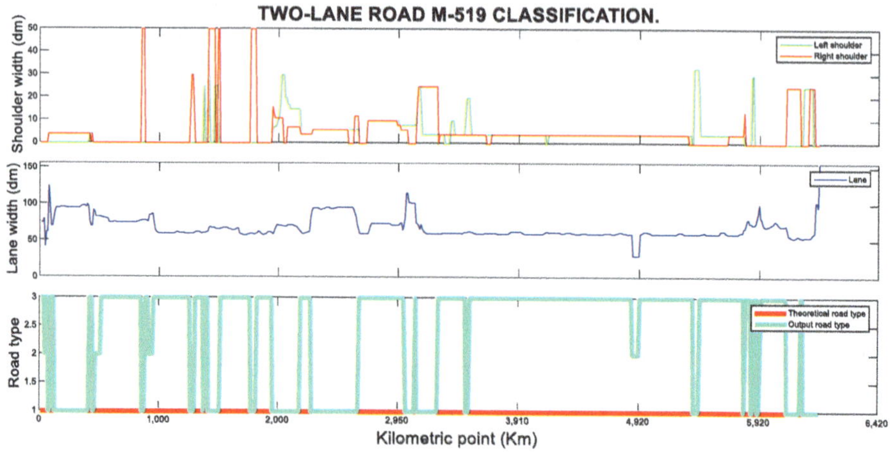

Fig. 7. Two-lane road M-519 classification.

Similarly (Fig. 7), M-519 road is considered class I (Table 1), but it has a section classified as class III (suburban). Most of the road section has a medium or wide lane with narrow shoulders along most of its length, thus, the proposed fuzzy system identifies it as class III. It is questionable whether M-519 is class I, since its geometry is not similar to other class I roads such as M-607 road that has wider shoulders. Therefore, M-519 should be considered as a medium speed road (class III).

To summarize, the fuzzy rule-based system is able to correctly identify most of the road classes for a particular road, and it is also able to classify sections of different classes of the same road. This tool can facilitate a more updated and realistic categorization of conventional roads, using current measurements of the geometric characteristics of the roads, which can lead to suggest a more suitable and appropriate speed for these road sections, improving driving safety.

4 Neuro-Fuzzy System

An ANFIS neuro-fuzzy system has also been tested [14, 15]. Before applying this strategy, repeated samples were removed in the pre-processing phase. Input data set was divided into two sets, one for training (50%) and another for validation (50%). A k-fold cross validation scheme, with $k = 5$, was used. The average of each measure on the different partitions was obtained (Table 4).

The neuro-fuzzy system identifies well the road sections of M-607and M-509. It also assigns the right class to M-852 road. But this system does not identify any road section of class III. Therefore, the performance of a neuro-fuzzy system is worse than the rule-based fuzzy system.

Table 4. Classification results (% correct) with the neuro-fuzzy system.

Road section	% correct class			% correct detection per section		
	Class I	Class II	Class III	Class I	Class II	Class III
M-607	**94.31**	5.69	0	**95.32**	0	0
M-519	38.66	**63.16**	0.18	**36.65**	0	0
M-852	9.72	**88.70**	1.58	0	**96.10**	**7.86**
M-618	**80.31**	19.69	0	**85.03**	0	**0**
M-305	**85.38**	14.09	0.53	0	0	**0.52**
M-509	**90.79**	9.17	0.04	**90.79**	0	0
M-601	6.24	**85.12**	8.64	0	0	**8.64**

5 Conclusions and Future Work

In this paper two fuzzy-based systems, Mamdani-type and neuro-fuzzy, have been designed and applied to classify conventional two-lane roads according to their geometric characteristics. The classifiers based on fuzzy logic use as inputs the current dimensions of the road: lane and shoulders width.

The results obtained with the fuzzy rule-based system are interesting and useful. On one hand, the class of two-lane roads assigned during the design phase is rightly identified. In addition, in some cases there are sections of the road where the geometric characteristics have changed and now they correspond to a different class of road, and this has also been detected.

This is important because, depending on the two-lane road class, the driving speed is determined. So a more updated and realistic classification allows increasing the safety and comfort while driving.

As future works, the design of a speed recommender system according to the real class of road section is proposed. This speed determination system can be applied to generate maps for checking lateral signaling, and to find "black spots" on certain sections of a road. Besides, a comparative study between different computational techniques is planned.

References

1. Díaz, J., Vuelvas, J., Ruiz, F., Patiño, D.: A Set-membership approach to short-term electric load forecasting. RIAI **16**(4), 467–479 (2019)
2. Santos, M.: One approach applied intelligent control. Rev. Iberoamericana de Automática e Informática Ind. RIAI **8**(4), 283–296 (2011)
3. Santos, M., López, V.: Fuzzy decision system for safety on roads. In: Handbook on Decision Making, pp. 171–187. Springer, Heidelberg (2012)
4. Martín, S., Romana, M.G., Santos, M.: Fuzzy model of vehicle delay to determine the level of service of two-lane roads. Expert Syst. Appl. **54**, 48–60 (2016)

5. Amini, J., Saradjian, M.R., Blais, J.A.R., Lucas, C., Azizi, A.: Automatic road-side extraction from large scale image maps. Int. J. Appl. Earth Obs. Geoinf. **4**, 95–107 (2002)
6. Tuncer, O.: Fully automatic road network extraction from satellite images. In: 2007 3rd International Conference on Recent Advances in Space Technologies, pp. 708–714. IEEE (2007)
7. Yan, X., Zhang, R., Ma, J., Ma, Y.: Considering variable road geometry in adaptive vehicle speed control. Math. Probl. Eng. **2013**, 12 p. (2013). Article ID 617879
8. Achwickart, T., Voos, H., Hadji-Minaglou, J.R., Darouach, M.: A novel model-predictive cruise controller for electric vehicles and energy-efficient driving. In: 2014 IEEE/ASME International Conference Advanced Intelligent Mechatronics, pp. 1067–1072. IEEE (2014)
9. Burrieza, A., Munoz-Velasco, E., Ojeda-Aciego, M.: A flexible logic-based approach to closeness using order of magnitude qualitative reasoning. Logic J. IGPL **28**, 121–133 (2019)
10. Leal, J.C.E., Angulo, J.R.M., Zambrano, J.H.B., Manriquez, A.D.: Using a microelectromechanical system to identifying roadway surface disruptions. IEEE Lat. Am. Trans. **16**(6), 1664–1669 (2018)
11. Spanish Ministry of Development: Standard 3.1- IC. Road tracing. Order FOM/273/2016, of February 19 (2016). www.fomento.gob.es/recursos_mfom/norma_31ic_trazado_orden_fom_273_2016.pdf
12. Coordination and information center. General Directorate of Roads. Department of Transportation, Housing and Infrastructure. Madrid Regional Government, Madrid, Spain (road data gathered in 2018, Unpublished)
13. Highway capacity manual, 6th edn., Chapter 15. In: Two-Lane Highways (2018)
14. Santos, M., López, R., de la Cruz, J.M.: A neuro-fuzzy approach to fast ferry vertical motion modelling. Eng. Appl. Artif. Intell. **19**(3), 313–321 (2006)
15. Santos, M., Dexter, A.L.: Temperature control in liquid helium cryostat using self-learning neurofuzzy controller. IEE Proc.-Control Theory Appl. **148**(3), 233–238 (2001)

Swarm Modelling Considering Autonomous Vehicles for Traffic Jam Assist Simulation

Javier Echeto[1], Manuel G. Romana[2], and Matilde Santos[3](✉)

[1] Computer Engineering Department, National University of Distance Education, Madrid, Spain
javierecheto@hotmail.com

[2] Civil Engineering Transportation Department, Universidad Politécnica de Madrid, Madrid, Spain
manuel.romana@upm.es

[3] Institute of Knowledge Technology, University Complutense of Madrid, Madrid, Spain
msantos@ucm.es

Abstract. Autonomous and connected cars are almost here, and soon will be an everyday reality. Driver desired comfort, road conditions, travel dynamics and communication requirements between vehicles have to be considered. Simulation can help us to find how to improve road safety and comfort in traveling. Traffic flow models have been widely used in recent years to improve traffic management through understanding how current laws, with human drivers, should change in this new environment. Early attempts to driving modelling were restricted to the macroscopic level, mimicking continuous physical patterns, particularly waves. However, extensive improvements in technology have allowed the tracking of individual drivers in more detail. In this paper, the Intelligent Driver Model (IDM) is used to examine traffic flow behavior at a vehicle level with emphasis on the relation to the preceding vehicle, similarly as it is done by the Adaptive Cruise Control (ACC) systems nowadays. This traffic model has been modified to simulate vehicles at low speed and the interactions with their preceding vehicles; more specifically, in traffic congestion situations. This traffic jam scenario has been analyzed with a developed simulation tool. The results are encouraging, as they prove that automatic car speed control can potentially improve road safety and reduce driver stress.

Keywords: Simulation · Traffic flow · Congested traffic · Traffic jam · Intelligent Driver Model · Vehicle swarm · Autonomous vehicles · Optimization

1 Introduction

Our roads will, in the near future, hold a continuous flow of autonomous and connected vehicles. Thus, it is becoming more and more important to explore how to optimize the use of road infrastructures, user driving comfort and the communication requirements between vehicles [1]. Simulation can help us to understand how to improve road safety and to make driving more comfortable.

Á. Herrero et al. (Eds.): SOCO 2020, AISC 1268, pp. 429–438, 2021.
https://doi.org/10.1007/978-3-030-57802-2_41

Congestion management on roads and city streets is usually approached in two different ways. Currently the preferred one is based on gathering and providing information on the actual road traffic conditions, advising all drivers on travel times in order to help them decide which route to follow, as well as managing the flows of vehicles coming into the highway in the on-ramps. The second approach focuses on the vehicles, developing intelligent systems that are able to adjust vehicle speed based on the behavior of the preceding vehicle, and modifying the dynamics of the traffic response [2].

Traffic flow models have been widely developed, studied and improved over the last years to better understand traffic management and to validate conceptual solutions that result in an improvement of traffic flow [3]. There models either consider time-space behavior of individual drivers under the influence of vehicles in their proximity (microscopic models), predicting driver behavior without explicitly distinguishing their individual time-space performance (mesoscopic model), or from the viewpoint of the collective vehicular flow (macroscopic models) [4].

Although early model driving behavior attempts were restricted to the macroscopic level, recent and continuous technology improvements have allowed the tracking of individual drivers in more detail. As a consequences, the number of microscopic models being explored has greatly increased in the last decade.

In this work we apply the microscopic Intelligent Driver Model (IDM) to examine traffic flow behavior at an individual level with emphasis on the relation to the following vehicle. This traffic model has been modified in order to simulate vehicles travelling at low speed and, more specifically, in traffic jams. A simulation tool has been developed to analyze this traffic scenario. The results are encouraging and prove that automatic car speed control can potentially improve road safety and reduce drivers stress.

The structure of the paper is as follows: in Sect. 2 a brief state of the art is presented. Section 3 is devoted to the description of the Intelligent Driver Model (IDM). The application of a modified version of this model as a traffic jam assistant is developed in Sect. 4. Results are discussed. The conclusions and future work end the work.

2 State of the Art

The Traffic Jam Assistant System helps to avoid rear-end collisions in dense traffic and traffic jams. Traffic Jam Assist works in combination with proven systems such as the Adaptive Cruise Control (ACC) and Lane Assist Systems (such as LDW, or Lane Departure Warning). The partially automated comfort function takes over the longitudinal and lateral control of the vehicle [5], meaning that ACC will be responsible of the longitudinal control and LDW will warn the driver when the vehicle begins to move out of its lane to keep the vehicle always on a safe trajectory.

The working principle of a Traffic Jam Assist function is to continuously analyze the speed of the surrounding vehicles, comparing it with its own driving speed [6]. If the system detects dense traffic or a traffic jam at speeds below 60 km/h (approx. 35 mph), the driver can activate the functionality by pressing a button. The vehicle will now automatically follow the vehicle in front, taking over driving, accelerating, braking and steering within the same lane.

Simulations can be used to, on one hand, prove the efficiency of these systems in traffic jams and, on the other, analyze traffic flows in these conditions. This is particularly useful when new systems need to be tested in an extensive set of complex scenarios, ensuring safety under all circumstances [7].

However, academic studies on this topic are scarce. Indeed, they use general traffic models that do not adapt well to any speed and different types of vehicles. In [8], systems designed to assist drivers in traffic jams are described. This development is based on the equations of the movement of a vehicle along a trajectory, taking into account only the vehicle's own movement. No use is made of simulation. In [7], authors simulate the dynamics of a vehicle using a multibody vehicle model to show the utility of a virtual platform they have developed, but they do not seem to address the problem of traffic jams. In [9] authors present a survey of the state-of-the-art related to vehicle platooning, swarm robotics concepts, swarm path planning and traffic simulators.

Authors in [10] present simulations of congested traffic in open systems with the IDM car following model. Microsimulations with identical vehicles on a single lane qualitatively agree with real traffic data.

Other authors, such as in [11], discuss modeling features for human and automated (ACC) driving by means of microscopic traffic simulations They conclude that a small amount of ACC equipped cars and, hence, a marginally increased free and dynamic capacity, may lead to a drastic reduction of traffic congestion.

3 Intelligent Driver Model (IDM)

The Intelligent Driver Model (IDM) is a deterministic car-following (time-continuous and autonomous) model of the Optimal Velocity Model (OVM) family with descriptive parameters to make it accident-free [12]. The main idea of this strategy is to combine the possibility to reach the desired speed in a free traffic situation with the ability to identify how much braking is necessary to stay clear of any collision. Figure 1 shows the graphical representation of the IDM model.

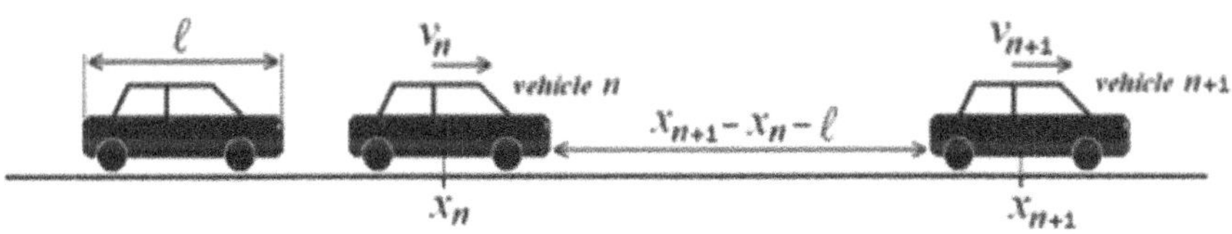

Fig. 1. Representation of the IDM model.

The IDM acceleration $\dot{v}$ of each vehicle n is a continuous function of the velocity v_n (m/s), the distance gap or space headway s_n (m), and of the relative speed Δv_n (m/s) of the leading vehicle. The following formula describes the acceleration in the case of a population of homogeneous vehicles:

$$\dot{v}_n(s_n, v_n, \Delta v_n) = a_{\propto}\left[1 - \left(\frac{v_n}{v_{0,n}}\right)^{\delta} - \left(\frac{S^*(v_n, \Delta v_n)}{s_n}\right)^2\right] \quad (1)$$

Where the desired gap s* is formulated as follows:

$$s^*(v_n, \Delta v_n) = s_{0,n} + s_{1,n}\sqrt{\frac{v_n}{v_{0,n}}} + T_n v_n + \frac{v_n \Delta v_n}{2\sqrt{a_n b_n}} \tag{2}$$

The IDM parameters of (1) and (2) can be described as:

- $s_{0,n}$: is the minimum bumper-to-bumper distance to the front vehicle (m)
- $s_{1,n}\sqrt{\frac{v_n}{v_{0,n}}}$: is the comfortable distance (m)
- T_n: is the desired safety time headway when following other vehicles (sec)
- $\frac{v_n \Delta v_n}{2\sqrt{a_n b_n}}$: is the anticipation (m)
- $v_{0,n}$: is the desired speed when driving on a free road (m/s)
- *a:* is the acceleration in everyday traffic (m/s^2)
- *b:* is the comfortable braking deceleration in everyday traffic (m/s^2)
- δ: is the acceleration exponent (m/s^2)

In (2), the first term $s_{0,n}$, aims at maintaining the desired distance. This term has the highest influence when traffic is at constant speed and gap. The second term depends on the speed of the vehicle, so that it gives the desired level of comfort to the trip. That is, it provides the addition of some extra distance to the "desired distance". Then, the driver has more time to react to changes in the speed of the preceding vehicle, and therefore, the driver feels safer and more comfortable. The distance that is added is determined by the jam distance parameter, $s_{1,n}$, together with the ratio between the actual speed and the desired speed.

As opposed to the term "comfortable distance", the safe time headway is the absolute minimum distance necessary to stop completely if the predecessor vehicle suddenly brakes. This distance becomes larger at higher speeds, since it is obtained by multiplying the speed by T, which is referred to as the safe time headway parameter.

The developed simulation tool was initially tested using Eq. (1) but this produced an unrealistic behavior in its results. It was required to update the limit of the last term to prevent it from taking negative values. Equation (2) was then re-written as follows:

$$S^*(v_n, \Delta v_n) = s_{0,n} + max\left(0, s_{1,n}\sqrt{\frac{v_n}{v_{0,n}}} + T_n v_n + \frac{v_n \Delta v_n}{2\sqrt{a_n b_n}}\right) \tag{3}$$

In the IDM model a driver considers only the first vehicle ahead. If this predecessor is found to be increasingly closer to the considered car, the simulated driver will respond by either releasing the gas pedal or braking directly, depending on the speed reduction desired. This is modelled by setting a higher desired distance. The relative speed will be positive in this case, since it is calculated as the speed of vehicle n minus the speed of the leading vehicle:

$$\Delta v_n = v_{n+1} - v_n \tag{4}$$

The anticipation term also contains the deceleration parameter, b, which controls the deceleration when breaking. Note, however, that the deceleration is theoretically not limited, as opposed to the acceleration.

For each vehicle, the acceleration is integrated over time to obtain the velocity, and then the velocity is again integrated over time to produce the current location x.

$$\dot{x}_n = v_n \tag{5}$$

The actual distance to the predecessor is calculated by the difference between the positions of the leading vehicle x_{n+1} and the position of the follower, x_n, adding the vehicle length (l), which is an initial parameter of the model (Fig. 1).

$$s_n = x_{n+1} - x_n - l_n \tag{6}$$

According to [9] a normal driving behaviour of a vehicle can be simulated with the parameters listed in Table 1:

Table 1. IDM parameters in the homogeneous implementation.

Parameters	Variable	Realistic value
Maximum acceleration	$a_n\left(m/s^2\right)$	0.73
Desired speed	$v_0(m/s)$	33.3
Linear jam distance	$s_n^0(m)$	2
Non-Linear jam distance	$s_n^1(m)$	3
Safe time headway	$\tau_n(s)$	1.3
Comfortable deceleration	$b_n\left(m/s^2\right)$	1.67
Acceleration exponent	δ_n	4

The IDM model parameters can be interpreted by considering the following three standard situations [12]:

- When accelerating on a free road from stopped, the vehicle has a maximum initial acceleration a. As speed grows, the acceleration decreases gradually, reaching zero as the speed reaches the desired speed v_0. The exponent δ controls the reduction rate: the higher its value, the larger the reduction of the acceleration when approaching the desired speed. The limit value as $\delta \rightarrow \infty$ corresponds to the acceleration profile of Gipps' model, while $\delta = 1$ reproduces the overly smooth acceleration behaviour of the Optimal Velocity Model.
- When following a leading vehicle, the spacing (distance gap) is approximately given by the safe distance $(s_0 + v_n T_n)$. This safe distance is determined by the time gap $\boldsymbol{\tau_n}$ plus the minimum spacing s_0.
- When approaching slower or stopped vehicles, the deceleration usually does not exceed the comfortable deceleration b_n. The acceleration function is smooth during transitions between these situations.

4 Traffic Jam Assistant

We have developed a traffic jam assistance function based on the IDM model which has been simulated with Matlab. This traffic model has been modified as described earlier. Any vehicle in a congested road shows the same behaviour, and therefore its parameters and behaviour are the same regarding the modelling but the length. We have differentiated two types of vehicles, cars (Type1) and bigger ones such as trunks or buses (Type 2). Table 2 shows realistic parameters for the main variables of the proposed traffic jam scenarios.

Table 2. Original IDM vs Traffic jam settings.

Parameter	Var	Veh Type 1	Veh Type 2	Traffic jam
Maximum acceleration (m/s^2)	a_n	0.73	0.7	0.7
Desired speed (m/s)	v_0	33.3	19	1.38
Linear jam distance (m)	s_n^0	2	1.53	0.1
Non-linear jam distance (m)	s_n^1	3	0.36	0.3
Safe time headway (s)	τ_n	1.6	2	0.2
Comfortable deceleration (m/s^2)	b_n	1.67	1.61	1.5
Acceleration exponent	δ_n	4	4	4
Car length (m)	l	4	12	4/12

The existence of a traffic jam condition is verified through monitoring the individual vehicle speed and distance to the preceding vehicle every sample time. Congestion is identified if the speed and distance between vehicles are below a predefined threshold. If traffic conditions are validated, the IDM car-following model adapts the parameters of Table 2 to this scenario.

The simulation considers different scenarios of vehicles in a traffic jam with relative low vehicle speeds (lower than 10 km/h) and inter-vehicle distance up to 2 m.

4.1 Single Lane with Heterogeneous Vehicles

In this first scenario, a convoy of forty different vehicles ($n = 40$) on a single lane driving at low speed and with inter-vehicle distance of 1.25 m were simulated, as shown in Fig. 2, left. The jam velocity is 5 km/h. Vehicles Types 1 (small blue squares) and Type 2 (red squares) are placed on the lane. Simulation time is 700 s.

Figure 2, right, shows the headway distance (m). According to the results, the initial inter-vehicle distance of 1.25 m is maintained by all the vehicles in the queue within an interval of around ± 0.1 m. The signal fluctuations are due to the addition of a limited random noise into the IDM model to produce a more realistic behavior.

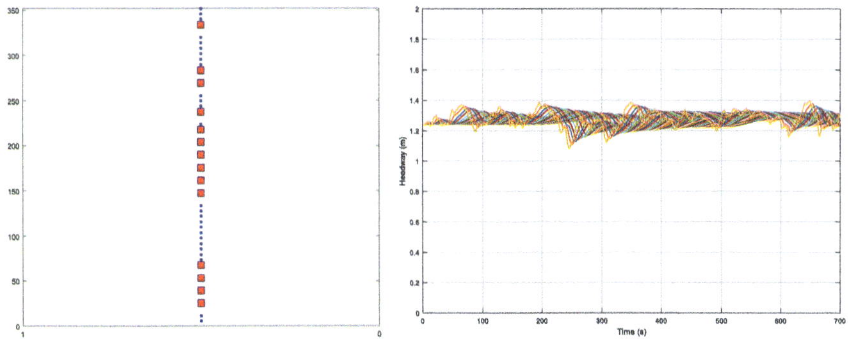

Fig. 2. Single lane traffic jam simulation (left) and headway distance (right)

Both speed (Fig. 3, left) and acceleration (Fig. 3, right) seem to be quite smooth and without large oscillations (speed, ±0.04 m/s; acceleration, ±0.005 m/s^2), which is most desirable under traffic jam conditions in order to avoid unintended crashes.

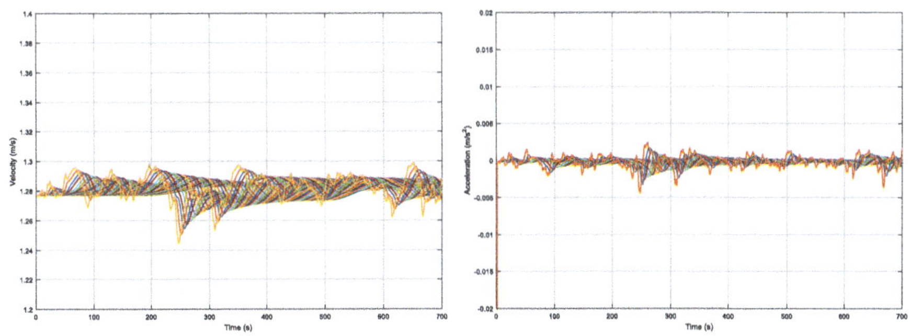

Fig. 3. Traffic jam vehicle speed and acceleration performance.

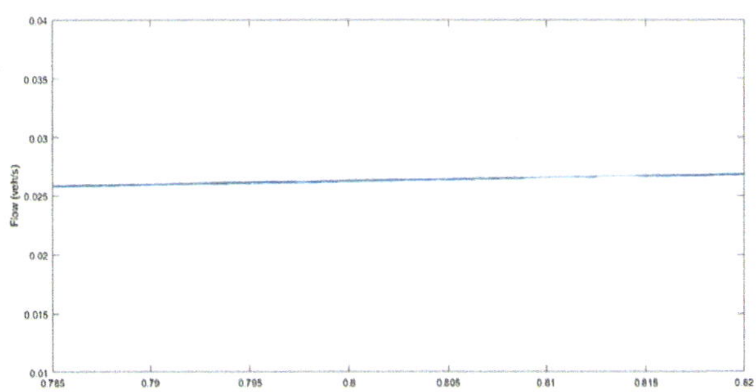

Fig. 4. Flow rate in traffic jam assistance.

As a consequence, the flow rate or volume (vehicles/s) is kept almost constant during the simulation time (Fig. 4).

4.2 Multiple Lanes with Heterogeneous Vehicles

Next, a convoy of forty different vehicles (n = 40) has been simulated on a 3-lane road driving at low speed, with inter-vehicle distance of 1.25 m (Fig. 5, left). The jam velocity is 5 km/h, vehicle Type 1 (blue squares) and Type 2 (red square) are placed on the three lanes; simulation time is also 700 s.

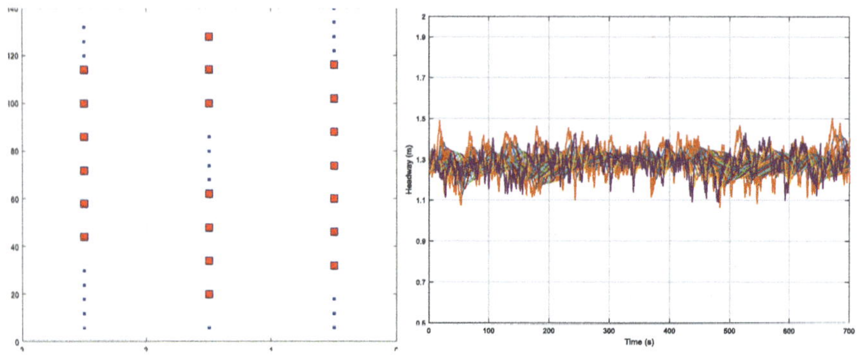

Fig. 5. Multiple lanes traffic jam simulation (left) and headway distance (right).

According to Fig. 5, right, the initial inter-vehicle distance of 1.25 m is closely kept by all vehicles in each ow with a small deviation of ±0.25 m. The fluctuations are due to the limited random noise included into the IDM model to produce a more realistic behaviour.

Vehicle speed and acceleration (Fig. 6, left and right, respectively) are kept free from sudden changes, and show no relevant oscillations (speed, ±0.5 m/s; acceleration, ±0.03 m/s^2).

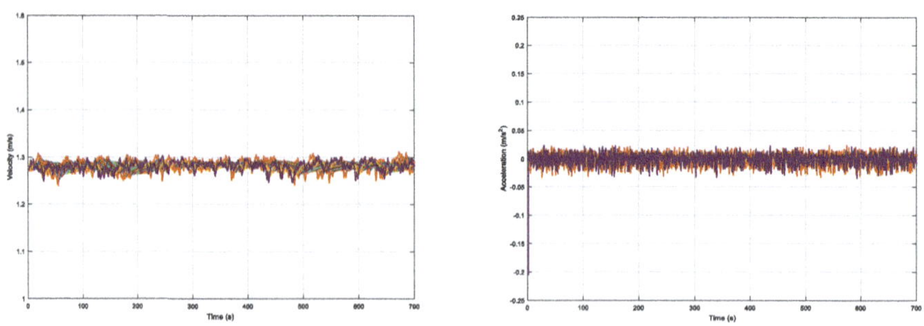

Fig. 6. Vehicle speed (left) and acceleration (right) in a three-lane congestion simulation.

In Fig. 7 the simulation of the flow rate is shown. The flow rate in the three lanes is kept stable and almost constant during the simulation time.

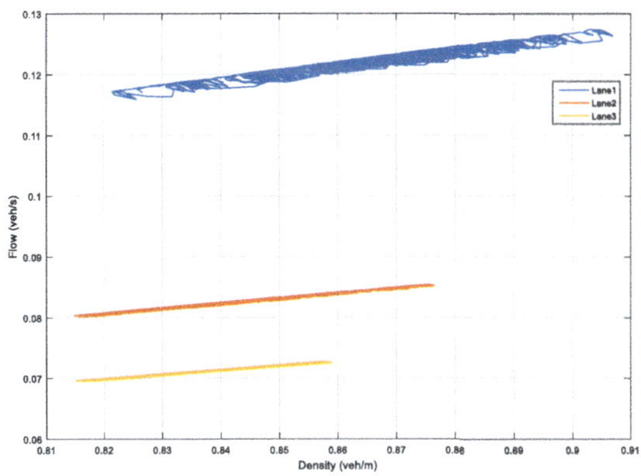

Fig. 7. Flow rate in traffic jam assistance

5 Conclusions and Future Work

This paper presents the results of a research that uses the simulation of different scenarios for a congested road, for which a modified version of the microscopic IDM traffic flow model was chosen. The aim of this car following simulation is to provide the longitudinal control in a traffic jam assistance function.

The IDM model, without the modification, has been widely used for the simulation of ACC systems for high speed vehicles. However, it has been modified here, and implemented for low speeds and, specifically, in congested situations (traffic jams). Simulations results show that the new model gives a stable and smooth behaviour regarding vehicle speed and acceleration while traffic flow is kept constant. This fact confirms that the use of longitudinal control systems like ACC could potentially improve road safety and reduce driver stress by taking over driving off and the acceleration.

Additionally, this traffic jam assist system could also help drivers to be more relaxed while getting to their destination, even in dense traffic or in traffic jams. It could be used by car makers to indicate advised acceleration and deceleration rates, and comfortable spacing with preceding vehicles to persons driving cars in congested situations, albeit, in this case, an additional buffer should be taken into account for safety.

Regarding future works, the following are proposed. First, to validate the proposed IDM model parameters in real congested traffic situations. Second, to integrate the traffic jam assistance functionality here proposed within a cooperative network, on which the real time status information of the vehicles is shared among them to get a more efficient traffic management strategy.

References

1. Santos, M., López, V.: Fuzzy decision system for safety on roads. In: Handbook on Decision Making, pp. 171–187. Springer, Heidelberg (2012)
2. Milanés, V., Shladover, S.E.: Modeling cooperative and autonomous adaptive cruise control dynamic responses using experimental data. Transp. Res. Part C Emerg. Technol. **48**, 285–300 (2014)

3. Martín, S., Romana, M.G., Santos, M.: Fuzzy model of vehicle delay to determine the level of service of two-lane roads. Expert Syst. Appl. **54**, 48–60 (2016)
4. Hoogendoorn, S.P., Bovy, P.H.: State-of-the-art of vehicular traffic flow modelling. Proc. Inst. Mech. Eng. Part I J. Syst. Control Eng. **215**(4), 283–303 (2001)
5. Pérez, J., Gajate, A., Milanés, V., Onieva, E., Santos, M.: Design and implementation of a neuro-fuzzy system for longitudinal control of autonomous vehicles. In: International Conference on Fuzzy Systems, pp. 1–6. IEEE (2010)
6. Malerczyk, J., Lerch, S., Tibken, B., Kummert, A.: Impact of intelligent agents on the avoidance of spontaneous traffic jams on two-lane motorways. In: MATEC Web of Conferences, vol. 308, p. 05003. EDP Sciences (2020)
7. Marcano, M., Matute, J. A., Lattarulo, R., Martí, E., Pérez, J.: Low speed longitudinal control algorithms for automated vehicles in simulation and real platforms. Complexity, 12 p. (2018). Articulo ID 7615123
8. Lüke, S., Fochler, O., Schaller, T., Regensburger, U.: Traffic-jam assistance and automation. In: Handbook of Driver Assistance Systems: Basic Information, Components and Systems for Active Safety and Comfort, pp. 1–13 (2014)
9. Caruntu, C.F., Ferariu, L., Pascal, C.M., Cleju, N., Comsa, C.R.: A concept of multiple-lane vehicle grouping by swarm intelligence. In: 2019 24th International Conference on Emerging Technologies and Factory Automation (ETFA), pp. 1183–1188. IEEE (2019)
10. Treiber, M., Hennecke, A., Helbing, D.: Microscopic simulation of congested traffic. In: Traffic and Granular Flow'99, pp. 365–376. Springer, Heidelberg (2000)
11. Kesting, A., Treiber, M., Schönhof, M., Kranke, F., Helbing, D.: Jam-avoiding adaptive cruise control (ACC) and its impact on traffic dynamics. In: Traffic and Granular Flow'05, pp. 633–643. Springer, Berlin, Heidelberg (2007)
12. Treiber, M., Kesting, A.: Car-following models based on driving strategies. In: Traffic Flow Dynamics, pp. 181–204. Springer, Heidelberg (2013)

Special Session: Soft Computing and Machine Learning in Non-linear Dynamical Systems and Fluid Dynamics: New Methods and Applications

Exploring Datasets to Solve Partial Differential Equations with TensorFlow

Oscar G. Borzdynski[1], Florentino Borondo[2,3(✉)], and Jezabel Curbelo[1,2]

[1] Departamento de Matemáticas, Universidad Autónoma de Madrid, Cantoblanco, 28049 Madrid, Spain
[2] Instituto de Ciencias Matemáticas (ICMAT), Cantoblanco, 28049 Madrid, Spain
[3] Departamento de Química, Universidad Autónoma de Madrid, Cantoblanco, 28049 Madrid, Spain
f.borondo@uam.es

Abstract. This paper proposes a way of approximating the solution of partial differential equations (PDE) using Deep Neural Networks (DNN) based on Keras and TensorFlow, that is capable of running on a conventional laptop, which is relatively fast for different network architectures. We analyze the performance of our method using a well known PDE, the heat equation with Dirichlet boundary conditions for a non-derivable non-continuous initial function. We have tried the use of different families of functions as training datasets as well as different time spreadings aiming at the best possible performance. The code is easily modifiable and can be adapted to solve PDE problems in more complex scenarios by changing the activation functions of the different layers.

Keywords: Deep learning · Partial derivative equations · TensorFlow · Keras · Neural Network

1 Introduction

The use of Machine Learning (ML) is spreading across many fields in Applied Science, often showing a very good performance in the resolution of many different practical tasks, such as weather forecasting [14], self driving cars [12], or translation [2], just to name a few. However, ML is not very popular in Mathematics or other theoretical sciences, despite the fact that strong evidence of its great potential has been recently reported in the literature [6]. Reservoir computing [11], for example, is one such method, which unfortunately is very demanding computationally.

In this paper we explore a more economic computationally alternative way of approximating the numerical solution of Partial Differential Equations using Deep Neural Networks (DNN) based on the Keras [4] and Tensorflow softwares [1]. This framework is widely used for its performance and versatility [5].

Á. Herrero et al. (Eds.): SOCO 2020, AISC 1268, pp. 441–450, 2021.
https://doi.org/10.1007/978-3-030-57802-2_42

Table 1. DNN Structure and activation functions

Layer	No. of neurons	Activation function
Entry layer	100	Linear
First hidden layer	1250	Linear
Second hidden layer	2500	Linear
Third hidden layer	5000	Linear
Exit layer	5000	Linear

Deep learning techniques are promising in solving PDEs because they are able to represent complex-shaped functions very effectively, specially when compared to other traditional methods which experience the "curse of dimensionality" difficulties. For instance, the experiments in [8] show that the artificial neural networks exhibit a better performance than finite element methods for several cases of PDEs.

Similar work to ours has been reported in the literature. In particular, Deep-XDE [9] is a code made to solve PDE using Tensorflow that allows the user to make an approximation without making a big effort in choosing the structure of the DNN. Good results have been obtained using this library. For example, it has been applied to the study of inverse problems in nano-optics and metamaterials [3], and space-time fractional advection-diffusion equations [10]. We decided to use plain Tensorflow to be capable of finely tuning the network for our problem.

To illustrate and analyze the feasibility and performance of our method we apply it to a well known PDE, as it is the heat equation [15] with Dirichlet boundary conditions for a non-derivable non-continuous initial function. We tried different families of particular solutions as training datasets, and check the use of different ways to span the time interval, seeking for the best performance. Excellent solutions are found for generic initial functions in all cases explored so far.

The paper is organized in four sections. In Sect. 2, we describe the Deep Neural Network structure, activation functions, and training dataset. Sect. 3 is devoted to briefly explain the heat equation and its possible theoretical solutions. In Sect. 4, we illustrate our method by presenting the results obtained in several numerical experiments. Finally, in Sect. 5 we summarized our conclusions, and discuss possibilities for future work.

2 Data and Methods

2.1 Our Deep Neural Network

A DNN is formed by a series of layers, each one consisting a certain number of neurons with a given activation function. The activation function defines the information flows along the network.

The parameters defining the structure of our DNN are given in Table 1. The entry layer receives 100 equidistant samples of the initial function. The hidden layers are incremental with 1250, 2500 and 5000 neurons, respectively. The exit layer has 5000 outputs which correspond to a matrix of 100×50 with the first dimension being the position and the second the time.

This structure is chosen for several reasons. The first one is the possibility of predicting non-bounded negative values, the linear activation function makes this possible since it is defined in the range $(-\infty, \infty)$. We consider that the neurons receive a vector X and they have a vector of weights W where W_i corresponds to the input X_i referring i neurons in the previous layer. A linear activation function means that the exit signal of the neurons is $W^{\intercal}X$, this implying a linear transformation of the input to the output data. Second, the behavior of the activation function near zero is not as steep as others functions, like for example the sigmoid [6]. The third is the growing effect obtained from an increasing number of neurons, adding information instead of removing or shuffling it.

As the last parameters of out network we need to specify an optimizer and a loss function. The loss function is the objective to minimize, it compares the exit of the DNN with the expected result, and returns a metric which indicates the distance between them. The optimizer is the algorithm that determines how the parameters of the network change to minimize the loss, fitting the data to the expected result.

We decided to use the root mean square error (RMSE) as loss function because it penalizes big errors, and we want a uniform fit to the solution. Also we used the well known ADAM optimizer [7] as it has been empirically shown [13] to work well, improving the performance of other alternative methods.

2.2 Our Training Datasets

The dataset that we use consists of 2000 equations, and we train our DNN with 1600 (80%) of them in batches of 100. After several tries we decided that 20 epochs were sufficient to achieve acceptable results.

3 Example: The Heat Equation

To illustrate the performance of our method we use the well known heat equation with Dirichlet boundary conditions

$$\begin{cases} u_t = \alpha u_{xx} & t > 0, 0 < x < L, \\ u(x,0) = f(x) & 0 < x < L, \\ u(0,t) = u(L,t) = 0 & t > 0, \end{cases} \tag{1}$$

which solution is

$$\begin{aligned} u(x,t) &= \sum_{n=1}^{\infty} \left\{ b_n \sin \frac{nx\pi}{L} \exp\left[-\left(\frac{n\alpha\pi}{L}\right)^2 t \right] \right\}, \\ b_n &= \frac{2}{L} \int_0^L f(x) \sin \frac{nx\pi}{L} dx. \end{aligned} \tag{2}$$

Table 2. Definition of datasets and testing ways in the different experiments. Linear and exponential time means, respectively, that the time steps are equally, or exponentially, separated times (see text for details).

Experiment	Training dataset	Testing
A	Family of functions $f(x)$ defined in Eq. (4) with different random intervals and linear time	$h(x)$ defined in Eq. (5) and linear time
B	Family of functions $f(x)$ defined in Eq. (3) with different random a parameter and linear time	$h(x)$ defined in Eq. (5) and linear time
C	Family of functions $f(x)$ defined in Eq. (3) with different random a parameter and exponential time	$h(x)$ defined in Eq. (5) with exponential time
D	Family of functions $f(x)$ defined in Eq. (3) with different random a parameter and linear time	$h(x)$ defined in Eq. (5) with linear time and double interval

assuming a one-dimensional rod of length $L = \pi$, $0 \leq t < 0.05$, and defining two different initial conditions:

$$f(x) = \sin(ax), \tag{3}$$

$$f(x) = \begin{cases} 1 \text{ if } x \in I \text{ and,} \\ 0 \text{ otherwise,} \end{cases} \tag{4}$$

where a is a characteristics parameter, and I is a given known set consisting of two non-overlapping intervals defined by four limits.

4 Results

In order to train the DNN a dataset is needed. We are going to explore four different ways of generating it, and one way of testing it, as summarized in Table 2.

In the first experiment A, the initial data $f(x)$ for Eq. (1) is given by (4) defined with random intervals I. For the second experiment B, $f(x)$ is given by (3) with random a. In both experiments the temporal grid is uniform, i.e. $t_i = 0.001i$ with $i = 0, \ldots, N-1$ where N is the number of temporal nodes. Experiment C is equal to experiment B but with the node distance following the expression $t_i = [-1 + \exp(i/N)]/20$ for $i = 0, \ldots, N-1$. Experiment D is the same as experiment B but with an extended (doubled) time interval.

In all the previous scenarios we use a test function $h(x)$, as initial condition of (1), which is the non-derivable non-continuous function:

$$h(x) = \begin{cases} 0 & \text{if} \quad x = 0 \\ 0.3 & \text{if} \quad 0 < x < \pi/2 \\ 0.8 & \text{if} \quad \pi/2 \leq x < \pi \\ 0 & \text{if} \quad x = \pi. \end{cases} \tag{5}$$

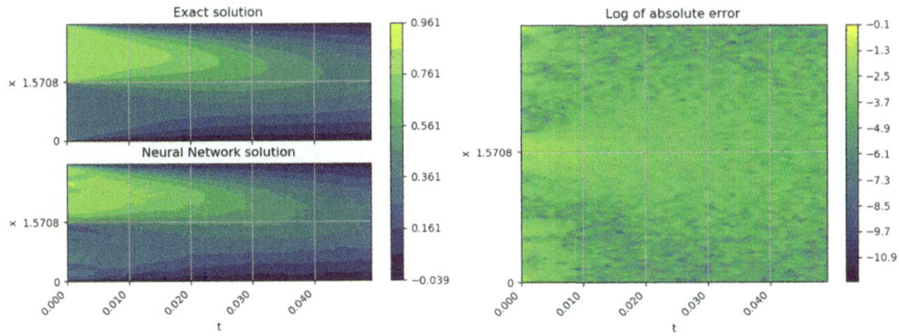

Fig. 1. Result of experiment A of Table 2. (Left) Theoretical and approximate solution obtained by the DNN. (Right) Logarithm of the error of the DNN approximation, where yellow/blue color means bigger/smaller errors. The maximum error occurs in the extremes of the rod at $t = 0$, where the model does not comply with the Dirichlet boundary condition. Its value is 0.8181 (top left corner). The maximun error for $t > 0$ is 0.2813 and the mean error is 0.0305.

Note that this function barely complies with the Dirichlet condition, and the solution is not easily computed as it need to be transformed to a Fourier series. Although $h(x)$ plays the same role as $f(x)$ in Eq. (1), we use different notation to easily differentiate the functions used for training and testing.

The hardware used in all the experiments is a very modest:

- i7-4790 8 threads 3.6 Ghz
- 16 GB of RAM
- 250 GB SSD

No use of the GPU (graphics processing card) was made, as it is customarily done, to test if a conventional computer was able to be trained and used to predict in a model like our. The typical time needed to generate the dataset was roughly 3 hours, and the training was performed in about 15 min.

In Figs. 1, 2, 3 and 4 we present the results obtained with the DNN specified in Table 1 for the four different scenarios described in Table 2.

In the first experiment A, we used a Dirac-delta shaped functions in random intervals. For testing, we used the function defined in Eq. (5). As can be seen in Fig. 1, the shape of the predicted and exact solutions are very similar, and the error is very uniform everywhere. The maximum error is 0.8181, which happens at the extreme of the rod, where the initial function has a very big gap. We will see that this effect happens in every test we have made. The mean error is 0.0305.

In the second experiment B we used a sin family of functions, while for testing the function defined in Eq. (5) is used. As it is seen in Fig. 2 errors mostly occurs at the initial time. We think that this is probably due to the big variation of $\sin(ax)$ when a is big. The maximum error takes a similar value as in experiment A, being this equal to 0.8134. The mean error is also similar to the previous case, and equal to 0.0330.

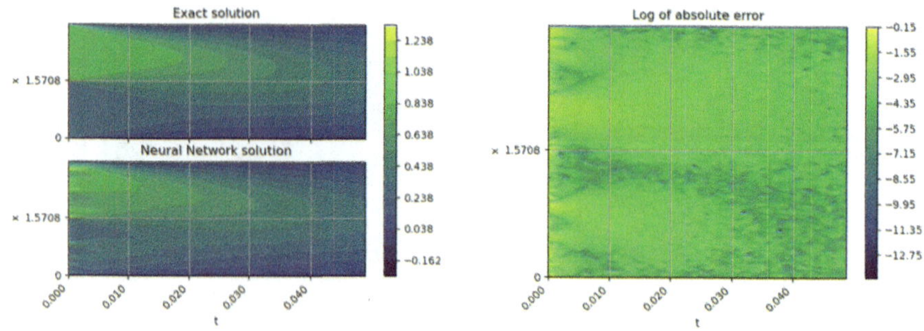

Fig. 2. Same as Fig. 1 for experiment B of Table 2. The corresponding maximums and mean error values are 0.8134, 0.3068, and 0.0330, respectively

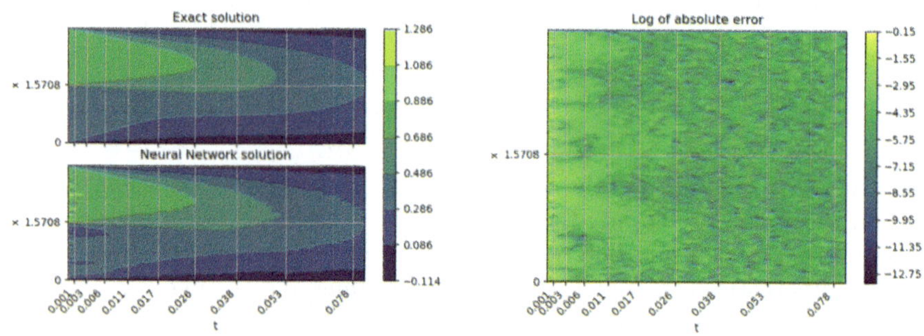

Fig. 3. Same as Fig. 1 for experiment C of Table 2. The corresponding maximums and mean error values are 0.8024, 0.2070, and 0.0179, respectively

In order to optimize the results obtained in the initial time portion, we bring closer the initial time steps and separate them a bit the further ones in experiment C (see results in Fig. 3). First, we see a big improvement in the mean error, being it reduced to 0.0179. The error at the end of the rod and the initial time is still the maximum obtained error, equal to 0.8024. We appreciate a smaller and more uniform error as time advances.

Finally, in the last experiment D, which results are shown in Fig. 4 we tried a new approach to see how the model works when the total time interval is extended (to twice the value used before in experiment C). To achieve this task, we reevaluated the last value obtained by the previous evaluation, obtaining twice the time. We see that the shape is similar, but the error gets a lot bigger, rising to 0.0506.

In all the experiments we have monitored the maximum error without considering the initial error at $t = 0$. The conclusion is that the error drastically decreases in all cases, but the maximum still happens as time goes to 0. Another conclusion that can be drawn from the previous results is that after those initial experiments, the best possible strategy is to stick to the exponential time approximation, since it renders the best results.

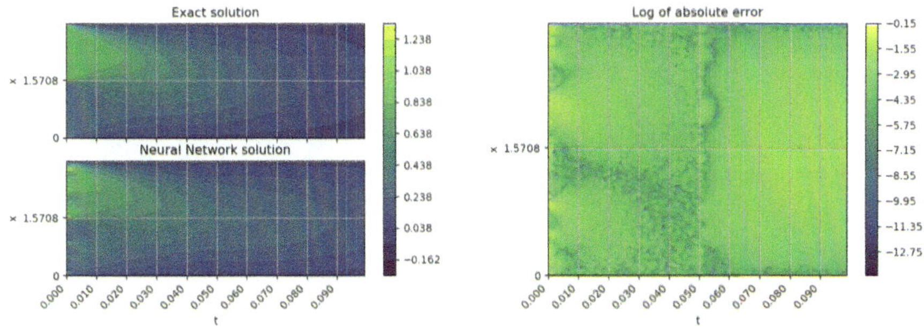

Fig. 4. Same as Fig. 1 for experiment D of Table 2. The corresponding maximums and mean error values are 0.8134, 0.3068, and 0.0506, respectively

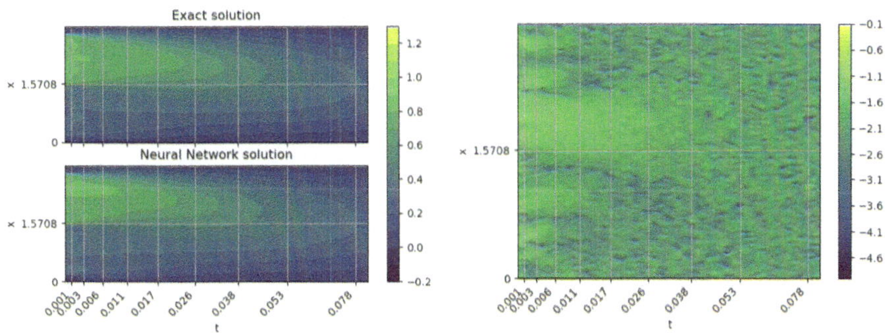

Fig. 5. (Left) Theoretical and approximate solution obtained by the DNN with exponential time for the function $\tilde{h}(x)$ defined in Eq. (6). (Right) Logarithm of the error of the DNN approximation, where yellow/blue color means bigger/smaller errors.

Therefore, we next try to evaluate smoother initial functions. Note that the used training dataset will be the same as before. In this new batch of numerical experiments, we try a function that is equal to $h(x)$ but not ending so close to the end of the rod, thus preventing the error at $t = 0$. For this purpose we use the following definition:

$$\tilde{h}(x) = \begin{cases} 0 & \text{if} \quad x \leq 0.2 \\ 0.3 & \text{if} \quad 0.2 < x < \pi/2 \\ 0.8 & \text{if} \quad \pi/2 \leq x < \pi - 0.2 \\ 0 & \text{if} \quad \pi - 0.2 \leq x. \end{cases} \tag{6}$$

The corresponding results are shown in Fig. 5, where we present the solution and the approximation made by our algorithm. Note that in this case the maximum error reduces to 0.0915, and that the mean error reduces to 0.0132. Also, it can be seen that the maximum error does not happen at $t = 0$. This result indicates that when the function does not fit the Dirichlet condition our method can not obtain a good approximation.

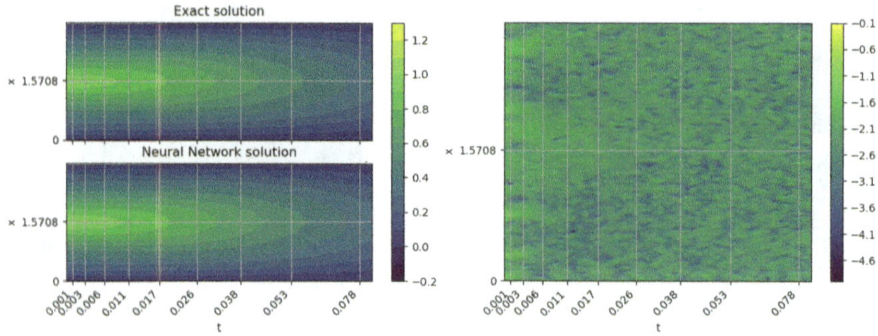

Fig. 6. Same as Fig. 5 for the function $\overline{h}(x)$ defined in Eq. (7).

We next try a new testing function that it is continuous, but non-derivable, defined in the following way:

$$\overline{h}(x) = \begin{cases} \dfrac{2x}{\pi} & \text{if} \quad x \leq \dfrac{\pi}{2}, \\ 2 - \dfrac{2x}{\pi} & \text{if} \quad x > \dfrac{\pi}{2}. \end{cases} \tag{7}$$

The corresponding results are presented in Fig. 6, where it is seen that the error is much smaller than in previous experiments; the maximum error is 0.0447, and the mean error 0.0055. Notice that this is the best case obtained along all our work, so that one can conclude that the use of continuous testing functions greatly improve the performance of our method.

As the last testing function, we try a continuous and derivable initial condition:

$$h^{*}(x) = -(x - 0)(x - \pi). \tag{8}$$

The corresponding results are presented in Fig. 7, where it is seen that a maximum error of 0.1177 and a mean error of 0.0156 are obtained. Notice that

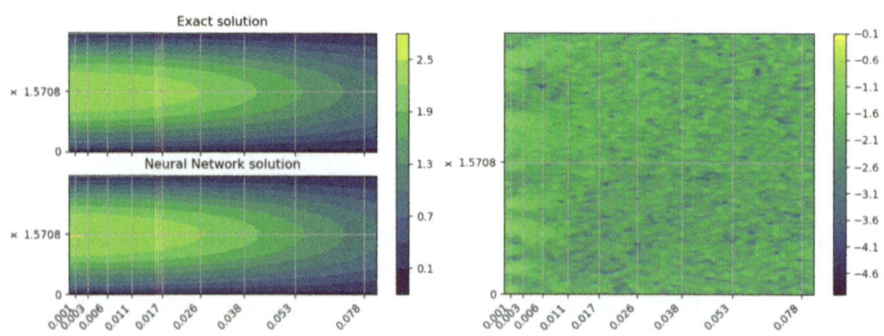

Fig. 7. Same as Fig. 5 for the function $h^{*}(x)$ defined in Eq. (8).

the maximum value of the solution is much larger than in the previous cases, so the percentage of error is more or less the same here.

5 Summary and Conclusions

In this work we have developed a simple DNN based on readily available software which is able to find accurate approximate numerical PDEs on modest laptop computers. We have use the well know heat equation to check the performance of the method. This represents a good alternative in terms of computational effort and cost to more sophisticated methods, such as the increasingly popular Reservoir Computing [11], whenever an extremely high accuracy is not required.

To optimize our DNN we have tried four different approaches, two families of functions and three different time spans, having obtained better results when compressing the time steps in the initial time and expanding them as time increases. Other initial functions where tested with this method, as the function smoothed the results improving them, achieving in this way a mean error of 10^{-3} in the best case.

The numerical experiments done in this paper show that deep learning may be used to approximate non-easily computable functions with a decent error in an everyday computer, even when the initial function does not fully comply with the boundary conditions. The only small problem of our approach is the generation of the training dataset, since a large number of solutions need to be computed. When the problem is theoretically solvable the required datasets can be easily obtained.

The main objective of our work was to develop a method able to run in an modest computer, thus making Deep Learning available to any researcher in computer science. Running it with a bigger dataset or more complex network structures will need bigger computational means that would improve the performance of our approximation. Also a combination of various families of functions has proven to improve the results, but we wanted to keep the dataset in this first paper as simple as possible.

Acknowledgments. This work has been partially supported by the Spanish Ministry of Science, Innovation and Universities, Gobierno de España, under Contracts No. PGC2018-093854-BI00, and ICMAT Severo Ochoa SEV-2015-0554, and from the People Programme (Marie Curie Actions) of the European Union's Horizon 2020 Research and Innovation Program under Grant No. 734557.

References

1. Abadi, M., et al.: TensorFlow: Large-Scale Machine Learning on Heterogeneous Systems (2015)
2. Bahdanau, D., Cho, K.H., Bengio, Y.: Neural machine translation by jointly learning to align and translate. In: 3rd International Conference on Learning Representations, ICLR 2015 - Conference Track Proceedings. International Conference on Learning Representations, ICLR (2015)

3. Chen, Y., Lu, L., Karniadakis, G.E., Dai Negro, L.: Physics-informed neural networks for inverse problems in nano-optics and metamaterials, December 2019
4. Chollet, F., et al.: Keras (2015). https://github.com/fchollet/keras
5. Gulli, A., Pal, S.: Deep learning with Keras (2017)
6. Han, J., Jentzen, A., Weinan, E.: Solving high-dimensional partial differential equations using deep learning. Proc. National Acad. Sci. (USA) **115**(34), 8505–8510 (2018)
7. Kingma, D.P., Ba, J.: Adam: a method for stochastic optimization. In: 3rd International Conference on Learning Representations, ICLR 2015 - Conference Track Proceedings. International Conference on Learning Representations, ICLR (2015)
8. Lagaris, I.E., Likas, A., Fotiadis, D.I.: Artificial neural networks for solving ordinary and partial differential equations. IEEE Trans. Neural Netw. **9**(5), 987–1000 (1998)
9. Lu, L., Meng, X., Mao, Z., Karniadakis, G.E.: DeepXDE: a deep learning library for solving differential equations, July 2019
10. Pang, G., Lu, L., Karniadakis, G.E.: FPINNs: fractional physics-informed neural networks. SIAM J. Sci. Comput. **41**(4), A2603–A2626 (2019)
11. Pathak, J., Hunt, B., Girvan, M., Lu, Z., Ott, E.: Model-free prediction of large spatiotemporally chaotic systems from data: a reservoir computing approach. Phys. Rev. Lett. **120**(2), 1 (2018)
12. Ramos, S., Gehrig, S., Pinggera, P., Franke, U., Rother, C.: Detecting unexpected obstacles for self-driving cars: fusing deep learning and geometric modeling. In: IEEE Intelligent Vehicles Symposium, Proceedings, pp. 1025–1032. Institute of Electrical and Electronics Engineers Inc., July 2017
13. Ruder, S.: An overview of gradient descent optimization algorithms. ArXiv e-prints. https://arxiv.org/abs/1609.04747 (2016)
14. Salman, A.G., Kanigoro, B., Heryadi, Y.: Weather forecasting using deep learning techniques. In: ICACSIS 2015 - 2015 International Conference on Advanced Computer Science and Information Systems, Proceedings, pp. 281–285. Institute of Electrical and Electronics Engineers Inc., February 2016
15. Salsa, S.: A Primer on PDEs : Models, Methods, Simulations. La Matematica per il 3+2, 1st edn. (2013)

Modeling Double Concentric Jets Using Linear and Non-linear Approaches

Juan A. Martín[1(✉)], Adrián Corrochano[1], Javier Sierra[2], David Fabre[2], and Soledad Le Clainche[1]

[1] School of Aerospace Engineering, Universidad Politécnica de Madrid, 28040 Madrid, Spain
juanangel.martin@upm.es

[2] Institut de Mécanique des Fluides de Toulouse, IMFT, Université de Toulouse, CNRS, 31400 Toulouse, France

Abstract. This article models the wake interaction between double concentric jets. The configuration is formed by a rounded jet surrounded by an external annular jet and is defined in a two-dimensional domain imposing axi-symmetric conditions. The flow is studied at laminar conditions (low Reynolds number) in three different cases based on the velocity of the two jets defined as U_i and U_e: case (i) $U_i = U_e$, case (ii) $2U_i = U_e$ and case (iii) $U_i = 2U_e$. Linear stability theory (LST) predicts the most unstable modes identifying a steady and an unsteady mode, both localized in the near field in the empty area between the two jets, forming a bubble. Neutral stability curves identify the critical Reynolds number for each test case, showing that this value is larger in case (iii) than in case (i), although the velocity of the inner jet in case (iii) is twice the velocity in case (i), suggesting that the flow bifurcation is delayed in case (iii). Finally, dynamic mode decomposition is applied to create a model for the non-linear solution of the concentric jets in case (i). The method retains the modes predicted by LST plus some other modes. Using these modes is possible to extrapolate the solution from the transitory of the numerical simulations to the attractor with error ∼2%, resulting in a reduction of the computational time in the numerical simulations of 50%.

Keywords: Double concentric jets · Linear stability analysis · Dynamic mode decomposition · Low-order model · Data forecasting

1 Introduction

Complex flows are found in a wide range of industrial and natural application, for this reason studying these types of flows is a research topic of high interest since the past [9,10]. Understanding the physical mechanism defining these flows is a starting point to create simple models that allows describing the flow complexity in a simplest and efficient manner. Then, it is possible to use these models to

Á. Herrero et al. (Eds.): SOCO 2020, AISC 1268, pp. 451–459, 2021.
https://doi.org/10.1007/978-3-030-57802-2_43

predict spatio-temporal patterns using computers, reducing in a notorious way the computational cost (memory and time).

Double concentric jets is a flow configuration, which represents a complex flow that models several industrial applications such as jet pumps, ejectors, industrial burners, et cetera, where the efficiency of such devices depend on the turbulent mixing between the two jets. Hence, creating a model of this flow that can be used for data forecasting is a research topic of high interest.

This article studies in detail the configuration of double concentric jets at low Reynolds number. Using a linear approach of the equations modeling the flow and linear stability analysis, or using a non-linear approach of such equations, and using the data-driven method dynamic mode decomposition, it is possible to identify the most relevant modes driving the flow motion. In this article we use these modes to create a low-order model of this complex flow. This is a starting point that intends to bring new ideas to create a soft computing model for future applications.

The article is organized as follows. Section 2 defines the problem of the double concentric jets. Sections 3 and 4 introduce the methodology used to study the problem: linear stability theory and dynamic mode decomposition (DMD). The main results for the linear stability analysis are presented in Sect. 5, while the model constructed using DMD and its performance is presented in Sect. 6. Finally, Sect. 7 summarizes the main conclusions.

2 Test Problem: Double Concentric Jets

We study the wake interaction of double concentric jets: a rounded jet defined in the inner part of the domain with radio $d/2 = 0.5$ and an annular jet surrounding the first jet with internal and external radii $D_i/2 = d/2 + L$ and $D_e/2 = 3d/2 + L$, respectively, where L is the distance between the two jets. At the studied flow conditions (laminar flow) the problem is axi-symmetric, thus the Navier-Stokes equations are solved in cylindrical coordinates in the two-dimensional mesh shown in Fig. 1. A 2D cartesian grid is employed to describe half of the computational domain, and axi-symmetric condition on the bottom is imposed. The origin of the Cartesian frame of reference is considered located on the inner jet, while the external jet is located at a distance $L = 1$ from the inner jet. The x-axis is chosen to be parallel to the incoming freestream velocity, while the y-axis with the cross-stream velocity.

Numerical simulations have been performed using two different numerical codes, StabFem [1] and Nek5000 [7]. On the one hand, StabFem is an open-source code that uses finite element method as spatial discretization and solves the linear (and in some cases non-linear) form of the (incompressible and compressible) Navier-Stokes equations (NSE) and is suitable to perform linear stability analysis in two-dimensional complex geometries (see more details in [6]). On the other hand, Nek5000 is an open source code that uses high-order spectral elements as spatial discretization, providing highly accurate solutions of the non-linear (and in some cases linear) form of the incompressible NSE in two-

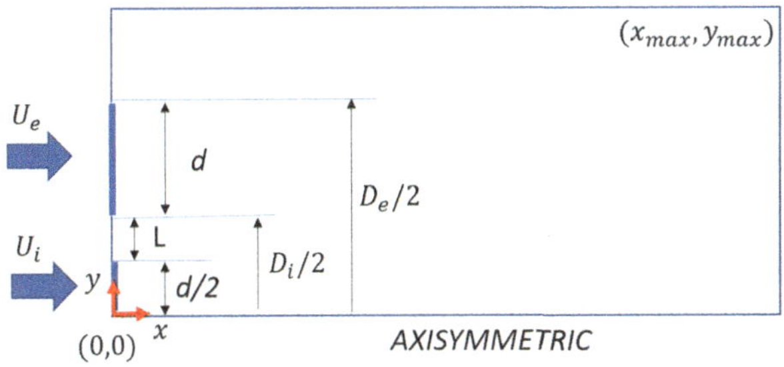

Fig. 1. Computational domain modeling two concentric annular jets.

and three-dimensional geometries. In both cases, NSE are non-dimensionalized and the Reynolds number is defined with the diameter of the internal jet, as $Re = Ud/\nu$.

In the numerical simulations performed with both numerical codes, the boundary conditions used at the entrance of the jets are freestream velocity U_i and U_e, for the inner and outer jets, respectively, and Neumann conditions for pressure. Zero velocity (wall boundary condition) is imposed on the remaining surfaces at the entrance of the computational domain. In the upper side and the outlet surfaces of the domain Neumann and Dirichlet boundary conditions are imposed for velocity and pressure, respectively. Three different test cases have been studied varying the value of the freestream velocity of the internal and external jet, named as cases (U_i, U_e): case $(1, 1)$, case $(1, 2)$ and case $(2, 1)$.

The size of the computational domain for the streamwise and normal coordinates are defined as $0 \leq x \leq 80$ and $0 \leq y \leq 60$, respectively. These dimensions have been set after carrying out a grid independence study (not shown for the sake of brevity) comparing the value of the streamwise and cross-stream velocities at several locations of the computational domain in four different cases with different dimensions and number of cells.

Firstly, to model the complex wake of two concentric jets we have used StabFem to perform the linear stability analysis, identifying two unstable modes (steady and unsteady), which define the first two flow bifurcations. Secondly, we have calculated the unsteady non-linear solution of NSE with Nek5000 and then we have constructed a model using the data-driven DMD method. The model is used to predict the attractor from a set of data collected in the transitory of the numerical simulations.

3 Linear Global Stability Analysis

Linear stability analysis theory considers as a path of transition the bifurcation process from laminar to turbulent flow. In fluid dynamics, Navier-Stokes equations (NSE) describe the movement of viscous Newtonian flows.

Linear stability theory (LST) studies the evolution of a small perturbation $\tilde{\mathbf{q}}(\mathbf{x}, t)$ imposed upon a base flow $\overline{\mathbf{q}}(\mathbf{x}, t)$, generally defined as a steady state for laminar flows [8,13] (although for unsteady solutions is considered as the mean flow [12]). Starting from the Reynolds decomposition of the instantaneous flow field $\mathbf{q}(\mathbf{x}, t)$ ($\mathbf{q}$ represents the velocity vector and pressure), defined as

$$\mathbf{q}(\mathbf{x}, t) = \overline{\mathbf{q}}(\mathbf{x}, t) + \tilde{\mathbf{q}}(\mathbf{x}, t), \tag{1}$$

it is possible to introduce this decomposition into the non-linear NSE and linearize these equation over the base flow, resulting in the linearized Navier-Stokes equations (LNSE).

These equations can be written as an initial-value-problem. Assuming that the perturbation can be separated between temporal and spatial coordinates, it is possible to introduce a Fourier decomposition in time as $\tilde{\mathbf{q}} = \hat{\mathbf{q}} e^{-i\lambda t}$, leading to the following generalized matrix eigenvalue problem (EVP)

$$\mathbf{B}\hat{\mathbf{q}} = \lambda \mathbf{A}\hat{\mathbf{q}}, \tag{2}$$

where the matrices $\mathbf{A}$ and $\mathbf{B}$ collect information regarding the boundary conditions of the problem. The eigenvalues λ, defined as $\lambda = \sigma + i\omega$, represent the frequency, ω, and growth rate, σ, of the most unstable modes, which are driving the flow motion. These are the modes with positive growth rates, called *flow instabilities*. The eigenvectors $\hat{\mathbf{q}}$ define the shape of the *unstable modes.*

4 Data Analysis Using Dynamic Mode Decomposition

Dynamic mode decomposition (DMD) [11] is a technique generally used for the analysis of non-linear dynamical systems and to identify coherent structures in complex flows. The method decomposes spatio-temporal data $\mathbf{v}_k = \mathbf{v}(t_k)$, where t_k is the time, as a temporal expansion of M Fourier-like modes $\mathbf{u_m}$, called as DMD modes, in the following way,

$$\mathbf{v}_k \simeq \sum_{m=1}^{M} a_m \mathbf{u_m} e^{(\sigma_m + i\omega_m)t_k}, \tag{3}$$

where a_m, σ_m and ω_m are the amplitudes, growth rates and frequencies of the modes. The data are equi-distant in time with time interval Δt.

For the analysis of complex data such as noisy experiments, transient or turbulent flows, an extension of the DMD algorithm is introduced, named as higher order dynamic mode decomposition (HODMD) [4]. This is the algorithm used for the analysis of the data presented in this article.

5 Creating a Simple Linear Model

Linear stability analysis has been carried out to analyze the three cases defined in Sect. 2 as $(U_i, U_e) = (1, 1)$, $(1, 2)$ and $(2, 1)$, of the double concentric jets. Figure 2 shows the base flow, where it is possible to follow the evolution of the

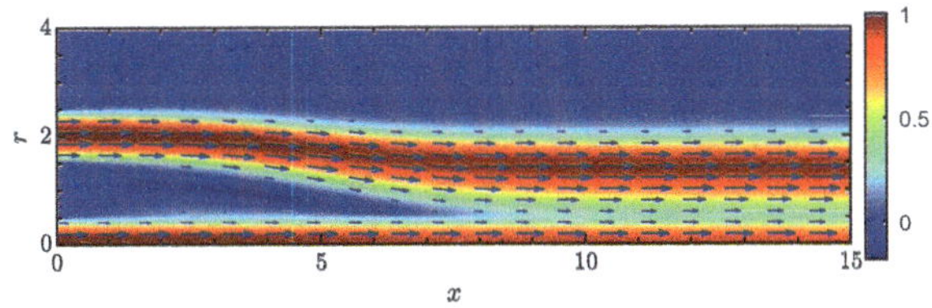

Fig. 2. Base flow in two concentric annular jets. Colormap of streamwise velocity. Arrows indicate the intensity of the streamwise velocity

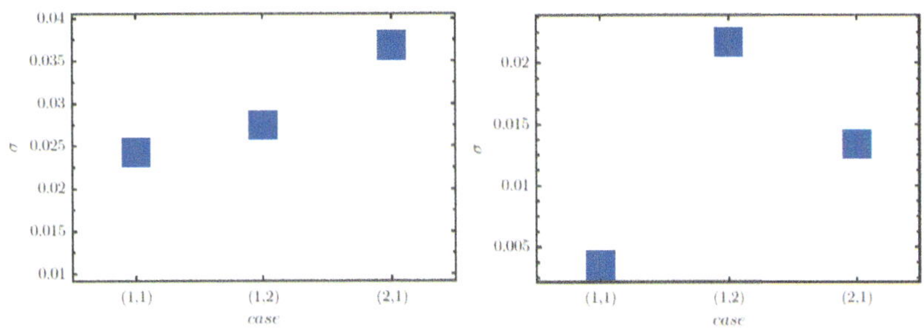

Fig. 3. Growth rate σ of the most unstable modes identified by linear stability analysis. Left: steady mode S_1. Right: unsteady mode F_1.

two jets starting at different positions and finally mixing up at $x \simeq 8$ forming a single thicker wake downstream this position. Upstream this point, we identify a region with zero velocity between the two jets.

Two different unstable modes have been identified in the linear stability analysis, the steady mode S_1 and the unsteady mode F_1, which is a low frequency mode, with $\omega \simeq 0.08$. Figure 3 shows the value of the modes S_1 and F_1, for the three different configurations studied. As expected, the mode becomes more unstable (largest growth rate) rising the velocity of one of the jets. The most unstable case in S_1 occurs in case $(2,1)$ when the velocity of the inner jet is twice the velocity of the external jet. On the contrary, the most unstable case in F_1 occurs with the highest velocity of the external jet, in case in case $(1,2)$. The frequency (figure not shown for the sake of brevity) also slightly increases with the rise in the jet velocity, finding the largest value in case $(2,1)$.

Neutral stability curves have been performed to identify the critical Reynolds number (when the growth rate is zero) of the previous modes with the aim of showing the flow conditions at which these modes become unstable. Figure 4 shows that the critical Reynolds number of mode S_1 for the case $(1,1)$ is $\sim$370, while for the mode F_1 is $\sim$1800. This fact suggests that the topology of the base flow changes at $Re \simeq 380$ but the flow remains steady until the second flow bifurcation, which is given at $Re \simeq 1800$. Similar behavior is found in the remaining two cases, where the order of magnitude of the critical Re identified for the modes S_1 and F_1 is similar. Regarding the differences between the three cases analyzed, both modes become unstable at lower Reynolds number in case $(1,2)$

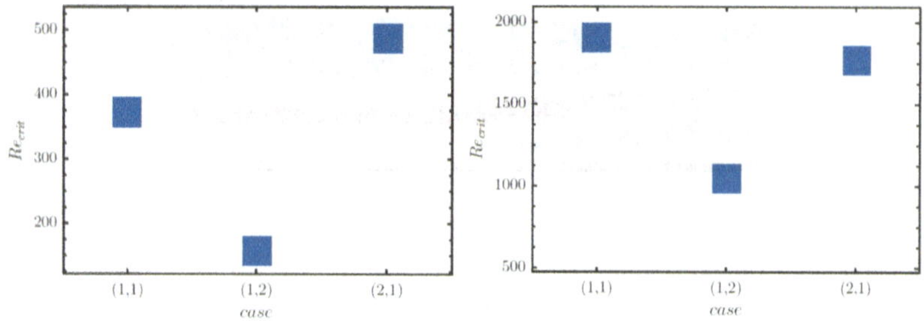

Fig. 4. Critical Reynolds number (Re_{crit}) at which the value of the growth rate in modes S_1 (left) and F_1 (right) is zero in the three cases analyzed.

compared to case $(1, 1)$. This is a consequence of the velocity rise, producing a more complex flow at lower Reynolds number, overtaking the flow transition. On the contrary, the critical Reynolds number in case $(2, 1)$ is larger than in case $(1, 1)$. This result is unexpected, since the velocity rise in the internal jet produces an effect contrary to the increase in flow complexity, but it delays the flow bifurcation. In other words, using an external annular jet with half velocity value than the internal jet produces a mechanism for flow control, increasing the critical Reynolds number and consequently delaying the flow transition.

The shape of the modes S_1 and F_1 is presented in Figure 5. The highest intensity of these modes is located in the region between the two jets. The steady mode is formed by a large size bubble, probably related to some changes in the topology of the base flow that needs to be studied more in detail in future works. The flow becomes unsteady due to the presence of the mode F_1, triggering the oscillations of the bubble.

6 Creating a Complex Non-linear Model

With the aim at modeling the complex interaction between double concentric jets, the non-linear solution of the flow is analyzed at flow conditions slightly above the critical Reynolds number. The prediction of the critical Reynolds number carried out by the linear theory presented in the previous section have been validated performing numerical simulations solving the non-linear solution of Navier-Stokes equations with the solver Nek5000. The simulations converge to a steady solution for values of $Re \leq 340$, while the solution is unsteady for $Re \geq 345$, thus the critical Re is approximated as $Re_{crit} \simeq 342$. This value is slightly smaller than the one obtained in the linear stability analysis, which is consistent with the non-linear approximation.

One of the main drawbacks of solving the non-linear equations is the large computational cost compared to the linear solution. The order of magnitude of the computational time of the linear analysis to identify global stability modes is minutes, while using Nek5000 it is necessary to use several hours of computational time to converge the solution to the attractor, which is approximated at

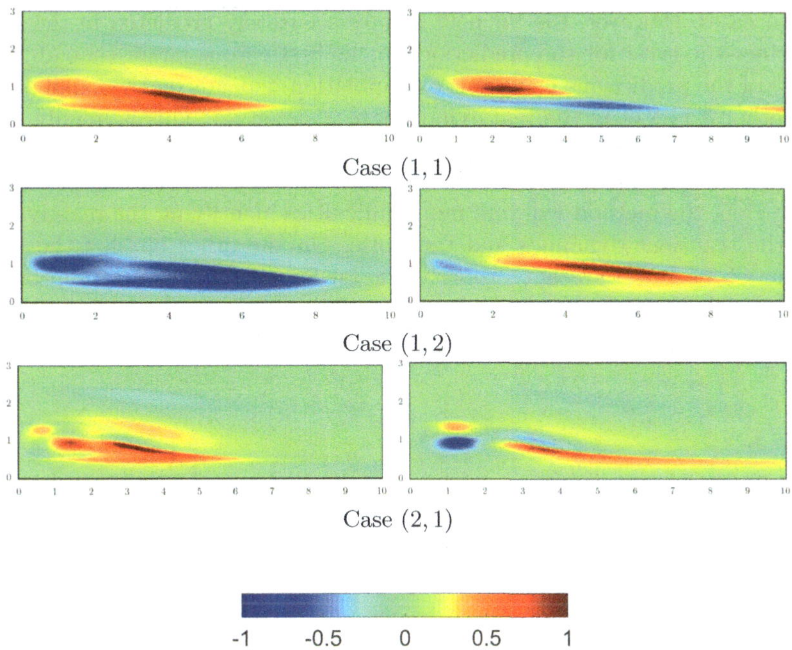

Fig. 5. Most unstable modes S_1 (left) and F_1 (right). Real component of streamwise velocity. The modes are normalized with their maximum value.

time ~1000. To overcome such issue, we propose a model for data forecasting using DMD in the case (1, 1), but this method could be extended to the two other cases presented in the previous section, cases (1, 2) and (2, 1).

DMD is applied to a group of data collected in the transitory of the numerical simulation, to calculate the DMD expansion (3). The growth rate of the modes is set to 0, and the temporal term, t_k, of the expansion is adjusted to a time interval representing the attractor. Two different test cases have been carried out. In the first case 40 snapshots have been collected in the time interval $t \in [105, 300]$, and in the second case 14 snapshots have been collected in the interval $t \in [340, 415]$. The model presented in Eq. (3) is constructed using $M = 20$ and $M = 12$ modes for the first and second case, respectively, and the solution is extrapolated to the attractor defined in the time interval $t \in [800, 1000]$. The speed-up factor representing the reduction of the computational time for the numerical simulations is $1000/300 \simeq 3.33$ (~75%) and $1000/415 \simeq 2$ (~50%), and the root mean square error of these predictions is $\sim 2.4 \cdot 10^{-1}$ and $\sim 2.2 \cdot 10^{-2}$ for the first and second cases, respectively. In both cases, the modes S_1 and F_1 predicted by the linear theory are included in the DMD expansion (3), but the larger complexity of this non-linear solution makes necessary retaining a larger quantity of modes to predict the attractor with such small error. Figure 6 shows that the method predicts with relatively high accuracy the near field of the double concentric jets in both cases. However, the far field is only accurately predicted in the second case, since the error for the predictions in the first case is larger than 20%. The quality of the predictions using this extrapolation depends on the capabilities of the method to identify the real dynamics in a signal [5]. This is mainly dependent on (i) the quality of the data and (ii) the setting parameters of the method [2,3]. On the one hand, if the data are noisy or they are representing a transient region, the method will find more difficulties identifying the real dynamics from the transient dynamics and the noise. On the other hand, f the setting parameters of DMD are not properly chosen for the analysis (minimizing the reconstruction error), the method will provide spurious information that will alter its good performance. See the references [2,3,5] for more information.

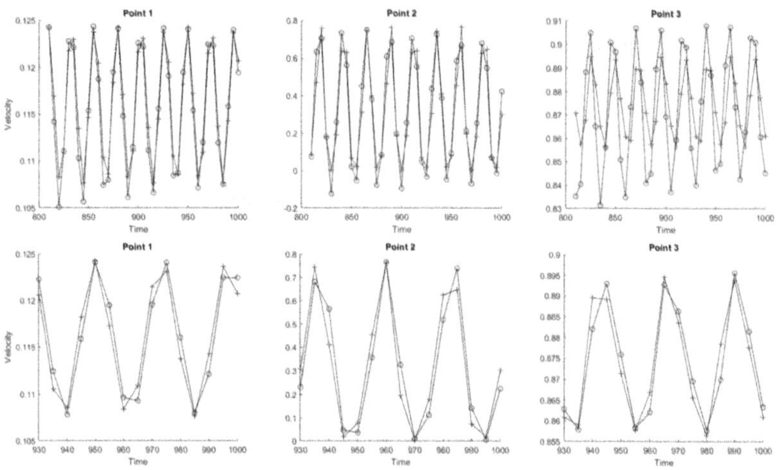

Fig. 6. Predictions of the attractor using the DMD expansion (3). From left to right: temporal evolution of streamwise velocity at points $(x, y) = (2, 0.5)$, $(2, 3.5)$ and $(6, 0.5)$. Data collected in the time interval $t \in [105, 300]$ (top) and $t \in [340, 415]$ (bottom).

7 Conclusions

This article models the wake of double concentric jets in laminar regime using axi-symmetric flow conditions. Depending on the velocity at the entrance of the jets, defined as U_i and U_e for the internal and external jets, respectively, three different test cases have been studied: cases $(U_i, U_e) = (1, 1)$, $(1, 2)$ and $(2, 1)$. Linear stability analysis has been applied to identify the main instabilities driving the flow motion, identifying a steady mode and an unsteady mode as first and second flow bifurcations. Neutral stability curves predict the critical Reynolds number for the transition from steady to unsteady flow, finding that, compared to the reference case $(1, 1)$, the presence of the flow bifurcation is overtaken in the case $(1, 2)$, in good agreement with the rise in flow complexity due to the higher velocity value used at the entrance of the external jet. However, the presence of the flow bifurcation is delayed in the case (2, 1), which is a result unexpected.

Finally, a model is constructed in the case (1, 1) applying DMD to a group of data collected in the transitory of the numerical simulations, which were solving the non-linear solution of this problem. The method identifies several modes, including the modes predicted by the linear theory. A DMD expansion is constructed using these modes, and the solution is extrapolated in time. This model predicts the attractor with error $\sim$2% for the near field, reducing the computational time for the numerical simulations $\sim$50%.

Acknowledgements. The work of Juan A. Martín been supported by the Spanish Ministerio de Ciencia, Innovación y Universidades under grant DPI2017-84700-R.

References

1. Stabfem. https://www.gitlab.com/stabfem/StabFem
2. Clainche, S.L.: Prediction of the optimal vortex in synthetic jets. Energies **12**(9), 1635–1661 (2019)
3. Clainche, S.L., Ferrer, E.: A reduced order model to predict transient flows around straight bladed vertical axis wind turbines. Energies **11**(3), 566–586 (2018)
4. Clainche, S.L., Vega, J.: Higher order dynamic mode decomposition. SIAM J. Appl. Dyn. Syst. **16**(2), 882–925 (2017)
5. Clainche, S.L., Vega, J.: Higher order dynamic mode decomposition to identify and extrapolate flow patterns. Phys. Fluids **29**(8), 084102 (2017)
6. Fabre, D., Citro, V., Sabino, D.F., Bonnefis, P., Sierra, J., Gianneti, F., Pigou, M.: A practical review on linear and nonlinear global approaches to flow instabilities. Appl. Mech. Rev. **70**(060802), 1–16 (2018)
7. Fischer, P.F., Lottes, J.W., Kerkemeier, S.G.: nek5000 Web page (2008). http://nek5000.mcs.anl.gov
8. Gomez, F., Clainche, S.L., Paredes, P., Hermanns, M., Theofilis, V.: Four decades of studying global linear instability: problems and challenges. AIAA J. **50**(12), 2731–2743 (2012)
9. Haller, G.: An objective definition of a vortex. J. Fluid Mech. **525**, 1–26 (2005)
10. Hunt, J.C.R., Wray, A., Moin, P.: Eddies, stream, and convergence zones in turbulent flows. Center for Turbulence Research Report CTR-S88 (1988)
11. Schmid, P.: Dynamic mode decomposition of numerical and experimental data. J. Fluid Mech. **656**, 5–28 (2010)
12. de Segura, G., García-Mayoral, R.: Turbulent drag reduction by anisotropic permeable substrates analysis and direct numerical simulations. J. Fluid Mech. **875**, 124–172 (2019)
13. Theofilis, V.: Advances in global linear instability analysis of nonparallel and three-dimensional flows. Prog. Aerosp. Sci. **39**, 249–315 (2003)

Unsupervised Data Analysis of Direct Numerical Simulation of a Turbulent Flame via Local Principal Component Analysis and Procustes Analysis

Giuseppe D'Alessio[1,2,3(✉)], Antonio Attili[4], Alberto Cuoci[2], Heinz Pitsch[4], and Alessandro Parente[1,3]

[1] Aero-Thermo-Mechanics Laboratory, Université Libre de Bruxelles, Bruxelles, Belgium
giuseppe.dalessio@ulb.ac.be
[2] CRECK Modeling Lab, Department of Chemistry, Materials and Chemical Engineering, Politecnico di Milano, Piazza Leonardo da Vinci 32, 20133 Milano, Italy
[3] Combustion and Robust Optimization Group (BURN), Université Libre de Bruxelles and Vrije Universiteit Brussel, Brussels, Belgium
[4] Institute for Combustion Technology, RWTH Aachen University, 52056 Aachen, Germany

Abstract. Direct Numerical Simulations (DNS) of reacting flows provide high-fidelity data for combustion model reduction and validation, although their interpretation is not always straightforward because of the massive amount of information and the data high-dimensionality.

In this work, a completely unsupervised algorithm for data analysis is investigated on a data-set obtained from a temporally-evolving DNS simulation of a reacting n-heptane jet in air. The proposed algorithm combines the Local Principal Component Analysis (LPCA) clustering algorithm with a variables selection algorithm via dimensionality reduction and Procustes Analysis. Unlike other data-analysis algorithms, it requires null or limited user expertise as all of its steps are unsupervised and solely entrusted to mathematical objective functions, without any hyperparameter tuning step required.

Keywords: Data analysis · Local variables selection · Principal Component Analysis · Direct Numerical Simulation · Turbulent flame

1 Introduction

Combustion data obtained from high-fidelity numerical simulations such as Direct Numerical Simulations (DNS) are routinely used for model development and validation, as well as for the understanding of chemical and physical processes. In any case, the first step is always the analysis of the massive amount

Á. Herrero et al. (Eds.): SOCO 2020, AISC 1268, pp. 460–469, 2021.
https://doi.org/10.1007/978-3-030-57802-2_44

of information that large-scale simulations produce, as they are usually characterized by a large number of statistical observations and several variables. Many data-driven approaches are available in literature and have been tested on combustion data, such as linear and non-linear dimensionality reduction techniques, i.e. Principal Component Analysis (PCA, also referred to as Proper Orthogonal Decomposition), Autoencoders (AE), Kernel Principal Component Analysis (KPCA), Isomap and Dynamic Mode Decomposition (DMD) [1–6], as well as techniques for high-dimensionality space exploration and visualization, such as Self Organizing Maps (SOMs) and t-SNE [7–10]. Although the effectiveness of these techniques is not questioned, as they have all proved to be effective in extracting information from data, their common limitation is related to the physical interpretation of the features, which is, in all the mentioned cases, not driven by objective criteria, but entrusted to the sole user experience, constituting a limitation to the algorithm analysis potential and extendibility. For some of these algorithms, such as Autoencoders and t-SNE, good performances can be obtained only after an accurate tuning of the hyperparameters, for which a thorough sensitivity analysis or a significant user expertise are required. For other algorithms, the applicability to combustion is limited because of their intrinsic linearity (PCA and DMD), or because of their CPU-intensive nature (KPCA and Isomap).

In this work, a local unsupervised algorithm for data analysis, which combines the effectiveness of the Local Principal Component Analysis (LPCA) algorithm input-space partitioning [11–13] and an automatic variables selection criterion via dimensionality reduction and Procustes Analysis [1,14], is tested on a dataset representing a 2D slice of a 3D temporally-evolving DNS simulation of a reacting n-heptane jet [15–18]. The algorithm's performances were assessed comparing the selected main local principal variables (LPVs) with the ones obtained by means of another well-known method used for data analysis and feature extraction, which exploits the rotation of the local principal components with the Varimax criterion [19]. The main advantage of the proposed local analysis is that both the partitioning and the LPV selection steps are accomplished according to mathematical criteria, not requiring any hyperparameter tuning, and no dependence from user expertise.

2 Theory

2.1 Variable Selection via Principal Component Analysis and Procustes Analysis

The PCA is a dimensionality reduction technique based on the eigenvalue-decomposition of a covariance matrix [1]. Given a matrix $\mathbf{X} \in \mathbb{R}^p$, consisting of n statistical observations of p variables, it is possible to compute the associated covariance matrix according to Eq. 1, which can be then decomposed by means of Eq. 2:

$$\mathbf{S} = \frac{1}{n-1}\mathbf{X}^T\mathbf{X}, \tag{1}$$

$$\mathbf{S} = \mathbf{A}\mathbf{L}\mathbf{A}^T. \tag{2}$$

The matrix $\mathbf{A}$ is an orthonormal basis of eigenvectors, the Principal Components (PCs), while $\mathbf{L}$ is a diagonal matrix of eigenvalues. The PCs are a linear combination of the original variables, and the dimensionality reduction is possible considering a subset of q eigenvectors, with $q < p$, associated to the q most powerful eigenvalues, such that the information loss is minimized.

In many applications, rather than reducing the dimensionality considering a new set of coordinates which are linear combination of the original ones, the main interest is to achieve a dimensionality reduction selecting a subset of m variables from the original set of p variables. One of the possible ways to accomplish this task is to couple the PCA dimensionality reduction with a Procustes Analysis [1,14]. To do that, PCA is firstly applied to the full data matrix $\mathbf{X} \in \mathbb{R}^p$, and a score matrix $\mathbf{Z} \in \mathbb{R}^q$ is obtained projecting the matrix $\mathbf{X}$ on the q-dimensional manifold spanned by the retained PCs:

$$\mathbf{Z} = \mathbf{X}\mathbf{A}. \tag{3}$$

After that, a subset consisting of m variables, with $q < m < p$, can be selected from the original matrix, thus obtaining the reduced matrix $\tilde{\mathbf{X}} \in \mathbb{R}^m$. At this point, PCA is applied to $\tilde{\mathbf{X}}$, and a scores matrix $\tilde{\mathbf{Z}} \in \mathbb{R}^q$ is obtained also in this case. If the choice of the m variables is done correctly, the discrepancies between the two scores matrices $\mathbf{Z}$ and $\tilde{\mathbf{Z}}$ are minimal, while there are significant differences otherwise [14]. A Procustes Analysis is thus carried out in order to quantitatively measure the similarity between the two matrices, calculating the sum of the squared differences between the points of $\mathbf{Z}$ and $\tilde{\mathbf{Z}}$. It consists in the computation of the M^2 coefficient:

$$M^2 = Tr(\mathbf{Z}\mathbf{Z}' + \tilde{\mathbf{Z}}\tilde{\mathbf{Z}}' - 2\mathbf{\Sigma}), \tag{4}$$

where $\mathbf{\Sigma}$ is the matrix of the singular values obtained from the decomposition of the square matrix $\tilde{\mathbf{Z}}'\mathbf{Y}$:

$$\tilde{\mathbf{Z}}'\mathbf{Z} = \mathbf{U}\mathbf{\Sigma}\mathbf{V}'. \tag{5}$$

By means of the minimization of M^2 as objective function, it is possible to build an iterative algorithm to select, in a totally unsupervised fashion, the best subset of m variables from the original set of p variables, as described in [14]:

1. The dimensionality of m is initially set equal to p.
2. Each variable is deleted from the matrix $\mathbf{X}$, obtaining p $\tilde{\mathbf{X}}$ matrices. The corresponding scores matrices $\tilde{\mathbf{Z}}$ are computed by means of PCA. For each of them, a Procustes Analysis is performed as in Eq. 4 with respect to the scores of the original matrix $\mathbf{X}$, and the corresponding M^2 coefficient is computed.
3. The variable which, once excluded, leads to the smallest M^2 coefficient is deleted from the $\tilde{\mathbf{X}}$ matrix.
4. Steps 2 and 3 are repeated until m variables are left, thus obtaining the reduced $\tilde{\mathbf{X}} \in \mathbb{R}^m$ matrix.

2.2 Unsupervised Data Analysis via Local Principal Variables

The coupling between PCA and the Procustes Analysis, proposed by Krzanovsky to select the main variables to preserve the multivariate data structure [14], can be easily extended to a local version by means of the LPCA clustering. The latter is an unsupervised algorithm to partition statistical observations in a high-dimensional space in clusters ($\mathbf{C}_i$, with $i \in [1, ..., k]$) via vector quantization (VQ), and after that the dimensionality reduction task is locally accomplished. This method has already been successfully applied in combustion for clustering purposes [20] as well as for model reduction [12,13]. The objective function for the unsupervised space partitioning is the PCA reconstruction error (ϵ_r), which is defined as:

$$\epsilon_r = ||\mathbf{x} - \tilde{\mathbf{x}}||, \tag{6}$$

where the vectors $\mathbf{x}$ and $\tilde{\mathbf{x}}$ in Eq. 6 represent the original and the reconstructed (from the reduced manifold) vectors, respectively. If data are partitioned in k clusters, and in each of them PCA is performed, it is possible to find k reduced basis of eigenvectors (LPCs) $\mathbf{A}^{(j)} \in \mathbb{R}^q$, with $j \in [1, ..., k]$. Thus, for each observation $\mathbf{x}$ of the data-set $\mathbf{X} \in \mathbb{R}^p$ it is possible to iteratively compute k reconstruction errors and assign it to a cluster $\bar{k}$, such that:

$$\bar{k} \mid \epsilon_{r,\bar{k}} = \min_{j=1,...k} \epsilon_{r,j}, \tag{7}$$

until the error variation for the reconstruction of the full data matrix $\mathbf{X}$ is below a fixed threshold. Considering k local sets of PCs ($\mathbf{A}^{(j)} \in \mathbb{R}^q$, with $j \in [1, ..., k]$), the errors arising from the dimensionality reduction are lowered with respect to the global PCA. The local method is piecewise-linear and not globally linear, thus being effective also for non-linear applications. Moreover, the possibility to select *locally* relevant variables can be more attractive from both data analysis and model development perspective. Locally optimized combustion reduced models have already proved to have several advantages with respect to global reduced models [20], as subsets of variables which are locally more coherent with the physics can be extracted from each group.

The algorithm has the following steps:

1. *Partitioning of the input space in clusters:* the thermochemical space is partitioned in k clusters via minimization of the reconstruction error.
2. *LPCs and local scores computation:* in each cluster $\mathbf{C}_i$ ($i \in [1, ..., k]$) found in the partitioning step, a local set of LPCs $\mathbf{A}^{(i)} \in \mathbb{R}^q$ is computed, and the corresponding local scores matrices $\mathbf{Z}_i$ are computed by projection of the clusters' points on the local reduced manifold.
3. *Local variables selection:* The variables needed to preserve the local multivariate structure are retained by means of the Krzanovsky algorithm [14].

3 Case Description

The data chosen to test the proposed algorithm were obtained from a 2D slice of a 3D temporally evolving DNS simulation of a n-heptane jet [15–18]. The fuel jet is nitrogen-diluted (85% in volume) at 400K, arranged in a coflow configuration with the oxidizer stream (air) at 800K. The turbulent jet is initialized with a Reynolds' number equal to 15,000, and a layer at stoichiometric composition is inserted in the region of smooth transition between the fuel and the oxidizer. Both the gas phase hydrodynamics and combustion were modeled using a reactive unsteady Navier-Stokes equation formulation within the low Mach number limit [21]. For the resolution of the gas velocity field, as well as for the reactive scalar fields, a finite-differences scheme was chosen [22], while the advection-reaction equations for soot moments were solved by means of a Lagrangian particle method [23,24]. Open boundary conditions were prescribed in the normal direction to the flame sheet in order to have a mass outflow for the combustion products, while periodic boundary conditions were imposed in the other two directions. The adopted kinetic mechanism for the n-heptane oxidation was reduced to 47 species and 290 reactions [25] from the detailed one developed by Blanquart et al. [26].

The 2D slice of the simulation considered for the analysis consisted of 1,048,576 grid points, each of them characterized by a thermochemical vector ϕ of temperature and 47 chemical species mass fractions. The data were organized as a matrix whose dimensions were $n \times p$, accounting for 1,048,576 observations of the 48 thermochemical variables.

4 Results

The local principal variables algorithm was applied to the data described in Sect. 3, with $k = 16$. In each cluster, the variables which were able to preserve the local multivariate structure (LPVs) were chosen according to a Procustes Analysis applied to the local scores matrices. In Fig. 1, the results obtained from the n-heptane jet clustering via LPCA are shown.

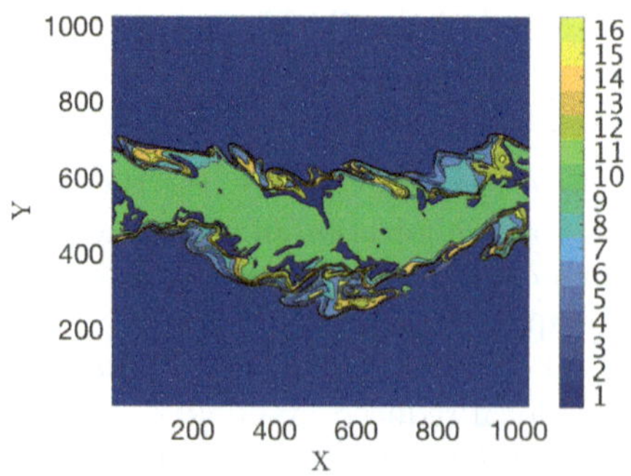

Fig. 1. LPCA unsupervised partitioning of the selected 2D slice of the 3D DNS simulation with 16 clusters.

In each cluster, a range from 4 to 9 chemical species was retained, according to the local manifold dimensionality, within the original 47 species implemented in the chemical mechanism. By means of this variable selection process, it was possible to easily interpret, from a physical point of view, the results obtained from the clustering process, as many subset of variables resulted to be chemically coherent. For example, a subset containing 4 variables (oxygen radical, hydroxyl radical, hydrogen radical, hydroperoxy radical) was identified in cluster number 2. All of these variables are involved in the oxygen branching reactions and are H-atom abstractors, a key step in the PAH formation. In several clusters, the selected LPVs were the ones involved in the soot formation as they consisted of mainly PAHs, such as in cluster number 5, 6 and 9. In Table 1, the LPVs selected by the algorithm in each cluster, according to the Procustes Analysis, are reported.

Table 1. Number of the cluster with the corresponding selected LPVs and coefficient of participation (ψ).

k	LPVs	ψ
1	CH_2O, CH_4, C_3H_6, C_4H_8, C_5H_6, A_1CHO	0.67
2	O, H, OH, HO_2	1
3	CH_2, HCO, C_2H_3, C_2H, HCCO, A_1^-, A_1CH_2	0.85
4	CH_2, O, CH, HCO	1
5	CH_3, A_2, A_1CH_2, $A_1C_2H^*$, A_2^-, A_1C_2H	0.67
6	CH_2, C_2H, A_1^-, $A_1C_2H^*$	0.75
7	A-C_3H_4, A_1, C_5H_6, A_2, $A_1C_2H_2$, A_2^-, A_1C_2H	0.71
8	CH_2O, C_2H_5, C_4H_8, C_5H_{11}, A_1CHO, C_7H_{15}	0.50
9	A_1^-, A_2^-, A_2, A_1CH_2, A_1C_2H	0.8
10	C_2H_6, C_4H_8, C_5H_{10}, A_1CHO	0.5
11	HO_2, HCO, CH_2O, CH_3, n-C_3H_7, C_7H_{15}	0.67
12	CH, HCO, C_2H, HCCO	1
13	CH_4, A-C_3H_5, C_4H_8, C_5H_6, C_5H_{11}, $A_1C_2H_2$, A_1CHO, C_7H_{15}	0.62
14	CH_2O, CH_3, C_2H_3, A-C_3H_5, n-C_3H_7, $A_1C_2H_2$, A_1CH_2	0.85
15	CH_2O, A-C_3H_5, n-C_3H_7, C_5H_{11}, $A_1C_2H_2$, C_7H_{15}	0.67
16	CH_2O, C_2H_3, A-C_3H_5, n-C_3H_7, A-C_3H_4, C_5H_6, $A_1C_2H_2$, A_1C_2H	0.75

A first, qualitative, assessment of the data analysis algorithm performances can be done comparing the maps of the local principal variables with the cluster shapes. In Fig. 2, the phenyl radical ($A1^-$) concentration map is compared with the shape of cluster number 9 (colored in yellow), where this species results to be a LPV. The maximum phenyl radical concentration values and gradients are placed in correspondence of the considered cluster, meaning that a correct variable was identified by the algorithm. Since a qualitative comparison by means of the contours shapes alone cannot be considered to be robust enough to evaluate the algorithm's performances in terms of data analysis, a quantitative assessment was carried out.

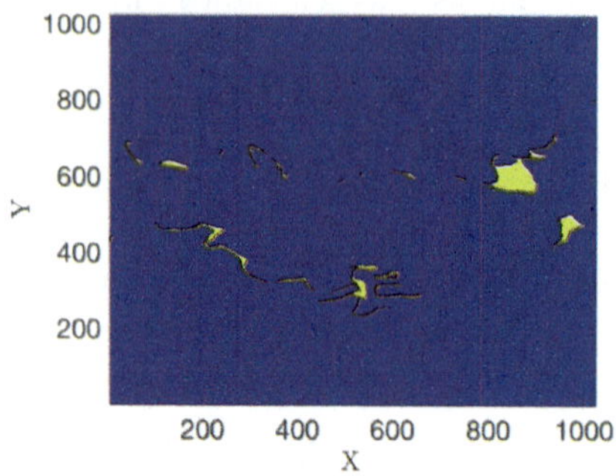

(a)

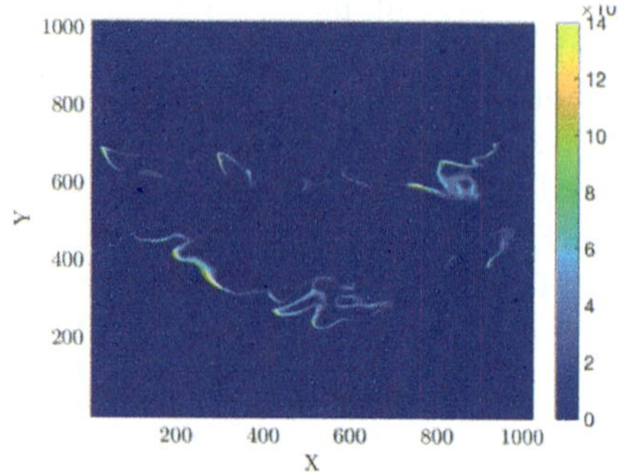

(b)

Fig. 2. (a) Cluster number 9 (in yellow) identified by means of the LPCA unsupervised partitioning algorithm applied to the DNS data, with $k = 16$; (b) phenyl radical ($A1^-$) map of concentration for the selected 2D slice of the 3D DNS simulation.

An assessment of the data analysis algorithm was done carrying out a comparison between the extracted LPVs and the variables which were considered important by another data analysis algorithm. The LPVs were compared with the variables having the highest weights on the LPCs, when rotated with the Varimax criterion. When PCA or LPCA are performed for data analysis tasks, the weights on the PCs must be visually inspected and interpreted, but it can easily happen that large weights are distributed on the eigenvectors over several variables, thus making impossible to associate the PC to a particular variable, nor a physical or chemical process. By means of rotation, instead, the PCs tend to align with only one or few variables, making their physical interpretation easier, as observed in [19]. A coefficient of participation ψ can be defined to represent the fraction of the LPVs having also the largest weight on the rotated LPCs, thus defined as the ratio between the number of LPVs found with largest weight on a rotated LPC in the considered cluster, and the total number of LPVs in that cluster:

$$\psi = \frac{N_{LPVs \in LPCs}}{N_{LPVs,tot}} \tag{8}$$

This coefficient can take values between zero and one, being equal to zero if the variables extracted by the two algorithms are completely different, and equal to one otherwise. Analyzing the ψ coefficients reported in Table 1 it is clear that, except for clusters number 8 and 10, all the PVs were found on the rotated LPCs. In particular, in clusters number 2, 4 and 12, all the selected LPVs were found to be important also by means of the rotation of the LPCs. Obtaining similar results by means of the two data analysis techniques is particularly relevant, as the analysis with the proposed local principal variables algorithm was achieved in an unsupervised fashion, without any visual inspection of the weights to be required. This is a considerable strength of the proposed algorithm, as it is possible to analyze massive data also using many clusters, a task which would result to be unfeasible if the visual inspection of the first q PCs in each cluster would be required.

5 Conclusions

In this work, an algorithm for local unsupervised data analysis was proposed and tested on a massive dataset obtained from a DNS simulation of a n-heptane reacting jet. The method consists of two steps. The first one is the data-set partitioning in different clusters, accomplished via the LPCA algorithm. After that, in each cluster the main variables are selected by means of an iterative algorithm for variables selection employing a Procustes Analysis.

A quantitative assessment of the algorithm' performances was carried out comparing the variables selected by means of the proposed algorithm with the ones selected by the rotation of the local principal components, and a satisfactory agreement was observed in all the clusters between the variables selected by the two algorithms. This result is particularly relevant, as it paves the way to the possibility to use a completely unsupervised tool to analyze the data, without any visual inspection nor interpretation of the weights.

The proposed algorithm for the local data analysis can constitute a functional tool aiding for the development and the validation of local reduced order models from DNS data. In fact, the formulation of local reduced order models has already shown to have several advantages over the global one, for example in the context of adaptive-chemistry simulations and the development of digital twins [20,27].

Acknowledgments. The first author acknowledges the support of the Fonds National de la Recherche Scientifique (FRS-FNRS) through a FRIA fellowship. A.A. and H.P. acknowledge funding from the European Research Council (ERC) under the European Union's Horizon 2020 research and innovation program under grant agreement No 695747. A.P. acknowledges funding from the European Research Council (ERC) under the European Union's Horizon 2020 research and innovation program, grant agreement No 714605.

References

1. Jolliffe, I.: Principal component analysis. In: Lovric, M. (ed.) International Encyclopedia of Statistical Science. Springer, Heidelberg (2011)
2. Sutherland, J.C., Parente, A.: Combustion modeling using principal component analysis. Proc. Combust. Inst. **32**(1), 1563–1570 (2009)
3. Sakurada, M., Yairi, T.: Anomaly detection using autoencoders with nonlinear dimensionality reduction. In: Proceedings of the MLSDA 2014 2nd Workshop on Machine Learning for Sensory Data Analysis, p. 4. ACM (2014)
4. Mirgolbabaei, H., Echekki, T., Smaoui, N.: A nonlinear principal component analysis approach for turbulent combustion composition space. Int. J. Hydrogen Energy **39**(9), 4622–4633 (2014)

5. Bansal, G., Mascarenhas, A.A., Chen, J.H.: Identification of intrinsic low dimensional manifolds in turbulent combustion using an Isomap based technique. Technical report, Sandia National Lab (SNL-CA), Livermore, CA (United States) (2011)
6. Grenga, T., MacArt, J.F., Mueller, M.E.: Dynamic mode decomposition of a direct numerical simulation of a turbulent premixed planar jet flame: convergence of the modes. Combust. Theory Model. **22**(4), 795–811 (2018)
7. Liukkonen, M., Hiltunen, T., Hälikkä, E., Hiltunen, Y.: Modeling of the fluidized bed combustion process and NOx emissions using self-organizing maps: an application to the diagnosis of process states. Environ. Model. Softw. **26**(5), 605–614 (2011)
8. Blasco, J.A., Fueyo, N., Dopazo, C., Chen, J.Y.: A self-organizing-map approach to chemistry representation in combustion applications. Combust. Theory Model. **4**(1), 61–76 (2000)
9. Fooladgar, E., Duwig, C.: Identification of combustion trajectories using t-distributed stochastic neighbor embedding (t-SNE). In: Salvetti, M., Armenio, V., Fröhlich, J., Geurts, B., Kuerten, H. (eds.) Direct and Large-Eddy Simulation XI, pp. 245–251. Springer, Cham (2019)
10. Fooladgar, E., Duwig, C.: A new post-processing technique for analyzing high-dimensional combustion data. Combust. Flame **191**, 226–238 (2018)
11. Kambhatla, N., Leen, T.K.: Dimension reduction by local principal component analysis. Neural Comput. **9**(7), 1493–1516 (1997)
12. Parente, A., Sutherland, J.C., Dally, B.B., Tognotti, L., Smith, P.J.: Investigation of the mild combustion regime via principal component analysis. Proc. Combust. Inst. **33**(2), 3333–3341 (2011)
13. Parente, A., Sutherland, J.C., Tognotti, L., Smith, P.J.: Identification of low-dimensional manifolds in turbulent flames. Proc. Combust. Inst. **32**(1), 1579–1586 (2009)
14. Krzanowski, W.J.: Selection of variables to preserve multivariate data structure, using principal components. J. Roy. Stat. Soc. Ser. C (Appl. Stat.) **36**(1), 22–33 (1987)
15. Attili, A., Bisetti, F., Mueller, M.E., Pitsch, H.: Formation, growth, and transport of soot in a three-dimensional turbulent non-premixed jet flame. Combust. Flame **161**(7), 1849–1865 (2014)
16. Attili, A., Bisetti, F., Mueller, M.E., Pitsch, H.: Effects of non-unity Lewis number of gas-phase species in turbulent nonpremixed sooting flames. Combust. Flame **166**, 192–202 (2016)
17. Attili, A., Bisetti, F., Mueller, M.E., Pitsch, H.: Damköhler number effects on soot formation and growth in turbulent nonpremixed flames. Proc. Combust. Inst. **35**(2), 1215–1223 (2015)
18. Attili, A., Bisetti, F.: Application of a robust and efficient Lagrangian particle scheme to soot transport in turbulent flames. Comput. Fluids **84**, 164–175 (2013)
19. Bellemans, A., Aversano, G., Coussement, A., Parente, A.: Feature extraction and reduced-order modelling of nitrogen plasma models using principal component analysis. Comput. Chem. Eng. **115**, 504–514 (2018)
20. D'Alessio, G., Parente, A., Stagni, A., Cuoci, A.: Adaptive chemistry via pre-partitioning of composition space and mechanism reduction. Combust. Flame **211**, 68–82 (2020)
21. Tomboulides, A.G., Lee, J.C.Y., Orszag, S.A.: Numerical simulation of low Mach number reactive flows. J. Sci. Comput. **12**(2), 139–167 (1997)

22. Desjardins, O., Blanquart, G., Balarac, G., Pitsch, H.: High order conservative finite difference scheme for variable density low Mach number turbulent flows. J. Comput. Phys. **227**(15), 7125–7159 (2008)
23. Cottet, G.-H., Koumoutsakos, P.D.: Vortex Methods: Theory and Practice. Cambridge University Press, Cambridge (2000)
24. Koumoutsakos, P.: Multiscale flow simulations using particles. Ann. Rev. Fluid Mech. **37**, 457–487 (2005)
25. Bisetti, F., Blanquart, G., Mueller, M.E., Pitsch, H.: On the formation and early evolution of soot in turbulent nonpremixed flames. Combust. Flame **159**(1), 317–335 (2012)
26. Blanquart, G., Pepiot-Desjardins, P., Pitsch, H.: Chemical mechanism for high temperature combustion of engine relevant fuels with emphasis on soot precursors. Combust. Flame **156**(3), 588–607 (2009)
27. Aversano, G., Bellemans, A., Li, Z., Coussement, A., Gicquel, O., Parente, A.: Application of reduced-order models based on PCA & Kriging for the development of digital twins of reacting flow applications. Comput. Chem. Eng. **121**, 422–441 (2019)

HODMD Analysis in a Forced Flow over a Backward-Facing Step by Harmonic Perturbations

José Miguel Pérez(✉), Soledad Le Clainche, and José Manuel Vega

School of Aerospace Engineering, Universidad Politécnica de Madrid, 28040 Madrid, Spain
josemiguel.perez@upm.es

Abstract. This work studies in detail the energy amplification produced by inflow excitation of a steady flow over a backward-facing step. The disturbances introduced in the inflow are composed of a convergent series of sine functions with different wavenumbers, but the same temporal frequency. The evolution of the perturbations in time is solved using a linear integrator of the Navier-Stokes equations. This information is stored in a group of snapshots and then is analyzed using a data-driven method, higher order dynamic mode decomposition. The method provides a modal decomposition of the data that is used to solve an optimization problem, which identifies the inflow condition giving the maximum energy growth. The results obtained using this novel technique are in qualitatively good agreement with the theory. This is the first step on a new method that could be used for the analysis of numerical and experimental data, without the technical restrictions given by the classical methods. It is possible to identify maximum energy growths without the need of previous knowledge of the adjoint operator or without imposing Dirichlet boundary conditions, generally used in non-modal analyses.

Keywords: Non-modal analysis · HODMD · Backward-facing step · Maximum energy growth

1 Introduction

The flow over a backward-facing step is a benchmark problem in fluid dynamics, generally used in the validation of numerical codes and methodologies. For this reason, it has been studied in detail by different authors over the past 20 years.

Blackburn et al. [2], showed that the flow in this problem presents sub-critical convective instability and transient growth. Mao [5] extended this analysis to the study of receptivity conditioned by the imposition of white noise at the inflow as an initial condition.

In this work, we consider the same parametric configuration as used in the previous references: similar geometry and Reynolds number, which is based on

Á. Herrero et al. (Eds.): SOCO 2020, AISC 1268, pp. 470–479, 2021.
https://doi.org/10.1007/978-3-030-57802-2_45

the step height. Similarly to Mao [5], we are interested in studying the response of the system to disturbances in the inflow. However, instead of introducing white noise, we introduce some perturbations given by an expansion of spatial sinusoidal functions.

This article presents a novel technique to study a system response, which uses a data-driven method, higher order dynamic mode decomposition. The main advantage of using this new tool is that it is possible to identify the energy growth of a system without any knowledge of the governing equations. This article presents a first step of a tool that can be potentially used for the analysis of any type of numerical and experimental data, without the technical restrictions generally imposed by classical methodologies.

This article is organized as follows. Section 2 introduces the description of the problem analyzed. Sections 3 and 4 introduce the classical methodology for the non-modal analysis and the data-driven method HODMD. Section 5 introduces the new methodology followed in this article. Finally, Sects. 6 and 7 introduces the main results and conclusions.

2 Problem Description

Figure 1 shows a sketch of the geometry studied in this problem. The domain is a two-dimensional channel with solid walls up and down, inflow at the left and outflow at the right. Regarding the dimensions, inlet and outlet heights are h and $2h$, respectively, being the height step equal to two. The length of the inlet and outlet channel is $L_i = 10h$ and $L_o = 50h$, respectively. These lengths are the same as those considered in Ref. [2], which were fixed after carrying out a grid convergence study. Finally (as the figure shows) the origin of the coordinate system was defined at the step edge.

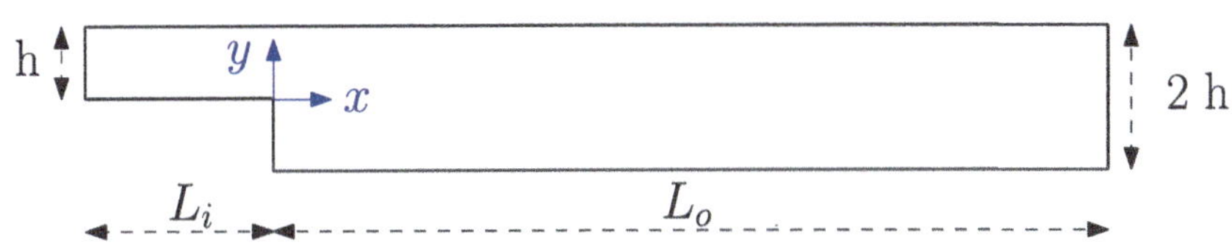

Fig. 1. Geometry of the backward-facing step with expansion ratio two and step height is equal to h. In this problem, flow goes from left to right

The base flow was calculated using Semtex [1,2], a numerical solver that solves the incompressible Navier–Stokes equation using high-order finite element methods. The boundary conditions used in the calculation of the base flow were: Non-slip boundary conditions at the solid walls (up and down), parabolic Poiseuille profile with centerline velocity equal to U_c at the inlet, and zero traction outflow boundary condition for velocity and pressure at the outlet. The Reynolds number is defined using the centerline velocity and the step height as $\mathrm{Re} = U_c h/\nu$, being ν the kinematic fluid viscosity, and is set to Re = 500. At these conditions, the flow is modally stable but unstable by transient growth.

The mesh is defined by using 563 rectangular macro-elements (see Fig. 2 top) with 7 nodes along the edges of these. These nodes are placed at the zeros of Legendre polynomials defined in a canonical square ($[-1, 1] \times [-1, 1]$) that is mapped to each macro-element.

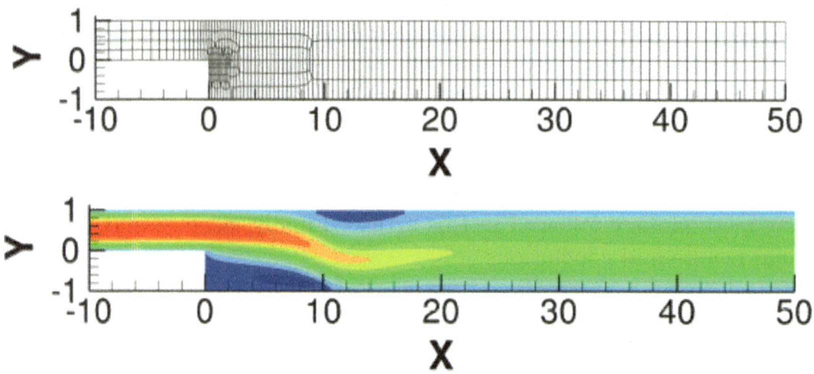

Fig. 2. Backward facing step. Top: mesh (macro-elements). Bottom: Streamwise velocity of the steady base flow at $Re = 500$.

The streamwise velocity obtained at Re = 500 is presented in Fig. 2 bottom. These results were converged numerically and validated. A classic non-modal stability analysis (transient growth) was applied to this base flow obtaining the optimal unstable mode and the growth rate documented in Ref. [2].

3 Non-modal Stability Analysis

Although technically it is possible considering general boundary conditions in non-modal analysis, generally this problem is solved assuming homogeneous Dirichlet boundary conditions in all the boundaries of the domain. On the one hand, using these conditions simplify the definition of the adjoint problem and its integration in time. On the other hand, the convective instabilities should have a negligible amplitude at the outflow.

In this work we will try to avoid these problems, solving this non-modal stability analysis using the data-driven method higher order dynamic mode decomposition. Therefore, the choice of boundary conditions will not be determined by any restriction.

4 Higher Order Dynamic Mode Decomposition

Higher order dynamic mode decomposition (HODMD) [3] is a method introduced to analyze flow fields and identify the spatio-temporal patterns. This algorithm is a generalization of dynamic mode decomposition [6] introduced for the analysis of complex flows, noisy data and non-linear systems [4].

HODMD algorithm lies in the following higher order Koopman assumption, which relates a set of temporal equidistant snapshots $\{v_k; k = 1, \dots K-1\}$ using a group of linear Koopman operators, $R_1, R_2, \cdots, R_d$ in the following way

$$\mathbf{v}_{k+d} \approx R_1\mathbf{v}_k + R_2\mathbf{v}_{k+1} + \dots + R_d\mathbf{v}_{k+d-1}, \quad k = 1, 2, \dots, K-d. \tag{1}$$

These operators contain the dynamics of the system. Thus, solving the eigenvalue problem in a matrix containing all these operators it is possible to represent the vector state $\mathbf{v}_k$ as an expansion of M DMD modes in the following way,

$$\mathbf{v}_k = \sum_{m=1}^{M} a_m \mathbf{u}_m e^{(\alpha_m + i\omega_m)(k-1)\Delta t}, \quad k = 1, 2, \dots K, \tag{2}$$

where α_m and ω_m are the temporal growth rate and frequency, obtained from the calculated eigenvalues, $\mathbf{u}_m$ are the DMD modes, obtained from the calculated eigenvectors, and a_m are the mode amplitudes, obtained by least squares fitting of the previous expression.

In the previous analysis, it is necessary to define two tolerances, ε_1 and ε_2, which allow filtering spatial redundancies and set the number of modes to retained in the previous DMD expansion. More details about this algorithm can be found in Ref. [3].

5 Studying the System Response Using HODMD

This section introduces the process carried out to study the response of a system using HODMD. The methodology is divided into three main steps.

5.1 Initial Data Processing

As a first step, we calculate the base flow, which is a steady flow. Then, the linearized Navier-Stokes equations are integrated assuming zero initial condition and the following boundary perturbation at the inflow,

$$\mathbf{u}_b = \tilde{\mathbf{u}}_b \left(1 - e^{-\sigma t^2}\right)\left(1 - e^{-\sigma(\tau - t)^2}\right)\cos(\omega t), \tag{3}$$

where τ is the final time, σ is a positive relaxation parameter, which allows a smooth transition at the beginning and end of the simulation, and ω is an angular temporal frequency. The term $\tilde{\mathbf{u}}_b$ is modeled so that each component is expressed as a convergent series of sine functions; $\sum_{i=1}^{500} \frac{1}{p^i} \sin(\pi x i)$, with $p = 2$.

In contrast to what is generally done in classical non-linear analysis, the outflow boundary conditions applied in the linear solver are the same as the one applied in the calculation of the base flow; Neumann boundary condition for velocity. The remaining boundary conditions are homogeneous Dirichlet, for all variables and boundaries.

When the numerical simulation is converged, we collect a group of snapshots to apply HODMD and create a DMD expansion (2). Such expansion can be rewritten by separating complex conjugate modes from real modes,

$$\mathbf{v}_k = 2\sum_{m=1}^{N} a_m e^{\alpha_m t_k}\left(\mathbf{u}_m^R \cos(\omega_m t_k) - \mathbf{u}_m^I \sin(\omega_m t_k)\right) + \sum_{m=N+1}^{N+M} a_m e^{\alpha_m t_k}\mathbf{u}_m\,, \tag{4}$$

where $\mathbf{v}_k$ is the result of the numerical simulation at time t_m. In this equation, $\mathbf{u}_m^R$ and $\mathbf{u}_m^I$ are the real and imaginary parts of $\mathbf{u}_m$. Note that in Eq. (4) the first sum is for complex conjugated modes and the second one is for real modes, ie. $\mathbf{u}_m$ is real for $m > N$.

To reduce the computational cost, the vector field was interpolated from the original mesh to a coarser mesh composed by the same number of macro-elements, which were discretized using only 3 nodes on each spatial direction. This interpolation was based on an spectral coarsening, retaining the same spectral element structure but reducing the interpolation order on each of them.

5.2 Creating a Matrix of Modes

Using the DMD modes presented in Eq. (4), we create a new matrix of modes as,

$$\mathbf{M}(0) = \left[\mathbf{u}_1^R, \mathbf{u}_1^I, \mathbf{u}_2^R, \mathbf{u}_2^I, \ldots, \mathbf{u}_N^R, \mathbf{u}_N^I, \mathbf{u}_{N+1}, \ldots, \mathbf{u}_{N+M}\right]\,, \tag{5}$$

composed by J rows (spatial dimension of the modes, which is equivalent to the number of grid points of the computational domain) and $2N + M$ columns (number of DMD modes). The temporal evolution of this matrix is given by,

$$\mathbf{M}(t) = \mathbf{M}(0)\mathbf{F}(t)\,, \tag{6}$$

where $\mathbf{F}(t)$ is a matrix of dimension $(2N + M) \times (2N + M)$ defined as,

$$\mathbf{F}(t) = \begin{bmatrix} \mathbf{F}_1(t) & 0 & \ldots & 0 & 0 \\ 0 & \mathbf{F}_2(t) & \ldots & 0 & 0 \\ \vdots & \vdots & \vdots & \vdots & \vdots \\ 0 & 0 & \ldots & \mathbf{F}_N(t) & 0 \\ 0 & 0 & \ldots & 0 & \mathbf{F}_M(t) \end{bmatrix}. \tag{7}$$

being

$$\mathbf{F}_m(t) = e^{\alpha_m t}\begin{pmatrix} \cos(\omega_m t) & \sin(\omega_m t) \\ -\sin(\omega_m t) & \cos(\omega_m t) \end{pmatrix}\,, \quad \text{with } m = 1, 2 \ldots N\,, \tag{8}$$

and

$$\mathbf{F}_M = \begin{pmatrix} e^{\alpha_{N+1} t}\mathbf{I}_2 & 0 & \ldots & 0 \\ 0 & e^{\alpha_{N+2} t}\mathbf{I}_2 & \ldots & 0 \\ \vdots & \vdots & \vdots & \vdots \\ 0 & 0 & \ldots & e^{\alpha_{N+M} t}\mathbf{I}_2 \end{pmatrix}, \tag{9}$$

with $\mathbf{I}_2 = \mathbf{F}(0)$. Note that $\mathbf{F}_m(t)$ is a contraction or dilatation (depending on the sign of α_m) plus a rotation of counter-clockwise angle equal to $-\omega_m t$.

Given a mesh composed by J grid points, it is possible to create a diagonal matrix $\mathbf{A}$ with dimension $J \times J$, whose elements are the square root of the areas of the cells, representing the weight for each element of the mesh. Applying a singular value decomposition (SVD) to matrix $\mathbf{AM}(0)$, we obtain

$$\mathbf{AM}(0) = \mathbf{V\Sigma W}^T , \tag{10}$$

where $()^T$ is the transpose conjugate, $\mathbf{V}$ and $\mathbf{W}$ are the left and right singular vectors and $\mathbf{\Sigma}$ is a diagonal matrix containing the singular values.

Introducing this decomposition in (6), gives

$$\mathbf{AM}(t) = \mathbf{V\Sigma W}^T\mathbf{F}(t) . \tag{11}$$

Finally, a generic perturbation at time t could be written as a function of the matrix of modes $\mathbf{M}(t)$ as follows,

$$\mathbf{u}(t) = \mathbf{M}(t)\mathbf{b} , \tag{12}$$

where $\mathbf{b}$ is a column vector that defines the amplitude of modes for this perturbation.

5.3 Prediction of Optimal Perturbation

This section defines the process of predicting the optimal perturbation for a given value of the angular temporal frequency, ω (see Eq. (3)). This is the procedure followed by the authors who study the receptivity problem, see for instance [5]. However, the methodology proposed in this work is more general, allowing to introduce a range of angular temporal frequencies at the inflow, thus this method is also able to optimize the value of the frequency. Although, optimizing the frequency is beyond the scope of this paper and remains as an open topic for future tasks.

Given a generic perturbation at time τ, its energy is defined as,

$$E(\tau) = (\mathbf{u}(\tau), \mathbf{u}(\tau)) , \tag{13}$$

where $(,)$ is the inner product defined as the integral defined in the whole domain Ω,

$$(\mathbf{u}(\tau), \mathbf{u}(\tau)) = \int_\Omega \mathbf{u} \cdot \mathbf{u} dv . \tag{14}$$

The maximum energy growth G at time τ is the maximum ratio between the final energy in the whole domain $E(\tau)$ and the energy at the inflow for a given frequency, E_b, given by

$$G = \max_{\mathbf{u}(0)} \frac{E(\tau)}{E_b} . \tag{15}$$

Introducing Eq. (12) in Eq. (15) and approximating the integral defined in Eq. (14) by a Gaussian quadrature, gives

$$\int_{\Omega} \mathbf{u} \cdot \mathbf{u} dv \approx \mathbf{u}^T \mathbf{A}^T \mathbf{A} \mathbf{u} = ||\mathbf{A}\mathbf{u}||_2^2 . \tag{16}$$

Subindex 2 (in the last term) means the Euclidean L_2 norm.

Combining Eqs. (12), (13), (15) and (16) we obtain

$$G(\tau) = \max_{\mathbf{b}} \frac{\mathbf{b}^T \mathbf{M}(\tau)^T \mathbf{A}^T \mathbf{A} \mathbf{M}(\tau) \mathbf{b}}{\mathbf{b}^T \mathbf{M}_b(0)^T \mathbf{M}_b(0) \mathbf{b}} . \tag{17}$$

This equation can be simplified using Eqs. (10) and (11) and taking into account that $\mathbf{V}^T\mathbf{V} = \mathbf{I}$. Note that, in general, $\mathbf{V}_b^T \mathbf{V}_b \neq \mathbf{I}$, where the number of rows and columns of matrix $\mathbf{V}_b$ are the number of spatial integration nodes at the boundary and temporal modes, respectively. That is, $\mathbf{V}_b$ is the restriction of $\mathbf{V}$ at the inflow. Then we have,

$$\begin{aligned} G(\tau) &= \max_{\mathbf{b}} \frac{\mathbf{b}^T \mathbf{F}(\tau)^T \mathbf{W} \boldsymbol{\Sigma}^2 \mathbf{W}^T \mathbf{F}(\tau) \mathbf{b}}{\mathbf{b}^T \mathbf{W} \boldsymbol{\Sigma}^T \mathbf{V}_b^T \mathbf{V}_b \boldsymbol{\Sigma} \mathbf{W}^T \mathbf{b}} \\ &= \max_{\hat{\mathbf{b}}} \frac{\hat{\mathbf{b}}^T \hat{\mathbf{F}}^T \boldsymbol{\Sigma}^2 \hat{\mathbf{F}} \hat{\mathbf{b}}}{\hat{\mathbf{b}}^T \boldsymbol{\Sigma}^T \mathbf{V}_b^T \mathbf{V}_b \boldsymbol{\Sigma} \hat{\mathbf{b}}} , \end{aligned} \tag{18}$$

where,

$$\hat{\mathbf{F}} = \mathbf{W}^T \mathbf{F} \mathbf{W} \tag{19}$$

and

$$\hat{\mathbf{b}} = \mathbf{W}^T \mathbf{b} . \tag{20}$$

Solving this optimization problem over all possible combinations of inflow boundary perturbations is equivalent to find the largest eigenvalue of the following generalized eigenvalue problem,

$$\hat{\mathbf{F}}^T \boldsymbol{\Sigma}^2 \hat{\mathbf{F}} \hat{\mathbf{b}} = G(\tau) \boldsymbol{\Sigma}^T \mathbf{V}_b^T \mathbf{V}_b \boldsymbol{\Sigma} \hat{\mathbf{b}} . \tag{21}$$

Once the amplitudes of the optimal mode have been calculated for a given time horizon (τ), is straightforward to obtain the shape of the optimal mode as,

$$\mathbf{u}(t) = \mathbf{A}^{-1} \mathbf{A} \mathbf{M}(t) \mathbf{b} = \mathbf{A}^{-1} \mathbf{V} \boldsymbol{\Sigma} \mathbf{W}^T \mathbf{F}(t) \mathbf{b} = \mathbf{A}^{-1} \mathbf{V} \boldsymbol{\Sigma} \hat{\mathbf{F}} \hat{\mathbf{b}} . \tag{22}$$

combining Eqs. (12), (11), (19) and (20).

6 Results

The HODMD analysis has been performed considering the following parameters: $\varepsilon_1 = 10^{-3}$, $\varepsilon_2 = 10^{-6}$ and $d = 1$. Figures 3 shows the maximum value of G as a function of time. As can be seen, G decreases when ω increases. Mao [5] observed similar behaviour in the evolution of the energy, although there are

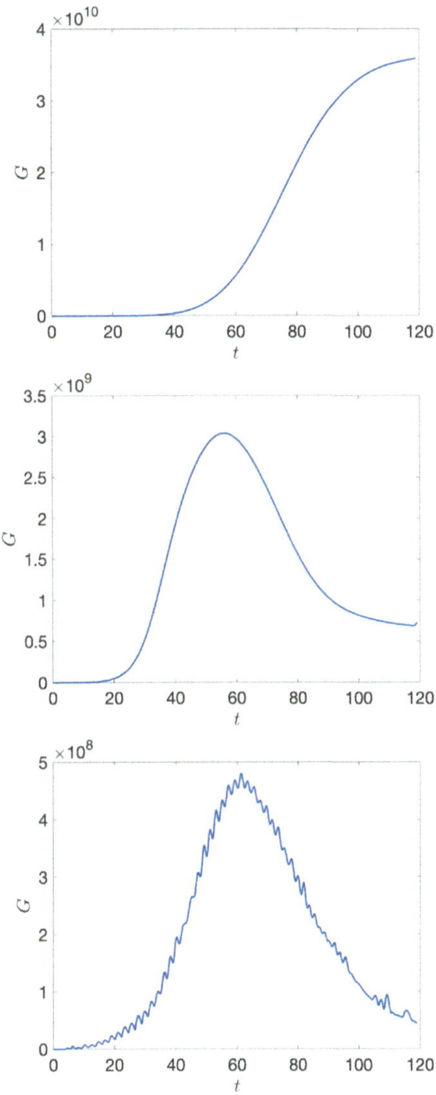

Fig. 3. Temporal evolution of the maximum value of the energy G for different values of ω (temporal angular frequency). From top to bottom: $\omega = 0.5$, $\omega = 1.0$ and $\omega = 1.5$.

some differences found in the value of energy growth. One of the main reasons could be related to the different boundary perturbations using in this problem and the different methodologies carried out.

Figures 4 shows the vorticity field in the z (spanwise) direction. These structures are similar to those obtained in Ref. [5], observing the same pattern in the case of $\omega = 0.5$ and recovering part of the activity of the mode in the region near the step in the case of $\omega = 1.5$.

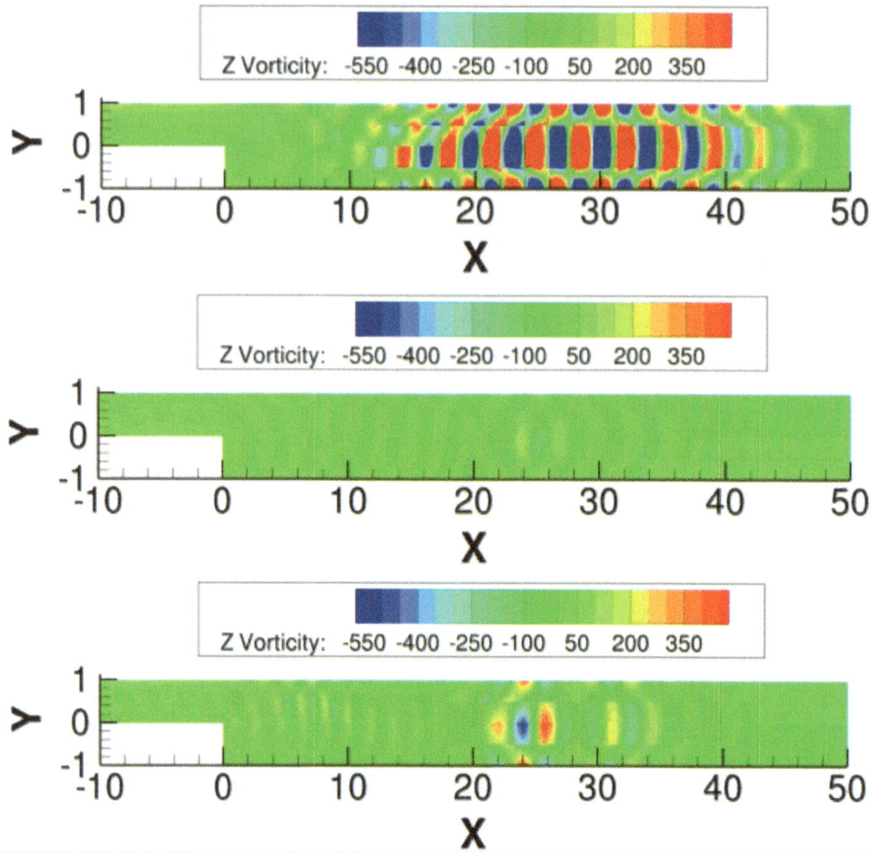

Fig. 4. Spanwise vorticity (vorticity normal to the XY plane) for different values of ω (temporal angular frequency). From top to bottom: $\omega = 0.5$, $\omega = 1.0$ and $\omega = 1.5$.

Although this method provides qualitative results in good agreement with the literature [5], the differences found in the maximum energy level encourage the authors to continue improving the method presented in this article for future research. Nevertheless, we have been able to present a new method for soft computing, efficient, to study the energy growth of a system.

7 Conclusions

This work introduces a novel optimization methodology for the analysis and prediction of optimal perturbations. The results obtained are promising, showing the expected trends in both the growth rates and the shape of modes. The disturbances introduced have been generated using a spatial multi-frequency perturbation at the inflow trying to excite the relevant modes necessary for the optimization process, which are evolved in time using a linear solver of the

Navier-Stokes equations. These modes are calculated applying the data-driven method HODMD to the database generated in the linear integration. Once the dominant modes are characterized, an eigenvalue problem is solved, being its leading eigenvalue the maximum growth rate and its corresponding eigenvector the coefficients to use in the reconstruction of the optimal perturbation from the HODMD modes. In this way, it is not necessary to use adjoint codes and therefore the problem related to the imposition of the boundary conditions of the adjoint problem is avoided. This methodology could also be applied to experimental data, extending the field of application of the classical analysis.

References

1. Blackburn, H.M.: Three-dimensional instability and state selection in an oscillatory axisymmetric swirling flow. Phys. Fluids **14**(11), 3983–3996 (2002)
2. Blackburn, H.M., Barkley, D., Sherwin, S.J.: Convective instability and transient growth in flow over a backward-facing step. J. Fluid Mech. **603**, 271–304 (2008)
3. Le Clainche, S., Vega, J.: Higher order dynamic mode decomposition. SIAM J. Appl. Dyn. Syst. **16**(2), 882–925 (2017)
4. Le Clainche, S., Vega, J.: Analyzing nonlinear dynamics via data-driven dynamic mode decomposition-like methods. Complexity **2018**, 6920783 (2018)
5. Mao, X.: Effects of base flow modifications on noise amplifications: flow past a backward-facing step. J. Fluid Mech. **771**, 229–263 (2015)
6. Schmid, P.: Dynamic mode decomposition of numerical and experimental data. J. Fluid Mech. **656**, 5–28 (2010)

An Application of Variational Mode Decomposition in Simulated Flight Test Data

Carlos Mendez[1,2](✉)

[1] School of Aeronautics, Universidad Politécnica de Madrid, Madrid, Spain
c.mendezg@alumnos.upm.es
[2] Facultad de Ciencias Químicas, Universidad Nacional de Asunción, San Lorenzo, Paraguay
cmendez@qui.una.py

Abstract. In this work, a new application for variational mode decomposition (VMD) is proposed to identify frequencies on an input signal, simulated numerically. The recent advances in the use of this technique have promoted its application in different areas of science. The present work applies VDM to identify the central dynamics in a non-linear dynamical system modeling the evolution of flutter in-flight testing, an aerodynamic instability that produces undesirable effects, and that can even lead to the aircraft destruction. This work presents a first step in the search of new methods that can reduce logistics costs in real flight testing experiments. The results of the present work are compared with the modal analysis obtained with the open-source solver NeoCASS. The method described in this article provides fast and accurate results compared to other methods already used in the field of flight testing.

Keywords: Variational mode decomposition · Flight test data · NeoCASS

1 Introduction

In the aeronautical industry, the fast and accurate detection of aeroelastic frequencies and damping is a research topic of major interest that brings the development of new numerical methods in flight flutter testing. All these methods are used to identify the dynamic of the system, which includes the detection of frequencies, damping rates, and modal shape using the information captures on flight tests by accelerometers. A comparison between the most effective flutter methodologies is presented by some authors in the literature [6,7]. Among the most popular techniques used to predict flutter it is possible to find non-linear autoregressive moving average exogenous (NARMAX) [8], auto-regressive moving-average method (ARMA) [9], least-squares curve-fitting method (LSCF) [10], moving-block approach (MBA) [11] and other like presented in [12]. The

Á. Herrero et al. (Eds.): SOCO 2020, AISC 1268, pp. 480–489, 2021.
https://doi.org/10.1007/978-3-030-57802-2_46

success of this method are related to the quantity of data available captured by accelerometers, and it is important to use other methods that eventually reduce the number o input data in order to obtain the same result.

Variational mode decomposition (VMD) is an adaptive and non-recursive signal decomposition method developed by Dragomiretskiy and Zosso in 2014 [1] a promising method using in [4] for comparing with traditional soft computing techniques and see the good performance. The VMD method transforms the mode decomposition problem into a variational solution problem. The signal is decomposed into a K discrete number of sub-signals which are near the corresponding center frequency. The method has been applied in signal decomposition for audio engineering, climate analysis, and various flux, respiratory, and neuromuscular signals found in medicine and biology [1], other studies used the method to predict damping rates in civil structures [2] and electrical application [3]. The main advantage of this method is that it only needs one input to extract information, and this premise is used to apply it in a real flight test experimental campaign. As the first step to implement the VMD in the detection, we used a software that includes the fluid interaction in order to use more realistic signals, for this work, we use the NeoCASS software.

This work presents a new application for VMD using a simulated signal that represents the flight test data. The main goal of this work is to detect the frequencies with accuracy and efficiently.

The algorithm was based on the work of [1]. This work is organized as follows. Section 2 briefly introduce the mathematics and parameters of the code, and the characteristics of the NeoCASS and the signal output are presented in Sect. 3. The main results are presented in Sect. 4, and finally Sect. 5 presents the main conclusions.

2 Variational Mode Decomposition

The frequency detection based on VMD as two main steps: 1) Modal decomposition assuming the input s signal can be decomposed in K modes u_k.

$$\sum_{K} u_k(t) = s(t) \tag{1}$$

and, 2) the detection of the modal frequency in each mode extracted, where $\{u_k(t)\} = \{u_1(t), ..., u_K(t)\}$ denotes the set of all sub-signals, and each sub-signal is compact around a center frequency $\{f_k\} = \{f_1, ..., f_K\}$, this set of frequencies represents the response of the system on certain conditions. K is the number of modes and needs to be specified. The decomposition is made solving the constrained variational problem:

$$\min_{\{v_k\},\{f_k\}} \left\{ \sum_{k=1}^{K} \left\| \partial_t \left[\left(\delta(t) + \frac{j}{\pi t} \right) * u_k(t) \right] e^{-j2\pi f_k t} \right\|_2^2 \right\}, \tag{2}$$

where ∂_t represents de partial derivative respect to time, $\delta(t)$ is the Dirac distribution, $j = \sqrt{-1}$, '$*$' is the convolution operator and '$\|\|_2$' the L2 norm.

The unconstrained problem (augmented Lagrangian) using the Lagrange multiplier λ and the quadratic penalty factor α is defined as follow:

$$L(\{u_k(t)\},\{f_k(t)\},\lambda(t)) = \alpha\sum_{k=1}^{K}\left\|\partial_t\left[\left(\delta(t)+\frac{j}{\pi t}\right)*u_k(t)\right]e^{-j2\pi f_k t}\right\|_2^2 + \left\|s(t)-\sum_{k=1}^{K}u_k(t)\right\|_2^2 + \left\langle\lambda(t), s(t)-\sum_{k=1}^{K}u_k(t)\right\rangle \quad (3)$$

where '⟨⟩' is the inner product of two vectors. The minimization problem Eq. 2 can be solved using a sequence of iterative sub-optimizations named alternate direction method of multipliers (ADMM). The ADMM procedure for searching u_k and f_k is given by:

$$\hat{u}_k^{n+1}(f) = \frac{\hat{s}(f)-\sum_{i\neq k}\hat{u}_i^n(f)+\hat{\lambda}(f)/2}{1+2\alpha(2\pi f-2\pi f_k^n)^2}, \quad (4)$$

$$f_k^{n+1} = \frac{\int_0^\infty 2\pi f|\hat{u}_k^n(f)^2|df}{\int_0^\infty|\hat{u}_k^n(f)|^2 df}, \quad (5)$$

where n is the number of iterations, and $\hat{s},\hat{\lambda}$ and $\hat{u}_k$ represents the Fourier transform of the signal. The criterion to stop the iterative process is the threshold ε (pre-specified number), which is the following:

$$\frac{\sum\left\|u_k^{n+1}-u_k^n\right\|_2^2}{\sum\left\|u_k^n\right\|_2^2} < \varepsilon \quad (6)$$

The final modes $u_k(t)$ can be obtained using the inverse Fourier transform $\hat{u}_k(f)$ and taking the real component, as;

$$u_k(r) = R\{ifft(\hat{u}_k)(f)\}. \quad (7)$$

The success of the VMD is related to the selection of K [1]; some research tries to explain the effect between the selection of K and the detected frequencies in two scenarios, K < actual mode number and $K >$ actual mode number. The first stage of this work is related to evaluate the effect of K and the penalty factor α over the accuracy of the detected frequencies compared to known modal frequencies of the input signal in two cases, the first case we assume $K =$ Rigid body + Aeroelastic modes and the second case $K =$ Aeroelastic modes and the values of α given for $1 \leq \alpha \leq 1.10^5$. In the second stage, we use the obtained values of the parameters K and α with three different input signals, and they correspond to different maneuvers on the airplane.

3 Simulated Signal Input Using NeoCASS

The input signal is generated using software NeoCASS, a numeric analysis tool developed by Politecnico di Milano, Dipartimento di Ingegneria Aerospaziale, mainly used in the design of aircraft, [14]. It is based on real insight; the algorithm provides structural information to build the aircraft investigating aeroelastic behavior using structural and aerodynamic numerical methods. The dynamic response analysis is made using NeoRESP, a second-order frequency domain formulation, which solves the dynamic response of the aircraft to some externally controlled inputs, using matrices directly in their original form. Frequency responses are evaluated over the frequency range of interest, and inverse transformations are finally applied to recover the time-histories of states and outputs.

3.1 NeoCASS Parameters

For this work, we use a Boeing 747 model (B747-100); this model includes the aerodynamic model, the structural model, and the aeroelastic model. The model consists of 512 nodes uniformly distributed with default characteristics in the software see Fig. 1 (for more details see [13]).

It is necessary to define some inputs like aircraft velocity (VREF), Mach number (MACH), air density (RHOREF), upper-frequency limit (300 Hz, this range is selected based on previous research [5]) and the number of modes included in the modal analysis (for this work we use 9). The parameters selected are $VREF = 170\,\mathrm{m/s}$, $MACH = 0.5$ and $RHOREF = 1.225\,\mathrm{kg/m^3}$.

Finally, we define the conditions at which the control surfaces will be varied and its time evolution (the characteristics of the maneuver). We chose the temporal variation of the surfaces as a sinusoidal function with a duration of 0.5 seconds (s) for the aileron, rudder, and elevator see Fig. 1. The NeoCASS output consists of a series of files corresponding to displacement, velocity, and acceleration for every node of the mesh and every one of the six degrees of freedom (DOF).

3.2 Analysis of the Signals

As a previous step before applying the VMD algorithm, we analyze the intensity of the accelerometers for every DOF and for the three different maneuvers.

The total time of simulation was 10 s with a $\Delta t = 0.005\,\mathrm{s}$, and the total length of the signal is the 2000 points. In the Fig. 2 we can see the intensity of the acceleration for every node in the entire time in the z-direction, which is the main direction (comparing the directions x and y) and some remarkable things we can note, (1) not all the nodes are activated after the maneuvers, there are some nodes in which the intensity of the vibrations remains stable in time (that are the nodes candidates for applying VMD) mainly between ~180 and ~280, and (2) different maneuvers excite the nodes in different ways (we expect some frequencies in only some of the maneuver), this is related to the type of mode

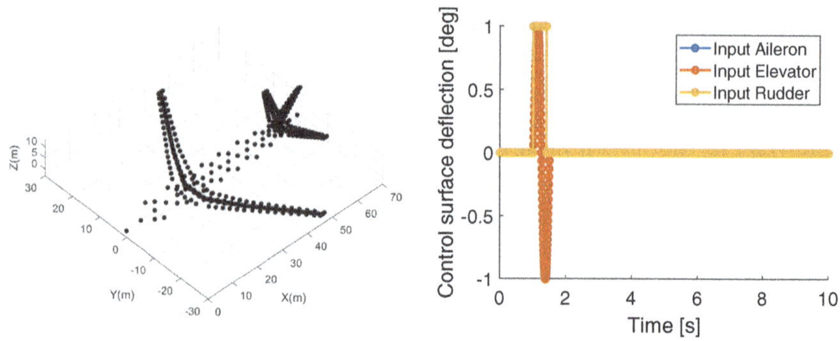

Fig. 1. Node representation of the airplane and signal input for the maneuvers for different control surface.

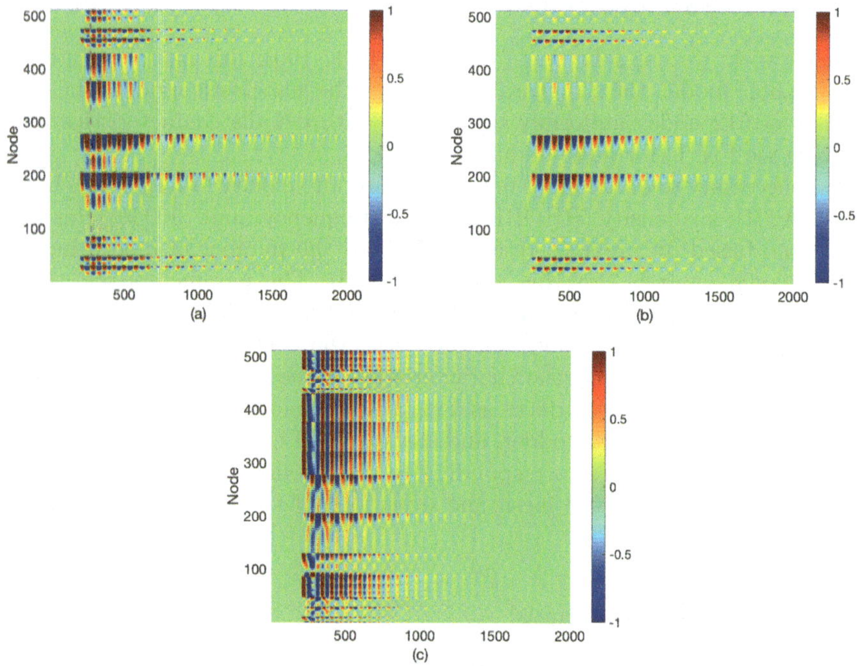

Fig. 2. Intensity of the acceleration for every node during the maneuvers modeling flight flutter testing with $\Delta t = 0.005$. Maneuvers carried out in (a) Aileron, (b) Rudder and (c) Elevator.

that we can find in aeroelastic problems, symmetric or antisymmetric modes, in this point we can note the inability of the method to differentiate them.

The VMD method use one input signal to decompose it in different modes. In other words, we only need one input signal to detect the frequencies. There is an advantage because we would only need one accelerometer to extract the information. However, if we only use one input we can lose information, since we can select one node that could not contain all the frequencies (or the amplitude of the frequencies is not sufficient to be detected). For this reason that we limit the original input signal to a region with high information (see Fig. 3). A good practice will be using various inputs (accelerometers) to extract the data, and that is part of the working line, which starts with this paper. The scope for this work is laying the groundwork for using this method as an alternative than the traditional, evaluating the algorithm parameters, and comparing it with de modal analysis results (knowing that the frequencies in the dynamic response are not the same as the modal analysis).

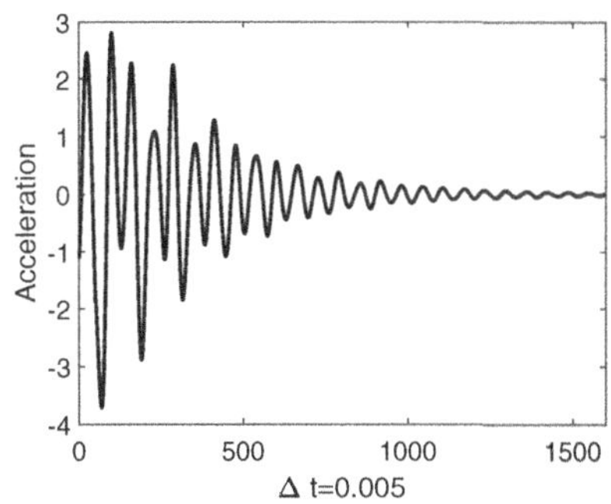

Fig. 3. Temporal evolution of the acceleration in one node in the z-direction.

The VMD is presented as an alternative to the traditional methods, evaluating the algorithm parameters, and comparing it with the modal analysis results (knowing that the frequencies in the dynamic response are not the same as the modal analysis).

The modal analysis is carried out in NeoCASS, and the results with the type of modes are presented in Table 1.

4 Application of VMD to Acceleration Output

The algorithm used in this work is based on [1] and adapted to use with our data. Initially, we use the aileron information (See Fig. 2(a)) in order to evaluate K as we mentioned in Sect. 2. We perform two groups of simulations, the first group uses $K = 11$ and the second uses $K = 5$. This selection of is based in the fact that NeoRESP has six modes related to body rigid and nine imposed (this represents 15 modes). For these nine modes, five only appear when we modify the aileron surface, this is because this maneuver activates only symmetric modes as expected. So, for the aileron, we have five possibles modes different that the rigid body modes. Respect to α we chose $1 \leq \alpha \leq 1.10^5$ (range based on [2]).

Table 1. Modal input

$Freq_{modal}$	Mode
1,3837	Symmetric
2,1058	Antisymmetric
2,6436	Antisymmetric
3,1505	Symmetric
3,3160	Symmetric
3,7014	Antisymmetric
4,4377	Symmetric
4,8167	Antisymmetric
5,1974	Symmetric

Table 2. Frequencies obtained using the number of modes $K = 11$

$\alpha = 1$	$\alpha = 10$	$\alpha = 100$	$\alpha = 1000$	$\alpha = 10000$	$\alpha = 100000$
0.8884	0.3725	0.6320	0.7542	0.7896	1.7296
1.7165	1.7079	1.7211	1.7277	1.7294	3.0363
2.8023	3.0537	3.0292	3.0118	3.0164	6.1897
4.9599	4.4417	3.9719	3.8127	3.8062	11.3289
9.3406	5.6491	6.4079	6.2578	6.2771	18.9464
18.9930	11.2109	10.2961	8.8491	8.9158	26.5063
27.7604	19.3879	16.5320	11.7525	16.0786	32.7032
42.2812	29.7074	25.2183	17.0420	31.0974	39.0454
46.0111	40.2986	35.9224	25.7844	38.9837	45.2828
52.6450	50.4267	47.3877	39.7502	48.9934	51.5436
61.1420	59.1675	58.2880	56.5433	56.7212	56.5394

Evaluating the intensity of the acceleration in Fig. 1 we can observe different nodes with high intensity on time, one of that node is the number 278 (we can expect more information in this node than others nodes with low intensity). The Tables 2 and 3 shown the results of the application of VMD algorithm over the acceleration (in z-direction) output of NeoCASS, as a prior observation, it is possible to see the number of potentially goods modes are around 3 and 4 ($\sim$1.5 Hz to $\sim$5.2 Hz), so this gives an idea to select the number K. Evaluating the of influence of α on the results, varying the frequencies from $\sim$0.372 Hz to higher frequencies $\sim$61.14 Hz. We are comparing the results of both tables with Table 1, and we note that the modal frequencies are between $\sim$1.5 Hz to $\sim$5.2 Hz and the best combination for achieving the best results are $K = 5$ and $\alpha = 1000$.

The modal decomposition of the input Fig. 3 are presented in Fig. 4 showing the excellent performance of the VMD with this kind of signals (simulation time was less than 1 s with a standard $i5$ notebook).

Table 3. Frequencies obtained using the number of modes number of modes $K = 5$

$\alpha = 1$	$\alpha = 10$	$\alpha = 100$	$\alpha = 1000$	$\alpha = 10000$	$\alpha = 100000$
1.7163	0.4542	0.6145	0.7536	0.7879	1.7295
3.1669	1.7091	1.7200	1.7276	1.7293	3.0365
6.6549	3.1834	3.0917	3.0122	3.0168	21.4672
23.2572	11.0047	6.9019	3.8288	3.8320	36.5388
51.7473	48.3084	43.7822	6.6361	41.3638	50.2820

4.1 Frequencies Detection for Different Maneuvers

Once we chose the algorithm parameters, we use this information to detect the aeroelastic frequencies in the signals for the three maneuvers, and the results are presented in Table 4. As seen, most of the frequencies are properly identified. However, the method also identifies some spurious frequencies in the three maneuvers. Nevertheless, this method presents an advantage compared to classical techniques (1) it is fast and (2) it is automatic. Improving the results presented imply comparing VMD with other methods remains an open topic for future works.

Table 4. Frequencies for every maneuver comparing to modal frequencies

$Freq_{Modal}$	Freq - $VMD_{Aileron}$	Freq - VMD_{Rudder}	Freq - $VMD_{Elevator}$
			0.9463
	0.7536	0.4028	0.9770
1.3837	1.7276	1.3608	
2.1058			1.9690
2.6436		2.1581	
3,1505		3.0375	3.1322
3.3160	3.0122		
3.7014	3.8288		
4.4377		4.6243	
4.8167			
5.1974	6.6361		5.1712

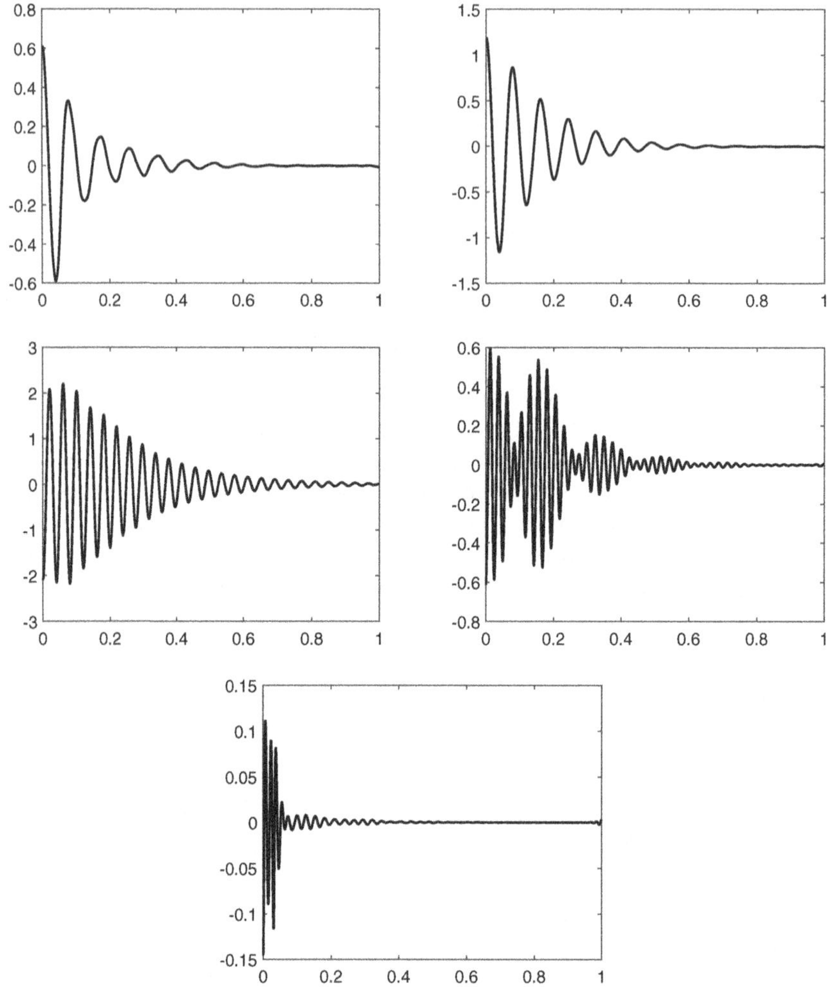

Fig. 4. Five mode decomposition of the signal input given in Fig. 3

5 Conclusions

In this paper, we have presented an application for Variational mode decomposition (VMD), to analyze the signal of and accelerometer which has been modeled using NeoCASS. VMD analyses the dynamic response over three different maneuvers modeling flight flutter testing. Comparing to other typical methods that are based on a linear approximation considering a high number of entrances (accelerometers), the VMD only needs one signal input to decompose it in their modes. A good selection of parameters gives satisfying results considered as a first approximation. In future works this application be coupled to other methods in order to reduce the time of simulation identifying the modes and their corresponding frequencies and damping rates, which is the main goal in flight flutter testing.

References

1. Dragomiretskiy, K., Zosso, D.: Variational mode decomposition. IEEE Trans. Signal Process. **62**(3), 531–544 (2014)
2. Zhang, M., Xu, F.: Variational mode decomposition based modal parameter identification in civil engineering. Front. Struct. Civ. Eng. **13**, 1082–1094 (2019). https://doi.org/10.1007/s11709-019-0537-3
3. Deng, W., Liu, H., Zhang, S.: Research on an adaptive variational mode decomposition with double thresholds for feature extraction. Symmetry **10**, 684 (2018)
4. Zounemat-Kermani, M., Seo, Y., Kim, S.: Can the decomposition approaches always enhance the soft computing models? Predicting the dissolved oxygen concentration in St. Johns River, Florida. Appl. Sci. (2019). https://doi.org/10.3390/app9122534
5. Follador, R., de Souza, C.E., da Silva, R.G.A., Góes, L.C.S.: Comparison of in-flight measured and computed aeroelastic damping: modal identification procedures and modeling approaches. J. Aerosp. Technol. Manag. **8**(2), 163–177 (2016)
6. Dimitriadis, G., Cooper, J.E.: Comment on "flutter prediction from flight flutter test data". J. Aircr. **43**, 862–863 (2006)
7. Lind, R.: Comment on "flight-test evaluation of flutter prediction methods". J. Aircr. **40**(5), 964–970 (2003)
8. Kukreja, S.L., Brenner, M.J.: Nonlinear black-box modelling of aeroelastic systems using structure detection: application to F/A-18 data. AIAA J. Guid. Control Dyn. **30**(2), 557–564 (2007)
9. Matsuzaki, Y., Ando, Y.: Estimation of flutter boundary from random responses due to turbulence at subcritical speeds. J. Aircr. **18**(10), 862–868 (1981)
10. Taylor, P.M., Moreno Ramos, R., Banavara, N., Narisetti, R.K., Morgan, L.: Flight flutter testing at Gulfstream Aerospace using advances signal processing techniques. In: Proceedings of 58th AIAA/ASCE/AHS/ASC Structures, Structural Dynamics, and Materials Conference, AIAA paper 2917-1823 (2017)
11. Hammond, C.E., Dogget Jr., R.V.: Determination of subcritical damping by moving-block/Randomdec Applications. In: Flutter Testing Techniques, NASA Scientific and Technical Information Oddice, Washington, D.C., pp. 59–76 (1975)
12. Mendez, C., Le Clainche, S., Vega, J.M., Moreno, R., Taylor, P.: Aeroelastic flutter flight test data analysis using a high order dynamic mode decomposition approach. In: Proceedings of AIAA Scitech 2019 Forum, AIAA paper 2019-1531 (2019)
13. Cavagna, L., Ricci, S., Riccobene, L.: Structural sizing, aeroelastic analysis, and optimization in aircraft conceptual design. J. Aircr. **48**(6), 1840–1855 (2011)
14. Cavagna, L., Ricci, S., Travaglini, L.: NeoCASS: an integrated tool for structural sizing, aeroelastic analysis and MDO at conceptual design level. Prog. Aerosp. Sci. **47**(8), 621–635 (2011)

Following Vortices in Turbulent Channel Flows

Jose J. Aguilar-Fuertes, Francisco Noguero-Rodríguez, José C. Jaen-Ruiz, Luis M. García-Raffi, and Sergio Hoyas(✉)

Instituto Universitario de Matemática Pura y Aplicada, Universitat Politècnica de València, 46022 València, Spain
sergio.hoyas@mot.upv.es

Abstract. The behaviors of individual vortices has become a matter of study in turbulent flows as the computational power to make their study feasible has become available. A new parallel vortex-tracking tool has been developed in this work, to track vortical coherent structures in all scales through the time steps of a DNS simulation. The tool first constructs a graph and decomposes it into their connected components to obtain the vortices in a given time step and then compares them to the ones present in a previous time step to obtain tracking information that can be reconstructed to obtain each vortex temporal evolution.

Keywords: DNS · Vortex tracking · Structures

1 Introduction

In the last decades, the computational power of computers has increased exponentially. In the 1990's, the biggest supercomputers reached computing powers of around 100 gigaflops. With the improvement of technology, the performance of computers has increased approximately one order of magnitude every five years. Nowadays, the fastest supercomputers reach computing powers of 200 petaflops. In addition, the use of thousands of processors is relatively easy and efficient. Moreover, to read and write massive database is more or less straightforward. These improvements have allowed researchers to use the Direct Numerical Simulations (DNS) for the study of turbulent flows during the last three decades. And, in fact, DNS has proven to be one of the most powerful tools to analyse them. However, it is also clear that we are still far away from reaching the large Reynolds numbers occurring in practical applications. An optimistic estimation is that it would be possible to run a DNS of a commercial jet plane around 2050. Given that an estimation from Jiménez [6] is that wall-bounded turbulence is responsible of the 5% of the CO2 dumped by mankind into the atmosphere every year, this gap is certainly too large. Thus the focus should be placed in the understanding of the internal mechanism of the flow to produce better models. Since the seminal work of Chong [3] and others, we are able to identify the

Á. Herrero et al. (Eds.): SOCO 2020, AISC 1268, pp. 490–496, 2021.
https://doi.org/10.1007/978-3-030-57802-2_47

basic coherent structures of turbulent flows. However, their behaviour or even the precedence (cause-effect) [9] of some of these structures is still an open problem. Moreover, to study the dynamics of these structures, a very large database of the flow, is needed to extract and follow the individual vortices. Typically, for medium Reynolds numbers this database is in the range of the decades to hundreds of TB.

In this article, we introduce a novel method to identify and track these vortices. The goal is to understand the particular life of every vortex and to identify the most energetic structures of the flow in an efficient way. The team acknowledges that machine learning algorithms bases on neural networks have proven very useful for tracking problems outside fluid dynamics, such as computer vision [12]. However, convolutional neural networks require a large amount of labeled data for training. This data can be generated by means of a *hard* computing algorithm such as the one introduced in this work.

In the next section, the numerical method used to obtain the data is briefly introduced, together with the mechanism used to identify the vortices. The coherent structure reconstruction and the temporal tracking is shown in Sects. 3 and 4, respectively. Results are shown in Sect. 5 and conclusions are outlined in Sect. 6.

2 Input Data

The flow can be described by means of the momentum and mass balance equations. They have been solved using the LISO code, which has successfully been employed to run some of the largest simulations of turbulence [4,5]. Briefly, the code uses the same strategy as [7], but using seven-point compact finite differences in y direction with fourth-order consistency and extended spectral-like resolution [8]. The temporal discretization is a third-order semi-implicit Runge-Kutta scheme [11]. The wall-normal grid spacing is adjusted to keep the resolution at $\Delta y = 1.5\eta$, i.e., approximately constant in terms of the local isotropic Kolmogorov scale $\eta = (\nu^3/\epsilon)^{1/4}$, where ϵ is the dissipation rate. In wall units, Δy^+ varies from 0.3 at the wall, up to $\Delta y^+ \simeq 12$ at the centerline. In the simulation used, the fluid is driven by a pressure gradient in the x direction, its characteristics are present in Table 1.

As a first step to follow the vortices in the flow, the decision of which points belong to vortices and which must not be taken. There is no strict definition for this transformation, but several criteria have been developed.

In this work the criterion introduced by Chong [3] has been used as modified by Del Álamo [1]. The first states that a point can be considered a vortex if the discriminant of the jacobian of the velocity in that point is nonzero, meaning that its jacobian matrix has complex eigenvalues that describe a swirling motion of the flow [3]. The modification introduced by Del Álamo normalizes the value of the discriminant relative to the distance to the wall and therefore produces a more homogeneous distribution of vortices along the domain: the dependence of the probability of a point being part of a vortex to the distance of that point to the wall is reduced. This allows for a clear vortex separation close to the wall while maintaining vortical structures in the centre of the channel [1].

Table 1. Main features of the input DNS simulations

Geometrical characteristics	
Number of points in x N_x	192
Number of points in y N_y	251
Number of points in z N_z	192
Domain length in x L_x	2π
Domain length in z L_z	π
Channel height	2

Flow characteristics	
Bulk Reynolds number Re	2500
Friction Reynolds number Re_τ	180
Kinematic viscosity ν	0.000308

Temporal resolution characteristics	
CFL	0.9

This homogeneity of the criterion, compared to other available local vorticity criterions, can produce a smoother evolution for vortices that move across the height of the channel during their evolution. As in other works [1,10], the threshold used for the vorticity criterion was set to the location of the percolation crisis, defined as the value of the threshold in which an increment increases the total number of vortices found, but does not reduce the proportion of the volume of all vortices contained just in the biggest one.

The application of this criterion to a velocity field produced by a DNS simulation assigns a boolean value to each point in the grid, which the tool developed in this work takes as input to reconstruct the coherent structures and track them through time.

3 Coherent Structure Reconstruction

Vortices are three-dimensional structures, and local vortex identification methods only produce information about the vorticity of a point with no relation to its neighbours. Therefore, a reconstruction algorithm is needed to obtain the vortices present in a flow field.

The condition that separates one vortex V_i from another one $V_{j\neq i}$, both formed by points in which the criterion has returned an affirmative result, is the possibility of tracing a continuous path between every pair of points in that vortex V_i only passing through the points in V_i by moving stepwise on any of the 3 cartesian directions. This can also be expressed by saying that every point contained in a vortex V_i must have at least one neighbour also contained in that vortex V_i and none contained in another $V_{j\neq i}$.

This three-dimensionality is however a problem for efficient computing, as computer memory organises data in only one dimension. Given a three-dimensional matrix such as the one containing the results of the vorticity criterion, only one of their indices will be contiguous in memory, making displacements through that direction efficient in memory but displacements in the second and especially third index inefficient.

The tool developed instead divides the fluid flow into planes perpendicular to the main flow velocity, and then reconstructs the 2D coherent regions of every plane in parallel. Each coherent region in 2D, which can be described as a *slice* of a vortex, is represented by a node on a graph **G**, which is formed by all the vortices present in the flow field. This process is detailed in Fig. 1.

Then, if the projection of any of the aggregated 2D *slices* on the next or previous planes overlaps the *slice* corresponding to any node from the projection plane, both will be part of the same vortex in 3D and an edge will be established between those two nodes. This evaluation can be performed in parallel for all nodes in the domain to construct all the set of edges in **G**.

Then, the individual vortices present in the field can be obtained by separating **G** into its connected components by means of the parallel component separation algorithm presented by Chin *et al.* [2]. Each connected component represents a set of 2D *slices* interconnected, in the third dimension by the edges of the graph, making a complete three dimensional coherent vortex (Fig. 2).

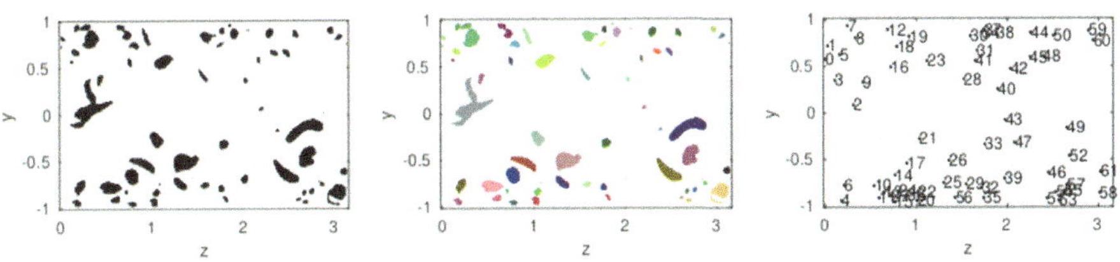

Fig. 1. (Left) Vorticity criterion result, where the black regions consist of the points where the criterion is fulfilled (Centre) Reconstruction of the coherent 2D vortex regions, each in a different colour (Right) Corresponding graph nodes to each region

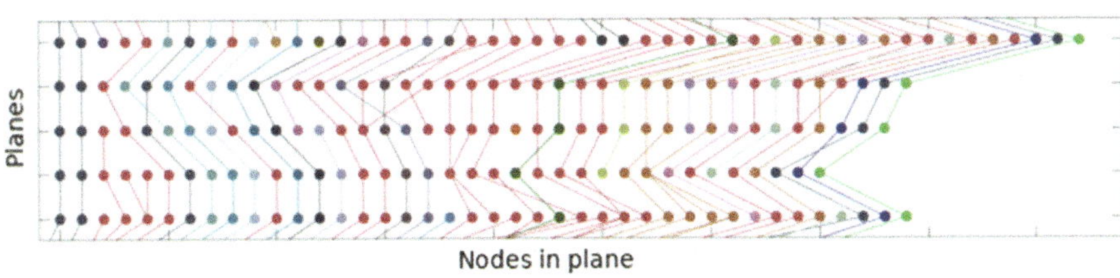

Fig. 2. Representation of a section of **G**, the vertical axis representing the plane a node belongs to and the nodes are spread through the horizontal axis. The colours represent the connected components in **G**.

4 Temporal Tracking

Once the vortices on a given field have been obtained, they can be compared to the ones present in the previous time step to assign a correspondence between the structures present in each time step. By repeating this process for a number of time steps and storing the correspondences, the life of a vortex can be followed through time. The computation of this correspondence in this work relies on the combination of two methods:

- Similarity checks: a vector of main features of every vortex is extracted in each time step, containing information about its centre of mass, bounding box dimensions and volume. This vector is compared by means of a weighted distance to the features of every vortex in the previous time step and the most similar is selected. If this most similar vortex is within a margin of error, thee tool establishes the correspondence that both are the same vortex in different time steps.
  The weight parameter for every feature and maximum relative error accepted are selected by means of a constrained nonlinear optimisation with the comparison success rate as the objective function and a set maximum failure rate as constrain.
- Point to point comparisons: this is the approach taken by other works tracking vortices such as [10], based on the evaluation of the intersection between vortices in consecutive time steps. This method does not have any accuracy constraints or calibration requirements, but requires access to a larger amount of data, making the computation more expensive. Also, this method produces result where the complex dynamics of vortex collisions and splits prevent the similarity comparisons from obtaining meaningful results.

As both of the methods are done in a vortex-by-vortex basis, both of them can be executed for all the vortices present in a flow field in parallel. In this work both methods are run, using the point to point comparison where similarity cannot provide information.

5 Preliminary Results

The algorithm presented has been tested on 200 time steps of the simulation presented in Sect. 2. In this short time, only the lifetimes of small vortices could be extracted. The distance to the wall and volume of one is represented in Fig. 3. This evolution shows how the vortex is born near the wall, starts growing and separating from the wall but dissipates before it can get escape the near-wall region. This is consistent with the expected results for a small vortex.

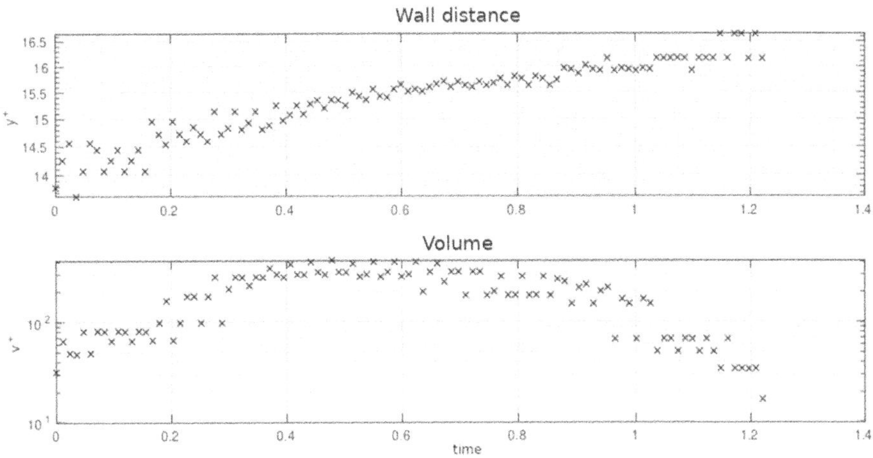

Fig. 3. (Up) Evolution of the distance between wall and centre of mass of a small vortex versus time. (Down) Evolution of the volume occupied by the same vortex versus time.

6 Conclusions and Future Works

A new vortex-tracking method has been implemented in this work, which allows for fluid structure tracking through parallel computing. This method is composed by a reconstruction part, abstracted into a graph problem that can be solved with available algorithms; and a tracking part, in which a combination of a point-to-point comparison and a similarity check are used. This implementation has been tested in a small set of fields to check its proper functioning.

In this work a initial test has been done following vortices as described by the criterion introduced by del Álamo, but the algorithm presented can be used to track any other aggregated phenomena occurring in fluid domains, such as intense Reynolds-stress events.

Acknowledgements. This work was supported by RTI2018-102256-B-I00 of MINECO/FEDER.

References

1. del Álamo, J.C., Jiménez, J., Zandonade, P., Moser, R.: Self-similar vortex clusters in the turbulent logarythmic region. J. Fluid Mech. **561**, 329–358 (2006)
2. Chin, F.Y., Lam, J., Chen, I.N.: Efficient parallel algorithms for some graph problems. Commun. ACM **25**(9), 659–665 (1982)
3. Chong, M., Perry, A., Cantwell, B.: A general classification of three-dimensional flow fields. J. Phys. A **2**(5), 765–777 (1990)
4. Gandía-Barberá, S., Hoyas, S., Oberlack, M., Kraheberger, S.: The link between the Reynolds shear stress and the large structures of turbulent Couette-Poiseuille flow. Phys. Fluids **30**(4), 041702 (2018). https://doi.org/10.1063/1.5028324

5. Hoyas, S., Jiménez, J.: Scaling of the velocity fluctuations in turbulent channels up to $Re_\tau = 2003$. Phys. Fluids **18**(1), 011702 (2006)
6. Jiménez, J.: Near-wall turbulence. Phys. Fluids **25**(10), 101302 (2013)
7. Kim, J., Moin, P., Moser, R.: Turbulence statistics in fully developed channels flows at low Reynolds numbers. J. Fluid Mech. **177**, 133–166 (1987)
8. Lele, S.K.: Compact finite difference schemes with spectral-like resolution. J. Comput. Phys. **103**(1), 16–42 (1992)
9. Lozano-Durán, A., Jiménez, J.: Effect of the computational domain on direct simulations of turbulent channels up to $Re_\tau = 4200$. Phys. Fluids **26**(1), 011702 (2014). https://doi.org/10.1063/1.4862918
10. Lozano-Durán, A., Jiménez, J.: Time-resolved evolution of coherent structures in turbulent channels: characterization of eddies and cascades. J. Fluid Mech. **759**, 432–471 (2014)
11. Spalart, P.R., Moser, R.D., Rogers, M.M.: Spectral methods for the Navier-Stokes equations with one infinite and two periodic directions. J. Comput. Phys. **96**(2), 297–324 (1991)
12. Wang, L., Ouyang, W., Wang, X., Lu, H.: Visual tracking with fully convolutional networks. In: Proceedings of the IEEE international conference on computer vision, pp. 3119–3127 (2015)

Special Session: Soft Computing Techniques and Applications in Logistics and Transportation Systems

Stable Performance Under Sensor Failure of Local Positioning Systems

Javier Díez-González[1](✉), Rubén Álvarez[2], Paula Verde[1], Rubén Ferrero-Guillén[1], David González-Bárcena[3], and Hilde Pérez[1]

[1] Department of Mechanical, Computer, and Aerospace Engineering, Universidad de León, 24071 León, Spain
{jdieg,hilde.perez}@unileon.es,
{pgarcv00,rferrg00}@estudiantes.unileon.es
[2] Positioning Department, Drotium, Universidad de León, 24071 León, Spain
ruben.alvarez@drotium.com
[3] IDR/UPM, Universidad Politécnica de Madrid, 28040 Madrid, Spain
david.gonzalez@upm.es

Abstract. Local Positioning Systems are an active topic of research in the field of autonomous navigation. Its application in difficult complex scenarios has meant a solution to provide stability and accuracy for high-demanded applications. In this paper, we propose a methodology to enhance Local Positioning Systems performance in sensor failure contexts. This fact guarantees system availability in adverse conditions. For this purpose, we apply a Genetic Algorithm Optimization in a five-sensor 3D TDOA architecture in order to optimize the sensor deployment in nominal and adverse operating conditions. We look for a trade-off between accuracy and algorithm convergence in the position determination in four (failure conditions) and five sensor distributions. Results show that the optimization with failure consideration outperforms the non-failure optimization in a 47% in accuracy and triples the convergence radius size in failure conditions, with a penalty of only 6% in accuracy during normal performance.

Keywords: CRLB · Genetic Algorithms · Localization · LPS · Sensor failure

1 Introduction

Autonomous navigation has meant a challenge for scientific development over the last few years. The high accuracy required has entailed the interest in Local Positioning Systems (LPS) where the positioning signal paths are reduced between targets and architecture sensors. This fact reduces noise and uncertainties trough the minimization of the global architecture errors with respect to Global Navigation Satellite Systems (GNSS).

LPS cover a defined and known space with architecture sensors where the capabilities of the system are maximized. LPS properties depend on the measurement of the physical

Á. Herrero et al. (Eds.): SOCO 2020, AISC 1268, pp. 499–508, 2021.
https://doi.org/10.1007/978-3-030-57802-2_48

magnitude used for the determination of the target location: time [1], power [2], frequency [3], angle [4], phase [5] or combinations of them [6].

Among these systems, the most extended are time-based models due to their reliability, stability, robustness and easy-to-implement hardware architectures. Time-based positioning computes the total or relative travel time of the positioning signals from the target to the receivers generating two different system conceptions: total time-of-flight-Time of Arrival (TOA) [7] - and relative time-of-flight -Time Difference of Arrival (TDOA) [8] systems-.

TDOA systems compute the relative time between the reception of the positioning signal in two different architecture sensors. Therefore, the synchronization of these systems is optional. Asynchronous TDOA architectures measures time differences in a single clock of a coordinator sensor [9] while in synchronous TDOA all architecture sensors must be synchronized to compute all together the time measurements.

Time relative measurements lead to hyperboloid surfaces of possible location of targets. Every two architecture sensors a hyperboloid equation is obtained while only $(n - 1)$ independent equations are achieved from n different sensors [10]. The required number of sensors to determine unequivocally the target location is 5 sensors for 3-D positioning in these methodologies. However, the intersection of three different hyperboloids in TDOA systems leads to two different potential solutions. Nevertheless, these solutions are not able to be discarded from a mathematical point of view.

In one of our previous works [11], we have shown that a reliable unique solution to the intersection of three hyperboloids or spheres can be obtained through the maximization of the distance between the two potential solutions by means of Genetic Algorithms (GA). We achieve this result by applying Taylor-based algorithms [12] from an initial iteration point which must be close enough to the final solution. Node deployment showed to have a direct impact for this achievement.

The sensor distribution has also relation with the accuracy of the LPS. Cramér-Rao Lower Bound (CRLB) [13, 14] derivations allow the characterization of the White Gaussian Noise (WGN) in the time measurements, estimating the minimum achievable error in positioning systems [15]. This has allowed us to study the node deployment optimization in TDOA systems by means of GA [16, 17]. The reason of the use of heuristic techniques relies on the NP-Hard problem solution of the 3D sensor deployment in LPS and it is widespread in the literature [18–20].

However, the consideration of sensor failures has not yet been considered for LPS sensor distribution optimizations. In this paper, we propose for this purpose a GA optimization for the 3D node deployment in a TDOA system with five architecture sensors that can suffer from sensor failures. We perform a multi-objective optimization in which we look for a trade-off between the accuracy of the system with five sensors and every combination of four nodes in a defined environment of an LPS. This methodology will ensure the availability of the system with acceptable accuracy in case of sensor failures in the architecture nodes.

The remainder of the paper is organized as follows: the algorithm for the target unequivocal location determination is introduced in Sect. 1, the CRLB modeling is presented in Sect. 2, the GA and the fitness function are detailed in Sect. 3 and Sects. 4 and 5 show the results and conclusions of the present paper.

2 Taylor-Based Positioning Algorithm in TDOA Systems

Relative time measurements in TDOA systems lead to hyperboloid equations of possible target locations. Numerical methods are needed to solve the intersection of these non-linear equations. Taylor-Based algorithms in TDOA systems are linearizations of the equation of the time difference of arrival:

$$\begin{aligned} R_{ij} &= d_{ij} = d_{Ei} - d_{Ej} = ct_{ij} = c\left(t_i - t_j\right) \\ &= \sqrt{(x_E - x_i)^2 + (y_E - y_i)^2 + (z_E - z_i)^2} \\ &\quad - \sqrt{\left(x_E - x_j\right)^2 + \left(y_E - y_j\right)^2 + \left(z_E - z_j\right)^2} \end{aligned} \tag{1}$$

where R_{ij} and d_{ij} represent the distance difference of the signal travel from the emitter to sensors i and j, d_{Ei} and d_{Ej} are total distance from the emitter (E) to sensors i and j, c is the speed of the radioelectric waves, t_{ij} is the time difference of arrival measured in the architecture sensors, t_i and t_j is the total time of flight of the positioning signal from emitter to receivers i and j respectively and (x_E, y_E, z_E), (x_i, y_i, z_i) and $\left(x_j, y_j, z_j\right)$ are the Cartesian coordinates of the target and the sensors i and j.

Taylor approximation truncated on first order is applied in Eq. 1 to linearize the equation from an initial iteration point (x_0, y_0, z_0):

$$R_{ij} = ct_{ij} = R_{ij_0} + \frac{\partial R_{ij}}{\partial x}\Delta x + \frac{\partial R_{ij}}{\partial y}\Delta y + \frac{\partial R_{ij}}{\partial z}\Delta z \tag{2}$$

where R_{ij_0} is the range difference of arrival in the initial iteration point, $\frac{\partial R_{ij}}{\partial x}$, $\frac{\partial R_{ij}}{\partial y}$ and $\frac{\partial R_{ij}}{\partial z}$ are the partial derivatives of the range differences measured in the i and j architecture sensors particularized in the initial iteration point. The application of Eq. 2 to every pair of sensors of the TDOA architecture leads to the following relation, that enables the obtainment of the target location.

$$\Delta \mathbf{P} = \left(\mathbf{H}^{\mathbf{t}}\mathbf{H}\right)^{-1}\mathbf{H}^{\mathbf{t}}\Delta \mathbf{R} = \begin{pmatrix} \Delta x \\ \Delta y \\ \Delta z \end{pmatrix} \tag{3}$$

where $\mathbf{H}$ is the partial derivative matrix, and $\Delta\mathbf{P}$ represents the incremental values from the last iteration point in each space direction which supposes the unknown of the equation.

3 CRLB Modeling in TDOA Systems

CRLB is an unbiased estimator of the lowest variance of a parameter. Its usage in the localization field is widespread [21–25] since it allows us to determine the minimum achievable error by the system. The uncertainties introduced in the measurements depend on the distance traveled by the positioning signal from the emitter to the architecture sensors in a heteroscedastic noise consideration. Recent studies [14] developed a matrix form of the CRLB considering heteroscedasticity in time measurements:

$$FIM_{mn} = \left[\frac{\partial h(TS)}{\partial x_m}\right]^T R^{-1}(TS)\left[\frac{\partial h(TS)}{\partial x_n}\right] + \frac{1}{2}tr\left\{R^{-1}(TS)\left[\frac{\partial R(TS)}{\partial x_m}\right]R^{-1}(TS)\left[\frac{\partial R(TS)}{\partial x_n}\right]\right\} \quad (4)$$

where **FIM** indicates the Fisher Information Matrix, m and n are the sub-indexes of the estimated parameters in the **FIM**, TS is the target sensor Cartesian coordinates, $\mathbf{h(TS)}$ is a vector that contains the travel of the signal in the TDOA architecture to compute a time measurement:

$$h_{TDOA_i} = \|TS - AS_i\| - \|TS - AS_j\| \qquad i = 1, \ldots, N_{AS} \quad j = 1, \ldots, N_{AS} \quad (5)$$

being AS_i and AS_j the coordinates of the architecture sensors i and j and N_{AS} the number of sensors involved in the position determination. $\mathbf{R(TS)}$ is the covariance matrix of the time measurements in the architecture sensors. The covariance matrix is built with a heteroscedastic noise consideration in the sensors modeled by a Log-normal path loss propagation model [17] obtaining the following variances:

$$\sigma^2_{TDOA_{ij}} = \frac{c^2}{B^2(P_t/P_n)}PL(d_0)\left[\left(\frac{d_{Ei}}{d_0}\right)^n + \left(\frac{d_{Ej}}{d_0}\right)^n\right] \qquad i = 1, \ldots, N_{AS} \quad j = 1, \ldots, N_{AS} \quad where\ i \neq j \quad (6)$$

where B is the signal bandwidth, P_T is the transmission power, P_n the mean noise level determined through the Johnson-Nyquist equation, n the path loss exponent, d_0 the reference distance from which the path loss propagation model is applied and $PL(d_0)$ the path-loss in the reference distance.

The trace of the inverse of the Fisher Information Matrix (**J**) provides the uncertainties associated with each variable to estimate, i.e. the three Cartesian coordinates of the target for a 3D positioning. The location accuracy is directly evaluated through the Root Mean Squared Error (RMSE), which is computed based on the trace of the **J** matrix.

4 GA Optimization

The strong influence of the sensor distribution in the LPS performance enables the maximization of their capabilities through the optimization of their sensor placement. This approach is especially critical in complex 3D environments, where the most important source of positioning error is promoted by the sensor distribution.

In our previous works [17], a GA for optimizing sensor distributions in 3D irregular environments is presented. The proposed methodology allows a modular definition of the optimization region and the reference surface for locating the sensors of the positioning architecture. In addition, the procedure allows the election of different selection techniques, percentage of elitism, crossover methodologies, mutation types, and convergence criteria. After the choice of the optimization method, the next step is the definition of the fitness function. In this case, we look for a multi-objective optimization for the combined maximization of the TDOA system accuracy in 3D environments when the whole architecture is available and when only four sensors are accessible, limited by the size of the convergence regions that allow the correct execution of the Taylor-based positioning algorithm. The fulfillment of these objectives guarantees the robustness of the TDOA architectures in adverse conditions of operation. The fitness function is detailed hereafter:

$$
\begin{aligned}
ff = \sum_{1}^{Comb} \left\{ \frac{C_1}{NT} \sum \left\{ 1 - \frac{\left[\left(\frac{1}{RMSE_{ref}} \right) - \left(\frac{1}{\mathbf{RMSE}_{4\,sensors}} \right) \right]^2}{\left(\frac{1}{RMSE_{ref}} \right)^2} \right\} \right. \\
\left. + \frac{C_2}{NT} \sum \left\{ \frac{\left[\left(\frac{1}{Dist_{ref}} \right) - \left(\frac{1}{\mathbf{Dist}} \right) \right]^2}{\left(\frac{1}{D_{ref}} \right)^2} \right\} \right\} \\
+ C_3 \sum \frac{\left\{ 1 - \frac{\left[\left(\frac{1}{RMSE_{ref}} \right) - \left(\frac{1}{\mathbf{RMSE}_{Ncs}} \right) \right]^2}{\left(\frac{1}{RMSE_{ref}} \right)^2} \right\}}{NT} - C_4 \frac{\sum_{i=1}^{N_{CS}} BL_i}{N_{CS}}
\end{aligned}
\tag{7}
$$

where *Comb* is the number of groups of four sensors which are obtainable based on the total number of architecture sensors, *NT* is the number of analyzed points, $RMSE_{ref}$ is the reference accuracy, $\mathbf{RMSE}_{4sensors}$ is the vector that contains the CRLB evaluation for each point at analysis with each combination of 4 sensors, $Dist_{ref}$ indicates the reference distance for the convergence criteria, **Dist** represents the vector that specifies the convergence evaluation in terms of the distance between the two possible solutions for each point at study, $\mathbf{RMSE}_{Ncs}$ is the vector that contains the CRLB analysis for each point at study when all architecture sensors are available, C_k are coefficients for calibration of the individual summands of the fitness function and BL_i is the penalization factor associated with the existence of sensors in banned regions (if they exist).

5 Results

In this section, the results of the optimization for sensor failure in TDOA architectures are detailed. Initially, a 3D complex scenario has been designed for carrying out the optimization, proving the adaptability of the proposed methodology in any environment.

In Fig. 1, the term TLE represents the Target Location Environment which defines the region where the targets are possibly located. For this simulation, the TLE region extends from 0.5 to 5 m of elevation from the base surface, based on a division of 10 m in x and y Cartesian coordinates, and 1 ms in z coordinate. This ensures the correct evaluation and continuity of the accuracy and convergence analysis.

The NLE area expresses the Node Location Environment, which indicates all possible sensor locations. In the case of the NLE region, the height of the sensors is limited in the

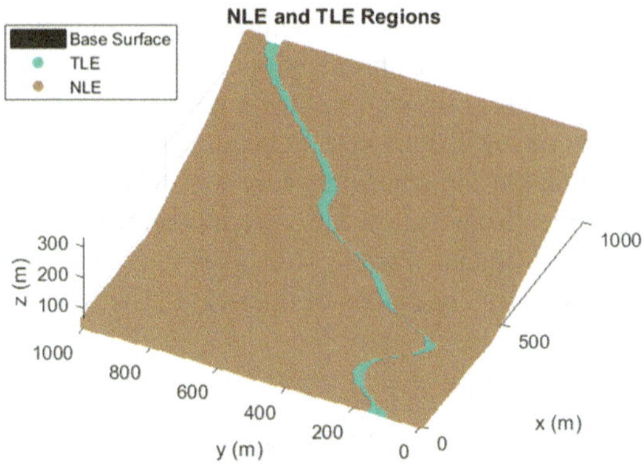

Fig. 1. The scenario of simulations. The reference surface is depicted is grey tones. NLE and TLE regions are respectively shown in orange and purple colors.

range 3 to 20 m from the base surface. The discretization of the NLE region depends on the codification of the individuals of the GA, precisely on the longitude of the chromosomes implemented. In this optimization, the resolution of the NLE area varies in the three Cartesian coordinates from 0.5 to 1 m, alluring a fine setting in the optimization of each sensor. Tables 1 and 2 show the principal parameters of configuration for the positioning system and the GA applied for the optimization.

Table 1. Parameters of configuration for the positioning system operation [15, 25, 26].

Parameter	Value
Transmission power	100 W
Mean noise power	−94 dBm
Frequency of emission	1090 MHz
Bandwidth	80 MHz
Path loss exponent	2.16
Antennae gains	Unity
Time-Frequency product	1

Values presented in Table 1 have been chosen in an attempt to stand for a generic positioning technology, expressed by the typical parameters of transmission power, frequency of emission and bandwidth. The GA configuration is based on the following aspects: population of 120 individuals with binary codification, Tournament 3 as selection procedure with 2% of elitism, single-point crossover, single-point mutation with a probability of 5%, and 90% of equal individuals as convergence criteria. This election allows the trade-off between fitness function maximization and processing time. For more information about the genetic operators and the design of the GA [17]. In addition,

C_k coefficients are defined as unity, searching for a solution with normal condition predominance in the final sensor deployment, but with good failure conditions performance. This GA was coded in the MATLAB software following every of these considerations. The results after the optimization process are shown for distributions of 5 sensors. The results for the optimized sensor placement with failure consideration, 5 sensors nominal operating conditions and convergence maximization (Conf. 1) are provided in Figs. 2 and 3 when two of the sensors are not available.

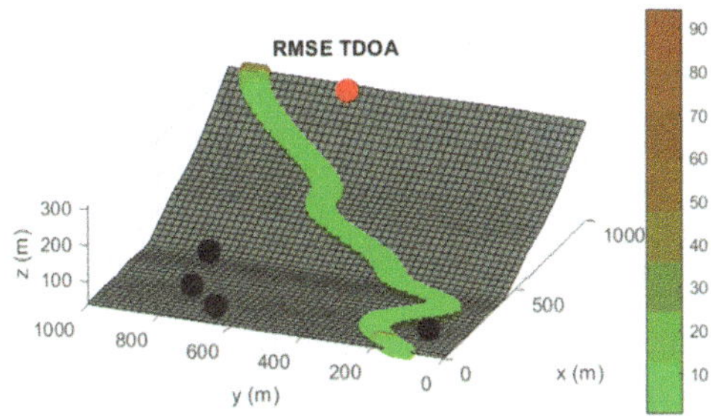

Fig. 2. Accuracy analysis in terms of CRLB for the optimized distribution of 5 sensors under possible failure of two arbitrary sensors of the architecture. Black spheres indicate active sensors and the red sphere symbolizes the failing sensor.

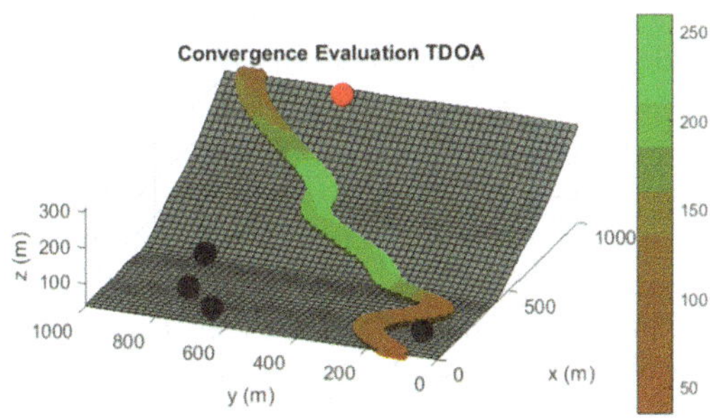

Fig. 3. Convergence radius analysis for the optimized distribution of 5 sensors under possible failure of two arbitrary sensors of the architecture.

In Table 2, a comparison between the optimized sensor distribution for sensor failure (Conf. 1) and the optimized sensor placement of 5 sensors without malfunction consideration and convergence maximization (Conf. 2) is supplied. It should be stressed that this last optimization is carried out through a fitness function with the direct evaluation of the CRLB for 5 sensors and the last term of the Eq. 7.

Results of Table 2 reveal that the optimization carried out in Conf. I not only minimizes the CRLB when only 4 sensors are accessible but also maximizes the region where the Taylor-based positioning algorithm is able to operate. The beauty of this combined multi-objective optimization is that the accuracy of the four-sensor combinations in failure conditions is increased by 47% while the accuracy of the normal operating

Table 2. Comparison between two optimized sensor distributions: with (Conf. 1) and without (Conf. 2) failure consideration.

	Operating condition	RMSE (m)		Convergence radius (m)	
		Mean	Max	Min	Mean
Conf. 1	Non-failure	1.154	3.234	40	171.687
	Failure	7.061	181.325	32	167.312
Conf. 2	Non-failure	1.085	7.114	4	145.807
	Failure	13.293	>300	0	37.841

five sensor distribution (Conf. 1) is reduced less than a 6% with regards to conventional node deployments (Conf. 2) that only consider the five-sensor optimization.

This new optimization procedure considering sensor failures does guarantee the robustness of the positioning system in complex conditions of operations, and the design of architectures considering these situations.

6 Conclusions

In this paper, a method to guarantee the system accuracy under sensor failure is proposed. We address the possible sensor malfunctioning or ineffective link between target and architecture sensors which are key factors in LPS actual deployments.

For this purpose, we have defined a 3D scenario in which a five-sensor distribution of a TDOA architecture is deployed in order to achieve practical results. The possible failure of two sensors in adverse operating conditions leads to the solution of the ambiguity in the target position determination with four receivers. We have proved that an unequivocal solution can be attained through the use of Taylor-Based positioning algorithms in combination with an optimized node location looking for a maximization of the distance between the two possible solutions in the four-sensor TDOA problem.

Accuracy analysis must be also carried out in both nominal and failure operating conditions. Therefore, we perform a multi-objective optimization of the node location by means of a Genetic Algorithm. This optimization looks for the maximization of the convergence of the positioning algorithms and the accuracy of the architecture to solve this NP-Hard problem.

Results show that both accuracy and convergence can be achieved under every possible sensor failure condition. The optimization considering only four effective links with the architecture sensors in failure conditions triples the values of the convergence region and increases the accuracy in 47% regarding to conventional optimizations that do not consider these adverse situations.

References

1. Shen, H., Ding, S., Dasgupta, S., Zhao, C.: Multiple source localization in wireless sensor networks based on time of arrival measurement. IEEE Trans. Signal Process. **62**(8), 1938–1949 (2014)
2. Yiu, S., Dashti, M., Claussen, H., Perez-Cruz, F.: Wireless RSSI fingerprinting localization. Sig. Process. **131**, 235–244 (2017)
3. Lindgren, D., Hendeby, G., Gustafsson, F.: Distributed localization using acoustic Doppler. Sig. Process. **107**, 43–53 (2015)
4. Rong, P., Sichitiu, M.L.: Angle of arrival localization for wireless sensor networks. In: 2006 3rd Annual IEEE Communications Society on Sensor and Ad Hoc Communications and Networks, Reston, VA, pp. 374–382 (2006)
5. Sackenreuter, B., Hadaschik, N., Faßbinder, M., Mutschler, C.: Low-complexity PDoA-based localization. In: Proceedings of the 2016 International Conference on Indoor Positioning and Indoor Navigation (IPIN), Alcalá de Henares, Spain, pp. 1–6 (2016)
6. Yin, J., Wan, Q., Yang, S., Ho, K.C.: A simple and accurate TDOA-AOA localization method using two stations. IEEE Signal Process. Lett. **23**(1), 144–148 (2016)
7. Shen, J., Molisch, A.F., Salmi, J.: Accurate passive location estimation using TOA measurements. IEEE Trans. Wireless Commun. **11**(6), 2182–2192 (2012)
8. Lanxin, L., So, H.C., Frankie, K.W., Chan, K.W., Chan, Y.T., Ho, K.C.: A new constrained weighted least squares algorithm for TDOA-based localization. Sig. Process. **93**(11), 2872–2878 (2013)
9. He, S., Dong, X.: High-accuracy localization platform using asynchronous time difference of arrival technology. IEEE Trans. Instrum. Meas. **66**(7), 1728–1742 (2017)
10. Priyantha, N.B., Balakrishnan, H., Demaine, E.D., Teller, S.: Mobile-assisted localization in wireless sensor networks. In: Proceedings IEEE 24th Annual Joint Conference of the IEEE Computer and Communications Societies, Miami, FL, pp. 172–183. IEEE (2005)
11. Díez-González, J., Álvarez, R., Sánchez-González, L., Fernández-Robles, L., Pérez, H., Castejón-Limas, M.: 3D TDOA problem solution with four receiving nodes. Sensors **19**(13), 2892 (2019)
12. Yang, K., Xu, Z.: A quadratic constraint total least-squares algorithm for hyperbolic location. Int. J. Commun. Netw. System Sci. **2**, 130–135 (2008)
13. Lanzisera, S., Zats, D., Pister, K.S.J.: Radio frequency time-of-flight distance measurement for low-cost wireless sensor localization. IEEE Sens. J. **11**, 837–845 (2011)
14. Kaune, R., Hörst, J., Koch, W.: Accuracy analysis for TDOA localization in sensor networks. In: Proceedings of the 14th International Conference on Information Fusion, Chicago, IL, USA (2011)
15. Rappaport, T.S.: Wireless Communications-Principles and Practice. Prentice Hall, Upper Saddle River (2002)
16. Álvarez, R., Díez-González, J., Alonso, E., Fernández-Robles, L., Castejón-Limas, M., Perez, H.: Accuracy analysis in sensor networks for asynchronous positioning methods. Sensors **19**(13), 3024 (2019)
17. Díez-González, J., Álvarez, R., González-Bárcena, D., Sánchez-González, L., Castejón-Limas, M., Perez, H.: Genetic algorithm approach to the 3D node localization in TDOA systems. Sensors **19**(18), 3880 (2019)
18. Peng, B., Li, L.: An improved localization algorithm based on genetic algorithm in wireless sensor networks. Cogn. Neurodyn. **9**(2), 249–256 (2015)
19. Domingo-Perez, F., Lazaro-Galilea, J.L., Wieser, A., Martin-Gorostiza, E., Salido-Monzu, D., de la Llana, A.: Sensor placement determination for range-difference positioning using evolutionary multi-objective optimization. Expert Syst. Appl. **47**, 95–105 (2016)

20. Zhang, Q., Wang, J., Jin, C., Ye, J., Ma, C., Zhang, W.: Genetic algorithm based wireless sensor network localization. In: Proceedings of the Fourth International Conference on Natural Computation, Jinan, China (2008)
21. Ruz, M.L., Garrido, J., Jiménez, J., Virrankoski, R., Vázquez, F.: Simulation tool for the analysis of cooperative localization algorithms for wireless sensor networks. Sensors **19**(13), 2866 (2019)
22. Kowalski, M., Willett, P., Fair, T., Bar-Shalom, Y.: CRLB for estimating time-varying rotational biases in passive sensors. IEEE Trans. Aerosp. Electron. Syst. **56**(1), 343–355 (2019)
23. Hu, D., Chen, S., Bai, H., Zhao, C., Luo, L.: CRLB for joint estimation of TDOA, phase, FDOA, and Doppler rate. J. Eng. **21**, 7628–7631 (2019)
24. Álvarez, R., Díez-González, J., Sánchez-González, L., Perez, H.: Combined noise and clock CRLB error model for the optimization of node location in time positioning systems. IEEE Access **8**(1), 31910–31919 (2020)
25. Álvarez, R., Díez-González, J., Strisciuglio, N., Perez, H.: Multi-objective optimization for asynchronous positioning systems based on a complete characterization of ranging errors in 3D complex environments. IEEE Access **8**(1), 43046–43056 (2020)
26. Yaro, A.S., Sha'ameri, A.Z.: Effect of path loss propagation model on the position estimation accuracy of a 3-dimensional minimum configuration multilateration system. Int. J. Integr. Eng. **10**(4), 35–42 (2018)

Solving the Two-Stage Supply Chain Network Design Problem with Risk-Pooling and Lead Times by an Efficient Genetic Algorithm

Ovidiu Cosma, Petrica Pop(✉), and Cosmin Sabo

Department of Mathematics and Computer Science,
North University Center at Baia Mare,
Technical University of Cluj-Napoca, Cluj-Napoca, Romania
{ovidiu.cosma,petrica.pop,cosmin_sabo}@cunbm.utcluj.ro

Abstract. Supply chain network design (SCND) nowadays represents an important part of Supply Chain Management (SCM) aiming to design a network such that to reduce the cost of the supply chain determined by the location of facilities and the flow of product between the selected facilities. In this paper we investigate a particular SCND, namely the two-stage supply chain network design problem with risk-pooling and lead times. We provide a mathematical model for this problem and as well a solution approach based on genetic algorithms for solving the problem. Computational experiments were performed on a set of instances and the obtained results prove that our proposed genetic algorithm provides good solutions within reasonable running times.

1 Introduction

Supply chains are part of our everyday lives. Almost everything that we purchase in a store comes to us as a part of a supply chain and managing and optimizing these networks is a complex, but important task. Designing a supply chain involves creating a network that incorporates all the facilities, means of production, products, and the transportation between the facilities. The design should also include details of the number and location of the facilities: plants, warehouses, and supplier base.

The two-stage supply chains involve manufacturers, distribution centers (DCs) and retailers and the aim of the supply chain network design (SCND) problem is to design a most efficient network possible such that to fulfill the demands of the retailers and ensure the lowest transportation cost. These problems have been intensively studied and several variants have been investigated as well.

The two-stage supply chain network design problems are referred to in the literature also as two-stage transportation problems. For these optimization

Á. Herrero et al. (Eds.): SOCO 2020, AISC 1268, pp. 509–519, 2021.
https://doi.org/10.1007/978-3-030-57802-2_49

problems Raj and Rajendran [11] proposed two scenarios: the first one, called Scenario 1, takes into consideration fixed costs associated to the routes in addition to unit transportation costs and boundless capacities of the DCs, for this variant of the problem we refer to [2,4,6,9,10]; while the second one, called Scenario 2, considers the opening costs of the DCs in addition to unit transportation costs. For more information on this variant of the problem, we refer to [1,5,11].

This work deals with a two-stage supply chain network design problem involving suppliers, distribution centers and retailers that takes into consideration the risk-pooling strategy that manages the demand uncertainty, see for more information Chen and Lin [3], and the lead time that is an important factor that affects the level of safety stock under customer request uncertainty. In the case of real applications, the lead times are determined by the pairs customer - supplier due to several aspects such as: distance, mean of transportation, production capacity, manufacturing technology, etc. For more information concerning lead times in supply chain management we refer to Yang and Geunes [12]. The objective of the investigated SCND problem with risk-pooling strategy and lead times is to determine and select the suppliers and the distribution centers fulfilling the demands of the customers under minimal transportation costs.

In the form considered in our paper, the problem was introduced by Park et al. [8]. They described a mathematical model of the problem based on nonlinear integer programming and as well a solution approach based on Lagrangian relaxation. The aim of this paper is to propose a novel solution approach for solving the investigated problem using genetic algorithms.

The rest of our paper is organized as follows: in Sect. 2, we define the considered SCND problem and present the mathematical model of the problem based on nonlinear integer programming. The developed genetic algorithm is described in Sect. 3 and the computational experiments and the obtained results are presented, analyzed and discussed in Sect. 4. Finally, in the last section, we conclude our work and present some future research directions.

2 Definition of the Problem

As considered by Park et al. [8], we make the following assumptions: the investigated SCND problem consists of a set of suppliers (manufacturers), distribution centers (warehouses) and retailers; the suppliers are not subject to production restrictions; the single item products are ordered from distribution centers to suppliers and then are shipped to retailers; every retailer gets products only from one distribution center and every distribution center is supplied only from a single supplier. The requests of the customer arrive daily at the retailer and these requests are supposed to be independent from each other and normally distributed. Every distribution center is supposed to have finite capacity and accomplishes immediate inventory in order to fulfill the requests of the retailers along with safety stock to deal with the uncertainty of the customer request to the retailers in order to reach the risk-pooling profits. Order lead times from distribution centers to suppliers are supposed to be deterministic, but dependent on the pairs DC's - suppliers.

The aim of the two-stage supply chain network design problem with risk-pooling and lead times is to find the routes to be opened and corresponding shipment quantities on these routes, and as well the inventory control decisions on the amount of products ordered, the amount of safety-stock at every distribution center such that the customer requests and all the shipment constraints are satisfied, and the total distribution costs are minimized.

In order to provide a mathematical model of the two-stage supply chain network design problem with risk-pooling and lead times, we will make use of the following notations of the involved parameters: l is the number of suppliers and k is the supplier identifier; m is the number of distribution centers (DCs) and j is the DC identifier; n is the number of retailers and i is the retailer identifier; g_k is the annual fixed setup cost for supplier k; f_{jk} is the annual fixed cost of locating DC j and assigning it to supplier k; p_{jk} is unit cost of transportation from supplier k to DC j; c_{ij} is unit cost of transportation from DC j to retailer i; A_j is the fixed inventory ordering cost at DC j; h_j is the unit per year inventory holding cost at DC j; B_j is the daily throughput capacity of the DC j; μ_i is the mean daily customer demand at retailer i; v_i is the variance of daily customer demand at retailer i; l_{jk} is the order lead time in days from supplier k to DC j; χ is the number of working days per year, α is the service level and z_α is the standard normal deviate such that $P(Z \leq z_\alpha) = \alpha$.

The considered decision variables are: the binary variables: z_k is 1 if the supplier k is used and 0 otherwise, u_j is 1 if the DC j is used and 0 otherwise, y_{jk} is 1 if the route from supplier i to DC j is used and 0 otherwise and x_{ij} is 1 if the route from DC j to retailer i is used and 0 otherwise, and the linear variable: Q_j representing the order quantity from DC j, r_j representing the reorder level at DC j and SS_j representing the safety stock level at DC j.

Then the investigated SCND problem with risk-pooling and lead times can be modeled as the following mixed integer problem described by Park et al. [8]:

$$\min \sum_{k=1}^{l} g_k z_k + \sum_{j=1}^{m}\sum_{k=1}^{l} f_{jk} y_{jk} + \sum_{i=1}^{n}\sum_{j=1}^{m}\sum_{k=1}^{l} \chi \mu_i p_{jk} x_{ij} y_{jk} + \sum_{i=1}^{n}\sum_{j=1}^{m} \chi \mu_i c_{ij} x_{ij}$$
$$+ \sum_{j=1}^{m} \sqrt{2\chi A_j h_j \sum_{i=1}^{n} \mu_i x_{ij}} + \sum_{j=1}^{m} z_\alpha h_j \sqrt{\sum_{i=1}^{n}\sum_{j=1}^{m} v_i l_{jk} x_{ij} y_{jk}}$$

$$s.t. \sum_{j=1}^{m} x_{ij} = 1, \ \forall\, i \in \{1, ..., n\} \tag{1}$$

$$x_{ij} \leq \sum_{k=1}^{l} y_{jk}, \ \forall\, i \in \{1, ..., n\}, \ j \in \{1, ..., m\} \tag{2}$$

$$\sum_{i=1}^{n} \mu_i x_{ij} \leq B_j, \ \forall\, j \in \{1, ..., m\} \tag{3}$$

$$\sum_{k=1}^{l} y_{jk} \leq 1, \ \forall\, j \in \{1, ..., m\} \tag{4}$$

$$y_{jk} \leq z_k, \forall\, j \in \{1, ..., m\},\ k \in \{1, ..., l\} \tag{5}$$

$$x_{ij}, y_{jk}, z_k \in \{0, 1\},\ \forall\, i \in \{1, ..., n\},\ j \in \{1, ..., m\},\ k \in \{1, ..., l\} \tag{6}$$

The objective function minimizes the total distribution cost: the fixed costs, transportation per-unit costs and on-hand/safety-stock inventory costs. Constraints (1) and (4) guarantee that the single sourcing assumption is satisfied for every retailer and every distribution center. Constraints (2) and (5) guarantee that every retailer and every opened distribution center must be assigned to exactly one of their possible providers. Constraints (3) guarantee that the storage capacities of the distribution centers are not surpassed. The last constraint ensure the integrality of the decision variables.

3 Description of the Proposed Genetic Algorithm

In this section, we propose a genetic algorithm for solving the two-stage supply chain network design problem with risk-pooling and lead times.

The chromosomes have been defined in such a way as to allow for a compact representation of the problem solutions and to allow the exploration of the entire space of potential feasible solutions. Each chromosome consists of two integer arrays. The first array has m genes that represent the links between DCs and suppliers. We will call this array DS. The value of the DS_j gene represents the supplier allocated to distribution center j. If $\sum_k y_{jk} = 0$, then there is no supplier allocated to DC_j and the DS_j gene is void. The second array hass n genes that represent the links between retailers and DCs. We will call this array RD. The value of the RD_i gene represents the distribution center allocated to retailer i. Because all retailers must be assigned to an exactly one DC, none of the genes in the RD array will be void.

An example of a chromosome is shown in Fig. 1a. The genes arrays of this chromosome are shown in Fig. 1b. The fitness of a chromosome is defined by the value of the objective function of the SCND calculated according to the chromosome genes.

The initial population consists of N chromosomes that are randomly constructed based on the following algorithm:

1. For each gene RD_i, choose a random number j, $j \in \{1, ..., m\}$, $RD_i \leftarrow j$.
2. If after choosing RD_i, $\sum_i \mu_i x_{ij} > B_j$, then another random value from the same set is chosen for RD_i, until the daily throughput capacities of all the DCs are respected.

This algorithm for generating the chromosomes may get stuck at step 2 in an infinite loop because of the limited daily throughput capacities of the DCs. For dealing with this aspect, if the selection of the RD_i gene fails after a certain

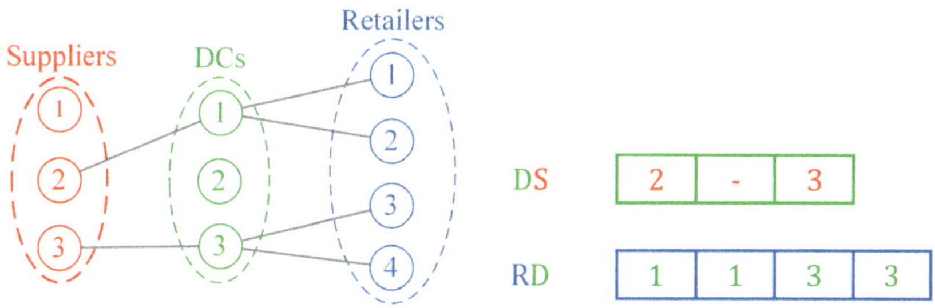

(a) An example of a chromosome (b) The genes arrays

Fig. 1. An example of a chromosome and its corresponding gene arrays

number of attempts, in our case m, then the entire operation is canceled, and the chromosome construction is restarted from scratch. After the completion of the RD array, we construct the DS array, as follows:

3. An integer k is randomly chosen, such that $k \in \{1, ..., l\}$. This will indicate the supplier used in the solution represented by the constructed chromosome.
4. If $\sum_i x_{ij} \geq 1$ then $DS_j \leftarrow k$, else DS_j will be void.

All the chromosomes generated by this algorithm will use a single supplier, but this does not represent a limitation of the solution search space, because the crossover operator can increase the number of suppliers up to l.

Two examples of random chromosomes generated using the described algorithm are shown in Fig. 2.

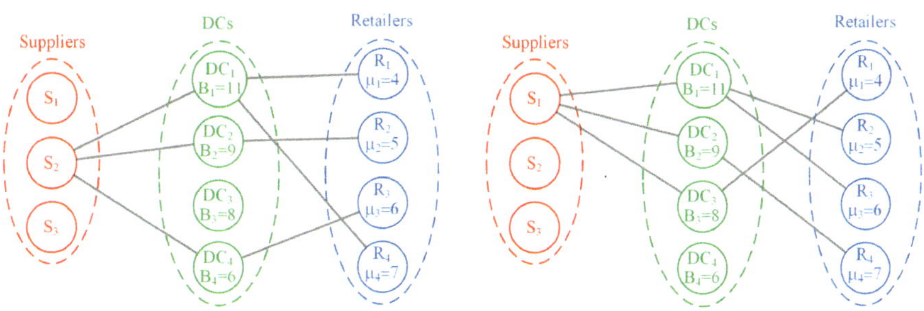

(a) First parent chromosome ($p1$) (b) Second parent chromosome ($p2$)

Fig. 2. Examples of random generated chromosomes

The initial population is processed by the population selection mechanism, resulting the first current population.

The crossover operator selects two parent chromosomes $p1$ and $p2$, from which it creates an offspring. The selection strategy is a combination between elitist and random selection. The first parent, $p1$, is always selected from the best 20% chromosomes in the current population. The second parrent, $p2$ is selected randomly from the entire current population. Each gene of the offspring is taken either from $p1$ or from $p2$, with equal probabilities.

An example of crossover operation is presented in Fig. 3. The parent chromosomes $p1$ and $p2$ are those illustrated in Fig. 2a and 2b, and the offspring chromosome o is shown in Fig. 3a.

The crossover operation begins with the retailers. They are processed in random order. The RD_i genes are taken with equal probabilities from $p1$ or from $p2$. If the selection of the RD_i gene exceeds the daily throughput capacity for DC_j, then the operator tries to take the gene from the other parent. If this operation is also not possible, then the DC allocated to retailer i is randomly chosen, such that $RD_i \in \{1, ..., m\}$, until the daily throughput capacity of each DC is respected. This processing could lead to an infinite loop, because of the DCs limited daily throughput capacities. If the processing does not finish after a certain number of retries, namely $m - 1$ if the RD_i genes of the parents are identical and $m - 2$ otherwise, than the whole crossover operation is abandoned, and the crossover operator restarts from scratch with the same two parents $p1$ and $p2$.

Figure 3b presents a situation that leads to an infinite loop when combining parents $p1$ and $p2$, none of the DCs can be allocated to retailer R_4 because of the previous allocations.

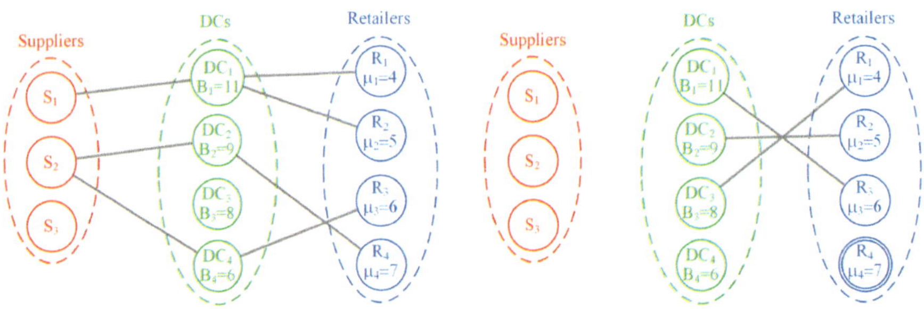

(a) Offspring chromosome (o) (b) Bad crossover

Fig. 3. Crossover examples

If the configuration of the RD array ends successfully, then the crossover operation continues with the configuration of the DS array. Each DS_j gene is taken either from $p1$ or from $p2$, with equal probabilities. The following types of problems may occur because of the random selection of the DS_j genes:

1. None of the suppliers is allocated to any DC,
2. DCs with allocated suppliers, but with no links to any retailer,
3. DCs allocated to retailers, that have no allocated supplier.

The last two types of problems that could appear when combining the two parrents $p1$ and $p2$ are shown in Fig. 4. Supplier S_2 is allocated to DC_4, but DC_4 is not allocated to any retailer. DC_3 is allocated to retailer R_1, but it has no allocated supplier.

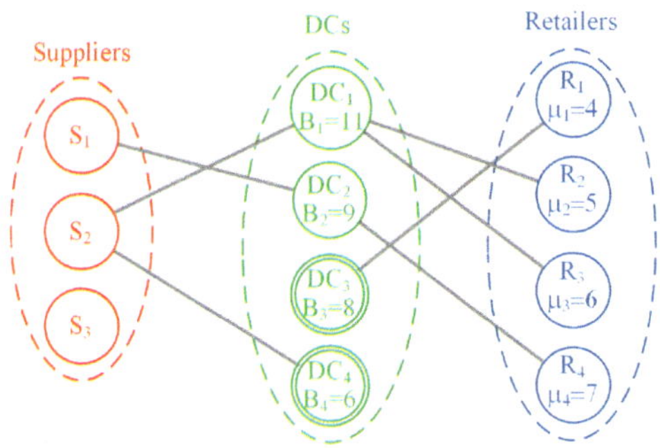

Fig. 4. An example showing the encountered problems by the crossover operator

The invalid offspring created by the crossover operator are corrected as follows:

1. All the DS_j genes corresponding to unallocated distribution centers are replaced with void genes. DC j is unallocated if $\sum_i x_{ij} = 0$,
2. If all the genes DS_j are void, then a retailer i and one of the two parents $p1$ and $p2$ are randomly chosen. Then for each allocated DC j, $DS_j \leftarrow p.DS_{RD_i}$, where p denotes the chosen parent and $p.DS_{RD_i}$ is the supplier allocated to the DC allocated to retailer i in the parrent chromosome p.
3. Randomly choose a supplier for each allocated DC j that does not have an allocated supplier. That is: $\sum_i x_{ij} \geq 1$, and $\sum_k y_{jk} = 0$. If there is a supplier k allocated to DC j in one of the two parrent chromozomes $p1$ and $p2$, then $DS_j \leftarrow k$. Else $DS_j \leftarrow random\ value \in \{1, ..., l\}$.

The new generation of chromosomes created using the crossover operator is processed by the population selection mechanism, resulting a new current population.

Each offspring created by the crossover operator can be altered by mutation with a probability of 0.1. Two mutation operators have been defined. The first mutation operator works as follows:

1. Randomly choose an allocated supplier k_1.
2. Randomly choose another supplier $k_2 \in \{1, ..., l\} \backslash \{k_1\}$.
3. Move to k_2 each DC that was allocated to k_1.

The second mutation operator works as follows:

1. Randomly choose an allocated DC j_1,.

2. Move each retailer a of j_1 to another allocated distribution center j_2 that has sufficient remaining daily throughput capacity. That is: $\sum_i \mu_i x_{ij_2} + \mu_a \leq B_{j_2}$. We name the retailers of DC j all the retailers DC j is allocated to. Retailer i is a retailer of DC j if $RD_i = j$.
3. If after step 2 DC j_1 becomes unallocated, then $DS_{j_1} \leftarrow void$.

The population selection mechanism merges the newly created generation with the current population (if one exists) and removes any duplicates. Then the resulting population is sorted by fitness value and only the best D chromosomes are retained for the new current population. All the rest are destroyed.

The genetic parameters have an important impact on the performance of the GAs. That is why in our developed GA the values of the parameters have been chosen based on preliminary computational experiments and statistical analysis as follows: the dimension of the current population D is 10000, the number of chromosomes in the initial population is $3 \times D$, the algorithm is stopped when the best known solution was not improved over the last 30 generations of chromosomes, the mutation probability is 0.1 and the number of crossover operations performed for completing each new generation of chromosomes is set to minimum $3 \times D$ and maximum $10 \times D$.

4 Computational Results

This section is dedicated to the achieved computational results with the aim of assessing the effectiveness of our developed approach for solving the two-stage supply chain network design problem with risk-pooling and lead times.

We performed our computational experiments for solving the investigated SCND problem on a set of 48 instances randomly generated with varying characteristics. Since the test instances used by Park et al. [8] could not be obtained in the literature, we generated new instances similar to that in Park et al. [8]: we considered six instance dimensions with the number of suppliers between 5 and 10, the number of DC's between 10 and 15 and the number of retailers between 20 and 40. All the other parameters of the problem have been chosen in the same way as Park et al. [8] did. All the instances used in our computational experiments are available at the address: https://sites.google.com/view/tstp-instances/.

We coded our algorithm in Java 8 and for each instance we carried out 30 independent trials, on a PC with Intel Core i3-8100 3.6 GHz, 8 GB RAM, Windows 10 Education 64 bit operating system.

Table 1 displays the computational results achieved by our genetic algorithm. The first two columns indicate the number of the instance and its size, the third and the fourth column show the cost of the best and average solutions achieved by our GA and fifth column displays the necessary average computational times in seconds in order to achieve the corresponding best solution in each run. The last columns contains the percentage gap calculated as follows: $\%gap = 100 \times (Best\ sol. - Average\ sol.)/Best\ sol.$, where *Best sol.* and *Average sol.* are the

Table 1. The computational results obtained by our proposed genetic algorithm

No.	Size (n × m × l)	Best solution	Average solution	Time [s]	Gap [%]
1	20 × 10 × 5	2483543.90	2483543.90	3.31	0.00
2	20 × 10 × 5	2169068.53	2169068.53	2.52	0.00
3	20 × 10 × 5	2462433.23	2462433.23	2.97	0.00
4	20 × 10 × 5	2828767.13	2828767.13	2.65	0.00
5	20 × 10 × 5	2688369.62	2688369.62	2.61	0.00
6	20 × 10 × 5	3792749.08	3792749.08	3.93	0.00
7	20 × 10 × 5	2517630.23	2517630.23	2.45	0.00
8	20 × 10 × 5	2132519.70	2132519.70	2.98	0.00
9	20 × 10 × 7	1809362.41	1809362.41	3.01	0.00
10	20 × 10 × 7	2410766.11	2410766.11	2.35	0.00
11	20 × 10 × 7	2458466.80	2458466.80	2.69	0.00
12	20 × 10 × 7	2408937.73	2408937.73	3.09	0.00
13	20 × 10 × 7	3134192.05	3134192.05	2.67	0.00
14	20 × 10 × 7	3676126.64	3676126.64	3.68	0.00
15	20 × 10 × 7	2914023.54	2914023.54	2.42	0.00
16	20 × 10 × 7	2802350.39	2802350.39	3.61	0.00
17	20 × 15 × 7	2149911.36	2149911.36	4.23	0.00
18	20 × 15 × 7	2559354.34	2559354.34	5.47	0.00
19	20 × 15 × 7	2089945.72	2089945.72	4.43	0.00
20	20 × 15 × 7	5208514.45	5208514.45	7.86	0.00
21	20 × 15 × 7	2870720.36	2870720.36	4.12	0.00
22	20 × 15 × 7	2978521.90	2978521.90	5.04	0.00
23	20 × 15 × 7	2564940.47	2564940.47	3.82	0.00
24	20 × 15 × 7	2513592.75	2513592.75	3.89	0.00
25	20 × 15 × 10	2393957.14	2393957.14	3.75	0.00
26	20 × 15 × 10	2232904.43	2232904.43	4.30	0.00
27	20 × 15 × 10	2210937.88	2210937.88	3.93	0.00
28	20 × 15 × 10	4141506.50	4141506.50	4.13	0.00
29	20 × 15 × 10	2153060.98	2153060.98	3.65	0.00
30	20 × 15 × 10	3018345.73	3018345.73	5.18	0.00
31	20 × 15 × 10	3197452.54	3197452.54	4.97	0.00
32	20 × 15 × 10	2632877.10	2632877.10	5.32	0.00
33	40 × 15 × 7	4635062.77	4635062.77	9.81	0.00
34	40 × 15 × 7	4454971.77	4456667.03	15.65	0.04
35	40 × 15 × 7	4063150.41	4064577.94	10.85	0.04
36	40 × 15 × 7	8271734.97	8271929.49	13.95	0.00
37	40 × 15 × 7	4446377.60	4446377.60	13.83	0.00
38	40 × 15 × 7	4485044.81	4485099.44	17.03	0.00
39	40 × 15 × 7	6129400.59	6129400.59	10.02	0.00
40	40 × 15 × 7	4925501.24	4925501.24	11.86	0.00
41	40 × 15 × 10	7082858.83	7090271.95	16.06	0.10
42	40 × 15 × 10	4669520.51	4673226.91	9.76	0.08
43	40 × 15 × 10	4710411.20	4712795.42	13.70	0.05
44	40 × 15 × 10	4408029.05	4416711.77	18.14	0.20
45	40 × 15 × 10	4924958.29	4929541.43	25.78	0.09
46	40 × 15 × 10	6124330.49	6133671.15	19.37	0.15
47	40 × 15 × 10	5651849.42	5663210.26	21.96	0.20
48	40 × 15 × 10	5456073.11	5470383.73	12.15	0.26

costs of the best respectively the average solutions achieved by our GA in the 30 runs of each instance.

Analyzing the results displayed in Table 1, we can remark that in 38 out of 48 instances, our GA provided the same best solutions in all the 30 runs, and for the other instances the percentage gap is at most 0.26%, fact that proves the stability of our proposed solution approach. The necessary average computational time value reported in seconds in order to achieve the corresponding solutions is at most 25.78 s.

5 Conclusions

In this paper an efficient genetic algorithm was developed in order to solve the two-stage supply chain network design problem with risk-pooling and lead times.

The results obtained through the use of our proposed approach are very promising, thus providing a reason to apply this kind of approach to other supply chain network design problems, with the aim of assessing the real practicality of the described method. Future research will focus on defining, detailing and adapting some other genetic operators (crossover, mutation and selection) to our GA and improving the developed solution approach by combining it with local search methods. In addition, our developed approach is going to be tested in the case of larger size instances of the problem.

References

1. Calvete, H., Gale, C., Iranzo, J.: An improved evolutionary algorithm for the two-stage transportation problem with fixed charge at depots. OR Spectr. **38**, 189–206 (2016)
2. Calvete, H., Gale, C., Iranzo, J., Toth, P.: A matheuristic for the two-stage fixed-charge transportation problem. Comput. Oper. Res. **95**, 113–122 (2018)
3. Chen, M.S., Lin, C.T.: Effects of centralization on expected costs in multi-location newsboy problem. J. Oper. Res. Soc. **40**(6), 597–602 (1989)
4. Cosma, O., Pop, P.C., Dănciulescu, D.: A novel matheuristic approach for a two-stage transportation problem with fixed costs associated to the routes. Comput. Oper. Res. **118**, 104906 (2020)
5. Cosma, O., Dănciulescu, D., Pop, P.C.: On the two-stage transportation problem with fixed charge for opening the distribution centers. IEEE Access **79**(1), 113684–113698 (2019)
6. Cosma, O., Pop, P.C., Pop Sitar, C.: An efficient iterated local search heuristic algorithm for the two-stage fixed-charge transportation problem. Carpathian J. Math. **35**(2), 153–164 (2019)
7. Holland, J.H.: Adaptation in Natural and Artificial Systems: An Introductory Analysis with Applications to Biology, Control and Artificial Intelligence. MIT Press, Cambridge (1992)
8. Park, S., Lee, T.-E., Sung, C.S.: A three level supply chain network design model with risk-pooling and lead times. Transp. Res. Part E **46**, 563–581 (2010)

9. Pop, P.C., Matei, O., Pop Sitar, C., Zelina, I.: A hybrid based genetic algorithm for solving a capacitated fixed-charge transportation problem. Carpathian J. Math. **32**(2), 225–232 (2016)
10. Pop, P.C., Sabo, C., Biesinger, B., Hu, B., Raidl, G.: Solving the two-stage fixed-charge transportation problem with a hybrid genetic algorithm. Carpathian J. Math. **33**(3), 365–371 (2017)
11. Raj, K.A.A.D., Rajendran, C.: A genetic algorithm for solving the fixed-charge transportation model: two-stage problem. Comput. Oper. Res. **39**(9), 2016–2032 (2012)
12. Yang, B., Geunes, J.: Inventory and lead time planning with lead-time-sensitive demand. IIE Trans. **33**(2), 439–452 (2007)

Genetic Algorithm Optimization of Lift Distribution in Subsonic Low-Range Designs

Rubén Ferrero-Guillén[1(✉)], Rubén Álvarez[2], Javier Díez-González[1], Álvaro Sánchez-Fernández[1], and Hilde Pérez[1]

[1] Department of Mechanical, Computer, and Aerospace Engineering, Universidad de León, 24071 León, Spain
{rferrg00,asancf03}@estudiantes.unileon.es, {jdieg,hilde.perez}@unileon.es

[2] Positioning Department, Drotium, Universidad de León, 24071 León, Spain
ruben.alvarez@drotium.com

Abstract. The optimization of the lift distribution is an essential analysis in the wing design segment of every aircraft project. Although it has been demonstrated that the optimal solution follows an elliptic distribution, there is no known relation between the parameters that define this distribution and its similarity to the elliptical one. Therefore, there is no direct approach for obtaining an exact solution, existing methodologies such as CFD simulations which require of a considerable amount of time and resources to offer accurate results. The methodology followed throughout this paper involves the application of metaheuristic techniques, such as genetic algorithms, in order to optimize the lift distribution obtained through the Prandtl lifting-line theory. Results show that the genetic algorithm proposed is able to obtain a satisfactory solution within a reasonable time.

Keywords: Genetic algorithm · Lift distribution · Wing design · Elliptical lift distribution

1 Introduction

Wing design stands as one of the most crucial analysis in every aircraft project, being the main contributor to the force that lifts the aircraft as well as playing a decisive role in the efficiency of the plane. Hence, it is critical that the wings provide the amount of lift required without deriving in other negative effects such as aerodynamic resistance, stall inception and lesser fuel capacity among others.

Therefore, companies undergoing the development of a new aircraft invest a substantial amount of resources for the R+D+i of the wing design especially the long-range models. Besides, due to the concurrent engineering fundamentals [1],

Á. Herrero et al. (Eds.): SOCO 2020, AISC 1268, pp. 520–529, 2021.
https://doi.org/10.1007/978-3-030-57802-2_50

the delay of a specific section of a project, such as wing design, may cause major consequences in other departments to the point of a complete setback of the project.

Moreover, the research and development of a specific airfoil is a rather demanding project, requiring severe research in both CFD (Computer Fluid Dynamics) simulations [2] and empirical experiments like wind tunnel testing [3]. Requiring these simulations of an extensive amount of time and resources to execute.

One of the most decisive analysis of the wing design is the optimization of the lift distribution. In an ordinary wing, the lift output usually does not remain constant and it varies from the distance from the root of the wing, due to the existence of variables such as the taper ratio λ, torsion angle α_t and the wing incidence α_{set} [4]. Hence, the lift output of every section of the wing varies, creating a lift distribution. It is concluded from multiple investigations that the optimal lift distribution is the elliptic one [5,6], and every deviation from this distribution result in negative consequences such as an increase in fuel consumption, or even develop the stall phenomenon [7] and its undesired consequences.

However, the optimization of this desired result is not easily achieved, being no known relation that could be drawn between the parameters that define the lift distribution of a wing and its similarity to an elliptical distribution. As a consequence, there is no direct approach available that could be used for obtaining an exact solution for this problem.

Nonetheless, the aeronautic industry have developed a series of methodologies [8,9] that could potentially obtain an exact solution. However, these techniques rely heavily on CFD simulations, which require of a considerable amount of resources when searching for a precise solution.

On the other hand, there are other techniques which do not require of CFD simulations and offer an approximated result [10], implementing numerical methods. However, the results of these methodologies may vary depending on the initial conditions of the problem.

In the endeavor to pursue a finer solution, we propose the application of metaheuristic techniques, such as genetic algorithms, as for finding a solution of this problem that does not rely on expensive simulations.

In the previous years, we have observed the rise of these methodologies over various disciplines, from economics and decision making [11] to driving optimization [12], positioning systems [13,14] and even aerodynamics in other aspects of wing design [15]. Hence, we propose the application of this algorithm in this particular problem with the intent of obtaining the combinations of parameters that optimizes the lift distribution of our wing in a reasonable time.

2 Description of the Problem

The wings are the main source of lift in an aircraft, this force is generated from the pressure difference from the static pressure in between the upper and lower surfaces of the airfoil as air flows through it, thus generating a force that pushes the wing upwards. The amount of force generated is heavily dependent on the geometry of the airfoil and does not remain constant along the chord or length of the airfoil.

When analysing the performance of an airfoil, it is preferred the term of lift coefficient of the airfoil C_l over its force of lift, which allow us to exclude all the environmental parameters out of the equation and adimensionalizate it by the airfoil's chord. This lift coefficient can be calculated in empirical test such as wind tunnels.

$$C_l = \frac{l}{qc} \tag{1}$$

where l is the lift force, q is the dynamic pressure and c is the chord of the airfoil.

The Eq. (1) provide the lift coefficient of an airfoil, a section of the wing, so in order to obtain the total lift coefficient of the wing C_L, more additional parameters are required as rarely the airfoil of a wing remains constant.

Therefore, given the airfoil in the root of the wing, in this case the NACA 23024, it is possible to define the shape of our wing as a function of a series of parameters, such as the wing surface S, the aspect ratio AR, the taper ratio λ, the twist angle α_t and the wing incidence α_{set}.

The aspect ratio, along the wing surface, provides the scope of the wing, and it is defined as the wingspan of the wing squared divided by the wing surface.

The taper ratio indicates the narrowing of the wing from root to tip. This narrowing serves multiple motives but mainly structural ones. Although its value depends on the project's specifications, we can obtain its value by dividing the chord's length at the tip by the chord's length at the root.

As for the twist angle, this parameter indicates the deviation of the angle of attack along the wingspan. The angle of attack of a wing is the angle formed between the mean aerodynamic chord of an airfoil and the incident flow. There is a direct relation between the angle of attack and the lift generated, however, over a certain value which depends on the airfoil, the airfoil no longer generates lift, knowing this phenomenon as stall [16]. The twist angle serves as a way to prevent this event from happening as well as adjusting the lift distribution to obtain its optimized value.

Finally, the wing incidence is the angle formed between the fuselage center line and the main aerodynamic chord. This parameter allows the wing to have a higher angle of attack above all, increasing the lift budget but compromising the stall of the wing.

All these parameters are the responsible for causing an irregular lift distribution along the wingspan, which usually tends to decrease from the distance from the root, mainly for structural purposes. Although there are multiple methodologies for obtaining this lift distribution, one of the most expanded and well rounded techniques is the Prandtl Lifting-Line Theory [17] from which we can obtain the value of the wing distribution. Despite being a traditional theory, it is still being used and codified in CFD simulations [18].

In conclusion, thanks to Prandtl's theory, it is possible to obtain the lift distribution of a wing as a function of the wing surface S, the aspect ratio AR, the taper ratio λ, the twist angle α_t and the wing incidence α_{set} as well as other aerodynamic parameters linked to the airfoil selected.

$$C_{L_\alpha} = \frac{4b\,\mu}{\overline{c}} \tag{2}$$

$$\mu = \frac{1}{\alpha_0 - \alpha} \cdot \sum_{n=1}^{N} A_n \sin(n\theta)\left(1 + \frac{\mu\, n}{\sin(\theta)}\right) \tag{3}$$

where b is the wingspan, $\overline{c}$ is the main aerodynamic chord, θ the polar coordinates, n the discretization, α the segment's angle of attack, α_0 the zero-lift angle of attack and A_n the coefficients of each point.

Following the Eqs. (2, 3) obtained from Prandtl's theory, it is possible to plot the lift distribution of a certain wing. As multiple studies have proved before [19], the optimal lift distribution of any sub-sonic wing design is always the elliptic distribution. Any deviation from this optimal distribution shall derive in undesired consequences such as an increase in the aerodynamic resistance, thus an increase in fuel consumption [20].

Nonetheless, there is no direct relation which could be drawn between these aerodynamic parameters and the likeness of the lift coefficient function to the ellipse distribution. Likewise, the most expanded methodology [8,9] to confront this problem relies on assumptions such as incompressible flow which is only valid on considerable low speed scenario. Besides, these approximations usually require a great deal of simulations in CFD software and real life experiments such as wind tunnels, increasing the global cost of the project.

Hence, we propose a different approach, relying on the application of heuristic algorithms such as genetic algorithms, as a way to achieve a more adequate solution than traditional methods.

3 Genetic Algorithm

Therefore, as a consequence of the lack of a viable exact solution that does not require the assumption of unfeasible conditions or the execution of laborious CFD simulations, we propose to approach this problem with metaheuristic methodologies. Although there are multiple algorithms that could prove suitable for this problematic situation, we propose the application of genetic algorithms as a result of their exploration and solution intensifying capabilities.

We have also observed the rise of genetic algorithms optimizations over the last years in a variety of disciplines, from economics and decision making [11], to optimizing driving routes [12], positioning [14] and even aerodynamic designs [15]. Therefore, their application to this problems seems feasible.

The genetic algorithm we propose will carry the parameters that defines the lift distribution, being these the aspect ratio, the taper ratio, the twist angle and the wing incidence. However, in this paper we are studying the lift distribution of a low range subsonic aircraft [21], hence not every value of these parameters can be considered acceptable. We can determinate from the design specifications as well as other similar projects that the parameters must be within a certain region, showed in Table 1.

Table 1. Parameters from the wing design

GA wing parameters			Aerodynamic constants	
Parameter	Max value	Min value	Parameter	Value
AR	13	11	S	$6.22\ m^2$
λ	0.7	0.3	α_0*	-1.25 rad
α_t	$-3°$	$-1°$	$\alpha_{2\pi}$*	2π rad
α_{set}	$3°$	$0°$		

**Values obtained from airfoil NACA 23024*

Furthermore, the proposed algorithm would carry all these variables in each and every individual of the population, coded in binary. From the difference in the range of these parameters we have created different length arrays for each variable, with a criteria for separating the digits from the whole number to the decimal part, as well as if it has a negative or positive value.

$$\alpha_{set} = \underbrace{1}_{\text{sign}} \quad \underbrace{010}_{\text{whole number}} \quad \underbrace{0110101101}_{\text{decimal number}} = 2.419°$$

These parameters define the lift distribution, hence, in order to optimize this distribution we must search the combination of parameters that generates the most likeness to the elliptical one. As a result, we can build a fitness function based on the difference of the lift distribution generated from these parameters and the optimal ellipse. It is possible to compute this difference with the MAE (Mean Absolute Error) or the RMSE (Root Mean Square Error).

The MAE is considered among some authors as generally the best method for evaluating a model performance [22,23], being the preferred methodology for evaluating uniform error distributions, nonetheless is a well rounded valid method.

On the other hand, the RMSE proves a better performance in normal error distributions, however, the bigger difference from the MAE is that the RMSE penalizes heavily large errors that deviate from the standard value [24].

Although both methodologies would prove suitable for this problem, the best approach is the RMSE, for a large singular error deviation may be less desirable than a low uniformed error distribution.

However, certain parameters such as the aspect ratio AR or surface of the wing S will define the dimensions of the wing, thus the scope of the lift distribution. Hence, the scope of the ellipse used to measure the elliptical likeness of the current lift distribution shall display similar dimensions with it. As a consequence, a new ellipse will be generated with each individual of the genetic algorithm.

Thence, it is possible to obtain the coordinates of the ellipse desired by adapting the ellipse equation so that it contains the lift coefficients at the root and the wingspan of the wing as they represent the intersection of the ellipse with the 2-D axis.

$$y_{\text{Ellipse}} = \sqrt{\left(1 - \frac{x^2}{C_{L_{\text{root}}}^2}\right)\frac{b^2}{2}} \quad (4)$$

where x is the discretization of the wing, b is the wingspan and $C_{L_{\text{root}}}$ the value of the lift coefficient at the root of the wing.

Nonetheless, following this approach, a more sizeable lift distribution might present a bigger RMSE than a smaller one due to its actual dimensions, even if it presents a much more suited likeness to the proposed ellipse. Still, this impediment could be easily arranged by adimensionalizating the RMSE, dividing it by the maximum value of the ellipse.

Furthermore, it is important to clarify that not every combination of these aerodynamic parameters is acceptable. Depending on the specifications of the aircraft project, these parameters should stay within certain limits. As a solution for this issue, we have created a correction factor κ which is a function of all these parameters, being its value bigger the farthest a variable stray from its expected value and null when it stays within the range specified in Table 1. Hence, the final value of κ would be added to the RMSE of the likeness of the lift distribution in order to penalize extreme and unfeasible combinations.

For the calculation of κ, we propose the following equations:

$$\kappa_{AR} = \max\left(1, \frac{|AR - AR_{\max}|}{|AR_{\max} - AR_{\min}|}, \frac{|AR - AR_{\min}|}{|AR_{\max} - AR_{\min}|}\right) \quad (5)$$

$$\vdots$$

$$\kappa = \left(4 - \kappa_{AR} - \kappa_{\lambda} - \kappa_{\alpha_t} - \kappa_{\alpha_{\text{set}}}\right) \cdot \varepsilon \quad (6)$$

where $AR_{\max}$ and $AR_{\min}$ are the maximum and minimum values of the interval AR specified in Table 1, and ε is the coefficient whose purpose is to determine the intensity of the κ penalization

Therefore, we can propose the following fitness functions, with MAE and RMSE error evaluation.

$$\text{ff}_{\text{RMSE}} = \frac{1}{C_{L_{\text{root}}}}\sqrt{\frac{\sum_{i=1}^{n}(y_{C_{L\alpha}} - y_{\text{Ellipse}})^2}{n}} + \kappa \quad (7)$$

$$\text{ff}_{\text{MAE}} = \frac{1}{C_{L_{\text{root}}}}\frac{\sum_{i=1}^{n}|y_{C_{L\alpha}} - y_{\text{Ellipse}}|}{n} + \kappa \quad (8)$$

Lastly, the algorithm shall rely on tournament based selection methodology [25] with 3 competing individuals. On the other hand, for the crossover methodology, we have implemented a multipoint based crossover. Likewise, the algorithms shall operate with small percentages of elitism and mutation, deduced in base of the results of previous simulations (Table 2).

Table 2. Genetic algorithm configuration

GA	Selection
Population size	60
Selection technique	Tournament 3
Elitism	5%
Crossover	Multi-point
Mutation	3%
Convergence criteria	50 generations or 80% individual equals
ε data validation	$5 \cdot 10^{-3}$

4 Results

Once set up and executed in the Python programming language, the algorithm showed a rapid convergence to an acceptable solution in a short interval of time. Due to the circumstances of this problem, a limited population had sufficed to reach the desired solution in an adequate number of generations, proving that this method could be considered as a viable alternative over long and resource-heavy CFD simulations. Therefore, the genetic algorithm proposed have obtained the following solution:

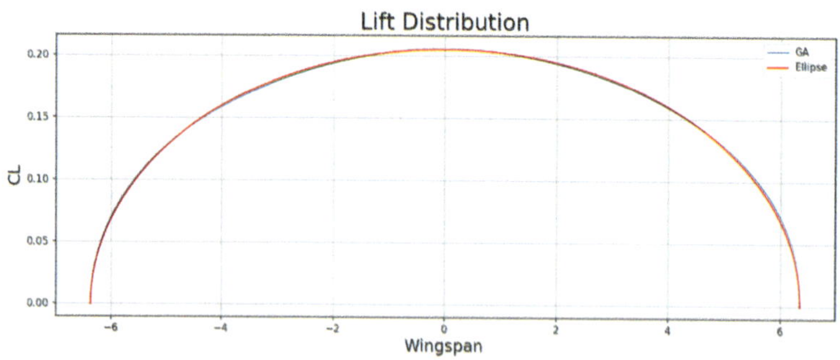

Fig. 1. Lift Distribution provided by GA. The blue curve represents the lift distribution through the wingspan (meters), provided by the RMSE variation of the genetic algorithm

As shown in Fig. 1, the lift distribution provided by the genetic algorithm proves a convenient likeness to the elliptical distribution desired, proving the suitability of this methodology.

Figure 2 shows the evolution of the RMSE along generations, thus we can appreciate the accelerated convergence to the final solution within a couple generations (Table 3).

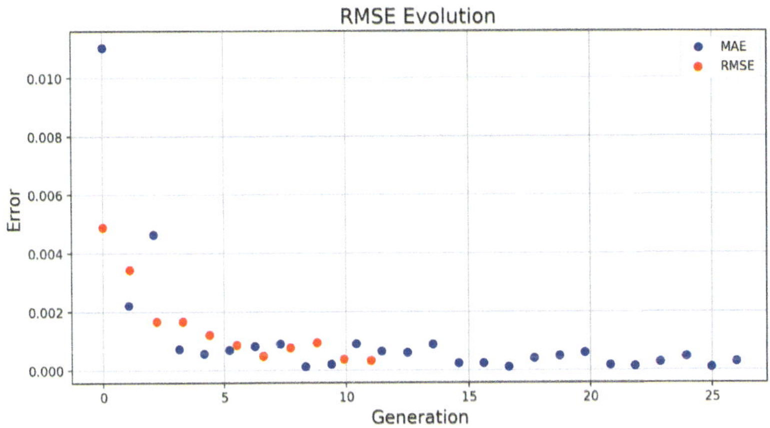

Fig. 2. Genetic Algorithm's lowest error for every generation with RMSE and MAE adaptations. The RMSE variation converged in generation 11 unlike the MAE where the convergence criteria was fulfilled in generation 26

Table 3. Results of the Genetic Algorithm

	RMSE variation	MAE variation
AR	11.5	10.8
α_t	$-1.738°$	$-1.684°$
α_{set}	$0.403°$	$0.234°$
λ	0.817	0.832
Lowest error	$4.949 \cdot 10^{-4}$	$\mathbf{3.88 \cdot 10^{-4}}$

Both variations of the genetic algorithm have proven to be satisfactory. The MAE variation showed a lower error in the best individual but the RMSE was rather stable and had a faster convergence.

5 Conclusion

Wing design represents a substantial analysis in every aircraft project, being one of the fields with the largest amount of resources invested in. One of the most important steps of the wing design is the optimization of the lift distribution, as the airfoil of the wing usually suffer a deviation from its original form in the root. It is concluded that the optimized lift distribution is the elliptical one, thus every deviation from this ideal distribution will result in undesired consequences such as an increase in fuel consumption.

However, there is no know relation between the aerodynamic parameters that define the wing and the likeness of the lift distribution to an ellipse. This problem has been confronted by numerous methodologies, from CFD computer simulations that could provide an exact solution, thought requiring of a considerable amount of time and resources to execute, to numerical methods that offer a close approximation.

In this paper we have proposed the application of metaheuristic techniques such as genetic algorithms to confront this problem in the pursue of an acceptable solution that does not require of any laborious simulations. We have discussed the different approaches for constructing the genetic algorithms with multiple fitness functions and we have made the adjustments required.

Results show that the genetic algorithm proposed is able to reach a robust solution in a reasonable time with both fitness functions designed, being thus fulfilled the main objective of this paper.

References

1. Prasad, B.: Concurrent Engineering Fundamentals, vol. 1. Prentice Hall PTR, NJ (1996)
2. Anderson, J.D., Wendt, J.: Computational Fluid Dynamics, vol. 206. Springer (1995)
3. Barlow, J.B., Rae, W.H., Pope, A.: Low-Speed Wind Tunnel Testing (1999)
4. DeYoung, J.: Theoretical additional span loading characteristics of wings with arbitrary sweep, aspect ratio, and taper ratio (1947)
5. Multhopp, H.: Methods for calculating the lift distribution of wings (subsonic lifting-surface theory). Aeronautical Research Council, London (1950)
6. Weissinger, J.: The Lift Distribution of Swept-Back Wings (1947)
7. McCroskey, W.J.: The phenomenon of dynamic stall. Technical report, National Aeronuatics and Space Administration Moffett Field Ca Ames Research ... (1981)
8. Albano, E., Rodden, W.P.: A doublet-lattice method for calculating lift distributions on oscillating surfaces in subsonic flows. AIAA J. **7**(2), 279–285 (1969)
9. Schrenk, O.: A simple approximation method for obtaining the spanwise lift distribution. Aeronaut. J. **45**(370), 331–336 (1941)
10. Yu, Y., Lyu, Z., Xu, Z., Martins, J.R.R.A.: On the influence of optimization algorithm and initial design on wing aerodynamic shape optimization. Aerosp. Sci. Technol. **75**, 183–199 (2018)
11. Metawa, N., Hassan, M.K., Elhoseny, M.: Genetic algorithm based model for optimizing bank lending decisions. Exp. Syst. Appl. **80**, 75–82 (2017)
12. Mohammed, M.A., Abd Ghani, M.K., Hamed, R.I., Mostafa, S.A., Ahmad, M.S., Ibrahim, D.A.: Solving vehicle routing problem by using improved genetic algorithm for optimal solution. J. Comput. Sci. **21**, 255–262 (2017)

13. Díez-González, J., Álvarez, R., Sánchez-González, L., Fernández-Robles, L., Pérez, H., Castejón-Limas, M.: 3D TDOA problem solution with four receiving nodes. Sensors **19**(13), 2892 (2019)
14. Díez-González, J., Álvarez, R., González-Bárcena, D., Sánchez-González, L., Castejón-Limas, M., Perez, H.: Genetic algorithm approach to the 3D node localization in TDOA systems. Sensors **19**(18), 3880 (2019)
15. Boutemedjet, A., Samardžić, M., Rebhi, L., Rajić, Z., Mouada, T.: UAV aerodynamic design involving genetic algorithm and artificial neural network for wing preliminary computation. Aerosp. Sci. Technol. **84**, 464–483 (2019)
16. Dickinson, M.H., Lehmann, F.O., Sane, S.P.: Wing rotation and the aerodynamic basis of insect flight. Science **284**(5422), 1954–1960 (1999)
17. Sivells, J.C., Neely, R.H.: Method for calculating wing characteristics by lifting-line theory using nonlinear section life data (1947)
18. Phillips, W.F., Snyder, D.O.: Modern adaptation of Prandtl's classic lifting-line theory. J. Aircr. **37**(4), 662–670 (2000)
19. Anderson, D., Graham, I., Williams, B.: Aerodynamics. In: Flight and Motion, pp. 14–19. Routledge (2015)
20. Browand, F.: Reducing aerodynamic drag and fuel consumption. In: Advanced Transportation Workshop, October, pp. 10–11 (2005)
21. Torenbeek, E.: Advanced Aircraft Design: Conceptual Design, Analysis and Optimization of Subsonic Civil Airplanes. Wiley (2013)
22. Willmott, C.J., Matsuura, K.: Advantages of the mean absolute error (MAE) over the root mean square error (RMSE) in assessing average model performance. Clim. Res. **30**(1), 79–82 (2005)
23. Chai, T., Draxler, R.R.: Root mean square error (RMSE) or mean absolute error (MAE)?-arguments against avoiding RMSE in the literature. Geosci. Model Dev. **7**(3), 1247–1250 (2014)
24. Taylor, K.E.: Summarizing multiple aspects of model performance in a single diagram. J. Geophys. Res. Atmos. **106**(D7), 7183–7192 (2001)
25. Miller, B.L., Goldberg, D.E., et al.: Genetic algorithms, tournament selection, and the effects of noise. Complex Syst. **9**(3), 193–212 (1995)

Hybrid Genetic Algorithms and Tour Construction and Improvement Algorithms Used for Optimizing the Traveling Salesman Problem

Vladimir Ilin[1(✉)], Dragan Simić[1], Svetislav D. Simić[1], and Svetlana Simić[2]

[1] Faculty of Technical Sciences, University of Novi Sad, Trg Dositeja Obradovića 6, 21000 Novi Sad, Serbia
{v.ilin,dsimic,simicsvetislav}@uns.ac.rs, dsimic@eunet.rs
[2] Faculty of Medicine, University of Novi Sad, Hajduk Veljkova 1–9, 21000 Novi Sad, Serbia
svetlana.simic@mf.uns.ac.rs

Abstract. The traveling salesman problem (TSP) aims at finding the shortest tour that passes through each vertex in a given graph exactly once. To address TSP, many exact and approximate algorithms have been proposed. In this paper, we propose three new algorithms for TSP based on a genetic algorithm (GA) and an order crossover operator. In the first algorithm, a generic version of a GA with random population is introduced. In the second algorithm, after the random population is introduced, the selected parents are improved with a 2-OPT algorithm and processed further with a GA. Finally, in the third algorithm, the initial solutions are obtained with a nearest neighbor algorithm (NNA) and a nearest insertion algorithm (NIA); afterwards they are improved with a 2-OPT and processed further with a GA. Our approach differs from previous papers for using a GA for TSP in two ways. First, every successive generation of individuals is generated based primarily on 4 best parents from the previous generation regardless the number of individuals in each population. Second, we have proposed the new hybridization between GA, NNA, NIA and 2-OPT. The overall results demonstrate that the proposed GAs offer promising results, particularly for large-sized instances.

Keywords: Traveling salesman problem · Genetic algorithm · Nearest neighbor algorithm · Nearest insertion algorithm · 2-OPT algorithm · Hybrid approach

1 Introduction

The traveling salesman problem (TSP) is a typical combinatorial optimization problem in the fields of computer sciences, operation research, logistics and transportation, and so on. The problem is to find the shortest tour that passes through a set of n vertices so that each vertex is visited exactly once. In logistics and transportation, the vertices are represented as cities. The TSP can be described as the minimization of the total distance traveled. The TSP can be classified into two classes based on the structure of distance

Á. Herrero et al. (Eds.): SOCO 2020, AISC 1268, pp. 530–539, 2021.
https://doi.org/10.1007/978-3-030-57802-2_51

matrix: symmetric and asymmetric. The TSP is symmetric if distance from city i to city j is the same as from city j to city i. Otherwise, TSP is asymmetric. For n cities, there are $(n-1)!/2$ possible ways to find a tour for a symmetric distance matrix and $(n-1)!$ possible ways to find a tour for an asymmetric distance matrix. Therefore, TSP belongs to the class of NP-hard problems, in which optimal solution to the problem cannot be obtained within a reasonable computational time for large size problems.

To address TSP, many exact and approximate algorithms have been developed. Exact algorithms for TSP include branch and bound [5], cutting planes [14], dynamic programming [19], and linear programming [3]. Nevertheless, exact algorithms can only address small scale TSP, as their complexity increases exponentially with the number of nodes. Heuristics, metaheuristics and hybrid algorithms are implemented when approximate solutions are sufficient and exact algorithms are computationally costly.

Heuristic algorithms for TSP include tour construction algorithms and tour improvement algorithms. The tour construction algorithms iteratively extend a partial tour or iteratively combine several partial tours into one tour. The tour construction algorithms include nearest neighbor algorithm [18], Clarke-Wright algorithm [6], insertion procedures [18], and so on. The tour improvement algorithms start with an initial tour and then replace two or more branches within the tour to obtain a shorter tour. Typical representatives of tour improvement algorithms are 2-OPT [13], 3-OPT [13], and k-OPT [12] algorithms.

Metaheuristic algorithms for TSP include ant colony optimization [4], neural networks [1], simulated annealing [24], and so on. Metaheuristic algorithms for TSP are often hybridized with other metaheuristics and with construction and improvement algorithms. For example, in [7], ant colony optimization is used for the path construction and bee colony optimization is used for the path improvements.

Genetic algorithms (GA) are typical representatives of evolutionary algorithms and metaheuristics as well. GAs are often used to solve TSP due to a large number of different crossover operators and various hybridizations with other metaheuristics and construction and improvement algorithms [4, 8]. A review of GA approaches for TSP was presented in [16]. In a recent paper [10], a review of crossover operators for TSP was introduced.

In this article, we propose three new algorithms for TSP based on a GA and an order crossover (OX) operator. In the first algorithm, a generic version of a GA with random population is introduced. In the second algorithm, after the random population is introduced, the selected parents are improved with a 2-OPT algorithm and processed further with a GA. Finally, in the third algorithm, the initial solutions are obtained with a nearest neighbor algorithm (NNA) and a nearest insertion algorithm (NIA); afterwards they are improved with a 2-OPT and processed further with a GA. Our approach differs from previous papers using GA for TSP in two ways. First, every successive generation of individuals is generated based primarily on 4 best parents from the previous generation regardless the number of individuals in each population. Second, we have proposed the new hybridization between GA, NNA, NIA and 2-OPT. The NNA and NIA algorithms are selected for generating a starting solution because both algorithms are relatively easy to implement and both algorithms run in proportional time. This paper continuous the authors' previous researches in transportation planning [11, 20–22].

The rest of the paper is organized in the following way. A different strategy of using the OX crossover operator in GA for solving the TSP is presented in Sect. 2. Section 3 overviews the use of NNA, NIA and 2-OPT for TSP. Section 4 introduces three new algorithms for TSP based on a GA and the OX crossover operator. Experimental results and discussion are presented in Sects. 5 and 6 provides concluding remarks.

2 A Genetic Algorithm and Crossover Operators for TSP

The genetic algorithm (GA) is a metaheuristic search algorithm based on the biological evolutionary process of "the survival of the fittest" – the principle proposed by Charles Darwin. The GA was proposed by Holland [9].

The purpose of using a GA is to find the individual from the search space with the best "genetic material". The quality of an individual is measured with a fitness function and the part of the search space to be examined is called the population. The GA is known for its good global search ability, high efficiency, and good scalability in solving TSP [8, 15, 23]. However, GA usually needs a number of iterations to obtain high quality solutions. The basic steps of GA are the following: initialization, encoding scheme, selection, crossover, mutation, evaluation, and termination (Fig. 1).

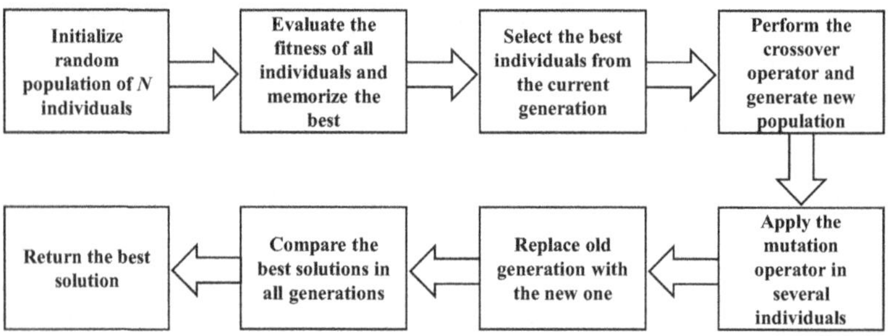

Fig. 1. The basic steps of GA

There are many representations to solve TSP using GAs. The binary, path, adjacency, ordinal, and matrix representations are often used. However, the most natural way to present a tour is using path representation. As an example, a tour can be represented simply as 1 → 4→8 → 2→5 → 9→3 → 6→7 → 1.

Since TSPs are combinatorial with path representation, the classical crossover operators such as one-point, two-point, and uniform crossovers are not suitable [10]. Frequently used path representations for TSP include partially mapped (PMX), order crossover (OX) and cycle crossover (CX) operators.

In this paper, we explore the use of the OX operator for TSP. The OX was proposed by Davis [2]. The OX method builds offspring by selecting a subtour of a parent and preserving the relative order of bits of the other parent. The subtour of a parent is generated by randomly selected two cut points. For example, parents P1 (1 → 7→9 → 2 || **3 → 4→6** || 5 → 8→1) and P2 (1 → 4→8 → 3 || **6 → 7→9** || 2 → 5→1) with randomly selected two cut points marked with „||", produce offspring in the following way. First, the selected bits between two cuts from P1 are added to O2 (1 → X→X → X || **3 → 4→6** || X → X→1) and selected bits between two cuts from P2 are added to O1 (1 → X→X → X || **6 → 7→9** || X → X→1). The first bit and the last bit in both parents and offspring are fixed as they represent the depot. The rest of the bits from the parents are transformed relative to the second cut point in both P1 and P2. The sequence of the bits in P1 from the second cut point (excluding bit 1) is: 5 → 8→7 → 9→2 → 3→4 → 6. After removing the bits 6, 7 and 9, which are already fixed in O2, the new sequence is added to O2 starting from the second cut point: O2 (1 → 2→3 → 4 || **6 → 7→9** || 5 → 8→1). In a similar manner, O1 is generated: O1 (1 → 8→7 → 9 || **3 → 4→6** || 2 → 5→1).

If we explore this mechanism further, we may notice that different cut points may be assigned to P1 and P2. Therefore, two parents can produce more than two offspring. This feature may be exploited to generate a new population in a different manner.

Example 2:
P1 (1→7→9→2 || **3→4** || 6→5→8→1), and
P2 (1→4→8→3 || **6→7** || 9→2→5→1)
produce
O3 (1→8→6→7 || **3→4** || 9→2→5→1) and
O4 (1→2→3→4 || **6 →7** || 5→8→9→1).

Example 3:
P1 (1→7→9→2→3 || **4** || 6→5→8→1) and
P2 (1→4→8→3→6 || **7** || 9→2→5→1)
produce
O5 (1→8→3→6→7 || **4** || 9→2→5→1) and
O6 (1→9→2→3→4 || **7** || 6→5→8→1).

3 Other Heuristic Algorithms for TSP

In this section, we briefly explain tour construction and tour improvement algorithms for TSP. The nearest neighbor algorithm, nearest insertion algorithm and 2-OPT algorithm are discussed.

3.1 A Nearest Neighbor Algorithm

The nearest neighbor algorithm (NNA) was proposed by Rosenkrantz et al. [18]. In this algorithm, a path is constructed as follows:

Step 1. Start with any node as the beginning of a path.
Step 2. Find the node closest to the last node added to the path. Add this node to the path.
Step 3. Repeat Step 2 until all nodes are included in the path. Then, join the first and the last node.

We selected the depot (the first node) as a starting and ending point for all routes.

3.2 A Nearest Insertion Algorithm

The nearest insertion algorithm (NIA) was also proposed by Rosenkrantz et al. [18]. In this algorithm, a path is constructed as follows:

Step 1. Start with a subgraph consisting of node i only.
Step 2. Find node k such that c_{ik} is minimal and form subtour $i - k - i$.
Step 3. Selection step. Given a subtour, find node k not in the subtour closest to any node in the subtour.
Step 4. Insertion step. Find the arc (i, j) in the subtour which minimizes $c_{ik} + c_{kj} + c_{ij}$. Insert k between i and j. c_{ij} is the distance from node i to node j.
Step 5. Go to Step 3 unless a Hamiltonian cycle is completed.

We selected the depot (the first node) as a starting and ending point for all routes.

3.3 A 2-OPT Algorithm

The 2-OPT algorithm was proposed by Lin [13]. In this algorithm, a path is constructed as follows:

Step 1. Find an initial tour randomly or by applying some other algorithm.
Step 2. Try to improve the tour using the two-branch exchange method.
Step 3. Continue Step 2 for all combinations and return the improved tour or the tour that is already 2-optimal.

4 The Variants of the Genetic Algorithm for TSP

In this section, we describe three proposed algorithms for TSP, named: (1) generic GA for TSP, (2) GA with 2-OPT for TSP, and (3) GA with NNA, NIA and 2-OPT for TSP.

4.1 A Generic Genetic Algorithm for TSP

First, we describe an approach in the generic GA for TSP (Algorithm 1). Starting from an initial random population (Lines 1–2) and other setup parameters (Lines 3–5), for each iteration, 4 best parents are selected and placed in the main solution pool (Line 9). Then, the other good potential parents are selected and placed in the auxiliary solution pool (Line 10). Both pools of solutions are used to generate a new population using a crossover OX operator (Line 11). Regularly, the crossover probability is set to be high (usually above 0.5) in order to introduce the majority of good individuals to the production process and to enable convergence towards better solution. In this paper, a new way of using the crossover OX operator is introduced, that is, two parents can produce a lot more than two offspring. Our goal is to explore a search space pertaining the selected 4 best parents in each iteration. Different strategies can be investigated on how to generate new subpopulations. We opted to produce 50% of offspring in the new generation from two best parents in the previous generation (Line 12). The other 50%

of offspring are generated from different crossover strategies between all 4 best parents in the previous generation (Lines 13-17). In the next step, randomly selected offspring from the new population (Line 18) are mutated in order to maintain variability of the population and to escape the search becoming trapped too quickly in a local optimum (Line 19). Finally, we compare the obtained solutions in all generations (Line 23) and return the best solution found (Line 24). The initial parameters of a GA are presented in Sect. 5.

Algorithm 1 GA (OX) for TSP

1: Initialize number of individuals (I) in each population (P): *parameter N*
2: Initialize random population (I_1-I_N): *Pstart*
3: Initialize maximal number of generations: *parameter Gmax*
4: Initialize crossover probability: *parameter Cprob*
5: Initialize mutation probability: *parameter Mprob*
6: Calculate the fitness and memorize the best in the *Pstart*
7: **while** *Gmax* is not reached **do**
8:     Calculate the fitness of all individuals in *Pstart*
9:     Select 4 best individuals in the main solution pool (*Mpool*): I_1 to I_4
10:     Select *Cprob* * *N* - (I_1 to I_4) individuals in the auxiliary solution pool (*Apool*)
11:     Perform the crossover (OX) on the *Mpool* and *Apool* to generate a new population:
12:         $Pa \leftarrow I_1$ OX I_2 (50% of offspring in the new population)
13:         $Pb \leftarrow I_1$ OX I_3 (10% of offspring in the new population)
14:         $Pc \leftarrow I_1$ OX I_4 (10% of offspring in the new population)
15:         $Pd \leftarrow I_2$ OX I_3 (10% of offspring in the new population)
16:         $Pe \leftarrow I_2$ OX I_4 (10% of offspring in the new population)
17:         $Pf \leftarrow I_3$ OX I_4 (10% of offspring in the new population)
18:         New population: $Pnew \leftarrow Pa + Pb + Pc + Pd + Pe + Pf$
19:     Perform mutation operator in randomly selected *Mprob* * *N* individuals
20:     Calculate the fitness and memorize the best in that generation
21:     $Pstart \leftarrow Pnew$
22: **end while**
23: Sort best solutions in all generations and select the best solution
24: **return** best solution in all generations

Parent Selection, Crossover and Mutation

Each iteration in the generic GA for TSP includes a selection of 4 best parents in the main solution pool and other potential parents in the auxiliary solution pool. The number of parents in the auxiliary solution pool depends on the crossover probability. The proposed strategy in this paper is to thoroughly explore a search space pertaining 4 best parents in each iteration.

The OX operator is selected for crossover (Algorithm 2). Starting from the input phase, two parents from the main solution pool are selected (Line 1) and the auxiliary solution pool is introduced (Line 2). Then, other setup parameters are defined (Lines 3–5). In case that two parents from the main solution pool are equal, one of them is replaced with a non-equal parent from the auxiliary solution pool with a strongest fitness function (Lines 6–9). In that manner, the subpopulation is produced by two non-identical parents. In the next step, until the number of offspring in the new subpopulation is not reached their production is performed (Lines 10–12). Finally, a new subpopulation is returned (Line 13).

Algorithm 2 Crossover OX for TSP

Input: Parents P_i and P_j from the main solution pool – *Mpool*
Input: All potential parents from the auxiliary solution pool – *Apool*
Initialize new empty matrix for the new subpopulation: *parameter NewSPop*
Initialize number of offspring produced by P_i and P_j: *parameter NewSPopMax*
Initialize random set of *NewSPopMax* / 2 two cut points: *parameter TwoCutsList*
if P_i and P_j are equal **then**
    Replace P_j with not equal parent with a strongest fitness function from the *Apool*
    Update *Apool*
end if
while *NewSPopMax* is not reached **do**
    Produce offspring in the new subpopulation and update *NewSPop*
end while
return *NewSPop*

The swapping mutation operation depends on the mutation probability. This defines how many new offspring need to be mutated. The candidates for mutation are selected randomly. Subsequently, two crossing points are randomly chosen and swapped. Regularly, the mutation probability is significantly lower than the crossover probability (usually below 0.2) as it represents the divergence strategy, i.e. an opportunity to escape from a local optimum.

4.2 A Genetic Algorithm with a 2-OPT Algorithm for TSP

In the next step, a GA (Algorithm 1) is modified with a 2-OPT algorithm. The 2-OPT algorithm is introduced after the fitness function is calculated and 4 best parents are selected and placed in the main solution pool. The 2-OPT algorithm is then applied on these 4 best parents. In case that, during the crossover process, 2 parents are equal, one of them is replaced with another non-equal parent (see Algorithm 2), and the 2-OPT algorithm is applied on that new parent. In that manner, improved parents will provide additional quality in genetic material for the offspring in each generation.

4.3 A Genetic Algorithm with a Nearest Neighbor Algorithm, a Nearest Insertion Algorithm and a 2-OPT Algorithm for TSP

In the final step, a GA (Algorithm 1) is modified with NNA, NIA and 2-OPT. The NNA and the NIA are introduced to generate good starting solutions. The obtained solutions are then improved with the 2-OPT algorithm which makes 4 "strong" individuals in the starting generation. That number corresponds to the applied methodology to generate the new population based on 4 best parents from the previous generation. The rest of individuals in the starting generation are introduced randomly. The use of the 2-OPT algorithm in further steps corresponds to the use described in the Subsect. 4.2.

The main idea of introducing the NNA and NIA is to implement relatively fast algorithms that produce a solution with a good quality. Hybridization with a GA should improve initial solutions through the iteration process.

5 Computational Results and Discussion

In this section, we perform experimental results to analyze the performance of different algorithms based on a GA for TSP. The algorithms are implemented in Matlab and the experiments are run on a desktop computer with an Intel Core i5-2400, 3.1 GHz processor. The authors conducted the experiments on eleven benchmark instances from TSPLIB [17]. The obtained results are presented in Table 1.

Table 1. Comparison results between three proposed GAs and other GAs with different crossover operators

Instance (N) Opt. val.	Results	GA (OX)	GA (OX) + 2-OPT	GA (OX) + NNA + NIA + 2-OPT	GA (PMX) [10]	GA (OX) [10]	GA (CX2) [10]
fri26 (N = 26) 937	Best	975	990	1005	1056	1051	1099
	Worst	1277	1297	1039	1294	1323	1278
	Average	1098	1109	1027	1133	1158	1128
ftv33 (N = 34) 1286	Best	1488	1532	1442	1708	1804	1811
	Worst	2011	1959	1504	2399	2366	2322
	Average	1727	1754	1496	2012	2098	2083
ftv38 (N = 39) 1530	Best	1809	1780	1589	2345	2371	2252
	Worst	2178	2287	1589	2726	2913	2718
	Average	1963	2112	1589	2578	2617	2560
dantzig42 (N = 42) 699	Best	823	800	777	1298	1222	699
	Worst	1225	1134	777	1606	1562	920
	Average	945	966	777	1425	1301	802
ft53 (N = 53) 6905	Best	8553	9078	8452	13445	13826	10987
	Worst	11332	12232	8506	16947	16279	13055
	Average	9894	10479	8457	14949	14724	12243
krol124p (N = 100) 36230	Best	57802	53138	40893	90231	97122	92450
	Worst	70534	66369	40893	118386	122497	121513
	Average	63254	60693	40893	100335	103457	101229
ftv170 (N = 171) 2755	Best	7965	6806	3422	13346	15202	6421
	Worst	9572	8157	3515	19314	19708	8416
	Average	8641	7495	3502	16775	17569	7019
rbg323 (N = 323) 1326	Best	2355	1603	1493	4123	3998	4212
	Worst	2625	1804	1504	5147	5385	5342
	Average	2494	1708	1497	4434	4602	4654
rbg358 (N = 358) 1163	Best	2572	1528	1346	5380	5630	5404
	Worst	2846	1711	1364	5915	5948	6004
	Average	2677	1613	1354	5532	5830	5622
rbg403 (N = 403) 2465	Best	3688	2669	2498	6231	6196	6257
	Worst	3949	2804	2503	6653	6629	6671
	Average	3801	2743	2502	6536	6386	6455
rbg443 (N = 443) 2720	Best	4143	2933	2747	6754	6932	6854
	Worst	4339	3115	2766	7209	7351	7388
	Average	4250	3044	2752	6905	7121	6981

In these eleven instances, ftv33, ftv38, ft53, kro124p, ftv170, rbg323, rbg358, rbg403, and rbg443, are asymmetric, while fri26 and dantzig42 are symmetric TSPs. The initial parameters of GAs are as follows: the population size is 200, maximum generation is 500, crossover probability is 0.8, and mutation probability is 0.1. Each experiment was executed 30 times independently.

In Table 1, the proposed GA (OX) algorithm is performing better, on average basis, than the GA (PMX) [10] and the GA (OX) [10] for all tested instances. The GA (CX2) [10] displays better results than the proposed GA (OX) only for 2 out of 11 instances: dantzig42 and kro124p. The GA (CX2) [10] provides the optimum value for the instance dantzig42 sixteen out of thirty times.

The GA (OX) + 2-OPT algorithm and GA (OX) + NNA + NIA + 2-OPT algorithm further improve the obtained results of the GA (OX) algorithm. The GA (OX) + 2-OPT algorithm exhibits better results than the GA (OX) for larger instances: kro124p, ftv 170, rbg323, rbg358, rbg403, and rbg443. The GA (OX) + NNA + NIA + 2-OPT algorithm outperforms both algorithms, GA (OX) and GA (OX) + 2-OPT algorithm, for all instances. For instances rbg403 and rbg443, the obtained average values reveal gaps of 1.50% and 1.18% from the optimal values.

The overall results demonstrate that the proposed GAs based on the new strategy of using the OX crossover operator outperform other comparable GAs.

6 Conclusion and Future Work

In this study, three GAs based on a different strategy of using the OX crossover operator are presented. In the first algorithm, a generic version of a GA with random population is introduced. In the second algorithm, a hybridization between the GA and 2-OPT is presented. Finally, in the third algorithm, a hybridization between GA, NNA, NIA, and 2-OPT is displayed.

The presented algorithms are tested using eleven benchmark instances from TSPLIB [17] ranging from 26 to 443 cities. The obtained results indicate that the proposed GAs outperform other comparable GAs [10]. The hybrid GA (OX) + NNA + NIA + 2-OPT algorithm exhibits promising results for larger benchmark instances.

The future work could focus on extending the research on different strategies of using the OX crossover operator and mutation operator. Hybridization with some other tour construction and tour improvement algorithms can also be investigated.

References

1. Creput, J.C., Koukam, A.: A memetic neural network for the Euclidean traveling salesman problem. Neurocomputing **72**(4), 1250–1264 (2009)
2. Davis, L.: Applying adaptive algorithms to epistatic domains. IJCAI **85**, 162–164 (1985)
3. Diaby, M.: The traveling salesman problem: a linear programming formulation. WSEAS Trans. Math. **6**(6), 745–754 (2007)
4. Dong, G.F., Guo, W.W., Tickle, K.: Solving the traveling salesman problem using cooperative genetic ant systems. Expert Syst. Appl. **39**(5), 5006–5011 (2012)
5. Finke, G., Claus, A., Gunn, E.: A two-commodity network flow approach to the traveling salesman problem. Congressus Numerantium **41**, 167–178 (1984)

6. Golden, B.: A statistical approach to the TSP. Networks **7**, 209–225 (1977)
7. Gunduz, M., Kiran, M.S., Ozceylan, E.: A hierarchic approach based on swarm intelligence to solve traveling salesman problem. Turk. J. Electr. Eng. Comput. Sci. **23**(1), 103–117 (2015)
8. Ha, Q.M., Deville, Y., Pham, Q.D., Hà, M.H.: A hybrid genetic algorithm for the traveling salesman problem with drone. J. Heuristics **26**(2), 219–247 (2019). https://doi.org/10.1007/s10732-019-09431-y
9. Holland, J.H.: Adaptation in Natural and Artificial Systems: An Introductory Analysis with Applications to Biology, Control, and Artificial Intelligence. University of Michigan Press, Oxford (1975)
10. Hussain, A., Muhammad, Y.S., Sajid, M.N., Hussain, I., Shoukry, M.A., Gani, S.: Genetic algorithm for traveling salesman problem with modified cycle crossover operator. Comput. Intell. Neurosci. **2017**, 1–7 (2017)
11. Ilin, V., Simić, D., Tepić, J., Stojić, G., Saulić, N.: A survey of hybrid artificial intelligence algorithms for dynamic vehicle routing problem. In: Onieva, E., Santos, I., Osaba, E., Quintián, H., Corchado, E. (eds.) HAIS 2015. LNCS (LNAI), vol. 9121, pp. 644–655. Springer, Cham (2015)
12. Lin, S., Kernighan, B.: An effective heuristic algorithm for the traveling salesman problem. Opns. Res. **21**(2), 498–516 (1973)
13. Lin, S.: Computer solutions of the traveling salesman problem. Bell Syst. Tech. J. **44**, 2245–2269 (1965)
14. Miliotis, P.: Using cutting planes to solve the symmetric travelling salesman problem. Math. Program. **15**(1), 177–188 (1978)
15. Victer Paul, P., Ganeshkumar, C., Dhavachelvan, P., Baskaran, R.: A novel ODV crossover operator-based genetic algorithms for traveling salesman problem. Soft. Comput. **2**, 1–31 (2020). https://doi.org/10.1007/s00500-020-04712-2
16. Potvin, J.-Y.: Genetic algorithms for the traveling salesman problem. Ann. Oper. Res. **63**(3), 339–370 (1996)
17. Reinelt, G.: TSPLIB. http://comopt.ifi.uni-heidelberg.de/software/TSPLIB95/. Accessed 13 Feb 2020
18. Rosenkrantz, D., Stearns, R., Lewis, P.: Approximate algorithms for the traveling salesperson problem. In: Proceedings of the 15th Annual IEEE Symposium of Switching and Automata Theory, pp. 33–42. IEEE (1974)
19. Salii, Y.: Revisiting dynamic programming for precedence-constrained traveling salesman problem and its time-dependent generalization. Eur. J. Oper. Res. **272**(1), 32–42 (2019)
20. Simić, D., Kovačević, I., Svirčević, V., Simić, S.: Hybrid firefly model in routing heterogeneous fleet of vehicles in logistics distribution. Log. J. IGPL **23**(3), 521–532 (2015)
21. Simić, D., Simić, S.: Evolutionary approach in inventory routing problem. In: Rojas, I., Joya, G., Cabestany, J. (eds.) IWANN 2013. LNCS, vol. 7903, pp. 395–403. Springer, Heidelberg (2013)
22. Simić, D., Simić, S.: Hybrid artificial intelligence approaches on vehicle routing problem in logistics distribution. In: Corchado, E., Snášel, V., Abraham, A., Woźniak, M., Graña, M., Cho, S.-B. (eds.) HAIS 2012. LNCS (LNAI), vol. 7208, pp. 208–220. Springer, Heidelberg (2012)
23. Xu, X., Yuan, H., Matthew, P., Ray, J., Bagdasar, O., Trovati, M.: GORTS: genetic algorithm based on one-by-one revision of two sides for dynamic travelling salesman problems Soft. Comput. **24**, 7197–7210 (2020)
24. Zhan, S.H., Lin, J., Zhang, Z.J., Zhong, Y.W.: List-based simulated annealing algorithm for traveling salesman problem. Comput. Intell. Neurosci. **2016**, 1–12 (2016)

Segmentation Optimization in Trajectory-Based Ship Classification

Daniel Amigo(✉), David Sánchez(✉), Jesús García, and José Manuel Molina

Group GIAA, University Carlos III of Madrid, Madrid, Spain
{damigo,davsanch,jgherrer,molina}@inf.uc3m.es

Abstract. An analysis over trajectory segmentation techniques is carried out by the study of the different algorithms and the experimentation over a ship classification problem, which use a data preparation and classification system used in previous works. With the data preparation, the system handles real-world Automatic Identification System (AIS) data, cleaning wrong measurements and smoothening the trajectories by the application of an Interacting Multiple Model (IMM) filter. Also applies some balancing algorithms to address the lack of an equal distribution among classes. To correctly evaluate the classification with the imbalanced data a multiple objective analysis is proposed to consider the minority class and the global accuracy. Over that multi-objective analysis, different segmentation algorithms and its variations are tested to analyze the influence of them into the classification problem. The results show a Pareto front with different viable solutions for the proposed multi-objective problem, without a dominant algorithm over rest of the tested segmentation algorithms.

Keywords: AIS data · Class imbalance · Kinematic behavior · Ship classification · Track segmentation

1 Introduction

The maritime surveillance systems are an essential element for the protection of the seas, ensuring the safety of maritime transport and security of citizens. Detecting and locating vehicles is a solved problem using multiple technologies, but classifying the type of vessel is more challenging, which is an essential element for decision-making in maritime surveillance systems. Technologies such as AIS [1] provide information that allows the target identification, however as they work collaboratively the information is not always reliable, as it is susceptible to manipulation.

The problem of this study is the classification of trajectories to obtain the type of ship based on kinematics data that model its behavior. This is an extension of a previous study [2, 3], where the problem was defined and main subprocesses identified. These first approaches concluded that it was necessary to specifically analyze the impact of each subprocess on the classification. Thus, the objective of this paper is to study the impact of segmentation on the final performance, observing the variation compared the fixed-size

Á. Herrero et al. (Eds.): SOCO 2020, AISC 1268, pp. 540–549, 2021.
https://doi.org/10.1007/978-3-030-57802-2_52

segmentation initially proposed. To achieve it, more complex segmentation techniques are studied and analyzed, allowing variable size segments that can be better adjusted to the ships' motion. To proceed from the sensor detections to the ship classification, it is required a system that performs different processes on the data. This system has been developed in previous works [2, 3], being necessary within this study the analysis of the segmentation process.

The system used has several processes, starting from the data preparation to clean real-world data problems, an IMM filter is used to reduce the noise by smoothing the target trajectory. The proposed step is the segmentation of trajectories, splitting the original track by applying different criteria (uniform length, shape or direction preserving...) and then a process handles of the data imbalance, since the ship types are not distributed in a homogeneous manner (neither in trajectories or segments). Finally, the last process is classification, by using different algorithms applied to track segments to predict the ship type. Specifically, the objective is to determinate the membership in the fishing class, which is the minority in the used dataset. This classification process requires a prior sub-process that computes representative features from each trajectory segment, these will be variables used to model the behavior of the ship. Although other variables related to the trajectories context could provide useful information to classify them, the proposed system seeks to avoid this type of information, because it aims to find a system based on as little information as possible, focusing only on the track kinematics, which could be improved later by including the context information.

The experiments compare various segmentation techniques with respect to the original segmentation (fixed length). The results show the trade-off between accuracy and imbalance of classification so there is not an absolute optimal solution, but makes it clear the multi-objective nature of the problem, and solutions show a Pareto front.

This paper is organized as follows: In Sect. 2 the state-of-art methods in segmentation of maritime vehicles tracks are analyzed. In Sect. 3 is explained the process necessary to the trajectory-based classification in Sect. 4 there is the explanation of all the texted segmentations while in Sect. 5 results of the work are shown. Finally, the conclusions and perspectives for future works are presented in Sect. 6.

2 State of the Art

The state of the art covers previous works on two main problems: trajectory classification and trajectory segmentation.

A basic problem for trajectory classification is the feature extraction to infer intelligence from the available information. For example, these recent studies [4–6] perform a feature extraction on the trajectory of the ship to determine its behavior. This feature extraction is not adequate for a problem where long-duration trajectories or very heterogeneous mixture of trajectories appear.

As an alternative, feature extraction can be applied on each segment instead of the whole track in order to extract more precise information for the classifier. There are researchers [7] who perform a segmentation before classification, but they use their own segmentation technique very specific to their problem. Alternatively, this paper experiments with both classical and recent segmentation techniques to analyze how they influence the problem of classification trajectories. Note also that all these papers use context information, making them incomparable with the present proposal.

The field of trajectory segmentation has several approaches [8], one of them is the compression algorithms, which identify the key-points of the trajectory and use them to generate the segments. Segments are generated according to different conditions, e.g. time gaps, trajectory shape or its context. Also, they can be categorized according to whether they need the entire track (offline) or they can run in real time (online).

The simplest approach to segmentation is uniform sampling, which cut the track into segments of uniform size [9] (the approach used by default in the previous works). This paper explores segmentation algorithms according to the trajectory shape, generating segments that minimize error with respect to the trajectory. In Fig. 1 illustrates several segmentation algorithms achieving different outputs on the same track.

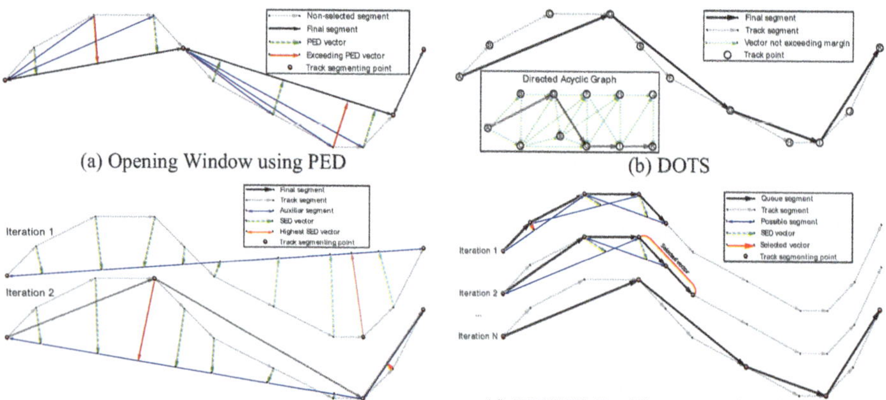

(a) Opening Window using PED (b) DOTS

(c) Top-Down using SED, also known as TD-TR (d) SQUISH-E, with a queue size of 4 points

Fig. 1. Example of how several track segmentation algorithms work

The classic algorithms for segmentation are:

- Opening Window (OPW) [10]: It generates variable size segments by setting the start of the track and including points in the window until an error threshold is exceeded. When this threshold is exceeded, as is shown in Fig. 1(a), the current segment is closed, and the window is restarted, following this process until the end.
- Top-down [11]: It starts with a segment that covers the entire trajectory and divides it recursively at the point where the error is highest, as shown happening twice in Fig. 1(c). This process continues until the selected error measurement is below the threshold for all points.
- Bottom-up: The inverse process to Top-Down. It starts with small segments unifying them when the error is the smallest, until cannot be unified anymore.

These algorithms calculate the segment error in relation to the trajectory by using the Perpendicular Euclidean Distance (PED) of each point. A big improvement is to use instead of PED, the Synchronized Euclidean Distance (SED) [10], which take into consideration track point timestamp with regard to the segment total time.

Based on the previous classic approaches there are many other algorithms that seek a better performance when performing the segmentation, like:

- SQUISH-E [12]: It works by using a queue of fixed size, adding points to it and in each iteration eliminating the one with the smallest SED error. Figure 1(d) shows this procedure, checking in trios the less relevant point and removing it from the queue. This algorithm uses two parameters for shaping the resulting segment: λ guarantees a compression ratio of the track, while μ indicates the maximum SED error.
- MRPA [13]: It works by approaching the track based on a bottom-up multiresolution approach, using an accumulated variation of the SED criterion (ISSED).
- DOTS [14]: This algorithm performs a variation to allow online running of the MRPA. It uses a DAG (Directed Acyclic Graph) to describe all potential segments of the trajectory, as can be shown in Fig. 1(b).

3 Ship-Type Determination Using Binary Classification

This section provides a brief explanation of the original system used on the general problem, summarizing its main subprocesses, starting from the input data up to the classification algorithms. The system was detailed in [2, 3].

The first step is cleaning the raw data from sensors. In this case, the available data is from AIS sensor. It provides kinematic data of ships integrated with additional information such as the ship type, which is used here to train the classifier. Specifically, the chosen repository is the one provided by the Danish Maritime Authority [15], in which there is a recompilation of daily AIS contacts since 2006. Dealing with real-world raw data requires a strong pre-processing which is critical for final performance, removing inconsistencies, null, wrong, and noisy values. These problems are generated by malfunction of AIS transmitters and human errors. The measurement noise taken by the sensor can either be outliers, directly detectable evaluating the offset in GPS coordinates, or small noises that can be smoothed by a filtering algorithm. In the proposed system an IMM filter has been implemented to smooth the noise, configured with two Extended Kalman Filters as modes of prediction for ship trajectories: the first one for linear movements and low prediction noise and the second one to model the movements that would be considered noisy (speed variations, turns, ...).

Prior to classification, its necessary a process to address the unbalance problem present in this domain due to the lack of an equal distribution among classes. For instance, long and frequent trajectories of cargo and passenger vessels populate the training data sets and bias the classification models towards these categories reducing the representation of other ones, like the fishing vessel category. To solve the problem, the system implements oversampling and undersampling techniques, which adjust the amount of data of each class by adding or removing instances [16]. The experimentation uses the original imbalanced dataset, and two balanced datasets: one using random undersampling, randomly removing instances of the majority classes, and another using the SMOTE algorithm [17], already used for track classification [5], oversampling the minority class by creating new artificial samples. The classification is based on the following features generated from the track points contained in the segments:

- Course variation: describing turnarounds between track points.
- Distance: characterizing movement range and complexity between track points.
- Speed: characterizing the movement velocity between track points.
- Time between measures: considering the time gaps between track points.
- Speed variation: describing acceleration and deceleration between track points.

Because the possible difference in the number of measures between segments, is necessary to make those kinematic variables suitable as a classification input. The following statistical measures are applied to aggregate all the segment track points: the mean, maximum, minimum, mode, standard deviation and three quartiles. Also, the total time of the segment is included to support the time gaps variables.

The classification problem considered in this work is predicting when a vessel is of fishing type and when it is not, i.e. a binary classification problem. Common classification algorithms in binary problems as the Support Vector Machine (SVM) and the decision tree algorithm are chosen, looking to keep the importance on the segmentation problem by using simple and well-known techniques but able to perform it.

To evaluate the results obtained by the classification we must consider two main factors, the accuracy of the general classification and the specific accuracy on the minority class (fishing), which is affected by the imbalance in the training process. Therefore, along with the classification accuracy, the F-measure metric [18] is considered to assess both effects. The simultaneous evaluation of both metrics prevents the domination of the classification accuracy by the effect of majority class. Besides, the presence of these two metrics makes the problem multi-objective, allowing to observe the Pareto's front when displaying the results from different algorithms and their parameters.

4 Trajectories Segmentation

This section presents the different experiments to be carried out using the track segmentation algorithms. Each algorithm has different parameters to set its functionality depending on the problem. In this case, as the configuration of each algorithm is not trivial with respect to its impact on the classification, different experiments are performed, varying from each of the parameters, allowing an analysis of the impact of each of them. A summary of the variations of each algorithm is shown in Table 1 and a detailed explanation of the 196 experiments tested in this paper is given below.

The base case used in previous works uses a uniform segmentation of 50 measures (around 9 min). For comparison, tests of 10 and 20 values are performed as well.

Opening window (OPW) has the following variants from its base implementation:

- The cut-off criterion: whether it occurs at the point where the window has exceeded the error (NOPW), and whether it is done at the previous point (BOPW) [10].
- Different error evaluation functions: PED or SED ("_TR", meaning Time-Ratio [10]). Three error values are tested to each function: 20, 30 and 50 m.

Table 1. Segmentation algorithms variations

<table>
<tr><th>Base algorithm</th><th>Variation (if any)</th><th>Error function</th><th>Error value (meters)</th><th>Minimum size</th><th>Compression rate</th></tr>
<tr><td colspan="2">Uniform Segment</td><td>PED</td><td>-</td><td rowspan="3">0
10
20
50</td><td rowspan="4">-</td></tr>
<tr><td>OPW</td><td>BOPW
NOPW
BOPW_TR
NOPW_TR</td><td rowspan="2">PED
SED</td><td rowspan="4">20
30
50</td></tr>
<tr><td>TopDown</td><td>DP
TD_TR</td></tr>
<tr><td colspan="2">BottomUp</td><td>PED</td><td rowspan="4">-</td></tr>
<tr><td colspan="2">SQUISH-E</td><td>SED</td><td>1, 5, 10</td></tr>
<tr><td colspan="2">DOTS</td><td rowspan="2">ISSED</td><td rowspan="2">100
500</td><td rowspan="2">-</td></tr>
<tr><td colspan="2">MRPA</td></tr>
</table>

- To ensure that the segments are generated with a minimum length, favoring the classification. A minimum segment size its tested with 0, 10, 20 and 50 points.

The Top Down algorithm has variations for the error evaluation function, marked as "DP" (Douglas Peucker algorithm [11]) when it uses PED and as "TD_TR" when it uses SED [10]. These variations use the same error and minimum segment size as OPW.

Bottom Up has no relevant variations according to the error function, as only the PED error function has been used in the literature.

SQUISH-E only uses the SED, with the same three error values already listed as μ value. In addition, it has the compression parameter λ, testing 1, 5 and 10 values.

Finally, both DOTS and MRPA only vary on the error values, using 100 and 500 as values for its accumulative SED variation.

5 Results Analysis

The performed experimentation is applied over three days in July 2017 from AIS contacts off the coast of Denmark. In total, more than 30 million contacts are available as system inputs. After the cleaning process, there are 7 million contacts, divided into 39077 different tracks. These trajectories are the inputs of the segmentation stage, which results in the number of segments shown in Fig. 2.

The figure also shows a demonstration of the imbalance problem, being possible to see the difference between the fishing class and the remaining instances (non-fishing).

As mentioned, to analyze the results of the different experiments carried out, the accuracy and F-measure are displayed together as a multi-objective problem, considering the total accuracy and the problem imbalance problem at the same time. In the Fig. 3, it can be seen the distribution of values of the accuracy and F-measure corresponding to different variations of the classification and balancing algorithms. The Pareto front is formed for those non-dominated solutions, i.e., those with no other solutions with higher values in the two metrics simultaneously. In the figure, this front is formed by the solutions appearing in the upper-right corner.

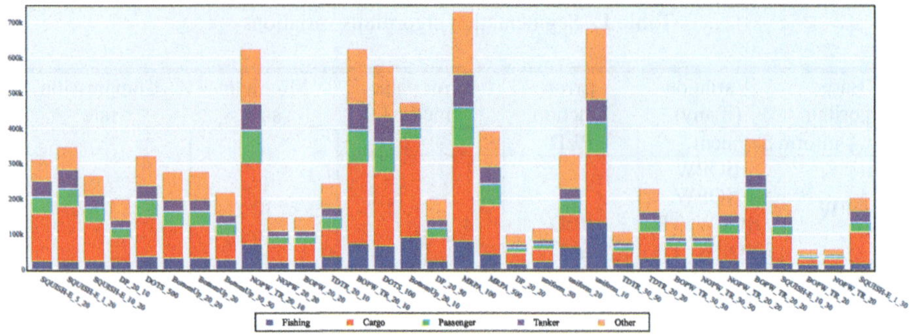

Fig. 2. Number of segments of the main ship types

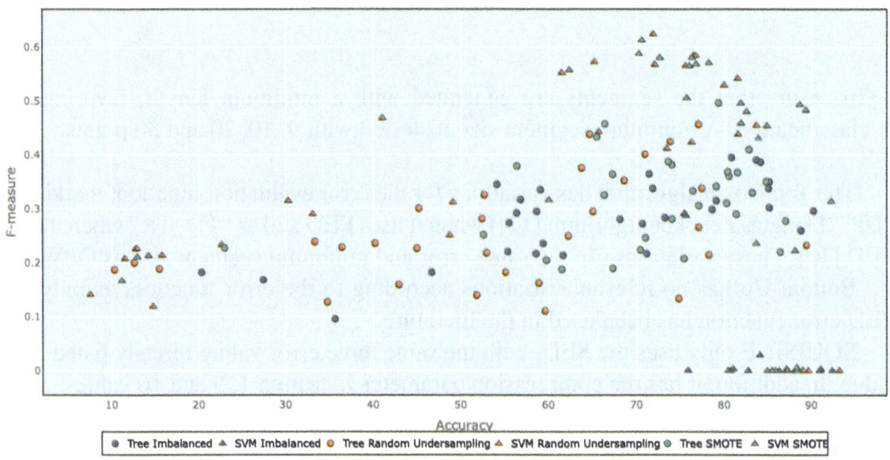

Fig. 3. Classification results for the different proposed variations

It can be appreciated how the SVM has results that are usually better with respect to accuracy, but in return it may have a worse performance when considering the class imbalance. That effect is produced because it is a boundary-based algorithm and has a trend to misclassify the minority class if it has a low impact in the total accuracy. This is especially noticeable in the imbalanced classification, which shows in many cases a zero value for F-measure (i.e., all samples of the minority class misclassified).

The decision trees have more moderate results, which do not stand out so much in the accuracy but in return they get better results in the F-measure. However, the front is clearly dominated by the SVM with balanced data sets, these although still have executions that demonstrate little success in the problem of the imbalance but also have the executions located in the front.

The most notable of these are the SVMs that operate on a balanced data set using SMOTE, although the random undersampling also have pareto front executions. To put the results in perspective, Fig. 4 shows all the segmentation algorithms executed by SVM applied on the SMOTE balanced data set.

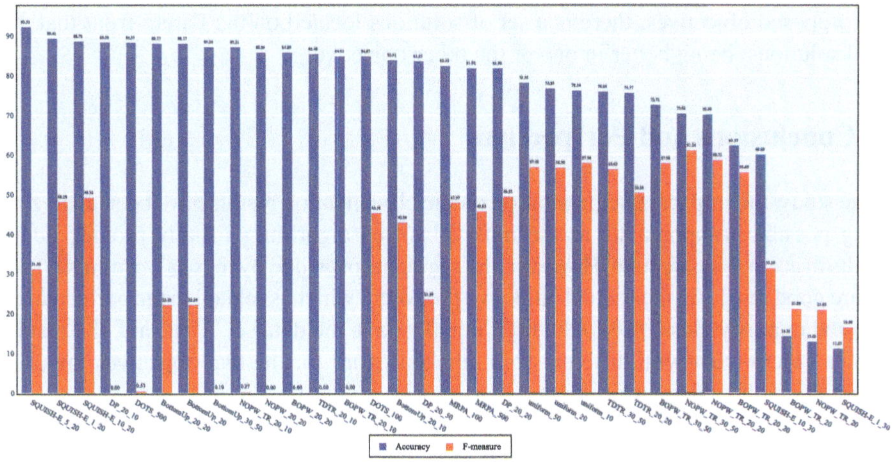

Fig. 4. SVM classification result for the segment variation in SMOTE balanced dataset

The figure not only shows the results of the accuracy but also the results for the F-measure which is not so positive since the most complex segmentations usually have slightly lower results in that metric.

There is no case that stands out especially from the rest, since when talking about a multi-objective problem between unbalance metrics and classification accuracy there is no algorithm that is especially good in both.

Being a point to emphasize that the best algorithms in one of the objectives clearly obtain their improvement when getting worse in the other one, an example would be the SQUISH-E with 20 error value and 5 compression parameter that obtains the best accuracy although its metrics are far below other algorithms. There is also the opposite case with the opening window algorithm, in which the best F-measure show an accuracy 20 points below that obtained by the specified SQUISH-E.

Regarding the higher complexity of the segmentation algorithms we can see how generally the segmentation algorithms that give better results when performing the compression of trajectories (SQUISH-E, MRPA, DOTS) do not ensure a better result within the proposed classification problem. Most of their executions seem to have good accuracy but not all of them good results in the F-measure used for the imbalance problem. In fact, one of the results belonging to the front and that therefore could be considered as one of the best, is obtained by the most basic segmentation algorithm, the uniform sampling with a size of 50.

Another aspect to consider is that the parameters introduced in the different segmentation algorithms influence the results variation, since the different executions of the same algorithm show very different results. For example, with the SQUISH-E algorithm, is possible to observe different results: one with the best accuracies, other with very poor results and another clearly within the Pareto front, achieving one of the best values within the two objectives with an accuracy close to 90% and balancing metrics only about 10 points below the best. Even if there is no absolute solution that meets the two proposed objectives, there is a set of solutions located on the Pareto front that are valid solutions, being better in one or the other objective.

6 Conclusions and Perspectives

In the study, the impact of segmentation on the classification results have been analyzed, being possible to appreciate as the most advanced algorithms usually provide better results in accuracy objective. However, the segments provided by these algorithms do not ensure good results in the second objective proposed, which is related to the performance with the minority class, due to the high imbalance in the data set. That said, the results show a Pareto front with different solutions that work for the two objectives imposed within the multi-objective problem

As a conclusion, it is very important the quality of the segments within the proposed process since there are trajectories with more measurements than others which create more segments with certain segmentation algorithms, affecting the classification. Also, by classifying segments it is possible to introduce noise with non-representative segments to its class (e.g. a ship departing from a port).

The SVM algorithm has demonstrated that it has the capacity to obtain good results for the classification, however it has a clear tendency towards the trivial solution, harming the minority class to obtain good results when maximizing the majority class.

Both classification algorithms are representative and responsive to the analyzed balancing algorithms. The main point of improvement is the testing of new segmentation or classification algorithms that achieve a better separation of instances, particularly those that can benefit most from the segments. Also, the application of the proposed method can approach other similar problems where classification is performed based on kinematic information of trajectories. For example, a classification oriented on pedestrian traffic could ensure safety (pickpocket identification), or the application in air traffic can allow flying mode identification thanks to the track segments adaptability.

Acknowledgement. This work was funded by public research projects of Spanish Ministry of Economy and Competitivity (MINECO), reference TEC2017-88048-C2-2-R.

References

1. Tu, E., Zhang, G., Rachmawati, L., Rajabally, E., Huang, G.B.: Exploiting AIS data for intelligent maritime navigation: a comprehensive survey from data to methodology. IEEE Trans. Intell. Transp. Syst. **19**, 1559–1582 (2018). https://doi.org/10.1109/TITS.2017.2724551
2. Amigo, D., Sánchez Pedroche, D., García, J., Molina, J.M.: AIS trajectory classification based on IMM data. In: 2019 22th International Conference on Information Fusion (FUSION), Ottawa, ON, Canada, pp. 1–8. IEEE (2019)
3. Sánchez Pedroche, D., Amigo, D., García, J., Molina, J.M.: Context information analysis from IMM filtered data classification. In: 1st Maritime Situational Awareness Workshop MSAW 2019, Lerici, Italy, p. 8 (2019)

4. Kraus, P., Mohrdieck, C., Schwenker, F.: Ship classification based on trajectory data with machine-learning methods. In: 2018 19th International Radar Symposium (IRS), Bonn, pp. 1–10. IEEE (2018)
5. Zhang, T., Zhao, S., Chen, J.: Research on ship classification based on trajectory association. In: Douligeris, C., Karagiannis, D., Apostolou, D. (eds.) Knowledge Science, Engineering and Management, pp. 327–340. Springer, Cham (2019)
6. Ichimura, S., Zhao, Q.: Route-based ship classification. In: 2019 IEEE 10th International Conference on Awareness Science and Technology (iCAST), Morioka, Japan, pp. 1–6. IEEE (2019)
7. Sheng, K., Liu, Z., Zhou, D., He, A., Feng, C.: Research on ship classification based on trajectory features. J. Navig. **71**, 100–116 (2018). https://doi.org/10.1017/S0373463317000546
8. Zheng, Y.: Trajectory data mining: an overview. ACM Trans. Intell. Syst. Technol. **6**, 1–41 (2015). https://doi.org/10.1145/2743025
9. Tobler, W.R.: Numerical map generalization. Michigan Inter-University Community of Mathematical Geographers (1966)
10. Meratnia, N., Rolf, A.: Spatiotemporal compression techniques for moving point objects. In: Lecture Notes in Computer Science (2004). https://doi.org/10.1007/978-3-540-24741-8
11. Douglas, D.H., Peucker, T.K.: Algorithms for the reduction of the number of points required to represent a line or its caricature. Can. Cartogr. **10**, 112–122 (1973). https://doi.org/10.3138/FM57-6770-U75U-7727
12. Muckell, J., Olsen, P.W., Hwang, J.-H., Lawson, C.T., Ravi, S.S.: Compression of trajectory data: a comprehensive evaluation and new approach. Geoinformatica **18**, 435–460 (2013). https://doi.org/10.1007/s10707-013-0184-0
13. Chen, M., Xu, M., Franti, P.: A fast O(N) multiresolution polygonal approximation algorithm for GPS trajectory simplification. IEEE Trans. Image Process. **21**, 2770–2785 (2012). https://doi.org/10.1109/TIP.2012.2186146
14. Cao, W., Li, Y.: DOTS: An online and near-optimal trajectory simplification algorithm. J. Syst. Softw. **126**, 34–44 (2017). https://doi.org/10.1016/j.jss.2017.01.003
15. Danish Maritime Authority: AIS Data. dma.dk/SikkerhedTilSoes/Sejladsinformation/AIS/Sider/default.aspx
16. Gosain, A., Sardana, S.: Handling class imbalance problem using oversampling techniques: a review. In: 2017 International Conference on Advances in Computing, Communications and Informatics (ICACCI), Udupi, pp. 79–85. IEEE (2017)
17. Chawla, N.V., Bowyer, K.W., Hall, L.O., Kegelmeyer, W.P.: SMOTE: synthetic minority over-sampling technique. jair **16**, 321–357 (2002)
18. Fernández, A., García, S., Galar, M., Prati, R.C., Krawczyk, B., Herrera, F.: Learning from Imbalanced Data Sets. Springer International Publishing, Cham (2018)

Bio-Inspired System for MRP Production and Delivery Planning in Automotive Industry

Dragan Simić[1(✉)], Vasa Svirčević[2], José Luis Calvo-Rolle[3], Vladimir Ilin[1], Svetislav D. Simić[1], and Svetlana Simić[4]

[1] Faculty of Technical Sciences, University of Novi Sad, Trg Dositeja Obradovića 6, 21000 Novi Sad, Serbia
dsimic@eunet.rs, {dsimic,v.ilin,simicsvetislav}@uns.ac.rs
[2] Lames d.o.o, 22000 Sremska Mitrovica, Serbia
vasasv@hotmail.com
[3] Department of Industrial Engineering, University of A Coruña, Avda. 19 de febrero s/n, 15405 Ferrol, A Coruña, Spain
jose.rolle@udc.es
[4] Faculty of Medicine, University of Novi Sad, Hajduk Veljkova 1–9, 21000 Novi Sad, Serbia
svetlana.simic@mf.uns.ac.rs

Abstract. Material requirement planning (MRP) can be characterized as an organized flow of material in the production process with the required sequence determined by the technological procedure. This paper presents biological swarm intelligence in general, as well as firefly optimization for modelling MRP and delivery planning in the production system of automotive industry. The aim of this research is to create a model which will satisfy production, deliver processes and optimize raw material inventory with real production constraints. The proposed system is tested in the data set collected in *Lames* Italian automotive company in the factory located in Serbia.

Keywords: Material requirement planning · Firefly optimization algorithm · Production · Delivery

1 Introduction

Supply chain management (SCM) has attracted increased attention and interest in the field of business logistics. The optimization of the supply chain is a major task. Different approaches have been developed to establish an efficient supply chain between companies. One of the most important enablers for efficient supply chain operations is *schedule stability*. Additionally, *stable schedule* has been listed as the seventh most important task recommended by the U.S. automotive industry to increase the U.S. competitiveness. In the field of production and *Material Requirements Planning* (MRP), the problems resulting from frequent plan revisions have been discussed in literature, for the past fifty years, since the middle of the nineteen seventies [1].

Á. Herrero et al. (Eds.): SOCO 2020, AISC 1268, pp. 550–559, 2021.
https://doi.org/10.1007/978-3-030-57802-2_53

MRP is a system for calculating the materials and components needed to manufacture a product. It consists of three primary steps: (i) taking inventory of the materials and components on hand, (iii) identifying which additional ones are needed, and then (iii) scheduling their production or purchase. There exist various negative aspects of frequent plan revisions. First, frequent re-planning leads to a general loss of confidence in planning. Then, production decisions that are continually altered generate confusion at an operational level and on the shop floors. Likewise, in multi-level production system, it propagates throughout the entire system, and disturbances may be amplified.

The modern concepts of materials management with a critical assessment of the MRP and the Japanese "Kanban" system are analysed in [2]. MRP proposes a centralized, multi-stage mechanism, which includes an analytic bill explosion as well as an optimum lot sizing procedure. On the other hand, Kanban describes a simple, yet effective possibility of decentralizing that process by a retrograde automatism. Different organizational issues of both concepts restrict their efficient applicability to a special class of material planning problems. Then, regarding the distinct coordination and information necessities, some most striking results are derived. Another modern concept of material handling, which sweeps through whole industries, is that of milk-run production [3, 4]. Material planning can be characterized as an organized flow of material in the production process with the required sequence determined by the technological procedure. It is a summary of operations presented by material conveying, storage, packaging and weighing, and technological manipulations and works directly related to the production process. Planning and dimensioning material flow challenges are difficult to overcome, especially in scenarios characterized by many hard constraints and by well-established processes [5].

This paper presents biological swarm intelligence in general, and particularly the model *firefly algorithm* (FFA) for modelling the optimized MRP system in automotive industry. The aim of this research is to create a model to minimize inventory raw material and inventory finish goods, when production is given on demand. This research continuous the authors' previous researches in supplier selection in supply chain management [6, 7], and inventory management system presented in [8, 9].

The rest of the paper is organized in the following way: Sect. 2 overviews the related work. Section 3 shows modelling MRP, and the firefly optimization algorithm implemented in MRP. This section also describes the used dataset. Experimental results are presented in Sect. 4 and finally, Sect. 5 provides concluding remarks.

2 Related Work

Based on the main MRP idea, many researchers, engineers and practitioners researched, developed, and applied different systems in production companies. The paper [10] discusses how to analyse, design, and develop a computer based and web based application system, for the raw material order planning system using the MRP method. Input of that system is, in the form of, Master Production Schedule (MPS) resulting from Production Scheduling Information System. MPS are generated from the demand forecasting results of sales transaction history data in *Online Transaction Processing*. That system produces output in the form of raw materials on the booking schedule, using MRP per week (period), though not using safety stock with the assumption that raw material goods arrive on time.

A review of literature on MRP implementation related to less developed countries in general, and Egypt in particular, revealed that no systematic study attempted to investigate how manufacturing companies have been implementing MRP systems. Thus, there are attempts to investigate the state of the art of MRP implementation in Egypt. The major mail survey findings were based on 93 responses received, of which 52 were MRP companies which operated in quite different business environments. The findings of that study [11] may enable MRP managers and users to obtain better understanding of MRP promoters, suggesting some ideas for further research on how manufacturing companies in Egypt are implementing new production management systems like the MRP system. The findings of that study suggest that MRP implementation in Egypt is relatively similar to the implementation in manufacturing companies in the newly industrialised countries in the west.

The research paper [12] deals with the problems of time, cost and optimal exploitation of available resources to achieve the project objectives and meet the required quality standards of implementing 5000 housing units in Benghazi, Libya. The problem concerning time evaluation and exploitation of available resources by the company concerned with project implementation was initially observed. Therefore, the researchers deal with that issue using the most common techniques in the operations research, based on MRP, in order to prepare and make the project timetable and control the implementation process.

Micro, small, and medium enterprises (MSME) are the largest business executors to the Indonesia's national industry. MSME in fashion industry is one of the most promising business ventures for outsourcing. Grooveline is a company which offers the service of T-shirt manufacturing that can be ordered according to customer's wishes in terms of design, colour, image to be printed, fabrics, and size. A calculation of MRP of each product produced is crucial to the business in order to design an effective purchasing order. The implementation of the aforementioned plan shall prevent the company from wasting materials, conduct a more effective production, and lead a more profitable business. The requirement that needs to be met in order to make a calculation of MRP, is the availability of product structure, MPS, Bills of Material, purchasing and production lead time, a time phased structure, Gross Material Requirement, a lot sizing, and net material requirement. The result of the total calculation has demonstrated that, when a company implements MRP from the beginning, that company can make savings in price by 11% [13].

The successful MRP implementation depends on SCM network design. Recently, the increasing need for sustainable freight transportation led to taking into account economic, environmental, and risk aspects. Greenhouse gas (GHG) emissions have a direct influence on the structure and behaviour of supply chains networks (SCN). The supply model consists of two-stage SCN: Secure & Green Supply Chain Network (SGSCN). In the SGSCN, a manufacturer is directly connected to several distribution centres, and each of them is connected to one or more customers. The objective of SGSCN is to minimize transportation costs whilst also maintaining a specified overall security level. A mathematical model for computing the risk and the applications for several SCN configurations and scenarios is illustrated in [14].

The two-stage supply chain problem, with manufacturers, distribution centres and, customers, with fixed costs associated to the routes and proposals an efficient heuristic algorithm for the minimisation of the total transportation costs is discussed in [15]. The algorithm starts with building several initial solutions by processing customers in a specific order and choosing the best available supply route for each customer. After each initial solution is built, a process of searching for better variants around that solution follows, restricting the way the transportation routes are selected.

A matheuristic approach for solving the two-stage transportation problem with fixed costs associated to the routes is depicted in [16]. The proposed heuristic algorithm is designed to optimize the transportation problem, which is obtained by incorporating a linear programming optimization within the framework of a genetic algorithm.

3 Modelling the Material Requirements Planning

There are two techniques that a manufacturing enterprise may employ for the purposes of inventory management. The first one is stock replenishment, popularly known as statistical inventory control or the order point system. Usually, it is applied for single item models. Under this approach, the depletion in the supply of each inventory item is monitored and a replenishment order is issued whenever the supply drops to a predetermined quantity. The second technique is MRP, which is used for a multi-level, multi-period production planning and inventory control system. It is commonly much more difficult to analyse and control than a single-stage method. MRP was introduced in the 1970's [17].

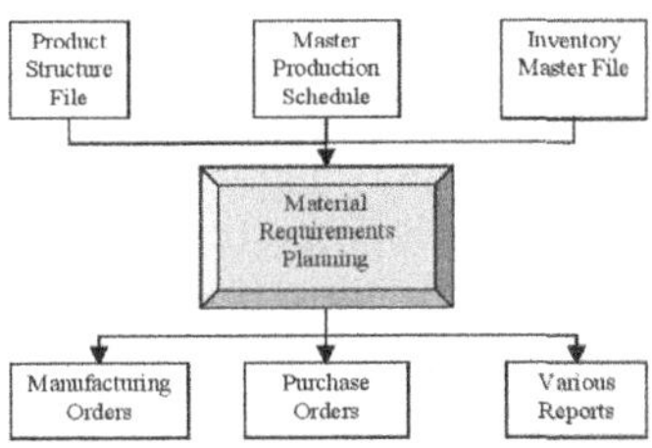

Fig. 1. An overview for material requirement planning system (Adopted from [18])

MRP is a system that controls inventory levels, plans production, helps in supplying management with important information, and supports the manufacturing control system with respect to the production of assembled parts [18] (Fig. 1). The MPS has to be feasible so that components can be produced within the capacity available in each time period, and the production-inventory system can be governed by the capacity constraints. Capacity constraints are considered in inventory planning for determining optimal target inventory positions.

3.1 Mainframe of the Firefly Algorithm

In this research to optimize MRP production and delivery planning in automotive industry in general, a bio-inspired optimization model, and particularly the firefly algorithm, is used. The firefly algorithm (FFA) is a relatively new swarm intelligence optimization method introduced in [19], in which the search algorithm is inspired by the social behaviour of fireflies and the phenomenon of bioluminescent communication. There are two critical issues in the firefly algorithm that represent a variation of light intensity referred as *cost value* and the formulation of attractiveness.

Algorithm 1 *The algorithm of* ***firefly algorithm***

Begin

Step 1: ***Initialization.*** *Set the generation counter* $\mathbf{G} = 1$*; Initialize the population of n fireflies* **P** *randomly and each firefly corresponding to a potential solution to the given problem; Define light absorption coefficient* γ*;*
Set controlling the step size α *and the initial attractiveness* β_0 *at* $\mathbf{r} = 0$*.*

Step 2: *Evaluate the cost function* ***I*** *for each candidate in* **P** *determined by* ***f(x)***

Step 3: ***While*** the termination criteria are not satisfied ***or*** **G** < MaxGeneration ***do***

```
    for i=1:n (all n candidate solution) do
        for j=1:n (n candidate solution) do
            if (I_j < I_i),
                move candidate i towards j;
            end if
            Vary attractiveness with distance r via exp[-γr²];
            Evaluate new solutions and update cost function;
        end for j
    end for i
    G = G+1;
```

Step 4: ***end while***

Step 5: *Post-processing the results and visualization;*

End.

Fireflies communicate, search for pray and find mates using bioluminescence with varied flashing patterns. Attractiveness is proportional to the brightness, which decreases with increasing the distance between fireflies. If there are no brighter fireflies than one particular candidate solution, it will move at random in the space [20]. The brightness of a firefly is influenced or determined by the objective function. For a *maximization/minimization* problem, brightness can simply be *proportional/inversely proportional* to the value of the cost function. More details about FFA and its variants are depicted in [21]. The basic steps of the FFA are summarized by the pseudo code revealed in Algorithm 1. The *light intensity* or attractiveness value $\boldsymbol{\beta}$ depends on the distance $\boldsymbol{r}$ between the fireflies and the media light absorption coefficient $\boldsymbol{\gamma}$. The *attractiveness* of each firefly is determined as monotonic decreasing function, where β_0 represents the attractiveness of the firefly at $r = 0$ and usually it is called *initial attractiveness*, using the equation:

$$\beta(r) = \beta_0 e^{-\gamma r^2} \tag{1}$$

The *movement* of a firefly f_j from position x_j to new position x_{j+1}, attracted to a brighter firefly f_i at position x_i is established by the equation:

$$x_{j+1} = x_j + \beta_0 e^{-\gamma r_{ij}^2} (x_j - x_i) + \alpha \varepsilon_i \tag{2}$$

where r_{ij} is distance between two fireflies, $\boldsymbol{\alpha}$ is the mutation coefficient and, ε_i is continuous uniform random numbers. In this experiment the following value of parameters: *maximum number of iterations* = 500; *number of fireflies* = 25; *media light absorption coefficient* $\boldsymbol{\gamma} = 0.4$; *initial attractiveness* $\boldsymbol{\beta_0} = 2$; *mutation coefficient* $\boldsymbol{\alpha} = 0.3$ are used in the firefly optimisation process.

3.2 Collection of Input Data

The first step of the method is the collection of input data. Generally, the input data come from the decisions the planner has made in the previous phases of identification of constraints and system design.

Table 1. Customer demand for products – **Pro.** 1 - left window regulator for front door – 5-door car; **Pro.** 2 - left window regulator for back door – 5-door car; **Pro.** 3 - right window regulator for front door – 5-door car; **Pro.** 4 - right window regulator for back door– 5-door car; **Pro.** 5 - left window regulator for front door – 3-door car; **Pro.** 6 - left window regulator for back door– 3-door car

Pro.	Days in 2013														Total
	Week 6							Week 7							
	1	2	3	4	5	6	7	1	2	3	4	5	6	7	
1	250	310	0	300	310	Weekend		325	325	0	450	0	Weekend		2270
2	250	310	0	300	310			325	325	0	450	0			2270
3	250	310	0	300	310			325	325	0	450	0			2270
4	250	310	0	300	310			325	325	0	450	0			2270
5	80	120	0	80	80			120	80	0	200	0			760
6	80	120	0	80	80			120	80	0	200	0			760

The data are collected from *Lames* Italian automotive company, and the factory is located in Serbia. *Lames* automotive company in Serbia produces door window regulators. The collected data set is from period 2013 and 2014; however, this research uses only two working weeks, 6 and 7, which are presented in Table 1. The ***first*** - initial production plan which is calculated as Demand for products + 2% industrial scrap is presented in Table 2.

The production process is organized in the following way. The factory works five days per week from Monday to Friday in two shifts, morning and afternoon. The factory has two production lines and both of them work in both shifts, with the *maximum production capacity* being 700 units per production line. That means that the factory can produce 2800 final items per day.

Also, what is important to mention is that it takes 15 min to change production tools in the production line in the moment when the production type is changing. In the production line, punctual item-related data are usually collected in a document, called

Table 2. The ***first*** - initial production plan (Demand + 2% predicted industrial scrap)

Pro.	Days in 2013														Total
	Week 6							Week 7							
	1	2	3	4	5	6	7	1	2	3	4	5	6	7	
1	255	318	0	306	318	Weekend		332	332	0	459	0	Weekend		2320
2	255	318	0	306	318			332	332	0	459	0			2320
3	255	318	0	306	318			332	332	0	459	0			2320
4	255	318	0	306	318			332	332	0	459	0			2320
5	82	123	0	82	82			123	82	0	204	0			778
6	82	123	0	82	82			123	82	0	204	0			778

Table 3. Plan *For Every Part* - PFEP - number of pieces in one final product

Type of final item	Raw material description	*BOM* for one FG	Production needs per hour	Type of final item	Material description	*BOMl* for one FG	Production needs per hour
	motor (L1)	1	700		motor (R2)	1	700
	slide bar (L1)	1	700		slide bar (R2)	1	700
Product 1	cable (L1)	1	700	**Product 4**	cable (R2)	1	700
	bumper	8	5600		bumper	8	5600
	plastic wheel	5	3500		plastic wheel	5	3500
	motor (L2)	1	700		motor (L3)	1	700
	slide bar (L2)	1	700		slide bar (L3)	1	700
Product 2	cable (L2)	1	700	**Product 5**	cable (L3)	1	700
	bumper	8	5600		bumper	8	5600
	plastic wheel	5	3500		plastic wheel	5	3500
	motor (R1)	1	700		motor (R3)	1	700
	slide bar (R1)	1	700		slide bar (R3)	1	700
Product 3	cable (R1)	1	700	**Product 6**	cable (R3)	1	700
	bumper	8	5600		bumper	8	5600
	plastic wheel	5	3500		plastic wheel	5	3500

Plan For Every Part (PFEP). In that document, information about every item or part needed for production, logistics and procurement can be found. The PFEP for these products is presented in Table 3. Product 1 refers to left window regulator for front door – 5-door car, shown as L1 type – left one, whose parts are presented in PFEP. It is important to notice that motors, slide bars and cables are different for every productions type. The calculations should also consider the following: stock quantity, safety stock, in-production plan, and in-raw-material order. The rest of the product types can be described in same manner, as presented in Table 3.

4 Experimental Results and Discussion

According to the nature of FFA, the experiment was repeated for 100 times with the data collection and experimental results for minimizing the inventory value, as presented in Table 4. For calculating the minimized inventory value, the following constraints were taken into account: *Lead time* from five different suppliers; *Type of packaging* for raw materials; *Minimal order quantity* for raw materials; and *Safety stock* for raw materials. The cost function of the FFA optimization inventory value is:

$$\min\,(inventory\ value) = \min \sum_{i=1\,day}^{14\,days} (demand\ for\ products_i - production\ plan_i)$$

and *production workload* is discrete uniform distribution function.

Table 4. Calculation for the production plan for *first* (I) and *second* (II) shift, and *production workload* in weeks 6 and 7

Product	Days in 2013														**Total**
	Week 6							*Week 7*							
Shift	*1*	*2*	*3*	*4*	*5*	*6*	*7*	*1*	*2*	*3*	*4*	*5*	*6*	*7*	
1 *(I)*	292	291	215	400	270	0	0	190	340	250	172	0	0	0	2320
2 *(I)*	276	307	350	90	270	0	0	325	265	85	352	0	0	0	2320
3 *(II)*	292	291	215	400	270	0	0	190	340	250	172	0	0	0	2320
4 *(II)*	276	307	350	90	270	0	0	325	265	85	352	0	0	0	2320
5 *(I)*	122	93	96	90	55	0	0	138	0	184	0	0	0	0	778
6 *(II)*	122	93	96	90	55	0	0	138	0	184	0	0	0	0	778
Shift I	690	691	561	580	595	0	0	653	605	519	524	0	0	0	
Shift II	690	691	561	580	595	0	0	653	605	519	524	0	0	0	
Workload %	*98.5*	*98.7*	*80.1*	*82.8*	*85.0*	*0*	*0*	*93.3*	*86.4*	*74.1*	*74.8*	*0*	*0*	*0*	*86.00*

Likewise, experimental results are based on MRP *Long term planning* and *Short term planning*. *Long term planning* is usually 6 months long and it is used to simulate the future demand and supply situation in all BOM levels. On the other hand, *Short term planning* is usually 4 weeks long, and presents the exact *Delivery Plan* to the customers. In order to make *Short term planning production* and *Delivery Plan* in automotive industry easier to understand, this paper presents only two-week *production planning*. Similarly, experimental results satisfy the *Short term forecast – delivery plan*.

The aim of this research is the need to optimize *Short term planning production* and *Delivery Plan* in the *Lames* factory in Serbia. Customer demand for products and delivery plan is satisfied and presented in Table 1 and Table 4. The production workload is between 74.1% and 98.7%. It can be observed that day 5 in the week 7 has the delivery value 0; nevertheless, one should not forget that automotive industry is the line production type of industry - when new customer's demand comes, finished products will be produced during the next working day, and for other delivery time.

It is not easy to compare implemented MRP systems. There are available varieties of different productions depending on type market environment systems; three of them are most common: (i) make-to-stock, (ii) make-to-order, and (iii) assemble-to-order. The company *Lames* is typical in the class of make-to-order. Therefore, some MRP systems can be compared with their efficiency qualitatively while the other companies can be compared quantitatively. For example, there are: (i) "the output of MRP is important since commands are issued through planning in order to launch the suggested orders with the required quantities and within the limited time period" [12]; and (ii) "result of the total calculation has shown that if company has implemented MRP from the beginning, company can make saving of 11%" [13].

5 Conclusion and Future Work

This paper presents a biological swarm intelligence model: the firefly optimization for modelling the material requirement planning production and delivery planning in automotive industry. MRP concept reduces overall inventory levels, creating positive economic effect on inventories and optimization of material handling in production. This model is tested on real-world dataset, collected in *Lames* Italian automotive company from the factory in Serbia, for short-term forecast – delivery plan.

Experimental results encourage further research. As the optimization method, the firefly has several parameters that determine its behaviour and efficacy. The future work could focus on extending research on good choice of parameters for various optimization scenarios which should help the production manager achieve different solutions and make better operational solutions and better production results with little effort. Then this model could be tested with the original very large real-world dataset obtained from existing different manufacturing companies.

References

1. Chu, C.-H., Hayya, J.C.: Buffering decisions under MRP environment: a review. Omega **16**(4), 325–331 (1988). https://doi.org/10.1016/0305-0483(88)90069-2
2. Fandel, G., Lundeberg, T.: Essays on Production Theory and Planning. Springer, Heidelberg (1988)
3. Simić, D., Svirčević, V., Corchado, E., Calvo-Rolle, J.L., Simić, S.D., Simić, S.: Modelling material flow using the milk run and Kanban systems in the automotive industry. Expert Syst. (2020). https://doi.org/10.1111/exsy.12546
4. Simić, D., Svirčević, V., Ilin, V., Simić, S.D., Simić, S.: Material flow optimization using milk run system in automotive industry. Advances in Intelligent Systems and Computing, vol. 950, pp. 411–421, Springer, Cham (2019). http://doi.org/10.1007/978-3-030-20055-8_39

5. Simić, D., Svirčević, V., Simić, S.: A hybrid evolutionary model for supplier assessment and selection in inbound logistics. J. Appl. Logic **13**(2), 138–147 (2015). https://doi.org/10.1016/j.jal.2014.11.007. Part A
6. Simić, D., Simić, S.: Hybrid artificial intelligence approaches on vehicle routing problem in logistics distribution. In: Hybrid Artificial Intelligence Systems. LNCS, vol. 7208, pp. 208–220. Springer, Heidelberg (2012). http://doi.org/10.1007/978-3-642-28942-2_19
7. Simić, D., Kovačević, I., Svirčević, V., Simić, S.: Hybrid firefly model in routing heterogeneous fleet of vehicles in logistics distribution. Logic J. IGPL **23**(3), 521–532 (2015). https://doi.org/10.1093/jigpal/jzv011
8. Ilin, V., Ivetić, J., Simić, D.: Understanding the determinants of e-business adoption in ERP-enabled firms and non-ERP-enabled firms: a case study of the Western Balkan Peninsula. Technol. Forecast. Soc. Change **125**, 206–223 (2017). https://doi.org/10.1016/j.techfore.2017.07.025
9. Simić, D., Svirčević, V., Ilin, V., Simić, S.D., Simić, S.: Particle swarm optimization and pure adaptive search in finish goods' inventory management. Cybern. Syst. **50**(1), 58–77 (2019). https://doi.org/10.1080/01969722.2018.1558014
10. Hasanati, N., Permatasari, E., Nurhasanah, N., Hidayat, S.: Implementation of material requirement planning (MRP) on raw material order planning system for garment industry. IOP Conf. Ser. Mater. Sci. Eng. **528** (2019). https://doi.org/10.1088/1757-899x/528/1/01206
11. Salaheldin, S., Francis, A.: A study on MRP practices in Egyptian manufacturing companies. Int. J. Oper. Prod. Manag. **18**(6), 588–611 (1998). https://doi.org/10.1108/01443579810209557
12. Imetieg, A.A., Lutovac, M.: Project scheduling method with time using MRP system – a case study: construction project in Libya. Eur. J. Appl. Econ. **12**(1), 58–66 (2015). https://doi.org/10.5937/EJAE12-7815
13. Iasya, A., Handayati, Y.: Material requirement planning analysis in micro, small and medium enterprise case study: grooveline – an apparel outsourcing company final project. J. Bus. Manag. **4**(3), 317–329 (2015)
14. Pintea, C.M., Calinescu, A., Pop Sitar, C., Pop, P.C.: Towards secure & green two-stage supply chain networks. Logic J. IGPL **27**(2), 137–148 (2019). https://doi.org/10.1093/jigpal/jzy028
15. Cosma, O., Pop, P.C., Sabo, C.: An efficient solution approach for solving the two-stage supply chain problem with fixed costs associated to the routes. Procedia Comput. Sci. **162**, 900–907 (2019). https://doi.org/10.1016/j.procs.2019.12.066
16. Cosma, O., Pop, P.C., Danciulescu, D.: A novel matheuristic approach for a two-stage transportation problem with fixed costs associated to the routes. Comput. Oper. Res. **118** (2020) https://doi.org/10.1016/j.cor.2020.104906. Article no. 104906
17. Orlicky, J.: Material Requirements Planning—The New Way of Life in Production and Inventory Management. McGraw-Hill, New York (1975)
18. Benton, W.C., Whybark, D.C.: Material requirements planning (MRP) and purchase discounts. J. Oper. Manag. **2**(2), 137–143 (1982). https://doi.org/10.1016/0272-6963(82)90029-8
19. Yang, X.-S.: Firefly algorithm, Lévy flights and global optimization. In: Bramer, M., Ellis, R., Petridis, M. (eds.) Research and Development in Intelligent Systems XXVI. Springer, London (2010). https://doi.org/10.1007/978-1-84882-983-1_15
20. Yang X.-S.: Cuckoo Search and Firefly Algorithm. Springer, Switzerland (2014). https://doi.org/10.1007/978-3-319-02141-6_1
21. Yang, X.-S.: Applications of Firefly Algorithm and Its Variants. Springer, Switzerland (2014). https://doi.org/10.1007/978-3-319-02141-6

Special Session: Soft Computing and Machine Learning in IoT, Big Data and Cyber Physical Systems

Time Series Data Augmentation and Dropout Roles in Deep Learning Applied to Fall Detection

Enol García González, José Ramón Villar(✉), and Enrique de la Cal

Computer Science Department, University of Oviedo, Oviedo, Spain
enolgargon@gmail.es, {villarjose,delacal}@uniovi.es

Abstract. Fall Detection is one of the most interesting and challenging research topics in the world today because of its implications in society and also because the complexity of processing Time Series (TS). Plenty of research has been published in the literature, several of them introducing Deep Learning (DL) Neural Network (NN) as the modelling element. In this study we analyse one of these contributions and address several enhancement using TS data augmentation and dropout. Moreover, the possibility of reducing the NN to make it lighter has been studied. The NN has been implemented using Keras in Python and the experimentation includes an staged fall publicly available data set. Results show the TS data augmentation together with dropout helped in learning a more robust and precise model. Future work includes introducing different types of cross-validation as well as introducing other types of DL models more suitable for TS.

Keywords: Fall detection · Neural network · Time series · Accelerometer · Wearables

1 Introduction and Related Work

The Fall Detection (FD) represents a challenge that, if overcome, could significantly improve the quality of life of people living alone, especially the elderly [1]. During the last few years, different approaches have been proposed to solve the problem of FD, but most methods face a major challenge when processing time series in which there is no window covering the whole time and data are obtained progressively. This study focuses on FD using wearable devices that includes an tri-axial accelerometer (3DACC) placed on a wrist; as shown in [2], this solution might be more usable for elder people.

The literature concerning the focused specific problem includes a wide variety of solutions. Machine learning (ML) is the main way to address the problems of fall detection and classification. Some examples of these methods are those presented In [3,4] feature extraction stage and Support Vector Machines are used

Á. Herrero et al. (Eds.): SOCO 2020, AISC 1268, pp. 563–570, 2021.
https://doi.org/10.1007/978-3-030-57802-2_54

in order to classify the TS windows. Thresholds could be used in FD [5–7] using instances that have been previously labeled as a function of acceleration predefined magnitude values. Thresholds could be used in FD in order to establish rules that drive the final decision [5,8,9]. [10] shows a comparison of these type of methods. In [11] an user centred solution is proposed, where a single model is build for a concrete individual, this model identifies the normal activities and every thing outside the normal activity could be a possible fall. Besides, a solution based on clustering the detected peaks and building specific models for each cluster was proposed in [12]. In [13], the authors show how all these studies can be useful in real life by improving the quality of people's lives through the advancement of technology with the diffusion and ease of access to wearables. However, the research published in [14] presents aspects such as power consumption and real-time data processing that make it necessary to work in lighter FD systems.

One of the main characteristics of all the previous approaches is that all of them developed models with a reduced computational cost. However, with the raise of Deep Learning (DL), several studies have proposed using DL as a FD service [15]. One of these is [16], where an open source dataset and LSTM deep learning model are used in order to detect falls and daily activities with a high success rate. Using DL the input change dramatically: instead of detecting peaks and producing a feature extraction from the TS, with DL a complete TS window is feed to the model without further pre-processing. Alternatively, a Convolutional network (convnet) solution is proposed in [17], where 4 blocks of convnet models are sequenced to produce the final TS label. Interestingly, the authors obtained a good model but some of the normal enhancements in DL were not considered.

This study focuses on extending the work presented in [17] to introduce two well known DL enhancements, such as data augmentation and dropout to avoid overfitting. Furthermore, several different configurations have been tested in order to reduce the dimension of the convnet. The main novelty of this study, however, is the designed TS data augmentation. The experimentation will compare all the solutions and shows the benefits of the designed TS data augmentation.

The structure of this manuscript is as follows. The next section describes the convnet proposed in [17] as it represents the basis of this research. Moreover, the section also give details of the TS data augmentation and the dropout introduced to the model. Finally, the data set and the experimentation set up are described at the end of this section. Section 3 includes the obtained results and the comparison of the different options analyzed in this research, together with a discussion on the results. Finally, the conclusions are drawn.

2 Materials and Methods

This section will be responsible for defining the motivation for the project, as well as the starting NN for the research. It also includes a description of the modifications that were made to the original NN in order to improve it and the data and experiments carried out.

2.1 Neural Network Models Used in This Study

This study is based on the network proposed in [17]. This NN is built by 4 levels where we found a Convolution layer, a Normalization layer, a ReLU layer and a Max Pooling layer in each level. Furthermore, every Convolution and max Pooling Layer has a filter size of 1×5. The first Convolution layer has 16 filters, the second 32, the third 64 and the fourth 128. Finally, there is a classification dense layer with Softmax activation which gives the output of the NN. Figure 1 depicts the structure of the NN. The authors studied this NN with the UMAFall data set mixing all the TS in a single bag and using 10 fold cross validation. From now on, this model is referred as CN_{CAS}.

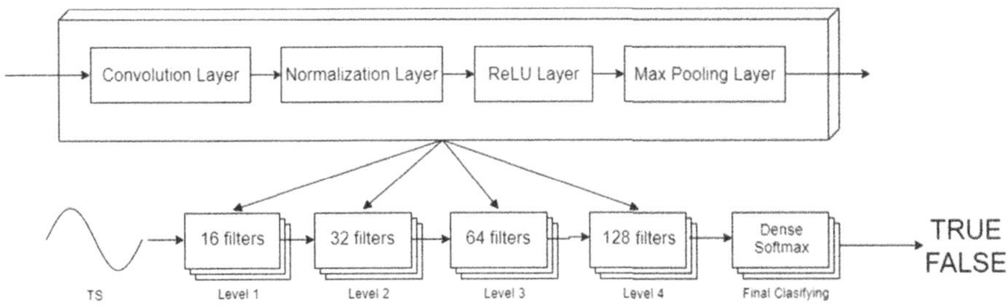

Fig. 1. Structure of the NN proposed in [17].

This NN model has been revisited, proposing different alternative configurations. The NN was considered to be too heavy and over-dimensioned for the size of the problem to be solved, therefore, networks were considered in which each level contained the same layers, but reducing the number of levels, in particular 2 and 3 levels were considered. These alternatives have been tested but their results were poorer than those from the CN_{CAS}, therefore their results have not been included in this research.

2.2 Enhancements in the Network Learning

Two main improvements have been included: introducing data augmentation in the training data feed and several over-fitting avoiding drop-out layers.

Data Augmentation. The first modification that was made to CN_{CAS} was to apply data augmentation to achieve a much wider and more varied set of training data. In this way, it is possible to eliminate the over-fitting that appears when all the falls are located at the same time in the data window or because they are too similar in magnitude. The NN that adds data augmentation on the CN_{CAS} NN will henceforth be referred to as CN_{CAS+DA}. The augmentation was done in two ways:

- A random number is generated to increase or decrease the difference between two consecutive values of the time series.
- In addition, the moment at which the fall within the time series occurs was modified. For this, the series was observed and it was seen that the magnitude at the beginning and at the end was identical, since it starts at rest and ends at rest, which allows moving the time series in time using rotation, that is, all the data are delayed a fixed number of milliseconds and those that exceed the end of the series are placed at the beginning of it.

An example of this data augmentation process can be seen in Fig. 2.

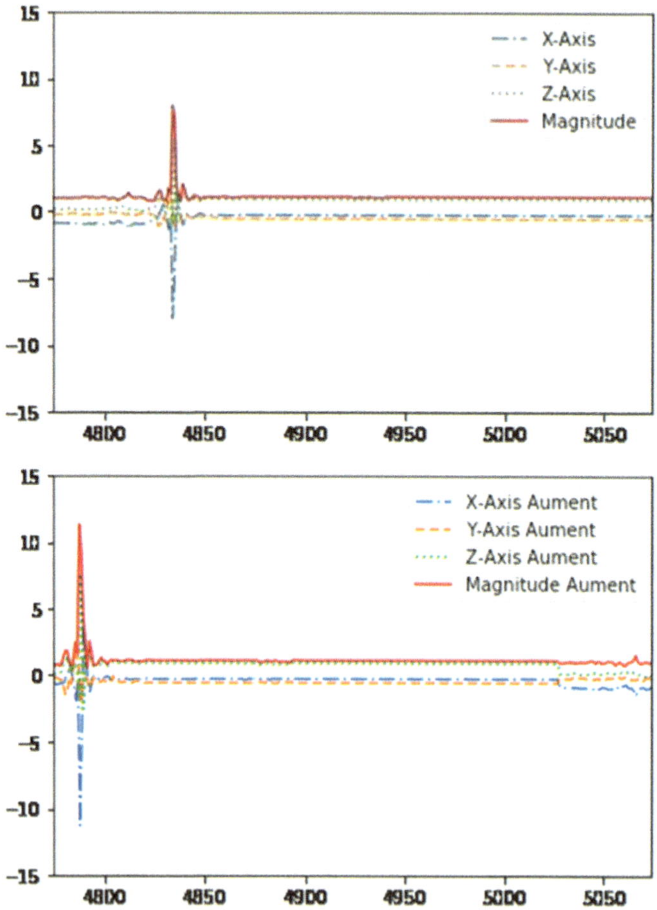

Fig. 2. Comparison of data before and after the DA process. The X-axis shows the sample's index, the Y-axis is multiple of $G = 9.8\,\mathrm{m/s^2}$. The scale and shift of the multivariate TS is clearly shown.

Drop-Out. A modification we made to the CN_{CAS+DA} NN to avoid over-adjustment during training was to add a drop-out layer between each level. We denote the NN that makes used of drop-out together with data augmentation as $CN_{CAS+DA+DO}$. Interestingly, the NN including drop-out only has also been evaluated but its results were not competitive.

2.3 Data Set and Cross Validation

In order to compare this results with the [17], we use the staged falls data set provided in [18]. In this data set, up to 19 participants performed several human activities of daily living plus staged falls. Three different types of fall were staged: forward, lateral and backwards fall. The data was gathered using inertial devices, including both 3DACC, magnetometer and gyroscope, placing the sensors on different body locations. In this research we consider only the TS from the 3DACC sensor placed on a wrist.

Each participant recorded several runs of each activity or staged fall, each run producing a TS including the acceleration components for each axis. All the TS have been introduced in a bag of TS with their corresponding label (either FALL or NOT_FALL), 20% of the TS are preserved for validation, while the remaining samples are kept for training and testing. A sliding window of size 650 ms with a shift of 1 sample is used to evaluate each interval within a TS.

We use 10 fold cross validation for the training and testing stage. In this cross validation configuration, TS belonging to any of the participants can be included in the train and test, there is no distinction by participant. We are going to compare the different options explained before. For each model and fold the Accuracy, Kappa factor, Sensitivity and Specificity are determined.

3 Results and Discussion

The obtained results are shown in several graphs and tables as detailed next:

- Table 1 shows the average of the different metrics and each network configuration CN_{CAS}, CN_{CAS+DA} and $CN_{CAS+DA+DO}$.
- Figure 3 shows the box plots obtained for each network configuration.

Table 1. Average of the metrics obtained during the experiments

Experiment	Accuracy	Specificity	Sensitivity	Kappa
CN_{CAS}	0.6967	0.7091	0.7091	0.3947
CN_{CAS+DA}	0.8915	0.8870	0.8870	0.7829
$CN_{CAS+DA+DO}$	0.9125	0.9142	0.9142	0.7712

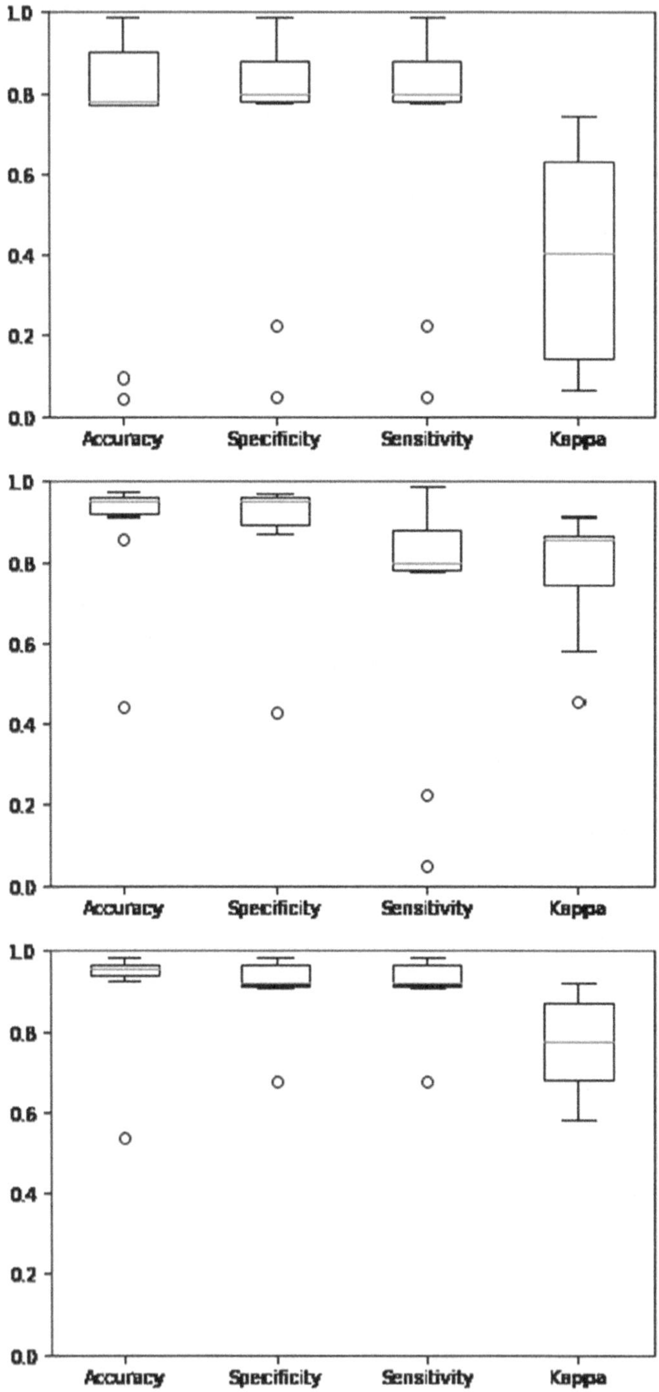

Fig. 3. Box plots obtained for each of the configurations. A box plot for each metric is included in each graph. Top, center and bottom correspond to CN_{CAS}, CN_{CAS+DA} and $\mathrm{CN}_{CAS+DA+DO}$.

The results of the experiments shows that both CN_{CAS+DA} and $CN_{CAS+DA+DO}$ obtaine better metrics than previous the proposed in CN_{CAS}; the most remarkable improvement happening with the CN_{CAS+DA} network. From this improvement we can see that CN_{CAS} NN took learned the peaks in an specific position within a TS window; when the peak came from an modified staged fall, the CNN_{CAS} was not able to identify the fall. By training the same network with a much larger data set, the performance with the test set has been greatly improved. Furthermore, if we look at the box plots of the networks we can see how the dispersion of metrics is reduced, since with the CN_{CAS} network both very good and very bad networks were obtained. However, with the CN_{CAS+DA} network the values taken from the metrics are more concentrated.

From the data reflected in the table, it seems that the $CN_{CAS+DA+DO}$ network offers an improvement over the CN_{CAS+DA} network, but this improvement is obtained at the cost of sacrificing the concentration of results discussed above. While the CN_{CAS+DA} network has the results of metrics very well concentrated, the new $CN_{CAS+DA+DO}$ network has more than one occasion in which it shows a malfunction. However, despite this dispersion, the results can be considered better.

Finally, it is worth mentioning that the Kappa and Accuracy results show how the balance in the data set is in compromise. Certainly, the data set has a big difference in the number of TS labeled as Fall or Not Fall. For the scope of this study we have not cope with this issue because we wanted to compare with the original method. However, future research includes a DA that copes with this particular issue, balancing the number of TS for each label.

4 Conclusion

In this study, a proposal for FD using DL NN has been refined with several elements: i) a 3DACC located on a wrist, ii) using TS data augmentation and iii) introducing dropout; these two latter to avoid the overfitting. A publicly available staged fall data set (UMA FALL) was used in the experimentation to evaluate and compare the options. With all the results obtained and compared in the previous topic it can be concluded that the CN_{CAS+DA} and $CN_{CAS+DA+DO}$ networks are options that significantly improve the performance of the initial network.

Future work includes analysing several different staged fall data sets, and designing other types of NN, such as LTSM and recurrent networks or CONV1D NN. Moreover, more interesting TS data augmentation designs can be introduced in order to get a good variation of the signals.

Acknowledgment. This research has been funded by the Spanish Ministry of Science and Innovation under project MINECO-TIN2017-84804-R and by the Grant FCGRUPIN-IDI/2018/000226 project from the Asturias Regional Government.

References

1. Jahanjoo, A., Naderan, M., Rashti, M.J.: Detection and multi–class classification of falling in elderly people by deep belief network algorithms. Ambient Intell. Human. Comput., 1–21 (2020)
2. Khojasteh, S.B., Villar, J.R., Chira, C., Suárez, V.M.G., de la Cal, E.A.: Improving fall detection using an on-wrist wearable accelerometer. Sensors **18**(5), 1350 (2018)
3. Zhang, T., Wang, J., Xu, L., Liu, P.: Fall detection by wearable sensor and one-class SVM algorithm. In: Huang, D.S., Li, K., Irwin, G. (eds.) Intelligent Computing in Signal Processing and Pattern Recognition. Lecture Notes in Control and Information Systems, vol. 345, pp. 858–863. Springer, Heidelberg (2006)
4. Wu, F., Zhao, H., Zhao, Y., Zhong, H.: Development of a wearable-sensor-based fall detection system. Int. J. Telemedicine Appl. **2015**, 11 (2015)
5. Bourke, A., O'Brien, J., Lyons, G.: Evaluation of a threshold-based triaxial accelerometer fall detection algorithm. Gait Posture **26**, 194–199 (2007)
6. Fang, Y.C., Dzeng, R.J.: A smartphone-based detection of fall portents for construction workers. Procedia Eng. **85**, 147–156 (2014)
7. Fang, Y.C., Dzeng, R.J.: Accelerometer-based fall-portent detection algorithm for construction tiling operation. Autom. Constr. **84**, 214–230 (2017)
8. Huynh, Q.T., Nguyen, U.D., Irazabal, L.B., Ghassemian, N., Tran, B.Q.: Optimization of an accelerometer and gyroscope-based fall detection algorithm. J. Sens. **2015**, 8 (2015)
9. Kangas, M., Konttila, A., Lindgren, P., Winblad, I., Jämsaä, T.: Comparison of low-complexity fall detection algorithms for body attached accelerometers. Gait Posture **28**, 285–291 (2008)
10. Hakim, A., Huq, M.S., Shanta, S., Ibrahim, B.: Smartphone based data mining for fall detection: analysis and design. Procedia Comput. Sci. **105**, 46–51 (2017)
11. Villar, J.R., de la Cal, E.A., Fáñez, M., Suárez, V.M.G., Sedano, J.: User-centered fall detection using supervised, on-line learning and transfer learning. Progress in AI **8**(4), 453–474 (2019)
12. Fáñez, M., Villar, J.R., de la Cal, E.A., Suárez, V.M.G., Sedano, J.: Feature clustering to improve fall detection: a preliminary study. SOCO **2019**, 219–228 (2019)
13. Godfrey, A.: Wearables for independent living in older adults: gait and falls. Maturitas **100**, 16–26 (2017)
14. Igual, R., Medrano, C., Plaza, I.: Challenges, issues and trends in fall detection systems. BioMedical Eng. OnLine **12**, 66 (2013)
15. Casilari-P'erez, E., Lagos, F.G.: A comprehensive study on the use of artificial neural networks in wearable fall detection systems. Expert Syst. Appl. **138** (2019)
16. Wu, X., Cheng, L., Chu, C.H., Kim, J.: Using deep learning and smartphone for automatic detection of fall and daily activities. In: Lecture Notes in Computer Science, vol. 11924, pp. 61–74 (2019)
17. Casilari, E., Lora-Rivera, R., García-Lagos, F.: A wearable fall detection system using deep learning. In: Advances and Trends in Artificial Intelligence, pp. 445–456 (2019)
18. Casilari, E.: Umafall: a multisensor dataset for the research on automatic fall detection. Procedia Comput. Sci. **110**, 32–39 (2017)

A Comparison of Multivariate Time Series Clustering Methods

Iago Vázquez[1], José Ramón Villar[2(✉)], Javier Sedano[1], and Svetlana Simić[3]

[1] Instituto Tecnológico de Castilla y León, Pol. Ind. Villalonquejar, 09001 Burgos, Spain
{iago.vazquez,javier.sedano}@itcl.es

[2] Computer Science Department, University of Oviedo, Oviedo, Spain
{villarjose,delacal}@uniovi.es

[3] Department of Neurology, Clinical Centre of Vojvodina Novi Sad, University of Novi Sad, Novi Sad, Republic of Serbia
svetlana.simic@mf.uns.ac.rs

Abstract. Big Data and the IoT explosion has made clustering Multivariate Time Series (MTS) one of the most effervescent research fields. From Bio-informatics to Business and Management, MTS are becoming more and more interesting as they allow to match events the co-occur in time but that is hardly noticeable. In this paper, we compare four clustering methods retrieved from the literature analyzing their performance on five publicly available data sets. These methods make use of different TS representation and distance measurement functions. Results show that Dynamic Time Warping is still competitive; APCA+DTW and Compression-based dissimilarity obtained the best results on the different data sets.

Keywords: Time Series · Clustering · Multivariate

1 Introduction

Multivariate Time Series (MTS) have regained the focus of the research community with the effervescence of Big Data, Internet of Things and Cyber-Physical Systems. In many cases, there is no information that introduce relationships among the MTS instances. Until recently, the problem was focused on univariate TS clustering; for instance, [1] proposed use Dynamic Time Warping (DTW) and k-means to cluster the performance of a photovoltaic power plant, so to predict the meteorological conditions. Similarly, k-means was used to cluster TS and then predict the weather conditions [2]. Interested readers can refer to [3] for a good review on this topic. Nevertheless, when more than one Time Series (TS) is involved the clustering problem becomes much more challenging. Additionally, it is possible to choose between unsupervised and semi-supervised methods to perform the clustering.

Á. Herrero et al. (Eds.): SOCO 2020, AISC 1268, pp. 571–579, 2021.
https://doi.org/10.1007/978-3-030-57802-2_55

Grouping MTS has been found interesting in order to perform complex event detection or to classify the current scenario. For instance, [4] proposed a Partitioning around Medoids and Fuzzy C-Medoids clustering for the problem of detecting high-value pollution records or alarms in the city of Rome. To group the instances, the similarity among the variables between two MTS instances is one of the most studied topics. As an example, the authors in [5] proposed Principal Component Analysis similarity factor combined with the average based Euclidean distance together with a fuzzy clustering scheme to group MTS instances. Discords have also been used in MTS instance clustering to identify anomalies [6]. Alternatively, hash functions have been proposed to index and to measure the similarities as well [7].

Interestingly, Machine Learning models have been also used in measuring the similarity between multivariate TS, i.e., Gaussian Mixture Models [8] or Recurrent Neural Networks [9,10]. A different approach is based on extracting features and then using these features to group the multivariate TS [11] or together with Self-Organized Maps [12], Hidden Markov Models [13] or Fuzzy Linear [14]. Still, this problem cannot be considered solved and a recent study found out that the combination of feature extraction and a classification stage performs better than the current approaches [15].

This paper shows a comparison among four MTS instance clustering methods. The MTS representation and the distance measurement are different from one method to the other. In all of them, hierarchical clustering is the algorithm responsible of the groupings according to the distance matrices; the obtained trees are cut to get the desired number of clusters k. In the experimentation, the 4 methods are compared using several published MTS data sets. Two different experiments are carried out: on the first hand, the best number of clusters is found using the elbow's rule; on the second hand, the number of groups are defined with the number of classes in each data set. These two experimentation set ups might provide some idea on the performance of the MTS clustering methods: the first one tackles the total ignorance of the problem (no knowledge in the number of classes) and how they behave with the elbow's rule; the second one represents the case of total knowledge, where the number of labels are known a-priori but not the grouping or the MTS patterns. The main goal of this study is to set the basis for a future research on merging the outcomes of different MTS data sets, giving some rules on how the different techniques perform and providing evidence on how to design the merging.

The structure of the paper is as follows. Next Section aims to give details of the 4 methods of this comparison, the data sets used in the comparison and the experimental set up. Section 3 discuss on the obtained results. Finally, the conclusions are drawn.

2 Materials and Methods

This section describes the 4 methods in this comparison first, then the MTS data sets are introduced and, finally, the experimental set up is detailed.

2.1 MTS Clustering Methods

Let us call raw MTS the temporal sequence of values for each of the variables gathered from a certain source. Each instance in this raw MTS data set (ts^i) can be written as $< x_1^i, x_2^i, \cdots, x_M^i >$, where M is the number of variables and $x_m^i = x_{m1}^i, x_{m2}^i, \cdots, x_{mN}^i >$, N is the number of samples, m is the variable and i is the index on the MTS data set. We call $x^i[t] =< x_{1t}^i, x_{2t}^i, \cdots, x_{Mt}^i >$ the sample at time t. We assume a MTS data set as a collection of instances of raw MTS with arbitrary length. Note that we can store MTS for which the variables have different sampling rate provided there are some timestamps where all the sampling of all the variables coincide in time [16,17] using polynomial interpolation. Besides, long MTS are expected to be split in different instances; automatic segmentation of MTS can be employed in these cases to produce the set of suitable instances [14,17].

The four methods in this comparison are included in the following listing. In all of them, the distance between each pair of MTS instances in the data set are stored in a matrix; then, the hierarchical clustering (hclust) is employed to group the MTS instances.

- **Adaptive Piecewise Constant Approximation (APCA) plus MINDIST and hclust** [18], denoted as APCA-MINDIST. In this study, each variable j in a raw TS is represented by M segments ($APCA(ts^{ij})$=$\{< v_1^{ij}, p_1^{ij} >, \cdots, < v_M^{ij}, p_M^{ij} >\}$). The coefficients v_k^{ij} are the mean of the values of variable j in the interval $[p_(k-1)^{ij}, p_k^{ij}]$, with $p_0^{ij} = 0$. The limits of the intervals are computed with the Haar Discrete Wavelet Transform [19]. The MINDIST, defined by the authors, is used as the distance measurement.
- **APCA plus DTW and hclust** [18], denoted as APCA-DTW. The main variation is that DTW [20] is used as the MTS instances distance measurement.
- **Fast Fourier Transform (FFT) combined with hclust** [21] and denoted as FFT-hclust. The FFT is computed on the z-scored raw data, limiting the coefficients to the 10 components. The distance between two univariate TS is measured with the Energy of the differences between them.
- **Raw data and measuring similarities with the Compression-based dissimilarity measure (CMD) on the raw data** [22] and denoted as CMD-hclust. To overcome with the problem of TS of different lengths, the longer TS is windowed and the CDM is averaged. Let lng be the length of the shorter TS instance, then we propose to use a sliding window of size lng with a shift of lng samples; padding the window with the last TS sample whenever needed to avoid incomplete sliding windows. We consider two TS of similar length whenever the differences in length do not surpass the 1.5 ratio.

We have used the rule of the elbow to select the number of clusters [23]. To do so, the sum of squares distances of each point to its cluster center as the measure of quality Q_k of the current number of clusters k. Thus, if Cl_k is the set of every clusters found for every possible number of clusters k used to feed the clustering algorithm, then $Q_k = \sum_{C \in Cl_k} \sum_{p \in C} d(p, c_C)^2$, where c_C is the center of the cluster C and d corresponds to the Euclidean distance.

2.2 Experimental Data Sets

To illustrate the performance of the different clustering methods we have used several MTS data sets from the Time Series Classification site [24]. All the instances of the proposed data sets are labelled, which allows to evaluate the performance of the different solutions. The following MTS data sets are included in the experimentation stage:

- ArticularyWordRecognition (AWR) [25,26]: 25 train and 25 test instances of 12 variables, each with 143 samples.
- Cricket (Cr) [26,27]: records the movements of the hands of 4 cricket umpires using accelerometers. A total of 12 classes, with 6 variables and 1197 samples each per instance. The data set includes 108 train instances and 72 test instances.
- Epilepsy (EP) [28]: this data set includes triaxial accelerometer data recorded for several Activities of Daily Living and simulated Epileptic seizures. The data set includes 137 train instances and 128 test instances. Each instance includes 3 variables and 206 samples.
- Finger Movements (FM) [29]: this dataset has a correspondence to Benjamin Blankertz for the BCI II competition (Data set IV). The data set includes 316 train instances and 100 test instances. Each instance includes 28 variables, 50 samples each.
- HeartBeat (HB) [30,31]: this dataset is derived from the PhysioNet/CinC Challenge 2016. The data set includes 61 instance for training and 61 for testing. Each instance has 61 variables and 405 samples.

2.3 Assessment of the Methods

We propose to use the following metrics to measure the performance of each method: Accuracy (ACC), Sensitivity (SEN), Specificity (SPE), and Kappa Factor (KPP). Therefore, we count the number of True Positives (TP), True Negatives (TN), False Positives (FP) and False Negatives (FN) on the set of pairs of instances of a data set. In the context of clustering, we define these measures the following way, based on the work exposed in [32]:

- If the two instances are in the same cluster and belong to the same class, the pair counts as a True Positive.
- If the two instances are in different clusters and belong to different classes, the pair counts as a True Negative.
- If the two instances are in the same cluster and belong to different classes, the pair counts as a False Positive.
- If the two instances are in different clusters but they belong to the same class, the pair counts as a False Negative.

3 Results and Discussion

Results are included in Table 1 and Table 2. The former includes the results for the best number of clusters in each case; the latter shows the figures when the number of clusters is set to the number of labels in the data set.

Table 1. Results for the **best number of clusters** found using the rule of the elbow.

Method	AWR					Cr				
	K	ACC	KPP	SEN	SPE	K	ACC	KPP	SEN	SPE
h-A-MIN	28	0.99	0.99	0.75	0.99	10	0.94	0.94	0.74	0.96
h-A-DTW	31	0.99	0.99	0.88	1.00	12	0.97	0.97	0.98	0.97
h-FFT	12	0.93	0.93	0.87	0.94	10	0.92	0.92	0.72	0.94
h-CMD	5	0.64	0.64	0.5	0.65	4	0.71	0.71	0.79	0.71
Method	EP					FM				
	K	ACC	KPP	SEN	SPE	K	ACC	KPP	SEN	SPE
h-A-MIN	5	0.70	0.68	0.26	0.84	5	0.70	0.68	0.26	0.84
h-A-DTW	6	0.71	0.70	0.37	0.83	3	0.50	0.37	0.41	0.509
h-FFT	5	0.64	0.61	0.32	0.74	5	0.64	0.61	0.32	0.74
h-CMD	5	0.80	0.79	0.64	0.86	5	0.80	0.79	0.64	0.86
Method	HB									
	K	ACC	KPP	SEN	SPE					
h-A-MIN	4	0.59	0.25	0.79	0.29					
h-A-DTW	3	0.61	0.23	0.87	0.23					
h-FFT	5	0.60	0.18	0.88	0.18					
h-CMD	5	0.47	0.37	0.28	0.75					

As it can be seen, there is no clear winner among the different data sets. AWR shows a high Accuracy and Kappa Coefficient for APCA-MINDIST and APCA-DTW, with a significantly better sensitivity for the second method in the two run experiments. With the Cr data set, the best performance is observed for the APCA+DTW. Nevertheless, all the methods performed rather well with these two data sets. In the case of the EP data set, however, CMD-hclust is the best clustering method, followed by APCA+DTW in both experiments.

The results obtained with the FM and HB data sets are clearly poorer. In FM, for the first experiment, each method shows an accuracy of 0.5, while the sensitivity is higher for APCA+MINDIST and APCA+DTW and the specificity is higher for FFT-hclust and CMD-hclust. However, as Kappa coefficient is higher for these last two methods, their performance is based on their ability to find relevant clustering rules, while the APCA based methods seem to get clusters with more differences among their quantity of elements. In the second experiment, we have also a similar accuracy for each method, but the low

Table 2. Results obtained when the **number of clusters (K)** is set to the number of classes in the data set.

Method	AWR				Cr			
	ACC	KPP	SEN	SPE	ACC	KPP	SEN	SPE
h-A-MIN	0.98	0.98	0.84	0.99	0.94	0.94	0.73	0.96
h-A-DTW	0.99	0.99	0.91	0.99	0.98	0.98	0.94	0.98
h-FFT	0.98	0.98	0.77	0.99	0.93	0.93	0.68	0.95
h-CMD	0.93	0.93	0.13	0.96	0.81	0.81	0.56	0.83
Method	EP				FM			
	ACC	KPP	SEN	SPE	ACC	KPP	SEN	SPE
h-A-MIN	0.64	0.62	0.31	0.75	0.50	0.27	0.62	0.37
h-A-DTW	0.69	0.67	0.37	0.79	0.50	0.33	0.501	0.49
h-FFT	0.62	0.59	0.33	0.72	0.50	0.15	0.82	0.18
h-CMD	0.79	0.78	0.71	0.82	0.50	0.33	0.501	0.48
Method	HB							
	ACC	KPP	SEN	SPE				
h-A-MIN	0.59	0.03	0.97	0.03				
h-A-DTW	0.61	0.23	0.87	0.23				
h-FFT	0.59	0.00	0.99	0.01				
h-CMD	0.51	0.29	0.52	0.49				

specificity and high sensitivity for FFT-hclust, along the low value of the Kappa factor. APCA+DTW and CMD perform similarly, while APCA+MIN shows a less balanced result than the two previous methods.

Finally, with the HB, the second experiment's results for FFT-hclust and APCA+MINDIST are the worst: the low Kappa Factor and specificity show that these two methods created two extremely imbalanced clusters, and their performance is similar to those obtained when clustering all the instances in the same cluster. APCA+DTW shows better performance, while CMD-hclust is the most balanced method considering the all the metrics. Overall, perhaps it can be concluded that the best two methods are APCA+DTW and CDM-hclust; however, what is really relevant is that the methods vary their performance according to the data set. More research is needed in obtaining MTS clustering methods that perform similarly among a wide variety of problems; perhaps an ensemble of techniques including some user feedback might help in driving the grouping process.

4 Conclusions

This study present a comparison of MTS clustering methods using publicly available MTS data sets. The aim of this research is to find which TS representation and distance measurements are more promising among APCA-DTW, APCA-MINDIST, FFT-hclust and CMD-hclust.

Results show that there is a strong variability in the results according to the data set, showing no clear winner method. Both APCA-DTW and CMD-hclust showed the best overall performance and were more balanced when considering all the metrics simultaneously. More research is needed in obtaining MTS clustering methods that perform similarly among a wide variety of problems; perhaps an ensemble of techniques including some user feedback might help in driving the grouping process.

Acknowledgment. This research has been funded by the Spanish Ministry of Science and Innovation under project MINECO-TIN2017-84804-R and by the Grant FCGRUPIN-IDI/2018/000226 project from the Asturias Regional Government.

References

1. Liu, G., Zhu, L., Wu, X., Wang, J.: Time series clustering and physical implication for photovoltaic array systems with unknown working conditions. Sol. Energy **180**, 401–411 (2019)
2. Lee, Y., Na, J., Lee, W.B.: Robust design of ambient-air vaporizer based on time-series clustering. Comput. Chem. Eng. **118**, 236–247 (2018)
3. Aghabozorgi, S., Shirkhorshidi, A.S., Wah, T.Y.: Time-series clustering - a decade review. Inf. Syst. **53**, 16–38 (2015)
4. D'Urso, P., Giovanni, L.D., Massari, R.: Robust fuzzy clustering of multivariate time trajectories. Int. J. Approximate Reasoning **99**, 12–38 (2018)
5. Fontes, C.H., Budman, H.: A hybrid clustering approach for multivariate time series - a case study applied to failure analysis in a gas turbine. ISA Trans. **71**, 513–529 (2017)
6. Hu, M., Feng, X., Ji, Z., Yan, K., Zhou, S.: A novel computational approach for discord search with local recurrence rates in multivariate time series. Inf. Sci. **477**, 220–233 (2019)
7. Yu, C., Luo, L., Chan, L.L.H., Rakthanmanon, T., Nutanong, S.: A fast LSH-based similarity search method for multivariate time series. Inf. Sci. **476**, 337–356 (2019)
8. Mikalsen, K.Ø., Bianchi, F.M., Soguero-Ruiz, C., Jenssen, R.: Time series cluster kernel for learning similarities between multivariate time series with missing data. Pattern Recogn. **76**, 569–581 (2018)
9. Vázquez, I., Villar, J.R., Sedano, J., Simic, S.: A preliminary study on multivariate time series clustering. In: 14th International Conference on Soft Computing Models in Industrial and Environmental Applications (SOCO 2019) - Seville, Spain, 13–15 May 2019, Proceedings, pp. 473–480 (2019)
10. Vázquez, I., Villar, J.R., Sedano, J., Simic, S., de la Cal, E.A.: A proof of concept in multivariate time series clustering using recurrent neural networks and SP-lines. In: Hybrid Artificial Intelligent Systems - 14th International Conference, HAIS 2019, León, Spain, 4–6 September 2019, Proceedings, pp. 346–357 (2019)
11. Ferreira, A.M.S., de Oliveira Fontes, C.H., Cavalcante, C.A.M.T., Marambio, J.E.S.: Pattern recognition as a tool to support decision making in the management of the electric sector. Part II: a new method based on clustering of multivariate time series. Int. J. Electr. Power Energy Syst. **67**, 613–626 (2015)

12. Salvo, R.D., Montalto, P., Nunnari, G., Neri, M., Puglisi, G.: Multivariate time series clustering on geophysical data recorded at Mt. Etna from 1996 to 2003. J. Volcanol. Geoth. Res. **251**, 65–74 (2013). Flank instability at Mt. Etna
13. Li, J., Pedrycz, W., Jamal, I.: Multivariate time series anomaly detection: a framework of hidden Markov models. Appl. Soft Comput. **60**, 229–240 (2017)
14. Duan, L., Yu, F., Pedrycz, W., Wang, X., Yang, X.: Time-series clustering based on linear fuzzy information granules. Appl. Soft Comput. **73**, 1053–1067 (2018)
15. Bode, G., Schreiber, T., Baranski, M., Müller, D.: A time series clustering approach for building automation and control systems. Appl. Energy **238**, 1337–1345 (2019)
16. Anstey, J., Peters, D., Dawson, C.: An improved feature extraction technique for high volume time series data. In: Proceedings of the Fourth IASTED International Conference on Signal Processing, Pattern Recognition, and Applications, pp. 74–81, January 2007
17. Keogh, E., Lonardi, S., Chiu, B.Y.c.: Finding surprising patterns in a time series database in linear time and space. In: Proceedings of the Eighth ACM SIGKDD International Conference on Knowledge Discovery and Data Mining, pp. 550–556 (2002)
18. Chakrabarti, K., Keogh, E., Mehrotra, S., Pazzani, M.: Locally adaptive dimensionality reduction for indexing large time series databases. ACM Trans. Database Syst. (TODS) **27**, 188–228 (2002)
19. Chan, K.P., Fu, A.W.C.: Efficient time series matching by wavelets. In: Proceedings of the 15th International Conference on Data Engineering, p. 126 (1999)
20. Bellman, R.: Adaptive Control Processes. Princeton University Press, Princeton (1961)
21. Singleton, R.: An algorithm for computing the mixed radix fast Fourier transform. IEEE Trans. Audio Electroacoust. **17**(2), 93–103 (1969)
22. Keogh, E., Lonardi, S., Ratanamahatana, C., Wei, L., Lee, S.H., Handley, J.: Compression-based data mining of sequential data. Data Min. Knowl. Disc. **14**, 99–129 (2007)
23. Öztürk, A., Lallich, S., Darmont, J.: A visual quality index for fuzzy C-means. In: Artificial Intelligence Applications and Innovations, June 2018
24. Bagnall, A., Lines, J., Bostrom, A., Large, J., Keogh, E.: The great time series classification bake off: a review and experimental evaluation of recent algorithmic advances. Data Min. Knowl. Disc. **31**(3), 606–660 (2017)
25. Wang, J., Balasubramanian, A., de la Vega, L.M., Green, J.R., Samal, A., Prabhakaran, B.: Word recognition from continuous articulatory movement time-series data using symbolic representations. In: ACL/ISCA Interspeech Workshop on Speech and Language Processing for Assistive Technologies, pp. 119–127 (2013)
26. Shokoohi-Yekta, M., HuHongxia, B., Wang, J., Keogh, E.: Generalizing DTW to the multi-dimensional case requires an adaptive approach. Data Min. Knowl. Disc. **31**(1), 1–31 (2017)
27. Ko, M., West, G., Venkatesh, S., Kumar, M.: Online context recognition in multisensor systems using dynamic time warping. In: Proceedings of the IEEE International Conference on Intelligent Sensors, Sensor Networks and Information Processing (ISSNIP), pp. 283–288 (2005)
28. Villar, J.R., Vergara, P., Menéndez, M., de la Cal, E., González, V.M., Sedano, J.: Generalized models for the classification of abnormal movements in daily life and its applicability to epilepsy convulsion recognition. Int. J. Neural Syst. **26**(06), 1650037 (2016)
29. Blankertz, B., Curio, G., Muller, K.R.: No Title. In: Advances in Neural Information Processing Systems 14 (NIPS 2001) (2011)

30. Goldberger, A.L., Amaral, L.A.N., Glass, L., Hausdorff, J.M., Ivanov, P.C., Mark, R.G., Mietus, J.E., Moody, G.B., Peng, C.K., Stanley, H.E.: PhysioBank, PhysioToolkit, and PhysioNet components of a new research resource for complex physiologic signals. Circulation **101**(23), E215–E220 (2000)
31. Liu, C., Springer, D., Li, Q., Moody, B., Juan, R.A., Chorro, F.J., Castells, F., Roig, J.M., Silva, I., Johnson, A.E.W., Syed, Z., Schmidt, S.E., Papadaniil, C.D., Hadjileontiadis, L., Naseri, H., Moukadem, A., Dieterlen, A., Brandt, C., Tang, H., Samieinasab, M., Samieinasab, M.R., SameniRoger, R., Mark, G., Clifford, G.D.: An open access database for the evaluation of heart sound algorithms. Physiol. Meas. **37**(12), 2181–2213 (2016)
32. Zakaria, J., Mueen, A., Keogh, E.: Clustering time series using unsupervised-shapelets. In: Proceedings of the 2012 IEEE 12th International Conference on Data Mining, pp. 785–794 (2012)

Synthesized A* Multi-robot Path Planning in an Indoor Smart Lab Using Distributed Cloud Computing

Morteza Kiadi[1], José Ramón Villar[1(✉)], and Qing Tan[2]

[1] University of Oviedo, Oviedo, Spain
Mkiadi2002@yahoo.com, villarjose@uniovi.es
[2] Athabasca University, Athabasca, Canada
qingt@athabascau.ca

Abstract. Finding the shortest path for an autonomous robot in static environments has been studied for many years and many algorithms exist to solve that problem. While path finding in the static setting is very useful, it is very limiting in real world scenarios due to collisions with dynamic elements in an environment. As a result, many static path planning algorithms have been extended to cover dynamic settings, in which there are more than one moving objects in the environment. In this research, we propose a new implementation of multi agent path finding setting through A* that emphasizes on the path finding through a centralized meta-planner that operates on the base of Bag of Tasks (BoT), running on the distributed computing platforms on the cloud or fog infrastructures and avoiding dynamic obstacles during the planning. We also propose a model to offer a "Multi-Agent A* path planning as-a-Service" to abstract the details of the algorithm to make it more accessible.

Keywords: A* algorithm · Pathfinding · Fog computing · Kubernetes

1 Introduction

Robot navigation is the process of finding and executing a path from the initial location towards a target position while avoiding obstacles [1]. Based on the availability and the knowledge of the environment, path planning is scoped at the local or global level [2]. While local level refers to modifications to a predefined path made by the robot based on information gathered from the available sensors [1], the global level is responsible for producing a valid path to each robot. When the obstacles are static and the start and goal cells are known beforehand, we calculate the path (based on an ideal criterion such as the "shortest" path), which is known as "global" path planning. However, this method cannot help for scenarios where the obstacles are moving, or the goal is not fixed. As will be seen in the next section, there are many existing solutions that combine local and global path planning [3].

Á. Herrero et al. (Eds.): SOCO 2020, AISC 1268, pp. 580–589, 2021.
https://doi.org/10.1007/978-3-030-57802-2_56

This study analyses the hybridization of local and global planning in multirobot environments, considering collision avoidance. From one viewpoint, our solution is similar to the offline path planning as we use the prior environmental knowledge while we also would like to consider other agents' moves during the "planning" phase. Our solution proposes implementing a central "meta-planner" module, running ultimately on the cloud computing facilities to minimize the communication overhead between agents while taking advantage of abundant processing power in the cloud computing to apply the heuristic algorithms to avoid collisions. This research shows a proof of concept for the meta-planner implementation as well as introducing the architecture of the cloud-based solution. We propose a cloud native (CN) multi-agent (MA) A* path planner (A*PP), in short (CNMA-A* PP).

The structure of the study is as follows. The next Section copes with the related work. Section 3 completely describes the hybridized global multi-robot path planning. The experimentation design and results are introduced in Sect. 4, including some discussion. Finally, the conclusions are drawn.

2 Related Work on Multi-robot Path Planning

Path planning is an important and essential issue for the navigation of autonomous robots among many other use cases [4]. The studies of path planning started in the late 60s and many different algorithms have been proposed [5]. In this path different frameworks to solve a multi-robot path planning have been proposed such as problem reduction through answer set programming [6] and SAT [7]. In the Optimal Multirobot Path Planning on Graphs [8], the problem of multi-robot path planning has been discussed over four minimization objectives: the makespan (last arrival time), the maximum distance (single-robot traveled), the total arrival time, and the total distance. These objectives are pairwise distinct and NP-hard to optimize, as a result we can see suggestions to solve optimal MPP by finding effective near-optimal algorithms [8]. The solution proposed in [8] to tackle such multi-objective minimization problem is to create a one-to-one mapping between MPP and that for a multicommodity network flow problem by translation of the MPP problem into an integer linear programming (ILP) model solvable using an ILP solver.

In operational research, these types of problems have been traditionally solved based on linear and/or integer programming. Such approaches inevitably place restrictions on the form of the objective function and/or the constraints by being linear. Other non-linear solutions rely on the differentiability of the problem in its nature. Furthermore, such traditional methods all assume deterministic quantities and have problems dealing with any stochastic effects [9]. In more complex settings, swarm intelligence optimization methods, such as Genetic Algorithms or Ant Colony Optimization among others, have been successfully to find the shortest path in more complex settings [10]. The ACO-A* is another proposal to use ACO metaheuristic to suggest the order of traveling and then use A* to find the shortest path pair-wise between two cells [11]. Our solution is similar to ACO-A* from this viewpoint that it has two phases however, we use heuristic than meta-heuristic in our method as in our setting we do not face with stochastic and unknown environment. Collision avoidance through danger immune algorithm [12] has presented how different information systems such as global positioning system (GPS), automatic identification system (AIS) and Automatic Radar Plotting Aid (ARPA) is widely used in collision avoidance system on most merchant ships. Our proposal is different from this viewpoint that we do not have positional systems or real time sensors to report the location of robot in real time. The proposed solution in the Concurrent Goal Assignment and Collision-Free Trajectory Generation for Multiple Aerial Robots [13] is very similar to our proposed solution with the exception that we use robots in the indoor settings that aerial robots flying in different altitudes. The similarity comes from the fact that Benjamin Gravell and et al. [13] suggest Constrained Collision Detection Algorithm(CCDA) and Constrained Collision Detection Algorithm with Delay Times(CCDA-DT) to resolve the collision which are similar to our approach to find the collision by creating a matrix of time-moves and introducing the "wait" action to avoid collision.

By far, graph-based algorithms are the most widely used methods in global path planning [1] in order to find the shortest path. Examples of these algorithms include [14]: breadth-first and depth-first search, the Dijkstra's algorithm, the Bellman-Ford algorithm or the Floyd-Warshall algorithm. Nevertheless, one of the most competitive algorithms is the A*, which solves the single-source shortest path problem for nonnegative edge costs. Our solution has extended the implementation of the A* algorithm in the Artificial Intelligent book [15] and implemented by the simpleai [16] library. The idea of simultaneous task assignment and planning (STAP) problem [17] sounds a promising approach to extend our solution to a more dynamic and unpredictable approach with randomly assigned costs to each path in a graph route. However, the approach in STAP different from our solution provided that, each robot has a local reactive collision detector to avoid collision with dynamic obstacles. In our proposed solution we do not use local sensors and we do have dynamic assignment of the robots to destinations, like the way STAP works.

2.1 A* Algorithms

In the informed search algorithms such as A*, we rely on a function called the "heuristic function", to help the algorithm to pick the next cell to explore based on its "closeness" to the goal state in the entire path-finding process. A heuristic function is all about the trade-off between its accuracy and its speed [18]. One example is to try to estimate the "best heuristic" and then incorporate that into the A*. This method works if the search process is not time sensitive. For example, one can make use of the output generated by the back-tracking techniques mentioned in [19] as heuristic values. The backtracking technique is useful when we do not have much knowledge about the topology of the environment, and we would like to find the state values by reinforcement learning and trials and errors. This technique is based on the optimization of the Markov decision-making process and tweaking the model's hyper-parameters. This technique approximates the real distance from the goal and as a result, the evaluation function produces the successors for the optimal path, obviating entirely the need for search [20].

A* Algorithm Pitfalls

The A* algorithm has the following shortcomings or limitations: 1) slow search in large scale path search. For example to get the optimal path in a 100 * 100 grid, at least 513 nodes need to be searched of [18] 2). The A* is only useful when there is some domain knowledge about the environment 3) Finding the right estimate for the heuristic function is tricky and it impacts the performance of the algorithm drastically 4) In large space searches, the algorithm needs lots of memory and 5) A* algorithm assumes one node is moving at a specific point in time. That is not a suitable algorithm for multi-node and dynamically changing environments.

The above issues, specifically the last one, motivates the researches to think about making a better version of A* algorithms, that it is the subject of the next section.

3 Solution Design Approach and Features

This study proposes utilizing the A* algorithm in a multi-agent setting in order to obtain a multi-robot path planning, that is, simultaneously obtaining a collision free path planning for each of the robots. We use the principle of Bag of Tasks (BoT) [21], where each agent runs the A* algorithm independently (the Agent's planning phase) and after all the agents are done with their planning, a module that we call it "meta-planner" starts modifying the results of independent tasks (the refinement phase) to create (synthesizing phase) a cohesive plan that works for all agents, in this case, a collision free path for each agent. Figure 1 shows the block diagram of this idea. Moreover, this procedure has been designed and implemented "as-a-service", finding a collision free path for multi-agent systems.

Fig. 1. The phases of a meta-planner to realize a multi-robot path with no collision.

Due to the need for a path planning for each robot, we need to perform A* for each of them and then analyze the results. Actually, the planning and the refine stages could be integrated by synchronizing the different A* running in parallel. To do so, several modifications to the A* algorithm are needed. Moreover, our implications to design a solution "as-a-service" suggested moving to a different solution path.

Alternatively, we opted to run each A* independently, merging their result and running the following stages afterwards This solution makes use of a distributed container scheduling open source project called Kubernetes, which has been successfully used in different initiatives like Cloud manufacturing [22] and distributed containerized serverless architectures [23]. Kubernetes is one of the well- adopted platforms when it comes to Cloud Native Applications (CNAs). It is the path to make a cloud based solution that is elastic, self-contained deployment, no lick-in to a cloud provider, cross platform, automated infrastructure management and containerization [24] We extend the idea of scheduling tasks in the fog computing by BoT [21] to run on Kubernetes and we suggest a new cloud based service for a multi-robot A* path planning based on the CNA principles. The categorization and taxonomy of distributed problem solving and planning detailed [25] have been considered in this research. Moreover, we have considered all the movements of the robots to take one slot of time and that all of them have the same speed. The basic movements can be configured to be the main cardinal directions or extended with the main diagonal as well. The complete solution steps to run in the Kubernetes platform have been shown in Table 1.

Table 1. Multi-agent A* path planning as a service

1 **Containerize** the A* algorithm
2 **Setup** the YAML manifest files to describe the run-time environment
2-1 **Set** the maze configuration
2-2 **Set** the number of agents and their valid movements
3 **Launch** the Kubernetes components (Pods. ConfigMaps, …) with A* containers in Step 1
4 **For each** agent run one A* algorithm
4-1 **Run** A* algorithm for each agent in a Pod
4-2 **Store** the shortest path for each agent in the shared storage
5 **While** the meta-planner has not done:
5-1 **Unify** the path lengths
5-2 **Map** the **wait** and **time** factors to the produced A* paths
6 **For each** item in the map:
6-1 **Run** the meta-planner **Heuristics** by:
6-1-1 **Detecting** the collisions by comparing t_n:(x, y) of each agent
6-1-2 **Introduce** the wait action and shift the next moves accordingly whenever a collision is detected
6-1-3 **If** the path is blocked by another agent **then** set the blocking agent to a lower priority
7 **Return** the final paths

3.1 Refinement and Synthesize Phases

The first step in the refinement phase is to unify the length of the path to the same number. To do so the length of the longest path is determined, then padding the shorter paths with "no move" (step 5-1 in Table 1). The next step is to represent the paths in terms of cells and time slots. A path will be represented as a sequence of pairs like **time: (cell x, cell y**) (step 5-2 in Table 1). Without losing generalization, the robots are considered having the dimension of one cell. After all, in the synthesize phase the collision detection tries (steps 6-1-1 and 6-1-2 in Table 1) to detect cells included in more than one path at the same time units. In addition, in this phase, we detect the agents that block other agents' paths (step 6-1-3 in Table 1). Both actions in the synthesizing phases are achieved by the heuristic logic, implemented in the meta-planner module.

3.2 The Heuristic of the Meta-planner

To better understand how heuristics of the meta-planner works, we need to elaborate on the details of its internal functions. The meta-planner injects three new elements to the path produced by a pure A* algorithm: 1) the wait action 2) temporal element 3) path-blocker detector. These tools are used in the meta-planner heuristic as they are described in the following paragraphs:

- If a cell is going to be taken by more than one agent at the same time, one of them must wait. We introduce the "time-step" concept to the solution in the refinement phase to make sure such a goal is achievable in the synthesize phase, by making each "time unit" equal to each move. So, at timeslot 1 (**t1**), we have n-move (where n is equal to the number of agents), and in **t2** we have another n-move, and so on. The agents need a different number of moves to reach their destinations (as they have different start and destination cells). When all paths are reported to the meta-planner, it unifies their sizes (practically by adding "no move" action to the end of shorter paths). The selection of the agent to wait in our solution is completely random but it could be based on a more advanced priority system.
- There is a possibility that the destination cell of an agent blocks other agents' paths. We do not manipulate or modify the decisions that are made by A*. The reason is A* already has proved itself as one of the most efficient path planners. We respect the A* quality in finding the shortest paths but we detect the blocking moves and delay those moves in favour of other agents that need those cells. So the path shapes in our solution are not changed.

3.3 Solution Architecture of CNMA-A* PP

The solution described in the previous subsections can be augmented using the Kubernetes platform by extending the meta-planner and A* executions to a cloud-based distributed service offering. We call this proposed solution as "CNMA-A*PP' to emphasize on its cloud-native nature, multi-agent A* path planning. The "CNMA-A*PP' converts a standalone A* single agent algorithm that works in static settings to a cloud-based, configurable, multi-agent A* global planner. To do so, each agent is mapped to one Kubernetes Pod to execute the A* algorithm independently (planning phase of the meta-planner). The Pods run in parallel and in a distributed manner, reducing the total service time. The results from each A* runs shall be saved in storage that is shared among the Pods. The meta-planner running the refinement and synthesizing processes also run in a Pod. Requesting such a service is realized entirely in the form of Kubernetes manifest YAML files describing the environment maze setup, number of agents, allowable moves, cost of moves, etc. The solution should launch a set of infrastructure components such as Pod(s) or ConfigMaps to realize a "multi-agent A* path finding-as-a-service". The architecture of this solution is shown in Fig. 3. The "CNMA-A*PP' agrees with the principle of composability that is about employing the same architecture to deploy self-managing service compositions or applications using the microservice architectural pattern [26].

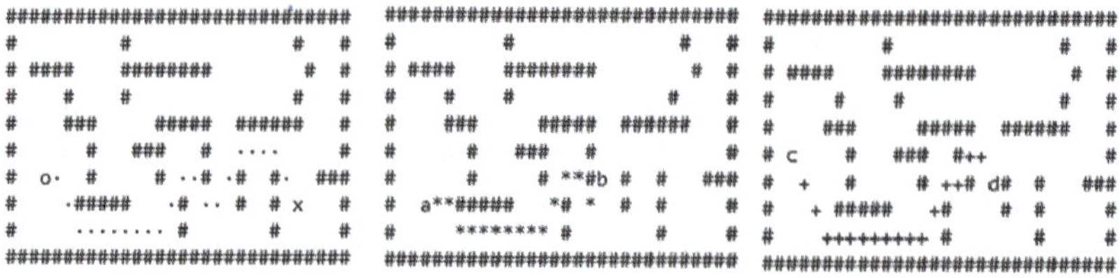

Fig. 2. Three agents generate their A* path independently in PoC.

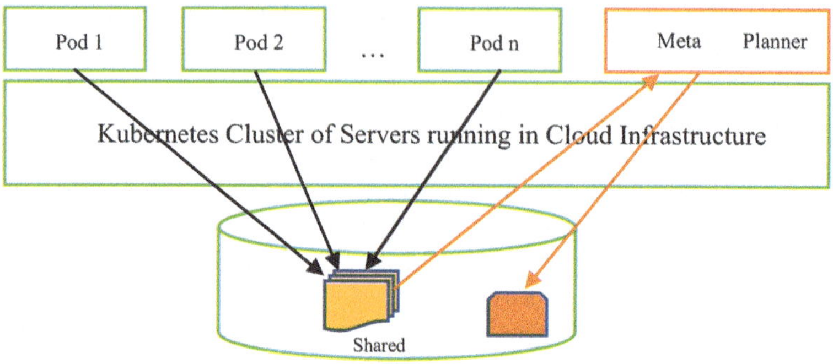

Fig. 3. A scheme of the Cloud Native-based design for a multi-robot path planner as-a-service

4 Experiment and Results

In this section, we present a proof of concept (PoC) implementation of our idea, i.e. CNMA-A*PP. The PoC is based on three agents, starting in different start points and targeting to different end points. We have intentionally positioned the start and end points to increase the chance of conflict to test our meta-planner performance and we have purposefully set the end cell of one agent in the middle of another agent's path to block the path. Figure 2 shows the maze, as well as the initial and goal points. Three agents are placed on it, each one with its starting points (**o**, **a** and **c**) and corresponding endings (**x**, **b**, and **d**) points. The paths found for each robot using A* are shown in Fig. 2. In the proposed path a few collisions exist.

The meta-planner instructs the priorities of agents if there is a blocker agent. As you can see in the following outputs, agent3 has been set to a "lower" priority by meta-planner due to the fact that the agent1 needs to pass through a cell that is the destination of the agent3 (that is where "d" is). Since the agent3 path has a lower number of moves (18 moves) it will reach the destination sooner than agent1, hence it will block agent1's move. To avoid this situation, the meta-planner suggests delaying its move.

```
The agent1 has no priority
The agent2 has no priority
The agent3 has low priority
--------------------------------
number of steps in path1 is: 23
number of steps in path2 is: 16
number of steps in path3 is: 18
```

The other meta-planner heuristic we have implemented is to detect the colliding cells. After executing the refinement and synthesizing phases of the meta-planner, the following moves are generated. As you can see in the following output, the agent1 has been set to wait(**w**) in t2 (**t2:w**) in favour of agent2 as both of them wanted to take cell (9,7) at t2.

```
Path for Agent 1 is:

t0:(7, 6) t1:(8, 6) t2:w t3:(9, 7) t4:(10, 8) t5:(11, 8) t6:(12, 8) t7:(13, 8) t8:(14, 8) t9:(15, 8) t10:(16, 8) t11:(17, 8) t12:(18, 7) t13:(19, 6) t14:(20, 6) t15:(21, 7) t16:(22, 7) t17:(23, 6) t18:(24, 5) t19:(25, 5) t20:(26, 5) t21:(27, 5) t22:(28, 6)

Path for Agent 2 is:

t0:(7, 7) t1:(8, 7) t2:(9, 7) t3:(10, 8) t4:(11, 8) t5:(12, 8) t6:(13, 8) t7:(14, 8) t8:(15, 8) t9:(16, 8) t10:(17, 8) t11:(18, 7) t12:(19, 6) t13:(20, 6) t14:(21, 7) t15:(22, 6) t16:0 t17:0 t18:0 t19:0 t20:0 t21:0 t22:0

Path for Agent 3 is:

t0:(6, 5) t1:(7, 6) t2:(8, 7) t3:(9, 8) t4:w t5:(10, 8) t6:(11, 8) t7:(12, 8) t8:(13, 8) t9:(14, 8) t10:(15, 8) t11:(16, 8) t12:(17, 8) t13:(18, 7) t14:(19, 6) t15:(20, 6) t16:(21, 5) t17:(22, 5) t18:(23, 6) t19:0 t20:0 t21:0 t22:0
```

5 Conclusions

This research is focused on collision avoidance multi-robot path planning. The aim of this study is to extend the outcome of A* with a simple heuristic to avoid the collisions, altogether designed and implemented in one of the latest state of the art distributed scheduling system in the cloud (i.e. Kubernetes) and adding meta-planner to augment A* to work in a multi-agent configuration.

The study represents a proof of concept and a standard maze used in path planning has been used to evaluate the heuristic proposed in this research. The performance of the heuristic has been found valid and the implementation with Kubernetes can be the next step to realize the CNMA-A*PP. Our proposal is aligned with a new trends in creating self-managed micro-services in the cloud [27]. In addition, in this paper we implemented a PoC along with two heuristics for meta-planner. This meta-planner heuristic can be upgraded to more advanced techniques such as the collision model that is proposed in [28]. The proposed solution in this paper is also aligned with the idea of Cloud4IoT which is containerizing IoT functions and optimize their placement and on the edge of network through fog computing [29].

Acknowledgement. This research has been funded by the Spanish Ministry of Science and Innovation, under project MINECO-TIN2017-84804-R, and by the Grant FC-GRUPIN-IDI/2018/000226 project from the Asturias Regional Government.

References

1. Mac, T.T., Copot, C., Tran, D.T., De Keyser, R.: Heuristic approaches in robot path planning: a survey. Rob. Auton. Syst. **86**, 13–28 (2016)
2. Xie, L., Xue, S., Zhang, J., Zhang, M., Tian, W., Haugen, S.: A path planning approach based on multi-direction A* algorithm for ships navigating within wind farm waters. Ocean Eng. **184**, 311–322 (2019)
3. Wang, L.C., Yong, L.S., Ang, M.H.: Hybrid of global path planning and local navigation implemented on a mobile robot in indoor environment. In: IEEE International Symposium on Intelligent Control - Proceedings (2002)
4. Han, S.D., Reliminaries, I.I.P.: Effective heuristics for multi-robot path planning in warehouse environments. In: 2nd IEEE International Symposium Multi-Robot Multi-Agent System, pp. 1–3 (2019)
5. Masuda, M., Wehner, N., Yu, X.: Ant colony optimization algorithm for robot path planning, vol. 3, no. 30, p. 30 (2010)
6. Erdem, E., Kisa, D.G., Oztok, U., Schüller, P.: A general formal framework for pathfinding problems with multiple agents. In: Proceedings 27th AAAI Conference Artificial Intelligence AAAI 2013, pp. 290–296 (2013)
7. Surynek, P.: Towards optimal cooperative path planning in hard setups through satisfiability solving. In: PRICAI 2012 Trends Artificial Intelligence, PRICAI 2012, pp. 564–576 (2012)
8. Yu, J., LaValle, S.M.: Optimal multirobot path planning on graphs: complete algorithms and effective heuristics. IEEE Trans. Robot. **32**(5), 1163–1177 (2016)
9. Andrew, A.M.: Modern heuristic search methods. Kybernetes **27**(5), 582–585 (1998)
10. Noreen, I., Khan, A., Asghar, K., Habib, Z.: A path-planning performance comparison of RRT*-AB with MEA* in a 2-dimensional environment. Symmetry (Basel) **11**(7), 945 (2019)
11. Yu, X., Chen, W.N., Gu, T., Yuan, H., Zhang, H., Zhang, J.: ACO-A∗: ant colony optimization plus A∗ for 3-D traveling in environments with dense obstacles. IEEE Trans. Evol. Comput. **23**(4), 617–631 (2019)
12. Xu, Q.: Collision avoidance strategy optimization based on danger immune algorithm. Comput. Ind. Eng. **76**, 268–279 (2014)
13. Gravell, B., Summers, T.: Concurrent goal assignment and collision-free trajectory generation for multiple aerial robots. IFAC-PapersOnLine **51**(12), 75–81 (2018)
14. Bruce: Heuristic Search Applications **53**(9) (2013)
15. Stuart, R., Peter, N.: Artificial Intelligence: A Modern Approach, Global Edition (2011)
16. Simpleai-team/simpleai. https://github.com/simpleai-team/simpleai/graphs/contributors
17. Yang, F., Chakraborty, N.: Multirobot simultaneous path planning and task assignment on graphs with stochastic costs. In: Proceedings IEEE MRS, pp. 1–3 (2019)
18. Mathew, G.E., Malathy, G.: Direction based heuristic for pathfinding in video games. In: 2nd International Conference Electronics and Communication Systems ICECS 2015, vol. 47, pp. 1651–1657 (2015)
19. Kiadi, M., Tan, Q., Villar, J.R.: Optimized path planning in reinforcement learning by backtracking, pp. 80–90 (2019)
20. Nilsson, N.J.: Problem-solving methods in artificial intelligence. McGraw-Hill Computer Science Series. McGraw-Hill, New York (1971)

21. Zhang, Y., Zhou, J., Sun, J.: Scheduling bag-of-tasks applications on hybrid clouds under due date constraints. J. Syst. Archit. **101**, 101654 (2019)
22. Dziurzanski, P., Zhao, S., Przewozniczek, M., Komarnicki, M., Indrusiak, L.S.: Scalable distributed evolutionary algorithm orchestration using Docker containers. J. Comput. Sci. **40**, 101069 (2020)
23. Soltani, B., Ghenai, A., Zeghib, N.: Towards distributed containerized serverless architecture in multi cloud environment. Procedia Comput. Sci. **134**, 121–128 (2018)
24. Kratzke, N., Quint, P.C.: Understanding cloud-native applications after 10 years of cloud computing - a systematic mapping study. J. Syst. Softw. **126**, 1–16 (2017)
25. Durfee, E.H.: Distributed problem solving and the DVMT, pp. 27–44 (1988)
26. Lewis, J., Fowler, M.: Microservices: a definition of this new architectural term
27. Toffetti, G., Brunner, S., Blöchlinger, M., Spillner, J., Bohnert, T.M.: Self-managing cloud-native applications: design, implementation, and experience. Futur. Gener. Comput. Syst. **72**, 165–179 (2017)
28. You, S.J., Ji, S.H.: Design of a multi-robot bin packing system in an automatic warehouse. In: ICINCO 2014 - Proceedings 11th International Conference on Informatics in Control, Automation and Robotics, vol. 2, pp. 533–538 (2014)
29. Dupont, C., Giaffreda, R., Capra, L.: Edge computing in IoT context: horizontal and vertical Linux container migration. In: GIoTS 2017 - Global Internet Things Summit, Proceedings, pp. 2–5 (2017)

Towards Fog-Based HiTLCPS for Human Robot Interactions in Smart Lab: Use Cases and Architecture Overview

Behzad Karim[1(✉)], Qing Tan[2], and Juan Carlos Alvarez[1]

[1] University of Oviedo, Oviedo, Spain
behzadk0@gmail.com, juan@uniovi.es
[2] Athabasca University, Athabasca, Canada
qingt@athabascau.ca

Abstract. This paper provides use case definitions and a high-level system architecture overview for human robot interaction in a fog computing-based Human in The Loop Cyber Physical System. Our focus is to develop a practical, natural, meaningful human robot interaction framework for single and multiple avatar (CPS) robots, and this paper outlines the research road ahead of us.

Keywords: Human-in-the-Loop CPS · Human-Machine interaction · Social robotics · Robotics · Fog computing · Cloud computing

1 Introduction

Safe and effective interaction is the key to operating multiple robots in a Human-Robot blended environment such as a smart laboratory in an educational setting [1]. In such an environment, humans and robots are co-existing, collaborating to participate in activities. This calls for robots with socially meaningful and acceptable behavior. In an academic setting, performing lab work in a smart lab environment remotely through an avatar robot while in presence of humans and other robots can be challenging. Remotely controlled avatar robots can be used to participate in lab work activities by anyone in need. This is one of the primary motivations behind our research. To perform the lab work, humans and robots need to communicate in a socially acceptable way while moving and working alongside each other. Social acceptability of mobile indoor robots in day to day educational facilities and in daily life, depends on practicality and efficiency of robots and *communication* plays a key role in this arena.

The ability of indoor robots to navigate autonomously, interact with humans, and act as a **member** of the team is of critical importance. Whether it is autonomous, semi-autonomous or an avatar robot controlled remotely by a human being, robots need to communicate and interact with entities around them and the smart environment in a dynamic and efficient manner. The ability for the robots to blend into the *social setting*

Á. Herrero et al. (Eds.): SOCO 2020, AISC 1268, pp. 590–599, 2021.
https://doi.org/10.1007/978-3-030-57802-2_57

requires this communication be natural and consistently based on acceptable conventions. This will not only foster intuitive action and response, but will also promote trust, and fruitful teamwork which will enhance the learning experience.

Success of smart labs in an educational setting depends greatly on the ability of the robots to provide a comfortable, intuitive and inspiring environment for learners to study and collaborate. Robots need to be socially acceptable as an integral part of the team and be not only supportive but also are trustworthy and reliable for the learners.

Cyber-physical systems (CPS) have been trending towards human-robot blended direction as well. In the past, technology and resources have been focused on making the robot perform and do the work while the human was acting as a knowledgeable operator who supervised the process. However, in recent years, rapid strides have been made towards building systems that need minimal user attention and are expected to use **human intuitive** clues to operate and adjust. This area of CPS has now been more commonly identified as, human-in-the-loop cyber-physical systems (HiTLCPSs), which takes **human's sensing, brain, and action** as an integral part of the CPS.

In recent years, robotics technology has significantly matured, and the industry is producing a variety of robots for specialized purposes. In the smart laboratory, **different kinds of robots** are expected to work together and interact with each other with learners as a part of the team in real-time. The interaction and cooperation between human and robot and among the robots must be **human understandable** and **human comprehensible**. Meaning that any human observer or collaborator who does not know anything about the specific robot technology should be able to perform work and interact with them.

Group collaboration, comprehension and evaluation is the fulcrum of being a functional team working towards a common goal. Our empathy and emotional intuition allow our interactions to encompass a broad range of social states and situations. As an example, if there is an earthquake drill going on in the workplace, the robots should immediately respond by guiding helping and assisting the team to follow the earthquake safety protocols. Indoor robots needs to gain and maintain a dynamic, sound, realistic, and humanly meaningful mental model which will be evolving as the group live, collaborate and get to know each other more. A higher level of collaboration would be for the robot to instill and promote the spirit of teamwork with positive influence on the team and be ready to operate under high pressure during emergencies or crisis situations.

The next section (Sect. 2) will review related research work. Section 3 will solution design approaches and features will be presented through the user cases. The architecture overview is introduced including some discussion in Sect. 4. Finally, the conclusions are drawn. Since this is the first step mapping the road ahead for our research project we do not have any numerical data to provide in support of our discussions other than the data provided in related research.

2 Related Research Work

In integrating the human element into CPS systems within an educational setting, specifically a smart lab environment one of the goals will be to establish a safe, understandable and efficient collaboration platform for smart lab robots. In such environments, both inspiring teamwork and safe operation in a crisis are important. Different levels of automation may change in relation to the type of cooperative partner in crisis management [2]. For example, in a normal setting in the smart laboratory, possible roles are professors, learners, and avatar robots (autonomous, semi-autonomous, or human manipulated). In a crisis, inside the same environment, firefighters, police officers and police robots could be added to the mix.

Human-Machine cooperative approach in driving has been studied without the intervention of assistance devices designed to improve lateral control. These studies suggest that driving assistance should be designed in a way that it is blended into drivers' actions [3].

Framework for rapid development and deployment of embedded HiTLCPS with assistive technology that augments human interaction and infers human intent has been developed in the research [4, 5]. A proactive social motion model (PSMM) that enables a service robot to navigate safely in crowded and dynamic environments was proposed and then combined with a path planning technique to generate a motion planning system [6]. In other research work, a closed-loop, sampling-based motion planner has been used for robot motion planning performing a learned task while reacting to the movement of obstacles and objects [7]. The task model is learned from expert demonstrations prior to task execution and is represented as a hidden Markov model.

Recent search has used neural network architecture for indoor robots to learn new navigation behavior [8]. By observing a human's movement in a room, a neural is built for spatial representations and path planning. Based on the human's motion, the robot learns a map that is used for path planning. In other research work on spatial cognition for navigation of an autonomous mobile robot in an indoor structured environment a fingerprint-based representation was used to create a compact environment model without relying on any maps and artificial landmarks [9]. Fog and cloud computing have opened new opportunities for provisioning and dynamic allocation of advanced robotic services including complicated Artificial Intelligence (AI) and Machine Learning (ML) algorithms [10]. Feasibility and efficiency of cloud robotics systems to provide provisioning location-based assistive services for active and healthy aging of elderly individuals [11]. A script based cognitive architecture for collaboration, incorporating Dynamic Bayesian Network (DBN), to detect user's intentions and goal, gain understanding of user initiatives, and govern robot action sequences has been tested for efficiency for real indoor robot task scenarios [12]. Interesting research work based on Grey systems theory which is a new method for studying problems of uncertainty with poor information suggests constructing the environmental information as manifestation of different cognition phase based on the different subsets of the grey hazy set produced by dynamic evolution [13]. Recent research work has demonstrated usage of Fog services to offload computationally expensive localization and mapping tasks without compromising operational reliability due to cloud connection issues [14].

3 Entities and Use Case Definitions

In this section, we outline the objectives and context of the cases we are going to study for Human-Robot Interaction in Smart Lab environment and provide use cases for each specific model. The context of this study is Human in The Loop Cyber Physical Systems (HiTLCPS), and we are only concerned with interactions related to motion and displacement within indoor smart lab environment.

Human-Robot interaction is a vast area of study and has been a popular area of research in academics. Our overall objective is to do a case study on human-robot interaction involving with movement and motion inside the laboratory. We focus on interactions, interpretations, decisions and actions related to movement and motion planning only and ignore all other cases.

3.1 Description of the Entities

The lab environment is smart. Meaning that devices and equipment can sense and may be able to transmit useful data for determining a robot's position and moving speed, and current status.

Entities moving inside the laboratory includes humans and robots moving autonomously, manually (remotely controlled), or in following mode. Humans inside the smart lab could are professors, lab assistants, students or lab workers. Avatar robots are remotely controlled in the following modes:

- ***autonomous robots*** move independently inside the lab environment. These robots plan and execute their own motion inside the lab but can also operate in *following mode* like avatar robots.
- ***manually-driven robots*** human operators through the Internet connection. All their movements are dictated to them through user controls in real time. However, there will be a latency involved in transmission of movement commands and receiving sensory data which needs to be considered.
- ***following mode robots*** are any robot (autonomous or avatar) following a human or another robot. The following mode robot could be one or multiple: a) One robot following a human or another robot, b) Multiple robots following a human or another robot.

3.2 Focus of the Study

The research focus will be on the interaction among **human and avatar (CPS) robots in the following mode**. The scope of these interactions will be limited to the ones related to motion and movement. When there is only one robot following, the interactions can be studied in one or two cases:

a) A robot is following a human.
b) A robot is following another robot.

While our primary goal is to study human-robot interaction, it will also be interesting to study *human understandable,* robot to robot interactions. It is our belief that robot to robot interaction and communications should always be human understandable for *Social Robots* if they are to be socially acceptable. Based on the above Categories of Interaction studied will be:

- Human to robot interaction
- Robot to robot interaction (human understandable)

Having avatar robots in following mode enables the user (learner) to focus on the task at hand, rather than manually driving the robot from point A to point B which may need extensive attention and expertise.

Robots in following mode need clear and concise instructions with regards to their movement to begin with how the movement starts. The robot should be capable to sense the changes of its following human or robot, to make decision, and to take proper action in order to adapt the changing situation. The interactions will not only cover verbal commands and directions given by humans, but will also cover cognitive decisions and responses based on changes in environmental conditions, human body motion and external factors such as emergencies.

The overall objective is to study and design the specifics of the interactions needed to perceive commands & directives, sense, gather useful information, think, and act accordingly on the field.

Commands and Directives

Using natural language, we would like to use a concise and practical verbal commands and directives for the robots in following mode. Natural language is efficient and understandable by humans. The same language set will be used in the case of a robot following another robot. Although it may not be the most efficient way for a robot to robot communication, it is an important feature for all the entities to communicate with the same language.

Data Collection, Sensory and Gathering Information

Every bit of useful information from every entity and intelligent device in the environment should be collected. An initial list of information sources may include: v A) verbal commands, B) body gestures, C) Sensory data collected from intelligent devices, D) movement and positioning data, E) emotional factors, F) emergency signals, G) global and local feeds through fog and/or cloud regarding important events (e.g. earthquakes, fire or other distress signals).

Cognition

Cognition is the processing of collected data into a digestible, realistic view of the environment to understand, act and/or make decisions, with consideration to the goals, when faced with diversity. This will probably be comprising most of the study as it a vast topic. This might sometimes mean letting go of initial directives and coming up with different options to achieve given goals.

Choosing and Decision Making

Making decisions and choosing between different options means making a proper assessment of the outcomes of each option, weighing the outcome against the goal at hand and comparing it to other options. This might seem trivial in ordinary situations however it could easily become more complicated when faced with exceptional cases.

In some cases, there might not be any viable options, which means the robot will have to go back to gathering more information and further cognitive effort.

Declaring

Once a decision is made, it must be declared (communicated) to other entities in the laboratory before execution. Depending on the situation and perception or need of other entities, the decision may be overruled or amended. Furthermore, some decisions may need consensus and/or approval. Depending on the importance of the decision, it may need multiple approvals.

This stage could become more interesting when other entities in the vicinity ask for clarification, modification, or alterations to robot's decision. This could mean going back to *sensing* and collecting information, understanding, coming up with new choices, and choosing again.

Acting

After communicating the "new" decision, action plan is prepared and will be carried out by the robot.

3.3 Use Cases for Robot in Following Mode

We created a preliminary list of basic use cases to capture different sequences of behavior, and unfold the scenarios based on situations. In the list below, use cases related to **Motion** are labeled as Mxx, use cases related to **Communicating Information** are labeled as Ixx, and use cases related to **Cognition** are labeled as Cxx, where xx is a serial number.

- **Use Case M01: Simply follow a lead** – Robot follows a point or a lead in the lab in this simple following mode scenario.
- **Use Case M02: Follow a Human to Get in Touch** – Robot follows a human because it needs to communicate, warn or just relay some information. This is particularly important when things are not going according to the plan (e.g. emergencies).
- **Use Case M03: Help Lead with Carrying Equipment** – Robot follows a human or another robot to help with carrying lab equipment or material.
- **Use Case M04: Follow Other Robots** – Robot needs to follow other robot(s) to move as a group from one point to another.
- **Use Case I01: Telepresence robot sends decisions/information** – Robots needs to communicate decisions and information to other entities in the lab.
- **Use Case I02: Lead Communicates Decision/Information** – Telepresence robot needs to communicate decisions and information to other entities in the lab.
- **Use Case I03: Telepresence robot receive information** – Telepresence robot receives information from other entities in the lab.
- **Use Case I04: Telepresence robot asks for approval** – Telepresence robot needs to get consensus or approval from other entities in the lab.
- **Use Case C01: Telepresence robot recognizes a natural language command** – Telepresence robot recognizes a natural language command or information from the lead that affects the motion plan.

- **Use Case C02: Telepresence robot recognizes a body language signal** – Telepresence robot recognizes a body language signal from the lead that affects the motion plan.
- **Use Case C03: Telepresence robot receives a user command** – Telepresence robot receives a user command from a remote user in control of it that affects the current motion plan.
- **Use Case C04: Telepresence robot realizes inconsistent action** – Telepresence robot realizes an inconsistent movement regarding the target point that affects the current motion plan.

Each one of these use cases have very detailed main success scenarios followed by several extensions. As an example, **Use Case M01: Simply follow a lead** is used. With respect to use cases, we follow three fundamental concepts of writing effective use cases [15]: **a) Scope**: What is the scope of the system being discussed, **b) Primary Actor**: What is the Actor's name and goal, and **c) Level**: How high or low level is this goal.

Use Case M01: Naturally following a lead

Scope: Preparation for remote lab work. **Primary Actor**: The Lab Lead. **Level**: Summary.

Telepresence robot follows a lead in the lab in this simple following mode scenario. The telepresence lab robots are always expected to be *"physically present"* when an instructor or team leader is inside the lab. To be *physically present* means *being* within a suitable distance to help and communicate with the lead.

The lead could be an instructor, a tutor, or a student in charge of lab work. This means that if the lead enters the room and the telepresence robots are not in the vicinity, they should move to displace themselves within a proper distance from the lead. And when the lead moves, the robots should follow by default. The exception is the telepresence robot user overriding the following mode navigation.

1. Lead is inside the lab at point A.
2. Telepresence robot is inside the lab at point C.
3. Lead declares moving to point B1.
4. Lead starts to move from point A to point B1.
5. Telepresence robot determines the destination point B2 as the new goal.
6. Telepresence robot starts moving from moves from point A to point B2 inside the lab.

4 Architecture Overview

The proposed system utilizes fog and cloud computing infrastructure and services for computation power and communication while delegating field specific execution, movement and improvisation. We do not intend to carry on all computation and decision-making power to cloud, quite the opposite, we intend to utilize the local computation power of robots as much as possible. However, for communication, group planning and tracking purposes, we engage scalable, flexible, and highly available cloud infrastructure.

In this paper, we focus on the cloud-based components of the architecture and their respective responsibilities (Fig. 1). Both autonomous and group robots will be relying on this framework. This design, relies on three main components or subsystems (all residing on cloud infrastructure).

A) ***Cloud Messaging Layer*** – the messaging layer is the subsystem responsible for delivering messages from source to destination, on a scheduled basis or on a publisher/subscriber model. This subsystem consists of a queueing system and a topic based publish subscriber system.
B) ***Robot Motion Planner*** – is the subsystem responsible for motion plan development for robots based on coordinate information and desired destination. This subsystem carries out the piece of processing related to groups. Groups could be comprised of humans in the lab and/or other moving robots heading towards the same destination. It can also act as the **control tower** for managing traffic between different groups and individual entities moving inside the lab environment.
C) ***Robot Motion Tracker*** – this subsystem acts as a continuous information gathering and coordinate recording server which receives and records information for Motion Planner and Reposting subsystems.
D) ***Admin Subsystem*** – This is the set of API and services used to update settings, maps, configurations and settings for the system.
E) ***Reporting Subsystem*** – This subsystem is used for reporting, monitoring and visualization.

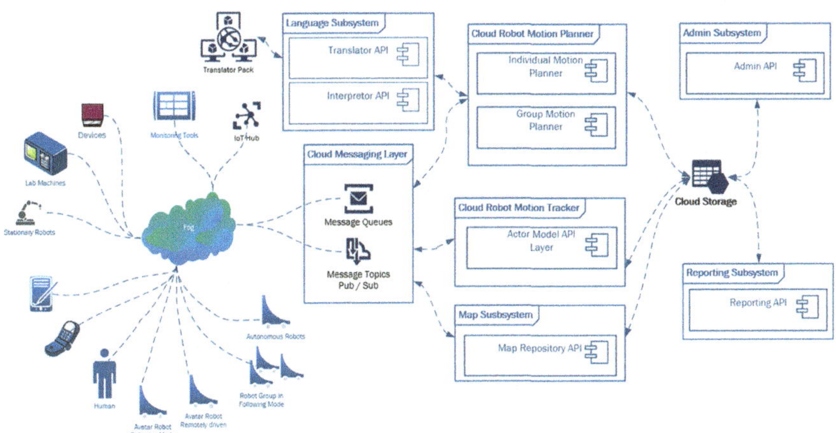

Fig. 1. High level view of fog & cloud-based components

F) ***Language Subsystem*** – This subsystem is responsible for interpretation and translation of messages from different sources into the Framework's language. Since we will be incorporating different model of robots in the system, the Translation and interpretation API is necessary. This layer decouples the framework from robots and messaging protocols.

G) ***Map Subsystem*** – Map subsystem listens for map request messages and responds accordingly. It also provides and API for adding, updating, reading and deleting maps.

5 Conclusions

Human comprehension, collaboration and evaluation is the fulcrum of being a functional team. Our initial goal is to add avatar robots to the smart lab environments for students who cannot be physically present in the lab as part of the team working on a project.

We have provided use cases and proposed a high level architecture to implement a robot in the following mode as the initial study of human robot interaction in HiTLCPS smart lab environment where human and robots coexist and collaborate.

In integrating the human element into CPS systems within an educational setting, specifically a smart lab environment one of the goals will be to establish a safe, understandable and efficient behavior collaboration platform for smart lab robots and avatar robots. In such environments, both inspiring teamwork and safe operation in a crisis is important.

Adding robots to these teams would only be meaningful if those robots could blend into the environment as reliable collaborators. This means, having robots as entities who interact with consideration to the social, and mental state of their team.

Utilizing fog and cloud computing to provision and dynamically allocate resources for robotic services and ML algorithms, we will enrich and clean collected data from multiple connected robons.

In future research, we will analyze different cognitive processes for indoor mobile robot navigation and design a suitable knowledge representation based on the latter. We will review navigational methodologies of robots with the help of various artificial intelligence techniques such as Deep Reinforcement Learning, Fuzzy Logic, Neural Network, Genetic Algorithm, Particle Swarm Optimization, etc. An important criteria is time-based dynamic evolution of the model through interaction, collaboration and building a memory model of individuals and personality traits of collaborators. Beyond being helpful and handling emergency situations such as fire or earthquake, we wish to explore modeling complicated cognitive activities such as curiosity and positive thinking patterns to support the learning experience for students while using the avatar robots.

References

1. Tan, Q., Denojean-Mairet, M., et al.: Toward a telepresence robot empowered smart lab. Smart Learn. Environ. **6**, 5 (2019). https://doi.org/10.1186/s40561-019-0084-3
2. Habib, L., Pacaux-Lemoine, M.-P., Millot, P.: Adaptation of the level of automation according to the type of cooperative partner. In: IEEE International Conference on Systems, Man, and Cybernetics, Banff, Canada, pp. 864–869, October 2017
3. Hoc, J.-M., Lemoine, M.-P.: Cognitive evaluation of human-human and human-machine cooperation modes in air traffic control. Int. J. Aviat. Psychol. **8**(1), 1–32 (1998)

4. Navarro, J., Mars, F., Hoc, J-M.: Lateral control support for car drivers: a human-machine cooperation approach. In: Proceedings of the 14th European Conference on Cognitive Ergonomics: Invent! Explore!, ECCE 2007, vol. 250, pp. 249–252. ACM (2007)
5. Feng, S., Quivira, F., Schirner, G.: Framework for rapid development of embedded human-in-the-loop cyber-physical systems. In: 2016 IEEE 16th International Conference on Bioinformatics and Bioengineering (BIBE), pp. 208–215, October 2016
6. Truong, X.-T., Ngo, T.D.: Toward socially aware robot navigation in dynamic and crowded environments: a proactive social motion. IEEE Trans. Autom. Sci. Eng. **14**(4), 1743–1760 (2017)
7. Bowen, C., Alterovitz, R.: Closed-loop global motion planning for reactive, collision-free execution of learned tasks. In: 2014 IEEE/RSJ International Conference on Intelligent Robots and Systems (2014)
8. Yan, W., Weber, C., Wermter, S.: A neural approach for robot navigation based on cognitive map learning. In: The 2012 International Joint Conference on Neural Networks (IJCNN). IEEE (2012)
9. Tapus, A., Siegwart, R.: A cognitive modeling of space using fingerprints of places for mobile robot navigation. In: Proceedings 2006 IEEE International Conference on Robotics and Automation, ICRA 2006 (2006)
10. Leite, I., Martinho, C., Paiva, A.: Social robots for long-term interaction: a survey. Int. J. Soc. Rob. **5**(2), 291–308 (2013)
11. Bonaccorsi, M., Fiorini, L., Cavallo, F., Saffiotti, A.: A cloud robotics solution to improve social assistive robots for active and healthy aging. Int. J. Soc. Rob. **8**(3), 393–408 (2016)
12. Park, H., Choi, Y., Jung, Y., Myaeng S.: Supporting mixed initiative human-robot interaction: a script-based cognitive architecture approach. In: 2008 IEEE International Joint Conference on Neural Networks, pp. 4107–4113 (2008)
13. Qu, W., Chen, Z.: A new cognitive approach based on dynamic evolution of the grey hazy set. In: 2014 19th International Conference on Methods and Models in Automation and Robotics (MMAR), Miedzyzdroje, pp. 572–577 (2014)
14. Sarker, V.K., Queralta, J.P., Gia, T.N., Tenhunen, H., Westerlund, T.: Offloading SLAM for indoor mobile robots with edge-fog-cloud computing. In: 2019 1st International Conference on Advances in Science, Engineering and Robotics Technology (ICASERT), Dhaka, Bangladesh, pp. 1–6 (2019)
15. Cockburn, A.: Writing Effective Use Cases. Addison-Wesley Professional, Boston (2000). ISBN 0-201-70225-8

Neural Models to Predict Irrigation Needs of a Potato Plantation

Mercedes Yartu[1], Carlos Cambra[2], Milagros Navarro[1], Carlos Rad[1], Ángel Arroyo[2], and Álvaro Herrero[2](✉)

[1] Composting Research Group (UBUCOMP), EPS-La Milanera, Universidad de Burgos, C/Villadiego s/n, 09001 Burgos, Spain
{minago,crad}@ubu.es

[2] Departamento de Ingeniería Informática, Grupo de Inteligencia Computacional Aplicada (GICAP), Escuela Politécnica Superior, Universidad de Burgos, Av. Cantabria s/n, 09006 Burgos, Spain
{ccbaseca,aarroyop,ahcosio}@ubu.es

Abstract. Reducing water consumption is an important target required for a sustainable farming. In order to do that, the actual water needs of different crops must be known and irrigation scheduling must be adjusted to satisfy them. This is a complex task as the phenology of plants and its water demand vary with soil properties and weather conditions. To address such problem, present paper proposes the application of time-series neural networks in order to predict the soil water content in a potato field crop, in which a soil humidity probe was installed. More precisely, Non-linear Input-Output, Non-linear Autoregressive and Non-linear Autoregressive with Exogenous Input models are applied. They are benchmarked, together with different interpolation methods in order to find the best combination for accurately predicting water needs. Promising results have been obtained, supporting the proposed models and their viability when predicting the real humidity level in the soil.

Keywords: Time series forecast · Neural networks · Autoregressive · Irrigation · Potato crop

1 Introduction

Originated and first domesticated in the Andes mountains of South America, the potatoes (*Solanum tuberosum*) belongs to the solanaceae family of flowering plants. In terms of agricultural production, potato crop is the third most important food crop in the world after rice and wheat. The EU produced 51.8 million tonnes of potatoes in 2018, with Germany, France, Poland and Netherlands as main producers [1]. In Spain, potato production reaches 2.24 million tonnes, and 40.3% of it is located in Castilla y León, mainly in Burgos (4%) occupying in 2017 around 2,400 ha of irrigated land [2]. In the Mediterranean context, irrigation supposes an extraordinary demand for available water, which constitutes an important problem in a context of water scarcity and climatic

Á. Herrero et al. (Eds.): SOCO 2020, AISC 1268, pp. 600–613, 2021.
https://doi.org/10.1007/978-3-030-57802-2_58

change. The application of innovative and an appropriate transfer of technologies to an adequate management of irrigation is a key factor to reach a sustainable crop production [3].

Monitoring weather variables and water status in soils are key factors to reach minimum water consumption without compromising crop production. The use of satellite or unmanned aerial vehicles (UAVs) imaginaries, automated weather stations and humidity or water potential probes are important tools to achieve precision irrigation adapted to crop phenology [4], which maximizes production avoiding water stresses, lixiviation of nutrients, or the incidence of crop pest and diseases.

In keeping with this idea, present paper proposes the application of IoT and Softcomputing to monitor a potato field crop, located in Cabia (Burgos, Spain), 42°16'57" N and 3°51'25" W, with sprinkler irrigation, to optimize water use efficiency. More precisely, a meteorological station, together with different sensors were placed in the crop in order to gather data in real time. Additionally, some measurements regarding crop development were taken and are analysed in present work. As there is no sensor to measure such features, these measurements must be taken manually. Thus, imaginary figures are not available on a daily basis and hence they must be interpolated in order to merge such data with those gathered through IoT. By taking into account all these data, neural networks for time series forecast are applied in order to predict water needs of the crop.

The remaining sections of this study will be structured as follows: previous work will be discussed in Sect. 2 while the methods applied in present study will be described in Sect. 3. Section 4 will introduce the real-life problem that is addressed, while the obtained results will be presented in Sect. 5. Finally, both conclusions and future work proposals will be discussed in Sect. 6.

2 Previous Work

Artificial Intelligence (AI) in general and Softcomputing in particular, have been previously applied to optimize irrigation systems. As stated in [5], different AI approaches and methods have been studied for smart controlling irrigation systems. More precisely, Neural Networks, Genetic Algorithms, and Fuzzy Logic could lead to optimum utilization of irrigation water resources.

Labbé et al. [6] modelled an irrigation decision process for limited water allocation, a very common pattern and challenge caused by climate change [7], and irrigation scheduling for corn plantations. The model consisted of irrigation management rules for different irrigation-related tasks that were derived from farmer surveys and based on the monitoring of their irrigation practices over a 2-year period. This model was incorporated into a simulator engine that, given the context of the decision, was able to predict irrigation schedules and irrigation volumes with an average error ranging from 6 to 13 mm for different farmers, reflecting an error below 6.7%. Instead of developing a model that captures the farmer's decision individually, using surveys and observations, in this study the Deep Learning and Artificial Intelligence AI were used to capture the agronomist's decision process in irrigation system [8].

Meanwhile, authors in [9] proposed a daily irrigation water demand calculation based on an Adaptive Neuro Fuzzy Inference System (ANFIS). This first-order Sugeno fuzzy model is combined with a backpropagation algorithm. It has a better performance (Root Mean Squared Error and Mean Absolute Percentage Error) predicting irrigation needs when compared to the Auto Regressive Moving Average models.

Khan et al. [10] compared different AI models and their error rates when it comes to irrigation prediction. It was found that among all the models, the 3-fold cross validation multiple decision trees SysFor model gave the best overall results. However, the actual amount of water required by the crop was accurately predicted by neural models. The difference in error percentage between ANNs and SysFor was almost 20%. The comparison concluded that SysFor, ANNs, and decision tree techniques are the most suitable ones for the task of irrigation prediction.

A neural network has been applied in [11] to model the temporal (surface) soil moisture fluxes. Only meteorological data and the soil moisture humidity itself are used as input data and no information from the crop is used. Present paper comprises a comprehensive study of different neural networks trained with different algorithms. Furthermore, it is the humidity level in the underground what is precisely predicted so that water demands could be more precisely calculated.

Similarly, non-linear time-series neural networks have been previously applied to some different problems ranging from workplace accidents [12] to road transportation [13] and fault detection [14]. Differentiating from previous works, present paper proposes time-series neural networks in order to predict the humidity in the underground.

3 Applied Methods

As previously stated, two kinds of methods have been applied in present paper; on the one hand, interpolation (described in Subsect. 3.1) has been applied to predict daily values of some features. On the other hand, neural networks (described in Subsect. 3.2) have been applied to predict the humidity level.

3.1 Interpolation

It is widely known that interpolation consists on generating new data points between a given range of values. In order to do that, several alternatives exist for one-dimensional problems. The following ones have been applied in present study:

- Cubic: this is a shape-preserving method for cubic interpolation. Based on the shape of the known data, new values are interpolated by piecewise cubic interpolation, taking into account the values at neighboring grid points.
- Spline: each new value calculated by this method is based on a cubic interpolation of the values at neighboring data in each respective dimension. The not-a-knot end conditions are applied.
- Makima: this modified version of the Akima cubic Hermite interpolation method [15]. Each new value calculated by this method is based on a piecewise function of polynomials (with degree smaller than or equal to 3). In the Akima formula, the value of the derivative at a given data point is a weighted average of nearby slopes. The weights are defined as:

$$w_1 = |\delta_{i+1} - \delta_i|;\ w_2 = |\delta_{i-1} - \delta_{i-2}| \tag{1}$$

Being δ_i the slope on the interval $[x_i x_{i+1})$. In the modified version, definition of weights is slightly different, as follows:

$$w_1 = |\delta_{i+1} - \delta_i| + \frac{|\delta_{i+1} + \delta_i|}{2};\ w_2 = |\delta_{i-1} - \delta_{i-2}| + \frac{|\delta_{i-1} + \delta_{i-2}|}{2} \tag{2}$$

Thanks to that, when two flat regions with different slopes meet, more importance is given to the side where the slope is closer to zero (horizontal), thus avoiding overshoot.

3.2 Neural Models

In order to predict the humidity level, once all data are available (i.e. after interpolation is carried out), 3 neural models for non-linear time-series forecast [16] have been applied, namely: Non-linear Input-Output (NIO), Non-linear Autoregressive (NAR) and Non-linear Autoregressive with Exogenous Input (NARX). These can be seen as feedforward networks in which the input weight has a tap delay line associated with it. Thanks to that, the network has a finite dynamic response to time series input data. The main differences between these 3 models are what data is given to the model in order to predict future values of humidity level. In the case of NIO, it is only the humidity level itself. In the case of NAR, all the other features (described in Sect. 4) except the humidity level are included. Finally, in the case of NARX, these two data sources are considered in the prediction. As a result, the NARX could be mathematically formulated as:

$$y(t) = f\left(y(t-1), \ldots, y\left(t-n_y\right), x(t-1), \ldots, x(t-n_x)\right) \tag{3}$$

Being *y(t)* the variable to be predicted in time instant t, $f\,()$ the function to be approximated by the neural model, *x(t)* an exogenous variable, n_{y} the maximum number of time delays in the output, and n_{x} the maximum number of time delays in the input. Consequently, the mathematical formulation for the NAR model is:

$$Y(t) = f(y(t-1), \ldots, y(t-ny)) \tag{4}$$

As it can be seen, in the case of the NAR model, the exogenous input (x) is not included in the formulation. Differentiating from this model, the predicted variable is replaced by this exogenous one in the NIO formulation:

$$y(t) = f(x(t-1), \ldots, x(t-n_x)) \tag{5}$$

4 Agronomic Setup

Field experiments were conducted from April 16th to October 10th 2019, in a potato field crop of 5 ha, located in Cabia (Burgos), 42°16'57" N and 3°51'25" W, with a semi-permanent sprinkler irrigation system. Soil was classified as *Calcic Luvisol*

(*LVk*) according to FAO, with loam texture, bulk density 1.26 kg L^{-1}, field capacity 0.31 (w/w), pH (1:5 w/v) 7.6, Electrical Conductivity (1:5 w/v, 25 °C) 0.65 dS m^{-1}, Organic Mater 3.33%, Total N 0.16% and lime 16.7%. Climate in this area is Attenuated Mesomediterranean, according to FAO.

As shown in Fig. 1, an agronomic IoT system was installed in the field, comprising an automatic weather station ATMOS 41 (METTER Group, USA) oriented to North. A soil humidity probe TEROS 10 (METTER) was installed at 15 cm depth, a soil water potential probe TEROS 21 at 30 cm depth and a rain gauge (ECRN 100) were connected to a EM60G data logger, remotely connected with ZENTRA Cloud System (METER Group, USA) that registered data each 30 min.

Fig. 1. Field map of the agronomic IoT system.

Potatoes (*Solanum tuberosum* L. Var. Agria) were planted in April 16th and from mid-June, phenological development was assessed according to BBCH-scale and four plants from the centre of the plot (20 × 20 m) were removed for laboratory analysis every 15 days. Morphological parameters such as length of aerial plant, number of stems and leaves, length of roots, number and weight of tubers, wet and dry biomass, chlorophyll content with SPAD, and N-content by a combustion autoanalyzer (TruSpec, LECO) were determined. Before harvesting, four sampling locations of 3 m^2 were chosen at random for yield estimation; tubers were classified by considering their diameter in different commercial classes: >80 mm, between 40–80 mm and <40 mm.

Public imaginary was captured from the satellite SENTINEL-2B under the scope of the EU Copernicus program. Nine images were obtained corresponding to day 11th to 171st, after plant emergency. From them, Normalized Difference Vegetation Index

(NDVI) was calculated according to the equation:

$$NDVI = \frac{(NIR - Red)}{(NIR + Red)} \tag{6}$$

Where Red and NIR are the spectral reflectance measurements acquired in the red (visible) and near-infrared regions, respectively. These data correspond to 4 and 8 of SENTINEL-2B bands, respectively. Raster layers were processed using the software QGIS v. 2.18 to obtain an NVDI vector layer. NVDI data were thereafter transformed into basal crop coefficients (K_{cb}) using equation:

$$K_{cb} = 1.44 \times NDVI - 0.1 \tag{7}$$

Crop evapotranspiration was calculated according to FAO Method 56 approach:

$$ET_C = (K_{cb} + K_s) \times ET_0 \tag{8}$$

Where K_s estimates soil evaporation, which is considered cero during the irrigation period as the crop development quickly cover soil surface.

As a result, the following features are available to apply the neural networks:

- Temperature: gathered from the temperature sensor (−40–50 °C) in the ATMOS 41 Weather Station (Meter Group, USA), Accuracy +/−0.5 °C.
- Precipitation: gathered from the precipitation sensor (0−400 mm/h) in the ATMOS 41 Weather Station (Meter Group, USA), Accuracy +/−5%. Daily
- CCM (Chlorophyl Content Index)[1]: CCM-200 plus Chlorophyll Content Meter (Opti-Sciences, UK) measures optical absorbance in two different wavelengths: 653 nm (Chlorophyll) & 931 nm (Near Infra-Red).
- Plant height[1]: a Carpenters meter (+/−1 mm) was used.
- Plant weight[1]: a weight scale (+/−1 mg) was used.
- % Humidity[1]: weight losses after 38 h at 70 °C (+/−1 °C).
- Aerial part length[1]: a Ruler lab (+/−1 mm) was used.
- Roots length[1]: a Ruler lab (+/−1 mm) was used.
- Plant Nitrogen content[1]: aerial part of plants was dried at 70 °C and thereafter, ground in a mill. Samples of 0.2 g were analysed by Dumas method in a TruSpec CN (LECO, USA) with IRD (Infra-Red Detector) and TCD (Thermal Conductivity Detector) for CO2 and N2, respectively.
- Tubers weight per plant[1]: a weight scale (+/−1 mg) was used.
- Number of tubers per plant[1]: tubers were visually counted.
- Tubers humidity[1]: weight losses after 38 h at 70 °C (+/−1 °C).
- Percentage of tubers in the 0–40 cm diameter range[1]: a squared measurement frame of 40 cm was used.
- Percentage of tubers in the 40–80 cm diameter range[1]: squared measurement frames of 40 and 80 cm were used.

[1] Interpolated by means of the methods described in Subsect. 3.1. All features are interpolated by means of same method each time.

- Percentage of tubers in the >80 cm diameter range[1]: a squared measurement frame of 80 cm was used.
- Tubers Nitrogen content[1]: crushed fresh tubers were dried at 70 °C and thereafter, ground in a mill. Samples of 0.2 g were taken.
- Underground humidity level: Teros 10 (Meter Group). It is a capacitance sensor that determines the dielectric permittivity of soil by measuring the charge time of a capacitor, which uses that medium as a dielectric. The sensor measures the time to charge a capacitor from a starting voltage, V_i to a voltage V_f with an applied voltage, V_f. Its working frequency (70-MHz) minimizes salinity and textural effects in the soil. This is the data feature to be forecast in the range [0, 1].

5 Experiments and Results

The results obtained through the different experiments are described in subsequent subsections. These results are presented by the applied interpolation method (Cubic, Makima, and Spline) and all the applied neural models (NAR, NIO, and NARX). During the experimental study, each one of these models has been tuned with different values of the appropriate parameters:

- Number of input delays: {1, 2, 3, 4, 5, 6, 7, 8, 9, 10}
- Number of output delays: {1, 2, 3, 4, 5, 6, 7, 8, 9, 10}
- Number of hidden neurons: {1, 5, 10, 15, 20}
- Training algorithm: {1 - Levenberg-Marquardt, 2 - Batch Gradient Descent, 3 - Gradient Descent with Momentum, 4 - Adaptive Learning Rate Backpropagation, 5 - Gradient Descent with Mo-mentum and Adaptive Learning Rate, 6 - Scaled Conjugate Gradient, 7 - Broyden–Fletcher–Goldfarb–Shanno Backpropagation}

As a result, 350 runs have been performed for the NIO and NAR models, and 3.500 for the NARX model. For each one of them, 10 executions have been carried out in order to obtain more statistically significant conclusions. Average Mean Squared Error (MSE) is provided in each case, calculated as the average MSE of all the included runs and executions. In each one of the tables, the lowest error value per column is in bold.

5.1 Results by Cubic Interpolation

Results (MSE) obtained when applying Cubic interpolation for the given features (listed in Sect. 4) are presented in this section. Firstly, results obtained by the neural models (NAR, NIO, and NARX) are presented per the number of input delays in Table 1.

Similarly, results obtained by the neural models (NAR, NIO, and NARX) are presented per the number of hidden neurons in Table 2.

Table 1. MSE of the results obtained by NAR, NIO, and NARX neural models after Cubic interpolation, averaged results are shown per the number of input delays.

N input delays	NAR	NIO	NARX
1	0.0014689	0.0009734	248.353764
2	0.0010200	0.0008537	293.284653
3	0.0012625	0.0007557	245.431746
4	0.0011707	0.0007253	2114.027857
5	0.0010601	0.0006607	250.708950
6	0.0010487	0.0005999	257.020836
7	0.0011634	0.0006057	513.295907
8	**0.0009762**	0.0005362	**28.8959075**
9	0.0010442	0.0005458	298.689252
10	0.0010452	**0.0005014**	283.937783

Table 2. MSE of the results obtained by NAR, NIO, and NARX neural models after Cubic interpolation, averaged results are shown per the number of hidden neurons.

N neurons	NAR	NIO	NARX
1	**0.00046848**	0.00050660	1096.76566
5	0.00085599	0.00170161	**188.035696**
10	0.00106059	0.00053109	282.978015
15	0.00149436	0.00032107	421.410713
20	0.00175055	**0.00031854**	406.349854

Finally, results obtained by the neural models (NAR, NIO, and NARX) are presented per the training algorithm in Table 3.

From the results obtained by Cubic interpolation, it can be said that NARX obtained, by far, the worst results in terms of error (MSE). When considering the number of input delays, the lowest error was obtained by the NIO model, with the highest number of delays (10). The lowest error for each one of the other neural models was also obtained with a high number of delays (8). After comparing the obtained results per number of hidden neurons, it is worth mentioning the best results in terms of MSE are obtained by NIO model comprising 20 neurons in the hidden layer. Finally, the training algorithm that outperforms all the other ones for the three neural models is Levenberg-Marquardt (LM). The lowest error when applying this algorithm is obtained by the NAR model.

Table 3. MSE of the results obtained by NAR, NIO, and NARX neural models after Cubic interpolation, averaged results are shown per the training algorithm.

Training algorithm	NAR	NIO	NARX
1	**0.000160087**	**0.000175692**	**8.623729042**
2	0.003173096	0.001686197	2215.944487
3	0.003385888	0.001619598	1016.390437
4	0.000293968	0.000286203	30.48441364
5	0.000353738	0.000360591	56.31164187
6	0.000202528	0.000249584	9.393541776
7	0.000312652	0.000352601	16.60766226

5.2 Results by Makima Interpolation

In a way like previous subsection, results (MSE) obtained when applying Makima interpolation are presented in this section. Firstly, results obtained by the neural models (NAR, NIO, and NARX) are presented per the number of input delays in Table 4.

Table 4. MSE of the results obtained by NAR, NIO, and NARX neural models after Makima interpolation, averaged results are shown per the number of input delays.

N input delays	NAR	NIO	NARX
1	**0.00016344**	0.00069287	0.00089254
2	0.00111653	0.00031333	0.00081513
3	0.00385679	0.00101315	0.00073656
4	0.00303471	0.00029949	0.00072213
5	0.00164207	0.00110668	0.00065707
6	0.00033552	**0.00029739**	0.00060338
7	0.00037262	0.00118705	0.00058923
8	0.00022522	0.00031471	0.00055197
9	0.00023622	0.00124011	0.00054781
10	0.00029200	0.00034244	**0.00053628**

Table 5 shows results obtained by the neural models (NAR, NIO, and NARX) presented per the number of neurons in the hidden layer of the models.

Table 5. MSE of the results obtained by NAR, NIO, and NARX neural models after Makima interpolation, averaged results are shown per the number of hidden neurons.

N neurons	NAR	NIO	NARX
1	**0.000463906**	**0.00050310**	**0.000488115**
5	0.000848853	0.00065632	0.000636506
10	0.001128539	0.00070203	0.000690203
15	0.001465014	0.00075088	0.000748559
20	0.001731256	0.00079128	0.000762679

Finally, results obtained by the neural models (NAR, NIO, and NARX) after Makima interpolation are presented per the training algorithm in Table 6.

Table 6. MSE of the results obtained by NAR, NIO, and NARX neural models after Makima interpolation, averaged results are shown per the training algorithm.

Training algorithm	NAR	NIO	NARX
1	**0.00016150**	**0.000180876**	**0.00014807**
2	0.00332818	0.000802008	0.00172820
3	0.00325209	0.000274582	0.00166450
4	0.00030008	0.000878147	0.00025697
5	0.00075088	0.000928166	0.00033010
6	0.00021049	0.000605822	0.00020037
7	0.00028627	0.000695771	0.00032827

After analyzing results in Tables 4, 5 and 6, it is worth mentioning that NAR and NARX models obtained the best results. When considering the number input delays, the minimum value (1) lead the NAR model to obtain the lowest error. In the case of NIO and NARX, 6 and 10 input delays respectively caused the models to reduce the error to the minimum. Regarding the number of hidden neurons, results are very consistent as the three models obtained the lowest MSE value when comprising only one hidden neuron. As it has been highlighted in the case of Cubic interpolation, LM is the training algorithm that let the models to obtain the minimum error when applied to Makima-interpolated data.

5.3 Results by Spline Interpolation

Finally, results (MSE) obtained when applying Spline interpolation are presented in this section. Firstly, Table 7 shows results obtained by the neural models (NAR, NIO, and NARX), presented per the number of input delays.

Table 7. MSE of the results obtained by NAR, NIO, and NARX neural models after Spline interpolation, averaged results are shown per the number of input delays.

N input delays	NAR	NIO	NARX
1	**0.00016068**	0.00098305	0.000937062
2	0.00095830	0.00083037	0.000805622
3	0.00394303	0.00083393	0.000741996
4	0.00274118	0.00067616	0.000675199
5	0.00167735	0.00064080	0.000646854
6	0.00031176	0.00062095	0.000613788
7	0.00035929	0.00060266	0.000595541
8	0.00029289	0.00057446	0.000548307
9	0.00024224	0.00054343	**0.000238717**
10	0.00031556	**0.00054079**	0.000538653

Similarly, results obtained by the neural models (NAR, NIO, and NARX) are presented per the number of hidden neurons in Table 8.

Table 8. MSE of the results obtained by NAR, NIO, and NARX neural models after Spline interpolation, averaged results are shown per the number of hidden neurons.

N neurons	NAR	NIO	NARX
1	**0.00044984**	**0.000512456**	**0.0004855**
5	0.00076514	0.000647422	0.0006236
10	0.00110439	0.000695201	0.0007031
15	0.00143401	0.000809941	0.0007577
20	0.00174776	0.000775962	0.0007533

Table 9 shows the results obtained by the neural models (NAR, NIO, and NARX), presented per the training algorithm.

It is worth mentioning that from the results obtained by Spline interpolation, as it happened in the case of Makima interpolation, best results have been obtained by NAR and NARX. Only one input delay was used by NAR to get the lowest error rate, while NIO and NARX employed high values (ten and nine respectively). As it happened when analyzing Makima-interpolated data, the three models obtained the lowest error when configured with one hidden neuron. NAR and NARX obtained the lowest error when they have been trained with the LM algorithm. Differentiating from these models, the lowest error was obtained by NIO on Spline-interpolated data when been trained with the Scaled Conjugate Gradient algorithm.

Table 9. MSE of the results obtained by NAR, NIO, and NARX neural models after Spline interpolation, averaged results are shown per the training algorithm.

Training algorithm	NAR	NIO	NARX
1	**0.00015990**	0.000947065	**0.0001458**
2	0.00328746	0.001717524	0.0017001
3	0.00305368	0.001668056	0.0016993
4	0.00029515	0.000302221	0.0002572
5	0.00039139	0.000350443	0.0003258
6	0.00020699	**0.000246865**	0.0001987
7	0.00030703	0.000355659	0.0003256

Finally, as in the case of the NARX model, both input and output delays are applied, Table 10 presents the results obtained by this model per interpolation method and number of output delays.

Table 10. MSE of the results obtained by NARX neural model, averaged results are shown per the number of output delays and interpolation method.

N output delays	Cubic	Makima	Spline
1	**253.2822868**	0.000678297	0.000677078
2	268.6515364	0.000667616	0.000670093
3	393.6519372	0.000653401	0.000674038
4	2120.995545	0.000682745	**0.000160276**
5	350.7752469	0.000667075	0.000669579
6	273.7858462	**0.000633245**	0.000666098
7	273.7598735	0.000656846	0.000656566
8	257.0172385	0.000666048	0.000658993
9	279.7134575	0.000684088	0.000666557
10	319.4469076	0.000662762	0.000667852

In this table it can be seen, as previously mentioned, that NARX obtained very bad results (high error rates) when applied to Cubic interpolation. On the contrary, acceptable results were obtained with medium values of output delays (6 and 4) in the case of Makima and Spline interpolation respectively.

6 Conclusions and Future Work

In general terms it can be said that present research has successfully addressed the initial targets. Different interpolation methods and time-series neural models have been combined and benchmarked in order to accurately predict the soil humidity level in a potato field. Thanks to the experimental validation, irrigation needs of the studied plantation could be adjusted.

After analyzing the presented results, conclusions can be derived from both the softcomputing and agricultural perspective. Taking into account the applied methods, it can be said that:

- The interpolation methods do not have a significant effect on the prediction except in one case. The NARX model when applied to Cubic-interpolated data obtained very high error rates.
- There is not a neural model that clearly outperforms the other ones; NIO obtained most best results when applied to Cubic-interpolated data while NAR and NARX outperformed it when applied to Makima and Spline-interpolated data. Furthermore, the parameter tuning of each model must be adjusted to each case as there is not a given combination of parameters that always leads to best results. The clearest conclusion about parameter tuning is that Levenberg-Marquardt is the best option when selecting the training algorithm. Except in one case (NIO applied to the Spline-interpolated data), it led the models to get the lowest error rates.

Actually, the activity of country-based institutional services involved in helping farmers to manage irrigation practices, are based only in forecast predictions, being more difficult to introduce in them predictions based in the available water content of the soil. The installation of non-expensive soil stations, with soil humidity probes located in reference soil profiles and covering wide irrigation areas, and the use of time series methods and neural networks for data analysis, would considerately improve soil water content monitoring and irrigation predictions.

As a proposal for future work, authors suggest applying some other softcomputing models to improve forecast. On the other hand, more input features may be considered and feature selection methods would be applied in order to identify those ones that are more important in order to predict the level of underground humidity.

Acknowledgements. : This work was financed by a grant agreement between Lab-Ferrer and UBUCOMP. Authors are grateful to the farmer Mr. José María Izquierdo for providing the experimental field and the monitoring of irrigation.

References

1. Agricultural Production Crops. https://ec.europa.eu/eurostat/statistics-explained/index.php/Agricultural_production_-_crops#Potatoes_and_sugar_beet. Accessed 02 Sept 2020
2. Yearly Statistics. https://www.mapa.gob.es/es/estadistica/temas/publicaciones/anuario-de-estadistica/2018/default.aspx?parte=3&capitulo=07&grupo=3&seccion=2. Accessed 02 Sept 2020

3. Pereira, L.S., Oweis, T., Zairi, A.: Irrigation management under water scarcity. Agric. Water Manag. **57**, 175–206 (2002)
4. Althoff, D., Alvino, F.C.G., Filgueiras, R., Aleman, C.C., da Cunha, F.F.: Evapotranspiration for irrigated agriculture using orbital satellites. Bioscience Journal **35**, 670–678 (2019)
5. Shitu, A., Tadda, M., Danhassan, A.: Irrigation water management using smart control systems: a review. Bayero Journal of Engineering and Technology **13**, 2449–2539 (2018)
6. Labbé, F., Ruelle, P., Garin, P., Leroy, P.: Modelling irrigation scheduling to analyse water management at farm level, during water shortages. Eur. J. Agron. **12**, 55–67 (2000)
7. Fry, A.: Water: facts and trends. World Business Council for Sustainable Development (2006)
8. Andriyas, S., McKee, M.: Recursive partitioning techniques for modeling irrigation behavior. Environ. Model Softw. **47**, 207–217 (2013)
9. Atsalakis, G., Minoudaki, C., Markatos, N., Stamou, A., Beltrao, J., Panagopoulos, T.: Daily irrigation water demand prediction using adaptive neuro-fuzzy inferences systems (anfis). In: Proceedings 3rd IASME/WSEAS International Conference on Energy, Environment, Ecosystems and Sustainable Development, pp. 369–374. WSEAS (2007)
10. Khan, M.A., Islam, M.Z., Hafeez, M.: Evaluating the performance of several data mining methods for predicting irrigation water requirement. In: AusDM, pp. 199–208 (2012)
11. Adeyemi, O., Grove, I., Peets, S., Domun, Y., Norton, T.: Dynamic neural network modelling of soil moisture content for predictive irrigation scheduling. Sensors **18**, 3408 (2018)
12. Contreras, S., Manzanedo, M.Á., Herrero, Á.: A hybrid neural system to study the interplay between economic crisis and workplace accidents in Spain. Journal of Universal Computer Science **25**, 667–682 (2019)
13. Alonso de Armiño, C., Manzanedo, M.Á., Herrero, Á.: Analysing the intermeshed patterns of road transportation and macroeconomic indicators through neural and clustering techniques. Pattern Anal. Appl. **23**(3), 1059–1070 (2020). https://doi.org/10.1007/s10044-020-00872-x
14. Taqvi, S.A., Tufa, L.D., Zabiri, H., Maulud, A.S., Uddin, F.: Fault detection in distillation column using NARX neural network. Neural Comput. Appl. **32**(8), 3503–3519 (2018)
15. Akima, H.: A method of bivariate interpolation and smooth surface fitting for irregularly distributed data points. ACM Trans. Math. Softw. **4**, 148–159 (1978)
16. Leontaritis, I.J., Billings, S.A.: Input-output parametric models for non-linear systems Part I: deterministic non-linear systems. Int. J. Control **41**, 303–328 (1985)

Special Session: Soft Computing Applied to Robotics and Autonomous Vehicles

Mathematical Modelling for Performance Evaluation Using Velocity Control for Semi-autonomous Vehicle

Khayyam Masood(✉), Matteo Zoppi, and Rezia Molfino

PMAR Robotics, DIME, University of Genova, Genova, Italy
khayyam.masood@edu.unige.it

Abstract. Freight Urban RoBOTic vehicle (FURBOT) is a semi autonomous vehicle for which it is desired that it could deliver freight autonomously from one destination to another. The vehicle is required to operate in Genova, Italy which in general has steep slopes. Additionally, safety of this vehicle and of the environment is of critical importance for urban autonomous driving thus the need for having a simulation model arises. Furthermore, the vehicle is expected to perform last mile freight delivery in European H2020 project SHOW for which highest autonomy is required. For these purposes, a mathematical model is constructed for autonomous velocity control over gradient varying hilly terrain. Autonomous traction and braking of the vehicle is introduced for catering for gradient varying terrain. The model built for this vehicle will serve as basis for embedding new sensors in future, tracking their performance and overall creating a safe environment for the vehicle to operate.

Keywords: Mathematical modeling · Performance evaluation · Mobile robots · Freight vehicle · Velocity control

1 Introduction

Freight Urban RoBOTic vehicle (FURBOT) is light weight, fully electronic vehicle designed for sustainable freight transport in urban areas. It is one of pioneering autonomous vehicles in freight delivery sector. The vehicle is expected to handle first and last mile freight delivery in an urban environment setting for the European H2020 project SHOW (SHared automation Operating models for Worldwide adoption). For the project SHOW, FURBOT is expected to attain maximum autonomy in its drive. Due to the autonomy requirements of SHOW project, it is essential for FURBOT to be modelled and simulated prior at length. For this purpose, it is very essential to build a custom-made simulation platform where automation testing and vehicle performance could be judged prior to experiments. This work is an effort to create such simulation platform in order to enhance the performance of the vehicle when integrated with new sensors and in general observing the performance anomalies if any.

Á. Herrero et al. (Eds.): SOCO 2020, AISC 1268, pp. 617–626, 2021.
https://doi.org/10.1007/978-3-030-57802-2_59

Dynamic modelling of new types of robots is becoming an essential research platform and is a required essential tool for developing better performing robots in current research. From Bi-pedal robots [1] to Hexa-slide robots [2] to parallel manipulators [3], dynamic modeling is becoming the key for better understanding the robots behavior in the natural environment.

Motion dynamics and custom control in robotics is essential for performance evaluation. Mobile robots specially wheeled robots which have vast human interaction and unknown environment requires simulation testing. Reference [4] uses sliding mode control for trajectory tracking for wheeled robot and [5] uses control to move their robot on slippery downhill. Mathematical modeling thus mainly revolves around modeling Newton-Euler equations [6,7] and using MATLAB as a tool for simulation.

This work, thus is essence of aforementioned researches in the field of mobile robotics. Due to need of modelling of vehicle, motion dynamics and controls are modelled for simulating our robot in MATLAB coupled with Simulink environment. As FURBOT is unique freight vehicle, modeling and simulating it is key for performance evaluation for vehicle and environmental safety.

2 Vehicle Dynamics

The current mathematical model of FURBOT is being developed for judging the performance of the vehicle before embedding any new hardware (sensors) or implementing any new automated technique which could result in an accident if not simulated prior to the experiments being conducted.

2.1 Constraints

The mathematical model is able to perform well on an empty straight road. Additionally, it is assumed that the vehicle does not have any inherent errors which can deviate it from the straight path e.g. tyre misalignment/balancing, lateral center of gravity shift etc. The road geometry consists of a straight road with null radius of curvature. The only varying feature of the road is the gradient of the road, which is extracted from Google Maps and simulated for traction power and braking control of the vehicle while driven on a hilly terrain.

2.2 Mathematical Modeling

An effort is made to make the mathematical model as precise as possible. The current coordinate system has its x-axis along the nose of the vehicle, y-axis towards left and z-axis is upwards [7]. It is considered that the vehicle is unable to roll, move along z-axis or pitch (unless due to road gradient change), making this mathematical model as a primarily three degree of freedom model. Both forces acting along x and y axis are modelled with moments acting along z-axis. However due to constraints, the motion of the vehicle is along a straight line.

Traction Force. Most of the definitions of forces acting along the body x-axis are taken from Ref. [8]. The forward force generated due to the torque acting on the driven wheels is given by Eq. 1

$$F_t = \frac{T_p \iota_g \iota_o \eta_t}{r_d} \tag{1}$$

Where F_t is the traction force and T_p is the torque output of the power plant and in our case is the output of the throttle controller. ι_g is the transmission gear ratio, ι_o is the final drive gear ratio, η_t is the final efficiency of the driveline from the wheels to the power plant and r_d is the radius of the wheels respectively.

Drag Force. The drag force calculation is straight forward and is estimated with the shape of FURBOT in mind. At present, drag coefficient (C_d) of 0.5 is selected which is in reference to usual drag coefficient of such shape vehicle. Equation 2 is the equation used for drag force calculation.

$$F_d = 0.5 \rho V^2 A_f C_d \tag{2}$$

Where F_d is the drag force acting on the body, ρ is the air density, A_f is the vehicle frontal area and V is the total velocity of the vehicle.

Gradient Force. Although the gradient force calculation is comparatively simpler i.e. dependent only on weight of the vehicle M_v and gradient angle α (Eq. 3, [8]), but the amplitude of this force can be significantly higher compared to other resistive forces. Also since in our case, there is no traffic or hurdle, this is the only force which cause braking to come in action (in case of negative gradient).

$$F_g = M_v g sin\alpha \tag{3}$$

Rolling Resistance Force. Rolling resistance force is due to the friction between tyres of the vehicle and the surface of road. The two main components attributing to this force are the normal force acting on the vehicle tyres and rolling resistance coefficient. Calculation of normal force is pretty straight forward, however there are number of different ways to calculate rolling resistance coefficient specially with varying velocity of the vehicle. The main equation for rolling force is given by Eq. 4 given in Ref. [8].

$$F_r = (M_v g cos\alpha) f_r \tag{4}$$

Where f_r is the rolling force coefficient and remaining is the normal force acting on the vehicle. For calculation of f_r numerous techniques are found in literature, however for this work, calculation of rolling force coefficient is taken from the work of Brian [9] which is also an extension of his work in [10] and is given in Eq. 5.

$$f_r = C_{sr} + \{3.24 C_{dr} (\frac{V}{100})^{2.5}\} \tag{5}$$

Where C_{sr} and C_{dr} represents the static and dynamic components of rolling resistance coefficient. In [11], variation of both, C_{sr} and C_{dr} are plotted against tyre pressure and Brian [9] used these graphs to extract polynomial expressions for C_{sr} and C_{dr} which are given in Eq. 6 and 7. These equations are thus taken from the work of Brian [9] and their validity is discussed in his work.

$$(C_{sr})_{ref} = -0.0000001687P_i^3 + 0.0000255349P_i^2 - 0.0012944847P_i + 0.0305104628 \quad (6)$$

$$(C_{dr})_{ref} = -0.0000002636P_i^3 + 0.0000404822P_i^2 - 0.0020812137P_i + 0.0381150798 \quad (7)$$

Where P_i denotes the tyre pressure. The above equations are considered for calculating the rolling resistance force coefficient for this work.

Forces Along y-Axis and Moment Along z-Axis. The Newton-Euler equations of motion for forces along y-axis and moment along z-axis are used for calculating respective forces and moments and are given by Eq. 8 and 9 [12].

$$\dot{v_y} = \frac{1}{M_v v_x}(-a_1 C_{af} + a_2 C_{ar})r - \frac{1}{M_v v_x}(C_{af} + C_{ar})v_y + \frac{1}{M_v}C_{af}\delta - r v_x \quad (8)$$

$$\dot{r} = \frac{1}{I_z v_x}(-a_1^2 C_{af} - a_2^2 C_{ar})r - \frac{1}{I_z v_x}(a_1 C_{af} - a_2 C_{ar})v_y + \frac{1}{I_z}a_1 C_{af}\delta \quad (9)$$

These equations are expressed in the body coordinate frame for the planar rigid vehicle [7]. C_{af} and C_{ar} are the cornering stiffness of the front and rear wheels respectively. δ is the steering angle and a_1/a_2 are the distances of the rear/front wheels from the CG of the vehicle. Since our steering angle for this work is considered zero, the forces acting along y-axis and moments acting along z-axis yield very negligible values which are not enough for moving the vehicle considerably in y-axis.

Longitudinal Equation of Motion Along x-Axis. The longitudinal equation of motion is relatively simple as the problem is reduced to forces along x-axis and is given by Eq. 10.

$$\dot{v_x} = \frac{\sum F_x}{M_v} = \frac{F_t + F_d + F_g + F_r}{M_v} \quad (10)$$

2.3 Velocity Control

Two separate controllers are designed for motion control for the autonomous vehicle FURBOT. One for traction power control and one for braking power control. The reason for using such cascade controller is because our requirements for braking and acceleration are different. For traction power control, we require a smooth robust controller whereas for braking power controller, we require sharp responses addition to smooth behavior. Traction power control is a PD controller with an error amplification factor, whereas braking power control is a simple proportional error control with a self-defined operational dead-band of 1 km/h speed, thus it is only initiated if there is a difference of at least 1 km/h speed between reference and actual speed. Reason for not including integral component for the controller designs was because of the overshoot integral values were causing in actuation values. The designing criteria for both the controllers was to keep the velocity error < 2 km/h. Inputs of both controllers is the difference between reference and actual velocity in km/h. Details of these controllers are given in Table 1.

Table 1. Velocity controllers

	Controller values		
Controller type	Error amplification	Proportional gain	Derivative gain
Traction power control	500	20	1
Braking power control	1	80	0

A switch is placed between traction controller and braking controller. If the signal of braking control is greater than 0 i.e. it is active, the traction controller is automatically turned off. The controller design is kept simple in order to build the mathematical model.

3 Simulation Setup and Results

The whole mathematical model was built in MATLAB and SimuLink and simulated for obtaining outputs for validating the model under genuine conditions.

3.1 Simulation Setup

Genova, Italy has one of the most unique terrain topology. Not only does it have beaches and mountain ranges, it is also quite dense in population. Driving an autonomous vehicle on such a terrain for the first phase of testing requires extensive simulation testing for environment and vehicle safety. A potential route is selected in Genova for FURBOT for its performance evaluation. The route selected for the FURBOT mathematical model verification and for elevation data is taken from Google Maps [13] and is shown in Fig. 1

For results evaluation, only the elevation profile is extracted. The current scope of work is to evaluate FURBOT capability over hilly terrain thus the selected path served as an ideal candidate for performance evaluation. The elevation profile extracted from the Fig. 1 is plotted in Fig. 2.

Fig. 1. Selected elevation route

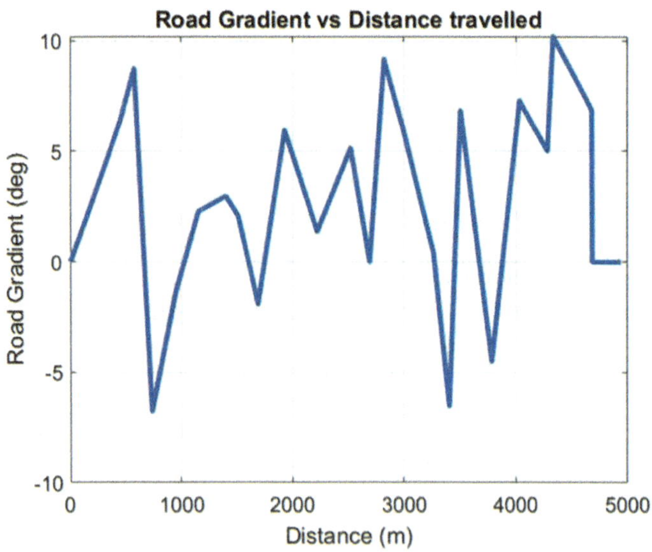

Fig. 2. Extracted elevation data

3.2 Results

In the current simulated scenario, the vehicle behaved comparatively well. The reference velocity was set to 40 km/h for the vehicle and velocity control contained the velocity error well within acceptable 1 km/h bound. Figure 3 shows the overall velocity of the vehicle and Fig. 4 shows the relative error in velocity in the whole simulation.

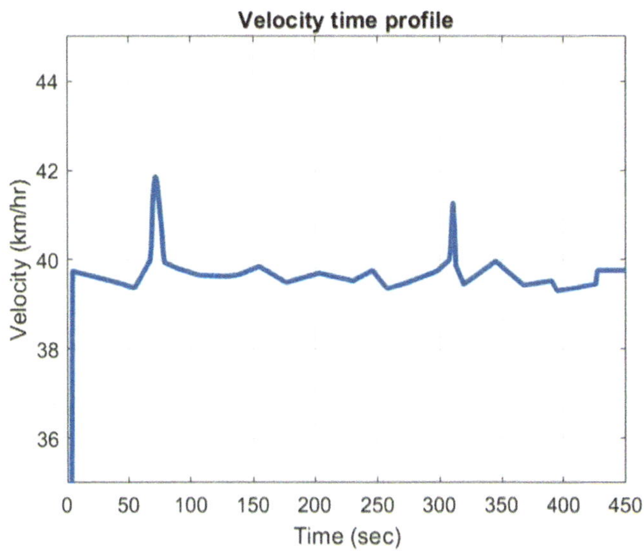

Fig. 3. Velocity profile over time

It is observed from the velocity profiles that it is easier for controlling the vehicle uphill compared to downhill as there are smooth transitions on positive gradient. The net effective traction force profile is given in Fig. 5. It is observed from the traction force profile that it follows the profile of the road gradient. Additionally, acceleration is zero in the negative gradient zone of the road which is also conceptually correct. Overall this shows satisfactory performance of the traction force controller.

The braking of the vehicle is triggered twice in the current scenario on the downhill journey of the vehicle. When compared with the gradient of the road, it shows coherence of braking with negative road gradient. A dead-band of 1 km/h is deliberately selected for avoiding any unnecessary use of braking. Figure 6 shows the comparison of braking force with the velocity error. This shows that braking is only triggered when higher than reference velocity is attained.

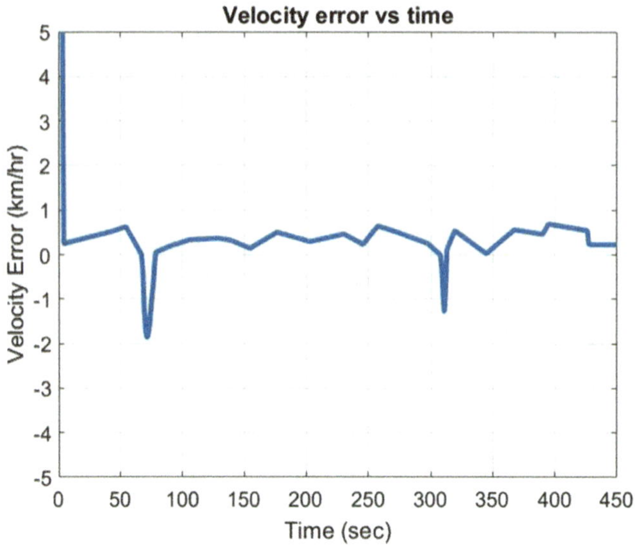

Fig. 4. Velocity error profile

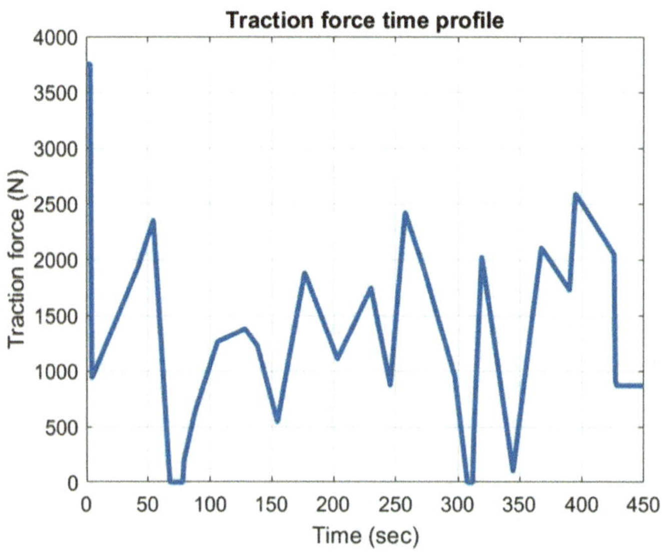

Fig. 5. Traction force profile

If we zoom-in Fig. 6 a) at one of the braking instances, as shown in Fig. 6 b), we can clearly observe the application of the velocity dead-band implemented for braking. Where green shows the velocity difference profile and red denotes the velocity difference profile input to the braking controller.

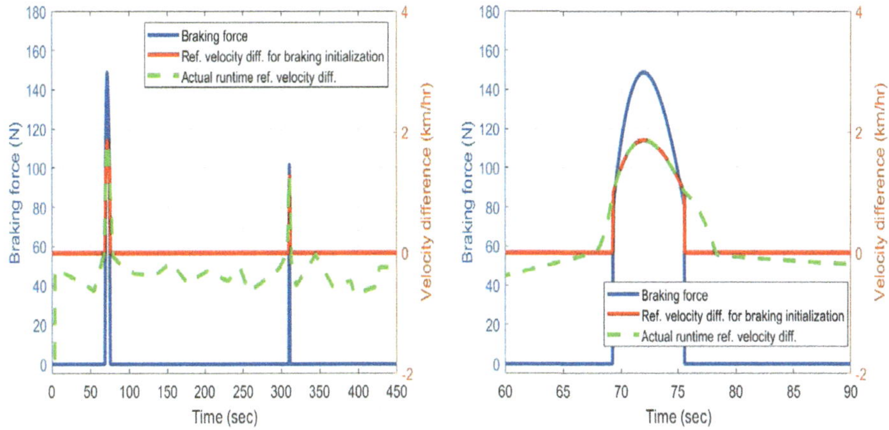

(a) Full profile

(b) Zoomed profile

Fig. 6. Braking force and velocity error profile

4 Conclusion

The mathematical model for FURBOT worked as per requirements. The velocity control for the vehicle created nominal errors which were within tolerable range of 2 km/h. The switching between the cascade controller for velocity control also behaved as per need. The vehicle was able to maintain its velocity over the uneven hilly terrain which was the goal of the research. Furthermore, the whole mathematical model generated realistic results.

After modeling complete road topology and embedding traffic data into the simulation, steering control can be incorporated. This can make the vehicle's mathematical model complete and autonomous which will be of critical importance for selecting and testing new sensors for the vehicle. This model will additionally serve as a platform for the future work on this autonomous vehicle. Number of safety enhancements can be incorporated in the vehicle after simulating its behavior. Some future outputs of this system include path planning, safe parking, cargo alignment and enhancing the safety of the vehicle and the environment which includes bounds on top speed, radial velocity, minimum safe distance and operational battery life before recharge is required.

Acknowledgment. This paper is supported by European Union's Horizon 2020 research and innovation programme under grant agreement No. 875530, project SHOW (SHared automation Operating models for Worldwide adoption).

References

1. Westervelt, E.R., Grizzle, J.W., Chevallereau, C., Choi, J.H., Morris, B.: Feedback control of dynamic bipedal robot locomotion. CRC Press, Boca Raton (2018)

2. Fiore, E., Giberti, H., Ferrari, D.: Dynamics modeling and accuracy evaluation of a 6-DoF Hexaslide robot. In: Nonlinear Dynamics. Conference Proceedings of the Society for Experimental Mechanics Series, vol. 1, pp. 473–479 (2016). https://doi.org/10.1007/978-3-319-15221-9_41
3. Pedrammehr, S., Qazani, M.R.C., Abdi, H., Nahavandi, S.: Mathematical modelling of linear motion error for Hexarot parallel manipulators. Appl. Math. Model. **40**(2), 942–954 (2016). https://doi.org/10.1016/j.apm.2015.07.004
4. Esmaeili, N., Alfi, A., Khosravi, H.: Balancing and trajectory tracking of two-wheeled mobile robot using backstepping sliding mode control: design and experiments. J. Intell. Robot. Syst. **87**(3–4), 601–613 (2017). https://doi.org/10.1007/s10846-017-0486-9
5. Asano, F., Seino, T., Tokuda, I., Harata, Y.: A novel locomotion robot that slides and rotates on slippery downhill. In: 2016 IEEE International Conference on Advanced Intelligent Mechatronics (AIM) (2016). https://doi.org/10.1109/aim.2016.7576804
6. Rodriguez, R., Ardila, D.L., Cardozo, T., Perdomo, C.A.C.: A consistent methodology for the development of inverse and direct kinematics of robust industrial robots. J. Eng. Appl. Sci. **13**(1), 293–301 (2018)
7. Marzbani, H., Khayyam, H., To, C.N., Quoc, D.V., Jazar, R.N.: Autonomous vehicles: autodriver algorithm and vehicle dynamics. IEEE Trans. Veh. Technol. **68**(4), 3201–3211 (2019). https://doi.org/10.1109/tvt.2019.2895297
8. Ehsani, M., Gao, Y., Longo, S., Ebrahimi, K.: Modern Electric, Hybrid Electric, and Fuel Cell Vehicles. CRC Press, Taylor & Francis Group, Boca Raton (2019)
9. Wiegand, B.P.: Estimation of the Rolling Resistance of Tires. SAE Technical Paper Series (2016). https://doi.org/10.4271/2016-01-0445
10. Wiegand, B.P.: Mass Properties and Advanced Automotive Design. SAWE Technical Paper 3602, 74th SAWE International Conference on Mass Properties Engineering; Alexandria, VA (2015)
11. Dixon, J.C.: Suspension Geometry and Computation. John Wiley & Sons Ltd., Chichester, UK (2009). ISBN 978-0-470-51021-6
12. Fu, C., Hoseinnezhad, R., Bab-Hadiashar, A., Jazar, R.N.: Electric vehicle side-slip control via electronic differential. Int. J. Veh. Auton. Syst. **6**, 1–26 (2014)
13. Google (n.d.): Google Maps directions for driving from Piazza del Portello, Genova to Righi, Genova. https://www.google.com/maps/dir/44.4114759,8.9345774/44.4241951,8.9379112/@44.4185213,8.9331592,15z/data=!4m2!4m1!3e0?hl=en. Accessed 12 Sept 2019

A Relative Positioning Development for an Autonomous Mobile Robot with a Linear Regression Technique

Daniel Teso-Fz-Betoño[1(✉)], Ekaitz Zulueta[1(✉)], Ander Sánchez-Chica[1(✉)], Unai Fernandez-Gamiz[2(✉)], Irantzu Uriarte[3(✉)], and Jose Manuel Lopez-Guede[1(✉)]

[1] System Engineering and Automation Control Department, University of the Basque Country (UPV/EHU), Nieves Cano, 12, 01006 Vitoria-Gasteiz, Spain
{daniel.teso,ekaitz.zulueta,ander.sanchez,jm.lopez}@ehu.eus

[2] Department of Nuclear and Fluid Mechanics, University of the Basque Country (UPV/EHU), Nieves Cano, 12, 01006 Vitoria-Gasteiz, Spain
unai.fernandez@ehu.eus

[3] Department of Mechanical Engineering, University of the Basque Country (UPV/EHU), Ingeniero Torres Quevedo, 1, 48013 Bilbao, Spain
irantzu.uriarte@ehu.eus

Abstract. Autonomous Mobile Robots (AMR) need a positioning function to move into unknown areas. These kinds of vehicles do not use a magnetic tape to guide into warehouses. Therefore, AMR use two different alternative techniques to solve the localization problem. First one is based on absolute positioning, and second one is established on relative localization. The absolute localization uses Simultaneous Localization and Mapping algorithms, in order to obtain a global position. However, the relative localization is based on odometry techniques. With the intention of developing a navigation system for an industrial mobile robot, which is being programmed in a structured text language, a relative localization is done utilizing LiDAR data acquisition. This novel concept analyzes two LiDAR datasets from different periods to calculate the AMR movement, by implementing Point matching and Linear Regression (LR) techniques. To understand the differences between conventional Iterative Closest Point (ICP) and LR a comparison is performed.

Keywords: Industrial mobile robots · Relative positioning · Iterative Closest Point · Linear regression · Point matching · Autonomous Mobile Robot · Automated mobile robots

1 Introduction

The industry is modifying the warehouses, by introducing new automation equipment, such us Automated Guided Vehicles (AGV) and Autonomous Mobile Robots (AMR). The difference between those reside in the navigation technique. AMR implementations

Á. Herrero et al. (Eds.): SOCO 2020, AISC 1268, pp. 627–635, 2021.
https://doi.org/10.1007/978-3-030-57802-2_60

are faster, smarter and more efficient, because AGV use fixed magnetic tapes as guide [1]. Thus, to transform an AGV into an autonomous vehicle, it requires an algorithm that controls the position. The localization is classified in two categories [2]: absolute localization and relative localization. Absolute navigation is known as Simultaneous Localization and Mapping (SLAM). The most common techniques are Particle Filter (PF) [3] and Extended Kalman Filter (EKF) [4]. However, this article is based in relative positioning. Thus, there are other techniques to implement as Wang et al. [5] explained, such us wheel odometry. This relative localization is used to estimate the movement of the robot. Nevertheless, Borestein et al. [6] mentioned, the odometry could not be a linear motion, due to wheel slippage. Therefore, this author mentioned different sensors to solve this problematic, such us inertial navigation sensors (gyroscopes and accelerometers), magnetic compasses, landmarks, etc.

Other authors try to find other techniques, such as Chambers et al. [7] analyzed. The pose estimation computes relative camera motion by comparing sequential images. Zheng et al. [8] proposed another Vision odometry to determinate the motion by parametrizing the robot pose. Apart from camera implementation, the LiDAR is also another instruments that it is used as a relative movement estimation. Chiella et al. [9] analyzed Light Detection and Ranging (LiDAR) sensors. This type of solution uses some spatial point in different time intervals to estimate the relative movement of the vehicle. Applying Iterative Closest Point (ICP) technique, it matches two different time interval datasets and calculates the translation and rotation matrices of the vehicle as Jinqiang et al. [10] mentioned. Moreover, there are some well-known ICP implementations for localization, such as Gressin et al. [11] or Yang et al. [12]. Thus, ICP is a conventional technique, which during the years has been improved, such us Du et al. [13] improvement, where apart from understanding the basic theory of ICP and how to improve it. In the end the ICP algorithm is based on lie group that converges monotonically to a local minimum. Moreover, this point matching it could be made applying singular value decomposition (SVD) as Oomori et al. [14] demonstrated, and SVD are used for Jacobian estimation as Papadopoulo et al. [15] presented. Other studies reveal that Jacobians can be calculated applying different methods, like multiple regression as Ferreira de Freitas et al. [16] presented or Therefore, Linear could be an implementation to estimate the position or motion. Chang et al. [17] develop a LR neural network to estimate the position of the robot and return faster to home position.

The aim of this publication is to make a LiDAR odometry. In order to use conventional ICP algorithm with SVD calculation, a linear regression is implemented to estimate the AMR rotation and translation. Moreover, a comparison between conventional ICP and LR will be performed in order to obtain a conclusion.

2 LiDAR Odometry Development

Using the same idea of ICP, the intention is to develop a linear regression function, which will estimate the rotation and translation matrices of the AMR. The intention of this development is to implement it on an industrial PC (IPC), by programming in a structured text (ST) language. Moreover, this computer uses the 80% of each capacity as a programable logic control (PLC). Normally these devices control logical operators and realize really simple mathematic operations, and do not execute an optimization function. Therefore, there are some limitations to program this device, such us there is no function to make an inverse of matrix, etc. The problem of ICP algorithm requires processing a singular value decomposition (SVD) function, which needs to compute a multiple matrix. The LR, however, works with two well-known matrices. Therefore, the motion of the AMR can be estimated doing an inversion, a transpose and a multiplication. Thus, to understand how this concept works, the section two is divided in two subsections. In the first one, the LR equations are analyzed to replicate the conventional ICP results. In the second one, a LiDAR Odometry pseudocode is presented.

2.1 Linear Regression Development

The LR will replicate the ICP function results. Therefore, the ICP function is represented in Eq. (1), where R and T are the rotation and translation matrices, *dataset* is the set of points that LiDAR detects, and t is the sample time.

$$[R, T] = f(dataset_{t-1}, dataset_t) \tag{1}$$

Moreover, it is important to consider the rotation matrix equation in Z-axis as analyzed in Eq. (2), and the rotation and translation functions are represented in Eq. (3).

$$R = \begin{bmatrix} \cos(\alpha) & -\sin(\alpha) \\ \sin(\alpha) & \cos(\alpha) \end{bmatrix} \tag{2}$$

$$\begin{bmatrix} X_2 \\ Y_2 \end{bmatrix} = \begin{bmatrix} \cos(\alpha) & -\sin(\alpha) \\ sin(\alpha) & cos(\alpha) \end{bmatrix} \cdot \begin{bmatrix} X_1 \\ Y_1 \end{bmatrix} + \begin{bmatrix} t_x \\ t_y \end{bmatrix} \tag{3}$$

Obviously X_1 and Y_1 are the coordinates of a single point from the $dataset_{t-1}$. X_2 and Y_2 represent the point location from $dataset_t$. The Eq. (3) is just for one point. Thus, this equation has to be modified for multiple points, considering that there will be i points in a dataset. This adaption is represented in Eq. (4).

$$\begin{bmatrix} X_{2i} \\ \vdots \\ Y_{2i} \\ \vdots \end{bmatrix} = \begin{bmatrix} -Y_{1i} & X_{1i} & 1\,0 \\ \vdots & \vdots & \vdots\,\vdots \\ X_{1i} & Y_{1i} & 0\,1 \\ \vdots\,\vdots & \vdots & \vdots\,\vdots \end{bmatrix} \cdot \begin{bmatrix} \sin(\alpha) \\ \cos(\alpha) \\ t_x \\ t_y \end{bmatrix} \tag{4}$$

The linear regression calculation is analyzed in Eq. (5).

$$\begin{bmatrix} \widehat{\sin(\alpha)} \\ \widehat{\cos(\alpha)} \\ \widehat{t_x} \\ \widehat{t_y} \end{bmatrix} = \left(\phi^t \cdot \phi\right)^{-1} \cdot \phi^t \cdot \vec{y} \tag{5}$$

$$\phi = \begin{bmatrix} -Y_{1i} & X_{1i} & 1 & 0 \\ \vdots & \vdots & \vdots & \vdots \\ X_{1i} & Y_{1i} & 0 & 1 \\ \vdots & \vdots & \vdots & \vdots \end{bmatrix}, \vec{y} = \begin{bmatrix} X_{2i} \\ \vdots \\ Y_{2i} \\ \vdots \end{bmatrix} \tag{6}$$

To improve the results of the motion estimation, it is important to perform a point matching technique, where the function will estimate which point, from the $dataset_t$, matches better with a point from the $dataset_{t-1}$. For this time point, matching technique analyzes all the distances between points and selects the minimum error as shown in Eq. (7).

$$\min\left(\sqrt{Point_{dataset_t} - Point_{dataset_{t-1}}}\right) \tag{7}$$

Once all the equations are seen, it will be described how the code works to obtain a motion estimation from the AMR.

2.2 LiDAR Odometry Pseudocode

The pseudo code is represented in the next line, where there are two essential functions.

```
While ≠ (dError < EPS)
   [ii, error]= FindNearestPoint(dataset_{t-1}, dataset_t)
   [R_1, t_1] = LRegresionMotionStimation(dataset_{t-1}, dataset_t, ii)
   dataset_t = R_1 · dataset_t
   dataset_t= [dataset_t(1,:) + t_1(1); dataset_t(2,:) + t_1(2)]
   R = R_1 * R
   T = R_1 * T + t_1
   dError = abs(preError − error)
   preError = error
end
```

FindNearestPoint is the function that calculates which points have the minimum distance between two datasets and *LRegresionMotionStimation* is the function that contains the Eq. (5). As it has been said, this code tries to minimize the error between points, in order to search the best matching position. That is the reason why $dataset_t$ is updated per each loop, until the dError minimizes to desired value, which is EPS.

This minimization is also made in a conventional ICP, the only part that differs LR pseudocode from ICP ones is the LRefressionMotionStimation function. In ICP code an SVD function is implemented. Therefore, the LiDAR odometry Test section reveal, the difference between how it works SVD and LR.

3 LiDAR Odometry Test

After developing the LiDAR Odometry function, the results will be compared to the conventional ICP. Thus, the comparison will reveal the difference of both algorithms.

In this case, it will be used a particular map from the school corridor to simulate the LiDAR data acquisition. In both cases, the relative pose estimation will be compared to the real relative pose, in order to obtain the error between both relative locations. The school corridor is represented in Fig. 1, and this map has been made using a LiDAR.

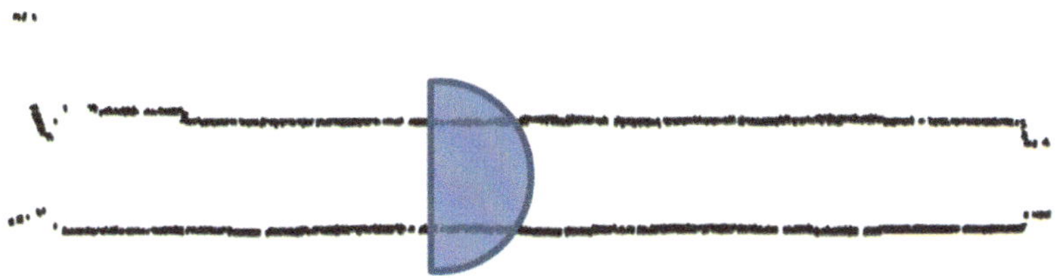

Fig. 1. School corridor map for LiDAR Odometry comparison

The black points represent the obstacles that the LiDAR has detected during the map building process. This corridor has 40 m from the start, which is in the left side line to end line that is in the right side, and the blue semicircle is the LiDAR limitations.

Therefore, in the simulation, the software will take a piece of map considering the localization of the LiDAR and measurements limitations. For this time, it can take measurements from 10 mm to 8000 mm and from $-\pi$ to π. Apart from that, this corridor seems to be simple; however, the problem of the most corridor comes, when the walls have not enough relevant information to known how much are the AMR moving, as Fig. 1 shown. The blue half circle revels how much this LiDAR can see. Thus, if the AMR starts moving, the SLAM has not got enough information to know where is. Therefore, in most cases only the wheel odometry is used. However, it is not always the best way to analyse displacement, due to the wheels can slip caused by the possible oils in the soil. This is the reason, why in this time a "simple" corridor is used, in which there is no extra information, such us landmarks, which helps to determinate displacement.

Once the map and the LiDAR limitations have been introduced, the comparison can be done. Using the LR LiDAR Odometry (LR-LO), the results of the execution are represented in Figs. 2 and 3. The error represents the difference between the movement and estimated movement. Therefore, the answers revels, how much millimeters the algorithm differs from the ideal estimation, and the idea of this comparation is to understand, if the Linear Regression works as a conventional ICP.

In Figs. 2 and 3, the maximum error oscillation for this execution test varies from −4,91 mm to 13,11 mm for X, from −14,53 mm to 11,85 mm for Y, and from 0 rad to 0,1468 rad for the orientation. The representations of the ICP LiDAR Odometry (ICP-LO) are illustrated in Figs. 4 and 5.

In Figs. 4 and 5, the maximum error oscillation for this execution test varies from −4,923 mm to 15,9 mm for X, from −1,823 mm to 15,9 mm for Y, and from 0 rad to 0,003 rad for the orientation. Matching LR-LO and conventional IPC-LO results in Table 1, some conclusions can be obtained.

Comparing the X error gaps, the LR-LO has smaller gaps than the conventional ICP-LO; however, the conventional ICP-LO method mean is 0,04 mm smaller, which

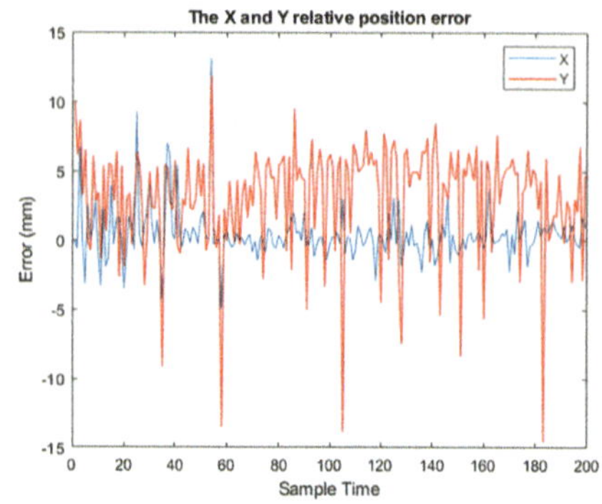

Fig. 2. LR-LO X and Y relative position error

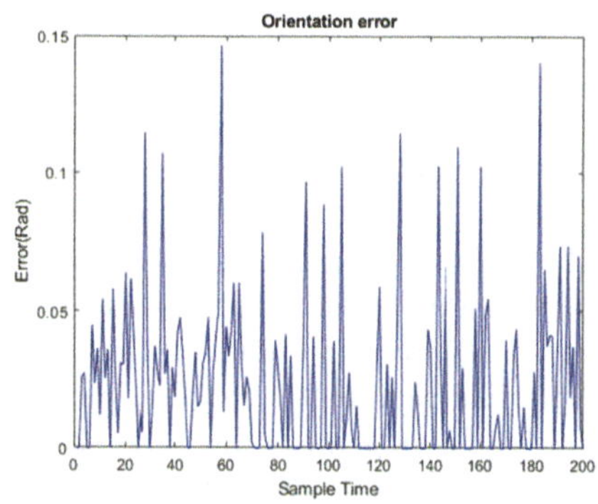

Fig. 3. LR-LO orientation error

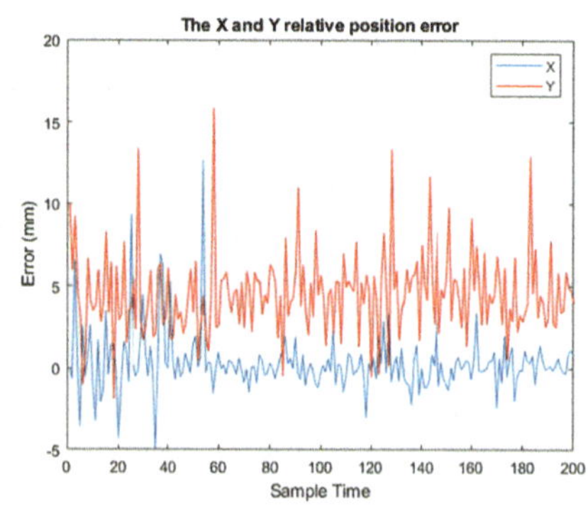

Fig. 4. ICP-LO X and Y relative position error

traduce on a 0,1231% difference. Thus, this difference can be considered negligible, as the commercial AMR makes ±1 cm mean error.

In the Y error gaps, the prediction changes. The LR-LO makes some picks during the prediction that increment the gap. Nevertheless, the LR-LO error mean values show

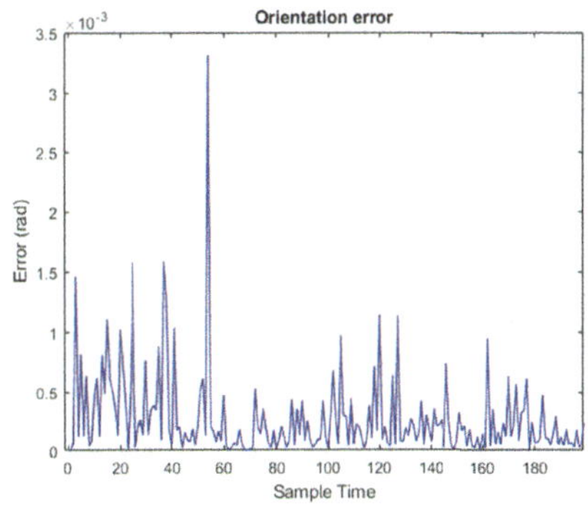

Fig. 5. ICP-LO Orientation error

Table 1. LR-LO and ICP-LO comparison

	X (mm)	Y (mm)	Orientation (rad)
LR-LO Error mean	0,3940	3,1426	0,0235
ICP-LO Error mean	0,3508	4,5596	2,7769e−04
LR-LO Error gap	−4,9 to 13,11	−14,53 to 11,85	0 to 0,1468
ICP-LO Error gap	−4,923 to 15,9	−1,823 to 15,9	0 to 0,003
LR-LO error mean respect to ICP-LO in %	0,1231%	−0,31%	83,62%

that conventional ICP-LO has higher values; in this case 1 mm higher, which revels a – 0,31% in comparison LR-LO with ICP-LO. In the orientation gap, the conventional ICP-LO predicts clearly better results and the different between both is 83,62%. However, this difference is insignificant, due to commercial AMR makes ±0,0349 rad, which is less than LR-LO mean error.

Over all, the prediction of the LR-LO during the time has more peaks. However, the mean calculation shows that this novel implementation obtains similar results in comparison with conventional IPC, which is the reference. Thus, this LR-LO works adequately to predict a relative positioning. Moreover, the proposal of this concept is to have other sensors which compare the wheel odometry with LiDAR, and detect sliding cases.

4 Conclusions

In this article, it can be affirmed that the novel relative positioning technique works satisfactorily and estimates the AMR relative movement with a little error. This error is smaller than 13 mm, and considering that the well-known ICP algorithm makes an error of 15,9 mm, the linear regression presents better results in some cases. In general, LR-LO and conventional IPC-LO have similar behavior, considering in X and Y positions the difference between both is less than 0,31%. The rotation is clearly worst, due to the difference is around 83%, however it maintains in correct parameters, because industrial AMR do not have a really precise localization. The mean error that makes LR-LO is 0,0235 rad, which is less than industrial robots present.

It is true that LR-LO has more error gaps, when the maximum and minimum values are analyzed. Nevertheless, the mean value affirms that the prediction is more stable, because LR-LO's value is lower than ICP's value. Apart from that, the linear regression is simpler to program in a ST language, and this is crucial for the vehicle that is being designed.

Clearly this development is the first step, and it is necessary to test on a real scenario. As a future work, this new algorithm will be implemented on an industrial AMR, which uses an IPC, to analyze the performance in a non-virtual scenario. This IPC uses each 80% of each capacity for PLC programing, and that is the reason why the LR-LO is programed in ST. Moreover, this technique will be compared with wheel odometry values to confirm which technique has better resolution, as the pose estimation of wheel odometry depends on floor sliding conditions.

Acknowledgments. Mercedes–Benz Vitoria is also acknowledged in especially to Emilio, Jose Carlos Velasco, the final assembly maintenance department of Mercedes-Benz Vitoria, Javier Loredo, Javier Gómez, Jose Antonio Hernando and Tomás Hernandez to give the opportunity to makes this research in intelligent production.

Funding. This research was financed by the plant of Mercedes-Benz Vitoria through PIF program to develop an intelligent production. Moreover, The Regional Development Agency of the Basque Country (SPRI) is gratefully acknowledged for economic support through the research project "Motor de Accionamiento para Robot Guiado Automáticamente", KK-2019/00099, Programa ELKARTEK. The authors are grateful to the Government of the Basque Country and to the University of the Basque Country UPV/EHU through the SAIOTEK (S-PE11UN112) and EHU12/26 research programs, respectively.

References

1. Cawood, G.J.; Gorlach, I.A.: Navigation and locomotion of a low-cost Automated Guided Cart, pp. 83–88. IEEE, November 2015
2. Cho, B., Seo, W., Moon, W., Baek, K.: Positioning of a mobile robot based on odometry and a new ultrasonic LPS. Int. J. Control Autom. Syst. **11**, 333–345 (2013). https://doi.org/10.1007/s12555-012-0045-x
3. Montemerlo, M., Thrun, S.: Simultaneous localization and mapping with unknown data association using FastSLAM, vol. 2, pp. 1985–1991. IEEE (2003)
4. Zhang, F., Li, S., Yuan, S., Sun, E., Zhao, L.: Algorithms analysis of mobile robot SLAM based on Kalman and particle filter, pp. 1050–1055. IEEE, July 2017
5. Wang, X., Li, W.: Design of an accurate yet low-cost distributed module for vehicular relative positioning: hardware prototype design and algorithms. TVT **68**, 4494–4501 (2019). https://doi.org/10.1109/TVT.2019.2901743
6. Borenstein, J., Everett, H.R., Feng, L., Wehe, D.: Mobile robot positioning: sensors and techniques. J. Robot. Syst. **14**, 231–249 (1997). https://doi.org/10.1002/(SICI)1097-4563(199704)14:43.3.CO;2-1

7. Chambers, A., Scherer, S., Yoder, L., Jain, S., Nuske, S., Singh, S.: Robust multi-sensor fusion for micro aerial vehicle navigation in GPS-degraded/denied environments. In: American Automatic Control Council, pp. 1892–1899, June 2014
8. Zheng, F., Tang, H., Liu, Y.: Odometry-vision-based ground vehicle motion estimation with SE(2)-constrained SE(3) poses. IEEE Trans. Cybern. **49**, 2652–2663 (2019). https://doi.org/10.1109/TCYB.2018.2831900
9. Chiella, A.C.B., Machado, H.N., Teixeira, B.O.S., Pereira, G.A.S.: GNSS/LiDAR-based navigation of an aerial robot in sparse forests. Sensors **19**, 4061 (2019). https://doi.org/10.3390/s19194061. https://search.proquest.com/docview/2296660065
10. Cui, J., Wang, F., Dong, X., Yao, K.A.Z., Chen, B.M., Lee, T.H.: Landmark extraction and state estimation for UAV operation in forest. In: TCCT, CAA, pp. 5210–5215, July 2013
11. Gressin, A., Mallet, C., Demantke, J., David, N.: Towards 3D lidar point cloud registration improvement using optimal neighborhood knowledge. ISPRS J. Photogramm. Remote Sens. **79**, 240–251 (2013). https://doi.org/10.1016/j.isprsjprs.2013.02.019
12. Yang, B., Chen, C.: Automatic registration of UAV-borne sequent images and LiDAR data. ISPRS J. Photogramm. Remote Sens. **101**, 262–274 (2015). https://doi.org/10.1016/j.isprsjprs.2014.12.025
13. Du, S., Zheng, N., Ying, S., Liu, J.: Affine iterative closest point algorithm for point set registration. Pattern Recogn. Lett. **31**, 791–799 (2010). https://doi.org/10.1016/j.patrec.2010.01.020
14. Oomori, S., Nishida, T., Kurogi, S.: Point cloud matching using singular value decomposition. Artif. Life Robot. **21**(2), 149–154 (2016). https://doi.org/10.1007/s10015-016-0265-x
15. Papadopoulo, T., Lourakis, M.I.A.: Estimating the Jacobian of the singular value decomposition: theory and applications. In: Computer Vision - ECCV 2000, pp. 554–570. Springer, Heidelberg (2000)
16. de Freitas, S.M.S.F., Scholz, J.P.: A comparison of methods for identifying the Jacobian for uncontrolled manifold variance analysis. J. Biomech. **43**, 775–777 (2010). https://doi.org/10.1016/j.jbiomech.2009.10.033
17. Chang, C., Chang, C., Tang, Z., Chen, S.: High-efficiency automatic recharging mechanism for cleaning robot using multi-sensor. Sensors (Basel, Switzerland) **18**, 3911 (2018). https://doi.org/10.3390/s18113911

Generating 2.5D Photorealistic Synthetic Datasets for Training Machine Vision Algorithms

Georgia Peleka(✉), Ioannis Mariolis, and Dimitrios Tzovaras

Centre for Research and Technology Hellas - CERTH, Information Technologies Institute, 6th km Charilaou-Thermi Rd., Thessaloniki, Greece
gepe@iti.gr

Abstract. The continued success of deep convolution neural networks (CNN) in computer vision can be directly linked to vast amounts of data and tremendous processing resources for training such non-linear models. However, depending on the task, the available amount of data varies significantly. Particularly robotic systems usually rely on small amounts of data, as producing and annotating them is extremely robot and task specific (e.g. grasping) and therefore prohibitive. Recently, in order to address the aforementioned problem of small datasets in robotic vision, a common practice is to reuse features that are already learned by a CNN within a large-scale task and apply them to different small scale ones. This transfer of learning shows some promising results as an alternative, but nevertheless it can not be compared with the performance of a CNN that is specifically trained from the beginning for that specific task. Thus, many researchers turned to synthetic datasets for training, since they can be produced easily and cost effectively. The main issue of such datasets that already exist, is the lack of photorealism both in terms of background and lighting. Herein, we are proposing a framework for the generation of completely synthetic datasets that includes all types of data that state-of-the-art algorithms in object recognition, and tracking need for their training. Thus, we can improve robotic perception without deploying the robot in time-consuming real-world scenarios.

Keywords: Robot vision · Machine learning · Neural networks · Synthetic data

1 Introduction

Recent advances in computer and robotic vision have been dominated by deep neural networks that have been trained in massive amounts of labeled data. State-of-the-art models appear to be extremely data-demanding since large amounts of training data are needed to optimize their variables. Acquiring such datasets is, however, a time-consuming task; thus, there has been a large increase in approaches where the model trains with a combination of real and synthetic, or exclusively on synthetic data.

Á. Herrero et al. (Eds.): SOCO 2020, AISC 1268, pp. 636–646, 2021.
https://doi.org/10.1007/978-3-030-57802-2_61

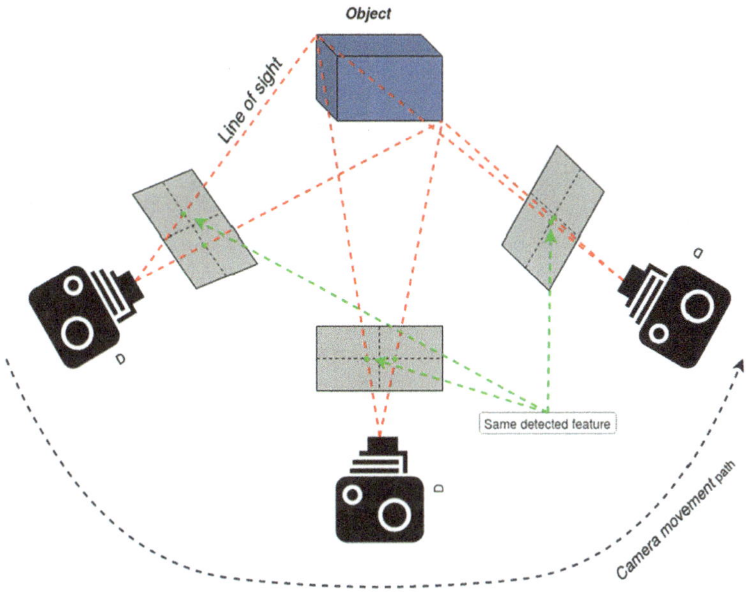

Fig. 1. Using structure from motion, photogrammetry can register camera positions during its movement. By detecting the location of detected and matched features, and by combining this information with the camera motion an estimation of every pixel's 3D coordinates can be made.

Apart from the acquisition of the data (in many cases millions of images), the usual annotation process consists of many hours of manual parsing, annotating, and labeling all these images. There are many labeling tools that make this process relatively user friendly, but nevertheless is exhaustively time-consuming. As an alternative to this, many researchers turn on synthetic datasets to bypass the barrier of manual annotation.

Synthetic data usually are produced utilizing a 3D model of the object of interest, and using rendering engines to create thousands images of it. As expected, since these images are produced in a simulation environment, every image is produced as an outcome of perfect conditions from a flawless sensor. Nevertheless, this is not the case in reality. No object is exactly a duplicate of another even within the same brand or genre (e.g. not all apples look the same), no sensor produces the perfect RGB or Depth image. In addition, the background behind the object isn't always the same, which can have an impact on the amount and color of the light that the environment casts around the object (e.g in case a cup is placed on a orange table cloth, the color of the light that hits the cup changes as it scatters over the tablecloth). So the main problem with models that are trained using solely synthetic data, is that they expect the input (e.g. RGB and/or depth images) to be flawless, with no artifacts or noise. Thus, arises the need for more photo-realistic synthetic datasets, as the ones we are proposing herein, that include noise and artifacts common in data acquired

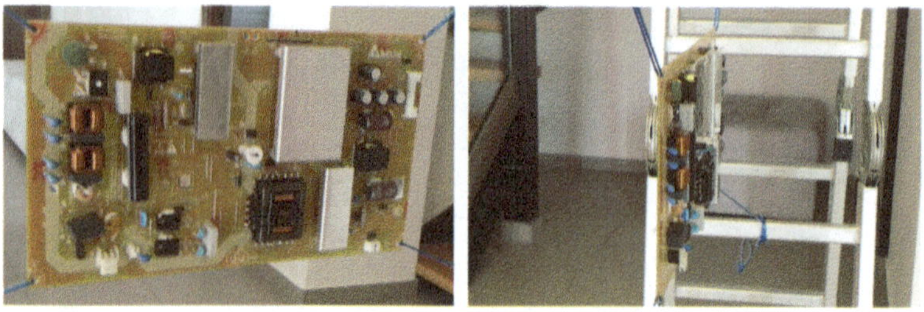

Fig. 2. During image acquisition for photogrammetry, there are cases where as the camera rotates around the object, the majority of it is visible (left), and others where only a slice of the object can be seen from the camera (right). In these cases the photogrammetry approach fails to match features from this image to the previous taken and the whole procedure fails.

from real camera sensors, incorporating changes in the luminance resulting from indirect light scattering from the environment. These types of datasets can ease the preparation for robotic grasping and object manipulation, and also enhance the perception of the environment from robotic platforms (Fig. 2).[1]

The following sections are structured as follows: Sect. 2 outlines work related to the use of the aforementioned types of datasets in computer and robotic vision. Section 3 explains our approach on ensuring photo-realistic 3D models. Section 4 analyzes our method for achieving photo-realistic lighting in synthetic data. In Sect. 5 we present our framework for synthetic dataset generation. In Sect. 6, we describe two datasets acquired with our framework. And Sect. 7 concludes with an outlook on future research, and discussion.

2 Related Work

During the last decade, the increased popularity of low-cost yet high-quality depth sensors like Microsoft Kinect [8], Intel Realsense depth sensors [9], Orbbec Astra depth cameras [4], has put the need for a complex level 3D object detection dataset to the spotlight. A number of previous efforts have been made to collect datasets with 2D and 3D observations for object detection, recognition, and tracking, both in terms of real and synthetic data.

Lai et al. in [10] proposed a dataset that includes 300 distinct objects from 51 classes. Every category comprised by 4–6 instances, and each object was densely photographed using a turntable. In total 207,920 RGB-D image pairs are provided, with roughly 600 images per object. For testing, each object is video recorded from three different angles. A total of 8 short video sequences are available, which allows only the evaluation of 4 categories (soda can, cap, bowl, and coffee mug) and 20 instances. However, this dataset does not appear to have noticeable viewpoint, background and lighting variability.

[1] An example dataset generated using the proposed framework will be publicly available upon the publication of the paper at hand.

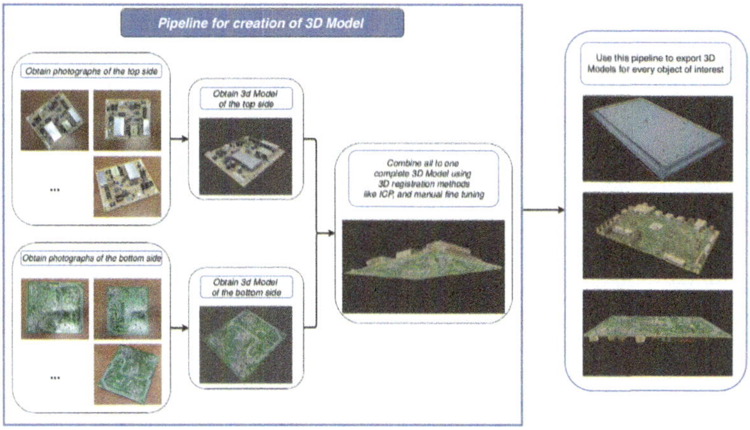

Fig. 3. The pipeline for creating photorealistic 3D models of flat and bulk objects using photogrammetry.

Browatzki et al. in [6] introduced a dataset of 18 types of items found typically in office and household environments, represented by 3–14 instances for each category (a total of 154 objects). For each object they obtained 36 images using a turntable (one image every 10°) leading to 5544 different views. For each view they provide 2 high resolution RGB images and a low resolution depth image. Despite the large amount of objects, there is a noticeable lack in variability of backgrounds and illumination (all objects were recorded on a turntable, in controlled luminance conditions).

A few researchers have explored training vision algorithms using synthetic computer generated data. Michels et al. [12] utilized 3D models of outdoor scenes to create synthetic images for autonomous driving. Agarwal et al. [5] superimposed human models on real backgrounds and used them to train models for pose estimation. Saxena et al. [14] used synthetic images for training, and result in robotic grasping of novel objects. Mariolis et al. [11] investigated cross-domain knowledge transfer utilizing a combination of synthetic and real depth data for training deep networks to recognize category and pose of highly deformable objects. They investigated the case of garments hanging from a single point, and using a hierarchy of deep convolutional neural networks they could estimate the pose and the category.

All aforementioned approaches, used either real of synthetic data to train vision algorithms for robotic applications. Nevertheless, there was a distinct absence of variation in backgrounds and lighting, that can be effectively assured using our framework for synthetic dataset generation.

3 Ensuring Photo-Realistic 3D Models

In order to simulate data acquisition of a real object of interest, the 3D model of this object is a necessity. The more photo-realistic the model, the more resemblance the resulting dataset will have to an equivalent real life acquired one. To create a photo-realistic model we employed photogrammetry [13].

Photogrammetry is the process used to create 3D models of objects or scenes from multiple overlapping images of them. The underlying principle is quite close to how many cameras today enable you to construct a panorama by combining overlapping images. Photogrammetry takes the principle further by using the structure from motion [15], using the camera position as it travels through 3D space to approximate the 3D (X, Y and Z) coordinates for each pixel of the original image (Fig. 1).

The aforementioned procedure produces really good results when the object used has some 3D volume, but it suffers with relatively flat objects. This is because, while the camera rotates around the object to capture the images, each image can overlap with the previous and the next one if the object has some volume. In the case of flat objects, there will be many photographs during the rotation of the camera that the object will be parallel to the line of sight and only a small part of it will be visible. In these cases, the photogrammetry algorithms have trouble matching the features of the object in this parallel position with the features previously observed in the images that the object is near perpendicular to the view of the camera (Fig. 4).

In this work we are focusing in perception for robotic assembly tasks, where PCB boards are manipulated by the robot for assembling an LCD TV. Thus, during the acquisition of this dataset, we mainly had flat objects to work with. To overcome this problem, but also ensure that we obtain a photo-realistic 3D model for each object, firstly, we placed the objects on a flat surface (e.g. the TV had to be placed on a table due to its weight), or had them hanging from a high point. Then, we obtained around 100 photos for each object, using different camera orientation and distances from them. We processed the images using the free version of photogrammetry tool 3D Zephyr [1], tuning the parameters of feature matching in order to ensure that all the acquired images can be registered successfully. In this way we extracted a 3D model for each side of a flat object. We then used a variant of Iterative Closest Point [7] to stitch this models together, creating a complete model. This process resulted in high detailed, photo-realistic textured 3D models for all the objects of interest (Fig. 3).

4 Ensuring Photo-Realistic Lighting Conditions

In the past years there have been many developments in the computer graphics industry. New tools, new algorithms, the use of Artificial Intelligence (AI), particularly neural networks, were some parts of the evolution that leads to photorealism in Computer Generated Imagery (CGI). For many years 3D artists had to manually add light sources and fine tune them in order to resemble true lighting conditions in their 3D scenes. Indoor scenes, where natural light coexists with artificial, are always more challenging. Since there is not a single source of illumination that casts light all over the scene, but multiple ones, with different intensities. Taking this into account, combined with the light dispersed

Fig. 4. HDRI of an indoor office taken from [3] (top row). A sample 3D object with various types of materials lighted by this HDRI (bottom row). The reflections of this environment are also visible when reflective materials are used (third material from the left).

and reflected on and from the surfaces inside the 3D scene, the illumination of a photorealistic indoor 3D scene is a very demanding task.

Recent developments introduced the use of High Dynamic Range Imaging (HDRI) for achieving accurate illumination in 3D scenes. HDRI is a panoramic image that incorporates all viewpoints from a single point. It incorporates a vast amount of data (usually 32 bits per pixel per channel) that can then be used to illuminate the CG scene. Creating high-quality HDRI from start is a challenging task that requires specialized tools and careful monitoring of the entire process. But currently there are many online repositories, used by the majority of 3D artists for photorealistic lighting, that can provide high quality HDRIs free of charge (e.g. [3]). To our knowledge, the proposed HDRI approach for generating photo-realistic synthetic data for training machine vision algorithms, has not yet been employed in the related literature.

5 Dataset Generation Framework Description

In order to decide the type of data that we need to include in these synthetic datasets, we took into account various machine learning algorithms and the types of data that the majority of them require for their training. Thus, in the proposed dataset generation framework, all known information needed in order

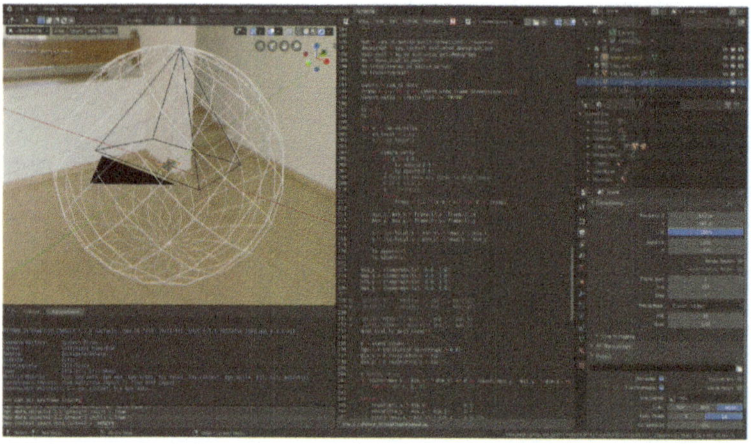

Fig. 5. The Blender setup for rendering the object using 3 different distances for the camera.

to train most common state-of-the-art neural networks and algorithms for object recognition are present. The generated datasets can be used to train deep neural networks for object recognition, object tracking, and object detection algorithms, and since they include ground truth information (such as object pose, bounding box etc.) it can also be used for testing. The proposed framework can also allow multiple objects of interest interacting in the same scene.

The synthetic dataset in our robotic assembly use case includes three objects (two PCB boards and a TV frame), 15 different HDRIs were used to render the objects in different lighting conditions, and 3 different camera distances for the object. The type of data that this dataset provides for every different camera position (for every image) is as follows: RGB image, Depth image, Depth image with added noise, Mask image, Bounding box, Bounding polygon, Object pose, and the Camera pose. Additionally, the camera intrinsic parameters, all the used HDRIs, and the detailed textured 3D models of the objects are provided.

6 Dataset Acquisition

6.1 Single Object Datasets for Training

To acquire single object datasets we utilized Blender [2] and it's Python API. All 3D models were individually imported into Blender, and since we chose 3 different camera to object distances, 3 spheres with these distances as a radius were created (Fig. 5). In addition, we created a python script, that used Blender's python API to keyframe the camera positions and iterate through the selected HDRIs.

As a first step we decided that the 3 distances that we were going to use was 2.0 m, 2.2 m, and 2.5 m. Three spheres were created, with these as radius

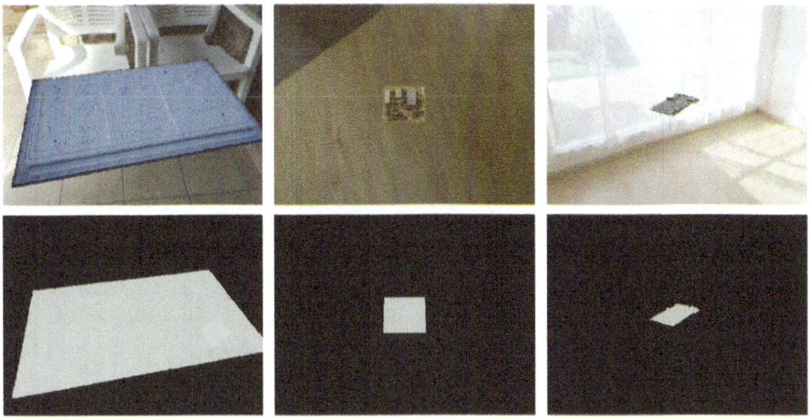

Fig. 6. Examples of RGB (top row) and Mask (bottom row) images from thesingle object datasets. From illumination with different HDRIs, there are noticeable differences in the type and color of light that these scenes have.

and 242 vertices each. For every vertex a key-frame was added to the camera position and rotation, so the camera, frame by frame would move from vertex to vertex and in each of them would look at the object (to ensure the object is always within camera's field of view). The camera intrinsic parameters were chosen to simulate the Orbbec Astra camera [4]. The whole scene was illuminated by iterating through 15 different indoor HDRIs. The HDRIs were also used to provide realistic backgrounds for the RGB images.

In every key-framed camera position we rendered: RGB, Depth, Mask images and also saved the camera pose and the object pose. In post processing we introduced noise to the depth image, using a Gaussian distribution with $\mu = 0$ and $\sigma = 30$, and extracted the bounding box and minimum bounding polygon from the mask images. So in total the main dataset includes $242 \times 3 \times 15 = 10890$ frames for each object, accompanied by the aforementioned extra data (Fig. 6).

6.2 Multiple Object Datasets for Testing

In order to evaluate object detection models that are trained with our single object dataset, we also created test datasets that include all multiple objects in the same scene, mimicking the real scenario of our use case, where the two PCB boards are mounted on the TV frame. The three test datasets were also created with Blender, by animations that included the PCBs moving one by one from a storage position to their position on the TV frame, until both are in contact with the TV frame. In these scenarios the camera was placed above the objects in a fixed position and orientation. The TV frame was fixed to a planar surface, and the two PCB boards were placed next to it.

In the first test dataset the camera view is not obstructed by anything, the PCB boards move one by one from their position to their final position on the

Fig. 7. Examples of RGB (top row) and Mask (bottom row) images from the multiple object datasets where the objects interact with each other and the view is: occluded by 2 beams hanging from the ceiling, with added self occlusion, and unobstructed.

TV frame. Their movement is such that the PCB is always perpendicular to the line on sight of the camera, thus, the PCB is always fully visible. In the second scenario we added a twist in the PCB movement, so they aren't always completely visible from the camera as they move to reach their final position. Finally, at the third test dataset we added two beams between the camera and the TV (the use case scenario implies an industrial environment) so the camera view is obstructed partially but the beams. In this scenario, the PCB boards also twist as they move. Thus, there are frames that the PCBs are partially occluded by a beam and also in an angle that makes them partially visible from the camera due to self-occlusion (Fig. 7).

7 Discussion and Future Prospects

The proposed synthetic dataset generation framework provides a complete pipeline for acquiring 2.5D photo-realistic data for training and testing object detection and tracking algorithms. It includes all possible types of information the current state-of-the-art machine learning methods need for training. Since currently deep neural network based methods are the majority of object detection algorithms, it is a necessity to explore methods that make the data acquisition for training a process less time and resource consuming. With the extreme improvement of computer graphics it comes as natural to exploit this for computer vision. Currently the majority of synthetic datasets with 2.5D data are confined to task specific data. Using the propose framework, one can produce large scale datasets about any object pool, and in any conditions. In that direction, we plan to incorporate the whole methodology to one specific tool which will enable the user to extract datasets using a graphical user interface. As a next step, we intend to augment these datasets with data captured in real conditions in order to investigate cross-domain knowledge transfer and get quantitative evaluation of the improvement of the trained algorithms due to the use of the HDRI approach.

Acknowledgement. This work has been supported by the European Union's Horizon 2020 research and innovation programme funded project namely: "Co-production CeLL performing Human-Robot Collaborative AssEmbly (CoLLaboratE)" under the grant agreement with no: 820767.

References

1. 3DFZephyr (2020). https://www.3dflow.net/3df-zephyr-pro-3d-models-from-photos/. Accessed 30 Apr 2020
2. Community, B.O.: Blender - a 3D modelling and rendering package. Stichting Blender Foundation, Amsterdam (2018). http://www.blender.org. Accessed 30 Apr 2020
3. HdriHaven (2020). https://hdrihaven.com/. Accessed 30 April 2020
4. Orbec: Orbec structured light camera (2020). https://orbbec3d.com/product-astra-pro/. Accessed 30 Apr 2020
5. Agarwal, A., Triggs, B.: A local basis representation for estimating human pose from cluttered images. In: Asian Conference on Computer Vision, pp. 50–59. Springer (2006)
6. Browatzki, B., Fischer, J., Graf, B., Bülthoff, H.H., Wallraven, C.: Going into depth: evaluating 2D and 3D cues for object classification on a new, large-scale object dataset. In: 2011 IEEE International Conference on Computer Vision Workshops (ICCV Workshops), pp. 1189–1195. IEEE (2011)
7. Chetverikov, D., Stepanov, D., Krsek, P.: Robust Euclidean alignment of 3D point sets: the trimmed iterative closest point algorithm. Image Vis. Comput. **23**(3), 299–309 (2005)
8. Freedman, B., Shpunt, A., Machline, M., Arieli, Y.: Depth mapping using projected patterns, 23 July 2013, US Patent 8,493,496
9. Keselman, L., Iselin Woodfill, J., Grunnet-Jepsen, A., Bhowmik, A.: Intel realsense stereoscopic depth cameras. In: Proceedings of the IEEE Conference on Computer Vision and Pattern Recognition Workshops, pp. 1–10 (2017)
10. Lai, K., Bo, L., Ren, X., Fox, D.: A large-scale hierarchical multi-view RGB-D object dataset. In: 2011 IEEE International Conference on Robotics and Automation, pp. 1817–1824. IEEE (2011)
11. Mariolis, I., Peleka, G., Kargakos, A., Malassiotis, S.: Pose and category recognition of highly deformable objects using deep learning. In: 2015 International Conference on Advanced Robotics (ICAR), pp. 655–662. IEEE (2015)
12. Michels, J., Saxena, A., Ng, A.Y.: High speed obstacle avoidance using monocular vision and reinforcement learning. In: Proceedings of the 22nd International Conference on Machine Learning, pp. 593–600 (2005)
13. Pollefeys, M., Gool, L.V.: From images to 3D models. Commun. ACM **45**(7), 50–55 (2002)

14. Saxena, A., Driemeyer, J., Ng, A.Y.: Robotic grasping of novel objects using vision. Int. J. Robot. Res. **27**(2), 157–173 (2008)
15. Schonberger, J.L., Frahm, J.M.: Structure-from-motion revisited. In: Proceedings of the IEEE Conference on Computer Vision and Pattern Recognition, pp. 4104–4113 (2016)

Control of Industrial AGV Based on Reinforcement Learning

Jesus Enrique Sierra-García[1] and Matilde Santos[2(✉)]

[1] University of Burgos, Burgos, Spain
jesierra@ubu.es
[2] Institute of Knowledge Technology, Complutense University of Madrid, Madrid, Spain
msantos@ucm.es

Abstract. Automatic Guided Vehicles (AGV) suffer degradation in their electro-mechanical components which affect the navigation performance over time. The use of intelligent control techniques can help to alleviate this issue. In this work a new approach to control an AGV based on reinforcement learning (RL) is proposed. The space of states is defined using the guiding error, and the set of control actions provides the reference for the velocities of each wheel. Two different reward strategies are implemented, and different updating policies are tested. Simulation results show how the RL controller is able to successfully track a complex trajectory. The controller has been compared with a PID obtaining better results.

Keywords: Reinforcement learning · Automatic Guided Vehicles (AGV) · Control · Q-learning

1 Introduction

Automatic Guided Vehicles (AGV) are unmanned transport vehicle mainly used in the industrial sector to replace manned trucks and conveyors. These autonomous vehicles can help to make industrial processes more efficient and to reduce human errors and operative costs. They have become very popular in recent years under the industry 4.0 approach [1]. The AGVs play a major role in the temporal and spatial flexibility requested by this new paradigm. For these and other reasons, the research on AGVs modelling and control is becoming more and more interesting and useful [2].

Industrial AGVs are usually controlled by conventional PID regulators. These control techniques, though effective, usually demand high efforts on calibration. Moreover, the parameters of the AGV are not constant, the size of the wheels are reduced by the friction and the payloads, and the electro-mechanical components suffer degradation over time. All these factors may worsen the navigation performance of these vehicles.

Therefore, adaptive controllers are necessary to address these issues. Artificial intelligent techniques in general, and reinforcement learning in particular, have been proved efficient with theses complex problems [3–6]. Reinforcement learning seems to be a good strategy to improve the guiding over time. In this work a new approach to control

Á. Herrero et al. (Eds.): SOCO 2020, AISC 1268, pp. 647–656, 2021.
https://doi.org/10.1007/978-3-030-57802-2_62

an AGV by reinforcement learning (RL) is proposed. The space of states is defined using the guiding error, and the set of control actions generates the speed reference for each wheel. Two different reward strategies are proposed, and several updating policies are tested and compared with a PID controller with encouraging results.

The works on AGV control found in the literature are usually focused on kinematics and control, although navigation algorithms, power storage, and charging systems are also dealt with. Since the work by Maxwell and Muckstadt [7], several control approaches have been presented. In [8], a kinematical and dynamical analysis of a tricycle mobile robot is shown. A comparison of control techniques for robust docking maneuvers of an AGV is presented in [9]. Espinosa uses event-based control approach for an indoor mobile robot [10]. A deep state-of-art about localization of AGVs is presented in [11]. In [12] non-linear observers are used to control the traction in an electrical vehicle. Other works are devoted to study related problems, such as manufacturing process scheduling for multi-AGV using RL [13] or the control of a fleet of AGVs [14], where a detailed state-of-art on the design and control of AGVs is also presented. But these papers do not apply reinforcement learning for controlling the AGV.

The rest of the paper is organized as follows. In Sect. 2 the equations that describe the AGV and the environment are presented. Section 3 describes the reinforcement learning approach that is applied. Simulation results are presented and discussed in Sect. 4. The document ends with the conclusions and future works.

2 System Description

2.1 AGV Model

The Easybot AGV model of the ASTI Mobile Robotics company is going to be used [15] (Fig. 1). The kinematics of the Easybot AGV is a combination of a differential vehicle and a tricycle. The traction unit is a differential robot, but the body is linked with the traction unit by a shaft and it revolves around it. The movement of the body around this pivot performs as a tricycle vehicle. In this work we will focus on the control of the traction unit by reinforcement learning techniques.

Fig. 1. Easybot model

The equations of the movement of the traction unit are given by the kinematic model of a differential robot [16]:

$$\dot{x}_h = \frac{V_l + V_r}{2}\cos(\Phi_h) \tag{1}$$

$$\dot{y}_h = \frac{V_l + V_r}{2}\sin(\Phi_h) \tag{2}$$

$$\dot{\Phi}_h = \frac{V_r - V_l}{L_h} \tag{3}$$

Where V_l and V_r are the longitudinal velocity of the left and right wheels (m/s); L_h is the distance between the wheels (m), and the set of variables $[x_h, y_h, \Phi_h]$ denotes the position (m) and attitude (rad) of the center of the imaginary line which links both wheels in the traction unit in a 2D inertia coordinate system (Fig. 2).

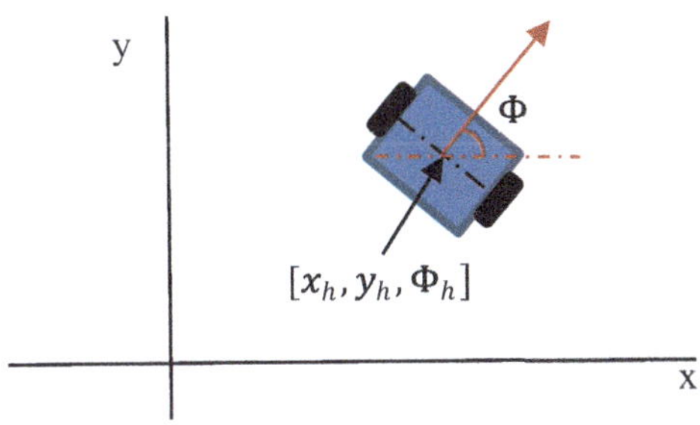

Fig. 2. Coordinate system of the traction unit

In order to control the movement of the traction unit the control signals V_c and w_c are used. V_c is the translational reference velocity (m/s) and w_c is the angular speed reference (rad/s). From this references the target wheel speeds, V_{l_c} and V_{r_c} (m/s), are obtained using Eqs. (4) and (5).

$$V_{l_c} = \frac{2V_c - w_c \cdot L_h}{2} \tag{4}$$

$$V_{r_c} = \frac{2V_c + w_c \cdot L_h}{2} \tag{5}$$

The guiding system of this AGV provides information about the deviation between the AGV and the current route in the working space. Different types of sensors can be used: optical sensors, to follow a line painted on the floor; magnetic sensors, to follow a magnetic tape placed on the floor; inductive sensors, to follow a buried cable, or even more advanced measurement systems based on SLAM technologies to follow a virtual line. In this work the Easybot robot is equipped with a magnetic sensor, but equivalent results could be obtained with any other sensor. The magnetic sensor gives the error guiding signal, err_{gui}, which is measured from the center of the magnetic tape to the center of the guiding sensor (Fig. 3).

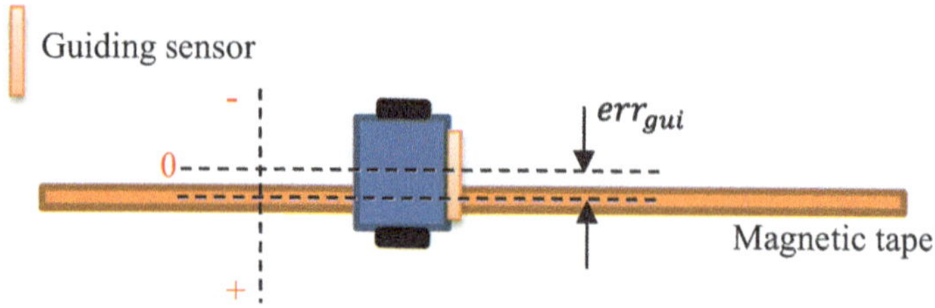

Fig. 3. Guiding error measurement

The translational reference velocity V_c is provided by the user, and it is normally defined by sections. Thus, the angular speed is going to be adjusted to correct the deviation from the path. In this work we apply a reinforcement learning control strategy to follow the reference, and it will be compared to a PID control action.

$$w_c = K_P \cdot err_{gui} + K_D \cdot d\frac{err_{gui}}{dt} + K_I \cdot \int err_{gui}dt \tag{6}$$

2.2 Workspace

The workspace scenario is a magnetic tape loop (in green) with a charging station and a traffic light (Fig. 4). The charging station recovers the energy consumption and this way the AGV does not need to leave the circuit. In the simulations the charging station and the traffic light will be obviated to focus on the guiding problem.

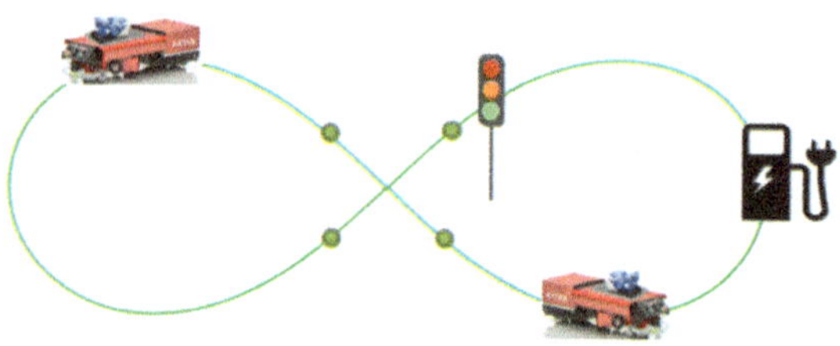

Fig. 4. Workspace of the AGV

The trajectory described by the AGV (Fig. 4) is a Bernouilli's lemniscate. It can be described as the set of P points where the product of the distance to two focus, $F1$ and $F2$, located at a distance $2a$ between them, is a^2 (7).

$$\left(x^2 + y^2\right)^2 - 2a^2\left(x^2 - y^2\right) = 0 \tag{7}$$

The intersection between the straight line projected by the guiding sensor and the lemniscate is used to calculate the guiding error. To calculate the crossing points, the straight line equation $y = mx + b$ is substituted in (7) and a fourth degree polynomial is obtained (8):

$$k_4x^4 + k_3x^3 + k_2x^2 + k_1x + k_0 = 0 \tag{8}$$

The solution of this polynomial provides the x coordinate of the intersection points. The constants k_0 to k_4 of the polynomial are given by the following expressions:

$$k_4 = m^4 + 2m^2 + 1 \tag{9}$$

$$k_3 = 4m^3 b + 4mb \tag{10}$$

$$k_2 = b^2\left(2 + 6m^2\right) - 2a^2\left(1 + m^2\right) \tag{11}$$

$$k_1 = 4m(b^3 + a^2 b) \tag{12}$$

$$k_0 = b^4 + 2a^2 b^2 \tag{13}$$

3 Reinforcement Learning Control Approach

The reinforcement learning approach consists of an environment, an agent and an interpreter. The agent, taking into consideration the current state of the environment and the previous rewards, selects the best action to carry out. This action produces an effect on the environment. This fact is observed by the interpret who provides information to the agent about the new state and the reward of the previous action, closing the loop [17]. Some authors consider that the interpreter is embedded in either the environment or the agent; anyway, the function of the interpret is always present.

Discrete reinforcement learning can expressed as follows [18]:

- S is a finite set of states
- A is a finite set of actions
- s_t is the state at t
- a_t is the action executed when the agent knows the environment at state s_t
- r_{t+1} is the reward received after action a_t is carried out.
- s_{t+1} is the state after action a_t is carried out.
- The environment or world is a Markov process:

$$MDP = \langle s_0, a_0, r_1, s_1, a_1, r_2, s_2, a_2 \ldots \rangle$$

- $\pi : S \times A \rightarrow [0, 1]$ is the policy; this function provides the probability of selection of action a for every pair (s, a).
- $p^a_{ss'} = \Pr\{s_{t+1} = s' | s_t = s \wedge a_t = a\}$ is the probability that a state changes from s to s' with action a
- $p^\pi(s', a')$ is the probability of selecting action a' at state s' under policy π
- $r^a_s = E\{r_{t+1} | s_t = s \wedge a_t = a\}$ is the expected one-step reward
- $Q^\pi_{(s,a)} = r^a_s + \gamma \sum_{s'} p^a_{ss'} \sum_{a'} p^\pi(s', a') Q^\pi_{(s',a')}$ is the expected sum of discounted rewards

The objective of the reinforcement learning algorithm is to find the best policy π^* that maximizes $Q^\pi_{(s,a)}$ for every state; formally:

$$\pi^* = \arg MAX_\pi \left[Q^\pi_{(s,a)}\right] \forall s \in S \tag{14}$$

3.1 Space of States and Set of Actions

From Eqs. (1–6) several state variables may be initially identified: x_h, y_h and Φ_h. However, this internal variables of the system are not accessible to the AGV, that is, they are not observable by the reinforcement learning agent and thus they are discarded.

The only observable variable by the AGV is the error signal err_{gui}, measured by the guiding sensor. Then the state variable s is obtained by the discretization of err_{gui} and its derivative $\dot{err}_{gui}$ (15–17):

$$err_{gui_S}(t) = MIN\left(err_{gui_{MAX}}, MAX\left(err_{gui_{MIN}}, err_{gui}(t)\right)\right) \tag{15}$$

$$\dot{err}_{gui_S}(t) = MIN\left(\dot{err}_{gui_{MAX}}, MAX\left(\dot{err}_{gui_{MIN}}, \dot{err}_{gui}(t)\right)\right) \tag{16}$$

$$s_t = \begin{bmatrix} s1_t \\ s2_t \end{bmatrix} = \begin{bmatrix} DIV\left(\left[err_{gui_S}(t) - err_{gui_{MIN}}\right] \cdot N_{s1}, err_{gui_{MAX}} - err_{gui_{MIN}}\right) \\ DIV\left(\left[\left[\dot{err}_{gui_S}(t) - \dot{err}_{gui_{MIN}}\right] \cdot N_{s2}, \dot{err}_{gui_{MAX}} - \dot{err}_{gui_{MIN}}\right)\end{bmatrix} \tag{17}$$

Where $[err_{gui_{MIN}}, err_{gui_{MAX}}, \dot{err}_{gui_{MIN}}, \dot{err}_{gui_{MAX}}]$ defines the range of the controller variables, being $err_{gui_{MIN}} < err_{gui_{MAX}}$ and $\dot{err}_{gui_{MIN}} < \dot{err}_{gui_{MAX}}$; DIV denotes the integer division, the first argument is the dividend and the second one is the divisor. The range of err_{gui} is divided by N_{s1} and the range of $\dot{err}_{gui}$ is divided by N_{s2}. This way $S \in [n \in \mathbb{N} \cup 0 | n \langle N_{s1}] \times [n \in \mathbb{N} \cup 0 | n < N_{s2}]$.

The control action can be designed using the speed reference of each wheel (low level control), or using the translational and angular speed references (medium level control). In our case we have implemented the low level control, with control action a given by the references $\left[V_{l_c}, V_{r_c}\right]$. The set A is designed to cover all the range of movements of the AGV, here reduced to:

$$A = \{[V_c, V_c], [V_c, -V_c], [-V_c, V_c]\} \tag{18}$$

Where V_c is the translational reference velocity (4–5). Therefore, the AGV can only go forward or rotate clockwise and anti-clockwise.

3.2 Design of the Reward Strategy

Since the only observable variable is err_{gui}, it will be used to design the reward strategy. Two different reward strategies have been tested. The first one does not punish any action but it only rewards. The reward decreases when the error grows, up to 0. We call it "not-punish" reward strategy. The equation to compute the reward is given by:

$$r_t = K_{r1} \cdot \left(err_{gui_{MAX}} - \left|err_{gui_S}(t)\right|\right) \tag{19}$$

where K_{r1} is a constant which can be used to adjust the weight of err_{gui_S} in the reward.

The second strategy takes into consideration if the AGV is either approaching or moving away from the magnetic tape line. If it is approaching the path line, it receives a reward otherwise it is punished (negative reward). The reward is proportional to the approaching and moving-away speed ("speed reward" strategy), given by

$$r_t = \begin{cases} -K_{r2} \cdot \dot{err}_{gui}(t) & err_{gui}(t) > 0 \\ K_{r2} \cdot \dot{err}_{gui}(t) & err_{gui}(t) \leq 0 \end{cases} \tag{20}$$

where K_{r2} is a constant which can be used to adjust the weight of the derivative of err_{gui} in the reward.

3.2.1 Methods to Update the Policy

Since the space of states is discrete, the policy is codified with a table $T^{\pi}_{(s,a)} : S \times A \rightarrow \mathbb{R}$, which relates each pair (s, a) with a measurement of the expected reward. At each time the action with greater long term expected reward is selected (21-22):

$$T^{\pi}_{(s_{t-1},a_{t-1})}(t) := f_{\pi}\left(T^{\pi}_{(s_{t-1},a_{t-1})}(t-1), r_t\right) \tag{21}$$

$$a_t = argMAX_a\left(T^{\pi}_{(s_t,a_t)}(s_t)\right) \tag{22}$$

where f_{π} is the method to update the policy, i.e., to update the long term expected reward of table $T^{\pi}_{(s,a)}$. Five different methods to update the policy have been tested. The first one only considers the last one step (OS) reward,

$$OS : T^{\pi}_{(s_{t-1},a_{t-1})}(t) = r_t \tag{23}$$

The second method considers all the previous rewards, "not-forgotten" (NF):

$$NF : T^{\pi}_{(s_{t-1},a_{t-1})}(t) = T^{\pi}_{(s_{t-1},a_{t-1})}(t-1) + r_t \tag{24}$$

The third method considers all the previous rewards, but only a percentage of each reward given by the learning rate parameter $\alpha \in \mathbb{R}[0, 1]$. We call this method OL-LR, "only learning with learning rate"

$$OL - LR : T^{\pi}_{(s_{t-1},a_{t-1})}(t) = T^{\pi}_{(s_{t-1},a_{t-1})}(t-1) + \alpha \cdot r_t \tag{25}$$

The fourth method uses the learning rate α and the value $(1-\alpha)$ to forget. Thus, the "learning-forgetting with learning rate" (LF-LR) is:

$$LF - LR : T^{\pi}_{(s_{t-1},a_{t-1})}(t) = (1-\alpha) \cdot T^{\pi}_{(s_{t-1},a_{t-1})}(t-1) + \alpha \cdot r_t \tag{26}$$

The last method is the Q-learning algorithm, where γ is the discount factor:

$$a_{max} = argMAX_a\left(T^{\pi}_{(s_t,a)}(t-1)\right) \tag{27}$$

$$QL : T^{\pi}_{(s_{t-1},a_{t-1})}(t) = (1-\alpha) \cdot T^{\pi}_{(s_{t-1},a_{t-1})}(t-1) + \alpha\left(r_t - \gamma \cdot T^{\pi}_{(s_{t-1},a_{max})}(t-1)\right) \tag{28}$$

4 Simulation Results

Simulation results have been obtained using Python/Spyder software. The simulation is run until the AGV arrives at the destination point. The origin coordinates are [2.15, 1.25] and the destination is [2, 1.24], so the path is almost a complete loop. It is important to remark that the AGV knows neither the origin nor the destination points, and neither its position during the trip. It only uses the guiding sensor information. The sampling time is set to 10 ms. During the simulation the maximum speed of each wheel is limited to 2 m/s and the values of $\left[err_{gui_{MIN}}, err_{gui_{MAX}}, \dot{err}_{gui_{MIN}}, \dot{err}_{gui_{MAX}}\right]$ are, respectively, $[-0.2, 0.2, -0.5, 0.5]$.

Figure 5 shows the trajectory followed by the AGVs when the controller based on reinforcement learning is applied. The reward strategy is "speed reward", and $(N_{s1}, N_{s2}) = (10, 5)$. The policy update method is Q-learning with $[\alpha, \gamma] = [0.5.0.1]$. The reference is represented in red, the trajectory followed by the AGV is the black line, the origin point is the green circle and the destination is blue circle. It is possible to observe how the AGV tracks the lemniscate successfully.

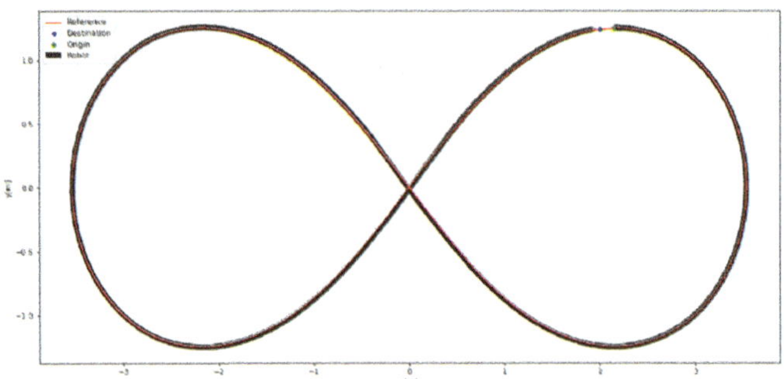

Fig. 5. Trajectory followed by the AGV with the reinforcement learning controller

Figure 6 (left) shows the trajectory followed by the AGVs (black) when a PID controller (6) with [KP, KD, KI] = [2, 5, 10] is applied. The track is good bad, but worse than with the reinforcement controller. Besides, the performance of the PID controller is very sensitive to the tuning. Figure 6 (right) shows the tracking of the same trajectory with the PID tuning parameters [10, 0, 39]. The response presents overshoot.

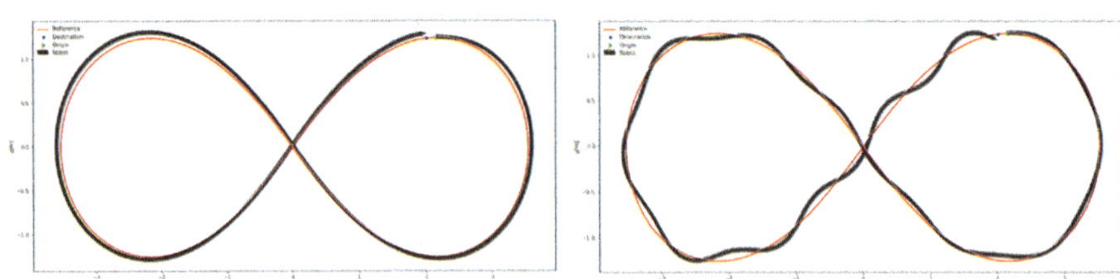

Fig. 6. Trajectory followed by the AGV with the PID controller

In addition, several experiments were carried out to measure the performance of the RL controller with different updating policies and reward strategies (Table 1). The parameter α is set to 0.5 and parameter $\gamma = 0.1$. The best results have been boldfaced. The column time indicates the time spent to reach the destination position.

Table 1. Results for different updating policies and reward strategies

Reward strategy	Policy update	Time	MSE	Time*MSE
"not–punish"	OS	20.83	2.26e−4	4,71e−03
	NF	–	–	–
	OL-LR	–	–	–
	LF-LR	21.9	1.86e−4	4,07e−03
	QL	20.29	2.46e−4	4,99e−03
"Speed reward"	OS	20.39	2.42e−4	4,93e−03
	NF	24.03	**1.25e−4**	**3,00e−03**
	OL-LR	24.03	**1.25e−4**	**3,00e−03**
	LF-LR	20.39	2.42e−4	4,93e−03
	QL	20.35	2.35e−4	4,78e−03
PID		**19.33**	4.01e−3	7.73e−02

One remarkable result is that NF and OL-LR do not converge with the "not-punish" reward strategy. This may be because the rewards with this strategy are always positive, and then when it is combined with either NF or OL-LR, the expected reward only can increase and never decreases.

The best time is obtained by the combination ("not-punish", QL) and the best MSE and Time*MSE is given by the combination ("speed reward", NF) and ("speed reward", OL-LR). All the configurations give smaller MSE and Time*MSE than with the PID, however time to destination is larger.

5 Conclusions and Future Works

Automatic guiding vehicles, AGVs, are very useful and widely used in logistics and production chains. However, their effectiveness may decrease due to the degradation of their electro-mechanical components over time, that negatively affects the guiding performance. Intelligent control techniques such as reinforcement learning can help to reduce these issues. The controller learns how to optimize the control actions over time and to modify the control law in order to adapt it to the changes in the system.

In this work a novel AGV control approach based on RL is implemented. The space of states and the set of actions are identified based on a mathematical model of the AGV kinematics and the environment. Two different reward strategies, based on the guiding error, are proposed, one of them without punishment. In addition, five different updating policies are described and the combination of them with the two reward strategies are tested and compared with a PID control. Simulation results validate the effectiveness of the RL controller to track a lemniscate trajectory.

Among other possible future works we may highlight consider dynamic trajectories, the modification of the controller to avoid obstacles and the control of a fleet of AGVs.

References

1. Bechtsis, D., Tsolakis, N., Vouzas, M., Vlachos, D.: Industry 4.0: sustainable material handling processes in industrial environments. In: Computer Aided Chemical Engineering, vol. 40, pp. 2281–2286. Elsevier (2017)
2. Theunissen, J., Xu, H., Zhong, R.Y., Xu, X.: Smart AGV system for manufacturing shopfloor in the context of industry 4.0. In: 2018 25th International Conference on Mechatronics and Machine Vision in Practice (M2VIP), pp. 1–6. IEEE, November 2018
3. Sierra, J.E., Santos, M.: Modelling engineering systems using analytical and neural techniques: hybridization. Neurocomputing **271**, 70–83 (2018)
4. Santos, M., López, V., Botella, G.: Dyna-H: a heuristic planning reinforcement learning algorithm applied to role-playing game strategy decision systems. Knowl. Based Syst. **32**, 28–36 (2012)
5. Martín-H, J.A., de Lope, J., Santos, M.: A method to learn the inverse kinematics of multi-link robots by evolving neuro-controllers. Neurocomputing **72**(13–15), 2806–2814 (2009)
6. Santos, M.: An applied approach of intelligent control. Revista Iberoamericana de Automática e Informática Industrial RIAI **8**(4), 283–296 (2011)
7. Maxwell, W.L., Muckstadt, J.A.: Design of automatic guided vehicle systems. IIE Trans. **14**, 114–124 (1982)
8. Bonilla, M., Reyes, F., Mendoza, M.: Modelling and simulation of a wheeled mobile robot in configuration classical tricycle. In: Proceedings of 5th WSEASA International Conference on Instrumentation, Measurement, Control, Circuits and Systems (2005)
9. Villagra, J., Herrero-Pérez, D.: A comparison of control techniques for robust docking maneuvers of an AGV. IEEE Trans. Control Syst. Technol. **20**(4), 1116–1123 (2011)
10. Espinosa Zapata, F., Lázaro Galilea, J.L., Olivares Bueno, J.: ALCOR project: contributions to optimizing remote robot guidance in intelligent spaces. Revista Iberoamericana de Automática e Informática Industrial **15**(4), 416–426 (2018)
11. Durrant-Whyte, H., Rye, D., Nebot, E.: Localization of autonomous guided vehicles. In: Robotics Research, pp. 613–625. Springer, London (1996)
12. Aligia, D.A., Magallán, G.A., De Angelo, C.H.: Traction control of an electric vehicle based on nonlinear observers. Revista Iberoamericana de Automática e Informática Industrial **15**(1), 112–123 (2018)
13. Xue, T., Zeng, P., Yu, H.: A reinforcement learning method for multi-AGV scheduling in manufacturing. In: 2018 IEEE International Conference on Industrial Technology (ICIT), pp. 1557–1561. IEEE, February 2018
14. Vis, I.F.: Survey of research in the design and control of automated guided vehicle systems. Eur. J. Oper. Res. **170**(3), 677–709 (2006)
15. ASTI Mobile Robotics 2020. https://www.astimobilerobotics.com/
16. Oriolo, G.: Control of nonholonomic systems (2019). https://www.dis.uniroma1.it/~oriolo/cns/cns_slides.pdf
17. Alvarez-Ramos, C.M., Santos, M., López, V.: Reinforcement learning vs. A* in a role playing game benchmark scenario. In: Computational Intelligence: Foundations and Applications, pp. 644–650 (2010)
18. Chen, C., Dong, D., Li, H.X., Chu, J., Tarn, T.J.: Fidelity-based probabilistic Q-learning for control of quantum systems. IEEE Trans. Neural Netw. Learn. Syst. **25**(5), 920–933 (2013)

Shared Control Framework and Application for European Research Projects

Mauricio Marcano[1,2(✉)], Sergio Diaz[1], Myriam Vaca[1,2], Joshué Pérez[1], and Eloy Irigoyen[2]

[1] TECNALIA, Basque Research and Technology Alliance (BRTA), 48160 Derio, Spain
{mauricio.marcano,sergio.diaz,myriam.vaca,joshue.perez}@tecnalia.com
[2] University of the Basque Country, 48013 Bilbao, Spain
eloy.irigoyen@ehu.eus

Abstract. Current commercial Advanced Driver Assistance Systems (ADAS) assist the driver indirectly through warning signals. However, a new generation of ADAS and Automated Driving applications, known as Shared Control, where driver and automation control the vehicle together, have the potential to influence upcoming functionalities, improving the driving performance and reducing the driver's physical and mental workload. The development of such a system has the attention of the European Commission, and different Research Innovation Actions (RIA) are developing new technologies for the human-centered design of partially and highly-automated vehicles. In particular, the PRYSTINE and HADRIAN projects are facing the challenge of sharing the authority of the dynamic driving task between driver and automation. In this sense, a common approach is shared between these projects to combine the necessary systems for a complete collaborative driver-automation framework. The integration of a Driver Monitoring System, a cooperative HMI, and a Shared Control System is part of their goals. In particular, the control system in charge of changing the control authority will be presented in this article for a collaborative overtaking scenario, analyzing two modalities: a collision-avoidance system, and a control transition system. Results, discussion, and future challenges are presented.

Keywords: Shared control · Autonomous vehicles · Driver-automation cooperation · Arbitration · Partially automated vehicles

1 Introduction

Automated Driving (AD) applications have increased in impact and maturity in the last years, due to the technological advances in perception sensors, processing hardware capabilities, artificial intelligent techniques, and new legal concessions

Á. Herrero et al. (Eds.): SOCO 2020, AISC 1268, pp. 657–666, 2021.
https://doi.org/10.1007/978-3-030-57802-2_63

to test automated vehicles on public roads [1]. However, despite the impressive demonstrators of automated driving functionalities, including commercial vehicles with partially automated driving features, the realization of such technology at a greater scale in our society is still a challenge [2], which could take decades to be achieved, while facing the technological, legal, and social barriers.

In parallel, the relevant advances achieved up to know can contribute to the development of human-centered vehicles that offer continuous control support during the driving task, reducing mental and physical workload, and ensuring a safer, more comfortable, and less demanding experience [3]. This collaborative driving strategy is suitable for inclusion as an especial mode of operation in partially automated vehicles (SAE Level 2 (L2) [4]). In these vehicles, automation has control over steering and pedals, but the driver has to monitor the environment and be ready to take full control in critical scenarios.

Nonetheless, current L2 vehicles, work under the on/off standard, with almost any cooperative control interaction with the driver. Furthermore, when the driver is out of the control loop, it leads to over-trust in automation, and consequently, increases the chance of a late take-over maneuver [5]. In this sense, ADAS with control cooperative components (or shared control ADAS) is a topic of interest in the AD research community. In these systems, the driver and the automation are guiding the vehicle together, with the proper authority that corresponds to the situation (e.g., driver distraction increases the authority of automation).

Shared control in the context of automated driving, is defined using the terminology presented by Abbink [6] as: "driver and automation interacting congruently in a perception-action cycle to perform a dynamic driving task that either the driver or the system could execute individually under ideal circumstances". Also, a joint effort with Flemisch [7] has included shared control in a cooperative framework at different task support levels: 1) operational, related to the control task, 2) tactical, for the maneuvers and decisions, and 3) strategical, which refers to the planning strategy of going from A to B.

The study of shared control systems has particular interest in steering applications, which is the most critical control interface in the driving task. Therefore, many European projects, as part of the mobility needs for a more safe and comfortable driving, have faced the challenge of human-machine cooperation in automated vehicles, aiming for a collaborative system that: 1) increase safety in dangerous maneuvers, such as lane change with a blind spot, 2) assist driver in authority transitions to ensure a smooth, progressive, fluid and safe control resuming, and 3) make the driving task comfortable and less demanding. These ADAS for partially automated vehicles have been studied in different EU research projects such as HAVEit [8], DESERVE [9], and the ABV Project [3].

Recently, two European projects continue this research line, looking for the implementation of collaborative human-centered vehicles using the shared control concept. First, PRYSTINE (Programmable Systems for Intelligence in Automobiles) project [2,10], studies shared control under the framework of fail-operational systems. Secondly, HADRIAN (Holistic Approach for Driver Role Integration and Automation Allocation for European Mobility Needs), makes emphasis on the dynamic adjustment of (fluid) human-machine interfaces (HMI) that take environmental, vehicle and driver conditions into account to provide adaptive signals and information, transfer control authority, and lead to safe transition between automated driving levels. These two projects have similarities and differences that will be highlighted in this article. Additionally, a common design framework will be presented, with an emphasis on the shared control system design that will be part of both approaches.

This article is structured as follows: Sect. 2 presents the description and objectives of the projects, together with the common framework, Sect. 3 describes the design of the shared control system. Section 4 analyzes the results of the system in an overtaking maneuver. Lastly, Sect. 5 closes with conclusions and future works.

2 Driver-Automation Framework for PRYSTINE and HADRIAN

The European Commission has granted funding for the development of Research Innovation Actions (RIA) in the context of automated driving. In this sense, PRYSTINE and HADRIAN are part of the ongoing projects that evaluate, design, and implement the human-centered concept in vehicles SAE Level 2, 3, and 4. PRYSTINE focuses the attention in fail-operational systems with an emphasis on the perception of the external environment using cameras, radar, and LiDAR, but also, considering in-cabin sensor fusion to detect the driver state. On the other hand, HADRIAN evaluates the human-centered design implementing fluid interfaces to improve driver automation-interaction not only at the operational level, but also from the human-acceptance perspective.

On the one hand, PRYSTINE intends to increase the Technology Readiness Level (TRL) to TRL 3–4, with validation in a Hardware and Driver in the Loop (HDiL) simulator. Conversely, HADRIAN push to take this technology to implementation in real vehicles and achieve demonstrations in relevant environments,

Table 1. Comparison of scopes of PRYSTINE and HADRIAN

	PRYSTINE	HADRIAN
Period	2018–2021	2020–2023
Objective	Fail-operational system	Fluid interfaces
Test platform	HWiL/DiL simulator	Experimental vehicle
DMS	Fusion of audio and vision-based sensor for driver distraction and drowsiness	Multisensor platform with driver model and RT-learning process
HMI	Visual HMI	Multi-sensory HMI Haptic, auditory, and visual
Scenario	Distraction in urban environment Authority transition in overtaking	Elderly driver assistance system
Acceptance	One cycle testing	Two iteration cycles

increasing the TRL index to 5–6, with more emphasis on driver acceptance tests. A more detailed comparison between these two projects is given in Table 1.

Additionally, in the context of these projects, a common control framework is proposed to integrate the driver and the automation in the collaborative and dynamic driving task. This integration requires interactions between different systems related to automated driving functionalities. Previously, a general architecture has been proposed for fully automated vehicles by Gonzalez et al. [9], with six high-level modules: acquisition, perception, communication, decision, control, and actuation. However, there are additional necessary modules to be included if the driver is sharing the authority of the vehicle with the automation: 1) a Driver Monitoring System (DMS), 2) a set of Human-Machine Interfaces (HMI), and 3) a Shared Control System (SCS). These systems are integrated into the original framework, and highlighted in green in Fig. 1, to indicate an addition to the original architecture.

Driver Monitoring System: It evaluates the driver's capability to execute the driving task by his/her own, and indicates the degree of assistance needed. According to the World Health Organization, most of the road traffic deaths are due to driver recognition errors, such as distraction or inattention [11]. That is the reason for the importance of taking into account the driver state as a variable of primary importance for decision making. The factors that can affect the driver's behavior can be assigned to a specific group taking into account outer (i.e. surrounding cars and vehicle density) and inner factors (i.e. distraction, drowsiness(fatigue), and medical conditions) [12]. Other measures indicating physical and mental workload are relevant in this module.

Human-Machine-Interfaces: They help the driver to understand the automation intention, state, and actions, increasing situation awareness and trust in the automated vehicle. In this sense, the system can communicate information to the driver by 1) a visual screen, through text or images, for example showing the representation of the environment with nearby vehicles, 2) haptic interfaces, using vibration in the pilot seat, at the steering wheel, or any other surface in contact with the driver, and 3) audio warnings, either by sound alerts or tutoring voice. The design of such strategies should follow the principles of comfort, usability, and avoid excessive information to not overwhelm the driver.

Shared Control System: It is the critical module of a human-centered vehicle framework where the decision and control actions are implemented. In the decision block, an arbitration sub-module is in charge of distributing the authority between the driver and the automation. This module calculates two relevant values: 1) the Level of Haptic Authority (LoHA), that represents the strength of intervention of the system when safety is compromised, (it is the stiffness of the controller around the optimal command [13]), and 2) the Level of Shared Authority (LoSA), a continuous value which indicates the mode of automation,

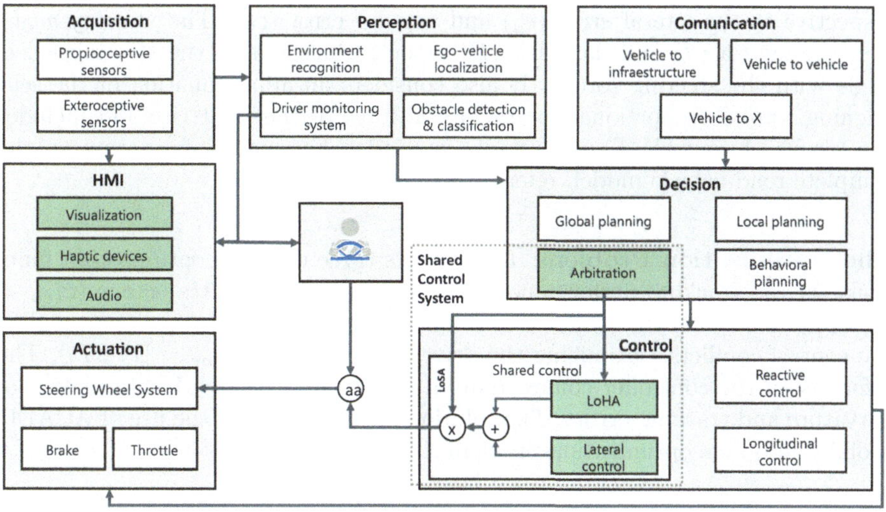

Fig. 1. Driver-automation framework

either fully automated or manual, to allow smooth, progressive, and comfortable transitions. This shared control system is explained in detail in the next section.

3 Shared Control System

The shared control system architecture is comprised of two subsystems. First, the lateral shared controller that assists the driver in the lane keeping task. Secondly, an arbitration system calculates the two levels of authority. On the one hand, the Level of Haptic Authority (LoHA) increases the default authority of the steering controller to avoid hazardous situations. On the other hand, the Level of Shared Authority (LoSA) manages the transitions of control from manual-to-automated and automated-to-manual, as a continuous value.

3.1 Lane-Keeping Controller

The lane-keeping system for shared control applications makes use of the optimization framework of Model Predictive Control (MPC) as in previous works [14,15], considering the torque at the steering wheel as the control input, to facilitate the driver-automation cooperation. The design of the controller has two considerations: 1) the representation of the system model through differential equations, and 2) the configuration of the optimization problem.

The System Model: It comprehends three sub-systems: the vehicle, the lane-keeping model, and the steering mechanism. This combination represents the road-vehicle model. The *vehicle* model uses dynamic bicycle system equations for a front steered vehicle. The *lane-keeping* model includes two differential equation respective to the lateral error (e_y) and angular error (e_Ψ). The *steering model* uses the inertia (J) and damping (B) model, which relates the steering wheel angle with the steering torque. It also considers an approximation of the self-aligning torque proportional to the lateral force of the front tire and includes the torque of control (T) as part of the model. For more information on the complete road-vehicle model, refer to [15].

The Optimization Problem: It considers three different optimization functions: 1) the tracking performance, to follow the reference trajectory ($\boldsymbol{z_{tra}} = [e_y, e_\Psi]$), 2) the driving comfort, minimizing speeds ($\boldsymbol{z_{com}} = [v_y, \dot{\psi}, w]$), and 3) the control conflicts, optimizing the driver control effort ($\boldsymbol{z_{tor}} = [T, \Delta T]$). The reference of the controller comes from an offline trajectory with information of curvature and tracking errors. The solution is obtained with the use of ACADO toolkit, an online optimization problem solver [16].

3.2 The LoHA Controller

Additionally to the lane-keeping controller, the shared control system needs a sub-module to increase the intervention of the controller to override the driver's intention in situations when safety is compromised. In this sense, a LoHA controller is added in cascade to the lane-keeping controller. The LoHA is the stiffness around the optimal steering angle. The higher the LoHA, the harder for the driver to override automation. This controller is defined as a proportional term to the difference between driver and automation command $T_{LoHA} = K_{LoHA}(\theta - \theta_d)$. It changes the original stiffness of the system K, to a new equivalent value $K_{eq} = K + K_{LoHA}$. Therefore, to keep the system stable, a new equivalent damping is found using the damping ratio formula $B_{eq} = B\sqrt{(K + K_{LoHA})/K}$ resulting in the following LoHA controller:

$$T_{LoHA} = K_{LoHA}(\theta - \theta_d) + (B_{eq} - B)w \tag{1}$$

3.3 The Arbitration System

The arbitration system is based in a Fuzzy Inference System, a powerful soft computing technique that allows to include human knowledge into the design of control and decision algorithms [17]. The fuzzy scheme comprehends four inputs and two outputs. The representative inputs are: 1) the driver's intention, considering the lateral error and its derivative, 2) the driver effort, measured as the equivalent torque at the steering wheel, and 3) the risk of collision, calculated as the time-to-collision with the vehicle in the left lane. These three variables allow us to calculate the following outputs: 1) the LoHA, which represents the need for a greater intervention of the system to avoid collisions, and 2) the LoSA, which is the variable authority for a progressive transition from automated-to-manual and manual-to-automation. The design rules are shown in Tables 2 and 3 for the conditions of a low and high risk of collision respectively.

Table 2. Rules for low collision risk

	e_y	Low			Med			High			
	de_y/dt	↓	–	↑	↓	–	↑	↓	–	↑	
T	$=0$	A	A	A	A	T	M	M	M	M	LoSA
	>0	T	T	T	T	M	M	M	M	M	
	$=0$	LOW									LoHA
	>0										

Manual (M) - Transition (T) - Auto(A)

Table 3. Rules for high collision risk

	e_y	Low			Med			High			
	de_y/dt	↓	–	↑	↓	–	↑	↓	–	↑	
T	$=0$	AUTOMATED									LoSA
	>0										
	$=0$	L	L	L	L	M	H	H	H	H	LoHA
	>0	L	M	M	M	M	H	H	H	H	

Low (L) - Medium (M) - High (H)

4 Use Case and Results

This section presents the tests of the shared control system in a collaborative overtaking maneuver, where the system assists the driver in three scenarios. First, if the driver plans to do a lane change, but there is a high risk of collision with the side vehicle, the system increases the LoHA to guarantee the safety and guides the vehicle towards the main lane, as shown in the top of Fig. 2. Secondly, when there is low collision risk, the system decreases the LoSA conceding the transition from automated-to-manual (see bottom of Fig. 2). Lastly, once the driver has passed the front vehicle and returns to the main lane, the system increases the LoSA to allow the transition from manual-to-automated and continue with the lane-keeping assistance.

The implementation of this maneuver is performed by one real driver in the HiL Automated Driving Simulator shown in Fig. 3. It is comprised of a high-performance computer, running Matlab/Simulink, and communicates with a steering wheel capable of a maximum torque of 15 N.m. The automated driving software simulator is based on Dynacar [18], a vehicle dynamic software based on a multi-body formulation.

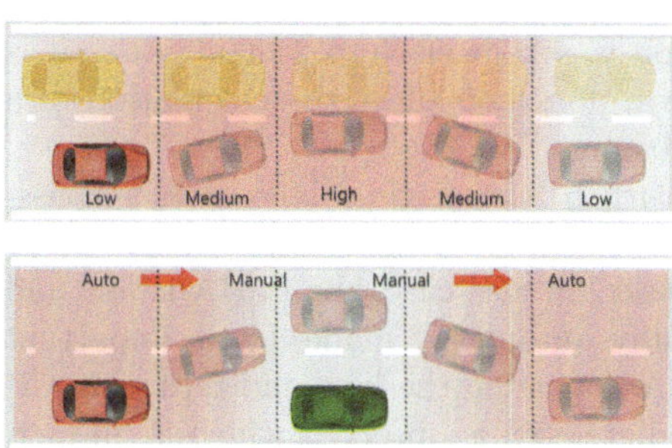

Fig. 2. Collaborative overtaking

Fig. 3. HDiL simulator platform

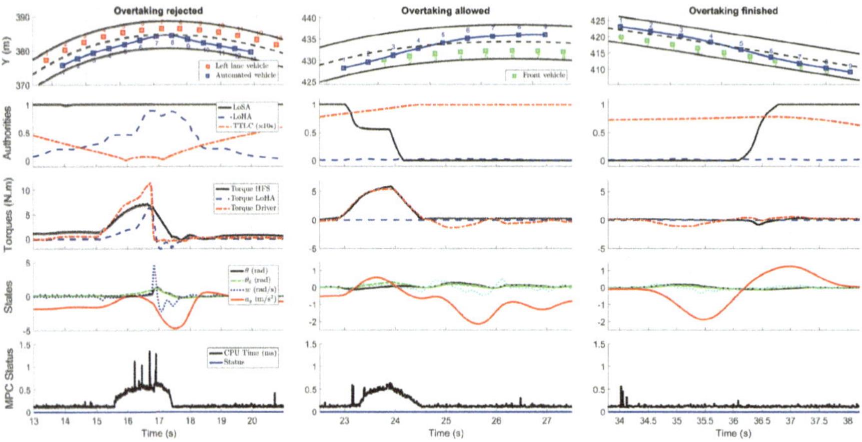

Fig. 4. Shared control system results in three scenarios

The results of the three scenarios are shown in Fig. 4. First, the collision avoidance system is tested with the vehicle starting in a fully automated mode. Initially, the LoHA is very low as there is no risk of collision. Then, in the second 16, the driver intends to make a lane change, but the system detects a low time-to-collision with the left lane vehicle. The arbitration system maintains the automated mode and increases the LoHA to strengthen the intervention of the system ensuring safety. The system achieves an assistance torque of 10 N.m, and the driver releases the steering wheel. In this case, the system can return to the lane without losing stability. On the one hand, safety was the priority, but also, the comfort was compromised with a lateral acceleration close to -5 m/s^2. The results also show that the MPC solver always found a feasible solution calculated in less than 1.5 ms.

In the second scenario, the driver intends to do a lane change again, but in this case, the system does not detect any collision risk and allows the transition from automated-to-manual. It is shown in the second column of Fig. 4, that the LoSA is changed smoothly and progressively, making the transition comfortable and understandable for the driver, with a maximum effort of 5 N.m in a short period. The maximum lateral acceleration was kept close to 2 m/s^2. Also, it is observed that the variation of authorities, does not affect the calculation of a feasible solution of the optimization problem.

Lastly, when the driver wants to return to the original lane after surpassing the front vehicle, the system changes from manual-to-automated and keeps assisting the driver in the lane-keeping task. In this case, the LoHA is low and the LoSA changes progressively to 1 (fully automated mode). It is important to mention that the behavior of the LoSA departing the lane and returning the lane is different. In the first, an intermediate step is observed which is, in fact, helpful for the driver to confirm the lane change intention. In the second one, the transition is performed without medium steps, allowing activation of the lane-keeping that is barely notable to the driver, as shown by the low lateral acceleration and steering wheel angular velocity.

5 Conclusions and Future Works

This article presents a shared control framework for implementations in two European RIA projects, PRYSTINE and HADRIAN, to improve the development of advanced control techniques for human-centered vehicles SAE Level 2, 3 and 4. The shared control system is comprised of an arbitration function that calculates the appropriate control authorities, based on fuzzy logic, a well-known soft computing technique, and a cascade architecture controller including an MPC and a PD controller for the lane-keeping task.

Results show the effectiveness of the system in a collaborative overtaking maneuver. When the risk of lateral collision is high, the automation overrides the driver's intention increasing the level of intervention (LoHA). Conversely, when there is no risk, and a lane change intention is recognized a transition of authority takes place in a fluid, progressive, and comfortable manner by changing the LoSA. The system is also able to reactivate the automated mode when returning to the lane.

In future works, the integration of the shared control system with other complementary modules such as the DMS and the cooperative HMI is necessary to prove the feasibility of the complete collaborative framework. The implementation of this approach will be tested with different drivers for a complete driver acceptance test and evaluates the utility of this cooperative control system in passenger vehicles.

References

1. Ertrac, E., Snet, E.: Ertrac automated driving roadmap. ERTRAC Working Group **7** (2017)
2. Druml, N., Macher, G., Stolz, M., Armengaud, E., Watzenig, D., Steger, C., Herndl, T., Eckel, A., Ryabokon, A., Hoess, A., Kumar, S., Dimitrakopoulos, G., Roedig, H.: Prystine - programmable systems for intelligence in automobiles. In: Proceedings 21st Euromicro Conference Digital System Design (DSD), pp. 618–626, August 2018
3. Sentouh, C., Popieul, J.C., Debernard, S., Boverie, S.: Human-machine interaction in automated vehicle: the abv project **47**, 6344–6349 (2014)
4. Committee, S.O.R.A.V.S., et al.: Taxonomy and definitions for terms related to on-road motor vehicle automated driving systems. SAE Standard J3016, pp. 01–16 (2014)
5. Saito, T., Wada, T., Sonoda, K.: Control authority transfer method for automated-to-manual driving via a shared authority mode. IEEE Trans. Intell. Veh. **3**(2), 198–207 (2018)
6. Abbink, D.A., Carlson, T., Mulder, M., de Winter, J.C., Aminravan, F., Gibo, T.L., Boer, E.R.: A topology of shared control systems-finding common ground in diversity. IEEE Trans. Hum. Mach. Syst. **99**, 1–17 (2018)

7. Flemisch, F., Abbink, D.A., Itoh, M., Pacaux-Lemoine, M.P., Weßel, G.: Joining the blunt and the pointy end of the spear: towards a common framework of joint action, human–machine cooperation, cooperative guidance and control, shared, traded and supervisory control. Cogn. Tech. Work, 1 (2019). https://doi.org/10.1007/s10111-019-00576-1
8. Hoeger, R., Amditis, A., Kunert, M., Hoess, A., Flemisch, F., Krueger, H.P., Bartels, A., Beutner, A., Pagle, K.: Highly automated vehicles for intelligent transport: Haveit approach. In: ITS World Congress, NY, USA (2008)
9. Gonzalez, D., Perez, J., Milanes, V., Nashashibi, F., Tort, M.S., Cuevas, A.: Arbitration and sharing control strategies in the driving process. In: Towards a Common Software/Hardware Methodology for Future Advanced Driver Assistance Systems, p. 201 (2017)
10. Marcano, M., Díaz, S., Pérez, J., Castellano, A., Landini, E., Tango, F., Burgio, P.: Human-automation interaction through shared and traded control applications. In: International Conference on Intelligent Human Systems Integration, pp. 653–659. Springer (2020)
11. Rolison, J.J., Regev, S., Moutari, S., Feeney, A.: What are the factors that contribute to road accidents? An assessment of law enforcement views, ordinary drivers' opinions, and road accident records. Accid. Anal. Prev. **115**, 11–24 (2018). https://www.sciencedirect.com/science/article/pii/S0001457518300873
12. Aksjonov, A., Nedoma, P., Vodovozov, V., Petlenkov, E., Herrmann, M.: Detection and evaluation of driver distraction using machine learning and fuzzy logic. IEEE Trans. Intell. Transp. Syst. **20**(6), 2048–2059 (2019). https://ieeexplore.ieee.org/document/8440785/
13. van Paassen, M.R., Boink, R.P., Abbink, D.A., Mulder, M., Mulder, M.: Four design choices for haptic shared control. In: Advances in Aviation Psychology, Volume 2: Using Scientific Methods to Address Practical Human Factors Needs, p. 237 (2017)
14. Guo, H., Song, L., Liu, J., Wang, F., Cao, D., Chen, H., Lv, C., Luk, P.C.: Hazard-evaluation-oriented moving horizon parallel steering control for driver-automation collaboration during automated driving. IEEE/CAA J. Automatica Sinica **5**(6), 1062–1073 (2018)
15. Ercan, Z., Carvalho, A., Tseng, H.E., Gökaşan, M., Borrelli, F.: A predictive control framework for torque-based steering assistance to improve safety in highway driving. Veh. Syst. Dyn., 1–22 (2017)
16. Houska, B., Ferreau, H.J., Diehl, M.: Acado toolkit-an open-source framework for automatic control and dynamic optimization. Optimal Control Appl. Methods **32**(3), 298–312 (2011)
17. Marcano, M., Matute, J.A., Lattarulo, R., Martí, E., Pérez, J.: Low speed longitudinal control algorithms for automated vehicles in simulation and real platforms. Complexity **2018** (2018)
18. Iglesias-Aguinaga, I., Martin-Sandi, A., Pena-Rodriguez, A.: Vehicle modelling for real time systems application. the virtual rolling chassis. DYNA **88**(2), 206–215 (2013)

A First Approach to Path Planning Coverage with Multi-UAVs

Alfredo Pintado[1] and Matilde Santos[2(✉)]

[1] Canal de Isabel II, Madrid, Spain
apintado@canaldeisabelsegunda.es
[2] Institute of Knowledge Technology, Complutense University of Madrid, Madrid, Spain
msantos@ucm.es

Abstract. In this paper, a multi-UAV system is applied to explore a searching area. The influence of the partition of the surface and the effects of varying the number of UAVs are analyzed. The covering of the area is based on small rectangular polygon area decomposition. Each sector is assigned to an UAV and efficient coverage algorithms are applied. The UAV follows a zig-zag navigation strategy to go through the way-points located at the center of the cells of the corresponding area. The performance of the multi-UAV system is discussed for different scenarios. Simulation results in terms of travel time are presented.

Keywords: Unmanned Aerial Vehicles (UAV) · Area coverage · Path planning · Polygon area decomposition · Multi-UAVs

1 Introduction

Unmanned aerial vehicles (UAV) have been proved an efficient technique to solve a great variety of problems of different fields. However, these systems still presents technological challenges such as security, reliability, robustness, etc. [1]. In addition, they present limitations that can become critical depending on the mission and are far from being successfully solved [2].

One of UAV main problems is limited flight autonomy. This is a crucial issue when dealing with search and rescue (SAR) missions [3]. In this case it is important to optimize the area coverage under study. The use of multiple UAVs to explore a scenario can help to better and faster cover the search map. But it is necessary to establish intelligent strategies that allow multiple vehicles to completely explore the area in the minimum time possible [4, 5].

In this work, the use of multiple UAVs for area coverage is addressed. Different scenarios are analyzed and small rectangular polygon area decomposition is applied to distribute the area between the UAVs. Increasing number of UAVs are tested, all of them with the same technical characteristics.

The paper is organized as follows. The polygon area decomposition strategy and the way-point vectors that will guide the navigation strategy are described in Sect. 2. Simulation results using different number of UAVs on different scenarios are discussed in Sect. 3. The paper ends with the conclusions and future works.

Á. Herrero et al. (Eds.): SOCO 2020, AISC 1268, pp. 667–677, 2021.
https://doi.org/10.1007/978-3-030-57802-2_64

2 Coverage Area Decomposition and Navigation Strategy

Most of the works that deal with the problem of area coverage with multiple UAVs are focused on the navigation strategy, and simplified models of the UAVs, work scenarios, weather conditions, loads, etc., are assumed. The majority of the studies try to solve the area coverage using way-points [6], so the problem is reduced to represent all possible paths as a graph with way-points as nodes and then use any path-following algorithm to efficiently compute the shortest path [7].

The area decomposition is thus transformed into the problem of dividing the area into sub-connected graphs connecting the centers of the cells [8]. A common simplified approach represents the scenario with regular convex polygons [9–11]. This makes easier the navigation as it is just to find the longest side of the polygon in order to set the flight orientation. For instance, in [12] a trapezoidal area decomposition is applied and the trapezoids are joined again according to different criteria. The work [9] proposes to reduce a convex polygon into smaller parts that correspond to the area assigned for each UAV using a divide-and-conquer algorithm. It starts with as many vertices as available UAVs, divides the polygon into linear segments and gradually changes the orientation until the area assigned to each UAV corresponds to the established area. For non-convex polygons, Nielsen [13] proposes a solution that applies trapezoidal decomposition to convert them into multiple convex polygons, but for a single UAV.

Models based on decomposition in cells are usually used when obstacles or prohibited areas are found in the area to be covered, since the polygons are divided taking into account those obstacles [13–16].

The UAV physical characteristics and navigation models usually consider only the flight speed and attitude, and do not include other real features such as weight and acceleration; similarly, the turning capacity is approached with linear trajectories and weighted turns (up to 90° or even up to 45°) [17, 18].

Path planning is usually designed to minimize the number of turns of the UAV during the flight. This may be relevant depending on the rotational speed of the aerial vehicle. This coverage path planning has been addressed with different approaches, including artificial intelligent techniques [19–21]. A classic solution that reduces the number of turns is the zig-zag trajectory, used in this work. Once the searching area has been decomposed into polygons, these are traveled starting from the largest side of the polygons. This solution guarantees that the number of turns is minimal if the surface to be explored is a regular convex polygon.

2.1 Area Decomposition Strategy

In this work we propose the searching area decomposition into small vertical rectangular polygons and their subsequent reallocation according to the number of UAVs (Fig. 1). This is a simplified variant of the typical area decomposition strategy for convex polygons, when the most suitable orientation is known (the largest side of the polygon).

This polygon side always gives the minimal width of the polygon (Fig. 2, left). An optimal decomposition according to this strategy of a convex rectangular polygon is shown in Fig. 2, right.

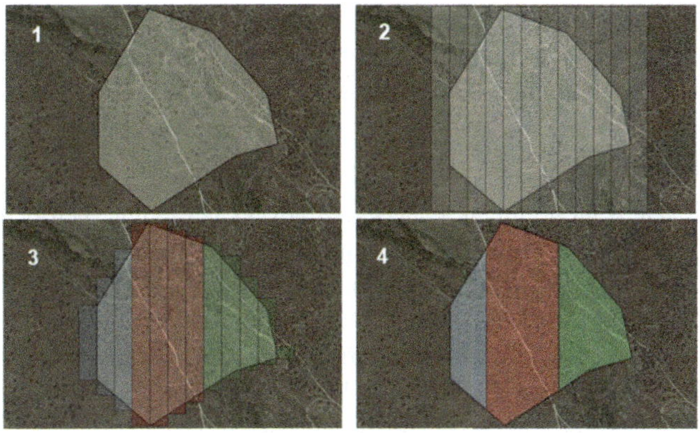

Fig. 1. Area decomposition in vertical rectangular polygons and reallocation for 3 UAVs.

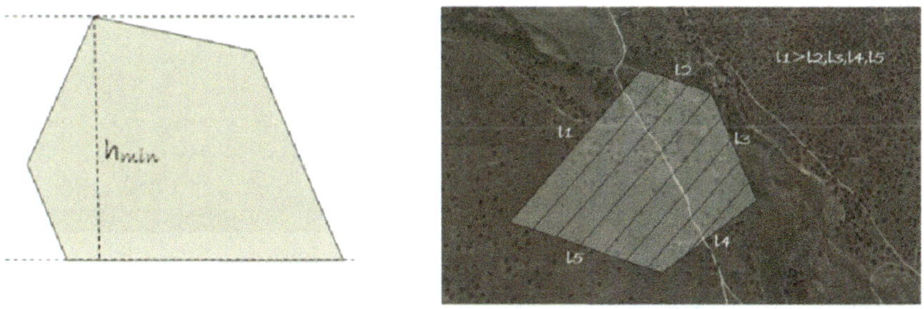

Fig. 2. Minimal width of the convex polygon (left) and optimal decomposition (right).

For other scenarios where the searching area is non-convex (regular concave areas or irregular areas), the optimal decomposition solution cannot always be obtained. For instance, Fig. 3 shows a vertical (left) and a horizontal (right) area decomposition of the same non-convex polygon for 3 UAVs. Depending on the orientation, very different results are obtained.

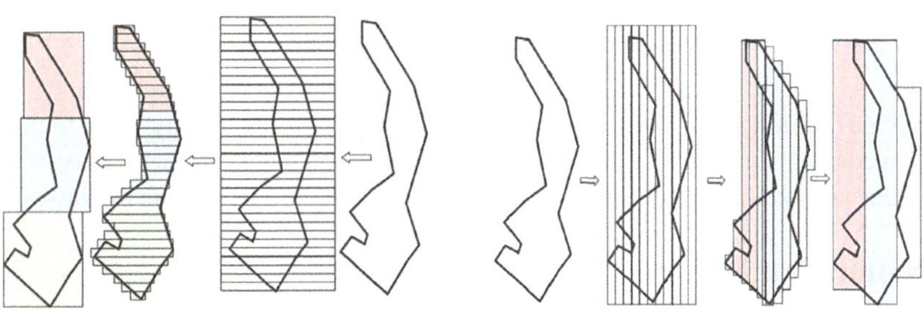

Fig. 3. Regular non-convex polygon decomposition with different orientations for 3UAVs

An alternative for highly irregular polygons or with isolated areas is to smooth the surface to be explored, but this sometimes results in longer time required to cover the area due to the greater number of turns.

2.2 Navigation Strategy

Once an UAV has been assigned a surface to cover, way-point vectors are generated according to the embedded camera scope and the flight attitude of the UAV in order to ensure the full coverage of the map (Fig. 4).

The surface is initially divided into square cells of size $C_i x C_i$, along the longest side of the assigned polygon, with the aim of minimizing turns. A way-point w_i is placed at the center of each cell. It is necessary to check that all the way-points are on the searching surface, as part of the cell may be outside the surface to be explored.

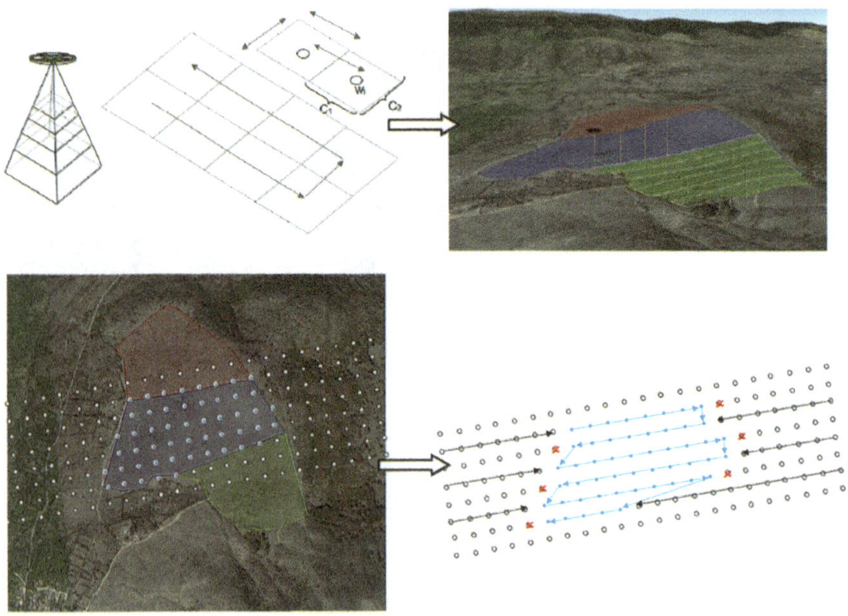

Fig. 4. UAV field of vision of UAV. Area decomposition. Way point vectors and possible turns

The navigation strategy consists of moving the UAV from one way-point to the next one at a constant speed. Note that outside the assigned area the cells are empty, so if the UAV reaches a cell without any way-point it turns to the following line of way-points.

3 Simulation Results

3.1 Performance Measurements

The final navigation cost of each UAV is obtained as the sum of the costs of moving it from way-point to the next one; it also includes an initial time, d_{Bi}, and a return time, d_{Bf}, that are defined as follows:

- d_{Bi}: time to reach the first way-point on the map from the base station
- d_{Bf}: time since the last way-point was visited until reaching back the base station

Take-off and landing times are not considered as they are the same for all the simulated scenarios.

The time spent in each turn is obtained by the following equation [17], which represents the simplified rotational model of the UAV:

$$\tau_{ri} = \tau_C[n_w - 1] + \sum[k\alpha] + \frac{1}{V_{UAV}}[d_{Bi} + d_{Bf}] \tag{1}$$

where k is the penalty factor for each angle, set to $k = 0.444$ [17], and α is the rotated angle. This model does not consider any constraints regarding the turns. In this equation, V_{UAV} is the travel speed, n_w the number of way-points, and τ_C the travel time between waypoints in a region C, that is calculated as,

$$\tau_C = \frac{C}{V_{UAV}} \tag{2}$$

3.2 Scenarios

The simulations have been run with the following characteristics. The discrete scenario size is 500 m^2. The UAVs are initially at the base station, outside the search area. The flight speed of all UAVs is 10 m/s. The distance between way-points is 5 m. The number of UAVs is 2, 4, 8 and 16. All the UAVs are considered to have the same technical characteristics (batteries, weight, etc.).

Three scenarios are considered:

- Regular convex (square)
- Irregular compact
- Irregular non-convex

The surface has been equally assigned to each UAV in terms of volume (area), regardless the shape. Indeed, once the decomposition of the area into rectangular polygons has been carried out, each polygonal sector is assigned consecutively to any available UAV. Thus, each UAV covers the same volume but the length of the routes can be very different.

Simulations with the same scenario and parameters have been carried out 5 times each, as random variables are involved. The best results are shown. It was not considered necessary to run more simulations as the variance was very small.

The procedure of area decomposition into polygons is shown in Fig. 5, and it is detailed as follows.

1. Divide the scenario into sectors (300 sectors).
2. Generate random clouds of points on the scenario (5000 points).
3. Classify the points that have been generated within the map to be covered. Discard the rest of the points.
4. Select sectors that contain any point of the map. Discard the rest of the sectors.
5. Generate random clouds of points on the sectors resulting from the previous process (8000 points).
6. For each sector, add the number of points included in the map in order to get an approximation of the area to be covered in each sector.
7. Join the sectors consecutively and proportionally to the number of points that each sector has, forming as many groups as UAVs are available. This results in that the surface assigned to each UAV is equitable.
8. Close the borders of the polygon to be covered by each UAV including all the points of the sectors. The more random points generated in step 5, the finer the polygon.

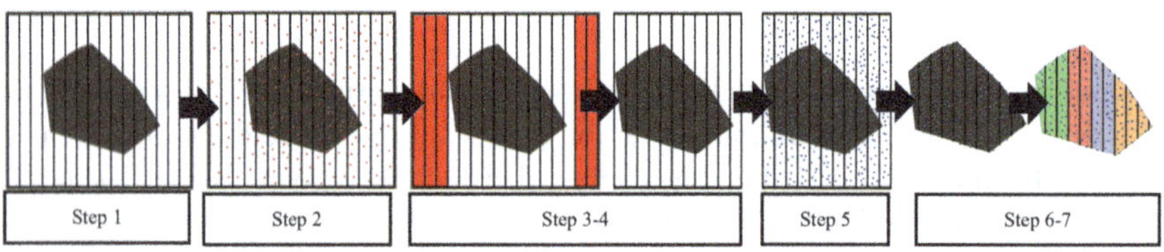

Fig. 5. Steps 1 to 7 of the polygon area decomposition strategy for multiple UAVs

The pseudo-code of the algorithm that implements that area decomposition is the following.

Algorithm 1: Area decomposition (Split andMmerge)

```
//STEP 1
Width Partition Sector = Size Scenario/Number of Sectors
//Split sectors generation
while (i < Number of Sectors)
i++
Pxf = Pxo + Width Partition Sector
Pyf = Pyo + Size Scenario
Create New Sector [i] = [(Pxo , Pyo), …, (Pxf , Pyf)]
//Move new origin point to next sector
(Pxo , Pyo) = (Pxf , Pyf)
end while
//STEP 2 to STEP 4
while (i < Number of Points to Classify Sectors)
  i++
  Point (Px , Py)= new random Point (Px , Py)
//Select sectors of interest
If (Point ∈ Polygon)
    add List SectorROI <> = Sector [j]
end while
//STEP 5 to STEP 6
while (i < Number of Points to Define Polygons)
  i++;
  Point (Px , Py)= new random Point (Px , Py)
//Select sectors and count number of points per sector
If (Point ∈ Polygon)
  //Select Sector
  while (classify Point== false)
If (Px , Py < Pxj , Pyj)
j++
classify Point = false
Else If
      number of points Sector [j] ++
      number of total Points ++
      classify Point = true
end while
//STEP 7
Number of Points per UAV = Total Points Number/Total UAVs Number
while (i < Number of Total UAVs)
  i++
  while (Number of Points per UAV > Total Points assigned UAV[i])
```

```
      Sectors total UAV[i] += Sector[j]
      Total Points assigned UAV[i] += number of points Sector[j]
      j++
    end while
end while
```

The results of applying this area decomposition into polygons strategy to three scenarios (rows), for 2, 4, 8 and 16 UAVs (columns), are shown in Fig. 6.

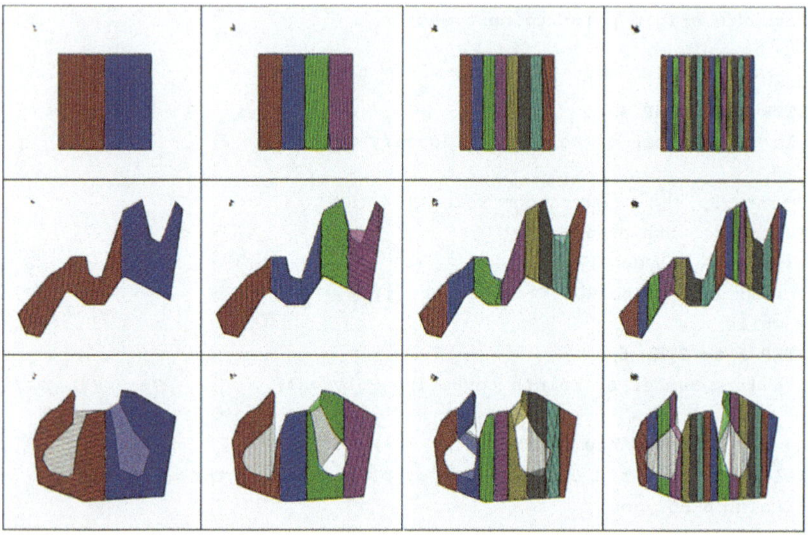

Fig. 6. Area decomposition for 2, 4, 8 and 16 UAVs (columns) of the three scenarios (rows)

Table 1 presents the simulation results. The regular convex (square) scenario has been taken as reference since it allows a straightforward interpretation of the influence of an increasing the number of UAVs. The percentage (%) of mission time saved represents the time saved using one more UAV than in the previous case (i.e., instead of using one UAV, using two or three UAVs to cover the area). Times are always given in seconds as a conventional unit for comparison purposes, though it is not real time.

Based on these results, a significant over cost of the total resources is observed when using more UAV systems. This is due to the fact that very little time is spent flying over the area assigned for each UAV with respect to the time needed to reach the area from the base station.

A priori, if all unmanned aerial vehicles were at the first way point of the area, regardless of the time to go and return to the base station, the zig-zag strategy in the square scenario would improve linearly in terms of travel time with the number of UAV, without affecting the total cost of the mission. To illustrate this fact, Table 2 shows a theoretical example of the loss of linearity in the performance as the number of UAVs increases on the square scenario. Variables T_i and T_f are the time it takes for each UAV to reach the assigned search area and return to the base station after completing the mission.

Table 1. Simulation results for different number of UAVs and different scenarios.

Scenario	No. UAVs	Time (slowest UAV)	Mission time saved	All UAVs time	% over-cost	Profit 50%–50%	Profit 90%–10%
Convex	2	971	–	1841	–	–	–
	4	530	45,42	2081	−13,04	32,38	39,57
	8	326	66,43	2337	−26,94	39,48	57,09
	16	195	79,92	2577	−39,98	39,94	67,93
Irregular convex	2	1212	–	2214	–	–	–
	4	839	30,78	2397	−8,27	22,51	26,87
	8	365	69,88	2364	−6,78	63,11	62,22
	16	280	76,90	2989	−35,00	41,89	65,71
Irregular non-convex	2	1388	–	2669	–	–	–
	4	1004	27,67	2831	−6,07	21,60	24,29
	8	660	52,45	3214	−20,42	32,03	45,16
	16	322	76,80	3237	−21,28	55,52	66,99

Table 2. Loss of the linearity in the performance when increasing the number of UAVs on the square scenario (time)

Convex square scenario	$T_i + T_f$ for each UAV	No. UAVs	Time UAV1	Time UAV2	Time UAV3	Time UAV4	Time (slowest UAV)	All UAVs time
On the area	0	1	2400	–	–	–	2400	2400
	0	2	1200	1200	–	–	1200	2400
	0	4	600	600	600	600	600	2400
From base station	90	1	2490	–	–	–	2490	2490
	90	2	1290	1290	–	–	1290	2580
	90	4	690	690	690	690	690	2760

This is just an example of the complexity of the problem due to high number of factors that influences the performance of an area covering strategy.

4 Conclusions and Future Works

This work presents an area decomposition strategy into polygons. The search surface, once divided, is assigned to multiple UAVs that travel it following a zig-zag strategy in order to cover it.

An interesting conclusion of this study is that there is a great unpredictability regarding the effects of increasing the number of UAVs on irregular scenarios. This may be because when the surface is divided among several UAVs, the corresponding areas are much more irregular and thus the distances traveled by each UAV also varies a lot. This is shown in the resulting over-cost with compact irregular scenario with 8 UAVs and the non-convex surface with 16 UAVs.

Beyond the influence of the different selected configurations, the weights given to the resources and costs are crucial. For example, if few resources are available cost savings are important, while if a rapid emergency action is required, the slowest UAV time reduction will be pursued.

As future works, different area assignment approaches and others navigation strategy could be tested. Some UAVs technical characteristics that strongly affect the results could be considered, such as the flight model, the embedded camera scope, etc.

References

1. Sierra, J.E., Santos, M.: Modelling engineering systems using analytical and neural techniques: Hybridization. Neurocomputing **271**, 70–83 (2018)
2. Pajares, G., Ruz, J.J., Lanillos, P., Guijarro, M., Santos, M.: Trajectory generation and decision making for UAVs. Revista Iberoamericana de Automática e Informática Industrial **5**(1), 83–92 (2008)
3. San Juan, V., Santos, M., Andújar, J.M.: Intelligent UAV map generation and discrete path planning for search and rescue operations. Complexity **2018**(1), 1–17 (2018)
4. Cabreira, T.M., Brisolara, L.B., Ferreira Jr., P.R.: Survey on coverage path planning with unmanned aerial vehicles. Drones **3**(1), 4 (2019)
5. García-Auñón, P., Santos Peñas, M.: Use of genetic algorithms for unmanned aerial systems path planning. In: Decision Making and Soft Computing: Proceedings 11th International FLINS Conference, pp. 430–435 (2014)
6. Almadhoun, R., Taha, T., Seneviratne, L., Zweiri, Y.: A survey on multi-robot coverage path planning for model reconstruction and mapping. SN Appl. Sci. **1**(8), 847 (2019).
7. Fernández, C., Pantano, N., Godoy, S., Serrano, E., Scaglia, G.: Parameters optimization applying Monte Carlo methods and evolutionary algorithms. Enforcement to a trajectory tracking controller in non-linear systems. Revista Iberoamericana de Automática e Informática Industrial **16**(1), 89–99 (2019)
8. Wu, Y., Zhu, J., Gao, K.: Multi-UAVs area decomposition and coverage based on complete region coverage. In: IOP Conference Series: Materials Science and Engineering, vol. 490, no. 6, p. 06. IOP Publishing (2019)
9. Maza, I., Ollero, A.: Multiple UAV cooperative searching operation using polygon area decomposition and efficient coverage algorithms. In: Distributed Autonomous Robotic Systems 6, pp. 221–230. Springer, Tokyo (2007)
10. Jiao, Y.S., Wang, X.M., Chen, H., Li, Y.: Research on the coverage path planning of UAVs for polygon areas. In: 2010 5th IEEE Conference on Industrial Electronics and Applications, pp. 1467–1472. IEEE (2010)
11. Choset, H., Pignon, P.: Coverage path planning: the boustrophedon cellular decomposition. In: Field and Service Robotics, pp. 203–209. Springer, London (1998)
12. Driscoll, T.M: Complete coverage path planning in an agricultúral environment. Theses Dissertations. Iowa State University (2011)

13. Nielsen, L.D., Sung, I., Nielsen, P.: Convex decomposition for a coverage path planning for autonomous vehicles: interior extension of edges. Sensors **19**(19), 4165 (2019)
14. Khan, A., Noreen, I., Habib, Z.: On complete coverage path planning algorithms for non-holonomic mobile robots: survey and challenges. J. Inf. Sci. Eng. **33**, 101–121 (2017)
15. Horvath, E., Pozna, C., Precup, R.E.: Robot coverage path planning based on iterative structured orientation. Acta Polytechnica Hungarica **15**(2), 231–249 (2018)
16. Hert, S., Lumelsky, V.: Polygon area decomposition for multiple-robot workspace division. Int. J. Comput. Geom. Appl. **8**(4), 437–466 (1998)
17. Santana, E., Moreno, R., Sánchez, C., Piera, M.À.: A framework for multi-UAV software in the loop simulations. Int. J. Serv. Comput. Oriented Manuf. **3**(2–3), 190–211 (2018)
18. Sierra, J.E., Santos, M.: Wind and payload disturbance rejection control based on adaptive neural estimators: application on quadrotors. Complexity **2019**, 20 (2019)
19. Santos, M.: An applied approach of intelligent control. Revista Iberoamericana de Automática e Informática Industrial RIAI **8**(4), 283–296 (2011)
20. García-Auñón, P., Santos-Peñas, M., de la Cruz García, J.M.: Parameter selection based on fuzzy logic to improve UAV path-following algorithms. J. Appl. Logic **24**, 62–75 (2017)
21. Fonnegra, R., Goez, G., Tobón, A.: Orientation estimating in a non-modeled aerial vehicle using inertial sensor fusion and machine learning techniques. Revista Iberoamericana de Automática e Informática Industrial **16**(4), 415–422 (2019)

Special Session: Soft Computing for Forecasting Industrial Time Series

Copper Price Time Series Forecasting by Means of Generalized Regression Neural Networks with Optimized Predictor Variables

Gregorio Fidalgo Valverde[1](✉), Alicja Krzemień[2], Pedro Riesgo Fernández[1], Francisco Javier Iglesias Rodríguez[1], and Ana Suárez Sánchez[1]

[1] University of Oviedo, 33004 Oviedo, Spain
gfidalgo@uniovi.es
[2] Central Mining Institute, 40-166 Katowice, Poland

Abstract. This paper presents a twelve-month forecast of copper price time series developed by means of Generalized regression neural networks with optimized predictor variables. To achieve this goal, in first place the optimum size of the lagged variable was estimated by trial and error method. Second, the order in the time series of the lagged variables was considered and introduced in the predictor variable. A combination of metrics using the Root mean squared error, the Mean absolute error as well as the Standard deviation of absolute error, were selected as figures of merit. Training results clearly state that both optimizations allow improving the forecasting performance.

Keywords: Copper price · Time series · Forecasting · Generalized regression neural networks · Lagged variable · Rolling window · Predictor variable

1 Introduction

Following the seminal work by Matyjaszek et al. [1], this paper develops a twelve-month forecast of copper price time series by means of Generalized Regression Neural Networks (GRNN) as described by Krzemień [2], using optimized predictor variables.

The optimization of the predictor variables was twofold: in first place, the optimum size of the lagged variable was calculated by trial and error method. After estimating an approximate optimum size, a range of values was selected for testing, in order to cover a period above and below this figure that allows including a multiple of twelve months, so any possible periodicity hidden in the time series will be considered. Second, the order of the lagged variables in the time series was included as an intrinsic signal in order to feed the neural network with additional information that will not be considered otherwise [3].

A combination of metrics using the Root mean squared error (RMSE), the Mean absolute error (MAE) and the Standard deviation of absolute error (STD of AE) were selected as figures of merit in order to determine the artificial neural network model that best fits the time series.

Á. Herrero et al. (Eds.): SOCO 2020, AISC 1268, pp. 681–690, 2021.
https://doi.org/10.1007/978-3-030-57802-2_65

2 Materials

The training data set to be used will be the monthly copper prices in \$/t from January 1960 until August 2018, totalizing 704 values. The validation data set will be the monthly copper prices in \$/t from September 2018 until August 2019.

Both data series are from the World Bank Pink Sheet [4] and are used under a Creative Commons Attribution 4.0 International License [5].

Figure 1 presents Copper (LME), grade A, minimum 99.9935% purity, cathodes and wire bar shapes, settlement prices in \$/t from January 1960 until August 2019, comprising both the training data set and the validation data set [4].

Fig. 1. Copper (LME), grade A, minimum 99.9935% purity, cathodes and wire bar shapes, settlement prices in \$/t from January 1960 until August 2019 [4].

On the other hand, the programs used within this paper were @RISK 7.5 and Neural-Tools 7.5 from Palisade Corporation (Ithaca, New York). Both the University of Oviedo and the Central Mining Institute have license of this software.

3 Method

3.1 Length of the Lagged Variables

In order to estimate the optimal number of time-delayed input terms that should form the length of the lagged variables, also known as rolling windows [6], Ren et al. [7] used the seasonal characteristic of the autocorrelation function plot (ACF).

In this case, and the same as described in Matyjaszek et al. [8], after achieving mean and variance stationarity using a logarithmic transformation and a second order differencing deseasonalization with a period of 28 months when representing a consistent genome of copper price time series (Fig. 2), it was not possible to extract any seasonal component (Fig. 3).

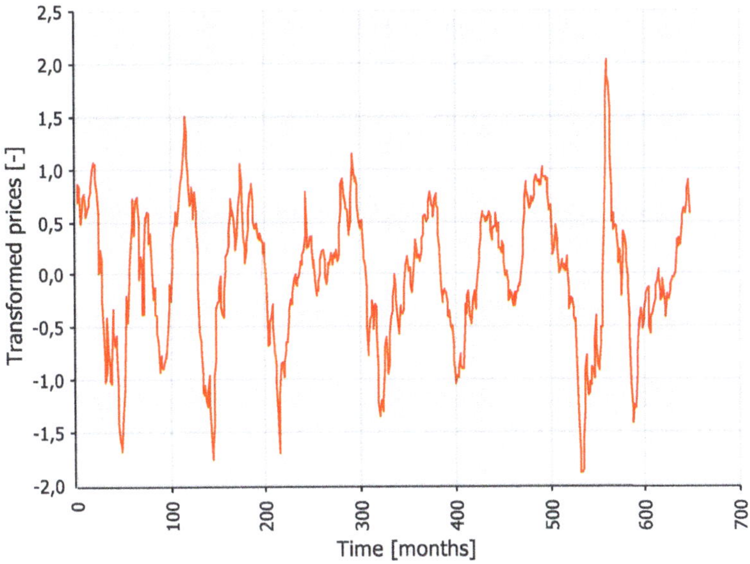

Fig. 2. Copper transformed prices after a logarithmic transformation and a second order differencing deseasonalization with a period of 28 months.

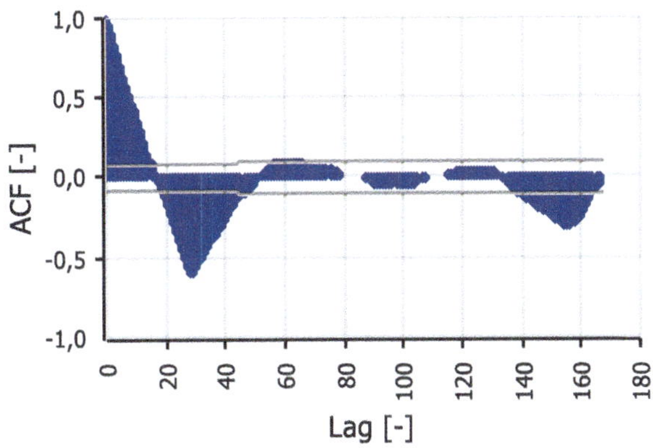

Fig. 3. Autocorrelation function plot with no seasonal component available

Among the alternative approaches to estimate this value apart from the minimum sample size requirement according to Turmon and Fine [9], the one presented in Matyjaszek et al. [8] was used: the number of time-delayed input terms will be coincident with the k value calculated by means of Eqs. (1) and (2):

$$\text{Total no of data} \leq n^2 + 2n + 1 \tag{1}$$

$$n = 1 + k + 1 \tag{2}$$

The value obtained with 704 monthly prices, corresponding to the period from January 1960 until August 2018 (training data set), is $k = 24$. Nevertheless, this value can only be considered as an approximation.

Thus, in order to test a wider range of values for the number of time-delayed input terms that should form the lagged variables, or the length of the rolling window, the GRNN will be trained starting from k = 12 until k = 36, exploring by doing this way any periodical features that may be hidden between a one and three year's period.

Table 1 presents as an example the first ten lagged variables of the GRNN model with 5 time-delayed input terms.

Table 1. First ten lagged variables for a rolling window size of $k = 5$ including the dependent variable to be estimated t.

Lagged variable	t–5	t–4	t–3	t–2	t–1	t
First	715.4	728.19	684.94	723.11	684.75	691.59
Second	728.19	684.94	723.11	684.75	691.59	702.39
Third	684.94	723.11	684.75	691.59	702.39	675.94
Fourth	723.11	684.75	691.59	702.39	675.94	647.06
Fifth	684.75	691.59	702.39	675.94	647.06	612.66
Sixth	691.59	702.39	675.94	647.06	612.66	642.13
Seventh	702.39	675.94	647.06	612.66	642.13	636.91
Eight	675.94	647.06	612.66	642.13	636.91	606.71
Ninth	647.06	612.66	642.13	636.91	606.71	616.63
Tenth	612.66	642.13	636.91	606.71	616.63	621.26

3.2 Figures of Merit

A combination of metrics are commonly used in order to evaluate different neural network models [11]. Following Lazaridis [12], RMSE and MAE were used, complemented with the STD of AE, in order to characterize the dispersion of absolute errors:

$$RMSE = \sqrt{\frac{\sum_{t=1}^{n}(A_t - F_t)^2}{n}}, \tag{3}$$

$$MAE = \frac{1}{n}\sum_{t=1}^{n}|A_t - F_t| \tag{4}$$

Where A_t and F_t are actual and forecasted values, and n is the number of forecasts.

4 Results and Discussion

4.1 Best Model Selection

Table 2 presents the training results of the GRNN using 12 to 36 time-delayed input terms in the lagged variable.

Table 2. Training results for the GRNN model using 12 to 36 time-delayed input terms (figures in bold correspond to the model that achieves better performance metrics).

Time-delayed input terms	No lagged variables	% Bad predictions with 5% tolerance	RMSE	MAE	STD of AE
12	693	21.7893%	77.96	50.65	59.27
13	692	23.8439%	81.89	49.35	65.35
14	691	20.5499%	74.61	42.17	61.55
15	690	24.9275%	86.52	52.74	68.58
16	689	19.3033%	69.45	40.27	56.58
17	688	24.5640%	89.41	51.75	72.92
18	687	21.9796%	84.83	48.20	69.81
19	686	21.1370%	78.64	44.49	64.84
20	685	19.4161%	73.88	41.03	61.44
21	684	21.0526%	73.49	40.39	61.40
22	683	19.3265%	68.21	37.16	57.19
23	682	18.7683%	66.96	35.43	56.81
24	681	25.8443%	88.67	51.36	72.29
25	680	24.5588%	85.31	48.89	69.91
26	679	23.7113%	82.92	47.01	68.31
27	678	20.0590%	75.12	40.57	63.22
28	677	20.6795%	73.74	41.17	61.18
29	676	29.1420%	105.68	62.65	85.11
30	675	17.9259%	70.10	38.06	58.87
31	674	16.6172%	67.57	36.25	57.02
32	673	16.3447%	66.94	35.01	57.06
33	672	15.3274%	65.10	33.73	55.68
34	671	14.4560%	62.67	32.31	53.70
35	670	12.9851%	60.48	30.75	52.07
36	**669**	**11.5097%**	**57.65**	**29.14**	**49.74**

Thus, the best model has 36 time-delayed input terms, with a percentage of bad predictions with a 5% tolerance of 11.5097%, a RMSE of 57.65, a MAE of 29.14 and a STD of AE of 49.74.

Now it was checked whether considering the order of the lagged variables in the time series improve or not the forecast. Table 3 presents the first three lagged variables of the GRNN model with 12 time-delayed input terms, including the order of the lagged variable in the time series. Table 4 presents the training results of the GRNN from 12 to 36 time-delayed input terms including the order of the lagged variable. It is interesting to highlight that even with the GRNN being fed with the order, it is not capable to detect that the dependent variable is the first number of the next lagged variable.

Table 3. First three lagged variables for a rolling window size of $k = 5$ plus the dependent variable t, and including the order of the lagged variables in the time series.

Lagged variable	Order	t–5	t–4	t–3	t–2	t–1	t
First	1	715.4	728.19	684.94	723.11	684.75	691.59
Second	2	728.19	684.94	723.11	684.75	691.59	702.39
Third	3	684.94	723.11	684.75	691.59	702.39	675.94

Thus, the best model in this case has 27 time-delayed input terms, with a RMSE of 55.29, a MAE of 26.51 and a STD of AE of 48.52, improving all the previous results.

Table 4. Training results for the GRNN model using 12 to 36 time-delayed input terms and including the order of the lagged variable in the time series (figures in bold correspond to the model that achieves better performance metrics).

Time-delayed input terms	No lagged variables	% Bad predictions with 5% tolerance	RMSE	MAE	STD of AE
12	693	15.4401%	62.88	39.49	48.94
13	692	12.4277%	53.66	29.90	44.56
14	691	18.9580%	72.45	42.68	58.54
15	690	17.6812%	68.97	40.22	56.04
16	689	22.3512%	84.26	51.28	66.85
17	688	24.8547%	90.88	53.84	73.21
18	687	14.4105%	63.25	34.69	52.89
19	686	24.0525%	87.35	50.80	71.06
20	685	23.7956%	84.26	48.54	68.87
21	684	12.2807%	56.91	29.91	48.41
22	683	21.0835%	79.01	44.58	65.23

(continued)

Table 4. *(continued)*

Time-delayed input terms	No lagged variables	% Bad predictions with 5% tolerance	RMSE	MAE	STD of AE
23	682	20.3812%	78.59	42.99	65.79
24	681	18.7959%	74.35	40.56	62.31
25	680	17.6471%	72.17	38.72	60.90
26	679	16.4948%	68.73	36.60	58.17
27	**678**	**11.6519%**	**55.29**	**26.51**	**48.52**
28	677	14.1802%	62.04	32.45	52.88
29	676	21.5976%	91.34	50.59	76.06
30	675	20.4444%	90.63	48.61	76.49
31	674	11.4243%	56.51	28.60	48.74
32	673	18.7221%	82.75	44.08	70.04
33	672	17.7083%	80.54	42.93	68.15
34	671	16.2444%	79.15	40.95	67.74
35	670	15.9701%	77.63	39.76	66.68
36	669	15.3961%	74.84	39.00	63.87

4.2 Copper Prices Forecast

Using a GRNN with 27 time-delayed input terms plus the order of the lagged variable in the time series, the copper price forecast for the period from September 2018 until August 2019 (validation data set) is presented in Table 5.

Table 5. Copper prices forecast.

Month	Price ($/t)
September 2018	6 051.05
October 2018	6 051.05
November 2018	6 051.05
December 2018	6 051.05
January 2019	6 051.05
February 2019	6 051.05
March 2019	6 051.05
April 2019	4 640.39
May 2019	5 147.12
June 2019	4 520.11
July 2019	4 611.31
August 2019	4 557.37

Table 6 presents the performance measures in order to evaluate this model.

Table 6. Performance measures.

Figure of merit	Value
RMSE	876.44
MAPE	0.1045
MAE	633.38

Finally, Fig. 4 presents the forecasted prices versus the validation data subset. As it can be clearly observed, during the first six months the forecasted prices, although being constant, are able to follow quite close the validation data subset. On the other hand, during the last six months, forecasted prices follow quite a different trend, with big differences compared with the validation data subset. Something similar to this happened to Krzemień et al. [13] when forecasting twelve months of European thermal coal spot prices with GRNN.

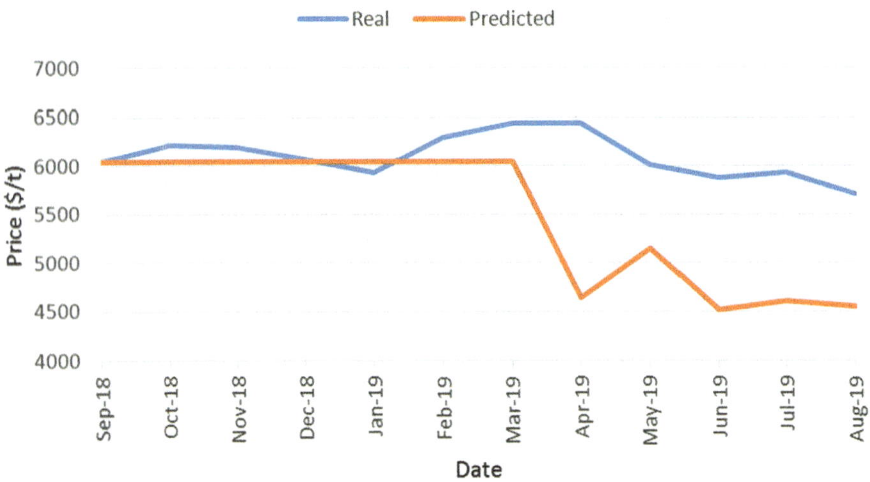

Fig. 4. Forecasted prices versus validation data subset.

Finally, Table 7 presents the performance measures considering only the six first forecasted prices that improve substantially the previous ones using 12 months.

Table 7. Performance measures for the first six months of the validation data subset.

Figure of merit	Value
RMSE	144.18
MAPE	0.0188
MAE	116.56

5 Conclusions

Firstly, in cases as the one analysed in this work, there is no proper method to select the optimum size of the rolling window so, until further research will be developed in this area, trial and error method over an estimated range should be executed.

Secondly, including the order in the time series of the lagged variables into the predictor variable helps improving the forecast accuracy. Nevertheless, this cannot be considered as an every-case rule. Again, further research is needed about this question.

Finally, the GRNN is able to achieve good figures of merit for the first six forecasting periods, while losing accuracy when the forecast is extended.

Despite all of these considerations, GRNN are generally able to improve other forecast methods [2] such as MARS models [14, 15], although with the disadvantage of being a 'black box'. Nevertheless, they are usually defeated when hybrid methods are applied [16].

References

1. Matyjaszek, M., Fidalgo Valverde, G., Krzemień, A., Wodarski, K., Riesgo Fernández, P.: Optimizing predictor variables in artificial neural networks when forecasting raw material prices for energy production. Energies **13**, 15 (2020)
2. Krzemień, A.: Dinamic fire risk prevention strategy in underground coal gasification processes by means of artificial neural networks. Arch. Min. Sci. **64**(1), 3–19 (2019)
3. Barabási, A-L.: Network Science. 1st ed., Cambridge University Press, Cambridge (2016)
4. World Bank. http://pubdocs.worldbank.org/en/561011486076393416/CMO-Historical-Data-Monthly.xlsx. Accessed 17 Apr 2020
5. Creative Commons Homepage (2008). https://creativecommons.org/licenses/by/4.0/. Accessed Jan 2020
6. Morantz, B.H., Whalen, T., Zhang, G.P.: A weighted window approach to neural network time series forecasting. In: Zhang, G.P. (ed.) Neural Networks in Business Forecasting. IRM Press (2004)
7. Ren, Y., Suganthan, P.N., Srikanth, N., Amaratunga, G.: Random vector functional link network for short-term electricity load demand forecasting. Inf. Sci. **367**, 1078–1093 (2016)
8. Matyjaszek, M., Riesgo Fernández, P., Krzemień, A., Wodarski, K., Fidalgo Valverde, G.: Forecasting coking coal prices by means of ARIMA models and neural networks, considering the transgenic time series theory. Resour. Policy **61**, 283–292 (2019)
9. Turmon, M.J., Fine, T.L.: Sample size requirements for feedforward neural networks. In: Advances in Neural Information Processing Systems, Denver, Colorado, USA, vol. 7, pp. 1–18 (1994)

10. Modaresi, F., Araghinejad, S., Ebrahimi, K.: A comparative assessment of artificial neural network, generalized regression neural network, least-square support vector regression, and K-nearest neighbor regression for monthly streamflow forecasting in linear and nonlinear conditions. Water Resour. Manag. **32**(1), 243–258 (2017). https://doi.org/10.1007/s11269-017-1807-2
11. Chai, T., Draxler, R.R.: Root mean square error (RMSE) or mean absolute error (MAE)?-arguments against avoiding RMSE in the literature. Geosci. Model Dev. **7**, 1247–1250 (2014)
12. Lazaridis, A.G.: Prosody modelling using machine learning techniques for neutral and emotional speech synthesis, Department of Electrical and Computer Engineering Wire Communications Laboratory, University of Patras, Greece (2011)
13. Krzemień, A., Riesgo Fernández, P., Suárez Sánchez, A., Sánchez Lasheras, F.: Forecasting European thermal coal spot prices. J. Sustain. Min. **14**, 203–210 (2015)
14. García Nieto, P.J., Alonso Fernández, J.R.R., Sánchez Lasheras, F., de Cos Juez, F.J., Díaz Muñiz, C.: A new improved study of cyanotoxins presence from experimental cyanobacteria concentrations in the Trasona reservoir (Northern Spain) using the MARS technique. Scien. Tot. Environ. **430**, 88–92 (2012)
15. Krzemień, A.: Fire risk prevention in underground coal gasification (UCG) within active mines: temperature forecast by means of MARS models. Energy **170**, 777–790 (2019)
16. Ordóñez, C., Sánchez Lasheras, F., Roca-Pardiñas, J., de Cos Juez, F.J.: A hybrid ARIMA–SVM model for the study of the remaining useful life of aircraft engines. J. Comput. Appl. Math. **346**, 184–191 (2018)

A Multivariate Approach to Time Series Forecasting of Copper Prices with the Help of Multiple Imputation by Chained Equations and Multivariate Adaptive Regression Splines

Fernando Sánchez Lasheras[1(✉)], Javier Gracia Rodríguez[2], Paulino José García Nieto[1], Esperanza García-Gonzalo[1], and Gregorio Fidalgo Valverde[2]

[1] Faculty of Sciences, University of Oviedo, c/ Federico García Lorca 18, 33007 Oviedo, Spain
{sanchezfernando,pjgarcia,espe}@uniovi.es

[2] School of Mining, Energy and Materials Engineering, University of Oviedo, c/ Independencia 13, 33004 Oviedo, Spain
{graciajavier,gfidalgo}@uniovi.es

Abstract. This research presents a novel methodology for the forecasting of copper prices using as input information the values of this non-ferrous material and the prices of other raw materials. The proposed methodology is based on the use of multiple imputation with chained equations (MICE) in order to forecast the values of the missing data and then to train multivariate adaptive regression splines models capable of predicting the price of copper in advance. The performance of the method was tested with the help of a database of the monthly prices of 72 different raw materials, including copper. The information available starts on January 1960. The prediction of prices from September 2018 to August 2019 showed a root mean squared error (RMSE) value of 318.7996, a mean absolute percentage error (MAPE) of 0.0418 and a mean absolute error (MAE) of 252.8567. The main strengths of the proposed algorithm are two-fold. On the one hand, it can be applied in a systematic way and the results are obtained without any human with expert knowledge having to take any decision; on the other hand, all the trained models are MARS. This means that the models are equations that can be read and understood, and not black box models like artificial neural networks.

Keywords: Copper price · Time series forecasting · Multiple Imputation by Chained Equations (MICE) · Multivariate Adaptive Regression Splines (MARS)

1 Introduction

Non-ferrous metals play a key role in the development of many products and technologies. The production and sales of this kind of metal are affected by crises and economic cycles [1]. Nowadays, one of the most important non-ferrous metals is copper. It is the non-precious metal which best conducts electricity. This, together with its ductility and

Á. Herrero et al. (Eds.): SOCO 2020, AISC 1268, pp. 691–701, 2021.
https://doi.org/10.1007/978-3-030-57802-2_66

resistance to corrosion, have made it the material most used for manufacturing electrical cables. Copper conductors are also used in various types of energy-efficient electrical equipment, such as generators, motors and transformers. Indeed, most telephone cables are copper, which also allows internet access. It should also be considered that all computers and telecommunications equipment contain copper to a greater or lesser extent in their integrated circuits, transformers and wires. Renewable energy sources will be crucial in meeting the growing demand for energy that will accompany the industrial development of the 21st century. A simple turbine contains more than a ton of copper. All these systems rely heavily on copper to transmit the energy they generate with maximum efficiency and minimum environmental impact. Despite the fact that aluminium is an alternative for copper in some applications, copper is one of the most important metals in the raw materials markets.

Copper, like zinc, platinum or the soya bean, is one of the raw materials traded in future commodity markets. Copper's demand is mainly linked with electrical, industrial and building economic sectors. The evolution of the price of copper depends on several factors related to the world economic situation and the price of the US Dollar, as it is usually traded in this currency. Concerning the evolution of the price of copper, it can be said that at the beginning of the 21st century and up to 2006 there was a significant upward trend. When the financial crisis of 2007 began, prices fell. The historical maximum was achieved in January 2011, with almost 9,900 US Dollars per metric ton. Nowadays, the price of copper remains stable, with an average value of 6,600 US Dollars per metric ton in the last year.

The aim of the present research is to forecast the future prices of copper considering its previous prices, as well as to forecast the prices of other raw materials. Due to the large amount of information available from different markets, the source of data for the present research will be the World Bank Commodity Price Data, with prices in nominal US dollars.

There have been many different attempts at predicting metal prices, not only in the case of copper [2, 3] but also of other metals such as iron ore [4], rare earths [5], thermal coal [6] and even the profitability of tungsten mining projects [7]. These references, and others like them, make use of a wide range of methodologies, such as general time series methods [8] or more specific ARIMA [2, 5, 9]; Artificial Neural Networks [2, 10].

2 Materials and Methods

2.1 The Database

The database employed for this study is the World Bank Commodity Price Data [11]. It is available online and used under a Creative Commons Attribution 4.0 International License (CC BY 4.0). The database has a total of 72 variables, after excluding those that are nowadays discontinued and so are not employed in the present research. Table 1 shows all the variables included in this work and the units in which their prices are measured. The information stored starts in January 1960. Prices of all the available variables from this month to August 2019 were retrieved. Please note that not all the information of all the variables for every month is available, as the record of some of them started after the date referred to. Also, it is possible that after starting to record

Table 1. Variables included in the present research and units in which their prices are measured.

Variable	Unit	Variable	Unit	Variable	Unit
Aluminum	($/mt)	Liquefied natural gas	($/mmbtu)	Sawnwood, Cameroon	($/cubic meter)
Banana, Europe	($/kg)	Logs, Cameroon	($/cubic meter)	Sawnwood, Malaysian	($/cubic meter)
Banana, US	($/kg)	Logs, Malaysian	($/cubic meter)	Shrimps, Mexican	($/kg)
Barley	($/mt)	Maize	($/mt)	Silver	($/troy oz)
Beef	($/kg)	Meat, chicken	($/kg)	Sorghum	($/mt)
Coal, Australian	($/mt)	Meat, sheep	($/kg)	Soybean meal	($/mt)
Coal, South African	($/mt)	Natural gas index	(2010 = 100)	Soybean oil	($/mt)
Cocoa	($/kg)	Natural gas, Europe	($/mmbtu)	Soybeans	($/mt)
Coconut oil	($/mt)	Natural gas, US	($/mmbtu)	Sugar, EU	($/kg)
Coffee, Arabica	($/kg)	Nickel	($/mt)	Sugar, US	($/kg)
Coffee, Robusta	($/kg)	Orange	($/kg)	Sugar, world	($/kg)
Copper	($/mt)	Palm kernel oil	($/mt)	Sunflower oil	($/mt)
Cotton, A Index	($/kg)	Palm oil	($/mt)	Tea, avg 3 auctions	($/kg)
Crude oil, average	($/bbl)	Phosphate rock	($/mt)	Tea, Colombo	($/kg)
Crude oil, Brent	($/bbl)	Platinum	($/troy oz)	Tea, Kolkata	($/kg)
Crude oil, Dubai	($/bbl)	Plywood	(cents/sheet)	Tea, Mombasa	($/kg)
Crude oil, WTI	($/bbl)	Potassium chloride	($/mt)	Tin	($/mt)
DAP	($/mt)	Rapeseed oil	($/mt)	Tobacco, US import u.v.	($/mt)
Fish meal	($/mt)	Rice, Thai 25%	($/mt)	TSP	($/mt)
Gold	($/troy oz)	Rice, Thai 5%	($/mt)	Urea	($/mt)

(continued)

Table 1. (*continued*)

Variable	Unit	Variable	Unit	Variable	Unit
Groundnut oil	($/mt)	Rice, Thai A.1	($/mt)	Wheat, US HRW	($/mt)
Groundnuts	($/mt)	Rice, Viet Namese 5%	($/mt)	Wheat, US SRW	($/mt)
Iron ore, cfr spot	($/dmtu)	Rubber, SGP/MYS	($/kg)	Woodpulp	($/mt)
Lead	($/mt)	Rubber, TSR20	($/kg)	Zinc	($/mt)

the values of a variable, data for certain months is missing. It is not frequent, but in this case, for example, it happens with the prices of sunflower oil.

2.2 Multiple Imputation by Chained Equations

In recent years, multivariate imputation by chained equations (MICE) has become one of the most appealing methodologies for missing data imputation [12]. It was originally developed by van Buuren and Groothuis-Oudshoorn [13]. The use of the multiple imputation means that more than one forecast of the missing values is performed, or, in other words, a large number of complete candidate data sets are created during the imputation process. The MICE algorithm is a Markov Chain Monte Carlo (MCMC) method, where the state space is the collection of all imputed values [13].

The MICE algorithm works on the assumption that missing information is missing at random [14].

In the MICE algorithm a set of regression models are run whereby each variable with missing data is modelled conditionally upon the other variables in the data [15]. From a practical point of view, the consequence is that each variable is modelled according to its distribution [12]. The MICE algorithm has four main steps, as follows:

- In a first step, missing values are imputed by a simple method such as replacing them by the mean of the variable that they belong to.
- Afterwards. imputed values of only one of the imputed variables are removed again.
- Next, the values of the missing data from the previous step are calculated with the help of a regression model.
- The missing values are then replaced by those obtained with the help of the regression model.

The cycle now starts again by removing the values of any of the other variables that have been imputed using the mean to calculate them with the help of a regression model. This process is repeated for a certain number of cycles, with the imputation results updated after each cycle. At the end of the cycles, the last imputation is considered as the final imputed data set.

As is well-known [13], in order to converge, any Markov chain has to fulfil the three following properties [13]: irreducible, aperiodic and recurrence. MICE fulfills these three properties.

Finally, it can be said that the MICE algorithm has already been employed in different problems such as the imputation of electrical variables [14] or missing answers in questionnaires [16], underlining the usefulness of this method.

2.3 Multivariate Adaptive Regression Splines

Multivariate adaptive regression splines (MARS) is a multivariate regression technique developed by Friedman [17]. This regression method is employed for the prediction of continuous variables, using a multivariate set as independent variables.

The MARS model can be represented as follows [18]:

$$y = f(x) + e$$

Where e is the model error, if the same length as the number of variables in x and y is the dependent variable. The function f represents a weighted sum of basis functions in a simplified way.

One of the main advantages of MARS is that it does not require any a priori assumptions about what the relationship between dependent and independent variables is [19].

MARS models make use of basis functions as elemental mathematical components that determine which variables will take part in it [17]. In order to know which basis functions are to be included in a model, these methods that determine their importance are generally employed.

The first of them is generalized cross-validation (GCV) [20]. Its formula is as follows:

$$GCV(M) = \frac{\frac{1}{n}\sum_{i=1}^{n}\left(y_i - \hat{f}_M(x_i)\right)^2}{1 - \frac{C(M)}{n^2}} \tag{1}$$

f_M represents a MARS model with M basis functions that forecast the value of y_i, $C(M)$ is a complexity penalty function that increases as the number of basis functions in the model grows. It can be expressed as follows:

$$C(M) = (M + 1) + d \cdot M \tag{2}$$

Where M is the number of basis functions and d a penalty value that in the present research, as in most cases [18] has been fixed at 2.

Another of the parameters employed to find out the importance of each variable in a MARS model is the saw residual sum of squares (RSS). The RSS can be expressed as follows:

$$RSS = GCV \cdot N \cdot \left(1 - \frac{ep}{N}\right)^2 \tag{3}$$

Where N is the number of observations and ep is the effective number of parameters, which in the context of MARS models is as follows:

$$ep = N + d \cdot \frac{(N-1)}{2} \tag{4}$$

Finally, the third criterion employed to establish the importance of a variable in a model is called n subsets. It represents the number of subsets that include each of the variables participating in the model [17].

Finally, the performance of the MARS model obtained will be performed with the help of r-squared $\left(R^2\right)$, root mean squared error [21, 22], mean absolute percentage error and mean absolute error [23, 24].

The MARS method has successfully been employed in the past by the authors in problems such as credit scoring [18] or the forecasting of the presence of cyanotoxin in continental waters [19].

2.4 The Algorithm

Figure 1 shows the flowchart of the proposed algorithm. After loading the available information, the first step consists of entering the missing data in all the variables with the help of the MICE algorithm. Once the data set is complete, a MARS model is trained that uses the price of copper as its output variable and the rest of variables as inputs. This MARS model can assess the importance of variables with the help of the parameter nsubsets, GCV and RSS. All the variables that are found to be of importance in this model will be employed for the training of the following models. In our case, no cutting points in these parameters are fixed, but all the variables employed for the MARS model of the second step of the flowchart are employed in the following models. Afterwards, a set of models is trained that is able to predict copper prices from 1 to 12 months ahead. Please note that, for example, the model that is trained to predict the value of copper in the i-th month makes use of the values of copper in the previous months, but in our case this does not mean that we are employing data for prediction beyond August 2018. Finally, these models are used to forecast copper prices from September 2018 to August 2019.

3 Results

3.1 Missing Data Imputation

12.52% of the data in the database employed for this study was missing. However, missing data was not equally distributed in all variables but was present in 22 of the 72 variables. This means that the average percentage of missing data per variable was 41.05%. Please note that the imputed database that we are employing corresponds to the information available from January 1960 to August 2018, as the prices to be forecasted are monthly values from September 2018 to August 2019.

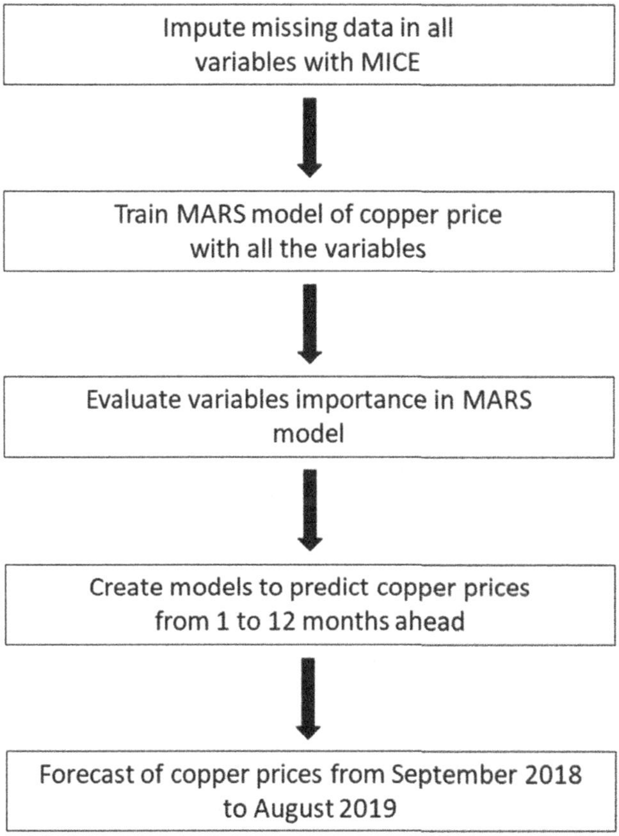

Fig. 1. Flowchart of the proposed algorithm.

3.2 Training of a Copper Price MARS Model with All the Variables

After the imputation of missing data, a MARS model was trained using the copper price as its output variable and the rest of variables as inputs. The R^2 of the model obtained, which has degree 6, was 0.9965. For the model training the threshold value was fixed at 10^{-8} and the maximum degree of variables interaction allowed was 9.

Of a total of 72 variables employed for training the MARS model, only 23 were found to be relevant, and therefore the rest were not included in the mathematical model obtained. The importance of those variables is shown in Table 2 where they are listed according to their nsubsets, GCV and RSS values. Please note that in this case, the order is the same for all three variables.

3.3 Training of MARS Models for Prediction from One to Twelve Months Ahead

The next step in the process involves training a set of twelve MARS models that are able to predict the copper price from one to twelve months in advance. These models use as training information the values from January 1960 to August 2018 of the 23 variables

Table 2. Variables importance measured by means of nsubsets, GCV and RSS.

Variable	nsubsets	GCV	RSS
Platinum	45	100	100
Zinc	44	35.5	35.5
Sawnwood, Malaysian	43	17.1	17.4
Rubber, SGP/MYS	43	17.1	17.4
Potassium chloride	42	14.9	15.1
Beef	41	13.7	14
Aluminum	40	12.8	13.1
Tin	39	11.9	12.1
Crude oil, Dubai	38	10.9	11.2
Sorghum	37	10	10.3
Lead	36	9.2	9.5
Fish meal	33	8.1	8.4
Plywood	33	8.1	8.4
Woodpulp	33	8.1	8.4
Tobacco, US import u.v.	29	6.7	7
Logs, Malaysian	27	6.4	6.6
Coconut oil	26	6	6.3
Logs, Cameroon	26	6	6.3
Groundnut oil	21	4.9	5.2
Soybeans	17	4	4.3
Shrimps, Mexican	16	3.7	4
Rice, Viet Namese 5%	11	2.4	2.8
DAP	8	1.6	2.1

found to be important for the previous model. In all the cases, the threshold employed was 10^{-8} with a maximum allowed degree for the model of 9. Please note that the results obtained showed that the degree of all models was from 3 to 5 with a minimum R^2 value of 0.9929.

The performance of the model is assessed with the help of root mean squared error (RMSE), mean absolute percentage error (MAPE) and mean absolute error (MAE) [25, 26]. Figure 2 shows a comparison of real and forecasted prices of copper from September 2018 to August 2019 expressed in dollars per metric ton. For this period the results show a RMSE value of 318.7996, MAPE of 0.0418 and MAE of 252.8567.

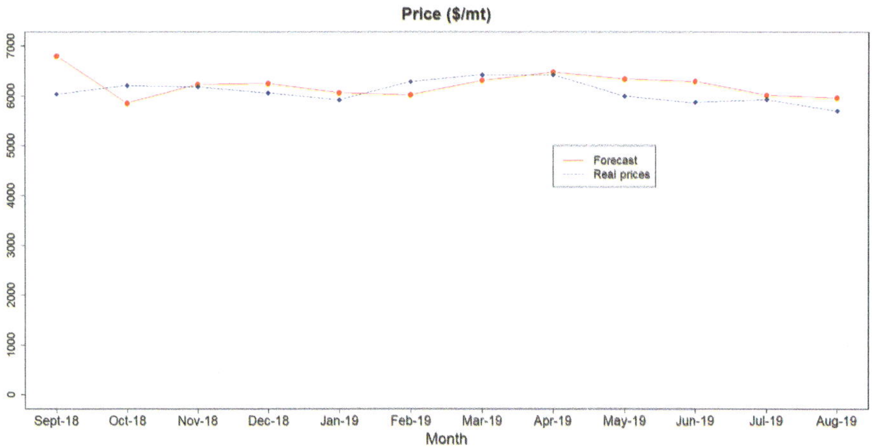

Fig. 2. Comparison of real and forecasted prices of copper from September 2018 to August 2019 expressed in dollars per metric ton.

4 Discussion and Conclusions

This research presents a novel method for the forecasting of monthly values of copper prices in a multivariate way that takes into account the values of copper in previous months and also the values of other raw materials. The main strengths of the proposed algorithm are that it can be applied in a systematic way, that the results are obtained without any decision that requires any kind of expert knowledge having to be taken and that all the models computed are MARS models in which relationships among variables are expressed by equations and not in the "blackbox" way as it happens in neural networks. In other words, the same methodology could be applied in an automatic way to any other time period or non-ferrous metal, or even any other variable expressed as a time series and also, a model that could be interpreted by the user would be trained.

We would like to remark that, as happens in the case of most of the forecasting methodologies applied to the stock market that try to forecast the evolution of either stocks or raw materials, one of the main weaknesses of the method is that there are many other exogenous variables (economical and financial news of the firms, political information and social movements, information given by the media, etc.) that are not taken into account and can have a greater influence over the future prices than those variables considered in the model.

Finally, from our point of view, it is also remarkable that this multivariate methodology would be employed in other fields. In general, it could be useful in any case where the evolution in time of data series is likely to depend on some other covariates. For example, in environmental research where the changes in the concentration of any pollutant would be affected by the concentration of the rest.

References

1. Iglesias García, C., Sáiz Martinez, P., García-Portilla González, M.P., Bousoño García, M., Jiménez Treviño, L., Sánchez Lasheras, F., Bobes, J.: Effects of the economic crisis on demand due to mental disorders in Asturias: data from the Asturias Cumulative Psychiatric Case Register (2000–2010). Actas Esp. Psiquiatr. **42**, 108–15 (2014)
2. Sánchez Lasheras, F., de Cos Juez, F.J., Suárez Sánchez, A., Krzemien, A., Riesgo Fernández, P.: Forecasting the COMEX copper spot price by means of neural networks and ARIMA models. Resour. Policy **45**, 37–43 (2015)
3. Tilton, J.E., Lagos, G.: Assessing the long-run availability of copper. Resour. Policy **32**, 19–23 (2007)
4. Ma, W., Zhu, X., Wang, M.: Forecasting iron ore import and consumption of China using grey model optimized by particle swarm optimization algorithm. Resour. Policy **38**, 613–620 (2013)
5. Riesgo García, M.V., Krzemień, A., Manzanedo del Campo, M.Á., Escanciano García-Miranda, C., Sánchez Lasheras, F.: Rare earth elements price forecasting by means of transgenic time series developed with ARIMA models. Resour. Policy **59**, 95–102 (2018)
6. Krzemień, A., Riesgo Fernández, P., Suárez Sánchez, A., Sánchez Lasheras, F.: Forecasting European thermal coal spot prices. J. Sustain. Min. **14**, 203–210 (2015)
7. Suárez Sánchez, A., Krzemień, A., Riesgo Fernández, P., Iglesias Rodríguez, F.J., Sánchez Lasheras, F., de Cos Juez, F.J.: Investment in new tungsten mining projects. Resour. Policy **46**, 177–190 (2015)
8. Dooley, G., Lenihan, H.: An assessment of time series methods in metal price forecasting. Resour. Policy **30**, 208–217 (2005)
9. Kriechbaumer, T., Angus, A., Parsons, D., Rivas Casado, M.: An improved wavelet–ARIMA approach for forecasting metal prices. Resour. Policy **39**, 32–41 (2014)
10. Khashei, M., Bijari, M.: An artificial neural network (p, d, q) model for timeseries forecasting. Expert Syst. Appl. **37**, 479–489 (2010)
11. World Bank Data. https://www.worldbank.org/en/research/commodity-markets Accessed 2 Jan 2020
12. Azur, M.J., Stuart, E.A., Frangakis, C., Leaf, P.J.: Multiple imputation by chained equations: what is it and how does it work? Int. J. Meth. Psy. Res. **20**(1), 40–49 (2011)
13. van Buuren, S., Groothuis-Oudshoorn, K.: mice: multivariate imputation by chained equations in R. J. Stat. Softw. **45**(i03) (2011)
14. Crespo Turrado, C., Sánchez Lasheras, F., Calvo-Rollé, J.L., Piñón-Pazos, A.J., de Cos Juez, F.J.: A new missing data imputation algorithm applied to electrical data loggers. Sensors **15**, 31069–31082 (2015)
15. de Cos Juez, F.J., Sánchez Lasheras, F., García Nieto, P.J., Álvarez-Arenal, A.: Non-linear numerical analysis of a double-threaded titanium alloy dental implant by FEM. Appl. Math. Comput. **206**, 952–967 (2008)
16. Ordóñez Galán. C., Sánchez Lasheras, F., de Cos Juez, F. J., Bernardo Sánchez, A.: Missing data imputation of questionnaires by means of genetic algorithms with different fitness functions. J. Comput. Appl. Math. **311**, 704–717 (2017)
17. Friedman, J.H.: Multivariate adaptive regression splines. Ann. Stat. **19**, 1–141 (1991)
18. de Andrés, J., Sánchez-Lasheras, F., Lorca, P., de Cos Juez, F.J.: A hybrid device of self organizing maps (som) and multivariate adaptive regression splines (mars) for the forecasting of firms' bankruptcy. J. Account. Manag. Inf. Syst. **10**, 351–374 (2011)
19. Garcia Nieto, P.J., Sánchez Lasheras, F., de Cos Juez, F.J., Alonso Fernández, J.R.: Study of cyanotoxins presence from experimental cyanobacteria concentrations using a new data mining methodology based on multivariate adaptive regression splines in Trasona reservoir (Northern Spain). J. Hazard. Mater. **195**, 414–421 (2011)

20. Sekulic, S., Kowalski, B.R.: MARS: a tutorial. J. Chemometr. **6**, 199–216 (1992)
21. García Nieto, P.J., Sánchez Lasheras, F., García-Gonzalo, E., de Cos Juez, F.J.: PM10 concentration forecasting in the metropolitan area of Oviedo (Northern Spain) using models based on SVM, MLP, VARMA and ARIMA: a case study. Sci. Total Environ. **621**, 753–761 (2018)
22. de Cos Juez, F.J., Lasheras, F.S., Roqueñí, N., Osborn, J.: An ANN-based smart tomographic reconstructor in a dynamic environment. Sensors **12**, 8895–8911 (2012)
23. Krzemień, A.: Fire risk prevention in underground coal gasification (UCG) within active mines: temperature forecast by means of MARS models. Energy **170**, 777–790 (2019)
24. Krzemień, A.: Dynamic fire risk prevention strategy in underground coal gasification processes by means of artificial neural networks. Arch. Min. Sci. **64**(1), 3–19 (2019)
25. Hyndman, R.J., Koehler, A.B.: Another look at measures of forecast accuracy. Int. J. Forecasting. **22**, 679–688 (2006)
26. Ordóñez Galan, C., Sánchez Lasheras, F., Roca Pardiña, J., de Cos Juez, F.J.: A hybrid ARIMA-SVM model for the study of the remaning useful life of aircraft engines. J. Comput. Appl. Math. **346**, 184–191 (2019)

Time Series Analysis for the COMEX Copper Spot Price by Using Support Vector Regression

Esperanza García-Gonzalo[1](✉), Paulino José García Nieto[1], Javier Gracia Rodríguez[2], Fernando Sánchez Lasheras[1], and Gregorio Fidalgo Valverde[2]

[1] Faculty of Sciences, University of Oviedo, c/Federico García Lorca 18, 33007 Oviedo, Spain
{espe,pjgarcia,sanchezfernando}@uniovi.es

[2] School of Mining, Energy and Materials Engineering, University of Oviedo, c/Independencia 13, 33004 Oviedo, Spain
{graciajavier,gfidalgo}@uniovi.es

Abstract. In this research work, support vector regression (SVR), a powerful machine learning tool that can identify nonlinear patterns effectively thanks to the introduction of a kernel function is used. Three different strategies (direct multi-step scheme, recursive multi-step scheme and direct-recursive hybrid scheme) for automatic lag selection in time series analysis are proposed. This article examines the forecasting performance of the three kinds of SVR models using published data of copper spot prices from the New York Commodities Exchange (COMEX). The numerical results obtained have shown a better performance of the direct-recursive hybrid scheme than the recursive multi-step scheme and direct multi-step scheme. The findings of this research work are in line of with some previous studies, which confirmed the superiority of SVR models over other classical techniques in relative research areas.

Keywords: Time series analysis · Support Vector Regression (SVR) · Copper price forecasting · New York Commodity Exchange (COMEX)

1 Introduction

Nonferrous metals are essential raw materials that are crucial for measuring the global economy. However, these materials, such as fossil fuels, are a limited resource. The production of nonferrous metals is strongly affected by several factors: supply, demand and share prices of non-ferrous metal companies. Copper is one of the main metal commodities and a nonferrous metal traded in the major physical futures trading exchanges [1–3]: the London Metal Exchange (LME), the New York Commodity Exchange (COMEX), and the Shanghai Futures Exchange (SHFE). Prices on these exchanges reflect the balance between copper supply and demand at a worldwide level, although they may be strongly influenced by currency exchange rates and investment flows, factors that may because volatile price fluctuations partially linked to changes in business cycle activity [4–6].

Á. Herrero et al. (Eds.): SOCO 2020, AISC 1268, pp. 702–708, 2021.
https://doi.org/10.1007/978-3-030-57802-2_67

Several methodologies have been used for metal price forecasting. Dooley and Lenihan [7] used two time-series forecasting techniques to conclude that ARIMA modelling provides marginally better forecast results than lagged forward price modelling. Cortazar and Eterovic [8] proposed multicommodity models to help estimate long term copper and silver futures prices. On the other hand, Khashei et al. [9] prefer artificial neuronal networks for time series forecasting. Ma et al. [10] proposed a grey model, optimized by particle swarm algorithm, to forecast iron ore import and consumption in China. Kriechbaumer et al. [11] decompose time series into its frequency and time domain to capture this cyclic behaviour dominant in the metal market. Finally, Sánchez Lasheras et al. [12] examine the forecasting performance of ARIMA model and two different neuronal networks to forecast the COMEX copper spot price.

In this article, a new methodology to foretell the COMEX copper spot price has been built and implemented. This paper introduces a novel methodology for estimating the COMEX copper spot price using support vector regression (SVR) in time series analysis [13, 14] with three different strategies: direct multi-step scheme, recursive multi-step scheme and direct-recursive hybrid scheme. The proposed method uses a kernel-penalized optimization of all hyperparameters in SVR identifying nonlinear input features with success.

2 Materials and Methods

2.1 Experimental Dataset

The present research used, as its main source of data, the monthly copper spot closing price from the COMEX from January 1960 till October 2019. The data set is the World Bank Commodity Price Data (The Pink Sheet): Copper (LME), grade A, minimum 99.9935% purity, cathodes and wire bar shapes, settlement price provided by: Bloomberg; Engineering and Mining Journal; Platts Metals Week; and Thomson Reuters Datastream; World Bank [15].

2.2 Support Vector Regression (SVR) for Time Series Analysis

This section presents $\varepsilon-$ SVR [16, 17] for time series analysis. Given a set of time series data, a training set consisting of a continuous dependent variable $y_i \in \Re,\ \forall i = 1, 2, \ldots, m$ and covariates $\mathrm{x}_i \in \Re^p,\ \forall i = 1, 2, \ldots, m$ can be constructed by taking p lags of y_i. The method $\varepsilon-$ SVR constructs a function $f(\mathrm{x}) = \mathrm{w}^T\mathrm{x} + b$, $\mathrm{w} \in \Re^n$, $b \in \Re$ that has at most a deviation of ε from y_i for all training instances x_i, and at the same time is as flat as possible [16–19]. Among a variety of kernel functions available, the radial basis function (RBF) kernel is chosen in many applications and in this research due to its superior performance [17–19].

2.3 Computational Procedure and Numerical Schemes

The training dataset comprises the data from January 1960 to August 2018 while the forecasted monthly prices start in September 2018 and end in August 2019. Thus, in this particular case, we must forecast twelve steps ahead. Thus, we will be performing multi-step forecasting. Three different strategies for the building of the training data will be used:

1. Direct multi-step forecast;
2. Recursive multi-step forecast; and
3. Direct-recursive hybrid forecast.

We have started using only one variable. The obvious variable is the copper price in previous years. Once this model is constructed, we have tried to improve the best model adding new variables from the dataset but no significant improvement was observed and thus, we have not included these other models in this study. Next, we are going to describe below the three different strategies for this problem of multi-step forecast.

Direct Multi-step Forecast

In this scheme, we construct different models for the different ahead forecasting:

$$\begin{aligned} pred(t+1) &= model1(obs(t), obs(t-1), \ldots, obs(t-s)) \\ pred(t+2) &= model2(obs(t), obs(t-1), \ldots, obs(t-s)) \\ \ldots & \\ pred(t+12) &= model12(obs(t), obs(t-1), \ldots\ldots, obs(t-s)) \end{aligned} \tag{1}$$

As we can see, the training set $(obs(t), obs(t-1), \ldots, obs(t-s))$ is the same for all the models but twelve different models have been constructed, one for each prediction. These models depend on five parameters: the first one is the lag, that is, the time period of observations used for each sample in the training set. In this case, we use s+1 observations per model. The observations in a given time can comprise one or more variables. We have started with only one variable, the copper price. The second parameter is the number of samples used. It depends on how much we go back in time taking samples into account to construct our model. Sometimes, the behavior of a variable changes with time and the model benefits from dropping samples during the first years. Finally, the last three parameters are those related with the method used, in this case, SVR technique with RBF kernel.

Recursive Multi-step Forecast

In this case, we construct only a model that could be the same as *model1* of the previous method. Then, at each step, we forecast only the next value. Then, we incorporate the predicted value, drop the oldest value and predict the next value. Thus, once the model has been constructed, the prediction process will be as follows:

$$\begin{aligned} pred(t+1) &= model(obs(t), obs(t-1), \ldots, obs(t-s)) \\ pred(t+2) &= model(pred(t+1), obs(t), obs(t-1), \ldots, obs(t-s+1)) \\ pred(t+3) &= model(pred(t+2), pred(t+1), obs(t), \ldots, obs(t-s+2)) \\ \ldots & \\ pred(t+12) &= model(pred(t+11), pred(t+10), \ldots, obs(t-s+12)) \end{aligned} \tag{2}$$

As we can see, we have a unique model. When we are predicting we move forward one step, incorporate the last prediction and drop the oldest observation. We have the same parameters as in the previous case.

Direct-Recursive Hybrid Forecast

This numerical scheme is a mixture of the two previous ones. We create a different model for each prediction but, in the predicting stage, the models are able to incorporate the predicted values one by one. In this case, the lag for each model increases as we advance in the prediction. That is, if we start with s + 1 observations for the first model, the second model will use one observation more, as it incorporates (in the forecasting stage) the newly predicted value.

$$\begin{aligned} pred(t+1) &= model1(obs(t), obs(t-1), \ldots, obs(t-s)) \\ pred(t+2) &= model2(pred(t+1), obs(t), obs(t-1), \ldots, obs(t-s)) \\ pred(t+3) &= model3(pred(t+2), pred(t+1), obs(t), obs(t-1), \ldots, obs(t-s)) \\ \ldots & \\ pred(t+12) &= model12(pred(t+11), pred(t+10), \ldots, obs(t-s)) \end{aligned} \tag{3}$$

In this case, we incorporate the predictions but we do not drop old observations as we advance in the prediction.

3 Results and Discussion

For the three numerical schemes, only a variable (copper price) has been used. All the available data has been used as training data. The available data set for training consist in the monthly copper prices between January 1960 and August 2017. The data between September 2017 and August 2018 has been used as validation set to optimize the hyperparameters with the grid-search method. Different models where created with the training data and the optimal hyperparameters were obtained with the grid-search method, using the validation set. The number of training samples varies with the lag. The shorter the lag, the greater the number of available samples, as a sample uses less observations and they span for a shortest period of time, allowing more samples with the same data. As the aim is to forecast monthly prices from September 2018 till August 2019, all the data related with this period of time (and the following one) have not been used during the training phase.

Table 1 indicates the goodness-of-fit parameters for the three different numerical schemes.

Table 1. Good-of-fitness parameters for the three different numerical schemes.

Numerical scheme	MAPE (%)	MAE	RMSE
Direct multi-step	9.4621	569.76	660.16
Recursive multi-step	5.7013	343.33	400.26
Direct-recursive hybrid	2.3647	144.21	170.15

Finally, Fig. 1 indicates observed and predicted COMEX copper spot price values using as predictor the SVR technique with a RBF kernel for the three different schemes.

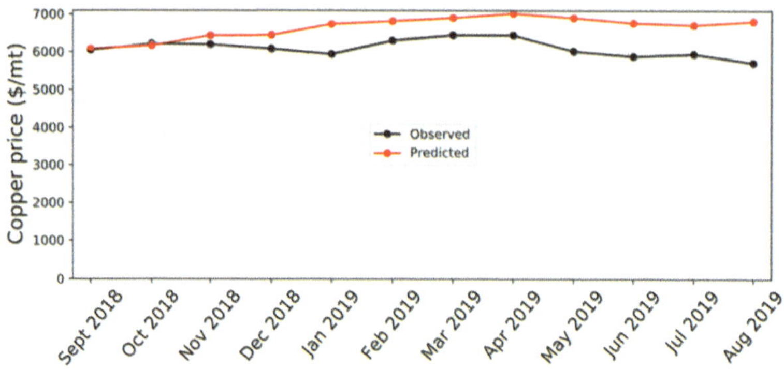

(a)

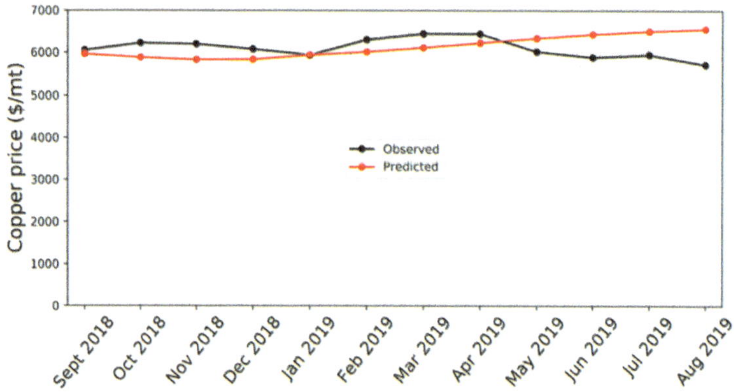

(b)

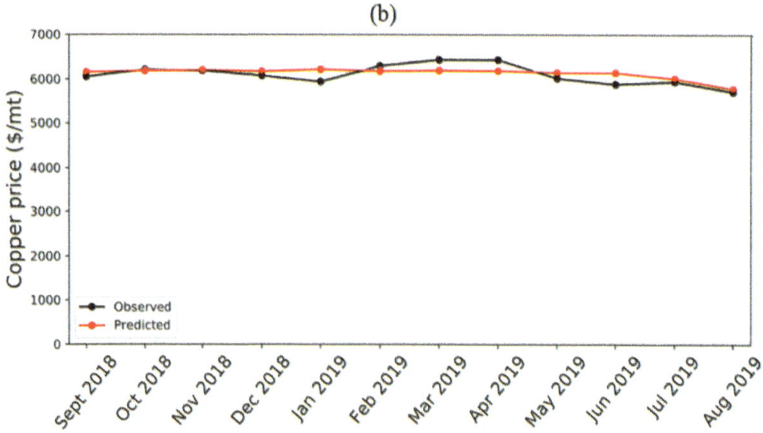

(c)

Fig. 1. Observed and predicted COMEX copper spot price values using as predictor the SVR technique with a RBF kernel for: (a) Direct multi-step scheme; (b) Recursive multi-step scheme; and (c) Direct-recursive hybrid forecast scheme.

4 Conclusions

According to the numerical results of the present research obtained with public data of copper in the COMEX market, it can be stated using as predictor the SVR technique that the performance level of the direct-recursive hybrid scheme is higher than those achieved by the recursive multi-step and direct multi-step schemes when analyzed in terms of statistics mean absolute percentage error (MAPE), mean absolute error (MAE) and root mean square error (RMSE). The direct multi-step method is the one that performs worst.

Finally, we believe there is a promising future for those lines of research combining hybrid models that are able to take full advantage of SVR models, creating models that combine machine learning techniques.

References

1. Streifel, S.: Impact of China and India on global commodity markets focus on metals & minerals and petroleum (2006)
2. Cuddington, J.T., Jerrett, D.: Super cycles in real metals prices? IMF Staff Pap. **55**, 541–565 (2008)
3. Roache, S.K.: China's impact on world commodity markets (2012)
4. Lahart, J.: Ahead of the Tape: Dr. Copper (2006)
5. Tilton, J.E., Lagos, G.: Assessing the long-run availability of copper. Resour. Policy. **32**, 19–23 (2007)
6. Gordon, R.B., Bertram, M., Graedel, T.E.: Metal stocks and sustainability. Proc. Natl. Acad. Sci. **103**, 1209–1214 (2006)
7. Dooley, G., Lenihan, H.: An assessment of time series methods in metal price forecasting. Resour. Policy. **30**, 208–217 (2005)
8. Cortazar, G., Eterovic, F.: Can oil prices help estimate commodity futures prices? The cases of copper and silver. Resour. Policy **35**, 283–291 (2010)
9. Khashei, M., Bijari, M.: An artificial neural network (p, d, q) model for timeseries forecasting. Expert Syst. Appl. **37**, 479–489 (2010)
10. Ma, W., Zhu, X., Wang, M.: Forecasting iron ore import and consumption of China using grey model optimized by particle swarm optimization algorithm. Resour. Policy **38**, 613–620 (2013)
11. Kriechbaumer, T., Angus, A., Parsons, D., Rivas Casado, M.: An improved wavelet–ARIMA approach for forecasting metal prices. Resour. Policy. **39**, 32–41 (2014)
12. Sánchez Lasheras, F., de Cos Juez, F.J., Suárez Sánchez, A., Krzemień, A., Riesgo Fernández, P.: Forecasting the COMEX copper spot price by means of neural networks and ARIMA models. Resour. Policy **45**, 37–43 (2015)
13. Brockwell, P.J., Davis, R.A.: Introduction to Time Series and Forecasting. Springer, Cham (2016)
14. Shumway, R.H., Stoffer, D.S.: Time Series Analysis and Its Applications: With R Examples. Springer, Cham (2017)
15. World Bank Commodity Price Data (The Pink Sheet). Bloomberg; Engineering and Mining Journal; Platts Metals Week; and Thomson Reuters Datastream; World Bank. http://pubdocs.worldbank.org/en/561011486076393416/CMO-Historical-Data-Monthly.xlsx
16. Steinwart, I., Christmann, A.: Support Vector Machines, Springer, New York (2008)

17. Schölkopf, B., Smola, A.J.: Learning with Kernels: Support Vector Machines, Regularization, Optimization, and Beyond. The MIT Press, Cambridge (2001)
18. Hamel, L.H.: Knowledge Discovery with Support Vector Machines. Wiley-Interscience (2011)
19. James, G., Witten, D., Hastie, T., Tibshirani, R.: An Introduction to Statistical Learning: with Applications in R. Springer, New York (2017)

Uncertainty Propagation Using Hybrid Methods

Juan Félix San-Juan[1(✉)], Montserrat San-Martín[2], Iván Pérez[1], Rosario López[1], Edna Segura[3], and Hans Carrillo[3]

[1] Scientific Computing Group (GRUCACI), University of La Rioja, 26006 Logroño, Spain
{juanfelix.sanjuan,ivan.perez,rosario.lopez}@unirioja.es

[2] Scientific Computing Group (GRUCACI), University of Granada, 52005 Melilla, Spain
momartin@ugr.es

[3] Department of Mathematics and Computer Science, University of La Rioja, 26006 Logroño, Spain
{edna-viviana.segura,hans-mauricio.carrillo}@alum.unirioja.es

Abstract. Small corrections in the argument of the latitude can be used to improve the accuracy of the SGP4 orbit propagator. These corrections have been obtained by applying the hybrid methodology for orbit propagation to SGP4, therefore yielding a hybrid version of this propagator. The forecasting part of the hybrid method is based on a state-space formulation of the exponential smoothing method. If the error terms that have to be considered during the model fitting process are taken as Gaussian noise, then the maximum-likelihood method can be applied so as to estimate the parameters of the exponential-smoothing model, as well as to compute the forecast together with its confidence interval. Finally, this hybrid SGP4 orbit propagator has been applied to data from Galileo-type orbits. This new propagator improves the accuracy of the classical SGP4, especially for short forecasting horizons.

Keywords: Time series · Hybrid methodology · Orbit propagation · Uncertainty

1 Introduction

The orbital motion of an artificial satellite, or space-debris object, is influenced by a variety of external perturbations, in addition to the Earth's gravity, which is the principal force that determines its orbit, such as the atmospheric drag, third-body influences, the solar radiation pressure, the Earth's tidal effects, and, in the case of an artificial satellite, also the small perturbing forces produced by its propulsion system. Numerical, analytical or semi-analytical methods can be

Á. Herrero et al. (Eds.): SOCO 2020, AISC 1268, pp. 709–717, 2021.
https://doi.org/10.1007/978-3-030-57802-2_68

used in order to solve the nonlinear equations of motion of this complex dynamical system. With the aim of simplifying it, some of the aforementioned external forces may be ignored depending on the intended purpose, for example the scientific requirements for the mission of an Earth's satellite, or the maintenance of a space-debris catalog. An orbit propagator is the implementation of one of the aforementioned solutions as a computer program.

The maintenance of a running catalog of space objects orbiting the Earth is an unavoidable duty in the management of the space environment close to the Earth, which requires the orbital propagation of tens of thousands of objects. Currently, these ephemerides are publicly available through the North American Aerospace Defense Command (NORAD) catalog, yet other organizations, like the European Space Agency (ESA), may make their own data, obtained from observations, also accessible.

Due to the huge number of objects to be propagated, a compromise between accuracy and efficiency must be established, depending on a variety of criteria. High-fidelity propagation models usually require step-by-step propagation by using numerical methods, which are computationally intensive because they rely on small step sizes. On the other hand, simplified models may admit analytical solutions, in this way notably alleviating the computational burden. In either case, the orbit propagation program relies only on the initial conditions, as well as on the propagation model, to make its predictions. However, the collection of past ephemerides provided by the catalog can be used to improve orbit predictions by taking non-modeled effects into account.

The main application of a space-debris catalog is the forecast of the future positions of all cataloged objects, since their extreme velocity converts them into uncontrolled projectiles that pose a real threat to operative satellites and space assets. As a result of this massive propagation activity, collision warnings have to be broadcast, so that satellite operators can perform collision-avoidance maneuvers. The assessment of the collision risk is strongly affected by all the uncertainties involved in the process of predicting the future positions of the cataloged objects.

The hybrid methodology for orbit propagation allows combining a classical propagation method, which can be numerical, analytical or semi-analytical, and a forecasting technique, based on either statistical time-series models [11] or machine-learning techniques, which is able to generate a compensation for the classical-propagation future errors from the time series of its former errors. This combination leads to an increase in the accuracy of the base propagator for predicting the future position and velocity of an artificial satellite or space-debris object, since it allows modeling higher-order terms and other external forces not considered in the base propagator.

In this work, we make use of a hybrid approach which combines the well-known analytical orbit propagator *Simplified General Perturbations-4* (SGP4), specially designed to be used with *Two-Line Elements* (TLE) as initial conditions [2,10,12,13,15], with a state-space formulation of the exponential smoothing method [4–6,14]. The consideration of the error terms as Gaussian noise during the model fitting process allows us to use the maximum likelihood method to estimate the parameters of the exponential smoothing model, as well as to compute the forecast and its confidence interval. Our goal in this study is to verify the capability of the hybrid orbit propagator to propagate the initial uncertainty.

The outline of the paper is structured around three sections. The hybrid methodology for orbit propagation is concisely summarized in Sect. 2. Then, the application of the hybrid SGP4 propagator to Galileo-type orbits is discussed in Sect. 3. Finally, Sect. 4 draws the conclusions of the study.

2 Hybrid Methodology

The hybrid methodology for orbit propagation is aimed at improving the estimation of the future position and velocity of any artificial satellite or space-debris object at a final instant t_f, expressed in some set of canonical or non-canonical variables, $\hat{\mathbf{x}}_f$. That improvement is performed on an initial approximation $\mathbf{x}_f^{\mathcal{I}}$, obtained by means of a base propagator that applies an integration method $\mathcal{I}$, which can be numerical, analytical or semi-analytical, to the system of differential equations that govern the behavior of the nonlinear dynamical system.

In order to enhance this initial approximation, it is necessary to somehow know the dynamics that the base propagator is missing. For that purpose, we can use the time series of its former errors, for which we need to know the real satellite ephemerides, either obtained by observation or simulated by high-fidelity slow numerical propagation, during a past *control interval*. For every epoch t_i in this control interval, we calculate the error $\boldsymbol{\varepsilon}_i$ as the difference between the real ephemeris $\mathbf{x}_i$ and the base-propagator approximation $\mathbf{x}_i^{\mathcal{I}}$:

$$\boldsymbol{\varepsilon}_i = \mathbf{x}_i - \mathbf{x}_i^{\mathcal{I}}. \tag{1}$$

This error $\boldsymbol{\varepsilon}_i$ is, in part, due to the fact that the base propagator implements a simplified model of the real system, although the intrinsic error in the initial conditions that we want to propagate can also contribute to it.

Once we have the time series of the base-propagator former errors, which embeds the dynamics that we want to reproduce, we can apply statistical time-series methods or machine-learning techniques in order to build a model. Later, we will use that model to predict an estimation of the base-propagator error at the final instant t_f, $\hat{\boldsymbol{\varepsilon}}_f$. Finally, the enhanced ephemeris at t_f, $\hat{\mathbf{x}}_f$, will be calculated by adding this estimated error to the base-propagator approximation $\mathbf{x}_f^{\mathcal{I}}$:

$$\hat{\mathbf{x}}_f = \mathbf{x}_f^{\mathcal{I}} + \hat{\boldsymbol{\varepsilon}}_f. \tag{2}$$

3 Application of the Hybrid SGP4 Propagator to Galileo-Type Orbits

3.1 SGP4 and AIDA Orbit Propagators

Two orbit propagators are involved in this study: SGP4, which is the base propagator whose accuracy we intend to improve, and AIDA [7], a high-precision numerical propagator that we use for generating the so-called *pseudo-observations* that represent the satellite real ephemerides.

SGP4 is an analytical orbit propagator originally based on Brouwer's theory [1] of satellite motion perturbed by the first five zonal harmonics of the Earth gravitational field. The description of the original Fortran code can be found in [2], although the complete documentation of all the mathematical equations was published in 2004 [3]. In this work, we use the most updated code developed by Vallado [15], which merges SGP4/SDP4 models, and is simply referred to as SGP4. This propagator includes the following force models: J_2 to J_4 zonal harmonics, air drag, and lunar and solar perturbations, as well as long-period resonant harmonics for the so-called deep-space satellites.

The input to the SGP4 propagator is the Two-Line-Element (TLE) set, which provides position and velocity vectors at a given time. The TLE includes information about the satellite and its orbit, such as the satellite number, orbit inclination, eccentricity, argument of perigee, derivatives of the mean motion, the BSTAR drag parameter, mean anomaly, and others.

The other orbit propagator, which we use for simulating observational data, is AIDA, the *Accurate Integrator for Debris Analysis*. It includes the following force models:

- geopotential acceleration computed using the EGM2008 model [8], up to an arbitrary degree and order for the harmonics;
- atmospheric drag, modeled using the NRLMSISE-00 air density model [9];
- solar radiation pressure with dual-cone shadow model;
- third body perturbations from Sun and Moon.

3.2 Numerical Results

This study has been conducted in the polar-nodal coordinates. The meaning of these variables is shown in Fig. 1. $Oxyz$ represents an inertial reference frame

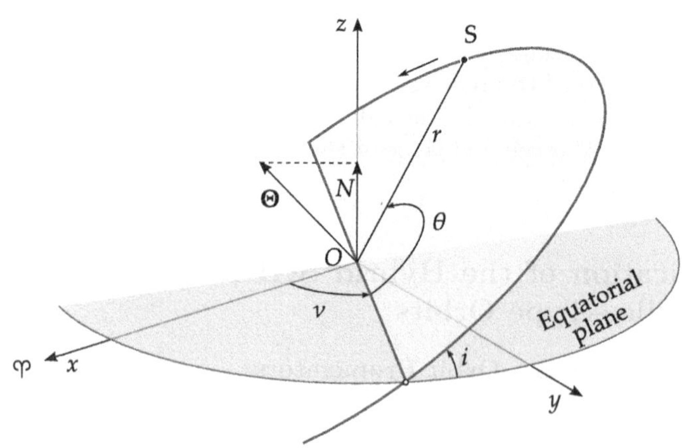

Fig. 1. Polar-nodal variables $(r, \theta, \nu, R, \Theta, N)$

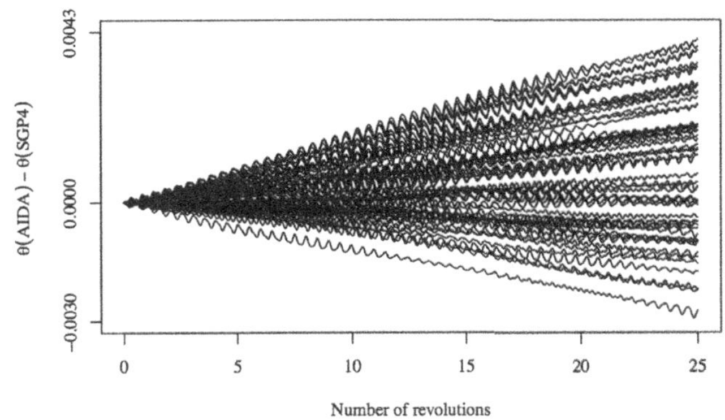

Fig. 2. $\varepsilon^{\theta} = \theta_{AIDA} - \theta_{SGP4}$ time series for several TLEs from the Galileo-8 satellite

centered at the center of mass of an Earth-like planet. The variable r denotes the distance from the center of mass of the Earth-like planet to the space object S, θ is the argument of the latitude of the object, ν represents the right ascension of the ascending node, R is the radial velocity, Θ designates the magnitude of the angular momentum vector $\boldsymbol{\Theta}$, whereas N represents the projection of $\boldsymbol{\Theta}$ onto the z-axis.

In this study, the hybrid methodology has been applied only to the argument of the latitude θ.

Figure 2 plots $\varepsilon^{\theta} = \theta_{AIDA} - \theta_{SGP4}$, the time series of the error in the argument of the latitude, for 53 different TLEs from the Galileo-8 satellite. TLE dates span from 28th March 2015 to 16th December 2016, including TLEs for every month between those two dates, with an approximately even distribution, although not completely regular. As can be seen, despite the fact that all these time series correspond to the same satellite, they do not seem to present a unique

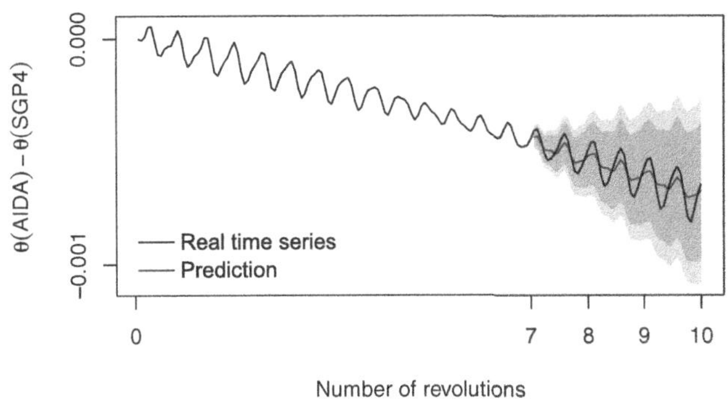

Fig. 3. Forecast of ε^{θ} for the next three satellite revolutions. The blue line represents the prediction and the shaded areas correspond to the 99% and 95% confidence intervals.

pattern. All of them show seasonal components, whose periods are approximately their Keplerian periods, and exhibit a high degree of variation in their trends.

Figure 3 displays the real and predicted values for one of the ε^{θ} time series shown in Fig. 2 for the following TLE,[1] which corresponds to 28th March 2015:

```
1 40545U 15017B   15087.10529976  .00000015  00000-0  00000+0 0  9997
2 40545 055.0895 094.8632 0005535 231.4671 034.4229 01.67457620    08
```

Predictions have been generated by applying the state-space formulation of the exponential smoothing method. This formulation considers the error component as Gaussian white noise, which allows applying formal estimation techniques, such as the maximum-likelihood method. Under reasonable conditions, this method provides consistent estimations with asymptotic efficiency. In essence, the state-space formulation allows representing the evolution of a set of components that underlie the time series, although they are not directly observable. This method is based on the recursive application of two equations: the *measurement equation*, which provides the estimation of the time-series values from the state vector, and the *transition equation*, which defines the evolution of the state vector. A detailed description of this formulation can be found in Ref. [6].

The first seven revolutions, which represent approximately 100 h, that is, about four days, constitute the control interval that we use for fitting the parameters of the model. Then, we use that model for predicting the time-series values during the next three revolutions. The line in black represents the real time series, whereas the line in blue corresponds to the forecast. As can be seen in the figure, the model works relatively well during the first forecasting revolutions. The use of the state-space formulation of the exponential smoothing method also allows determining the confidence interval of the prediction. The dark- and light-gray halos surrounding the forecast represent the 99% and 95% confidence intervals, respectively. It is worth noting that the size of the confidence

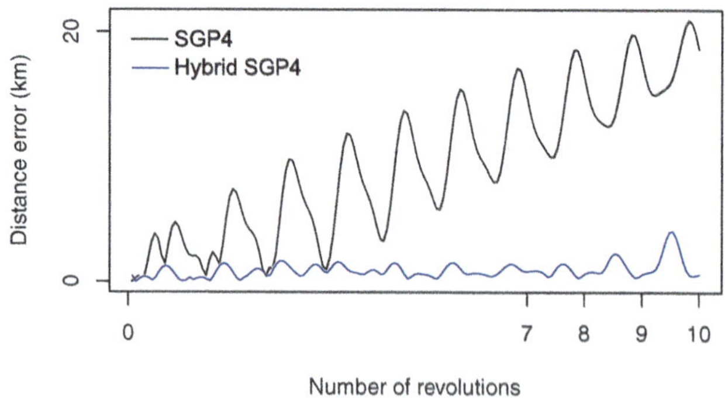

Fig. 4. Distance error of SGP4, in black, and the hybrid propagator, in blue, after a three-satellite-revolution propagation span

[1] TLEs can be downloaded from https://www.space-track.org.

interval is an indicator of the uncertainties associated to the fitting process of the statistical model.

A convenient way to evaluate how good a propagation is consists in translating the propagation errors into distance errors between the calculated position of the satellite and its real position, in this case determined through the AIDA numerical propagator. Figure 4 displays the distance error of SGP4 in black, and the hybrid SGP4 in blue. After the three revolutions during which we predict SGP4 errors, which correspond to approximately two propagation days, the maximum distance error for SGP4 is 20.92 km, whereas the hybrid propagator, in which the error correction has been applied, has a maximum distance error of only 5.98 km.

As can be expected, when we analyze the three orthogonal components of the position error, we verify that the main deviation takes place in the direction tangential to the orbit, namely the *along-track error*. Figure 5 displays the along-track error for both SGP4, in black, and the hybrid SGP4, in blue, with respect to the real satellite position, accurately computed with the numerical propagator AIDA. This figure also shows the 99% and 95% confidence intervals, as dark- and light-gray halos, respectively. These confidence intervals have been calculated from the confidence intervals of the prediction of ε^{θ}, shown in Fig. 3.

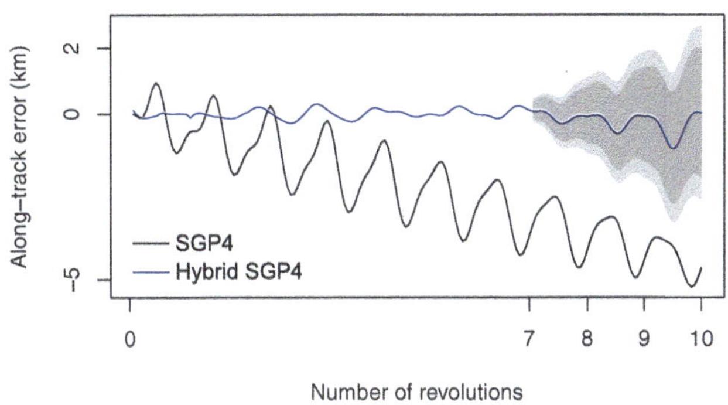

Fig. 5. Along-track error for SGP4, in black, and for the hybrid propagator, in blue. Shaded areas represent the 99% and 95% confidence intervals. Predictions start at revolution number 7.

Finally, this study has been extended to the 53 different TLEs from the Galileo-8 satellite showed in Fig. 2. The same procedure has been followed in all the cases: the time series of the argument-of-the-latitude error ε^{θ} during the first seven satellite revolutions has been used for fitting the parameters of the model, and then, future errors have been predicted for the next three revolutions. Table 1 presents some statistics for the distance errors for both SGP4 and the hybrid propagator HSGP4. As can be noticed, not only are HSGP4 errors smaller, but they also show a lower dispersion. The family of hybrid orbit propagators improves the accuracy of the classical SGP4, and is particularly good for short forecasting horizons.

Table 1. Statistics for the distance errors of SGP4 and the hybrid propagator HSGP4 (km)

	SGP4 error	HSGP4 error
Minimum	5.98	5.98
1st quartile	13.17	5.98
Median	20.89	5.98
Mean	23.72	6.95
3rd quartile	31.45	7.52
Maximum	51.65	12.06

4 Conclusions

The hybrid methodology for orbit propagation consists in complementing the approximate solution of a base propagator with a correction based on the time series of the propagator past errors, generated by means of statistical methods or machine learning techniques. It allows improving the accuracy of any base propagator, irrespective of its type, with a very light increment in the computational burden.

One of the most convenient statistical techniques for this purpose is the exponential smoothing method. We use it in order to create a model from the base-propagator past errors, and later to predict future errors.

In this study, we make use of the state-space formulation of the exponential smoothing method. Its main advantage lies in the fact that it allows applying the maximum likelihood method, which, by considering the error terms as Gaussian noise during the fitting process of the exponential-smoothing model parameters, allows determining the confidence interval of the predictions.

Knowing the confidence interval of the predictions allows propagating the uncertainties, which is necessary in order to determine the collision probabilities of space objects.

The study has been performed taking the well-known SGP-4 as the base propagator, and applying it to the propagation of Galileo-type orbits.

Acknowledgments. This work has been funded by the Spanish State Research Agency and the European Regional Development Fund under Project ESP2016-76585-R (AEI/ERDF, EU).

References

1. Brouwer, D.: Solution of the problem of artificial satellite theory without drag. Astron. J. **64**(1274), 378–397 (1959). https://doi.org/10.1086/107958
2. Hoots, F.R., Roehrich, R.L.: Models for propagation of the NORAD element sets. Spacetrack Report #3, U.S. Air Force Aerospace Defense Command, Colorado Springs, CO, USA (1980)

3. Hoots, F.R., Schumacher Jr., P.W., Glover, R.A.: History of analytical orbit modeling in the U.S. space surveillance system. J. Guidance Control Dyn. **27**(2), 174–185 (2004). https://doi.org/10.2514/1.9161
4. Hyndman, R.J., Koehler, A.B., Ord, J.K., Snyder, R.D.: Prediction intervals for exponential smoothing using two new classes of state space models. J. Forecast. **24**(1), 17–37 (2005). https://doi.org/10.1002/for.938
5. Hyndman, R.J., Koehler, A.B., Ord, J.K., Snyder, R.D.: Forecasting with Exponential Smoothing. The State Space Approach. Springer Series in Statistics, 1st edn. Springer, Berlin (2008). https://doi.org/10.1007/978-3-540-71918-2
6. Hyndman, R.J., Koehler, A.B., Snyder, R.D., Grose, S.: A state space framework for automatic forecasting using exponential smoothing methods. Int. J. Forecast. **18**(3), 439–454 (2002). https://doi.org/10.1016/S0169-2070(01)00110-8
7. Morselli, A., Armellin, R., Di Lizia, P., Bernelli-Zazzera, F.: A high order method for orbital conjunctions analysis: sensitivity to initial uncertainties. Adv. Space Res. **53**(3), 490–508 (2014). https://doi.org/10.1016/j.asr.2013.11.038
8. Pavlis, N.K., Holmes, S.A., Kenyon, S.C., Factor, J.K.: The development and evaluation of the Earth Gravitational Model 2008 (EGM2008). J. Geophys. Res. Solid Earth **117**(B4) (2012). https://doi.org/10.1029/2011JB008916
9. Picone, J.M., Hedin, A.E., Drob, D.P., Aikin, A.C.: NRLMSISE-00 empirical model of the atmosphere: Statistical comparisons and scientific issues. J. Geophys. Res. Space Phys. **107**(A12), 1–16 (2002). https://doi.org/10.1029/2002JA009430
10. San-Juan, J.F., Pérez, I., San-Martín, M., Vergara, E.P.: Hybrid SGP4 orbit propagator. Acta Astronaut. **137**, 254–260 (2017). https://doi.org/10.1016/j.actaastro.2017.04.015
11. San-Juan, J.F., San-Martín, M., Pérez, I.: An economic hybrid J_2 analytical orbit propagator program based on SARIMA models. Math. Prob. Eng. **2012**, 1–15 (2012). https://doi.org/10.1155/2012/207381. Article ID 207381
12. San-Juan, J.F., San-Martín, M., Pérez, I.: Application of the hybrid methodology to SGP4. Adv. Astronaut. Sci. **158**, 685–696 (2016). Paper AAS 16-311
13. San-Juan, J.F., San-Martín, M., Pérez, I., López, R.: Hybrid SGP4: tools and methods. In: Proceedings 6th International Conference on Astrodynamics Tools and Techniques, ICATT 2016. European Space Agency (ESA), Darmstadt, Germany, March 2016
14. Snyder, R.D., Koehler, A.B., Ord, J.K.: Forecasting for inventory control with exponential smoothing. Int. J. Forecasting **18**(1), 5–18 (2002). https://doi.org/10.1016/S0169-2070(01)00109-1
15. Vallado, D.A., Crawford, P., Hujsak, R., Kelso, T.S.: Revisiting spacetrack report #3. In: Proceedings 2006 AIAA/AAS Astrodynamics Specialist Conference and Exhibit, vol. 3, pp. 1984–2071. American Institute of Aeronautics and Astronautics, Keystone, August 2006. https://doi.org/10.2514/6.2006-6753. Paper AIAA 2006-6753

Special Session: Machine Learning in Computer Vision

Multidimensional Measurement of Virtual Human Bodies Acquired with Depth Sensors

Andrés Fuster-Guilló[1](✉), Jorge Azorín-López[1], Juan Miguel Castillo-Zaragoza[1], Cayetano Manchón-Pernis[1], Luis Fernando Pérez-Pérez[1], and Ana Zaragoza-Martí[2]

[1] Department of Computer Science and Technology, University of Alicante, Carretera Sant Vicent del Raspeig s/n, 03690 Alicante, Spain
fuster@ua.es

[2] Nursing Department, University of Alicante, Carretera Sant Vicent del Raspeig s/n, 03690 Alicante, Spain

Abstract. Obtaining 3D measurements of the human body requires precise scanning of the body, as well as methods for extracting these 1D/2D and 3D measurements from the selected volumes. The analysis of these 3D measurements and their monitoring over time (4D) in patients undergoing dietary treatment is a field that poses multidisciplinary challenges such as obtaining precise body models, automating the measurement process and analysing the data from a medical point of view. In this work, we propose a framework to acquire 3D models of patients and obtain measurements on these models. This framework incorporates computational methods for extracting 3D models that faithfully represent the human body, as well as methods for obtaining accurate measurements from those 3D models. An analysis of the accuracy of the proposed methods for obtaining measurements with both synthetic and real objects has been carried out. The low level of error observed in the experimentation on synthetic objects allows to attribute most of it to the scanning module. Experiments with real objects and body models show an error level comparable to other scanning systems based on RGB-D technologies. The main contribution of the work is to provide a framework to obtain in a selective and automatic way the 3D measurements of the human body, allowing the analysis of its evolution (4D) during the treatment of obesity.

Keywords: RGB-D · Human body measures · Dietetic treatment

1 Introduction

Measuring the volume of the human body with the aim of analyzing fat concentration as a symptom of overweight and obesity is a task often addressed in the health sector with traditional techniques and single-dimensional measurements. The study of anthropometric measurements and their variation over time in relation to fat accumulation presents multidisciplinary challenges of interest in the fields of information technology and health. The use of RGB-D devices can help to address the tasks of 3D scanning of the human body and later automatically obtaining 3D and 4D measurements of the

Á. Herrero et al. (Eds.): SOCO 2020, AISC 1268, pp. 721–730, 2021.
https://doi.org/10.1007/978-3-030-57802-2_69

selected human body volumes, with the inherent advantages of this kind of consumer oriented technologies [1].

The prevalence of overweight and obesity has increased worldwide, tripling over the last three decades in the countries of the European Union [2]. In the field of health, some pioneering work carried out in recent years has begun to incorporate the use of 3D models to analyse health parameters related to the volume or shape of the human body in obese patients [3, 4]. Classic treatments based on the body mass index (BMI) are beginning to be enriched with new anthropometric indices, such as the body volume index (BVI) based on the 3D shape of the human body [5]. However, these studies have focused on the measurement of static variables, without considering the temporal evolution of the body (4D) in response to dietetic treatment.

The 3D scanning of the human body was largely developed for the textile industry [6]. Today, 3D modelling of the human body is transforming our ability to accurately measure and visualize it, showing great potential for healthcare applications; from epidemiology to diagnosis and patient monitoring [7]. Recently, several 3D body scanning systems based on RGB-D technologies have appeared oriented to the fitness market, achieving 3D models not always with realistic visualizations incorporating texture (Naked.fit, Fit3D, Shapescale). Very recent works [8, 9] addresses the acquisition of human body models from RGB-D cameras and video footage providing textured models and avatars, but not oriented to the accuracy needed in healthcare applications. There are other acquisition systems focused on extracting 3D models of the body for avatar purposes, 3D printing, etc. but they consist of bulky devices.

There are different types of 3D sensors with different characteristics. Devices based mainly on lasers, such as Lidar or Time of Flight (ToF), are sensors that have good accuracy, but only obtain depth information and do not provide colour data. Stereo sensors use two colour cameras to infer the depth, which is usually a high cost, and are also difficult to transport since both cameras must be calibrated. Recently RGB-D cameras (Microsoft Kinect or Intel RealSense) use different technologies (structured light, ToF) to integrate colour and depth in one device. The characteristics of these RGB-D general purpose devices, accuracy, portability, capture frequency, etc., are causing its popularization and integration in mobile consumer devices [10]. For these reasons, in the present work these RGB-D devices are used to capture the 3D model, which implies an important scientific challenge.

In this context, this work focuses on the development of a framework for 3D reconstruction and measuring of the human body, using RGB-D devices. For the development of this research it is necessary to address the following issues:

Obtaining the 3D model: Acquisition of the body by capturing 3D images from several points of view simultaneously. These views must be statically aligned through the transformations obtained from the calibration to obtain the complete model.

Measuring selected volumes of the human body: Selection of different parts of the human body to obtain 1D, 2D and 3D measurements.

The rest of the paper is structured as follows: Sect. 2 details the set of methods used to obtain the 3D body model, Sect. 3 explains the methods proposed for the calculation of 2D and 3D measurements, Sect. 4 details the experiments and finally Sect. 5 summarizes the contributions and conclusions of the work.

2 3D Body Acquisition and Modelling

In order to obtain the 3D model of the human body, a network of RGB-D cameras has been used. The network is composed by 12 Intel RealSense RGB-D cameras located in a cabin with 4 aluminium masts and panels of 2200 × 800 × 800 mm distributed around the capture area. Figure 1 shows the cabin with RGB-D cameras to capture human body (left) depth and RGB colour images obtained by the system (right) [11].

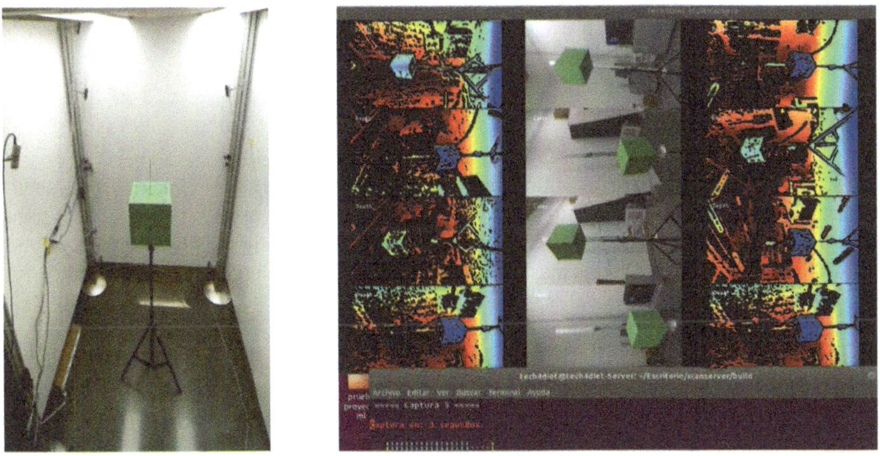

(a) (b)

Fig. 1. (a) Experimental set-up 12 RGB-D cameras. (b) Acquisition results for the cameras (colour and depth images). The green cube has been used for extrinsic calibration.

The pipeline used to obtain the 3D textured model from different RGB-D sensors is composed by five stages Fig. 2: acquisition, pre-processing, registration, mesh generation and texture projection. Calibration is not included as it is part of the set-up process.

To correct the distortions of the images caused by the lens, an intrinsic calibration is carried out using the Intel RealSense SDK. Since we are using a network of RGB-D cameras, it is necessary to carry out an extrinsic calibration to unify the different point clouds in the same coordinate space, obtaining the corresponding transformation matrixes. We use an extrinsic calibration based on 3D markers, spherical and cubic [12].

The network is composed by 12 Intel RealSense RGB-D cameras D435. Intel's SDK for RealSense has been used as the basis for the development. The acquisition process (Fig. 2 (a)) requires the synchronization of all the cameras in the network to perform the capture. Semaphore management has been used to address the synchronization.

At pre-processing stage, some noisy point clouds from the different RGB-D sensors have been obtained, so it is necessary to apply different methods to improve their quality (Fig. 2 (b)). First, the point cloud is truncated in the z-axis (depth) to remove the points beyond the centre of the capture area. After that, three filters are applied: median [13], bilateral [14] and statistical outlier removal (SOR) [15]. Finally, the normal vector for each point in the cloud is calculated [16].

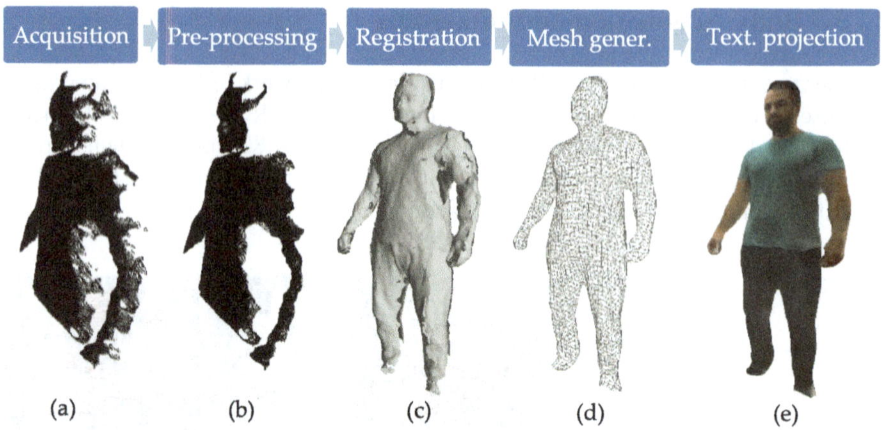

Fig. 2. Pipeline of 3D body reconstruction: The system can acquire several images from cameras (a) that are pre-processed in order to improve the quality of the acquisition (b). The set of points are registered in a unique origin of coordinates (c). In order to obtain the 3D model of the body, the 3D points are converted into a mesh (e) and, finally, the images are projected on it (d).

In order to align the different point clouds in a single 3D coordinate system, the transformation matrices T obtained from the extrinsic calibration are applied for registration (Fig. 2 (c)). We assume one camera as reference and transform the rest of point clouds to this one to obtain a unified dataset [17].

Different methods such as Greedy Projection or Marching Cubes were tested for mesh generation (Fig. 2 (d)), obtaining the best result with the Poisson surfacing algorithm [18]. It can reconstruct a triangle mesh from a set of oriented 3D points.

The present work is part of a project where objectives are addressed that require the texture projection (Fig. 2 (e)) for realistic visualization of the body model. Although obtaining measurements of the body only requires a 3D mesh model, the results of the textured model are presented to give it greater realism. The method proposed by Callieri et al. [19] has been used to carry out the raster projection and texture generation.

3 3D Body Measuring Method

From the mesh-based 3D models generated by the pipeline previously described, a method is proposed for obtaining perimeter, area and volume measurements of the selected 3D model sections. Given that the software developed for the visualization of the textured models makes use of the Unity graphic engine, the ray simulation system provided by this tool has been used as the core for the calculation of the points in space that determine the sections of the 3D model to be measured.

The following sections detail the method for obtaining perimetral measurements based on Unity, the method for the positioning of circles that allows the selection of the sections to be measured and finally the method for obtaining measurements of areas and volumes.

3.1 Perimetral Measurements Method

A method based on ray simulation has been used to measure the perimeter. This method consists of placing a circle cutting the entire perimeter that the user wants to measure. Thousands of coplanar orbiting rays are simulated on the chord all around the circle. The coplanar rays are launched along the contour to be measured, perpendicularly to tangent lines, storing the points of collision with the mesh. This process allows to obtain the set of points of the mesh on the perimeter at the height of the circle Fig. 3.

(a) (b)

Fig. 3. Rays perpendicularly launched along the circumferential chord of the circle storing the points of collision with the mesh (a). Rays perpendicularly launched along the circumferential chord of the circles inside the cylinder storing the points of collision with the mesh (b).

As the rays are launched, the points impacted are stored and the distance between the current point and the previous one is accumulated. The sum of all the distances forms the desired perimeter. The order in which the rays are projected generates the order of impacted points of the mesh so that the accumulation of the distances between these consecutive points provides the estimation of the perimeter of the 3D model intersected by the circle. The number of points impacted is related to the number of rays projected. The greater the number of impacted points, the greater the accuracy and the computational cost, as we can see in the experimental section. This allows the number of points to be adjusted according to the accuracy required.

3.2 Selection of Body Parts to Be Measured

Since the system has been designed to be used by dietetics specialists, the selection of the perimeters and volumes to be measured must be done by the user. An interface has been designed that allows a circle to be placed at the height of the 3D model of the body to be measured. The centre of the circle can be placed anywhere in the space. In addition, the circle can be angled to be properly oriented for measuring especial body parts such as arms… Finally, the size of the circle can be automatically adjusted to the surrounding figure. In Fig. 3, a circle can be seen sectioning parts of the body. For the selection of volumes, a similar scheme has been chosen in which a cylinder is located and sized around the part of the body whose volume is to be measured.

3.3 Estimation of Area and Volume

For the estimation of the area enclosed in the perimeter of a section of a 3D model, the set of points obtained in the calculation of the perimeter have been used. From this set of points, a pivot point has been selected for triangulating the area enclosed in the perimeter Fig. 4, that point will be the middle of the circle. The sum of the areas of the enclosed triangles provides the estimation of the sectioned area. The use of a single point as the pivot ensures that these areas do not overlap since the figure is not convex. This method is valid for this application since all body sections used are convex. As with the relationship between the number of projected rays and the precision of the perimeter estimation, the number of triangles used is directly related to the precision of the area measurement and its temporal cost, as shown in the experimental section.

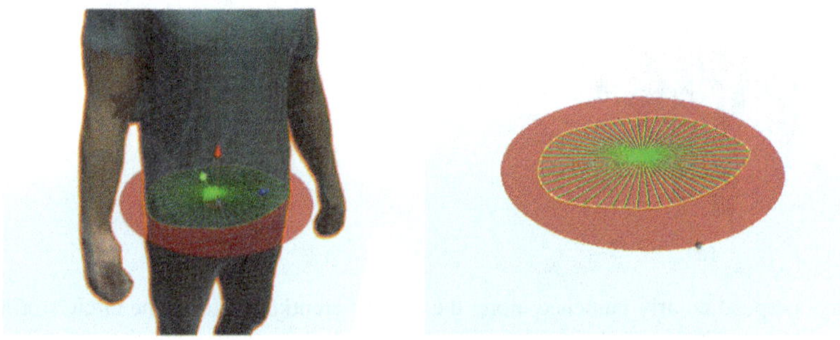

Fig. 4. Triangulation of the area enclosed in the perimeter

As mentioned in Sect. 3.2, for the calculation of volumes selected from 3D mesh based models, a cylinder is used that intersects the volume to be measured Fig. 3. The upper and lower circles of the cylinder determine the 3D volume to be measured. The method used to estimate the volume is based on the use of the triangulation method for the area calculation of a section, by iterating from the upper to the lower circle assuming a pre-set height "h" for each of the sections to transform to volume measurements.

4 Quantifying the Accuracy of the Method for Measuring Scanned 3D Models

The following section proposes a battery of experiments to quantify the accuracy of the methods. As explained previously, the complete process includes both the scanning of objects/bodies to obtain the corresponding mesh-based 3D models, and the methods for obtaining 1D, 2D and 3D measurements of selected sections of these models. Given the complexity of the complete process and the interest in differentiating which part of the error corresponds to the scanning and which to the obtaining of measurements, it has been proposed to carry out both experimentation with synthetic and real objects. In this way, in the experimentation with synthetic objects, the error attributable to the methods for obtaining measurements is estimated, while in the experimentation with real objects, the error of the complete process is measured, incorporating both the scanning error and the error of the methods for extracting measurements.

4.1 Experimentation with Synthetic 3D Models

The objective of the experimentation on synthetic models is to quantify the error corresponding to the methods for obtaining measurements explained in Sect. 3. These 3D models have been designed with pre-determined measurements whose estimation is the objective of the methods whose error is quantified.

Synthetic experimental setup: The following 3D models have been designed using Blender. Table 1 shows the measurements of the objects in cm:

Table 1. Synthetic objects and their 1D, 2D and 3D measurements in cm.

Object	Perimeter (1D)	Area (2D)	Volume (3D)
Cube 1	60,00	225,00	3.375,00
Cube 2	200,00	2.500,00	125.000,00
Cylinder 1	300,00	5.000,00	196.349,54
Cylinder 2	200,00	2.500,00	98.174,77
Cone	161,80	1.250,00	32.724,92
Pyramid	97,00	450,00	9.000,00

In Table 2 we can see the perimeter estimations of the different objects using the method described in Sect. 3 and the relative error. The estimation has been calculated by varying the number of rays projected onto the mesh, from 10^2 to 10^5. It is observed that the average relative error decreases as the number of projected rays increases. It is observed that from 10^4, the increase in the number of rays does not improve the estimation.

Table 2. Perimeter estimations (cm) (E.) for different number of rays and relative error (%).

Object	E. 10^2	Rel. ε	E. 10^3	Rel. ε	E. 10^4	Rel. ε	E. 10^5	Rel. ε
Cube 1	58,47	0,025500	59,95	0,000833	59,99	0,000167	59,93	0,001167
Cube 2	194,86	0,025700	199,84	0,000800	199,99	0,000050	199,87	0,000650
Cylinder 1	293,71	0,020967	299,11	0,002967	299,86	0,000467	299,44	0,001867
Cylinder 2	194,72	0,026400	199,57	0,002150	199,84	0,000800	199,62	0,001900
Cone	157,59	0,026020	160,14	0,010260	160,49	0,008096	159,35	0,015142
Pyramid	93,71	0,033918	95,81	0,012268	96,00	0,010309	95,88	0,011546
Average ε		**0,026417**		**0,004880**		**0,003315**		**0,005379**

In Table 3 we can see the area estimations for the different objects and the relative error. We provide the estimation varying the number of rays projected onto the mesh, from 10^2 to 10^5. It is observed that the average relative error decreases as the number of projected rays increases. As occurs with the perimeter estimations, it is observed that from 10^4, the increase in the number of rays does not improve the estimation.

Table 3. Area estimations (cm^2) (E.) for different number of rays and relative error (%).

Object	E. 10^2	Rel. ε	E. 10^3	Rel. ε	E. 10^4	Rel. ε	E. 10^5	Rel. ε
Cube 1	224,58	0,00187	225,01	0,000044	225,03	0,00013	225,13	0,000578
Cube 2	2495,36	0,00186	2500,04	0,000016	2.500,07	0,00003	2500,61	0,000244
Cylinder 1	4980,81	0,00384	4992,94	0,001412	4.993,92	0,00122	4990,69	0,001862
Cylinder 2	2491,79	0,00328	2496,41	0,001436	2.496,54	0,00138	2494,01	0,002396
Cone	1247,78	0,00178	1246,22	0,003024	1.247,16	0,00227	1245,64	0,003488
Pyramid	448,45	0,00344	449,78	0,000489	450,15	0,00033	451,43	0,003178
Average ε		**0,00268**		**0,00107**		**0,00089**		**0,00196**

Table 4 shows the volume estimations for the different objects and the relative error. We provide the estimation varying the number of rays projected onto the mesh, from 10^2 to 10^5. It is observed that the average relative error decreases as the number of projected rays increases. The value of "h" used for the sections of each circle was 1 cm.

Table 4. Volume estimations (cm^3) (E.) for different number of rays and relative error (%).

Object	E. 10^2	Rel. ε	E. 10^3	Rel. ε	E. 10^4	Rel. ε	E. 10^5	Rel. ε
Cube 1	3368,75	0,00185	3375,07	0,000021	3.375,15	0,00004	3376,2	0,000356
Cube 2	124768,49	0,00185	125002,1	0,000017	125.006,82	0,00005	125033,98	0,000272
Cylinder 1	194287,47	0,01050	194832,98	0,007724	194.716,61	0,00832	195587,54	0,003881
Cylinder 2	97385,65	0,00804	97623,71	0,005613	97.526,31	0,00661	97431,41	0,007572
Cone	32379	0,01057	32507,18	0,006654	32.511,76	0,00651	32671,93	0,001619
Pyramid	8970,9	0,00323	9003,78	0,000420	9.006,89	0,00077	9009,91	0,001101
Average ε		**0,00601**		**0,00341**		**0,00372**		**0,00247**

As a conclusion of the experiments carried out using synthetic objects, we can state that the relative error attributable to the estimation methods is very low, in the order of 0.005. Furthermore, we can affirm that the increase in the number of rays projected decreases the error committed, although given the increase in the temporary cost using more rays and the low error, it seems desirable to use a number of rays not too high.

4.2 Experimentation with Real 3D Models

The aim of the experimentation with real objects/bodies is to measure the error for the entire scanning and measurement system. The measurements of the real objects/bodies are known by manual procedures and their 3D models have been obtained by the scanning system detailed in Sect. 2. Since the error introduced by the measurement methods (Sect. 3) have been estimated very low, the error studied in this section will be mostly due to the scanning system. Since in the previous section the better number of rays has been estimated as 10^4, we will use it in this section.

Real experimental setup: The following 3D models have been used for the real experimentation (Table 5 shows the measurements of the objects in cm). Table 5 also shows the estimations obtained from 1D, 2D and 3D measurements for the cube and different parts of the body (perimeters, sections and volumes). The relative error of these measurements in relation to the real measurements obtained by manual procedures is also observed. The relative error is not calculated for 2D and 3D measurements of the body since their real values are not available. It is observed that the average relative error for perimeter is 0,036 and the average relative error for 2D and 3D measurements is 0,011. The average absolute error of the perimetral measurements is 2.4 mm. Although the comparison with other works such as [1] is not simple since the RGB-D capture method is not the same, we can affirm that comparable and even lower error levels are achieved.

Table 5. Real objects measurements (R.1D, R.2D, R.3D) their estimations (E.1D, E.2D, E.3D) in cm and their relative errors (Rel. ε). R = Real; E = Estimation; 1D = perimeter; 2D = area; 3D = volume

Object	R.1D	E.1D	Rel. ε	R.2D	E. 2D	Rel. ε	R.3D	E. 3D	Rel. ε
Cube 1	100	99,68	0,0032	625,00	632,09	0,011	15.625	15.805,6	0,012
Body head	56,10	55,88	0,0039	–	237,01	–	–	4.613,1	–
Body arm	29,65	29,53	0,004	–	66,63	–	–	1005,83	–
Body abdomen	101,4	101,71	0,0031	–	778,99	–	–	12.512,5	–
Average ε			**0,0036**			**0,011**			**0,012**

5 Conclusions

Obtaining 1D, 2D and 3D measurements of human body parts from scanned 3D models and the evolution of these measurements over time (4D) during dietetic treatment processes is a problem that poses interesting multidisciplinary challenges. There are few medical works that address the analytical study of data from morphological evolution in patients undergoing dietary treatment. The problem of 3D scanning of the human body with general-purpose RGB-D devices has been studied in various application contexts, although there are few studies that analyze the accuracy achievable by these low-cost devices for the extraction of body measurements. The main contribution of this work is to provide a framework to address both the scanning of 3D models of the human body and the selective and automatic extraction of 1D, 2D and 3D measurements from these models, reaching lower error levels (2,4 mm) than other works used as reference. As future work, the implantation of the framework in health centers is planned to assist specialists in the automatic extraction of body measurements. Moreover, we will develop methods to learn measurements from the experience to provide estimations using soft computing-based techniques of synthetic indices representative of these 3D measurements and their 4D temporal evolution.

Funding. This work has been partially funded by the Spanish Government TIN2017-89069-R grant supported with Feder funds.

References

1. He, Q., Ji, Y., Zeng, D., Zhang, Z.: Volumeter: 3D human body parameters measurement with a single Kinect. IET Comput. Vis. **12**(4), 553–561 (2018)
2. World Health Organization: World Health Organization. Estrategia mundial sobre régimen alimentario, actividad física y salud: marco para el seguimiento y evaluación de la aplicación. World Health Organization (2012)
3. Stewart, A.D., Klein, S., Young, J., Simpson, S., Lee, A.J., Harrild, K., Crockett, P., Benson, P.J.: Body image, shape, and volumetric assessments using 3D whole body laser scanning and 2D digital photography in females with a diagnosed eating disorder: preliminary novel findings. Br. J. Psychol. **103**(2), 183–202 (2012)
4. Giachetti, A., Lovato, C., Piscitelli, F., Milanese, C., Zancanaro, C.: Robust automatic measurement of 3D scanned models for the human body fat estimation. IEEE J. Biomed. Heal. Inform. **19**(2), 660–667 (2015)
5. Tahrani, A.A., Bolaert, K., Palin, S., Field, A., Redmayne, H., Barnes, R., Aytok, L., Rahim, A.: Body volume index: time to replace body mass index? (2008)
6. Apeagyei, P.R.: Application of 3D body scanning technology to human measurement for clothing Fit. Int. J. Digit. Content Technol. Appl. **4**(7), 58–68 (2010)
7. Treleaven, P., Wells, J.: 3D body scanning and healthcare applications. Comput. (Long. Beach. Calif.) **40**(7), 28–34 (2007)
8. Alldieck, T., Magnor, M.A., Xu, W., Theobalt, C., Pons-Moll, G.: Detailed human avatars from monocular video (2018). undefined
9. Yu, T., Zheng, Z., Guo, K., Zhao, J., Dai, Q., Li, H., Pons-Moll, G., Liu, Y.: DoubleFusion: real-time capture of human performances with inner body shapes from a single depth sensor (2018). undefined
10. Villena-Martínez, V., Fuster-Guilló, A., Azorín-López, J., Saval-Calvo, M., Mora-Pascual, J., Garcia-Rodriguez, J., Garcia-Garcia, A.: A quantitative comparison of calibration methods for RGB-D sensors using different technologies. Sensors (Switzerland) (2017)
11. Fuster-Guilló, A., Azorín-López, J., Zaragoza, J.M.C., Pérez, L.F.P., Saval-Calvo, M., Fisher, R.B.: 3D technologies to acquire and visualize the human body for improving dietetic treatment. Proceedings **31**(1), 53 (2019)
12. Saval-Calvo, M., Azorin-Lopez, J., Fuster-Guillo, A., Mora-Mora, H.: μ-MAR: multiplane 3D marker based registration for depth-sensing cameras. Expert Syst. Appl. **42**(23), 9353–9365 (2015)
13. PCL Team: Point Cloud Library (PCL): pcl::MedianFilter< PointT > Class Template Reference (2013). http://docs.pointclouds.org/1.7.1/classpcl_1_1_median_filter.html
14. PCL Team: Point Cloud Library (PCL): pcl::BilateralFilter< PointT > Class Template Reference (2019). http://docs.pointclouds.org/trunk/classpcl_1_1_bilateral_filter.html
15. PCL Team: Point Cloud Library (PCL): pcl::StatisticalOutlierRemoval< PointT > Class Template Reference (2013). http://docs.pointclouds.org/1.7.1/classpcl_1_1_statistical_outlier_removal.html
16. Radu Bogdan Rusu: Documentation - Point Cloud Library (PCL). http://pointclouds.org/documentation/tutorials/normal_estimation.php
17. Saval-Calvo, M., Azorín-López, J., Fuster-Guilló, A.: Model-based multi-view registration for RGB-D sensors. In: Rojas, I., Joya, G., Cabestany, J. (eds.) IWANN 2013. LNCS, vol. 7903, pp. 496–503. Springer, Heidelberg (2013). https://doi.org/10.1007/978-3-642-38682-4_53
18. Kazhdan, M., Bolitho, M., Hoppe, H.: Poisson Surface Reconstruction (2006)
19. Callieri, M., Cignoni, P., Corsini, M., Scopigno, R.: Masked photo blending: mapping dense photographic data set on high-resolution sampled 3D models. Comput. Graph. **32**(4), 464–473 (2008)

Event-Based Conceptual Architecture for the Management of Cyber-Physical Systems Tasks in Real Time

Henry Duque Gómez, Jose García Rodríguez, and Jorge Azorin-Lopez(✉)

Computer Technology Department, University of Alicante, Alicante, Spain
jazorin@ua.es

Abstract. Given the application domains and challenges presented to cyber-physical systems (CPSs), it is necessary to design a CPS system able to deal with temporal constrains. There are various software architecture models to meet this challenge. Models have been developed under three types of software structural units, such as: Component-based architecture, service-based architecture and agent-based architecture. These models are analyzed under the compliance of adaptability, autonomy and interoperability properties. Our proposal provides a conceptual architecture model driven by events for the real-time management of CPS, which is proposed under the rigor of software engineering based on a service-oriented architecture (SOA 2.0 - Service Oriented Architecture) and directed by EDA events (Event Driven Architecture), using event processing technology CEP (Complex Event Processing).

Keywords: CPS · IoT · SOA 2.0 · EDA · CEP

1 Introduction

CPS systems are devices that integrate computing, storage and communication capabilities in order to control and interact with a process in the physical world. CPSs are connected among them and to the virtual world and global digital networks [1, 2]. A CPS is a mechanism controlled or monitored by computer software-based algorithms and linked through the Internet, in which physical components and software are deeply integrated, where each element operates at different spatial and temporal scales [3]. The emergence of large-scale, highly distributed intelligent CPSs in the framework of the Internet of Things (IoT), cloud computing, mobility, big data, networks of interconnected devices and sensors. It involves that software architecture models have to work in an open and highly dynamic world driven by real-time CPS decision making [4, 5].

There are three main aspects to classify the structural units of CPS tasks: implementation of CPS based on a component architecture model, implementation of CPS based on the architecture model by services and implementation of CPS based on the agent-based architecture model [6]. These structural units are analysed in terms of their adaptability, autonomy and interoperability properties [6, 7]. These non-functional properties were

Á. Herrero et al. (Eds.): SOCO 2020, AISC 1268, pp. 731–740, 2021.
https://doi.org/10.1007/978-3-030-57802-2_70

proposed as critical in the challenges identified at the National Science Foundation (NSF) Cyberphysical Systems Summit [8].

In this paper, we propose a conceptual architecture model based on events for the management of CPS tasks in real time, given that currently there is no such architecture model by events. This architecture is Event Driven Architecture (EDA) and integrated to a Service Oriented Architecture (SOA), which evolves to the SOA 2.0 concept [9, 10]. EDA is an architecture in which the software executes an action when it receives one or more event notifications [9, 10]. It is designed to react and to make the CPS system devices interact with the environment by means of events and processed by the complex event processing technology (CEP) [11, 12].

The sections that compose the paper are organized as follows: Sect. 2 describes the state of the art and background of various software architecture models for real-time CPS tasks. Section 3 explains our event-driven architecture approach, while Sect. 4 describes the different components. Finally, the conclusions and further works are presented.

2 Related Works

This section summarizes the research works in the field organizing them in three main approaches of software engineering for CPS: components, services and agents.

2.1 Software Engineering for Component-Based CPS

Real-time distributed managed systems such as DREMS (Distribute Real-times Managed Systems) are typical component-based CPS runtime support models or platforms. It is conceived for distributed and mobile scenarios, e.g. groups of satellites or swarms of unmanned aerial vehicles (UAVs) [13]. The DREMS architecture is composed of two subsystems: design-time development platform and run-time support platform [6, 13].

In the CPS component model in DREMS, two non-functional elements are highlighted that are as important as the main functional characteristics of a system: real-time features of CPS components and independence and decoupling of the CPS components.

The RTCCM (Real Time Container Component Model) [14] and the ARINC-653 (ACM - Arinc Component Model) [15] support real-time operations. Functionally, it is a software specification for space and time partitioning in safety-critical real-time air vehicle operating systems [6, 14, 15].

On the other hand, the independence of the components and the fact that they are not coupled to the system play a fundamental role in a model for CPS based on components. DEECo (Dependable Emergence Ensembles of Components) [16] is a model aimed at working on the difficulties of large-scale distributed CPS- like dynamics or autonomy. A component system such as DEECo is characterised by its great independence.

Critical technical problems in this type of model usually exist in aspects such as the construction and deployment of CPS tasks based on components or the reconfiguration of CPS tasks.

The construction of complex CPS systems is implemented through the composition of simple blocks (components), previously developed independently of the application in which they will be used [15]. In fact, the separation between interface and implementation proposed for CPS tasks [16] allows a component to be conceived as a black box that encapsulates services. In this way, it is not necessary to know its internal details to use it. It is only necessary to characterize its interface.

Furthermore, the CPS that operate in dynamic and restricted environments are composed of multiple communication networks, controllers, sensors and actuators that involve constant and dynamic changes given the behavior of the physical scenario in which they act. That is why aspects such as the reconfiguration of CPS tasks is so important and acquires greater relevance in the those based on components [17].

2.2 Software Engineering for Service-Based CPS

Some researchers have designed and implemented models, platforms or frameworks for real-time support of service-based CPS, these models capture all the characteristics of SOA [18]. They created the OWL-S model - Web ontology Language for services [19]. With this model, they were able to develop context- and resource-sensitive CPS services. Below, we mention other models and real-time platforms for service-based CPS, highlighting the most important characteristics to be taken into account: CSSM - Context Sensitive service model [20], Physicalnet [21, 22] and RI - MACS - Radically innovative mechatronics and advanced control systems [23, 24].

The Context Sensitive service model is an extended model of the OWL-S model [20]. This service model (CSSM) based on the ontology of physical entities is context sensitive. In this service model, physical entities are organized hierarchically according to their relationships. In terms of context introduces two new constraints: the precondition of context and the effect of context. These are incorporated and treated as a traditional complement to the constraints of service provision [20].

Physicalnet [21] is a generic model or framework for the management and programming of embedded and distributed sensor and actuator resources in a multi-user and multi-network environment. It was designed and implemented under a four-level or layered service-oriented architecture [21, 22]:

A. Service provider Tier: Contains service providers and location nodes. A provider registers its services with one and only one negotiator. It receives configuration messages (commands) and periodically sends control messages (sample values or status reports) to its negotiator through the Communication Gateway.
B. Gateway Tier: It contains the gateway that provides the connection and translation between the service providers and the negotiators. The gateway integrates different types of network interfaces to communicate with various service providers. In this way, service providers can communicate with each other through the gateway even without common network interfaces.
C. Negotiator Tier: A negotiator is a registry of services, a database of service states and application requirements, and a center of authority to resolve requirement conflicts for multiple concurrent applications.
D. Applications Tier: It contains applications that periodically generate and cancel remote service requirements. Multiple applications can simultaneously access the same negotiator and a single application can involve multiple negotiators to access resources from different administrative domains.

RI - MACS (Radically innovative mechatronics and advanced control systems) [23, 24]. It is a model designed as an industrial automation proposal based on service-oriented architectures - SOA and Web services. It is used in systems of new factories that automate processes under CPS concepts, which require real time responses (Embedded Real Time Responses) [23, 24]. This model presents a hardware and software infrastructure for industrial automation, which takes advantage of open technologies such as SOA, Ethernet-based communications and real-time technologies. The implemented architecture is supported in communications that add network capacities of TCP/IP protocols with real time traffic management [23, 24].

2.3 Software Engineering for Agent-Based CPS

Several architectures designed for agent models use the JAVA library framework for the development of a set of agents called JADE (Java Agent Development Framework). The objective of JADE is to simplify the implementation of multi-agent systems through middleware that complies with the specifications of the Foundation for Intelligent Physical Agents (FIPA). It has as objective the definition of standards for the interaction of agents [25].

Providing real-time support for CPS is a major challenge, many models that provide tasks for CPS have time constraints and some low-level control tasks can be executed only on dedicated hardware. We will analyze some models of agents that guarantee the real time of CPS: the Holonic Agent Model - HLA [26] and the Rainbow Model- RM [27]. The HLA is a multi-agent platform composed by three main modules:

A. HLC – High Level Control Module: It is an autonomous software unit capable of making intelligent decisions, communicating and cooperating with other agents. Therefore, HLC and "agent" are synonymous. The complexity of the agent's behavior requires the use of some of the high-level programming languages such as C++ or Java for its implementation. The HLC makes high-level decisions to comply with the general agent functional logic.
B. LLC – Low Level Control Module: It is a common control program that is executed in a classic scan-based PLC controlled logic program, when the PLC reads the input values from the sensors in a controlled process through the analog or digital I/O cards; the control programs perform a calculation that includes the calculation of new output values; and finally, the output values are propagated to the actuators in the control system.
C. Control Interface: Provides the communication links between the HLC and the LLC. The agent captures notifications from the LLC about important events that require agent attention. Example: Status completion tasks, sensor activation, diagnosis of resource failure or insufficient resources in the CPS system.

The RM is a platform that allows the development of relatively easy applications for Smart Cities. The Rainbow architecture consists of three layers, designed to make calculations nearing the physical part:

1. **Physical Layer:** Rainbow is based on the adoption of a multi-agent layer distributed over the physical part, which is embedded in virtual objects (VOs) that run in the network and are assisted by cloud services.
2. **Distributed Middleware:** Sensors and actuators of the physical layer are represented by VOs, these offer agents transparency and ubiquitous access to the physical layer due to the exposure of an API interface. The CPS systems functionalities can be used by different types of VOs that can be combined with more sophisticated agents based on event management rules that can affect high-level applications and end users.
3. **Level Cloud:** In this layer, Rainbow nodes are configured. These nodes are virtualized and deployed over the existing Cloud infrastructure. The communication between the nodes connected to the physical layer is conducted by exchanging messages. The Rainbow Cloud Layer is a platform of services Paas - Platforms as a services.

3 Software Engineering for Event-Based CPS

Our proposal of conceptual architecture and software engineering for real-time CPS tasks based on events is proposed under the rigor of software engineering based on a service-oriented architecture (SOA 2.0 - Service Oriented Architecture). The architecture is directed by EDA events (Event Driven Architecture) in which actuators and sensors are integrated as services and CPS services are in platforms completely independent of the physical or virtual worlds. However, it is totally interoperable through interfaces that encapsulate and hide the particularities of each implementation. In this way, the services developed are independent of the manufacturer, operating system and development technology of each platform [6, 7, 9, 28]. Here we describe the main architecture concepts that support our proposal:

1. **Event-Driven Architecture EDA:** It is a pattern of software architecture designed to consume, produce, detect and react to a flow of events. These are emitters and producers of events, in which each consumer must react when they receive an event. It has become a paradigm of computational design that accelerates the implementation of software architectures for IoT and CPS [28]. It is an architecture in which the software executes an action when it receives one or more event notifications. It is not the data which is at the center of the software operation logic. The core idea is the concatenation of events that require a specific response. It is a completely new and different paradigm from the sequential logic of software that is massively used nowadays [28, 30, 31].
2. **Complex Event Processing (CEP):** It is an emerging technology that allows to capture, analyze and correlate a large number of heterogeneous events in order to detect critical or relevant situations in real time. An event is something that occurs or is expected to occur. Likewise, a situation is an occurrence of an event or succession of events that requires some immediate reaction. This technology is based on the filtering of minor events and the recognition of those events that are relevant to a particular domain. All events can be aggregated, formed and analyzed within the same context, using complex event processing technologies CEP [29, 30, 31].

3. **The SOA 2.0 architecture. ("Event-Driven SOA" - Event-Oriented Architecture):** This is an evolution of traditional SOA, in which communication between users and services is carried out through events. It allows business users to monitor, analyse and enrich events to make the connections between disparate events that at first do not seem to be intuitively obvious. This makes these rich events visible to others, especially business analysts or marketing managers, and also allows the SOA 2.0 system to possibly automate the actions to be taken to address some unique pattern. EDA and SOA are not exclusive architectures, they are simply complementary, which are integrated through an ESB (Enterprise Service Bus) integration layer. This layer is interoperable between different communication protocols and is used as an integration layer where applications are exposed as services [28, 32].

4 Components of the Conceptual Architecture

The core of the architecture for event-based CPS proposed in this document is based of five main components: the event producer, the event sender, the event bus or ESB (Enterprise Service Bus), the event manager and the event consumer. We will now briefly describe the components of the proposed conceptual architecture (Fig. 1).

Event Producer in 1. These are the components of the architecture from which information is obtained with the intention of detecting possible critical or relevant situations for the system (CPS - IoT) [32]. It emits an event when something of interest occurs. Some event producers are:

- Event sensors: They detect situations and generate or originate raw events from data or business flows. (Temperature transmission).
- Monitors and sounders: They produce events about the availability and problems of the systems that conform the CPS platform (Failures in the IT networks or sensor, actuator, communications).
- Business processes: They produce events in significant points of the processing or when a task of a specific process is accomplished.
- Services and applications: They produce events in key points of the processing.
- State machines: Produce events when the state is changed.

Event Emitter in 2. This logically couples with the event producer and is responsible for converting and packaging raw events from the producers for delivery to the event bus. It is conformed by:

- Event trigger: Takes events from the producer and does all the necessary to make it available for task of processing or delivery, which can include event aggregation, caching and serialization.
- Simple event processing services: Such as filtering and mediating events issued by one single producer, which enriches the event with information available at the time the event occurs.

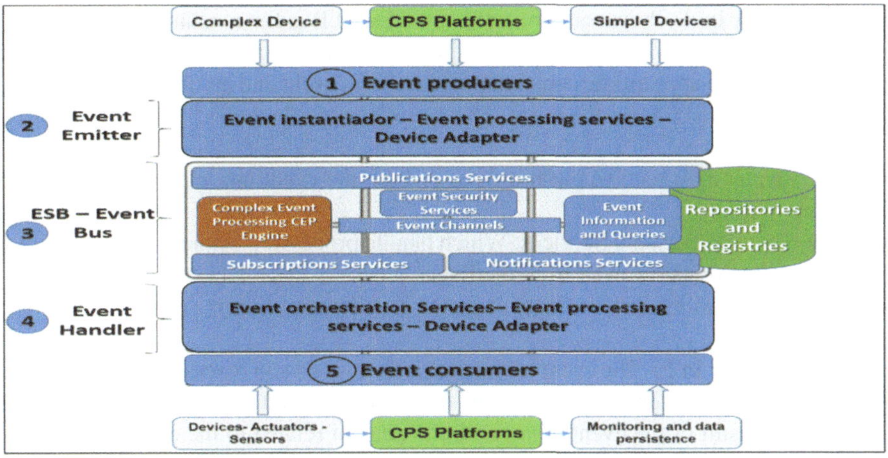

Fig. 1. Components of the conceptual architecture.

- Event adapters: Can offer formatting and protocol conversion of the event to create something that will be received by the event processing network.

The ESB Event Bus in 3. Receives events from the event senders and invokes consumers through event managers as a result of the events. Among the capabilities of the event bus, we can mention processing to produce a lower volume of more informative events using the input events. The event bus includes:

- Event channels: Which transmit events from the Event Transmitters to the Event Bus, between components of the Event Bus and to the Event Managers.
- Publishing Services: To enable producers to send events to the appropriate channels.
- Subscription services: To allow dynamic registration of producers and consumers of events.
- Notification Services: To notify Subscribed Event Managers when events are available.
- Query services: To allow the consultation of a repository in search of events.
- Event security services: To control access and authority relating to events.
- Event Processing and CEP Services: Which provide filtering, transformation and enrichment of events, and can also offer pattern comparison and event derivation. This includes complex event processing (CEP), which processes events from multiple sources and can perform pattern comparison that runs for a long period of time between events.
- Event Information Services: Enable administrators to add, remove and organize channels in order to organize event type metadata (syntax and semantics).
- Event logging: To offer a taxonomy of event types and an ontology of relationships between events.
- Event repository: To store events and so offer a persistence of events in the medium or long time.

Event Handler in 4. Prepares the events of the Event Bus for consumption of the consumers of events, receiving events and deciding how to react to them. Event Managers can also determine the appropriate consumer to react to an event and invoke the consumer(s) with a context derived from the event. The event manager includes:

- Event adapters: To receive event messages from the event bus and separate them to obtain event logs.
- Simple Event Processing Services: Which handle processing by the consumer to filter and mediate events received from the Event Bus.
- Event orchestration services: To manage the distribution of events among consumers.

Event Consumer in 5. The event consumer performs tasks in reaction to an event. The event consumer is not concerned with the origin of the event and only knows that it is invoked as a result of the event along with the context related to the event in question. The event consumer includes:

- Event activators: They are invoked to perform physical tasks inherent to CPS platforms (operation of valves, switches or alarms).
- Operator Control Panels: They display information about the behavior of the affected IT systems and services.
- Business control panels: They visualize information about the behavior of business processes.
- Business processes: Can be started or restarted in response to an event.
- Services and applications: Can be invoked in reaction to an event and can include external content management systems or event repositories.
- Status machines: The status of which can be changed in reaction to an event.

5 Conclusions

This paper proposes a theoretical architecture framework that reuses and integrates the concepts of EDA, SOA 2.0 and CEP event-driven architecture applied to support real-time CPS tasks. Each one of the architecture modules described facilitate the path to implement a network of processing events generated in real time from CPS platforms or infrastructure. This model allows to target the CPS devices as services and the communications are conducted through the integration of events in the ESB integration bus.

It is provided a useful framework for understanding the transition to be followed in an event model. Describing the event producers connected to the CPS infrastructure, from complex and simple devices to monitoring and data persistence. It is also described how events are prepared for consumption by event consumers. It also presents the modules that conform the event service bus, this last one is a vital component in the proposed architecture. From the event channels, the processing of complex CEP events, the security and information services to the subscription and notification of services. It summarizes the processing capabilities that may be required by event producers and consumers. The objective is to integrate all the modules that may be needed to implement event processing. Note that not all modules of the described conceptual architecture are necessarily required to implement a particular use case. This architecture provides the motivation to improve CEP engines to detect complex CPS events in real time execution.

The future direction of the work includes the investigation of a method for the detection of interaction events to improve the efficiency of the recovery of large volumes of compound and complex events. The case of use in study is the application of the proposed architecture using video surveillance camera networks as sensor devices in charge of capturing events from the physical environment for the CPS system.

References

1. Lee, E.A.: The past, present and future of cyber-physical systems: a focus on models. Sensors **15**, 4837–4869 (2015)
2. Ringert, J.A., Rumpe, B., Wortmann, A.: Architecture and behavior modeling of cyber-physical systems with MontiArcAutomaton. Aachener Informatik-Berichte, Software Engineering, Band 20. 2014, 27 February 2015
3. Lee, E.A.: Cyber physical systems: design challenges. In: 2008 11th IEEE International Symposium on Object Oriented Real-Time Distributed Computing (ISORC), 5–7 May 2008, Orlando, Florida, USA, pp, 363–369 (2008)
4. Perera, C., Liu, C.H., Jayawardena, S.: The emerging internet of things marketplace from an industrial perspective: a survey. EEE Trans. Emerging Top. Comput. **3**, 585–598 (2015)
5. Hamdaqa, M., Tahvildari, L.: Cloud computing uncovered: a research landscape. Adv. Comput. **86**, 41–85 (2012)
6. Sun, Y., Yang, G., Zhou, X.-S.: A survey on run-time supporting platforms for cyber physical systems. Frontiers Inf. Technol. Electron. Eng. **18**(10), 1458–1478 (2017)
7. Monostori, L.: Cyber-physical production systems: roots, expectations and R&D challenges. Procedia CIRP **17**, 9–13 (2014)
8. National Science Foundation: "Cyber-physical systems summit report", Missouri, USA, 24–25 April 2008. http://iccps2012.cse.wustl.edu/_doc/CPS_Summit_Report.pdf
9. Boubeta-Puig, J., Ortiz, G., Medina-Bulo, I.: MEdit4CEP: a model-driven solution for real-time decision making in SOA 2.0. Knowl. Based Syst. **89**, 97–112 (2015)
10. Service component architecture – unifying SOA and EDA: Technical report, Fiorano Software Technologies (2010)
11. Ollesch, J.: Adaptive steering of cyber-physical systems with atomic complex event processing services: doctoral symposium. In: Proceeding DEBS 2016, 20–24 June 2016
12. Boubeta-Puig, J., Ortiz, G., Medina-Bulo, I.: A model-driven approach for facilitating user-friendly design of complex event patterns. Exp. System with Apps. **41**(2), 445–456 (2014)
13. Levendovszky, T., Dubey, A., Otte, W.R., et al.: Distributed real-time managed systems: a model-driven distributed secure information architecture platform for managed embedded systems. IEEE Softw. **31**(2), 62–69 (2014)
14. Martínez, P.L., Cuevas, C., Drake, J.M.: RT-D&C: deployment specification of real-time component-based applications. In: Proceedings 36th EUROMICRO Conference on Software Engineering and Advanced Applications, pp. 147–155 (2010)
15. Dubey, A., Karsai, G., Mahadevan, N.: A component model for hard real-time systems: CCM with ARINC-653. Softw. Pract. Exper. **41**(12), 1517–1550 (2011)
16. Bures, T., Gerostathopoulos, I., Hnetynka, P., et al.: DEECO: an ensemble-based component system. In: Proceedings 16th ACM Sigsoft Symposium on Component-Based Software Engineering, pp. 81–90 (2013)

17. Martínez, P.L., Barros, L., Drake, J.M.: Design of component-based real-time applications. J. Syst. Softw. **86**(2), 449–467 (2013)
18. Huang, J., Bastani, F., Yen, I.L., et al.: Extending service model to build an effective service composition framework for cyber-physical systems. In: Proceedings IEEE International Conference on Service-Oriented Computing and Applications, pp. 1–8 (2009)
19. Martin, D., Paolucci, M., McIlraith, S., et al.: Bringing semantics to web services: the OWL-S approach. In: Cardoso, J., Sheth, A. (eds.) Semantic web services and web process composition, pp. 26–42. Springer, Heidelberg (2005)
20. Huang, J., Bastani, F., Yen, I.L., et al.: Toward a smart cyber-physical space: a context-sensitive resource-explicit service model. In: Proceedings 33rd Annual IEEE International Computer Software and Applications Conference, pp. 122–127, 125 (2009)
21. Vicaire, P.A., Hoque, E., Xie, Z., et al.: Bundle: a group-based programming abstraction for cyber-physical systems. IEEE Trans. Ind. Inform. **8**(2), 379–392 (2012)
22. Vicaire, P.A., Hoque, E., Xie, Z, Hoque, E., Stankovic, J.A.: Physicalnet: a generic framework for managing and programming across pervasive computing networks. In: RTAS 2010, Proceedings of the 2010 16th IEEE Real-Time and Embedded Technology and Applications Symposium, pp. 269–278, April 2010
23. Radically innovative mechatronics and advanced control systems (RIMACS)—Deliverable D1.2—Report on industrial requirements analysis for the next generation automation systems
24. Cucinotta, T., Mancina, A., Anastasi, G.F., Lipari, G., Mangeruca, L., Checcozzo, R., Rusina, F.: A real-time service-oriented architecture for industrial automation. IEEE Trans. Industr. Inf. **5**(3), 267–277 (2009)
25. Java Agent Development Framework (JADE): an Open Source platform for peer-to-peeragent based applications
26. Vrba, P., Radakovič, M., Obitko, M., et al.: Semantic technologies: latest advances in agent-based manufacturing control systems. Int. J. Prod. Res. **49**(5), 1483–1496 (2011)
27. Giordano, A., Spezzano, G., Vinci, A.: A smart platform for large-scale cyber-physical systems, pp. 115–134. Springer, Cham (2016)
28. Boubeta-Puig, J., Ortiz, G., Medina-Bulo, I.: Approaching the internet of things through integrating SOA and complex event processing. In: IGI Global Book Series Advances in Web Technologies and Engineering (AWTE). IGI Global (2014)
29. Luckham, D.: Event Processing for Business: Organizing the Real-Time Enterprise. Wiley, Nueva Jersey (2011)
30. Sosinsky, B.: Cloud Computing Bible. Wiley, Estados Unidos (2011)
31. He, M., Zheng, Z., Xue, G., Du, X.: Event driven RFID based exhaust gas detection services oriented system research. In: 4th International Conference on Wireless Communications, Networking and Mobile Computing, pp. 1–4 (2008)
32. Boubeta-Puig, J., Cubo, J., Nieto, A., Ortiz, G., Pimentel, E.: Proposal for a device architectures services with event processing. IGI Global (2013)

A Preliminary Study on Deep Transfer Learning Applied to Image Classification for Small Datasets

Miguel Ángel Molina[1], Gualberto Asencio-Cortés[1], José C. Riquelme[2], and Francisco Martínez-Álvarez[1](✉)

[1] Data Science & Big Data Lab, Pablo de Olavide University, 41013 Seville, Spain
mamolcab@alu.upo.es, {guaasecor,fmaralv}@upo.es
[2] Department of Computer Science, University of Seville, Seville, Spain
riquelme@us.es

Abstract. A new transfer learning strategy is proposed for image classification in this work, based on an 8-layer convolutional neural network. The transfer learning process consists in a training phase of the neural network on a source dataset of images. Then, the last two layers are retrained using a different small target dataset of images. A preliminary study was conducted to train and test the transfer learning proposal on Malaria cell images for a binary classification problem. The methodology proposed has provided a 6.76% of improvement with respect to other three different strategies of training non-transfer learning models. The results achieved are quite promising and encourage to conduct further research in this field.

Keywords: Transfer learning · Deep learning · Classification · Pattern recognition

1 Introduction

Deep learning has become quite popular in the field of big data and, in particular, in some applications such as remote sensing [1] or time series [2,3]. Transfer learning is a discipline suitable in situations in which there is a small amount of data to be mined (target data). The adequate training of deep neural network typically requires many data and much time. Nonetheless, a vast majority of real-world problems are not characterized by such amount of data and, therefore, models are not as accurate as expected. The integration of deep learning with transfer learning is called deep transfer learning and it makes the most of both paradigms. Thus, deep learning is used to model problems within big data contexts and, afterwards, re-purposed to transfer the knowledge to models with insufficient data [4]. There is a major flaw in transfer learning, which is the lack of interpretability of its models because pretrained models are applied to the new data without any prior information or understanding of the model [5].

Á. Herrero et al. (Eds.): SOCO 2020, AISC 1268, pp. 741–750, 2021.
https://doi.org/10.1007/978-3-030-57802-2_71

A new transfer learning strategy is proposed in this work, based on the application of a convolutional neural network (CNN). In particular, a 8-layer CNN is trained with the source dataset. Then, the last two layers are retrained with a training set from the target dataset. Different training sets, as explained in Sect. 3, are created in order to validate the robustness of the method. In short, New methodologies have been used, such as the differential study in four experiments and the analysis of similarities between source and target subsets through dendrograms.

To assess the performance of the proposal, the malaria cell images dataset [6], available along with the work in [7], is tested. This dataset is formulated as a binary classification problem, in which cells are either parasited or uninfected. Three additional strategies are also evaluated to compare the performance in terms of accuracy. The results achieved are quite promising.

The rest of the paper is structured as follows. Section 2 overviews recent and relevant papers in the field of deep transfer learning and its application to image classification. Section 3 describes the proposed methodology and how deep transfer learning can be applied to improve classification performance in datasets with few samples. Section 4 reports the results achieved and discusses their goodness. Finally, Sect. 5 summarizes the conclusions drawn from this work.

2 Related Works

Deep transfer learning is becoming one of the research fields in which much effort is being put into [8]. In fact, many applications can be found in the literature currently. Thus, Talo et al. [9] proposed a novel approach based on deep transfer learning to automatically classify normal and abnormal brain magnetic resonance images. Data augmentation, optimal learning rate finder or fine-tuning were the strategies used to infer the model.

A wide variety of applications in remote sensing problems are also available. In 2017, Zhao et al. [10] proposed a transfer learning model with fully pretrained deep convolution networks for land-use classification of high spatial resolution images. The authors claimed that the method accelerated the training process convergence with no loss of accuracy, as shown in the comparative analysis they report. The classification of Synthetic Aperture Radar (SAR) images through deep transfer learning was proposed in [11]. Given that labelling SAR images is quite challenging, the authors proposed to transfer learning from the electro-optical domain and used a deep neural network as classifier.

Another approach to range underwater source was recently introduced in [12]. In this case, the source domain was a set of large synthetic historical environmental data, which was transferred to the source domain (a deep-sea area). which migrates the predictive ability obtained from synthetic environment (source domain) into an experimental sea area (target domain). Reported results outperformed those of CNNs.

Another deep neural network model was proposed in [13] for plant classification. In particular, four different deep transfer learning models were applied to four public datasets, improving the performance of other methods.

Li et al. [14] linked emotions during conversations, by means of acoustic signals to behaviors through deep transfer learning as well, in an effort of explicitly quantifying the existing relationship. Hybrid architectures of both convolutional and recurrent neural networks to achieve the goal were explored.

3 Methodology

3.1 Image Preprocessing

The first step in the image preprocessing is to rescale all the images to the same dimensions, because it is necessary to have the same number of input pixels passed to the neural network. For the image rescaling process, the function *resize*() of the OpenCV library [15] was applied using a bilinear interpolation algorithm. The second step of the preprocessing is to encode the image labels, in order to have as many outputs of the neural network as image labels. Thus, a predicted probability is returned for each label.

3.2 Creation of Source and Target Subsets

Disjoint source and target subsets of images were extracted from the original set of images. The source subset is the dataset from which the initial model was trained. The target subset is the dataset used both to update such model (transfer process) and to test the updated model.

To extract the source and target subsets, it has been tried that such datasets were as different as possible. Additionally, the source subset is larger than the target one. The idea underlying this strategy is to check if the transfer learning is effective when the source and target subsets contain dissimilar images and the target set is a smaller one.

For this purpose, each image was firstly encoded using the Google Inception V3 deep neural network trained on ImageNet [16], which is available online. As a result of this process, for each image a vector of 2,048 real values were obtained from the weights of the last layer of such network. This process has been conducted using the Orange's Image Embedding node from the Orange-ImageAnalytics package (version 0.4.1) [17].

A table formed by the 2,048 real values of the image along with its label were generated for all the original images. Next, a hierarchical clustering was applied to each different label using such table as input. A dendrogram was generated for each label (U and P) after applying the hierarchical clustering.

Finally, the first two nodes of the first level of each dendrogram were selected. As example, for binary classification, two dendrograms were generated and four nodes were selected from them (U_A, U_B, P_A, P_B,). The node which contains the largest number of images for each label was added to the source subset. Similarly, the node which contains the smallest number of images for each label was added to the target subset.

It can be concluded that source and target subsets of images were generated in such a way that they contain dissimilar images and the source subset is larger than the target subset.

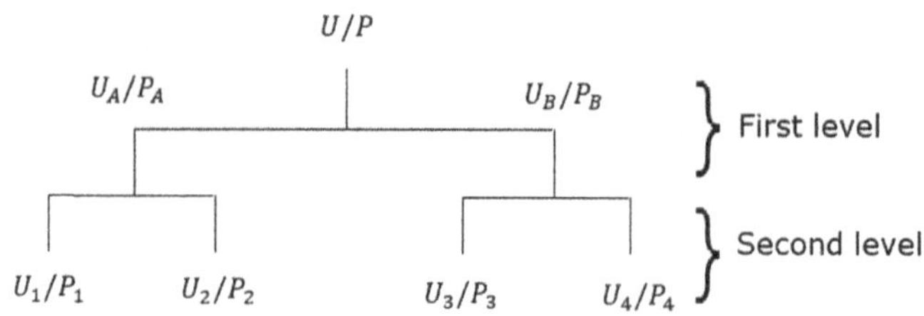

Fig. 1. Dendrogram scheme for each label U and P

3.3 Deep Neural Network Architecture

The next step consists in training a convolutional neural network and testing it using the subsets described in the previous section. The way these subsets are divided to validate the methodology will be explained in the next subsection.

The deep convolutional neural network is composed of three layers of 2D-based convolution using a kernel of size 3×3 and performing 32, 32 and 64 filters, respectively. Moreover, two layers of MaxPooling were added to the network, with a 2×2 size for the two of them. Finally, two dense flatten and fully-connected layers were added as the last layers of the network. The neural network proposed has $848,226$ parameters to be adjusted. The detailed network used is shown in Table 1. To implement the neural network architecture, Keras 2.2.4 over TensorFlow 1.14 was used [18].

Table 1. Deep neural network architecture used for transfer learning.

Layer (type)	Output shape	Params	Updateable
Conv2D	(None, 48, 48, 32)	896	No
Conv2D	(None, 46, 46, 32)	9,248	No
MaxPooling2D	(None, 23, 23, 32)	0	No
Conv2D	(None, 21, 21, 64)	18,496	No
MaxPooling2D	(None, 10, 10, 64)	0	No
Flatten	(None, 6400)	0	No
Dense	(None, 128)	819,328	Yes
Dense	(None, 2)	258	Yes

3.4 Four Validation Schemes

The target subset is randomly divided into two parts: training (70%) and test (30%). Freezing the same test part (30%) of the target subset, for a fair comparison, four different validation schemes have been proposed:

1. The model is generated using the training part (70%) of the target subset, and it is tested by evaluating its predictions over the test part (30%) of the target subset.
2. The model is generated using the whole source subset, and it is tested by evaluating its predictions over the test part (30%) of the target subset.
3. The model is generated using the whole source subset along with the training part (70%) of the target subset, and it is tested by evaluating its predictions over the test part (30%) of the target subset.
4. In this scheme the transfer learning procedure is carried out. The steps are the following:
   - The model is trained using the whole source subset.
   - Then, such model is updated using the training part (70%) of the target subset. This updating process only optimizes the weights within the two last layers of the neural network, maintaining the rest of its layers without changes.
   - The updated model is tested by evaluating its predictions over the test part (30%) of the target subset.

For each scheme, the methodology has been tested up to 10 times, having each execution a different random distribution of samples.

3.5 Source-Target Similarity Analysis

In order to check how the similarity between source and target subset images affects to the transfer learning effectiveness, the creation of these subsets was extended including the first two levels (instead of only the first level) of the dendrograms extracted from the hierarchical clustering process described in Sect. 3.2.

Extracting the images from each node of the second level of dendrograms, more combinations are available. Specifically, since dendrograms used are binary trees, there were 4 nodes taken from the second level of each dendrogram (U_1, U_2, U_3, U_4, P_1, P_2, P_3, P_4). As example, for the image binary classification, $4 + 4 = 8$ nodes were extracted, as it can be seen in Fig. 1.

To carry out the source-target similarity analysis, all combinations among extracted nodes are proved and the effectiveness achieved by transfer learning was analyzed in Sect. 4.

3.6 Class Imbalance Analysis

Finally, an analysis has been conducted to prove how the effectiveness of the proposed transfer learning methodology varies depending on the ratio between image classes (labels) in source and target subsets.

For such purpose, both source and target subsets derived from dendrograms were ranked according to the ratio between the minority and majority classes. Such ratio was expressed by a percentage and it was ranged from 50% (the number of images labeled with the minority class is half of the number of images labeled with the majority class) to 100% (same number of images for each class).

4 Experimentation and Results

4.1 Image Dataset

The set of images used to perform the methodology explained in previous section have been taken from Kaggle. Exactly, the chosen dataset is a set of images of cells which can be infected by the Malaria parasite or not. The challenge of these mages is to provide a complete data set of images in order to reduce the burden from microscopists in resource-constrained regions and improve diagnostic accuracy. The original source of images can be consulted in [6].

A set of 5000 images were randomly selected from the 13780 of the Kaggle challenge for each label; in total 10000 images were used to train and test our methodology. In order to work with similar type of images, all of them have been rescaled. This rescaling was to 50×50 pixeles.

4.2 Evaluation Metrics

In order to quantify the effectiveness of the methodology proposed, Binary Cross Entropy and Accuracy were computed.

The Binary Cross Entropy is a loss function that is applicable for binary classification. This is the most common loss function when working with this type of data sets. In this case, the output layer has one node. The typical activation function is a sigmoid and the formula is the following: $CE = -(y_i \cdot \log \hat{y_i}) + (1 - y_i)log(1 - \hat{y_i})$

The metric used for Accuracy is the Binary Accuracy, which calculates the mean accuracy rate across all predictions for binary classification problems. The formula is: $Acc = \frac{1}{n} \sum_{i=1}^{n} y_i = \hat{y}_i$

4.3 Experimental Settings

The experimental settings established to execute the experiments were:

Batch size: with a value of 128, it defines the number of samples that will be propagated through the network.

Epochs: one epoch is when an entire data set is passed forward and backward through the neural network only once. The number of epochs used was 5.

Optimizer: The optimizer used is the RMSprop. This optimizer recommends to leave the parameters at their default values, except the learning rate, which, in this case, has been set to $e_r = 1 \cdot 10^{-4}$.

4.4 Results and Discussion

The results obtained applying the methodology through the four validation schemes described in the previous section are shown in Table 2.

In Table 2, the four proposal schemes can be observed. In each sub table, *SelectedImages* explains how the training and test subsets are built. *Executios* indicated the number of executions of each scheme. *Loss* and *Accuracy* are the metrics used to evaluate the results of each scheme and which have been defined previously. *Average* and *SD* are the average and standard deviation of the accuracy of the ten executions

Table 2. Effectiveness achieved for each validation scheme with no transfer learning (Schemes 1, 2 and 3) and with our transfer learning proposal (Scheme 4).

Scheme 1					
Selected images	Execution	Loss	Accuracy	Average	SD
	1	0.5650	71.34%		
	2	0.5826	73.04%		
Train set:	3	0.5617	75.21%		
70% Target Domain	4	0.5782	71.73%		
	5	0.6076	65.92%	69.06%	0.04
	6	0.6135	64.45%		
Test set:	7	0.6046	66.46%		
30% Target Domain	8	0.5722	73.66%		
	9	0.6311	63.21%		
	10	0.6083	65.61%		

Scheme 2					
Selected images	Execution	Loss	Accuracy	Average	SD
	1	0.5994	65.61%		
	2	0.5997	69.48%		
Train set:	3	0.6132	61.74%		
Source Domain	4	0.6322	60.73%		
	5	0.6378	60.81%	60.58%	0.04
	6	0.6733	56.24%		
Test set:	7	0.6379	59.49%		
30% Target Domain	8	0.6316	59.18%		
	9	0.6963	55.92%		
	10	0.6702	56.55%		

Scheme 3					
Selected images	Execution	Loss	Accuracy	Average	SD
	1	0.6736	58.33%		
	2	0.5622	72.19%		
Train set:	3	0.5371	72.73%		
Source Domain and	4	0.7207	53.29%		
70% Target Domain	5	0.8145	53.29%	63.83%	0.10
	6	0.5489	74.13%		
Test set:	7	0.5259	73.04%		
30% Target Domain	8	0.8140	53.29%		
	9	0.6597	53.29%		
	10	0.4900	74.75%		

Scheme 4					
Selected images	Execution	Loss	Accuracy	Average	SD
	1	0.5065	76.92%		
Train set:	2	0.5153	75.45%		
Source Domain and	3	0.4701	78.23%		
	4	0.5157	76.14%		
Retrain with the	5	0.5065	78.23%	75.82%	0.02
70% Target Domain	6	0.5184	77.38%		
	7	0.5818	69.95%		
Test set:	8	0.5290	75.76%		
30% Target Domain	9	0.5345	71.88%		
	10	0.4908	78.23%		

As it can be seen in Table 2, the fourth scheme, which is the transfer learning one, is the scheme with the best results of all of them obtaining a better average accuracy, with an improvement of 6.76%, which is a very remarkable performance.

Another important feature is the robustness that the transfer learning technique brings to the results. The standard deviation of the transfer learning (Scheme 4) is smaller than the other schemes. Such result demonstrates that, with this technique, the learning is more robust and the dependence of the random train and test subsets is lower.

For the Source-Target similarity analysis, the four clusters obtained by the second level of the dendrogram for each class (image label) are used in order to make different combinations for constructing the source and the target subsets. The number of images of the second level for the class Uninfected are U1: 1883, U2: 2213, U3: 408 and U4: 496 images ($total = (U1 + U2) + (U3 + U4) = (1883 + 2213) + (408 + 496) = 4096 + 904 = 5000$ images). The number of images of the second level for the class Parasitized are P1: 377, P2: 2041, P3: 936 and P4: 1646 images ($total = (P1 + P2) + (P3 + P4) = (377 + 2041) + (936 + 1646) = 2418 + 2582 = 5000$ images).

With these clusters, the schemes 1 and 4 have been carried out again. The improvement for each group is shown in Table 3, where, the clusters obtained for the uninfected cells of Malaria set have been named as Target U and those obtained for the parasitized ones as Target P. The number of images obtained from the sum of them from the two previous clusters is the Target Dim. The sum of the rest of clusters is the Source Dim. Scheme1 Acc and Scheme4 Acc show the accuracy obtained from each scheme. The column named Improvement shows the percentage of improvement using transfer learning techniques. Finally, Cosine Distance indicates the cosine distance between the source and target subsets, where values close to 0 indicate very similar data sets. The formula has the following expression:

$$Cos\theta = \frac{\boldsymbol{a} \cdot \boldsymbol{b}}{\|\boldsymbol{a}\| \, \|\boldsymbol{b}\|} = \frac{\sum_1^n a_i b_i}{\sqrt{\sum_1^n a_i^2}\sqrt{\sum_1^n b_i^2}}$$

To facilitate the understanding of the graphs, the values indicated in the table will be 1-Cosine Distance.

Figures 2 and 3 show the improvements caused by the transfer learning technique depending on the distance between source and target subsets (Fig. 2) and the ratio between classes in source and target subset (Fig. 3). In Fig. 2, the relationship between the distance of the two subsets against the improvement using transfer learning can be observed. The distances obtained from the different combinations give a narrow range of values due to the own characteristics of the set of images. This causes that the improvements produced by transfer learning techniques are not noticeable. However, if the linear regression line of the curve obtained is drawn, a worsening of the results is observed as the distance between the two subsets is greater. In Fig. 3, the X axis shows the ratio of the minority class in each subset, and the Y axis the improvement between scheme 1 and scheme 4. As the ratio of the minority class grows, an effectiveness improvement of the scheme 4 is observed (particularly with higher ratios of minority class in the source subset). Only in the two last cases this effect is not appreciable. These two cases are, precisely, those related with the two subsets with bigger distances between them. Other aspects to be studied in future works are the influences of the number of samples in each cluster in order to get more information to learn general behaviour. It is possible that some limitations in the results can be associated with these aspects besides the architecture of the neural network. Also, the linear regression line is drawn to show the trend of the transfer learning improvement.

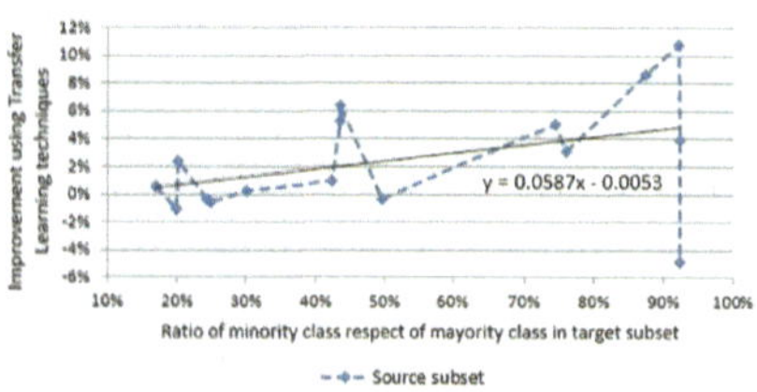

Fig. 2. Relationship between source-target subset distances and the transfer learning accuracy improvement.

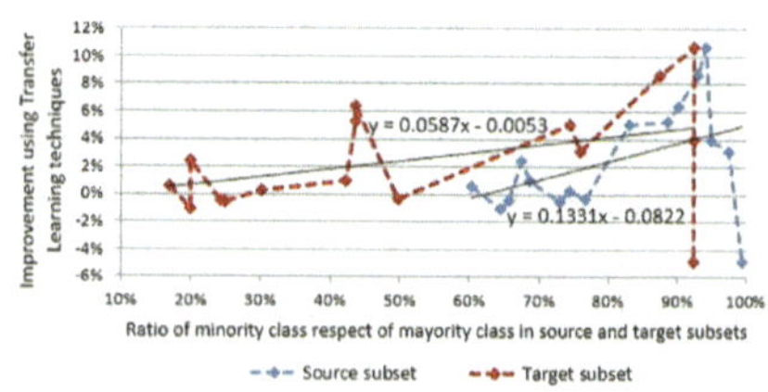

Fig. 3. Relationship between class imbalance ratio and the transfer learning accuracy improvement.

Table 3. Image classification accuracy achieved by transfer learning (Scheme 4).

Target U	Target P	Target Dim.	Source Dim.	1-Cosine distance	Scheme1 Acc.	Scheme4 Acc.	Improvement
U2	P2	4254	5746	0.1742	58.21%	68.96%	**10.75%**
U1	P4	3529	6471	0.1774	69.45%	78.13%	**8.68%**
U4	P3	1432	8568	0.2017	67.93%	74.30%	**6.37%**
U3	P3	1344	8656	0.2034	67.33%	72.60%	**5.27%**
U2	P4	3859	6141	0.1809	60.89%	65.98%	**5.09%**
U1	P2	3924	6076	0.1885	69.42%	73.33%	**3.91%**
U4	P1	873	9127	0.2118	88.24%	91.41%	**3.17%**
U1	P1	2260	7740	0.1980	86.99%	89.40%	**2.41%**
U2	P3	3149	6851	0.1774	80.31%	81.29%	**0.98%**
U2	P1	2590	7410	0.1749	92.68%	93.26%	**0.58%**
U4	P4	2142	7858	0.1909	74.66%	74.90%	**0.24%**
U1	P3	2819	7181	0.1853	89.34%	89.04%	**−0.30%**
U4	P2	2537	7463	0.1739	83.03%	82.64%	**−0.39%**
U3	P4	2054	7946	0.1859	80.75%	80.20%	**−0.55%**
U3	P2	2449	7551	0.1740	87.62%	86.57%	**−1.05%**
U3	P1	785	9215	0.2132	91.99%	87.12%	**−4.87%**

5 Conclusions

In this paper the benefits of transfer learning have been empirically demonstrated using a dataset of images of cells parasitized, or uninfected, by the Malaria disease. First, comparing the fourth validation schemes proposed, the use of transfer learning techniques has provided a 6.76% of improvement with respect to different ways to train non-transfer learning models. Also, transfer learning has provided more robustness, reflected in the smaller standard deviations obtained, bringing more general knowledge of the treated data sets. According to the analysis of improvements, similarities of images and class imbalance ratios, no clear improvements have been observed. However, some relationship has been found between the class ratio and the improvement of transfer learning, in such a way that more balanced datasets produce higher improvement using transfer learning. These works are a starting point to continue exploring the benefits and limitations of transfer learning, like the number of samples, distances and neural network structure. In future works, the results will be tested previously applied strategies with those being proposed here.

Acknowledgements. The authors would like to thank the Spanish Ministry of Economy and Competitiveness for the support under the project TIN2017-88209-C2-1-R.

References

1. Bui, D.T., Hoang, N.-D., Martínez-Álvarez, F., Ngo, P.-T.T., Hoa, P.V., Pham, T.D., Samui, P., Costache, R.: A novel deep learning neural network approach for predicting flash flood susceptibility: a case study at a high frequency tropical storm area. Sci. Total Environ. **701**, 134413 (2020)

2. Torres, J.F., Galicia, A., Troncoso, A., Martínez-Álvarez, F.: A scalable approach based on deep learning for big data time series forecasting. Integr. Comput. Aided Eng. **25**(4), 335–348 (2018)
3. Torres, J.F., Troncoso, A., Koprinska, I., Wang, Z., Martínez-Álvarez, F.: Big data solar power forecasting based on deep learning and multiple data sources. Expert Syst. **36**(4), e12394 (2019)
4. Deng, Z., Lu, J., Wu, D., Choi, K., Sun, S., Nojima, Y.: New advances in deep-transfer learning. IEEE Trans. Emerg. Top. Comput. Intell. **3**(5), 357–359 (2019)
5. Kim, D., Lim, W., Hong, M., Kim, H.: The structure of deep neural network for interpretable transfer learning. In: Proceedings of the IEEE International Conference on Big Data and Smart Computing, pp. 1–4 (2019)
6. Tatman, R.: R vs. Python: The Kitchen Gadget Test, Version 1 (2017). https://www.kaggle.com/iarunava/cell-images-for-detecting-malaria. Accessed 29 Jan 2020
7. Rajaraman, S., Antani, S.K., Poostchi, M., Silamut, K., Hossain, M.A., Maude, R.J., Jaeger, S., Thoma, G.R.: Pre-trained convolutional neural networks as feature extractors toward improved malaria parasite detection in thin blood smear images. PeerJ **6**, e4568 (2018)
8. Tan, C., Sun, F., Kong, T., Zhang, W., Yang, C., Liu, C.: A survey on deep transfer learning. In: Proceedings of the International Conference on Artificial Neural Networks, pp. 270–279 (2018)
9. Talo, M., Baloglu, U.B., Yıldırım, Ö., Acharya, U.R.: Application of deep transfer learning for automated brain abnormality classification using MR images. Cogn. Syst. Res. **54**, 176–188 (2019)
10. Zhao, B., Huang, B., Zhong, Y.: Transfer learning with fully pretrained deep convolution networks for land-use classification. IEEE Geosci. Remote Sens. Lett. **14**(9), 1436–1440 (2017)
11. Rostami, M., Kolouri, S., Eaton, E., Kim, K.: Deep transfer learning for few-shot SAR image classification. Remote Sens. **11**(11), 1374 (2019)
12. Wang, W., Ni, H., Su, L., Hu, T., Ren, Q., Gerstoft, P., Ma, L.: Deep transfer learning for source ranging: deep-sea experiment results. J. Acoust. Soc. Am. **146**, EL317 (2019)
13. Kaya, A., Keceli, A.S., Catal, C., Yalic, H.Y., Temucin, H., Tekinerdogan, B.: Analysis of transfer learning for deep neural network based plant classification models. Comput. Electron. Agric. **158**, 20–29 (2019)
14. Li, H., Baucom, B., Georgiou, P.: Linking emotions to behaviors through deep transfer learning. Comput. Electron. Agric. **6**, e246 (2020)
15. Bradski, G.: The OpenCV library. Dr. Dobb's J. Softw. Tools **25**, 120–125 (2000)
16. Szegedy, C., Liu, W., Jia, Y., Sermanet, P., Reed, S., Anguelov, D., Erhan, D., Vanhoucke, V., Rabinovich, A.: Going deeper with convolutions. In: 2015 IEEE Conference on Computer Vision and Pattern Recognition (CVPR), pp. 1–9, June 2015
17. Demšar, J., et al.: Orange: data mining toolbox in Python. J. Mach. Learn. Res. **14**, 2349–2353 (2013)
18. Chollet, F., et al.: Keras (2015). https://github.com/fchollet/keras

Burr Detection Using Image Processing in Milling Workpieces

Virginia Riego del Castillo, Lidia Sánchez-González(✉), Laura Fernández-Robles, and Manuel Castejón-Limas

Departamento de Ingenierías Mecánica, Informática y Aeroespacial, Universidad de León, 24071 León, Spain
vriegc00@estudiantes.unileon.es,
{lidia.sanchez,l.fernandez,manuel.castejon}@unileon.es

Abstract. Manufacturing processes require to satisfy quality standards in the produced parts. In particular, the edge finishing must be burr-free, avoiding that it yields different problems such as wasting time removing them what increases the production cost and time. A burr can be noticed microscopically, but it can contain imperfections or evidence of poor piece design. In order to detect automatically this imperfections and to evaluate the quality of the edge finishing, this paper proposes a complete vision based method using image processing and linear regression. With the calculated function, the slope is isolated and compared to obtain quality assessment thresholds. Results validate the good performance of the proposed method to differenciate three types of burrs.

Keywords: Quality estimation · Milling machined parts · Burrs in workpiece · Burr classification

1 Introduction

Actual technologies allow us to automate a wide range of processes, specifically in industry. In this sense, it is commonly assumed the use of collaborative robots due to the support they provide to operators during the decision-making process. In order to do so, robots hold certain intelligence that is achieved by the use of intelligent systems. Regarding manufacturing, there is an important requirement to improve the edge finishing of machined pieces to achieve the quality and price desired [5]. Traditionally, this analysis is made by visual inspection of operators what yields subjectivity and criteria changing across the different operators. The presence of burr in the edges of the parts is not desired and for that reason several approaches have been carried out to study this phenomenon.

On one hand, some researchers are focused in study the problem analytically. So, in [1] burr is predicted by considering the process parameters or in [13] burr is modeled by using the finite element method. Otherwise, other works search an explanation of how and why they are formed. [7] concludes that five types of

Á. Herrero et al. (Eds.): SOCO 2020, AISC 1268, pp. 751–759, 2021.
https://doi.org/10.1007/978-3-030-57802-2_72

burr can appear at the edge exit by carrying out different experiments in milling under certain conditions. The influence of cutting conditions on the formation of burr is analysed in [4]. Another feature, such as acoustic emission and cutting forces signals are considered to predict entrance and exit burrs size [9]. In [15], researchers analyse which is the effect of the accumulated remnant burr between passes on the burr size. A study of which exit angle should be used in order to reduce the burr is presented in [11]. Additional aspects like low uncut chip thickness or material microstructure heterogeneity are also found to have an effect in burr formation [12].

In some reviews [8], they control the formation of burrs by studying the parameters of the machines, such as machining direction and the tool engagement angle. While in [6], different contact and contactless solutions, like lasers and sensors.

There are works that use image processing in order to detect burr formation. In this sense, there is a method that searches the best-fitting rectangle to the position of the burr [3], while other searches over the horizontal axis [14]. This work proposes a new method based on the use of a vision system in order to detect the burr formation on machined workpieces.

This paper is structured as follows. Section 2 explains the computer vision method and how the functions describing the image are computed. Section 3 presents the experiments carried out to validate the method. Finally, Sect. 4 gathers the achieved conclusions and future work.

2 Inspection Method

In [2] a method to identify wear areas on machined workpieces is proposed; such computer vision is based on the use of an industrial boroscope that allows us to obtain images since it is linked to a microscope camera. The illumination system employs white LED. Using the same vision system, images of the considered pieces along their edge are acquired. Such views make possible to identify and classify automatically burrs by processing the obtained image. A flow chart of the entire process including the inputs and outputs of each phase is shown in Fig. 1.

By using the microscope camera, RGB images of 2592×1944 pixels with a resolution of 300 ppp, which concerns pixel par pouce, are acquired and saved in TIF format. The three manufactured parts used in this experiment are presented in Fig. 2. For each piece, a set of images have been captured specifically in the end of the piece. Some examples of the acquired images are shown in Fig. 3.

2.1 Image Processing

This step is divided in several stages in order to convert the input image into a binary image, which allows us to differentiate the background and the workpiece. The complete process is shown in Fig. 4.

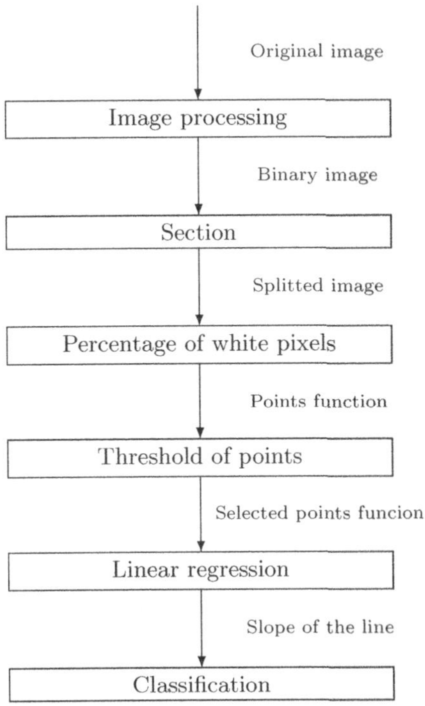

Fig. 1. Scheme of the proposed method followed to identify and classify images

Fig. 2. Parts considered in experiments.

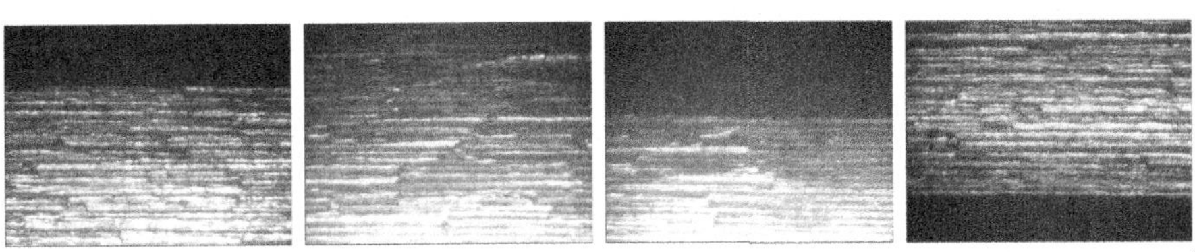

Fig. 3. Samples of acquired images.

1. Image Preprocessing: it consists of a image smoothing with a 5 × 5 kernel followed by a denoising with a template window size of 21 pixels and a search window size of 7 pixels.
2. Contrast-limited adaptive histogram equalisation (CLAHE) [10,16] that enhances the image contrast quality.
3. Otsu segmentation: that produces a binary image.
4. Morphological operations: dilate (with a 5 × 5 kernel) and erode (with a 25 × 25 kernel).

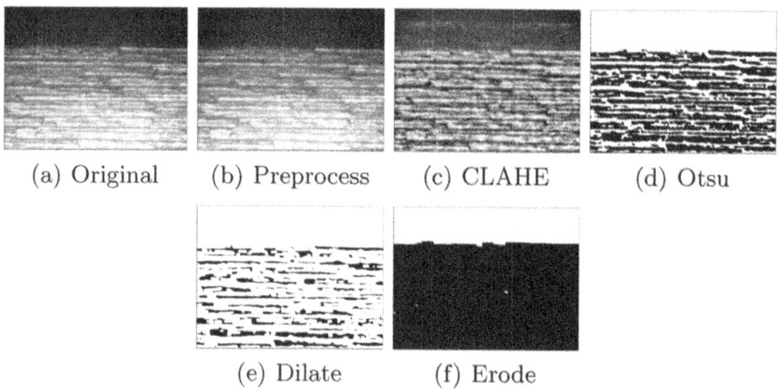

Fig. 4. Complete image processing procedure.

2.2 Section Image and Percentage of White Pixels

The binary image is split into 100 sections over its height, and for each section is computed the percentage of white pixels. These values form a feature vector that will be converted into a function that represents the blurr in order to compare with others.

2.3 Threshold of Points

The main idea of this stage is to delete the points that are irrelevant to be studied, such as the ones that are next to the higher and lower area of the part (see blue points in Fig. 5). To decide the relevance, the difference between each point and its previous point is computed; then those points whose difference with its previous point is higher than a 5-valued threshold are selected. These points are shown in Fig. 5 green-coloured crosses.

As light adjusting can make noise in some parts of the image, pixels whose position over the x axis is higher than 10 pixels from the previous two points are discarded, obtaining the definitive points to study (shown with red color stars in Fig. 5).

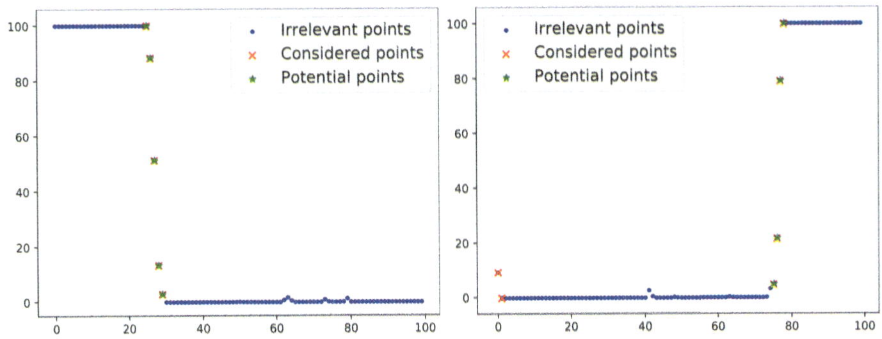

Fig. 5. Criteria for point selection.

2.4 Linear Regression

Once the points are selected, they are used to calculate a linear function. From the function, the slope is isolated to compare with others and makes possible to analyse the blurr presence.

The equation of regression line is defined as $h(x_i) = \beta_0 + \beta_1 x_i$ where $h(x_i)$ represents the predicted response value and β_0, β_1 are the regression coefficients.

Besides that, the residual error, ε_i, can be obtained as $y_i = \beta_0 + \beta_1 x_i + \varepsilon_i = h(x_i) + \varepsilon_i \rightarrow \varepsilon_i = y_i - h(x_i)$. Then, the cost function to attempt to minimize is $J(\beta_0, \beta_1) = \frac{1}{2n} \sum_{i=i}^{n} \varepsilon_i^2$

2.5 Criteria Selection

Researchers consider different number of burr types in the literature [7,12]. In this work, 3 categories are considered and an example of each category is shown in Fig. 6.

- The knife-type burr (K) does not chip, so there is no imperfections.
- The saw-type burr (S) is like a knife, but with small splinters.
- The burr-breakage (B) has large deformations due to fractures in manufacturing.

The representation of function slope with the associated class in Fig. 7 leads to establish a threshold to each class. By analysing the training data, the proposed criteria to determine the burr type as a function of the slope are:

- If it is higher than 10, there is knife-type burr (no burr formation).
- If it is lower than 1, there is burr-breakage.
- Otherwise, there is saw-type burr.

Fig. 6. Considered categories: knife-type burr (K) image on the left, saw-type burr (S) image on the center and burr-breakage (B) image on the right.

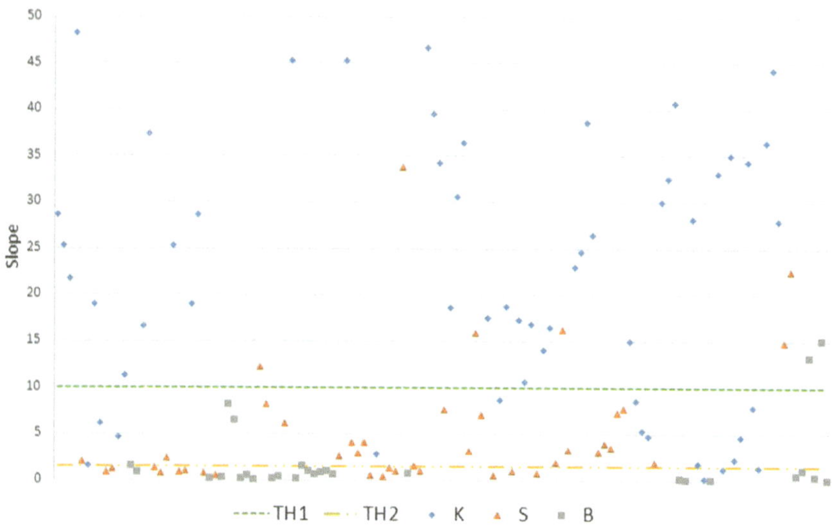

Fig. 7. Slope comparison for the different classes in training set

3 Experimental Results

The aim of the following experiment was to automatically detect whether the machined parts have a clean edge or have imperfections. Since imperfections do not allow the desired quality to be achieved.

In order to validate the proposed method, a dataset formed by 126 images is considered. These images are acquired with the described vision system and have the characteristics explained in Sect. 2. Each image was evaluated individually and classified visually by an expert according to their experience in three categories depending on if it has a clear edge finishing, little imperfections or an important lack of quality. This set is divided into training and testing subsets. In order to determine the parameters of the model proposed in Sect. 2.5, a set formed by 88 images is considered (the training set). The remaining 38 images are used in order to validate the model (the test set).

Let consider FP (False Positives) as those images that the proposed method identifies as they present burr formation but they actually do not and FN (False Negatives) those results that the method determine the burr is knife-type but the workpiece presents other type of burr. The confusion matrices for the training and test set are shown in Table 1 and Table 2.

Table 1. Confusion matrix of training set

	K	S	B
K	28	3	2
S	12	15	4
B	1	10	13

Table 2. Confusion matrix of test set

	K	S	B
K	12	4	2
S	2	9	1
B	0	2	8

The following performance metrics are calculated for each category. Precision is the fraction of results which are relevant and is given by $TP/(TP+FP)$, Recall by $TP/(TP+FN)$ is the fraction of total relevant results correctly classified and F1-score mixed both metrics by $2 * (precision * recall)/(precision + recall)$.

Table 3. Evaluation of each category using the test set

Class	TP	FP	FN	Precision	Recall	F1-score
K	12	4	2	0.7500	0.8571	0.8000
S	9	3	6	0.7500	0.6000	0.6667
B	8	2	1	0.8889	1.000	0.8421
Total	29	9	9	0.7632	0.7632	0.7632

According to the results shown in Table 3, machined workpieces that fulfill the quality standards and present defined end finishing are quite well identified (80%). Regarding blurr detection, when it shows big deformations it is well assessed (84.21%) but when it is not so significant, the perfomance is lower (66.67%).

4 Conclusions

In milling, manufactured parts must present clear edges in order to avoid costs and waste time removing them. Burr detection is a key aspect that guarantees that the machined workpiece satisfies certain quality standards. In this paper a method based on computer vision and linear regression is proposed in order to classify burrs from the images of the pieces. Using image processing techniques, the original image is converted into a image binary and the edge is analysed. By choosing the points near the ending, a function is defined and comparing them, a classification criterion is established. A proof of concept is presented that validates the method since it detects properly more than 80% of the burr formed on the workpiece of categories knife-type and burr-breakage. Future work involve different threshold selection and study its generalization for different parts as well as improving the detection of saw-type burr.

References

1. Bu, Y., Liao, W.H., Tian, W., Shen, J.X., Hu, J.: An analytical model for exit burrs in drilling of aluminum materials. Int. J. Adv. Manuf. Technol. **85**(9–12), 2783–2796 (2016)
2. Castejón-Limas, M., Sánchez-González, L., Díez-González, J., Fernández-Robles, L., Riego, V., Pérez, H.: Texture descriptors for automatic estimation of workpiece quality in milling. In: Pérez García, H., Sánchez González, L., Castejón Limas, M., Quintián Pardo, H., Corchado Rodríguez, E. (eds.) Hybrid Artificial Intelligent Systems, pp. 734–744. Springer, Cham (2019)
3. Chen, X., Shi, G., Xi, C., Zhong, L., Wei, X., Zhang, K.: Design of burr detection based on image processing. J. Phys: Conf. Ser. **1237**, 032075 (2019)
4. Chern, G.L.: Experimental observation and analysis of burr formation mechanisms in face milling of aluminum alloys. Int. J. Mach. Tools Manuf **46**(12–13), 1517–1525 (2006)
5. Dornfeld, D., Min, S.: A review of burr formation in machining. In: Aurich, J.C., Dornfeld, D. (eds.) Burrs - Analysis, Control and Removal, pp. 3–11. Springer, Heidelberg (2010)
6. Jin, S.Y., Pramanik, A., Basak, A.K., Prakash, C., Shankar, S., Debnath, S.: Burr formation and its treatments-a review. Int. J. Adv. Manuf. Technol. **107**(5), 2189–2210 (2020). https://doi.org/10.1007/s00170-020-05203-2
7. Lin, T.R.: Experimental study of burr formation and tool chipping in the face milling of stainless steel. J. Mater. Process. Technol. **108**(1), 12–20 (2000)
8. Niknam, S.A., Songmene, V.: Milling burr formation, modeling and control: a review. Proc. Inst. Mech. Eng. Part B J. Eng. Manuf. **229**(6), 893–909 (2015)
9. Niknam, S.A., Tiabi, A., Zaghbani, I., Kamguem, R., Songmene, V.: Milling burr size estimation using acoustic emission and cutting forces. In: ASME 2011 International Mechanical Engineering Congress and Exposition, pp. 901–909. American Society of Mechanical Engineers Digital Collection (2011)
10. Park, G.H., Cho, H.H., Choi, M.R.: A contrast enhancement method using dynamic range separate histogram equalization. IEEE Trans. Consum. Electron. **54**(4), 1981–1987 (2008)
11. Póka, G., Mátyási, G., Németh, I.: Burr minimisation in face milling with optimised tool path. Procedia CIRP **57**, 653–657 (2016). https://doi.org/10.1016/j.procir.2016.11.113. http://www.sciencedirect.com/science/article/pii/S2212827116312690. Factories of the Future in the digital environment - Proceedings of the 49th CIRP Conference on Manufacturing Systems
12. Régnier, T., Fromentin, G., Marcon, B., Outeiro, J., D'Acunto, A., Crolet, A., Grunder, T.: Fundamental study of exit burr formation mechanisms during orthogonal cutting of ALSi aluminium alloy. J. Mater. Process. Technol. **257**, 112–122 (2018). https://doi.org/10.1016/j.jmatprotec.2018.02.037, http://www.sciencedirect.com/science/article/pii/S0924013618300931

13. Régnier, T., Marcon, B., Outeiro, J., Fromentin, G., D'Acunto, A., Crolet, A.: Investigations on exit burr formation mechanisms based on digital image correlation and numerical modeling. Mach. Sci. Technol. **23**(6), 925–950 (2019). https://doi.org/10.1080/10910344.2019.1636274
14. Sharan, R., Onwubolu, G.C.: Measurement of end-milling burr using image processing techniques. Proc. Inst. Mech. Eng. Part B J. Eng. Manuf. **225**(3), 448–452 (2011)
15. Silva, L., Mota, P., Bacci Da Silva, M., Ezugwu, E., Machado, A.: Study of burr height in face milling of PH 13-8 Mo stainless steel–transition from primary to secondary burr and benefits of deburring between passes. CIRP J. Manuf. Sci. Technol. **10** (2015). https://doi.org/10.1016/j.cirpj.2015.05.002
16. Zuiderveld, K.: Contrast limited adaptive histogram equalization. In: Heckbert, P.S. (ed.) Graphics Gems IV, pp. 474–485. Academic Press Professional, Inc., San Diego (1994). http://dl.acm.org/citation.cfm?id=180895.180940

A Deep Learning Architecture for Recognizing Abnormal Activities of Groups Using Context and Motion Information

Luis Felipe Borja-Borja[1], Jorge Azorín-López[2(✉)], and Marcelo Saval-Calvo[2]

[1] Facul. de Ingen., Ciencias Físicas y Matemática, Universidad Central del Ecuador, Av. Universitaria, 170129 Quito, Ecuador
jazorin@ua.es

[2] Computer Technology Department, University of Alicante, 03080 Alicante, Spain

Abstract. Currently, the automation of activity recognition of a group of people in closed and open environments is a major problem, especially in video surveillance. It is becoming increasingly important to have computer vision architectures that allow automatic recognition of group activities to make decisions. This paper proposes a computer vision architecture capable of learning and recognizing abnormal group activities using the movements of the group in the scene. It is based on the Activity Description Vector, a descriptor capable of representing the trajectory information of a sequence of images as a collection of local movements that occur in specific regions of the scene. The proposal is based on the evolution of different versions of this descriptor towards the generation of images that will be input of a two-stream classifier capable of robustly classifying abnormal group activities. Moreover, it includes context information to provide extra information to classify the activities including it as the third stream of the classifier resulting in a robust architecture for one class classification problems. The architecture has been evaluated and compared with other approaches using Ped 1 and Ped 2 datasets, obtaining a high performance in abnormal group activity recognition.

Keywords: One-Class Classification · Human Behaviour Analysis · Activity Description Vector

1 Introduction

Automatic Human Behaviour Analysis (HBA) refers to the field of study in artificial intelligence that studies and analysis human actions and activities using machine learning techniques. Despite the large trajectory in this research field for its many applications [8,11], there is still challenges to solve. The large amount of CCTV cameras along with the improvements in computation capabilities,

Á. Herrero et al. (Eds.): SOCO 2020, AISC 1268, pp. 760–769, 2021.
https://doi.org/10.1007/978-3-030-57802-2_73

boosted the use of Deep Learning (DL) to solve and also open new HBA problems. One of the main current challenges is the study of multiple individuals conforming a group in the scene [6].

This paper focuses on group HBA in the case of one-class classification. The new HBA Deep Learning (DL) based approaches [16] require large sets of data to train the system for the whole spectrum of classes. However, in surveillance, the classes are mainly normal and abnormal, where normal is the large majority. Hence, some of the datasets have only normal behaviour in the training set making impossible to train binary classifiers. In order to cope with this, one-class classification techniques have been proposed [15].

It has been proved that using trajectory descriptors improves the quality of the actual behaviour estimation as it reduces some noise effects from the segmentation and tracking. The Activity Description Vector (ADV) [3,4] showed very good performance in description of trajectories for classic HBA analysis and prediction using neural networks and other classifiers. Furthermore, ADV was also used to analyse group behaviour (GADV) in [2] showing good results.

All this context led us to propose, as main objective of this work, an architecture for group activity recognition combining a variant of the ADV descriptor and machine learning/deep learning techniques, along with context or scene information. This architecture takes into account the problem of one-class classification. From it, the contribution of this paper improves the performance and generality in HBA classification tasks for one-class datasets. The variant of the ADV reduces the search space of all possible solutions, helping the afterwards classifier to faster and better perform its task.

The remaining of the paper is structured as follows: Sect. 1.1 describes the problem of one-class classification and provides an state-of-the-art of main works of HBA in this context; Sect. 2 introduces the Deep ADV proposal with a detailed explanation of the different components in the architecture for group action recognition; Sect. 3 shows a set of experiments that prove the performance of the proposal; and finally, Sect. 4 concludes the paper summarizing the main contributions and achievements, as well as future works.

1.1 One-Class Classification (OCC)

For the purposes of this study we assume that most elements of the training dataset are "normal" data (for this analysis the term "normal" is defined as not being anomalous). The goal is to learn from the data the meaning of "normal". Deviations or data different from this definition are considered as anomalies or "abnormal". In machine learning, if we want to have a binary classification we need samples of both classes. In many situations (e.g. survillance, fraudulent transactions...) we can hardly have samples of abnormal behaviours. For this cases, there is a common technique called one-class classification (OCC) [15]. It is based on the Minimum Coating Circle Problem [24], that tries to find the sphere with the smallest radius containing a set of points, for a finite set of points. In this case, it consists of learning the boundary that defines the single known normal class, and assuming that everything else is abnormal.

The combination of OCC and neural networks has been previously carried out in several works, such as the case of Chalapathy et al. [9] that proposed a model of a one-class neural network (OC-NN) to detect anomalies in complex datasets. The OCC is found in many real-world computer vision applications such as: anomaly detection [9], deep classification [17,20], novelty detection [1,22], and others.

About the use of one-class classification there are several works where researchers propose different approaches, in [26,27] Xu et al. present Appear DeepNet (AMDN) where multiple one-class SVM models are used to predict the anomaly scores of each input. A different proposal based on Generative Adversarial Nets (GANs) is presented in [19], or a variant called Conditional Generative Adversarial Networks (CGAN) [25] and combine with Denoising Autoencoders using multi level representations of both intensity and motion data. Another approaches based on local and global descriptors [21], and methods to integrate one-class Support Vector Machine (SVM) into Convolutional Neural Network (CNN), called Deep One-Class (DOC) [23]. In addition, proposal to colorize images for precise object detection [13] and to detect anomalies.

2 Deep Learning Architecture for Abnormal Classification

In this section, the deep learning architecture for recognizing abnormal activities of groups using context and motion information is presented. The context is related to any information of the scene that, although it is not the objective of the classification system, facilitates the achievement of it, in this case the recognition of activities. This is, for example, how the same activity in different places or times could be normal or abnormal, like a running in a train station could be abnormal whilst running in a playground is normal. Regarding the motion information, it is the main feature that the recognition system is built on. It is calculated from a variant of ADV [3–5]. Specifically, the deep learning variant of the ADV, coined D-ADV. This descriptor is enclosed in an architecture, named D-ADC-OC, that uses one-class classification.

The main stages of the pipeline are presented in Fig. 1. The D-ADV-OC is able to determine an abnormal activity carried out by groups from a sequence of images. It can be divided into two steps. First, two images called LRF and UDF are calculated by the D-ADV representation stage. After that, two modules are working in parallel to provide the classification of the activity in the scene and in the context.

2.1 Activity Description Vector

The first stage of the pipeline calculates a representation of the image sequence as a deep variant of the original Activity Description Vector (ADV). It is a trajectory-based feature initially presented in ADV [3] for representation trajectory data with classification purposes. For the sake of completeness, a brief summary of the ADV is shown but we refer you to [3,4] for further details.

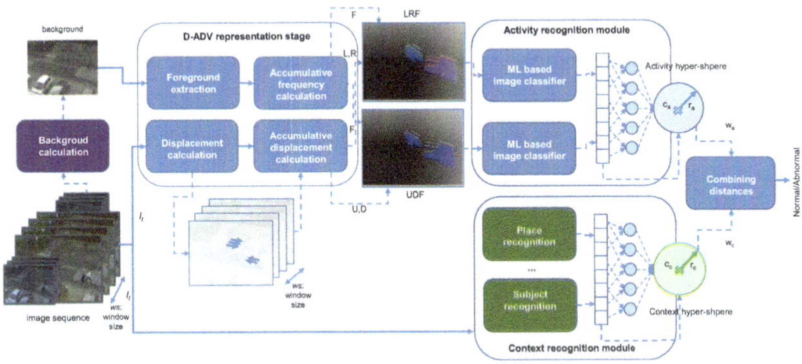

Fig. 1. Pipeline of the D-ADV-OC architecture. It is mainly divided into two parts: D-ADV representation stage were the descriptor is calculated and the second stage that defines the classifier to detect the activity and the context recognition.

ADV uses the number of occurrences of a person in a specific point of the scenario and its local movements in it. This method tessellates the ground scenario, G, in cellular regions as a grid, C, to discretize the environment. It is important to mention that, in order to have a more accurate result, G should be flatten using, for instance, homography. Each cell of the grid has information of the movements in the region including up (U), down (D), left (L), right (R) and frequency (F) data. The four former values are extracted from the single displacement between two consecutive points. If we focus on the U movement, it is calculated as follows:

$$U(p_i) = \begin{cases} (p_i - p_{i-1}) \cdot \begin{bmatrix} 0 \\ 1 \end{bmatrix} & if \frac{(p_i - p_{i-1}) \cdot \begin{bmatrix} 0 \\ 1 \end{bmatrix}}{\|(p_i - p_{i-1})\|} > 0 \\ 0 & otherwise \end{cases} \tag{1}$$

where p_i and p_{i-1} are two consecutive locations of the trajectory of an individual in G, and knowing that U is assumed to be a displacement in the positive vertical y axis. This formula is similar to the other displacements. On the other hand, frequency, F, is the number of occurrences of a person in a specific point.

Finally, the ground plane G is spatially sampled in a matrix C of $m \times n$ cells, so that the transformed points p_g and the functions of frequency and movements of it are in one of the cells of the matrix C. Each cell will describe the activity happened in that region of the scene considering the vector of relevant values, called *Activity Description Vector* (ADV_C). This vector will be composed by the frequency and the U, D, L and R movements of all points inside a cell :

$$ADV_C = \langle F, U, D, L, R \rangle \tag{2}$$

Therefore, within a particular cell, the accumulative histograms of the movements U, D, L, R and F for the points on G of the cell $C_{i,j}$ of C are calculated.

Let $u \times v$ the actual size of the scenario, split in $m \times n$ cells, and $p_{k,l}$ the point located in the position k and l of the G space, each ADV in a cell is:

$$\begin{aligned} \forall c_{i,j} \in C \wedge \forall p_{k,l} \in G / i = \left\lfloor \tfrac{kxm}{u} \right\rfloor \wedge j = \left\lfloor \tfrac{kxn}{v} \right\rfloor \\ ADV_{i,j} = \left(\sum F(p_{k,l}), \sum U(p_{k,l}), \sum D(p_{k,l}), \sum L(p_{k,l}), \sum R(p_{k,l}) \right) \end{aligned} \quad (3)$$

With this feature, the trajectory is described by dividing the scene into regions and compressing the data in cumulative values. It is interesting to highlight that Activity Description Vector integrates the trajectory information without length and sequential constraints.

2.2 D-ADV: Activity Descriptor for Deep Learning Purposes

The D-ADV uses a sequence of images as input. In contrast to ADV, the D-ADV is not based on the specific and individual movements of a person in the scene and the occurrences in it (i.e. Frequency). It considers the apparent motion of the subjects in the visual scene and the appearance of them assuming a specific background. For the former, the optical flow calculation is the starting stage of the process. It calculates the optical flow between two consecutive frames $(t, t + \delta t)$ of the sequence by using the differential method as the most widely used method [12]. It is based on the assumption of image brightness constancy: given a video sequence, the intensity of the pixel (x, y) of the frame t, $I_t(x, y)$, remains the same despite small changes of position and time period. Let (δx, δy, δt) the small change of the movement, and assuming the brightness constancy and expanding as Taylor series, it can be expressed and approximated as (more details can be found in [7,12]):

$$I_{t+\delta t}(x + \delta x, y + \delta y) \approx I_t(x, y) + \frac{\partial I}{\partial x}\delta x + \frac{\partial I}{\partial y}\delta y + \frac{\partial I}{\partial t}\delta t,$$

solving and dividing the second term throughout by δt, it is possible to obtain:

$$\frac{\partial I}{\partial x}\frac{\delta x}{\delta t} + \frac{\partial I}{\partial y}\frac{\delta y}{\delta t} + \frac{\partial I}{\partial t} = \frac{\partial I}{\partial x}U + \frac{\partial I}{\partial y}V + \frac{\partial I}{\partial t} \approx 0$$

where $U = \frac{\delta x}{\delta t}$ and $V = \frac{\delta y}{\delta t}$ are the two components of the optical flow in t.

In this case, the points p_i used to calculate the components of ADV as in Eq. 1 for the component Up (U) were those extracted from consecutive points in a trajectory on a plane. If we assume the image as a plane of the ground and a static camera (i.e. the apparent motion is only generated by the subjects in the scene, not for the observer – camera), the difference in the trajectory $(p_i - p_{i-1})$ could be approximated as the derivatives of pixels in x and y for the frame t as $(p_i - p_{i-1}) \approx (\frac{\delta x}{\delta t}, \frac{\delta y}{\delta t}) = (U, V)$. Moreover, as the movements are considered in each axis, the movements U and D are closely related to V component of the optical flow, and the components L and R related to the U. In consequence, the components could be calculated as:

$$U(I_t) = \begin{cases} -V_t \; if \; V_t < 0 \\ 0 \;\; otherwise \end{cases} \qquad D(I_t) = \begin{cases} V_t \; if \; V_t > 0 \\ 0 \; otherwise \end{cases} \tag{4}$$

With respect to the component F, it is estimated as:

$$F = |I - B| > 2 * std(I - B), \tag{5}$$

where B is the background calculated from a sequence of images, and std is the standard deviation of the difference between a frame and the background. The foreground is extracted in order to obtain the subjects that appear in the scene independently if the are moving.

This accumulative stage is responsible for calculating the ADV in a cell as presented in Eq. 2. On the one hand, accumulative displacement is responsible for the L, R, U and D parameters and the accumulative foreground is for the F component. The accumulation is considered for a set of consecutive frames of size, ws (see Fig. 1). In this case, the components are not concatenated all together, they are separated conforming two images LRF compose by the components L, R and F, and, similarly UDF combines the U, D and F components. Figure 1 shows an example of the UDF and LRF images where the accumulated data is shown in cyan and magenta.

2.3 D-ADV-OC: Abnormal Sequence Classifier Based on Two-Stream Activity Recognition and Context Information

The last stage of the proposal is composed by two modules that are combined to provide the normal or abnormal output of the model. The first module is the activity recognition stage based on deep neural networks. The proposed architecture D-ADV-OC for one-class problem considers a two-stream machine learning techniques (e.g. CNN, SVM...) able to classify the previously calculated single images: LRF and UDF. The proposal is experimented using various ML networks, and in particular, the CNNs are open to any existing one and any architecture could be used (VGG, ResNet, AlexNet, LeNet, etc.). This kind of networks usually uses a fully connected layer at the output with softmax activation in order to decide the class to which the image corresponds (e.g. objects, places, poses, etc.). The D-ADV-OC architecture does not take into account the individual dense layers. However, the previous layers in the convnet are concatenated into a late fusion way using in a concatenation layer from the two streams. Finally, a fully connected layer with linear activation is used to connect the concatenation layer to predict the abnormal activity in the group. It is based on the recent work proposed by Ruff et al. [20] that provides a deep model to train a neural network by minimizing the volume of a hypersphere that encloses the network representations of the data. Our approach differs as the work of Chalapathy et al. [10] does by combining the ability of networks to learn progressively rich representation of input data along with the one-class objective. Unlike the latter work, which uses auto-encoders to establish the representation of the input, defining the center of the hypersphere, in our work some layers of the

CNN based network are trainable allowing to keep learning both the center and the radius of the hypersphere. In order to avoid the problems of large datasets to train our model and with the objective that the model could be used for small datasets, we propose transfer learning from models trained with ImageNet.

The second module is related to the context information in the scene. Unlike the previous module, CNN-based networks are used to make predictions of objects, places, etc. that appear in the scene. In the training phase, the maximum values of the input patterns are calculated to normalize the output per object, place, etc. The average value of the performed normalization establishes the centre of the hypersphere, optimizing the radius of it through a fully connected layer at the end of the network.

Finally, the combining distances module uses the weights w_a and w_c for the activity and the context loss functions in order to train the network and calculates the distance from an input pattern to the normal class according in prediction stage using the following function:

$$dist = \frac{1}{n} w_a \sum_i ||i_a - c_a||^2 + w_c||i_c - c_c||^2,$$

being i_a the calculated representation for the activities using the motion; i_a, the calculated representation of the context in the scene; and, finally, c_a and c_c the centers of the hyperspheres.

3 Experiments

The experiments have been carried out using different data sets in order to assess the capabilities of the proposal. Additionally, comparisons have been made with other works where alternative solutions to the same problem are proposed. Specifically, we evaluated the effectiveness of our proposed architecture in two reference datasets, UCSD Ped 1 and Ped 2 [14] of scenes with groups of people. The datasets use one class defined as "normal", and anything different from this is considered "abnormal". For each dataset and architecture, the metrics used includes the Area Under Curve (AUC) and the Equal Error Rate (EER) from the Receiver Operating Characteristic (ROC).

Regarding the tested architectures, for the activity recognition a ResNet51 model and single fully connected layers have been used. For the context stage, a YOLO trained with VOC has been used. The window size (ws) as the number of consecutive frames considered in the accumulative process (see Fig. 1) is 10 for the different tests.

The experimental results are showed and compared with other state-of-the-art methods in Table 1 at frame-level. Results for the UCSD Ped 1 dataset show that the lowest value of EER and the highest value in AUC is provided in the work by Ravanbakhsh et al. [19] with 7% and 97.4% respectively. For the UCSD Ped 2 dataset, the best results are provided in the work of Vu et al. [25] achieving 2.49% for EER and, almost the perfect results for AUC achieving to the 99.21%. Our D-ADV-OC proposal has achieved very good results with and without context information. The performance is in accordance with those obtained in other works. However, for the UCSD Ped 1 dataset, our performance is lower some of than the state-of-the-art proposals.

The main characteristics of the methods with which our proposal has been compared to are: [13] is based on patches, [21] propose a cubic-patch-based method based on a cascade of classifiers, [18,19] propose GAN-based, [25] propose representation learning using Denoising Autoencoders (DAEs), [26,27] propose based on multiple Stacked Denoising AutoEnocders (SDAEs), and [23] integrate the one-class Support Vector Machine (SVM) into Convolutional Neural Network (CNN), named Deep One-Class (DOC) model.

Table 1. Classification of analyzed proposals

	Ped1 (frame-level)		Ped2 (frame-level)	
Reference	EER	AUC	EER	AUC
[18]	8%	97.4%	14%	93.5%
[19]	7%	96.8%	11%	95.5%
[26]	16%	92.1%	17%	90.8%
[27]	16.0%	92.1%	17.0%	90.8%
[21]	8.4%	93.2%	7.5%	93.9%
[25]	23.5%	82.34%	2.49%	99.21%
[23]	15.6%	91.4%	16.1%	91.1%
[13]	11.6%	90.0%	9.7%	92.0%
D-ADV-OC	25.7%	83.4%	18.2%	90.5%
D-ADV-OC+Context	**24.4%**	**84.4%**	**16.4%**	**91.4%**
D-ADV-OC+CNN	31.1%	76.8%	26.8%	80.5%
D-ADV-OC+CNN+Context	**30.2%**	**77.0%**	**25.7%**	**80.7%**

4 Conclusions

In this paper a novel group activity recognition architecture, D-ADV-OC based on trajectory descriptor context and machine learning or deep learning with One-Class Classification, has been proposed. The trajectory descriptor is a variant of the Activity Description Vector (ADV) named D-ADV serving as input of a classification stage. The variant considers any motion in the image, calculated by optical flow, instead of making use of specific trajectories of individual or the group, providing generality at the input, allowing its usage in many different situations and scenes. The apparent motion is accumulated in cells spatially distributed according to the input image of the sequence. It allows us to generate two images containing the description of the motion and the occurrence of subjects in the scene. The classification stage is fed by the previous images using two streams, context information, and using late fusion with a dense layer. Finally, the loss function to train the network is in charge of minimizing the volume of a hypersphere that encloses the network representations of the data.

Experiments have been carried out using the Ped 1 and Ped 2 datasets. The experimental results show the capacity of the architecture to classify the abnormal activities of the groups presented in the scene. Moreover, it is shown that the architecture is able to have good results using small datasets due to the use of the representation as the and input allow to the network to develop a hierarchy of higher understanding concepts from simpler ones. In this case, not from the image but from the motion representation.

Two main comparisons are made in the experiments, one without the use of a convolutional neural network (CNN), and the other with the use of the CNN, in both cases an additional comparison with the use of context is included. It can be verified by comparing D-ADV-OC+Context and D-ADV-OC+CNN+Context, that the values of AUC and EER where context is used are better. If we compare D-ADV-OC+CNN D-ADV-OC+CNN+Context, including an additional factor such as CNN, in the experiments we can conclude that the results of AUC and EER where it is not used are better, that is the case of D-ADV-OC+Context is the best result obtained from all the experiments performed.

References

1. Abati, D., Porrello, A., Calderara, S., Cucchiara, R.: Latent space autoregression for novelty detection. In: Proceedings of the IEEE Conference on Computer Vision and Pattern Recognition, pp. 481–490 (2019)
2. Azorin-Lopez, J., Saval-Calvo, M., Fuster-Guillo, A., Garcia-Rodriguez, J., Cazorla, M., Signes-Pont, M.T.: Group activity description and recognition based on trajectory analysis and neural networks. In: 2016 (IJCNN), pp. 1585–1592, July 2016
3. Azorín-López, J., Saval-Calvo, M., Fuster-Guilló, A., García-Rodríguez, J.: Human behaviour recognition based on trajectory analysis using neural networks. In: The 2013 (IJCNN), pp. 1–7. IEEE (2013)
4. Azorin-Lopez, J., Saval-Calvo, M., Fuster-Guillo, A., Garcia-Rodriguez, J., Orts-Escolano, S.: Self-organizing activity description map to represent and classify human behaviour. In: 2015 (IJCNN), pp. 1–7. IEEE (2015)
5. Azorin-López, J., Saval-Calvo, M., Fuster-Guilló, A., Oliver-Albert, A.: A predictive model for recognizing human behaviour based on trajectory representation. In: 2014 (IJCNN), pp. 1494–1501. IEEE (2014)
6. Borja, L.F., Azorin-Lopez, J., Saval-Calvo, M.: A compilation of methods and datasets for group and crowd action recognition. Int. J. Comput. Vis. Image Process. (IJCVIP) **7**(3), 40–53 (2017)
7. Bour, P., Cribelier, E., Argyriou, V.: Crowd behavior analysis from fixed and moving cameras. In: Alameda-Pineda, X., Ricci, E., Sebe, N. (eds.) Multimodal Behavior Analysis in the Wild, Computer Vision and Pattern Recognition, pp. 289–322. Academic Press, Cambridge (2019)
8. Chaaraoui, A.A., Climent-Pérez, P., Flórez-Revuelta, F.: A review on vision techniques applied to human behaviour analysis for ambient-assisted living. Expert Syst. Appl. **39**(12), 10873–10888 (2012)

9. Chalapathy, R., Chawla, S.: Deep learning for anomaly detection: A survey (2019). arXiv preprint arXiv:1901.03407
10. Chalapathy, R., Menon, A.K., Chawla, S.: Anomaly detection using one-class neural networks (2018)
11. Gowsikhaa, D., Abirami, S., Baskaran, R.: Automated human behavior analysis from surveillance videos: a survey. AI Rev. **42**(4), 747–765 (2014)
12. Ke, Q., Liu, J., An, S., Bennamoun, M., Sohel, F., Boussaid, F.: Computer vision for human–machine interaction. In: Leo, M., Farinella, G.M. (eds.) Computer Vision for Assistive Healthcare, Computer Vision and Pattern Recognition, pp. 127–145. Academic Press, Cambridge (2018)
13. Li, X., Li, W., Liu, B., Nenghai, Y.: Object and patch based anomaly detection and localization in crowded scenes. Multimedia Tools Appl. **78**(15), 21375–21390 (2019)
14. Mahadevan, V., Li, W., Bhalodia, V., Vasconcelos, N.: Anomaly detection in crowded scenes. In: 2010 IEEE Computer Society Conference on Computer Vision and Pattern Recognition, pp. 1975–1981. IEEE (2010)
15. Moya, M.M., Koch, M.W., Hostetler, L.D.: One-class classifier networks for target recognition applications. In: NASA STI/Recon Technical Report N, 93 (1993)
16. Nigam, S., Singh, R., Misra, A.K.: A review of computational approaches for human behavior detection. Archieves Comput. Methods Eng. **26**(4), 831–863 (2019)
17. Perera, P., Patel, V.M.: Learning deep features for one-class classification. IEEE Trans. Image Process. **28**(11), 5450–5463 (2019)
18. Ravanbakhsh, M., Nabi, M., Sangineto, E., Marcenaro, L., Regazzoni, C., Sebe, N.: Abnormal event detection in videos using generative adversarial nets. In: 2017 IEEE International Conference on Image Processing (ICIP), pp. 1577–1581. IEEE (2017)
19. Ravanbakhsh, M., Sangineto, E., Nabi, M., Sebe, N.: Training adversarial discriminators for cross-channel abnormal event detection in crowds. In: 2019 IEEE (WACV), pp. 1896–1904. IEEE (2019)
20. Ruff, L., Vandermeulen, R., Goernitz, N., Deecke, L., Siddiqui, S.A., Binder, A., Müller, E., Kloft, M.: Deep one-class classification. In: International conference on (ML), pp. 4393–4402 (2018)
21. Sabokrou, M., Fathy, M., Moayed, Z., Klette, R.: Fast and accurate detection and localization of abnormal behavior in crowded scenes. Mach. Vis. Appl. **28**(8), 965–985 (2017)
22. Sabokrou, M., Khalooei, M., Fathy, M., Adeli, E.: Adversarially learned one-class classifier for novelty detection. In: Proceedings of the IEEE Conference on Computer Vision and Pattern Recognition, pp. 3379–3388 (2018)
23. Sun, J., Shao, J., He, C.: Abnormal event detection for video surveillance using deep one-class learning. Multimedia Tools Appl. **78**(3), 3633–3647 (2019)
24. Sylvester, J.J.: A question in the geometry of situation. Q. J. Pure Appl. Math. **1**(1), 79–80 (1857)
25. Vu, H., Nguyen, T.D., Le, T., Luo, W., Phung, D.: Robust anomaly detection in videos using multilevel representations. In: Proceedings of the AAAI Conference on Artificial Intelligence, vol. 33, pp. 5216–5223 (2019)
26. Xu, D., Ricci, E., Yan, Y., Song, J., Sebe, N.: Learning deep representations of appearance and motion for anomalousevent detection (2015). arXiv preprint arXiv:1510.01553
27. Dan, X., Yan, Y., Ricci, E., Sebe, N.: Detecting anomalous events in videos by learning deep representations of appearance and motion. Comput. Vis. Image Underst. **156**, 117–127 (2017)

Implementation of a Low-Cost Rain Gauge with Arduino and Thingspeak

Byron Guerrero Rodríguez[1(✉)], Jaime Salvador Meneses[1(✉)], and Jose Garcia-Rodriguez[2(✉)]

[1] Central University of Ecuador, Quito, Ecuador
{bvguerreor,jsalvador}@uce.edu.ec
[2] Computers Technology Department, University of Alicante, Alicante, Spain
jgarcia@dtic.ua.es

Abstract. Recent studies determine that one of the triggers for landslides is torrential rains. This paper proposes the application of Arduino technology, and the platform for IoT Thingspeak to build a low-cost rain gauge equipment that allows the remote measurement of the variable's rainfall, temperature, soil moisture, relative humidity, longitude and GPS latitude, determined by the standard values of the sensors used. These data are processed in the Arduino card, and through a Wi-Fi communication the data is stored and visualized in real time in the Thingspeak platform, for its monitoring and interpretation. This data, combined with other geological, meteorological and satellite parameters, will make it possible to develop an artificial intelligence system to establish the threshold band where the landslides are triggered.

Keywords: Arduino · Thingspeak · Raing gauge · IoT

1 Introduction

Several studies determine that one of the triggers of the Mass Removal Phenomena in mountainous areas, and even more in tropical areas with abrupt morphologies, are the torrential rains [1]. For this reason, we intend to implement a prototype to obtain the measurement of the variables: rainfall, temperature, soil moisture, relative humidity, longitude and latitude GPS, using in the rainfall measurements containers with collection area [2]. This data, as well as others provided by different meteorological stations and satellite images, will be crucial information to identify landslide activity and characterize spatial and temporal patterns [3]. They will make it possible to develop an artificial intelligence system to determine the range of precipitation thresholds that may trigger these phenomena.

Recent advances in microcontroller technology have encouraged some research teams to develop and implement their own custom low-cost equipment, some based on highly customized WSNs [4]. Integrated systems have expanded into new application areas such as healthcare, automotive industry, robotics, home automation and smart cities, leading to the development of the Internet of Things (IoT) [5].

Á. Herrero et al. (Eds.): SOCO 2020, AISC 1268, pp. 770–779, 2021.
https://doi.org/10.1007/978-3-030-57802-2_74

In recent years the scientific community has begun to use electronic hardware and free software platforms such as Arduino to monitor, control and develop experimental hardware. Many studies have been carried out in this sense that have shown the capacity of the Arduino system to solve specific needs in different research fields [4]. In our case, we propose to implement a prototype based on an Arduino Nano card, which will allow to record the measurement of the variables using an "open-source" platform [6]. It incorporates a reprogrammable microcontroller and allows connections between the microcontroller and the different sensors [7].

Arduino has an IDE (Integrated Development Environment) which integrates a set of software tools to develop and record all the code needed to make it interpret the signals coming from the sensors and transmit them to the Thingspeak platform.

Thingspeak is an IoT analytical platform service, which allows to store and collect data using the HTTP protocol, through the Internet or through a local area network [6]. It provides instant visualizations of the data published by your devices in Thingspeak [8]. This platform is suitable for interacting with programs and math packages such as MatLab®, with hardware platforms such as Freescale®, Arduino® and other mobile devices [6].

The rest of the article is structured as follows: Sect. 2 describes the proposed system by describing all its components, Sect. 3 describes the software architecture, while Sect. 4 introduce the calibration and monitoring of key variables and we summarize our main conclusions and further research.

2 System Overview

In this work, an open source wireless acquisition system is developed to capture rainfall, relative humidity, temperature, soil moisture, longitude and latitude measurements. The following features are important for the implementation:

- A Wi-Fi wireless connection, since the accessibility to the location of the prototype in the measurement of the variables is remote. Depending on the location, the use of other mobile communication technologies such as 4G could be considered.
- Visualization of the data from any web browser or mobile device, for this process of data acquisition the Thingspeak platform is used.
- Self-supporting, the system includes a 100 W solar panel and a solid-state battery 12 V 12 Ah (see Fig. 1).

The prototype interacts with the Thingspeak platform and is structured by the following components: a Nano Arduino, four sensors; digital rain gauge TFA, DHT11 temperature-relative humidity sensor, the FC-28 soil moisture, the GPS module for Arduino NEO-GM-0-001, Node MCU Wi Fi, 16 × 4 LCD screen, solar panel, 12 VDC solid state battery.

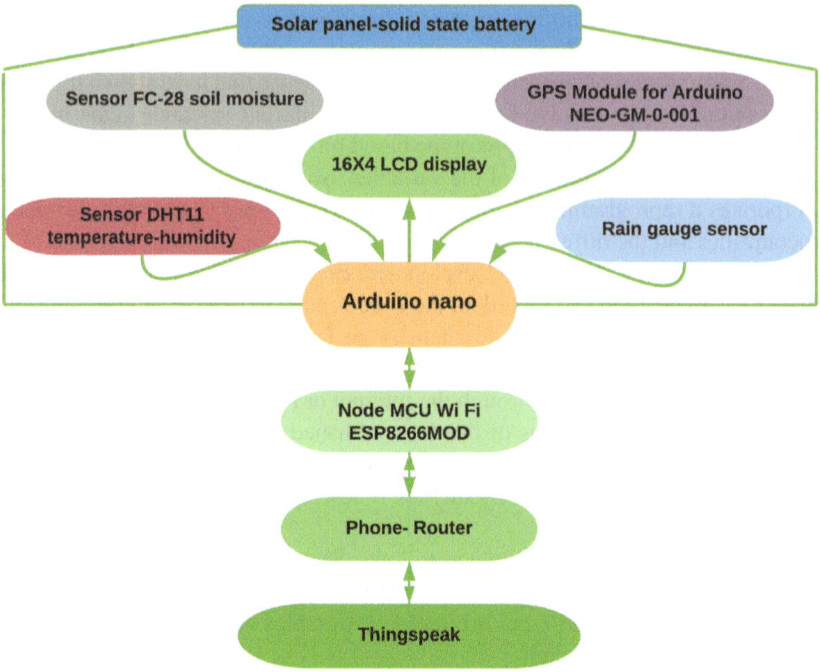

Fig. 1. Diagram for the implementation of the pluviometer.

2.1 Thingspeak[1]

Thingspeak is a MathWorks® IoT analysis platform service, with the ability to provide visualization and analysis of live data streams in the cloud. Thingspeak permits instant visualizations of the data published by your devices or computers, data loaded from the web or data sent from devices to a channel, the prototype sends the data to the Thingspeak channel. This platform accelerates the development of proof-of-concept IoT systems, especially those requiring analysis. It facilitates build IoT systems without configuring servers or developing web software [8].

The API Key allows to write data to a channel or read data from a private channel. These are automatically generated when a new channel is formed on the platform and they are used in the programming of the Arduino for pairing.

Thingspeak automatically records the data sent by the computer in a central location in the cloud. This way the data can be viewed from any web browser or mobile device for online or offline analysis. The API allows an easy visualization of the collected data through the use of spline graphs [8].

2.2 Arduino Nano

The development of the prototype is based on the small and user-friendly ATmega328-based Nano Arduino board [9]. The general specifications are shown in Table 1.

[1] https://thingspeak.com.

Table 1. Technical specifications of the Arduino Nano board.

Description	Value/Range
Operating Voltage	5 V
Recommended Input Voltage for Vin pin	7–12 V
Analog Input Pins	6 (A0 – A5)
Digital I/O Pins	14 (Out of which 6 provide PWM output)
DC Current on I/O Pins	40 Ma
DC Current on 3.3 V Pin	50 Ma
SRAM	2 KB
EEPROM	1 KB
Frequency (Clock Speed)	16 MHz
Communication	IIC, SPI, USART

Arduino Nano works with a programming environment that has been packaged as an application program; that is, it consists of a code editor, a compiler, a debugger, and a graphical interface builder. In addition, Arduino incorporates the tools to load the compiled program into the hardware's flash memory [10].

2.3 TFA Rain Gauge

The rain gauge is the scientific instrument for measuring rainfall, uses a rocker mechanism capacity of 10,000 l/mt2 resolution of 0.4 tested in the laboratory, this factor is used when programming the Arduino.

The dimension of the indoor unit is 122 × 94 × 25 mm. (height × front × depth) and the outer cup is 132 × 160 mm (see Fig. 2).

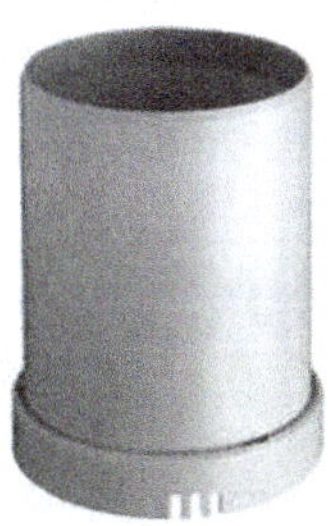

Fig. 2. TFA rain gauge.

Fig. 3. DTH11 temperature-relative humidity sensor.

2.4 DTH11 Temperature-Relative Humidity Sensor

The DHT11 sensor requires a power supply unlike many single-cable peripherals. The data sheet states that the DHT11 can be powered by a range of 3.3 to 5.5 V [11] (see Fig. 3).

The sensor allows the reading of humidity at temperatures of 20–80% and 0–50 °C, each with an error of 5% and ± 2 °C respectively [12].

2.5 Sensor FC-28 Soil Moisture

It is a soil moisture sensor made up of two exposed pads that work as probes for the sensor, acting together as a variable resistance [13]. The operating voltage is 3.3 V–5 V, its output ranges from 0 submerged in water, to 1023 in air (see Fig. 4).

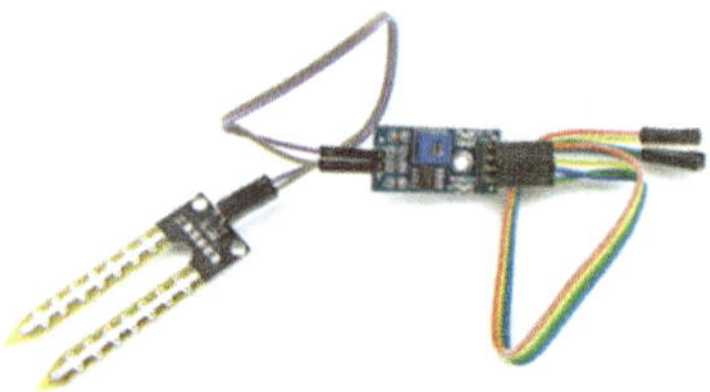

Fig. 4. Sensor FC-28.

Fig. 5. GPS module.

2.6 GPS Module for Arduino NEO-GM-0-001

The GPS module comes with a factory-configured EEPROM, a button battery to maintain the configuration data, an LED indicator and a ceramic antenna. It also has the Vcc, Rx, Tx and Gnd pins or connectors [14] (see Fig. 5).

2.7 MCU Wi Fi Node ESP8266MOD for Arduino

It is a stand-alone SOC with a built-in TCP/IP protocol stack that can provide access to any microcontroller in your network [15]. The 32-bit RISC CPU module features are: Tensilica Xtensa LX106 running at 80 MHz, 64 KiB RAM for instructions and 96 KiB RAM for data, IEEE 802.11 b/g/n Wi-Fi, 16-pin GPIO, SPI and I2C (see Fig. 6).

Fig. 6. Node MCU Wi Fi ESP8266MOD.

Fig. 7. LCD display.

2.8 LCD Display

An LCD (Liquid Crystal Display) is a device designed to show information in graphic form (see Fig. 7). The LCD display module is integrated in an LSI controller, the controller has two 8-bit registers, an instruction register (IR) and a data register (DR) [16].

2.9 Solar Panel

Photovoltaic systems consist of a set of elements, called solar cells or photovoltaic cells, arranged in panels, which directly transform solar energy into electrical energy [17]. It has the following characteristics: 100 W 12 V, polycrystalline, rigid, its dimensions $1014 \times 676 \times 35$ mm short circuit current ISC 5.79 A and its maximum output amps 5.79 A.

3 Software Architecture

The equipment includes an Arduino that is the central control unit, it processes the signals emitted by the sensors, these data are visualized in the LCD of the board in the place of location of the prototype, in turn they are transmitted by the module MCU of communication Wi-Fi to a mobile phone that is used as router, the parameters of the wireless communication to the Thingspeak platform are defined. This process is performed under programming conditions downloaded into the Arduino (see Fig. 8).

Each of the elements of the prototype rain gauge is mounted on the plate, and the measurement of the variable(s) is displayed on the LCD. The final cost of the prototype is around 250 EUR, being the cheapest equipment, we found in the market (see Fig. 9).

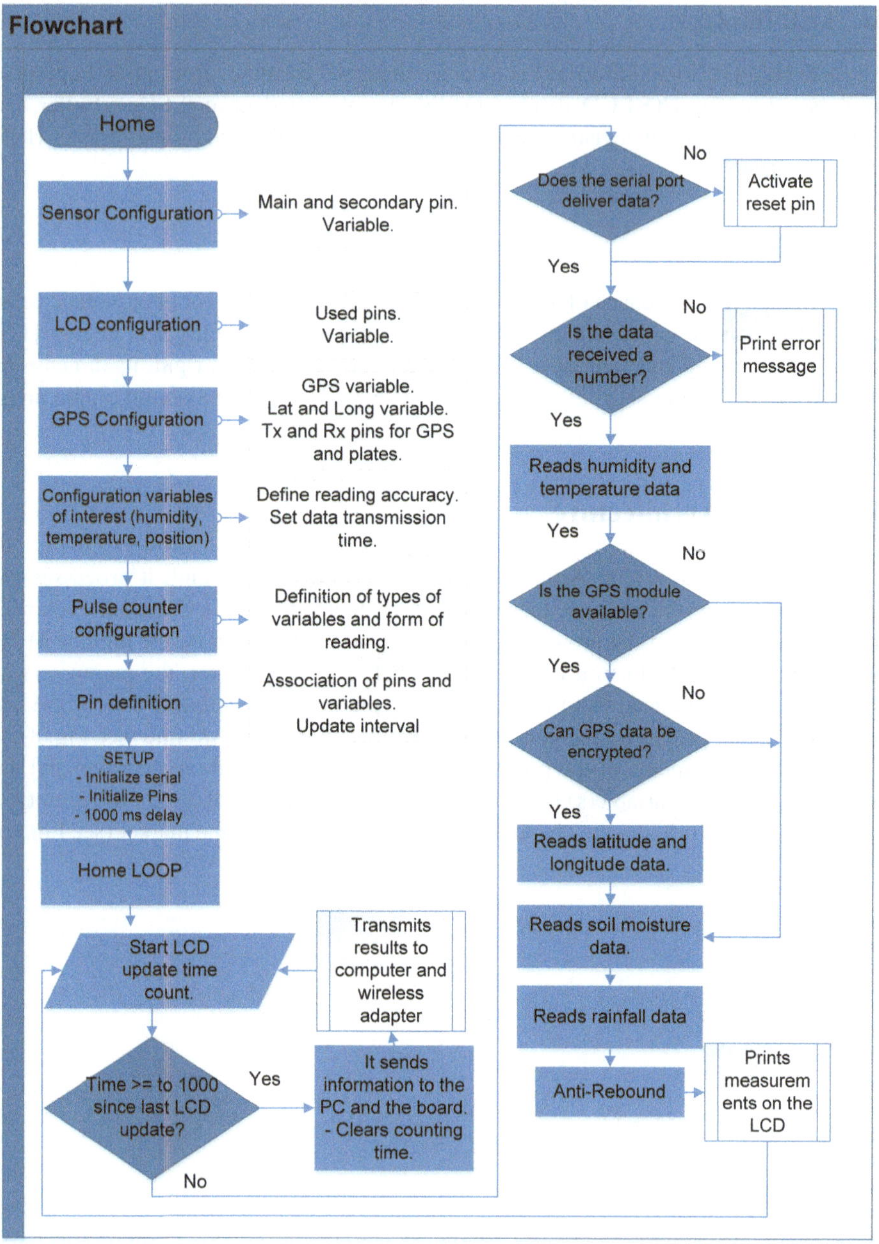

Fig. 8. Flowcharte of Arduino software system.

Fig. 9. Prototype implementation.

4 Calibration and Monitoring of Variables

In the Thingspeak platform the channels are created to visualize the data of the variables emitted by the sensors, which were conditioned in the Arduino device, (see Fig. 10). In addition, you can register the channel in the Thingview[2] Free application, which allows you to visualize the data on mobile devices.

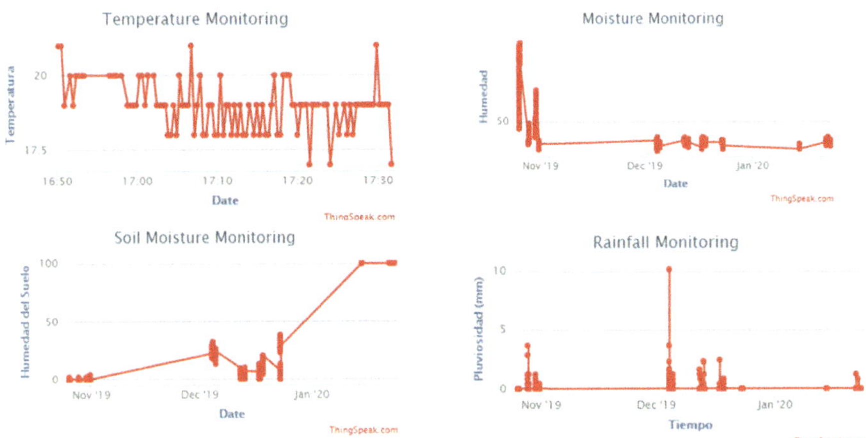

Fig. 10. Channel Statistics.

For the calibration of the equipment, the interruptions generated by the rain gauge's beam were analyzed. Each time the beam crosses the magnet, a digital signal is generated (see Fig. 11), these are processed by the Arduino. The measures delivered by the device were established under trial and error criteria. The device has provided results within the range of values offered by commercial equipments.

Finally, a metal support was built, where all the parts that were used for the implementation of the prototype were placed (see Fig. 12).

[2] https://play.google.com/store/apps/details?id=com.cinetica_tech.thingview&hl=es_49.

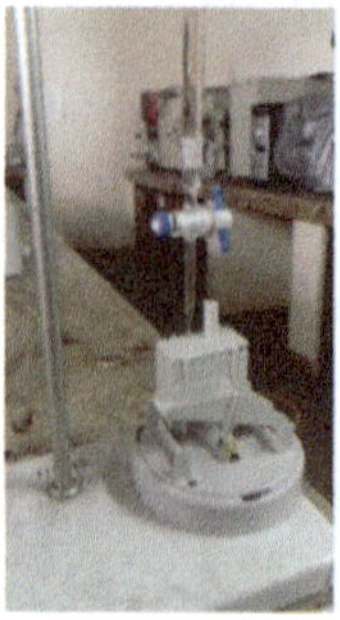

Fig. 11. Equipment testing and trials.

Fig. 12. Implementation of the field prototype.

5 Conclusions

The application of Arduino technology allowed the implementation of a low-cost device, allowing the data collection of variables like: temperature, rainfall, soil moisture, and relative humidity, which will serve as input for further studies.

The prototype connects to the Thingspeak platform, saving the data in the cloud, obtaining statistics and graphics of the measurements of the variables, which can be viewed from any browser or mobile device for interpretation.

By using a solar panel and a storage battery connected to the equipment, it allows it to be autonomous, thanks to this condition it can be installed in any area where it is required to obtain the measurements, without worrying about having a nearby power supply.

The data obtained with the developed system, in combination with information related to geological, meteorological and satellite hyperspectral images obtained from different sources, will make it possible to develop an artificial intelligence system that will establish the threshold band where landslides are triggered and will make it possible to generate a warning system to be used by government and social institutions.

Acknowledgement. I would like to express my gratitude to the Central University of Ecuador and FIGEMPA, which in the framework of the inter-institutional agreement with the University of Alicante, made this research work possible.

References

1. Gariano, S.L., Guzzetti, F.: Landslides in a changing climate. Earth Sci. Rev. (2016). https://doi.org/10.1016/j.earscirev.2016.08.011
2. Savina, M., Schäppi, B., Molnar, P., Burlando, P., Sevruk, B.: Comparison of a tipping-bucket and electronic weighing precipitation gage for snowfall. Rainfall Urban Context Forecast Risk Clim. Chang. **103**, 45–51 (2012). https://doi.org/10.1016/j.atmosres.2011.06.010
3. Rangnekar, A., Hoffman, M.: Learning representations to predict landslide occurrences and detect illegal mining across multiple domains. In: Proceedings of the 36th International Conference on Machine Learning, Long Beach, California, PMLR 97 (2019)

4. Soler Llorens, J.L.: Diseño, implementación y validación de sistemas de adquisición de datos sísmicos basados en Arduino (2019)
5. Mora, H., Peral, J., Ferrandez, A., Gil, D., Szymanski, J.: Distributed architectures for intensive urban computing: a case study on smart lighting for sustainable cities. IEEE Access **7**, 58449–58465 (2019). https://doi.org/10.1109/ACCESS.2019.2914613
6. Rodriguez Sotelo, J., López Londoño, A., Vega Botero, C., Darío Flórez Hurtado, R.: Sistema de monitoreo y control remoto usando IOT para un regulador de presión. Sci. Tech. **22**(4), 391–397 (2017). https://doi.org/10.22517/23447214.13291
7. ¿Que es Arduino?| Arduino.cl-Compra tu Arduino en Línea (2020). https://arduino.cl/que-es-arduino/. Accessed 23 Apr 2020
8. Gómez Maureira, M.A., Oldenhof, D., Teernstra, L.: ThingSpeak – an API and web service for the internet of things. In: World Wide Web (2014)
9. Arduino Nano| Arduino Official Store (2020). https://store.arduino.cc/usa/arduino-nano. Accessed 24 Apr 2020
10. Arduino Cloud| Aprendiendo Arduino (2020). https://aprendiendoarduino.wordpress.com/category/arduino-cloud/. Accessed 24 Apr 2020
11. Gay, W., Gay, W.: DHT11 sensor. In: Advanced Raspberry Pi, Berkeley, CA: Apress, pp. 399–418 (2018)
12. Adiono, T., Fathany, M.Y., Fuada, S., Purwanda, I.G., Anindya, S.F.: A portable node of humidity and temperature sensor for indoor environment monitoring. In: IGBSG 2018–2018 International Conference on Intelligent Green Building Smart Grid, pp. 1–5 (2018). https://doi.org/10.1109/igbsg.2018.8393575
13. FC-28 Soil Moisture Sensor Analog and Digital Outputs-Art of Circuits (2020). https://artofcircuits.com/product/fc-28-soil-moisture-sensor-analog-and-digital-outputs. Accessed 24 Apr 2020
14. Tutorial Módulo GPS con Arduino (2020). https://naylampmechatronics.com/blog/18_Tutorial-Módulo-GPS-con-Arduino.html. Accessed 24 Apr 2020
15. Baig, I., Muzamil, C., Dalvi, S., Campus, K.T.: Home automation using Arduino wifi module ESP8266, p. 8 (2016)
16. Vatronix Holdings Limited, LCD Module User Manual, 10, pp. 1–18
17. Montoya Rasero, C.: La energía solar fotovoltaica en el CTE. Electra, 137, pp. 94–100 (2006)

Functional Networks for Image Segmentation of Cutaneous Lesions with Rational Curves

Akemi Gálvez[1,2], Iztok Fister[2,3], Iztok Fister Jr.[3], and Andrés Iglesias[1,2](✉)

[1] Department of Information Science, Faculty of Sciences, Toho University, 2-2-1, Miyama, Funabashi 274-8510, Japan

[2] Department of Applied Mathematics and Computational Sciences, University of Cantabria, 39005 Santander, Spain
{galveza,iglesias}@unican.es

[3] Faculty of Electrical Engineering and Computer Science, University of Maribor, Maribor, Slovenia
{iztok.fister,iztok.fister1}@um.si

Abstract. This paper considers the problem of image segmentation for medical images, in particular, cutaneous lesions. Given a digital image of a skin lesion, our goal is to compute the border curve separating the lesion from the image background. This problem can be formulated as an optimization problem, where the border curve is computed through data fitting from a set of points lying on the lesion boundary. Some recent papers have applied artificial intelligence techniques to tackle this issue. However, they usually focus on the polynomial case, ignoring the more powerful (but also more difficult) case of rational curves. In this paper, we address this problem with rational Bézier curves by applying functional networks, a powerful extension of the classical neural networks. Experimental results on some benchmark medical images show that this method performs well and can be successfully applied to this problem.

Keywords: Cutaneous lesion · Image segmentation · Rational curves · Data fitting · Functional networks

1 Introduction

Image segmentation is an important problem in fields such as image processing, artificial vision, pattern recognition, virtual and augmented reality, and many others. This technology is also increasingly used in medical applications. Illustrative examples include computer tomography, magnetic resonance imaging, infrared imaging, and positron emission tomography, to mention just a few. An interesting medical application arises in dermatology, for early diagnosis and treatment of melanoma and other skin lesions. This is a critical issue in current medical and healthcare systems. Skin cancer is one of the most frequent types

Á. Herrero et al. (Eds.): SOCO 2020, AISC 1268, pp. 780–789, 2021.
https://doi.org/10.1007/978-3-030-57802-2_75

of cancer for both men and women. And melanoma is the most frequent and dangerous type of skin cancer.

Early detection is critical for an efficient treatment of melanoma and other malignant skin lesions. It has been reported that the five-year survival rate is about 99% for stage 0 melanoma (in situ), when the tumor is confined to the epidermis, while it is only 7% $\sim$ 20% for stage 4 melanoma, when the cancer spreads to other parts of the body. The most common diagnostic procedure is visual inspection by a specialist. However, it requires time and resources, and it is difficult to distinguish the melanoma from other skin lesions. Other diagnosis procedures include the ABCDE method, the Menzies scale, the 7-point checklist, and different types of biopsy [7,20]. All these procedures rely on human intervention. To overcome this limitation, image-based methods are gaining popularity in the field in recent years. They require image segmentation to identify the area of the lesion and separate it from the background. An important step in image segmentation is the border detection of the skin lesion from the image.

Until recently, the border detection was handled manually by dermatologists. However, some recent papers show that this border curve can be computed automatically (see, for instance, [21]). Popular segmentation approaches include thresholding methods [4,12], edge-based methods [1], clustering methods [22,24], level set methods [17], swarm intelligence methods [10,11], and active contours [16]. These methods work well and provides satisfactory results for the polynomial case. However, their accuracy can be improved by using more powerful and sophisticated functions.

In this work, we follow this approach by replacing the polynomial basis functions by rational ones. The resulting parametric curve is no longer a polynomial but a rational function. This procedure makes it possible to reduce the degree of the curve significantly without penalizing the approximation accuracy. Unfortunately, using rational curves is by far much more difficult than the polynomial case, because some extra variables (the weights) have also to be computed. In addition, the different free variables (data parameters, poles, and weights) are related to each other in a nonlinear way [5], leading to a difficult continuous multivariate nonlinear optimization problem. In this work, we address this problem with rational Bézier curves by applying functional networks, a powerful extension of the classical neural networks.

The structure of this paper is as follows: Sect. 2 describes the problem to be solved. Functional networks are discussed in Sect. 3. The proposed method is presented in Sect. 4. Then, it is applied to our optimization problem in Sect. 5. The paper closes in Sect. 6 with our conclusions and future work in the field.

2 The Problem

Suppose that we have a sorted collection of feature points $\{\mathbf{\Delta}_i\}_{i=1,\ldots,\kappa}$ in $\mathbb{R}^2$ obtained from medical images, and corresponding to the boundary curve between a skin lesion or tumor and the skin background. Note that in this paper vectors are denoted in bold. Such feature points might be subjected to measurement noise, irregular sampling, and other artifacts. Our goal is to compute a rational parametric curve $\mathbf{\Phi}(\tau)$ performing discrete approximation of the feature points $\{\mathbf{\Delta}_i\}_i$ in the least-squares sense. A *free-form rational Bézier curve* $\mathbf{\Phi}(\tau)$ *of degree* η is given by [6]:

$$\mathbf{\Phi}(\tau) = \frac{\sum_{j=0}^{\eta} \omega_j \mathbf{\Lambda}_j \phi_j^{\eta}(\tau)}{\sum_{j=0}^{\eta} \omega_j \phi_j^{\eta}(\tau)} \quad \text{with} \quad \phi_j^{\eta}(\tau) = \binom{\eta}{j} \tau^j \, (1-\tau)^{\eta-j} \tag{1}$$

where $\mathbf{\Lambda}_j$ are vector coefficients called the *poles*, ω_j are their scalar weights, $\phi_j^{\eta}(\tau)$ are the *Bernstein polynomials of index* j *and degree* η, and τ is the *curve parameter*, defined on the finite interval $[0, 1]$. By convention, $0! = 1$.

Now, our optimization problem consists of computing all parameters (i.e. poles $\mathbf{\Lambda}_j$, weights ω_j, and parameters τ_i associated with the $\mathbf{\Delta}_i$, for $i = 1, \ldots, \kappa$, $j = 0, \ldots, \eta$) of the rational Bézier curve $\mathbf{\Phi}(\tau)$ approximating the feature points better in the least-squares sense. This means minimizing the least-squares error, Υ, defined as the sum of squares of the residuals:

$$\Upsilon = \underset{\substack{\{\tau_i\}_i \\ \{\mathbf{\Lambda}_j\}_j \\ \{\omega_j\}_j}}{\text{minimize}} \left[\sum_{i=1}^{\kappa} \left(\mathbf{\Delta}_i - \frac{\sum_{j=0}^{\eta} \omega_j \mathbf{\Lambda}_j \phi_j^{\eta}(\tau_i)}{\sum_{j=0}^{\eta} \omega_j \phi_j^{\eta}(\tau_i)} \right)^2 \right]. \tag{2}$$

Now, taking:

$$\varphi_j^{\eta}(\tau) = \frac{\omega_j \phi_j^{\eta}(\tau)}{\sum_{k=0}^{\eta} \omega_k \phi_k^{\eta}(\tau)} \tag{3}$$

Eq. (2) becomes:

$$\Upsilon = \underset{\substack{\{\tau_i\}_i \\ \{\mathbf{\Lambda}_j\}_j \\ \{\omega_j\}_j}}{\text{minimize}} \left[\sum_{i=1}^{\kappa} \left(\mathbf{\Delta}_i - \sum_{j=0}^{\eta} \mathbf{\Lambda}_j \varphi_j^{\eta}(\tau) \right)^2 \right], \tag{4}$$

which can be rewritten in matrix form as $\mathbf{\Omega}.\mathbf{\Lambda} = \mathbf{\Xi}$, called the *normal equation*, where:

$$\mathbf{\Omega} = [\Omega_{i,j}] = \left[\left(\sum_{k=1}^{\kappa} \varphi_i^{\eta}(\tau_k) \varphi_j^{\eta}(\tau_k) \right)_{i,j} \right], \mathbf{\Xi} = [\Xi_j] = \left[\left(\sum_{k=1}^{\kappa} \mathbf{\Delta}_k \varphi_j^{\eta}(\tau_k) \right)_j \right],$$

$\mathbf{\Lambda} = (\mathbf{\Lambda}_0, \ldots, \mathbf{\Lambda}_\eta)^T$, for $i, j = 0, \ldots, \eta$, and $(.)^T$ means the transposition of a vector or a matrix. In general, $\kappa >> \eta$ meaning that the resulting system is over-determined. If values are assigned to the τ_i, our problem can be solved as a classical linear least-squares minimization, with the coefficients $\{\mathbf{\Lambda}_i\}_{i=0,\ldots,\eta}$ as unknowns. This problem can readily be solved by standard numerical techniques. On the contrary, if the values of τ_i are treated as unknowns, the problem becomes a very difficult over-determined, multimodal, multivariate, continuous, nonlinear optimization problem, unsolvable with the classical mathematical optimization methods. In this work, we address this problem through functional networks, explained in next section.

3 Functional Networks

Functional networks were firstly introduced in [2] as a generalization of the standard artificial neural networks, in which the scalar weights are replaced by neural functions. Since then, they have been applied to several problems in science and engineering; see, e.g., [3,9,14]. Functional networks share several common features with neural networks, including their graphical representation. Figure 1 shows the functional network used in this work, which is associated with the function in Eq. (1), but expressed in terms of the rational functions in Eq. (3). Following this figure, the main components of a functional network become clear:

1. Several layers of storing units: we have a first layer of input units containing the input information. In Fig. 1, it consists of the unit τ. We also have a set of intermediate layers of storing units. They are not neurons but units storing intermediate information. This set is optional and allows more than one neuron output to be connected to the same unit. In Fig. 1, we have two layers, each with $\eta + 1$ intermediate units, represented by small circles in black. Finally, we have an output layer, consisting only of the unit $\mathbf{\Phi}(\tau)$.
2. One or more layers of neurons or computing units. A neuron is a computing unit which evaluates a set of input values, coming from the previous layer, and gives a set of output values to the next layer. Neurons are represented by circles with the name of the corresponding neural function inside. For example, in Fig. 1 we have two intermediate layers of $\eta + 1$ neurons each, comprised of the neural functions $\varphi_j^\eta(\tau)$, $j = 0, \ldots, \eta$, and the $\times$ operator, respectively.
3. A set of directed links. They connect the input or intermediate layers to its adjacent layer of neurons, and neurons of one layer to its adjacent intermediate layers or to the output layer. Connections are represented by arrows, showing the flow direction, from the input layer to the output layer.

All these elements together form the *network architecture* or *topology* of the functional network, which defines the functional capabilities of the network. The main differences between neural networks and functional networks are:

1. In neural networks each neuron returns an output $y = f(\sum w_{ik} x_k)$ that depends only on the value $\sum w_{ik} x_k$, where $x_1, x_2, \ldots, x_n$ are the received inputs. Therefore, their neural functions have only one argument. In contrast, neural functions in functional networks can have several arguments.

2. In neural networks the neural functions are *univariate*: neurons can show different outputs but all of them represent the same values. In functional networks, the neural functions can be *multivariate*.
3. In functional networks the neural functions can be *different*, while in neural networks they are *identical*.
4. In neural networks there are weights, which must be learned. They do not appear in functional networks, where neural functions are learned instead.
5. In neural networks *the neuron outputs are different*, while in functional networks *neuron outputs can be coincident*. This leads to a set of functional equations, which have to be solved [2,3]. This means that neural functions can be reduced in dimension or expressed as functions of lower dimension.

All these features show that functional networks are more general and exhibit more interesting possibilities than neural networks.

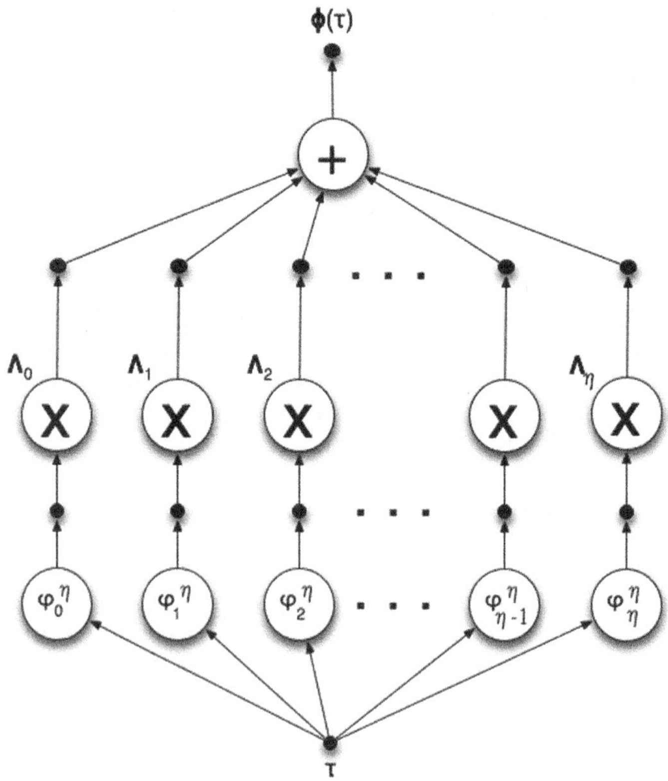

Fig. 1. Functional network used in this work.

4 Our Method

Our method consists of applying the functional network described in previous section to the optimization problem in Eq. (4). This process requires learning the function $\mathbf{\Phi}(\tau)$, which in turn, requires learning the rational functions $\varphi_j^\eta(\tau)$, $j = 0, \ldots, \eta$, and the poles $\mathbf{\Lambda}_j$. Note also that these rational functions do depend on both the polynomial functions $\phi_j^\eta(\tau)$, and the weights ω_j, which have also to be learned. The input of our optimization problem is the number of neurons in the intermediate layers, defined by η, and an initial collection of weights, ω_j^0, and parameters, τ_i^0. These two sets of parameters and initialized randomly on the intervals $(0, 100)$ and $[0, 1]$, respectively. Then, the method proceeds iteratively according to the following steps: using the parameters τ_i^0, the polynomial functions $\phi_j^\eta(\tau_i^0)$ are computed according to Eq. (1). With this output and weights ω_j^0, we compute $\varphi_j^\eta(\tau_i^0)$ according to Eq. (3). Then, we compute the poles $\mathbf{\Lambda}_j^0$ by solving the least-squares normal equations by Gaussian elimination or singular value decomposition (SVD). The values of the poles are then fixed, and used in next iteration to compute the new weights ω_j^1, using the previous parameters τ_i^0. Then, we perform parametric learning of the functional network using the data points through the error function in Eq. (4) to compute new parameters τ_i^1 using the new weights ω_j^1. Solving the least-squares normal equations again, we obtain new poles $\mathbf{\Lambda}_j^1$. This process is iteratively repeated until convergence, when no further improvement can be achieved. Finally, we compute the fitting error. We remark however, that this fitting error does not take into account the number of data points. To overcome this drawback, we also compute the $RMSE$ (root-mean squared error), given by: $RMSE = \sqrt{\frac{\Upsilon}{\kappa}}$. Accordingly, the fitting errors in next section will be given in terms of the RMSE instead of the functional error Υ.

5 Experimental Results

Our method has been applied to a benchmark of twelve skin lesion images obtained from a medical repository of digital images publicly available for research purposes. The whole set of images in our benchmark is not included here because of limitations of space. However, Fig. 2 shows four examples of the digital images for illustration. As you can see from the images, the skin lesions can be extremely varied in terms of shape, size, color, roughness, and other geometrical and visual features. As a result, it is very difficult to discriminate between benign and malignant tumors and even determine the specific type of skin lesion under analysis. Clearly, this makes any automatic procedure a very useful tool for medical diagnosis and treatment.

We have applied the method introduced in this paper to this benchmark and carried out several computational experiments. The corresponding results for the RMSE fitting error are reported in Table 1. The different examples are arranged in rows and labelled from I to XII. For each example, the table reports

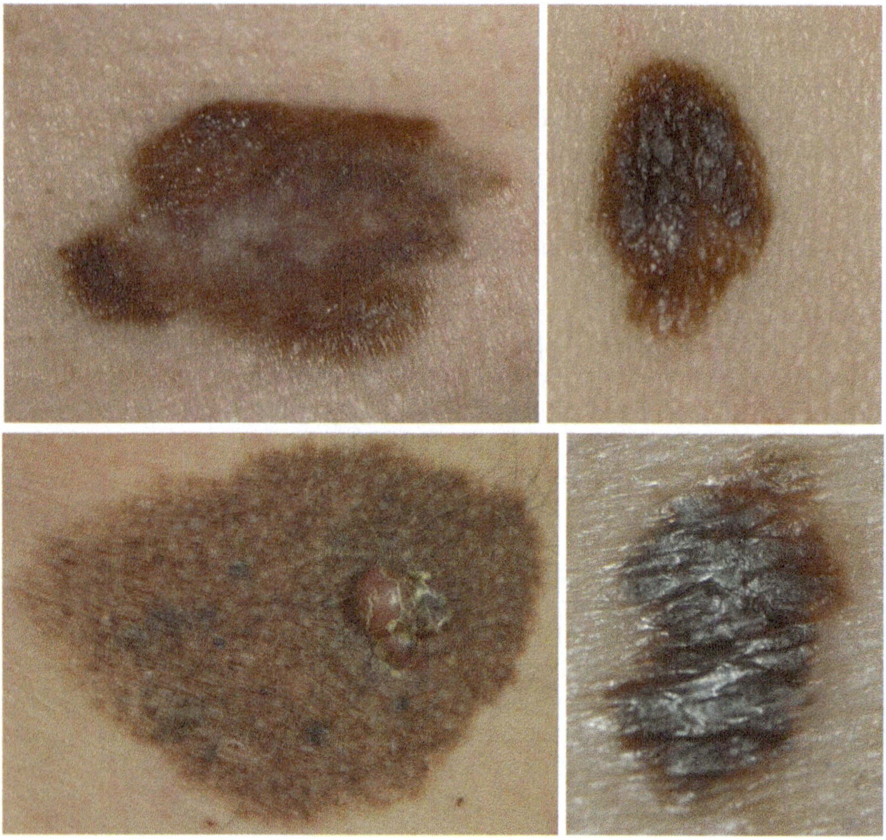

Fig. 2. Four examples of digital images of skin lesions in our benchmark (the images are not equally scaled; some of them have been resized for better visualization).

its RMSE fitting errors for five methods, arranged in columns 2–6. The results for the method introduced in this paper are shown in column 6. From the results there, we can see that the RMSE fitting error is of order $10^{-2} \sim 10^{-4}$, depending on the example. This is a very good value in the context of medical imaging. To support this assertion, we have carried out a comparison with other popular methods described in the literature. Our comparative work includes two of the most popular state-of-the-art approaches in medical imaging: thresholding [23] and clustering [4] (shown in columns 2 and 3 of Table 1, respectively). Finally, we also consider two other standard approaches: polynomial curve fitting and artificial neural networks. The former is applied through the `polyfit` Matlab command [19]. For the latter, we consider a a deep, artificial neural network called multilayer perceptron (MLP), which is well-known to be a universal function approximator [8,13]. The MLP in this comparison includes 15 neurons in a single hidden layer and uses the back propagation algorithm of Levenberg–Marquardt for training [15,18]. The best method for each example is highlighted

Table 1. Comparative results of the RMSE fitting error for the twelve examples in our benchmark (arranged in rows). For each example, the different methods (arranged in columns) are analyzed. Best results are highlighted in bold for prompt identification.

Example	Thresholding	Clustering	Polynomial Fitting	MLP NN	Our method
I	3.1746E−2	5.8617E−2	8.5503E−2	9.0641E−2	**1.3592E−2**
II	7.5762E−3	6.4809E−3	7.3882E−3	1.0370E−2	**4.6031E−3**
III	**2.1812E−2**	5.7384E−2	7.9217E−2	5.1166E−2	3.3814E−2
IV	1.8446E−2	3.6514E−2	4.2597E−2	2.1946E−2	**8.5917E−3**
V	1.1311E−2	**8.7226E−3**	5.7391E−2	1.0668E−2	1.2904E−2
VI	9.0433E−3	2.7756E−2	2.5814E−2	3.8353E−2	**6.1733E−3**
VII	6.0213E−3	4.5830E−3	5.7219E−3	3.6005E−3	**9.8092E−4**
VIII	**4.5785E−2**	5.3111E−2	2.1638E−1	6.6514E−2	6.6247E−2
IX	4.2017E−2	2.1636E−2	5.1422E−2	9.0988E−3	**6.1906E−3**
X	4.3018E−3	6.5639E−3	1.0026E−3	4.1377E−3	**8.3482E−4**
XI	3.9037E−2	**2.1173E−2**	6.2847E−2	4.0119E−2	2.5651E−2
XII	5.6317E−3	7.7664E−3	8.1565E−3	8.3026E−3	**4.9264E−3**

in bold for prompt identification. Regarding the implementation, all computational work has been performed on a personal PC with a 3.4 GHz Intel Core i7 processor and 16 GB of RAM. The source code has been implemented by the authors in the programming language of the popular numerical program *Matlab*, version 2018b.

Our comparative work shows that the method introduced here outperforms the other four approaches for most of the instances in our benchmark. In particular, our method is the best for 8 of the 12 examples analyzed. The exceptions are the examples III and VIII (with thresholding as the winner) and examples V and XI (where clustering is the winner). But even in those examples for which our method is not the best, the fitting error is the same order as the best, and usually, very close in value. This disparity in the winner indicates that there is no universal choice for this problem, as it might happen that state-of-the-art techniques become superior for a given example. Still, our functional network-based method yields very good results for almost all cases and outperforms other methods for most of them. Furthermore, the examples in our benchmark were chosen from a larger collection at random, without following any specific criteria, meaning that these results can likely be extrapolated to the whole set. Note also that thresholding and clustering also perform generally well, while the polynomial fitting and the MLP neural network do provide slightly worse results in most cases. This last observation is not really surprising, since the rational fitting is actually a generalization of the polynomial fitting, and similarly, the functional networks are a generalization of the neural networks.

6 Conclusions and Future Work

This paper addresses the problem of image segmentation for medical images of cutaneous lesions. Given a digital image of a skin lesion, we seek to compute the border curve separating the lesion from the image background. Such a border curve is usually constructed through data fitting from a collection of points assumed to lie on the lesion boundary. In our formulation, the fitting curve is assumed to be rational, extending the more common (but less powerful) case of polynomial fitting. This leads to a difficult continuous multivariate nonlinear optimization problem. In this paper, we address this problem by applying functional networks, a powerful extension of the classical neural networks. Experimental results on a benchmark of twelve digital images show that this method performs well, yielding approximating shapes with an acceptable accuracy for practical applications. We also carried out a comparison on our benchmark with four alternative methods: two state-of-the-art techniques (thresholding and clustering), polynomial fitting and multiplayer perceptron neural networks. The experimental results show that our functional network-based method yields very good results for almost all cases and outperforms the other methods for most of the instances in our benchmark. Our future work includes carrying out a deeper comparative work involving more methods (e.g., convolutional NNs) and a larger benchmark as well as measuring the time required to train each method in order to analyze the extensibility of the approach. Also, we aim at developing a procedure to determine automatically the number of neurons in the intermediate layers of the functional network (now assumed to be an input of the method).

Acknowledgments. Akemi Gálvez and Andrés Iglesias thank the financial support from the project PDE-GIR of the European Union's Horizon 2020 research and innovation program under the Marie Sklodowska-Curie grant agreement No. 778035, and from the Spanish Ministry of Science, Innovation and Universities (Computer Science National Program) under grant #TIN2017-89275-R of the Agencia Estatal de Investigación and European Funds FEDER (AEI/FEDER, UE). Iztok Fister Jr. thanks the financial support from the Slovenian Research Agency (Research Core Funding No. P2-0057). Iztok Fister acknowledges the financial support from the Slovenian Research Agency (Research Core Funding No. P2-0042).

References

1. Abbas, A.A., Guo, X., Tan, W.H., Jalab, H.A.: Combined spline and B-spline for an improved automatic skin lesion segmentation in dermoscopic images using optimal color channel. J. Med. Syst. **38**, 80–80 (2014)
2. Castillo, E.: Functional networks. Neural Process. Lett. **7**, 151–159 (1998)
3. Castillo, E., Iglesias, A., Ruiz-Cobo, R.: Functional Equations in Applied Sciences. Elsevier, Amsterdam (2005)
4. Celebi, M.E., Iyatomi, H., Schaefer, G., Stoecker, W.V.: Lesion border detection in dermoscopy images. Comp. Med. Imaging Graph. **33**(2), 148–153 (2009)
5. Dierckx, P.: Curve and Surface Fitting with Splines. Oxford University Press, Oxford (1993)

6. Farin, G.: Curves and Surfaces for CAGD, 5th edn. Morgan Kaufmann, San Francisco (2002)
7. Friedman, R.J., Rigel, D.S., Kopf, A.W.: Early detection of malignant melanoma: the role of physician examination and self-examination of the skin. Cancer J. Clin. **35**(3), 130–151 (1985)
8. Funahashi, K.I.: On the approximate realization of continuous mappings by neural networks. Neural Netw. **2**(3), 183–192 (1989)
9. Gálvez, A., Iglesias, A., Cobo, A., Puig-Pey, J., Espinola, J.: Bézier curve and surface fitting of 3D point clouds through genetic algorithms, functional networks and least-squares approximation. Lectures Notes in Computer Science, vol. 4706, pp. 680–693 (2007)
10. Gálvez, A., Fister, I., Fister Jr., I., Osaba, E., Ser, J.D., Iglesias, A.: Automatic fitting of feature points for border detection of skin lesions in medical images with bat algorithm. Stud. Comput. Intell. **798**, 357–368 (2018)
11. Gálvez, A., Fister, I., Osaba, E., Ser, J.D., Iglesias, A.: Hybrid modified firefly algorithm for border detection of skin lesions in medical imaging. In: Proceeding of IEEE Congress on Evolutionary Computation, IEEE CEC 2019, pp. 111–118. IEEE Computer Society Press, Los Alamitos (2019)
12. Garnavi, R., Aldeen, M., Celebi, M.E., Varigos, G., Finch, S.: Border detection in dermoscopy images using hybrid thresholding on optimized color channels. Comput. Med. Imaging Graph. **35**(2), 105–115 (2011)
13. Hornik, K., Stinchcombe, M., White, H.: Multilayer feedforward networks are universal approximators. Neural Netw. **2**(5), 359–366 (1989)
14. Iglesias, A., Gálvez, A.: Hybrid functional-neural approach for surface reconstruction. Math. Prob. Eng. **2014**, 13 (2014). Article ID 351648
15. Levenberg, K.: A method for the solution of certain non-linear problems in least squares. Q. Appl. Math. **2**(2), 164–168 (1944)
16. Ma, Z., Tavares, J.M.: A novel approach to segment skin lesions in dermoscopic images based on a deformable model. IEEE J. Biomed. Health Inform. **20**, 615–623 (2016)
17. Machado, D.A., Giraldi, G., Novotny, A.A.: Multi-object segmentation approach based on topological derivative and level set method. Integr. Comput. Aided Eng. **18**, 301–311 (2011)
18. Marquardt, D.: An algorithm for least-squares estimation of nonlinear parameters. SIAM J. Appl. Math. **11**(2), 431–441 (1963)
19. The MathWorks polyfit web page. https://www.mathworks.com/help/matlab/ref/polyfit.html. Accessed 20 Feb 2020
20. Nachbar, F., Stolz, W., Merkle, T., Cognetta, A.B., Vogt, T., Landthaler, M., Bilek, P., Braun-Falco, O., Plewig, G.: The ABCD rule of dermatoscopy: high prospective value in the diagnosis of doubtful melanocytic skin lesions. J. Am. Acad. Dermatol. **30**(4), 551–559 (1994)
21. Pathan, S., Prabhu, K.G., Siddalingaswamy, P.C.: Techniques and algorithms for computer aided diagnosis of pigmented skin lesions - a review. Biomed. Signal Process. Control **39**, 237–262 (2018)
22. Schmid, P.: Segmentation of digitized dermatoscopic images by two-dimensional color clustering. IEEE Trans. Med. Imaging **18**(2), 164–171 (1999)
23. Sezgin, M., Sankur, B.: Survey over image thresholding techniques and quantitative performance evaluation. J. Electron. Imaging **13**, 146–165 (2004)
24. Zhou, H., Schaefer, G., Sadka, A., Celebi, M.E.: Anisotropic mean shift based fuzzy c-means segmentation of dermoscopy images. IEEE J. Sel. Top. Sign. Proces. **3**(1), 26–34 (2009)

Manufacturing Description Language for Process Control in Industry 4.0

Mauricio-Andrés Zamora-Hernández[1(✉)], Jose Andrez Chaves Ceciliano[1(✉)], Alonso Villalobos Granados[1(✉)], John Alejandro Castro Vargas[2(✉)], Jose Garcia-Rodriguez[2(✉)], and Jorge Azorín-López[2(✉)]

[1] University of Costa Rica, San José, Costa Rica
{mauricio.zamorahernandez,jose.chavesceciliano, alonso.villalobos15}@ucr.ac.cr
[2] University of Alicante, Alicante, Spain
{jcastro,jgr,jazorin}@ua.es

Abstract. In factories, the assembly of products or components by operators is a complex task that is not free of recurring problems. In this process, operators often make mistakes that can lead to defective products. Therefore, they need to be inspected later to verify their correct assembly. The main problems are caused by several reasons, including high employee turnover due to a lack of experience in manufacturing specific products or confusion of instructions for similar components. In this paper, a novel structured language aimed to describe the required actions to manufacture a product in industrial assembly environments is presented. The main contribution is to provide a formal language that can help in the future to an automatic system can verify through visual control, whether the actions performed by the operator are carried out in accordance with the standard described by this language. It will allow to minimize the negative impact of errors during assembly.

Keywords: Assembly representation · Automatic inspection · Control language · Industry 4.0 · Recommendation system

1 Introduction

This is an investigation which focuses on improving the work in the industries during the product assembly process, to control that the actions are been carried out by the operators accordingly to the defined standards. The aim is to achieve homogeneity in the assembly of final products, minimizing losses due to manufacturing problems or waste of time and money due to reprocessing assemblies.

A problem that has always been present in industry is controlling manufacturing processes. Since the beginning of Industrial Engineering, the concept of the study of the method has been present, Kanawaty defines it as "The study or engineering of methods is the registry and systematic critical examination of the ways of carrying out activities, in order to make improvements" [7].

Á. Herrero et al. (Eds.): SOCO 2020, AISC 1268, pp. 790–799, 2021.
https://doi.org/10.1007/978-3-030-57802-2_76

Studying the method allows us to analyze processes from their most basic elements, such as the movement sequences necessary to complete tasks. In this way, improvements can be made in production processes, determining changes in sequences or reducing unnecessary movements.

It is, therefore, a crucial topic for research in Industrial Engineering, since it permits a correct production planning and provides an adequate analysis of operations, this becomes a way to establish more precise calculations in the capacity and production in industries. In addition, promotes the search for improving quality in the processes.

Along with advances in technology, alternatives have been proposed to solve the problem of assembly control, for which the techniques of automatic inspection systems are commonly used. These compare the assemblies against measurement standards. Due to the growing need for high quality and personalized products, new needs have been defined in quality control systems. Systems need to be able to learn to identify or process parts that are created for particular solutions. These quality inspection processes can be carried out making use of simple sensors such as those that measure weight, color, size [1]. Another technique used is Computer Vision (CV), which allows quality validation based on a standard through visual control mechanisms at workstations [4].

Some of the common applications of CV in manufacturing are: quality control (shapes, sizes, colors), collision detection [15], navigation [6] and augmented reality [11]. However, this work proposes to use CV beyond measurement of characteristics, in other words, to apply this technology into visually identifying the actions executed by the operator and compare them with the specified standard, which is to be design with our language. This language defines the sequence of product manual construction in order to confirm that a product meets the quality standard by the assurance of a correct assembling of the final product.

There are already research proposals based on Artificial intelligence (AI) systems, Image captioning, a technique that seeks to automatically generate an image description [10], which can be used to describe what happens in assembly environments, considering verification of the necessary manufacturing steps, but in certain circumstances more than a description with a simple label is wanted, meaning, a complete description of actions is required. For this reason, researchers such as Wang et al. [16], have carried out work focused in captioning for video, where they use Hierarchical Reinforcement Learning techniques to generate the descriptions. Krishna et al. [8], are using long short-term memory (LSTM) techniques in a dense-captioning model for event detection. In both cases, descriptive narratives in natural language are used. Yao et al. [18], proposes to create narratives in different ways: template-based methods, where structures are created, and seeks to generate narratives in fixed forms. Search for visual elements approaches are used and these texts and language-based models using k-nearest neighbor retrieval models are copied.

For it to be useful it is necessary to have a technique that makes it possible to make sense of the words used in the description, it is convenient to have grammatical systems to structure the instructions in such a way that they are able to express rules applicable to the industry. Authors like Nguyen et al. [13]. Are making proposals to understand and imitate human actions, without defining objectives or validations in the actions. Therefore, a grammar that allows structuring instructions and facilitates adequate communication of the actions to be carried out, without creating ambiguities between the parties involved is required.

To mitigate these deficiencies, Yang et al. [17], propose a system of convolutional neural networks, which, through video analysis constructs grammar trees of the observed actions. These are non-restricted general use grammars for a particular use, so it could generate unstructured actions for the strict verification of what is captured. Researchers such as Mancini et al. [12], are already working on object detection in specific domains of industry, but without the definition of a grammar to describe the sequence of actions. And so, the proposal of creating an assembly specific-domain grammar in Industry 4.0 is an novel idea.

The idea is to create a simple grammar that describes the daily activities in the product manual assembly. In order to increase its usability, the principles of the "Therbligs" theory were applied [2,3,14], this allows to represent all the entire assembly sequences using micro-movement primitives; the proposed language is based on the analysis of movements.

The Language allows describing the actions of the operators in a production cell, supported by the grammar that will be detailed in the next section. A new way of representing assembly instructions is designed; which will be the basis of a visual control system implemented for the quality control of the manual assembly process. All this, in conjunction with human collaborators, who will form a common environment with the machines that enable control by CV, operate synergistically to improve the final results of the product [4,5,9].

The rest of the document is structured as follows: Sect. 2 presents a general description of the proposed language, the general structure of its main elements, as well as particular syntax elements. In Sect. 3, the general validation of the proposal is executed using an example that shows how an assembly is developed, and ends with the conclusions and future works.

2 Manufacturing Description Language

The objective of this language is to evaluate if an operator is assembling a component or product the right way, according to the specifications. This inspection is carried out in several phases. The first phase is for an expert, usually a process or quality engineer, to utilize the language to describe the actions necessary to build the selected assembly. In the next phase, a operator performs his assembly tasks, while an artificial vision system records as inputs what is happening in the work area, then processes it and converts it into a textual description using video captioning techniques. Subsequently, a comparison is made between the system's output and the language description to determine if the operator followed the instructions.

In addition, the system is capable of determining the step or action that is currently being developed, in order to suggest the next steps to the operator and reduce the chance of him making a mistake when carrying out the assembly. The complete language proposal can be seen at GitHub[1].

The language is structured in system symbols (code 1.1), parameterization and assembly actions, for which each of these elements is presented in detail.

2.1 Parameterization

In the language design, two scenarios were considered, one where the CV system that describes the initial working environment will be auto-configured, through the identification of the present elements and their locations. In the other scenario, an operator or process engineer is required to describe the work area. In case of requiring a manual configuration, the following instructions are provided, which can be seen in detail in (code 1.2):

1. Product: The ID-code or name of the product to be registered is written in the assembly instructions.
2. Setup: Initial locations of parts, components, and tools are defined. In addition, the location of the existing components (assemblies and subassemblies) is indicated; as well as the quantity of components that are going to be generated during the manufacturing process. Finally, the dominant hand for the operator is selected, so that the system will configure the instructions according to the characteristics of each operator.

Within the language grammar these elements can be found in the setup-begin <sets>setup-end section. On the other hand, the sets can be each of the following options:

- assembly: Sets the location of an assembly to be used during the manufacturing process; if to _create is indicated it means that the assembly will be created from the union of 2 or more assemblies during execution. So the corresponding blocks for these assemblies are defined in the system.
- hand: Used to indicate to the system which is the operator's dominant hand, thus adjusting the instructions according to each individual.
- bin: Defines the location of a container and it's content, so that when instructions are established, the system knows where the operator should take or place the supplies for the assembly.
- accessory: Some tools use accessories. This indicates it's position to the system so that when indicated on the instructions, you can verify if the correct accessories were used.

2.2 Assembly Actions

This is the executable block, each basic execution instruction is called "step" and they can be found on the language grammar within the start steps end section, where each "step" can take two forms, which are detailed below:

[1] https://github.com/mazamorahdez/manufacturing_language.

Individual steps. These refer to an action that is defined as individual, among these are actions of type:

- hand: These operations are related to movements or actions with the hands, it can be seen in the code 1.3.
- tool: They are related to actions that require or apply the use of tools or their accessories, as it is exemplified in the 1.4 code.
- move: It allows to define displacements of the elements of the assembly, for example in the 1.4 code.

Block steps is the executable block, code 1.5, this type of step is considered special because they are atomic units of execution. It is made up of sequential steps, which in turn can also be of the block type, each of which has special characteristics. Among those blocks are:

- make-assembly: This is one of the most relevant blocks of the language, since it allows us to describe the concept of assembly (or sub-assembly), which are milestones during the process; defines the construction of complex elements from the union of more basic ones. A relevant element of this block is that it defines named units, so that they can be referenced in other "assemblies".
- repetition: There is a set of steps (including blocks), which are repeated several times, in these cases the set of steps to repeat and the number of times to be executed are defined.
- any-order: A basic premise of the language is that everything is executed in the written order, unless otherwise indicated, for this the any-order block is used, which tells the system that all instructions can be executed in any order.

Listing 1.1. Language symbols

```
<offset> := <digits>
<unit> := mm | cm | mm
<coordinate> := <sign><digits>,<sign><digits>
<sign> := <void> | <positive> | <negative>
<void> := ''
<positive> := +
<negative> := -
<position> := <x-position> | <y-position> | <y-position>-<x-position>
<x-position> := rigth | left
<y-position> := upper | lower
<identifier> := <char> | <char><word> | #bytes#
<word> := <alpha><word>
<alpha> := <char> | <digit> | <void>
<char> := a | b | ... | z | A | B |...| Z | _ | - | & | ' ' | . | , | @
<digits> := <digit> | <digit><digits>
<digit> := 0 | 1 | 2 | 3 | 4 | 5 | 6 | 7 | 8 | 9
```

Listing 1.2. Parameterization

```
<set> := assembly #<identifier> [to-create] [in <coordinate>:<offset>:<unit>];
<set> := hand <x-position>;
<set> := bin <part> [in <coordinate>:<offset>:<unit>];
<set> := tool <tool> [in <coordinate>:<offset>:<unit>];
```

Listing 1.3. Manual steps

```
<step> := <hand-action>:(<part>|<tool>) with <handused> [in assembly #<identifier>];
<step> := <hand-action>:<handused> [in <coordinate>:<offset>:<unit>];
<step> := <hand-action>:assembly #<identifier> with
(assembly #<identifier> | <handused>);
<handused> := hand | hand-nondominant | hand-any | hand-both
<hand-action> := put | hold | take | grip | release | push | spin | turn | join|move
```

Listing 1.4. Operations with tools and movements

```
<step> := move:assembly #<identifier> with <handused> from <coordinate>:<offset>
to <coordinate>:<offset>:<unit>;
<step> := <substep> [in <coordinate>:<offset>:<unit>];
<substep> := <hammer-action>:<hammer> | <wrench-action>:<wrench>
| <screwdriver-action>:<screwdriver> | <pliers-action>:<pliers>
| <driller-action>:<driller> with <accessory> | <clamp-action>:clamp
| <ratchet-action>:ratchet with (socket|none) | <screwdriver-action>:nut_driver
```

Listing 1.5. Step blocks

```
<make-assembly> := assembly-start #<identifier>:<steps> assembly-end;
<repetition> := repeat:<steps> until <digits> times;
<in-any-order> := any-order-begin <steps> any-order-end;
```

Other crucial elements involved in the definition of the language, and which are in general use in the parameterization and executable code part, are those described below:

- hand-action: These are the definition of all possible actions to be carried out with the hands. Some of these were taken from the Therbligs, others were incorporated according to the current reality.
- tool: This defines the list of tools with which you can define transformations or works on the elements in the product assembly. Basic and common families were created that were determined with a study of industrial operators. However, the language is designed to incorporate more tools mainly due to the constant development of new equipment. You can see the basic list of tools in the code 1.6.
- tool-action: Are the actions that can be performed on, or through the tools. Since not all tools share the same range of actions, there is an element that allows this association to be made. Like the tools, this section can be updated to represent the actions of the new tools that come on the market. The system has the capacity of extension in the actions, but a basic set is defined that can be seen in the code 1.6.
- substep: This is the section where the union of the own actions of a tool with the respective family of tools are made.
- part: Parts are the simplest and most common elements used in assemblies, like the rest of the elements, these can be extended. The system already incorporates a basic list that contains: screws, nuts, washers, among others.
- accessory: An accessory is defined as a complement for a particular tool, just like tool-actions, they are particular to each tool, so their relationship must also be established, in this case it is done in substep. An example of the accessories are drill bits and hubs, among others.
- coordinate: Allows to locate an ordered pair to locate the elements on the table, where the centroid of the artboard is assumed as the point (0,0). It is assumed that the visual control camera is located over the work table, which makes it possible to interpret the workspace as a plane.
- offset: To set the coordinate, the offset is defined, which is set as the length of each of the sides of a square, where the centroid of the square is the coordinate.
- unit: Are the units with which the offset and coordinates will work.

Listing 1.6. Tools and their actions

```
<tool> := <hammer> | <wrench> | <screwdriver> | <pliers> | <driller> | ratchet|clamp
| nut_driver
<hammer> := hammer_ball_pein | hammer_claw
<wrench> := wrench_adjustable | wrench_allen | wrench_combination
<screwdriver> := screwdriver_electric | screwdriver_phillips | screwdriver_slotted
<pliers> := pliers_diagonal | pliers_lineman | pliers_locking | pliers_long_nose
<driller> := drill_gun | drill_screw
<hammer-action> := nail | hammer_out | hit
<wrench-action> := pull | tight | locknut
<screwdriver-action> := screw
<pliers-action> := loosen | cut | hold | tighten
<driller-action> := drilling
<clamp-action> := loosen | tighten
<ratchet-action> := turn
```

3 Validation of the Proposal

For the validation of this work, we had the help of a group of industries whose work is made up of manufacturing cells. They were interviewed to determine how they represent their instructions for manual assemblies; as well as the set of tools and parts that they usually use. A series of stages was designed to carry out the validation, with the purpose to generate in the proposed language an equivalent to these instructions. The contrast of both formats for the assembly instructions are shown in the code 1.7 for the current format, and in the Fig. 1 its equivalent in proposed language. The steps used in the validation are as follows:

1. Identify at least one example of production sequences by industry.
2. For each sequence example the instructions were translated into the proposed language, to determine thus, if it has the capacity to express the instructions provided by each industry in an equivalent representation.
3. Take a group of operators from the industries in question and give them a brief introduction to the grammar. In this way they will be able to execute the instructions generated by the proposed language in their respective manufacturing cells.
4. In the industries where the collaboration of at least two operators was achieved, the second participant was asked if he was able to verify the actions carried out by the first, following the language instructions. This point is an approximation to what is expected in the future lines of research proposed in this work in the Sect. 4.

The example shown in this article was created from one of the videos generated by the team. Since in its original version, the complete example is too long to be presented in this paper, a simplified version was produced.

To generate the example presented below, the team carried out the following steps:

1. The time duration should not to exceed five minutes, including at least three different actions and two different tools.
2. A video player with the playback speed at 50% was used.
3. Each action, the playback was paused to write the action displayed in the code. If required, object coordinates was written down. Distances were measured with a tool that converts the distance from pixels to centimeters.

4. To reduce the size of the generated example, a repetitive sequences were searched. It was simplified by reducing the sequences that appeared several times.

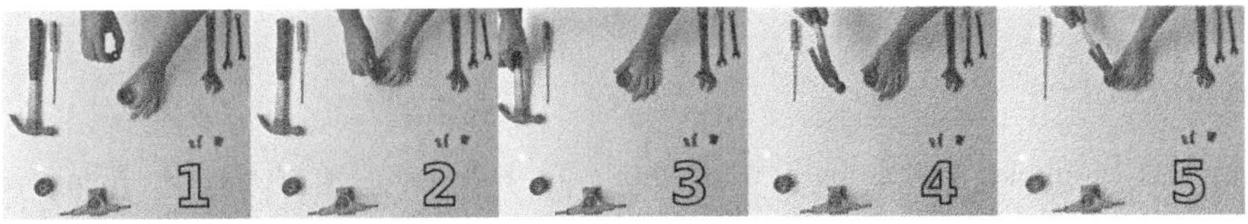

Fig. 1. Frames for code example

Listing 1.7. Simplified example of generated code

```
assemble-begin
skateboard;
setup-begin
hand right;
bin washer in -5,20 : 3 : cm;
bin screws in -5,25 : 3 : cm;
bin nut;
tool screwdriver_phillips in -30,35 : 3 : cm;
tool wrench_adjustable in 30,35 : 3 : cm;
tool wrench_combination in 33,35 : 3 : cm;
tool wrench_combination in 36,35 : 3 : cm;
tool hammer_claw in -33,35 : 3 : cm;
assembly #1 in -10,-35 : 7 cm;
assembly #2 in -30,-32 : 2 : cm;
assembly #3 in -10,-10 : 2 : cm;
setup-end
start
take : washer with hand;
move : assembly #1 with hand-nondominant
       from -10,-10 : 3 to 0,0 : 4 : cm;
hold : assembly #1 with hand-nondominant;
put : washer with hand in 0,0 : 1 : cm; // Frame #1
push : washer with hand in assembly #1; // Frame #2
release : assembly #1 with hand-nondominant;
take : hammer_claw with hand;            // Frame #3
take : assembly #1 with hand-nondominant;
hold : assembly #1 with hand-nondominant;
hit : hammer_claw in 0,0 : 4 : cm;       // Frame #4
spin : assembly #1 with hand-nondominant;
hit : hammer_claw in 0,0 : 4 : cm;       // Frame #5
turn : assembly #1 with hand-nondominant;
release : hammer_claw with hand in 0,0 : 4 : cm;
end
assemble-end
```

4 Conclusions

This paper proposes the design of a new structured language for the description of the activities of manufacturing operations. This language is part of a computer vision control system that allows determining the basic level of quality of a product, and defines the steps of how it was built according to the specifications of production in each organization. The language also allows transforming different process description systems in industries and works as a suggestion system for operators to minimize errors, creating a "poka yoke" system for assemblies.

As future lines of research, it is proposed to use this language in conjunction with video analysis systems to formalize the instructions carried out and verify if the instructions described in the proposed language are satisfied. In addition, it is very useful in the area of job design for job measurement and standard time calculation, along with Operations Engineering to calculate production capacity. The system can promote occupational safety by signaling to the operator that there are items that should not be present.

References

1. Fast-Berglund, Å., Fässberg, T., Hellman, F., Davidsson, A., Stahre, J.: Relations between complexity, quality and cognitive automation in mixed-model assembly. J. Manuf. Syst. **32**(3), 449–455 (2013)
2. Ferguson, D.: Therbligs: The Keys to Simplifying Work (2000). http://web.mit.edu/allanmc/www/Therblgs.pdf
3. Groover, M.P.: Work Systems and the Methods, Measurement, and Management of Work. Pearson Education Inc, Boston (2007)
4. Hedelind, M., Jackson, M.: How to improve the use of industrial robots in lean manufacturing systems. J. Manuf. Technol. Manag. **22**(7), 891–905 (2011). https://doi.org/10.1108/17410381111160951
5. Hermann, M., Pentek, T., Otto, B.: Design principles for industrie 4.0 scenarios. In: Proceedings of the Annual Hawaii International Conference on System Sciences, pp. 3928–3937, (March 2016). https://doi.org/10.1109/HICSS.2016.488
6. Hornung, A., Bennewitz, M., Strasdat, H.: Efficient vision-based navigation. Auton. Robots **29**(2), 137–149 (2010). https://doi.org/10.1007/s10514-010-9190-3
7. Kanawaty, G.: Introducción al estudio del Trabajo. Editorial Limusa S.A de C.V., 11 edn. (2008)
8. Krishna, R., Hata, K., Ren, F., Fei-Fei, L., Niebles, J.C.: Dense-captioning events in videos. In: Proceedings of the IEEE International Conference on Computer Vision, pp. 706–715, (October 2017). https://doi.org/10.1109/ICCV.2017.83
9. Lee, J., Bagheri, B., Kao, H.A.: Recent advances and trends of cyber-physical systems and big data analytics in industrial informatics. In: International Conference on Industrial Informatics (INDIN 2014), October 2014
10. Luo, R.C., Hsu, Y.T., Wen, Y.C., Ye, H.J.: Visual image caption generation for service robotics and industrial applications. In: Proceedings - 2019 IEEE International Conference on Industrial Cyber Physical Systems. (ICPS 2019), pp. 827–832 (2019). https://doi.org/10.1109/ICPHYS.2019.8780171
11. Makris, S., Karagiannis, P., Koukas, S., Matthaiakis, A.S.: Augmented reality system for operator support in human-robot collaborative assembly. CIRP Ann. Manuf. Technol. **65**(1), 61–64 (2016). https://doi.org/10.1016/j.cirp.2016.04.038
12. Mancini, M., Karaoguz, H., Ricci, E., Jensfelt, P., Caputo, B.: Kitting in the wild through online domain adaptation. In: IEEE International Conference on Intelligent Robots and Systems, pp. 1103–1109 (2018). https://doi.org/10.1109/IROS.2018.8593862
13. Real, F., Batou, A., Ritto, T., Desceliers, C.: Stochastic modeling for hysteretic bit–rock interaction of a drill string under torsional vibrations. J. Vib. Control (X)**25**(10), 1663–1672 (2019). https://doi.org/10.1177/1077546319828245
14. Universidad Politécnica deValencia: Therbligs (2018). http://evaluador.doe.upv.es/wiki/index.php/Therbligs
15. Wang, L., Schmidt, B., Nee, A.Y.C.: Vision-guided active collision avoidance for human-robot collaborations. Manuf. Lett. **1**(1), 5–8 (2013)

16. Wang, X., Chen, W., Wu, J., Wang, Y.F., Wang, W.Y.: Video captioning via hierarchical reinforcement learning. In: Proceedings of the IEEE Computer Society Conference on Computer Vision and Pattern Recognition, pp. 4213–4222 (2018). https://doi.org/10.1109/CVPR.2018.00443
17. Yang, Y., Li, Y., Fermüller, C., Aloimonos, Y.: Robot learning manipulation action plans by watching unconstrained videos from the World Wide Web. In: Proceedings of the National Conference on Artificial Intelligence, vol. 5, pp. 3686–3692 (2015)
18. Yao, T., Pan, Y., Li, Y., Qiu, Z., Mei, T.: Boosting image captioning with attributes. In: Proceedings of the IEEE International Conference on Computer Vision, pp. 4904–4912, (October 2017). https://doi.org/10.1109/ICCV.2017.524

ToolSet: A Real-Synthetic Manufacturing Tools and Accessories Dataset

Mauricio-Andres Zamora-Hernandez[1(✉)], John Alejandro Castro-Vargas[2(✉)], Jorge Azorin-Lopez[2(✉)], and Jose Garcia-Rodriguez[2(✉)]

[1] University of Costa Rica, San Jose, Costa Rica
mauricio.zamorahernandez@ucr.ac.cr
[2] University of Alicante, Alicante, Spain
{jcastro,jgr,jazorin}@ua.es

Abstract. The use of intelligent systems to improve manufacturing processes is the basis for the development of robotic solutions in Industry 4.0. Monitoring operators manipulating tools and objects is one of the key tasks. Deep learning methods are obtaining state-of-the-art results to solve this problem but large amounts of labelled data should be provided to these networks. However, no specific manufacturing tools datasets exist. For this purpose, we proposed a new dataset for this type of environment. An hybrid dataset of 29550 images has been proposed for network training that combines real and synthetic images of tools and components commonly used in manufacturing cells. This project is part of a set of proposed modules of a solution that allows us to evaluate in real-time the execution of assembly instructions of the operators throughout the production process.

Keywords: Dataset · Industry 4.0 · Toolset · UnrealEngine · YOLO

1 Introduction

In the industry there are different manufacturing phases in which robots are used to automate tasks and improve productivity. These types of machines are used in repetitive tasks or those with high precision requirements that mainly require mechanical actions. When a certain adaptation or creativity is required throughout the task, robots are limited to a rigid scheduled actions. In this cases human operators are more flexible to perform this kind of processes. In Industry 4.0 [10], artificial intelligence plays an important role improving productivity, quality and safety in the different stages of production [3]. This new time requires autonomous machines with a certain degree of intelligence, which are capable of adapting efficiently to different levels of production as well as safely collaborating in the process with human operators.

Among the different approaches used over the years, those using machine learning methods adapted to the application domain are especially remarkable.

Á. Herrero et al. (Eds.): SOCO 2020, AISC 1268, pp. 800–809, 2021.
https://doi.org/10.1007/978-3-030-57802-2_77

This is due to the fact that in recent years great advances have been made within these methodologies, which have overcome traditional approaches.

Methodologies based of the use of Deep Learning have obtained great relevance. These architectures learn features from input data with different levels of abstraction from multiple layers, and great improvements have been demonstrated in fields related to speech recognition, object recognition and object detection [11] to cite a few.

These approaches require a large amount of tagged data to obtain a relevant performance. This is a task that requires a lot of human effort, manually tagging images or videos that the network will use to learn to extract robust characteristics after a training process.

This work is part of a full intelligent architecture whose purpose is to assist operators throughout the different manufacturing phases, using an assembly description language that establishes the instructions to define the assembly process. One of the requirements for its development is to detect the objects that the operator needs and those that he is already using.

The proposed architectures evaluate the quality and accuracy of the manufacturing processes developed by human operators or recommend next action and necessary tools to complete the current task. Since most public datasets do not classify specific manufacturing tools and accessories, we proposed the creation of a dataset consisting of several objects that are used throughout the different manufacturing phases of a manual assembly. For data augmentation purpose we used a mixture model. A significant amount of data was generated synthetically through samples of real objects. A baseline based on YoloV3 [20] is provided to analyze its performance onto the dataset.

The rest of the paper is organized as follows: Sect. 2 is a review of works related with the topic. In Sect. 3 the proposed dataset of synthetic and real tools and accessories for manual manufacturing processes is described in detail. Section 4 is devoted to test the dataset with a well known baseline, Yolo3. Finally in Sect. 5 we present our conclusions and further lines of research.

2 Related Works

In this section, we review the basics of Industry 4.0 and YOLO, which is a cutting edge deep learning 2D object localization architecture. Also, we review the most widely used public datasets in the field.

2.1 Industry 4.0

Industry 4.0 is causing an update in companies through a transformation in their productive processes. Automation of processes and data exchange are the core of manufacturing technologies, working with human collaborators who formed a common environment with machines to work synergistically [8,9,12].

In the fourth industrial revolution, the concept of intelligent automation is one of its main axes. These production environments must consider robotics as a basis for their technologies, as well as Artificial Intelligence, Cyberphysical Systems, Big Data, the Internet of Things, in collaborative environments of robots and people as a unit [8,9,12].

With the high level of global competition, manufacturing has to be well planned to respond quickly with high quality products [6]. In order to do this, you must put pressure on each of the production engineering processes; from the design stages [18], process planning, complex calculations or modifications of production cells [6].

There is a need to improve production tools to adapt them to new challenges. Also, new trends and changes in the creation of customized products require operators to improve or learn new assembly skills, which are supported by intelligent machines [1,24].

2.2 You Only Look Once

You Only Look Once (YOLO) is an architecture for rapid detection, and precise tracking of multiple objects in real time, generating location coordinates for e detected object with a very high level of accuracy. To cite a couple of application examples: driving vehicles without specialized sensors or vehicles for people with disabilities [2,16,19].

YOLO architectures use a typical end-to-end network structure. This type of structure is more concise compared to the two-stage networks of R-CNN type. It integrates candidate area detection mechanisms, making the network faster than its counterparts with R-CNN type architectures [23].

The network that forms YOLO's backbone is based on Darknet-53 to extract features from images. The entire network mainly uses residual layers as building blocks. A total of five residual layers with different scales and weights. These only run between the residual layers and the output layers [13]. The convolutional layer uses alternating 1×1 and 3×3 convolutional cores to extract more abstract features [5].

The anchor box concept was introduced by Faster RFCNN and k-means which is used by YOLO v3 to determine the radius size for the anchor box that locates the searched object. Instead of directly mapping the coordinates into the bounding box, the parameters are relative to the anchor box that was predicted [13].

2.3 Relevant Datasets

In the field of object detection, the selection of the dataset to be used for network training is a key success-factor. Since it will determine the level of certainty that the network will have after being trained. To carry out this work, an exhaustive search was carried out to determine if within the available dataset, there was one that meets the requirements for the investigation. Among the most relevant datasets found are the following:

The Pascal Visual Object Classes (VOC). The Pascal Visual Object Classes (VOC)[1], is a publicly available dataset and an annual competition along with workshops since 2006. The dataset consists of 500,000 images in 20 categories that were retrieved from flickr[2] [7].

ImageNet Large Scale Visual Recognition Challenge. The ImageNet Large Scale Visual Recognition Challenge[3] contains 14,197,122 images, 21841 synsets indexed, organized according to the WordNet hierarchy. The challenge has been run annually from 2010 and it has become the standard benchmark for large-scale object recognition. The publically released dataset contains a set of manually annotated training images. A set of test images is also released, with the manual annotations [21].

COCO Dataset. The COCO Dataset (Common Object in Context)[4] is a large-scale object detection, segmentation, and captioning dataset. It contains 330 K photos of 91 objects types with a total of 2.5 million labeled instances. It has considerably more object instances per image as compared to ImageNet and PASCAL VOC (5 captions per image)[15].

SUN Database. The SUN (Scene UNderstanding) Database[5] is a scene categorization dataset. It contains 131 067 images in 908 Scene categories and 313 884 Segmented objects in 4479 Object categories. This dataset is based in WordNet [22], it has annotated images covering a large variety of environmental scenes, places and the objects within.

3 Toolset Dataset

Neural networks require large amounts of data to obtain acceptable levels of generalization. Make the annotation of the images is a task that consumes a large amount of time and human effort in the labeling process.

We revised relevant public datasets used to test object recognition and location objects. However those datasets are general and include only few examples and classes related with manufacturing tools and accessories. in order to obtain enough relevant and variable data to feed our deep architectures, we proposed the generation of a dataset made up of both real and synthetic data.

The real data is intended to contribute with noise and quality that are detectable in cameras to the learning process. It is intended that the network does not fall into overfitting because of the perfection in the captures obtained

[1] http://host.robots.ox.ac.uk/pascal/VOC/.
[2] https://www.flickr.com/.
[3] http://www.image-net.org/.
[4] http://cocodataset.org/.
[5] https://vision.cs.princeton.edu/projects/2010/SUN/.

through synthetic environments. Furthermore, given the difficulty in the process of obtaining and processing the images, the data has been increased through filters applied to transform the images, such as rotations and deformations. The objects were selected considering their use in different production processes. It consists of a total of 24 tools and materials such as: clamps, hammers and screws, examples of these objects are seen in Figs. 1 and 3.

3.1 Real Images

The dataset is made up of real pictures with a total of 591 images obtained from the Internet. To obtain a quantity of data that allows the network to be trained properly, data has been increased with data augmentation techniques.

Fig. 1. Objects which compose the real data with different random backgrounds.

This process has been carried out by segmenting the objects of interest from the original images to perform transformations on them. Such as varying the background of the images and randomly adding different transformations to the objects like: rotations, translations, deformations and noise.

A total of 50 transformations were performed for each object, using 1000 images to establish random backgrounds.

This method allows to generate a total of 29550 new samples correctly labelled from the size of the segmented objects (Fig. 1). The location of the bounding boxes can be estimated after applying the different transformations.

3.2 Synthetic Images

Real data requires a lot of effort in terms of collection and labeling. Moreover, they are limited to the perspective on which the images have been taken.

Synthetic data has been generated from 3D meshes for the different objects. In order to obtain a greater variety of the points of views of the objects and to increase the amount of data that will be available.

Fig. 2. Twelve cameras spread out around the spawn zone of the objects. In this area the background and orientation of the objects will vary.

This process was carried out using UnrealEngine, a video game engine, that include multiple plugins, as for example: UnrealROX [17], which facilitates the generation of synthetic datasets.

$$[label] = [bbox_{xcenter}/W][bbox_{ycenter}/H][bbox_w/W][bbox_h/H] \qquad (1)$$

This plugin allows to generate different types of data from the simulations executed in the engine, such as RGB images, depth and segmentation masks. The most relevant data types are color and segmentation masks, which are used to generate the labels in the format defined for YoloV3.

To generate the labels, we generate a bounding box with the maximum and minimum pixels of the segmentation mask, including the name of the tool. With these values the center of the bounding box is calculated, the width and height of the image are normalized as shown in the Eq. 1, where W and H represent the width and height of the image.

To obtain a relevant variability we deployed 12 cameras to represent different points of view of the object (Fig. 2). Moreover, to prevent the network from memorizing the working background and improving its testing performance, the background of the different captures was randomly varied with 50 different samples.

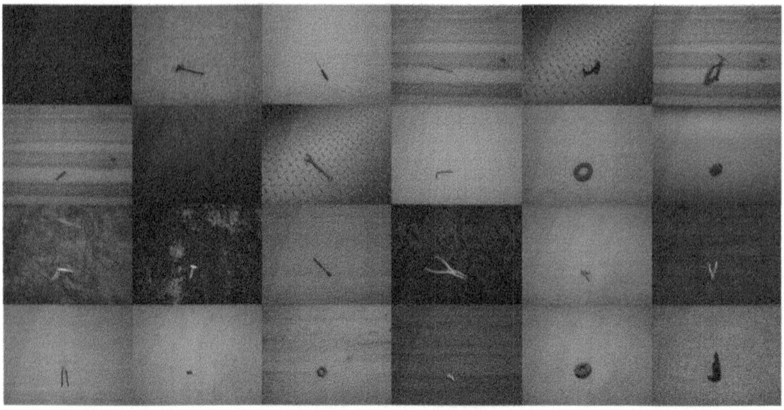

Fig. 3. 24 meshes used to generate the synthetic dataset. Four of them were generated using meshes obtained from YCB dataset [4], which are the screwdrivers, the adjustable wrench and the drill screw.

The system was prepared to take 100 captures with the cameras deployed for each object. Along this process, rotations were randomly applied to the objects, allowing variability to be obtained on the samples and generating a total of 28800 images3 with their corresponding bounding box label for YoloV3. The result of this process can be observed in the Fig. 3 for each mesh sample.

4 Experiments

YoloV3 was the network used as a baseline to detect the different objects contained in our dataset. The reason was that it combines high accuracy with a minimal impact on the run-time [20].

The experiments carried out with this network consisted on using a subset of the objects available in the dataset and training the network up to a maximum of 50200 epochs. There were used 18 different categories of objects, as hand tools, screwdrivers and hammers. The network was not trained by mixing hand tools together with materials such as screws and washes.

Table 1. The Mean Average Precision (mAP), Precision, Recall, F1 score and average in Intersection over Union (IoU) obtained as result of our training.

mAP	P	R	F_1	IoU
94.6	0.96	0.98	0.97	83.7

To perform the training, the real data was combined with the synthetic one and separated in a 20/80 proportion to generate a validation set and a training set. The training was then adjusted so that in addition to the increased offline data, additional transformations in training time were performed, such as rotations up to 40 degrees, variations in the HSVs channel of up to 50% and scale variations up to 30%, starting from an image size of 416. In addition, the hyper-parameters used to train the network were a learning rate of 0.001, momentum of 0.9 and a burn in of 1000.

The results obtained through the training can be seen in Table 1. Here we can see high values for accuracy, recall and f1 score, which may be an indication of overffiting in our training model. Therefore, we tested with additional objects to those used in our validation and test sets, where we obtained the results showed in Fig. 4. We noticed that the first two samples were detected correctly, although in the third one the net detects the drill screw as a drill gun, which is due to the fact that both tools share similar visual features.

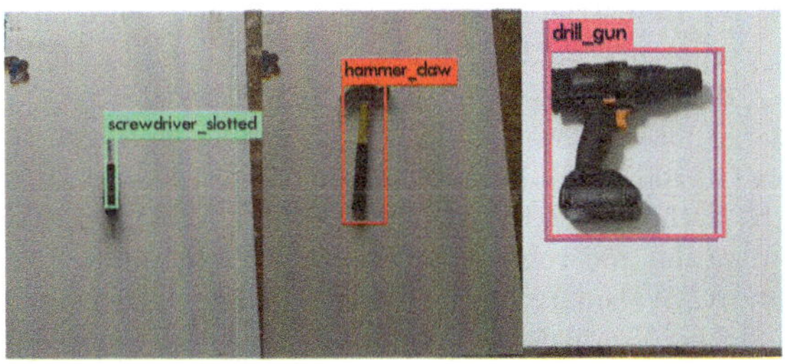

Fig. 4. Qualitative prediction results with objects not used in training nor validation sets.

5 Conclusions

We designed a dataset specifically focused to detect tools and materials in industrial environments of manual assembly. The aim of this process is to design algorithms to create a smart production system, with all the benefits this may offer, as for example, more safety for workers and improvements in the productivity.

This dataset was successfully prepared mixing real and synthetic data and it is publicly available[6] to the use of research community.

In the experiments developed using our dataset, can be deduced some degree of overfitting. However, it have shown certain tolerance working with YoloV3 to detect different tools. Also, as some of the objects which compose our set share similar visual features as drill gun and drill screw, in some points of view prediction can be confused, although it recognized successfully the existence of an object in the region.

As we mentioned before, this project is one of the proposed modules to assist workers with robotic instructions along the production process. So, this module

[6] https://drive.google.com/open?id=1VXTvh-AMyff9vCRG4JqKGfmLNzQHzB85.

will take part in the complete pipeline of future works, where the rest of the modules are being investigated.

Also, the results observed in this paper are sufficient for our proposals, but it will need to be improved to get best results in classification. One of the options which have been evaluated is the use of network architectures with best accuracy but slowest performance such as RetinaNet [14]. Another improvement pointed is to increase the amount of our data with new samples taken from real tools and more meshes in synthetic ones. This is specially important to face the overfitting.

Acknowledgements. Experiments were made possible by a generous hardware donation from NVIDIA.

References

1. Abdelhameed, W.: Industrial revolution effect on the architectural design. In: 2019 International Conference on Fourth Industrial Revolution. ICFIR 2019, pp. 1–6 (2019). https://doi.org/10.1109/ICFIR.2019.8894774
2. Aggarwal, C.C.: Neural Networks and Deep Learning. Springer, Heildelberg (2018). https://doi.org/10.1007/978-3-319-94463-0
3. Bahrin, M.A.K., Othman, M.F., Azli, N.N., Talib, M.F.: Industry 4.0: a review on industrial automation and robotic. J. Teknol. **78**(6–13), 137–143 (2016)
4. Calli, B., Singh, A., Walsman, A., Srinivasa, S., Abbeel, P., Dollar, A.M.: The YCB object and model set: towards common benchmarks for manipulation research. In: 2015 international Conference on Advanced Robotics (ICAR), pp. 510–517. IEEE (2015)
5. Cao, C.Y., Zheng, J.C., Huang, Y.Q., Liu, J., Yang, C.F.: Investigation of a promoted you only look once algorithm and its application in traffic flow monitoring. Appl. Sci. **9**(17), 3619 (2019). https://doi.org/10.3390/app9173619. https://www.mdpi.com/2076-3417/9/17/3619
6. Erdin, M.E., Atmaca, A.: Implementation of an overall design of a flexible manufacturing system. Procedia Technol. **19**, 185–192 (2015). https://doi.org/10.1016/j.protcy.2015.02.027, http://linkinghub.elsevier.com/retrieve/pii/S2212017315000286
7. Everingham, M., Eslami, S.M., Van Gool, L., Williams, C.K., Winn, J., Zisserman, A.: The pascal visual object classes challenge: a retrospective. Int. J. Comput. Vis. **111**(1), 98–136 (2014). https://doi.org/10.1007/s11263-014-0733-5
8. Hedelind, M., Jackson, M.: How to improve the use of industrial robots in lean manufacturing systems. J. Manuf. Technol. Manage. **22**(7), 891–905 (2011). https://doi.org/10.1108/17410381111160951
9. Hermann, M., Pentek, T., Otto, B.: Design principles for industrie 4.0 scenarios. In: Proceedings of the Annual Hawaii International Conference on System Sciences **2016-March**, pp. 3928–3937 (2016). https://doi.org/10.1109/HICSS.2016.488
10. Lasi, H., Fettke, P., Kemper, H.G., Feld, T., Hoffmann, M.: Industry 4.0. Bus. Inform. Syst. Eng. **6**(4), 239–242 (2014)
11. LeCun, Y., Bengio, Y., Hinton, G.: Deep learning. Nature **521**(7553), 436–444 (2015)
12. Lee, J., Bagheri, B., Kao, H.A.: Recent advances and trends of cyber-physical systems and big data analytics in industrial informatics. In: International Conference on Industrial Informatics (INDIN) 2014 October 2014

13. Li, J., Gu, J., Huang, Z., Wen, J.: Application research of improved YOLO V3 algorithm in PCB electronic component detection. Appl. Sci. (Switzerland) **9**(18) (2019). https://doi.org/10.3390/app9183750
14. Lin, T.Y., Goyal, P., Girshick, R., He, K., Dollár, P.: Focal loss for dense object detection. In: Proceedings of the IEEE International Conference on Computer Vision, pp. 2980–2988 (2017)
15. Lin, T.Y., Maire, M., Belongie, S., Hays, J., Perona, P., Ramanan, D., Dollár, P., Zitnick, C.L.: Microsoft COCO: common objects in context. In: Lecture Notes in Computer Science (including subseries Lecture Notes in Artificial Intelligence and Lecture Notes in Bioinformatics), 8693 LNCS(PART 5), pp. 740–755 (2014)
16. Lv, X., Dai, C., Chen, L., Lang, Y., Tang, R., Huang, Q., He, J.: A robust real-time detecting and tracking framework for multiple kinds of unmarked object. Sensors (Switzerland) **20**(1), 2 (2020). https://doi.org/10.3390/s20010002
17. Martinez-Gonzalez, P., Oprea, S., Garcia-Garcia, A., Jover-Alvarez, A., Orts-Escolano, S., Garcia-Rodriguez, J.: Unrealrox: an extremely photorealistic virtual reality environment for robotics simulations and synthetic data generation. Virtual Real. **24**, 271–288 (2020)
18. Puik, E., Telgen, D., van Moergestel, L., Ceglarek, D.: Assessment of reconfiguration schemes for reconfigurable manufacturing systems based on resources and lead time. Robot. Comput. Int. Manuf. **43**, 30–38 (2017). https://doi.org/10.1016/j.rcim.2015.12.011
19. Redmon, J., Divvala, S., Girshick, R., Farhadi, A.: You only look once: unified, real-time object detection. In: Proceedings of the IEEE Computer Society Conference on Computer Vision and Pattern Recognition **2016-December**, pp. 779–788 (2016). https://doi.org/10.1109/CVPR.2016.91
20. Redmon, J., Farhadi, A.: Yolov3: an incremental improvement. arXiv preprint, arXiv:1804.02767 (2018)
21. Russakovsky, O., Deng, J., Su, H., Krause, J., Satheesh, S., Ma, S., Huang, Z., Karpathy, A., Khosla, A., Bernstein, M., Berg, A.C., Fei-Fei, L.: ImageNet large scale visual recognition challenge. Int. J. Comput. Vis. **115**(3), 211–252 (2015). https://doi.org/10.1007/s11263-015-0816-y
22. Xiao, J., Ehinger, K.A., Hays, J., Torralba, A., Oliva, A.: SUN database: exploring a large collection of scene categories. Int. J. Comput. Vis. **119**(1), 3–22 (2016). https://doi.org/10.1007/s11263-014-0748-y
23. Xu, Q., Lin, R., Yue, H., Huang, H., Yang, Y., Yao, Z.: Research on small target detection in driving scenarios based on improved Yolo network. IEEE Access **8**, 27574–27583 (2020). https://doi.org/10.1109/ACCESS.2020.2966328. https://ieeexplore.ieee.org/document/8957514/
24. Zhou, L., Cao, S., Liu, J., Tan, T., Du, F., Fang, Y., Zhang, L.: Design, manufacturing and recycling in product lifecycle: new challenges and trends. In: 4th IEEE International Conference on Universal Village 2018, UV 2018, pp. 1–6 (2018). https://doi.org/10.1109/UV.2018.8709326

Special Session: Computational Intelligence for Laser-Based Sensing and Measurement

Robust 3D Object Detection from LiDAR Point Cloud Data with Spatial Information Aggregation

Nerea Aranjuelo[1,2](✉), Guus Engels[1], Luis Unzueta[1], Ignacio Arganda-Carreras[2,3,4], Marcos Nieto[1], and Oihana Otaegui[1]

[1] Vicomtech, Basque Research and Technology Alliance (BRTA), San Sebastian, Spain
naranjuelo@vicomtech.org
[2] Basque Country University (UPV/EHU), San Sebastian, Spain
[3] Ikerbasque, Basque Foundation for Science, Bilbao, Spain
[4] Donostia International Physics Center (DIPC), San Sebastian, Spain

Abstract. Current 3D object detectors from Bird's Eye View (BEV) LiDAR point cloud data rely on Convolutional Neural Networks (CNNs), which have originally been designed for camera images. Therefore, they look for the same target features, regardless of the position of the objects with respect to the sensor. Discarding this spatial information makes 3D object detection unreliable and not robust, because objects in LiDAR point clouds contain distance dependent features. The position of a group of points can be decisive to know if they represent an object or not. To solve this, we propose a network extension called FeatExt operation that enables the model to be aware of both the target objects features and their spatial location. FeatExt operation expands a group of feature maps extracted from a BEV representation to include the distance to a specific position of interest in the scene, in this case the distance with respect to the LiDAR. When adding the proposed operation to a baseline network in an intermediate fusion fashion, it shows up to an 8.9 average precision boost in the KITTI BEV benchmark. Our proposal can be easily added to improve existing object detection networks.

Keywords: 3D object detection · LiDAR · Feature extraction

1 Introduction

One of the keys to the success of Convolutional Neural Networks (CNNs) is their weight sharing property. The capability of identifying features anywhere in a given input has been an important factor to solve different challenges, specially in computer vision tasks, such as image classification, object detection or semantic segmentation [1,2]. Being able to detect complex patterns anywhere in

Á. Herrero et al. (Eds.): SOCO 2020, AISC 1268, pp. 813–823, 2021.
https://doi.org/10.1007/978-3-030-57802-2_78

the input data, no matter the location of the object, is an advantage in many fields. However, depending on the nature of the data and properties, the features of an object's class and its position may be strongly related. This is the case for point clouds obtained from a LiDAR sensor.

The LiDAR sends out high-speed pulses of laser-light and gets distances to surrounding objects based on the reflection time of the beams. The collision between an object and a laser beam is represented as a point in a 3D point cloud. Depending on the model, it can cover more than 100 m and up to 200 m. Its good performance in adverse weather and lighting conditions, and the precise 3D view that can be generated around the sensor, make the LiDAR a strong candidate to become a key component in advanced driving systems.

Consequently, and driven by the importance of 3D scene understanding in automotive field, diverse works have emerged with different deep learning based proposals for 3D object detection. Advances have been made rapidly and show promising results, but it is still an open question what the best way to process point cloud data with CNNs is. Many works try to adapt mature networks commonly used for camera images to this task [3–5], but they usually do not consider the special properties of LiDAR point clouds. An object present in a LiDAR point cloud at a distance of 5 m far from the sensor and the same object at 40 m does not have the same distribution and quantity of points. This is because the distance between points increases over the distance due to the properties of the LiDAR. Methods that convert the point cloud to a Bird's Eye View (BEV) representation in their pipeline tend to discard this information when they apply CNNs to localize the objects of interest in the BEV image.

In this work, we propose an effective way to add spatial location information to BEV-based methods to guarantee a more reliable and robust object detection.

Our main contributions could be summarized as follows:

- We propose a novel solution for including spatial location information in existing 3D BEV object detection networks for LiDAR.
- We introduce FeatExt, an operation which enriches the feature space by adding information regarding a specific location.
- We train a baseline network without FeatExt and compare it to two alternatives that integrate it by early fusion and intermediate fusion.
- We evaluate our proposal on the KITTI BEV benchmark and show that it boosts performance remarkably for all difficulty categories.

2 Related Work

2.1 BEV Object Detection Methods

Several 3D object detection approaches, specially the earlier ones, use image-based feature extraction networks [3–5]. The main idea of these methods is to project the point cloud to a BEV representation that can be used as input to a mature 2D CNN. Features that are encoded in the BEV map vary but often include the height of points, reflectance intensity or the density of points [5–7].

When this information is stored as 3 channels, a network architecture for RGB images is directly applicable. Some works store points height information in more than 3 channels to retain more information [6,7], but the architectures they use do not vary much from networks designed for image processing. BEV-based methods have shown promising results, but most of them have not payed much attention to the different properties of LiDAR data compared to image data. LiDAR point clouds are sparse and irregular and contain distance dependent features. Not adapting the network designs properly to the processed data nature may be one of the reasons for these methods to be lately surpassed by works processing directly the raw point cloud. Because of this, this work proposes how to better adapt these models and increase their performance.

2.2 Methods Learning from Raw Point Cloud

Some other works propose to process the point cloud data directly. One of the most influential works on applying deep learning to point clouds is PointNet [8], which processes the unstructured point cloud by a CNN and maps 3D points to a higher dimension feature space. Similarly, VoxelNet [9] proposes to learn point-wise features directly from point clouds. VoxelNet extracts features from point voxels by introducing a voxel feature encoding (VFE) layer. Then, the output is connected to a Region Proposal Network (RPN). Based on these ideas, diverse works have emerged [10–13]. For example, PointPillars [13] extracts point features with PointNet and then transforms them to a BEV representation to apply a FPN-inspired backbone [14]. PointRCNN [12] extracts point-wise features for 3D proposals generation, which are later on refined by a second stage. In general, most recent methods that learn features from point clouds achieve better results than BEV-based ones in the KITTI benchmark [15] but show some extra challenges such as the data quantity that needs to be processed in real time.

3 Methodology

3.1 Baseline Pipeline

We propose a common pipeline inspired by state-of-the-art 3D BEV object detection methods [3,5–7], shown in Fig. 1, to which we add later on our solution. The pipeline has as input a raw point cloud and outputs 3D bounding boxes containing the objects of interest. In order to do that, the point cloud is converted to a BEV image and fed to the object detection CNN.

Point Cloud to BEV Representation. Point clouds are unstructured and in order to apply standard CNNs to them, data need to be structured similarly to images. To do that, we can represent a point cloud from the BEV perspective (top view) by discretizing the cloud into a 2D grid with a specific resolution. We divide each cell vertically in three same-size voxels. In each voxel we store the maximum height of the contained points. This is encoded as a 3 channel image.

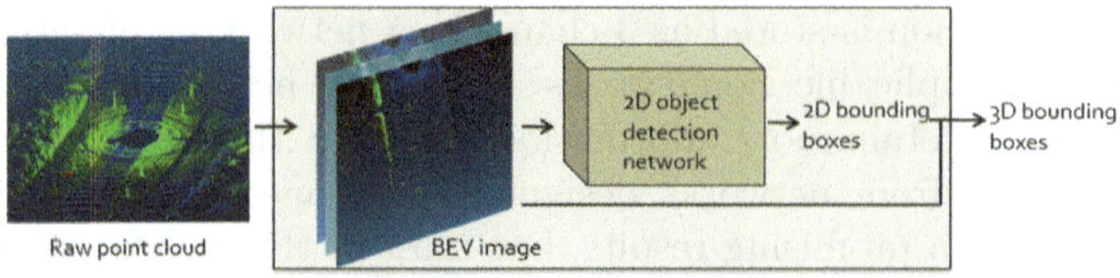

Fig. 1. Common baseline pipeline for 3D BEV object detection. Input point cloud is converted to a BEV representation. This is fed to a 2D object detection network. Height of the detected objects is estimated for final 3D object detections.

Object Detection Network. The BEV image is fed to a two-stage object detection network based on Faster R-CNN [16], but with the addition of a rotation regression branch in the head network to estimate oriented boxes.

The first stage of the network is the RPN, which extracts features from the input data using the three first blocks of the ResNet-50 [17] network. As target objects are small, we use a stride of 1 in the first block's last convolution to avoid downsampling the feature maps too early and losing detailed information fast.

The second stage of the network, takes as input the object proposals and the output feature maps of the RPN. After region of interest (ROI) pooling, each candidate's feature maps are fed to three fully connected (FC) layers with 1024 weights each. Output feature maps are processed by an horizontal and a rotational branch, each one containing a FC layer for object classification and a FC for bounding box regression. The horizontal branch considers bounding boxes without rotation, defined by the center of the box coordinates (x, y) and its dimensions (h, w), whereas the rotational branch estimates same objects with their corresponding orientation (x_r, y_r, h_r, w_r, θ).

The training loss of the RPN is the same as Faster R-CNN [16]. The head network contains a regression and a classification loss for each branch (horizontal and rotational). The regression loss is a summation of the absolute difference between the ground truth and the network prediction for each bounding box parameter when the box is axis-aligned (x, y, h, w) and oriented (x_r, y_r, h_r, w_r, θ). Based on that value, smooth L1 loss is computed.

The object classification loss is a softmax cross-entropy loss between the considered classes. All losses are combined as shown in the following equation:

$$\begin{aligned} Loss = {} & \lambda_{rpn_r} * L_{rpn_r} + \lambda_{rpn_c} * L_{rpn_c} + \\ & \lambda_{head_r} * (L_{head_{hr}} + L_{head_{rr}}) + \\ & \lambda_{head_c} * (L_{head_{hc}} + L_{head_{rc}}) \end{aligned} \tag{1}$$

where L_{rpn_r} and L_{rpn_c} are the RPN regression and classification losses, $L_{head_{hr}}$ and $L_{head_{hc}}$ are the regression and classification losses of the horizontal head network and $L_{head_{rr}}$ and $L_{head_{rc}}$ are the regression and classification losses of the rotational head network. λ_{rpn_c}, λ_{rpn_r}, λ_{head_r} and λ_{head_c} are the balancing parameters that control the trade-off between the losses.

3D Bounding Boxes. The object detection network outputs oriented 2D bounding boxes. In order to estimate the height of an object, maximum height of the points contained in the 2D bounding box is considered. This height is extracted directly from the BEV representation. To avoid wrong height estimations due to some extra points that may be on top of an object, a maximum height is defined for the object's class.

3.2 FeatExt Operation

Once the point cloud is transformed to the BEV representation, the information of the position of the objects in the scene, including the distance to the capturing sensor, is preserved in the form of pixel coordinates. However, most times this information is not used. When a fully convolutional network is applied, weight sharing guarantees that the same filters are applied to different input locations. This way, a network looks for the same specific features that represent an object in all those positions. Then, thanks to the translation equivariance property of the convolutions (a translated input produces a translated output) and the partial translation invariance that the pooling operations guarantee, an object can be found anywhere in the input. This means, that we look for specific features that are representative for an object class no matter where the object is in the input data. However, this is not valid for LiDAR point clouds. This can be appreciated in Fig. 2, where 3 car representations extracted from a point cloud are displayed. Each car is captured at a different distance from the LiDAR, and that affects the point distribution, density and distance between points.

In addition, operations like ROI pooling crop part of the feature maps to keep processing them, after discarding all spatial location reference. This information is decisive for the network to estimate if a group of points contains an object of interest or not.

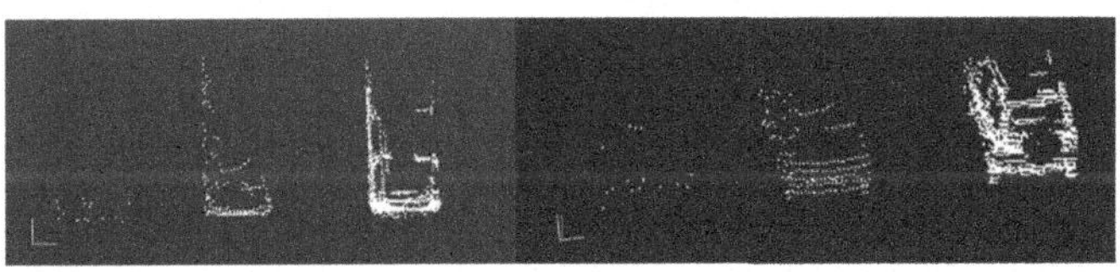

Fig. 2. Example of 3 car representations in LiDAR (Velodyne HDL-64E) point cloud at 41 m, 22 m and 8 m (from left to right) from the sensor. Left image shows cars in top view and right image same cars with certain perspective.

Inspired by [18], we introduce the FeatExt operation. FeatExt allows the filters to know how far a region is from a reference point. Adding this, allows the network to decide if the translation equivariance property should be kept or discarded. FeatExt is implemented by extending a group of feature maps with an extra channel that contains the distance to a specific position of interest in the scene, in this case the LiDAR sensor. Figure 3 depicts FeatExt (left) and the distance channel (right) for a BEV representation that assumes the LiDAR at the top middle. Each pixel in this distance matrix contains the radial distance d (meters) to the sensor. This is computed with the following equation:

$$
\begin{aligned}
x &= \frac{X_{max} - X_{min}}{W} * (i - \frac{W_f}{2}) \\
y &= \frac{Y_{max} - Y_{min}}{H_f} * j \\
d &= \sqrt{x^2 + y^2}
\end{aligned} \tag{2}
$$

where distances in lateral and longitudinal axis (x, y) are computed based on the pixel coordinate (i, j), the considered maximum and minimum point cloud range in meters (X_{max} and X_{min} laterally, Y_{max} and Y_{min} longitudinally) and the feature maps dimension to which the distance channel is added (H_f and W_f). This matrix is concatenated to the group of feature maps channel-wise.

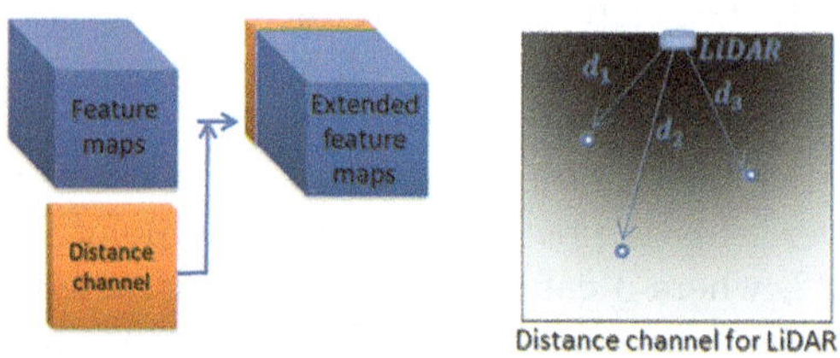

Fig. 3. FeatExt operation (left) and distance channel (right) from BEV perspective.

3.3 Fusion Design

Where to apply FeatExt operation is not a trivial issue. FeatExt fuses information from different feature spaces. These features need to be combined in a way that guarantees that the model is able to learn the relation between them, rather than discard the aggregated spatial information. Networks using multi-modal data fusion have explored different levels of abstraction where data can be fused for different tasks [19,20]. Following this, our work analyses the effect of adding FeatExt in an early fusion and an intermediate fusion stage.

Early fusion schemes integrate all the data into the input feature vector before feeding it to the neural network. For this integration we concatenate the distance channel to the BEV representation channels. Therefore, the input matrix for the model is W_{bev} x H_{bev} x 4, where W_{bev} and H_{bev} are the width and height of the BEV representation respectively. The pipeline for this low-level fusion is shown in Fig. 4 (left).

Intermediate fusion is about learning a shared representation of the data gradually. Based on this idea, our second proposal is to insert the FeatExt operation in a more progressive way. Figure 4 (right) shows the proposed architecture.

Features extracted by the backbone network are fed to the RPN, as explained in Sect. 3.1. Once the object candidates are computed, FeatExt extends the feature maps before applying the ROI pooling operation. This way, the pooled features of all the proposals contain the distance information before being fed to the second stage of the network.

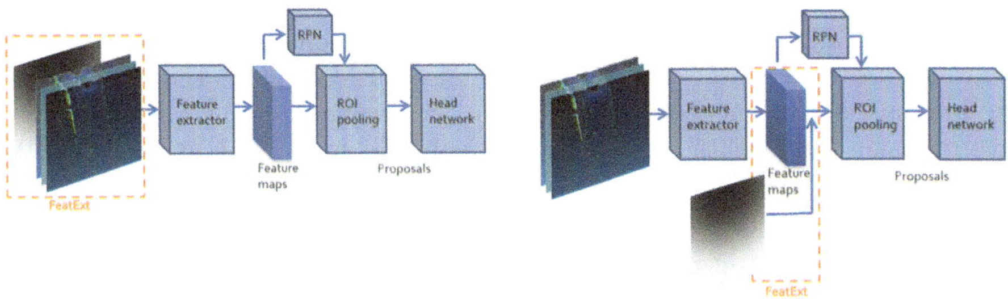

Fig. 4. FeatExt operation integrated in the baseline network in an early fusion (left) and intermediate fusion (right) manner.

4 Implementation

For the BEV representation we use point cloud within the range of $[-1.75, 1.25] \times [0, 70] \times [-35, 35]$ m, along Z, Y, X axis respectively, being the LiDAR installed at 1.73 m from the floor. We use a discretization resolution of 0.1 m laterally and longitudinally and 1 m vertically, which results in a BEV representation of $700 \times 700 \times 3$. Anchors of 45×20 pixels with sixteen orientations are used for cars, which is based on their real size in the BEV representation.

Pretrained weights on ImageNet data [21] are used to train the feature extractor. In the early fusion approach, as input depth contains an extra channel, the mean of the pretrained weights has been computed and concatenated so it can be applied to the fourth channel in the first convolutional layer. The head network is trained from scratch.

Regarding the loss functions, for the smooth L1 loss we use $\sigma = 3$ for the RPN network and $\sigma = 1$ for the head networks, as in Faster R-CNN [16]. λ_{rpn_c}, λ_{rpn_r}, λ_{head_c} and λ_{head_r} are empirically set to 2, 0.15, 4 and 2 respectively.

Networks are trained with a learning rate of 0.0003 and a decay factor of 3 is applied at $190k$ and $230k$ steps. Stochastic gradient descent with momentum of 0.9 is used for the optimization. Weight decay of 0.0001 is applied to prevent overfitting. Networks are trained on a Nvidia Tesla V100 GPU.

5 Experiments and Results

We test our proposal's performance on the BEV object detection task of the KITTI benchmark [15]. Each point cloud comes in a binary file containing the captured points, which are represented by their cartesian coordinates (x, y, z) and intensity value. We split the dataset, which contains 7481 samples, in training (50%), validation (25%) and testing (25%). The KITTI BEV benchmark considers three main classes (cars, pedestrians and cyclists). However, the car class contains more than half of all objects. In our experiments only cars are considered, as it is the only class with enough data to validate our proposal. For evaluation, we follow the easy, moderate and hard difficulty classification proposed by KITTI. We evaluate and compare three versions of the object detection network: baseline, FeatExt on baseline by early fusion and FeatExt on baseline by intermediate fusion (Sect. 3.3). Results are evaluated using the BEV average precision (AP) at 0.7 and 0.5 intersection over union (IoU) thresholds. We compare our proposals to the baseline network on the test set (Table 1).

Table 1. Results of trained models (baseline, FeatExt by early fusion, FeatExt by intermediate fusion) on KITTI BEV benchmark for 0.7 and 0.5 IoU thresholds.

Method	Easy	Moderate	Hard
Baseline (0.7)	79.5	73.1	66.6
Baseline+FeatExt (early) (0.7)	75.9	73.2	66.5
Baseline+FeatExt (intermediate) (0.7)	**84.4**	**75.7**	**75.5**
Baseline (0.5)	89.5	88.4	80.3
Baseline+FeatExt (early) (0.5)	89.4	88.2	80.3
Baseline+FeatExt (intermediate) (0.5)	**96.8**	**89.4**	**81**

It can be seen that adding the FeatExt operation in any abstraction level does not guarantee an improvement. Indeed, early fusion model provides a similar result to the baseline, even slightly worse in some cases. The reason may be that first layers are looking for basic patterns on the input data and are not able to relate the information added in a so low level stage. However, when FeatExt is added in an intermediate fusion fashion, it provides an important boost for all difficulties. The improvements in AP with IoU threshold of 0.7 (easy: 4.9, moderate: 2.6, hard: 8.9) and 0.5 (easy: 7.3, moderate: 1, hard: 0.7) indicate that the model is able to detect much more accurately cars, no matter the difficulty, and that the model is able to distinguish more robustly if a group of points is a car, specially for the easy category.

Figure 5 shows a qualitative comparison of the baseline and the baseline with FeatExt as intermediate fusion models. A car extracted from a point cloud at 8 m from the LiDAR is synthetically located in the point cloud at 8 m, 18 m, 28 m, 38 m, 48 m and 58 m. The images correspond to the inference result on the BEV representation (top row). The bottom row shows the detected cars in the 3D point cloud. Figure 5 (left) shows that baseline model detects the car in all the positions, even if it is not possible to have that point distribution in so a far distance as 58 m. The model with FeatExt, detects the car as far as 38 m, but farther points are discarded.

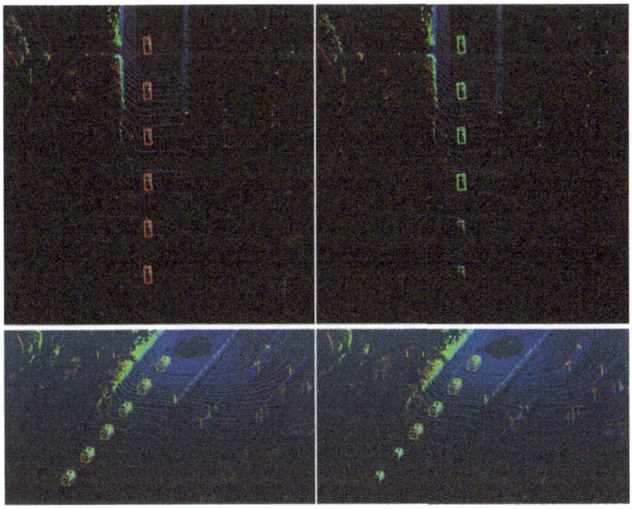

Fig. 5. Inference results on a BEV image (top row) that contains a car synthetically positioned at 8 m, 18 m, 28 m, 38 m, 48 m and 58 m from LiDAR. Left image corresponds to baseline result and right image to baseline with FeatExt as intermediate fusion. Bottom row shows the detected cars in the 3D point cloud.

6 Conclusions

In this work we introduce FeatExt operation, which aggregates information regarding the LiDAR position to BEV object detection networks. We evaluate two fusion alternatives for its integration in a baseline network and their effect on the model performance. Our experiments on the KITTI car detection task show the importance of the way information from different feature spaces is handled. Adding the distance information to any depth level does not guarantee the model learns to relate it to the features extracted from the BEV image. The results also show that when FeatExt is added by intermediate fusion, the AP boost is very remarkable (from 0.7 up to 8.9 AP improvement). Consequently, the model is able to use the aggregated data in a complementary manner. Our proposal can be easily added to existing BEV object detectors to boost their performance. In addition, the same idea can be extended to other tasks and fields, where data features are strongly related to already known specific spatial locations.

Future work includes further improvement of the robustness and reliability of the 3D object detector. To this end, we plan to extend the input data sources to exploit the complimentary data that different sensor modalities offer. RGB cameras could provide complementary object texture data, but other sensors such as Radars would also be suitable to boost the robustness of the detections in a safety-critical scenario such as automotive. As shown in this work, the way data are fused from different feature spaces will be crucial to learn the intermodality relationships.

References

1. Guo, Y., Liu, Y., Oerlemans, A., Lao, S., Wu, S., Lew, M.S.: Deep learning for visual understanding: a review. Neurocomputing **187**, 27–48 (2016)
2. Zhao, Z.Q., Zheng, P., Xu, S.T., Wu, X.: Object detection with deep learning: a review. IEEE Trans. Neural Netw. Learn. Syst. **30**(11), 3212–3232 (2019)
3. Li, B., Zhang, T., Xia, T.: Vehicle detection from 3d lidar using fully convolutional network. arXiv preprint, arXiv:1608.07916 (2016)
4. Wu, B., Wan, A., Yue, X., Keutzer, K.: Squeezeseg: convolutional neural nets with recurrent crf for real-time road-object segmentation from 3d lidar point cloud. In: 2018 IEEE ICRA, pp. 1887–1893. IEEE (2018)
5. Beltrán, J., Guindel, C., Moreno, F.M., Cruzado, D., Garcia, F., De La Escalera, A.: Birdnet: a 3d object detection framework from lidar information. In: 2018 21st International Conference on ITSC, pp. 3517–3523. IEEE (2018)
6. Chen, X., Ma, H., Wan, J., Li, B., Xia, T.: Multi-view 3d object detection network for autonomous driving. In: Proceedings of the IEEE Conference on Computer Vision and Pattern Recognition, pp. 1907–1915 (2017)
7. Yang, B., Luo, W., Urtasun, R.: Pixor: real-time 3d object detection from point clouds. In: Proceedings of the IEEE conference on Computer Vision and Pattern Recognition, pp. 7652–7660 (2018)
8. Qi, C.R., Su, H., Mo, K., Guibas, L.J.: Pointnet: deep learning on point sets for 3d classification and segmentation. In: Proceedings of the IEEE Conference on Computer Vision and Pattern Recognition, pp. 652–660 (2017)
9. Zhou, Y., Tuzel, O.: Voxelnet: end-to-end learning for point cloud based 3d object detection. In: Proceedings of the IEEE Conference on Computer Vision and Pattern Recognition, pp. 4490–4499 (2018)
10. Qi, C.R., Liu, W., Wu, C., Su, H., Guibas, L.J.: Frustum pointnets for 3d object detection from RGB-D data. In: Proceedings of the IEEE Conference on Computer Vision and Pattern Recognition, pp. 918–927 (2018)
11. Yan, Y., Mao, Y., Li, B.: Second: sparsely embedded convolutional detection. Sensors **18**(10), 3337 (2018)
12. Shi, S., Wang, X., Li, H.: Pointrcnn: 3d object proposal generation and detection from point cloud. In: Proceedings of the IEEE Conference on Computer Vision and Pattern Recognition, pp. 770–779 (2019)
13. Lang, A.H., Vora, S., Caesar, H., Zhou, L., Yang, J., Beijbom, O.: PointPillars: fast encoders for object detection from point clouds. In: Proceedings of the IEEE Conference on Computer Vision and Pattern Recognition, pp. 12697–12705 (2019)
14. Lin, T.Y., Dollár, P., Girshick, R., He, K., Hariharan, B., Belongie, S.: Feature pyramid networks for object detection. In: Proceedings of the IEEE conference on computer vision and pattern recognition, pp. 2117–2125 (2017)
15. Geiger, A., Lenz, P., Urtasun, R.: Are we ready for autonomous driving? The kitti vision benchmark suite. In: 2012 IEEE Conference on Computer Vision and Pattern Recognition, pp. 3354–3361. IEEE, June 2012
16. Girshick, R.: Fast r-cnn. In: Proceedings of the IEEE ICCV, pp. 1440–1448 (2015)
17. He, K., Zhang, X., Ren, S., Sun, J.: Deep residual learning for image recognition. In: Proceedings of the IEEE CVPR, pp. 770–778 (2016)

18. Liu, R., Lehman, J., Molino, P., Such, F. P., Frank, E., Sergeev, A., Yosinski, J.: An intriguing failing of convolutional neural networks and the coordconv solution. In: Advances in Neural Information Processing Systems, pp. 9605–9616 (2018)
19. Liu, J., Zhang, S., Wang, S., Metaxas, D. N.: Multispectral deep neural networks for pedestrian detection. arXiv preprint, arXiv:1611.02644 (2016)
20. Ngiam, J., Khosla, A., Kim, M., Nam, J., Lee, H., Ng, A. Y.: Multimodal deep learning. In: Proceedings of the 28th ICML-11, pp. 689–696 (2011)
21. Deng, J., Dong, W., Socher, R., Li, L. J., Li, K., Fei-Fei, L.: Imagenet: a large-scale hierarchical image database. In: 2009 IEEE CVPR, pp. 248–255. IEEE (2009)

A Comparison of Registration Methods for SLAM with the M8 Quanergy LiDAR

Marina Aguilar-Moreno(✉) and Manuel Graña(✉)

Computational Intelligence Group, University of the Basque Country (UPV/EHU), San Sebastian, Spain
marina.aguilar@ehu.eus, manuel.grana@ehu.es

Abstract. LiDAR based SLAM is becoming affordable by new sensors such as the M8 Quanergy LiDAR, but there is still little work reporting on the accuracy attained with them. In this paper we report on the comparison of three registration methods applied to the estimation of the path followed by the LiDAR sensor and the registration of the overall cloud of points, namely the iterated closest points (ICP), Coherent Point Drift (CPD), and Normal Distributions Transform (NDT) registration methods. In our experiment, we found that the NDT method provides the most robust performance.

Keywords: Point cloud registration · LiDAR · SLAM

1 Introduction

The simultaneous localization and mapping (SLAM) aims to estimate a reconstruction of the environment along with the path traversed by the sensor has become an integral part of the robotic operating system (ROS) [13,14]. One of the most widely used kinds of sensors used for SLAM are laser based depth measurement sensors, or light detection and ranging (LiDAR) sensors, which have been used for scanning and reconstruction of indoor and outdoor environments [3], even in underground mining vehicles [12]. Fusion of LiDAR with GPS allows for large scale navigation [4] of autonomous systems.

New affordable LiDAR sensors, such as the M8 from Quanergy that we are testing in this paper, allow for further popularization of LiDAR based SLAM applications. Due to its specific innovative characteristics, the M8 sensor still needs extensive testing by the community in order to assume its integration in the newly developed systems [9]. The work reported in this paper is intended partly to provide such empirical confirmation of the M8 sensor quality. We have not carried out any precise calibration process of the sensor [5,6]. Instead, we are assessing the sensor through the comparison of three standard point cloud registration methods over experimental data gathered in-house.

This paper is structured as follow: A brief presentation of the environment where experiment was carried out and the LiDAR sensor used in it, Quanergy

Á. Herrero et al. (Eds.): SOCO 2020, AISC 1268, pp. 824–834, 2021.
https://doi.org/10.1007/978-3-030-57802-2_79

M8. Next, the mathematical description of the three 3D registration methods used in the paper: Iterative Closest Point (ICP), Coherent Point Drift (CPD) and Normal Distribution Transform (NDT). Then, the algorithm developed to register LiDAR data with the three methods and reconstruct an indoor surface. Finally, experimental results are presented for each registration method and a comparative between them in terms of root mean square error of the Euclidean distance, path obtained and resulting surface.

2 Materials

Both the time sequence of M8 captured point clouds and the Matlab code used to carry out the computational experiments has been published as open data and open source code[1] in the Zenodo repository for reproducibility.

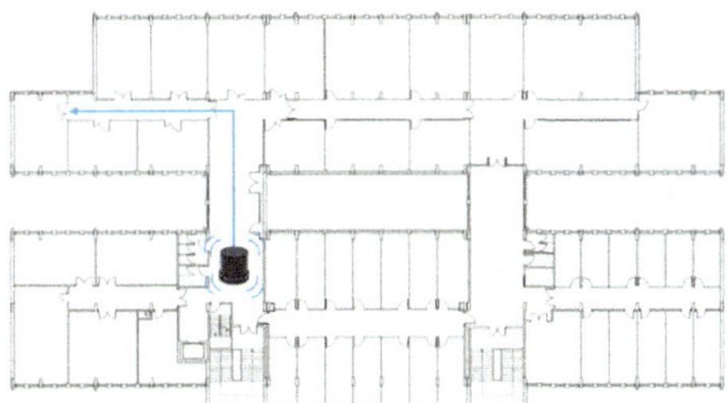

Fig. 1. Nominal path followed during the LiDAR recording.

Location and Experiment Setting. The experiment was carried out in the third floor of the Computer Science School of the UPV/EHU in San Sebastian. Figure 1 shows the nominal path followed by the M8 LiDAR on a manually driven mobile platform. The actual path shows small perturbations around the nominal path. We do not have a precise actual path measurement allowing to quantify the error in the trajectory.

LiDAR M8 Quanergy. The Quanergy M8 LiDAR sensor is a multi-laser system with 8 2D line scanners located on a spinning head. The Fig. 2 shows the M8 Quanergy LiDAR physical aspect and some of its specifications. This system is based on Time-of-Flight (TOF) technology whose spin rate is between 5 Hz and 20 Hz and its maximum range is 100 m. The Table 1 shows the M8 LiDAR main parameters. Besides, M8 LiDAR comes with 2 desktop applications to manage and visualize point clouds, a SDK to record and show data in real time, and a SDK in framework ROS.

[1] http://doi.org/10.5281/zenodo.3633727.

Table 1. Quanergy M8 sensor specifications

Parameter	M8 sensor specifications
Detection layers	8
Returns	3
Minimum range	0.5 m (80% reflectivity)
Maximum range	100 m (80% reflectivity)
Spin rate	5 Hz–20 Hz
Intensity	8-bits
Field of view	Horizontal 360° – Vertical 20° (+3°/−17°)
Data outputs	Angle, Distance, Intensity, Synchronized Time Stamps

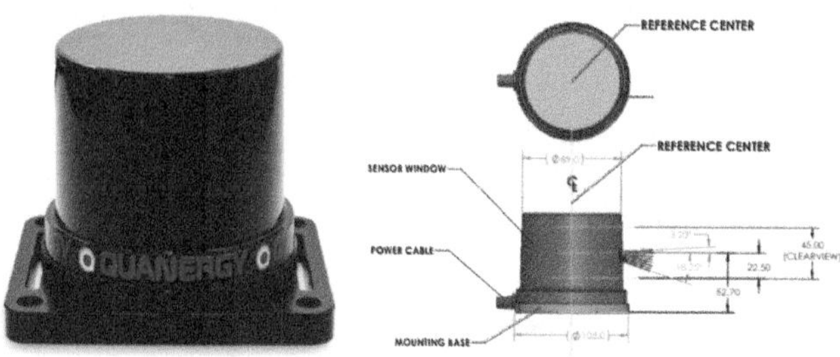

Fig. 2. The M8 Quanergy LiDAR and diagrammatic specs.

3 Point Cloud Registration Methods

Point cloud registration methods are composed of two steps: (a) finding the correspondence between points in one cloud (the moving) to the points in the other cloud (the reference), and (b) the estimation of the motion parameters that achieve optimal match of the moving points to the reference points after correcting for the motion. If the motion is modeled by a rigid body or an affine transformation, then a matrix transformation common to all points is estimated. If the motion is some non linear deformation, then we have to estimate a flow field. In this paper we are restricted to rigid body transformations, which are compositions of a translation and a rotation. The transformation estimation process takes the form of a minimization problem where the energy function is related to the quality of the correspondence achieved. Next we recall the basics of the three point cloud registration methods.

3.1 ICP

The most popular and earliest point cloud registration method is the Iterative Closest Point (ICP) proposed by Besl in 1992 [1]. This technique has been exploited in many domains, giving rise to a host of variations whose relative merits are not so easy to assess [11]. Given a point cloud $P = \{\mathbf{p}_i\}_{i=1}^{Np}$ and a shape described by another point cloud $X = \{\mathbf{x}_i\}_{i=1}^{Nx}$ (The original paper includes the possibility to specify other primitives such as lines or triangles with well defined distances to a point, but we will not consider them in this paper.) the least squares registration of P is given by $(\mathbf{q}, d) = \mathcal{Q}(P, Y)$, where $Y = \{\mathbf{y}_i\}_{i=1}^{Np}$ is the set of nearest points from X to the points in P, i.e. $\mathbf{p}_i \in P; \mathbf{y}_i = \arg\min_{\mathbf{x}\in X} \|\mathbf{x} - \mathbf{p}_i\|^2$, denoted $Y = \mathcal{C}(P, X)$, and operator $\mathcal{Q}$ is the least squares estimation of the rotation and translation mapping P to Y using quaternion notation, thus $\mathbf{q} = [\mathbf{q}_R \mid \mathbf{q}_T]^t$ is the optimal transformation specified by a rotation quaternion $\mathbf{q}_R$ and a translation $\mathbf{q}_T$, and d is the registration error. The energy function minimized to obtain the optimal registration is $f(\mathbf{q}) = \frac{1}{N_p}\sum_{i=1}^{Np} \|\mathbf{y}_i - \mathbf{R}(\mathbf{q}_R)\mathbf{p}_i - \mathbf{q}_T\|^2$, where $\mathbf{R}(\mathbf{q}_R)$ is the rotation matrix constructed from quaternion $\mathbf{q}_R$. The iteration is initialized by setting $P_0 = P$, $\mathbf{q}_0 = [1, 0, 0, 0, 0, 0, 0]^t$, and $k = 0$. The algorithm iteration is as follows: (1) compute the closest points $Y_k = \mathcal{C}(P_k, X)$, (2) compute the registration $(\mathbf{q}_k, d_k) = \mathcal{Q}(P_0, Y_k)$, (3) apply the registration $P_{k+1} = \mathbf{q}_k(P_0)$, and (4) terminate the iteration if the results are within a tolerance: $d_k - d_{k+1} < \tau$.

3.2 CPD

The Coherent Point Drift (CPD) [7,10] registration method considers the alignment of two point sets as a probability density estimation problem. The first point set $X = \{\mathbf{x}_i\}_{i=1}^{N}$ is considered the data samples generated from the Gaussian mixture model (GMM) whose centroids are given by the second point set $Y = \{\mathbf{y}_i\}_{i=1}^{N}$. Therefore, the CPD registration tries to maximize the likelihood X as a sample of the probability distribution modeled by Y after the application of the transformation $T(Y, \theta)$, where θ are the transformation parameters. The GMM model is formulated as $p(\mathbf{x}) = \omega\frac{1}{N} + (1-\omega)\sum_{m=1}^{M}\frac{1}{M}p(\mathbf{x}\,|m)$ assuming a uniform distribution for the *a priori* probabilities $P(m) = \frac{1}{M}$, and adding an additional uniform distribution $p(\mathbf{x}\,|M+1) = \frac{1}{N}$ to account for noise and outliers. All Gaussian conditional distributions are isotropic with the same variance σ^2, i.e. $p(\mathbf{x}\,|m) = \left(2\pi\sigma^2\right)^{-D/2}\exp\left(\frac{\|\mathbf{x}-\mathbf{y}_m\|^2}{2\sigma^2}\right)$. The point correspondence problem is equivalent to selecting the centroid $\mathbf{y}_m$ with maximum *a posteriori* probability $P(m\,|\mathbf{x}_n)$ for a given sample point $\mathbf{x}_n$. The CPD tries to minimize the negative log-likelihood $E\left(\theta, \sigma^2\right) = -\sum_{n=1}^{N}\log\sum_{m=1}^{M}P(m)\,p(\mathbf{x}\,|m)$ by an expectation-maximization (EM) algorithm. The E step corresponds to solving the point correspondence problem using the old parameters, by computing the *a posteriori* probabilities with the old parameters $P^{old}(m\,|\mathbf{x}_n)$. Let $p_{n,m}^{old} = \exp\left(-\frac{1}{2}\left\|\frac{\mathbf{x}_n - T\left(\mathbf{y}_n, \theta^{old}\right)}{\sigma^{old}}\right\|\right)$, then $P^{old}(m\,|\mathbf{x}_n) = p_{n,m}^{old}\left(\sum_{k=1}^{M}p_{k,m}^{old} + c\right)^{-1}$. The M step is the estimation of the new parameters minimizing the complete negative log-likelihood $Q = -\sum_{n=1}^{N}\sum_{m=1}^{M} P^{old}(m|\mathbf{x}_n)\log\left(P^{new}(m)\,p^{new}(\mathbf{x}|m)\right)$. For rigid transformations, the objective function takes the shape: $Q\left(\mathbf{R},\mathbf{t},s,\sigma^2\right) = \frac{1}{2\sigma^2}\sum_{n,m=1}^{N,M} P^{old}(m|\mathbf{x}_n)\left\|\mathbf{x}_n - s\mathbf{R}\mathbf{y}_m - \mathbf{t}\right\|^2 + \frac{N_p D}{2}\log\sigma^2$ such that $\mathbf{R}^T\mathbf{R} = \mathbf{I}$ and $\det(\mathbf{R}) = 1$. Closed forms for the transformation parameters are given in [10].

3.3 NDT [2]

The key difference of this method is the data representation. The space around the sensor is discretized into regular overlapped cells. The content of each cell having more than 3 points is modelled by a Gaussian probability distribution of mean $\mathbf{q} = \frac{1}{\mathrm{n}}\sum_i \mathbf{x}_i$ and covariance matrix $\Sigma = \frac{1}{n-1}\sum_i(\mathbf{x}_i - \mathbf{q})(\mathbf{x}_i - \mathbf{q})^t$, so that the probability of a LiDAR sample falling in the cell is of the form: $p(\mathbf{x}) \sim \exp\left(-\frac{1}{2}\right)(\mathbf{x}-\mathbf{q})\Sigma^{-1}(\mathbf{x}-\mathbf{q})$. Given an initial rigid body transformation $T(\mathbf{x};\mathbf{p}_0)$, where $\mathbf{p}$ is the vector of translation and rotation parameters, a reference point cloud $\{\mathbf{x}_i\}$ modelled by the mixture of the cells Gaussian distributions, and the moving point cloud $\{\mathbf{y}_i\}$, the iterative registration process is as follows: the new laser sample points $\mathbf{y}_i$ are transformed into the reference frame of the first cloud $\mathbf{y}'_i = T(\mathbf{y}_i;\mathbf{p}_{t-1})$, where we find the cell where it falls and use its parameters $(\mathbf{q},\Sigma)$ to estimate its likelihood $p(\mathbf{y}'_i)$. The score of the transformation is given by $score(\mathbf{p}) = \sum_i p(\mathbf{y}'_i)$. The maximization of the score is carried out by gradient ascent using Newton's method, i.e. $\mathbf{p}_t = \mathbf{p}_{t-1} + \triangle\mathbf{p}$. The parameter update is computed solving the equation $\mathbf{H}\triangle\mathbf{p} = -\mathbf{g}$, where $\mathbf{H}$ and $\mathbf{g}$ are the Hessian and the gradient of the $-score(\mathbf{p}_{t-1})$ function, respectively. Closed forms of $\mathbf{H}$ and $\mathbf{g}$ are derived in [2] for the 2D case. An extension to 3D is described in [8].

4 Registration and SLAM Algorithm

Figure 3 presents a flow diagram of the general algorithm that we have applied to obtain the registration of the LiDAR point clouds recorded at each time point $t = \{1,\ldots,T\}$ while the sensor is being displaced manually in the environment according to the approximate path in Fig. 1. The final result of the process is a global point cloud $M(T)$ that contains all the recorded 2D points registered relative to the first acquired point cloud $N(0)$, and the estimation of the LiDAR recording positions relative to the initial position. These recording positions are given by the composition of the point cloud registration transformations estimated up to this time instant. The trajectories displayed below all start from the XY plane origin for this reason. The process is as follows: For each acquired point cloud $N(t)$ at time t, firstly we remove the ego-vehicle points denoting $N(1)(t)$ the new point cloud. Secondly we remove the ground plane applying a threshold on the height, obtaining $N(2)(t)$. Thirdly, we downsample the point cloud to decrease the computation time and improve accuracy registration, obtaining $N(3)(t)$. For the initial point cloud at $t = 0$, $N(3)(t)$ becomes the global merged cloud $M(0)$. For subsequent time instants $t > 0$, the fourth step is to estimate the transformation T_t of the acquired data $N(3)(t)$ to the previous global point cloud $M(t-1)$. For this estimation, we use any of the registration algorithms described above to register $T_{t-1}(N(3)(t))$ to $M(t-1)$ obtaining T_t. We then apply this transformation to the acquired point cloud previous to downsampling $N(4)(t) = T_t(N(2)(t))$, which is used to obtain the new global registered point cloud by merging $M(t) = merge(M(t-1), N(4)(t))$.

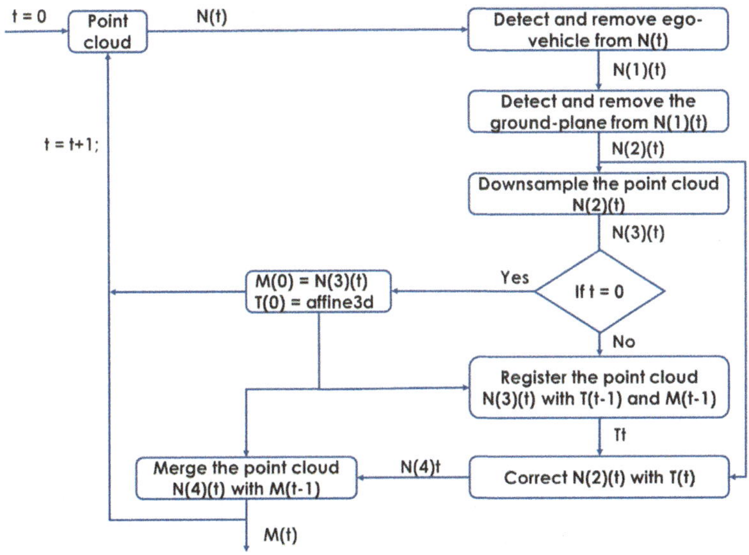

Fig. 3. Flow diagram of the registration algorithm. N(i)(t) is the point cloud at time t after the i-th step of processing. M(t) is the overall point cloud up after merging all the registered point clouds processed up to time t.

5 Results

Figure 4 shows the evolution of the registration error of the SLAM algorithm described in this article for different registration method: ICP, CPD and NDT. The point clouds used are recorded along the path shown in Fig. 1. The plot is logarithmic scale in order to be able to represent the three error plots in the same scale. The NDT algorithm gives the minimal error all along the path. The

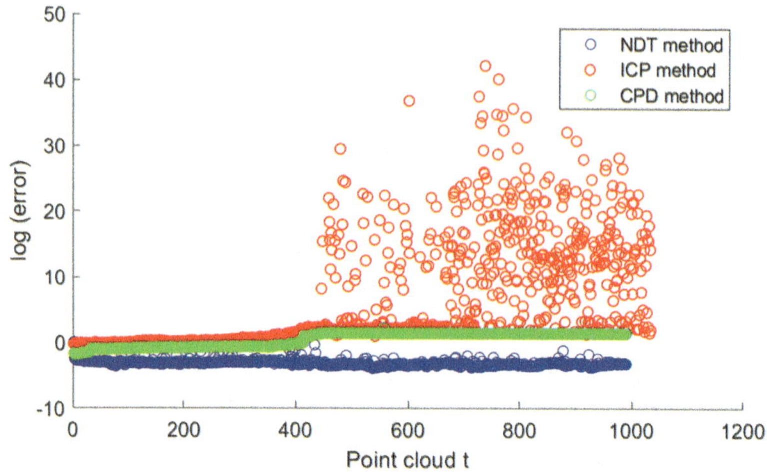

Fig. 4. Evolution of the registration error (log plot) for NDT (blue dots), CPD (green dots), and ICP (red dots).

error of both NDT and CPD registration methods remains bounded, however the error of the ICP method explodes after a point in the trajectory, specifically the turning point at the end of the main hallway in Fig. 1. Figure 5(right) shows the overall point cloud obtained at the end of the SLAM process, and the estimated trajectory (white points). After some point in the trajectory, the ICP registration loses track and gives random looking results. Figure 5(right) shows the results of the ICP registration up to the turning point, which are comparable with the results of the other algorithms. Figure 6(right) shows the results of the CPD algorithm in terms of the registered and merged overall cloud of points and the trajectory estimation (white points). It can also be appreciated that the SLAM process gets lost after the path turning point, however the registration of point clouds does not become unwieldy. Finally, Fig. 7(up) shows the results of the NDT algorithm. The trajectory (white points) is quite accurate to the actual path followed by the sensor. The trajectory turning point was in fact as smooth as shown in the figure. The overall registered and merged point cloud has a nice fit of the actual hallway walls, as can be appreciated in Fig. 7(bottom), including a communication switch closet signaled in the figure with an arrow, that is not present in the original floor plan.

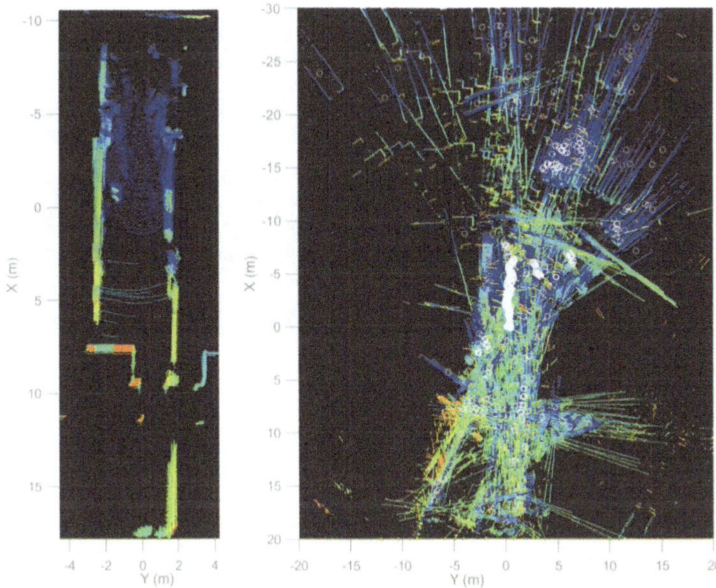

Fig. 5. Estimated trajectory (white points) and registered cloud of points using ICP (right). Registration of the cloud points before reaching the turning point (left).

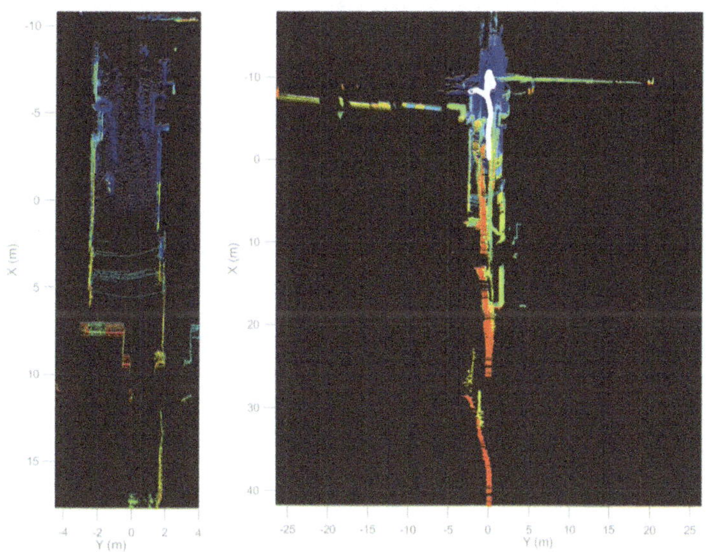

Fig. 6. Estimated trajectory (white points) and registered cloud of points using CPD (right). Registration of the cloud points before reaching the turning point (left).

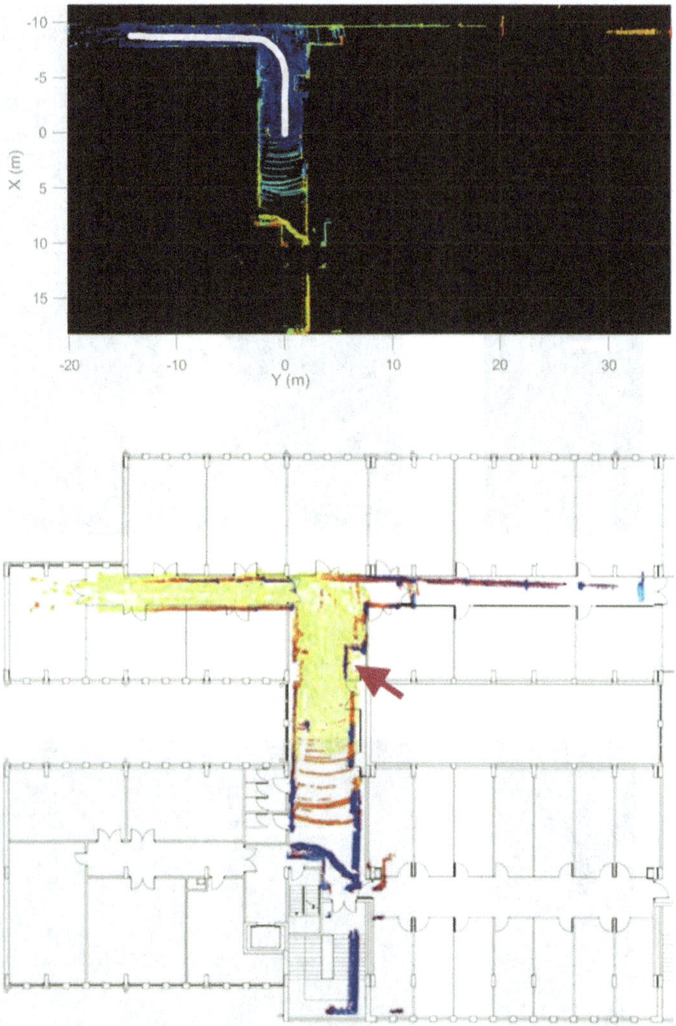

Fig. 7. Estimated trajectory (white points) and registered cloud of points using NDT (Above). Projection of the NDT registered point cloud on the plan of stage 3 of the building.

6 Conclusion

In this paper we report a comparison between three registration methods for 3D point clouds, namely the Iterative Closest Point (ICP), Coherent Point Drift (CPD) and Normal Distributions Transform (NDT). To collect point sets, we have located the M8 Quanergy LiDAR sensor on a manually driven mobile platform through the third floor of the Computer Science School of the UPV/EHU in San Sebastian. The registration algorithm followed in this paper includes preprocessing (detect and remove ego-vehicle and floor, and downsample), registration, transformation and merger point cloud. For each method described in this paper, we have obtained the registration error, the estimation of the path traversed by the sensor, and the reconstructed point cloud. For the ICP and CPD methods, the error is larger than for the NDT method. Besides, after the turning point in the nominal path, ICP and CPD obtained path and resulting point cloud are incorrect. NDT registration obtains coherent experimental results and an accurate trajectory compared with the nominal path followed.

Future works would be to combine the three methods described in this paper to obtain a better result than obtained separately.

Acknowledgments. This work has been partially supported by FEDER funds through MINECO project TIN2017-85827-P, and grant IT1284-19 as university research group of excellence from the Basque Government.

References

1. Besl, P.J., McKay, N.D.: A method for registration of 3-D shapes. IEEE Trans. Pattern Anal. Mach. Intell. **14**(2), 239–256 (1992)
2. Biber, P., Straßer, W.: The normal distributions transform: a new approach to laser scan matching, vol. 3, pp. 2743–2748, November 2003
3. Caminal, I., Casas, J.R., Royo, S.: SLAM-based 3D outdoor reconstructions from LIDAR data. In: 2018 International Conference on 3D Immersion (IC3D), pp. 1–8, December 2018
4. Deng, Y., Shan, Y., Gong, Z., Chen, L.: Large-scale navigation method for autonomous mobile robot based on fusion of GPS and lidar SLAM. In: 2018 Chinese Automation Congress (CAC), pp. 3145–3148, November 2018
5. Levinson, J., Thrun, S.: Robust vehicle localization in urban environments using probabilistic maps. In: 2010 IEEE International Conference on Robotics and Automation, pp. 4372–4378, May 2010
6. Levinson, J., Thrun, S.: Unsupervised Calibration for Multi-beam Lasers, pp. 179–193. Springer, Heidelberg (2014)
7. Lu, J., Wang, W., Fan, Z., Bi, S., Guo, C.: Point cloud registration based on CPD algorithm. In: 2018 37th Chinese Control Conference (CCC), pp. 8235–8240, July 2018
8. Magnusson, M., Lilienthal, A., Duckett, T.: Scan registration for autonomous mining vehicles using 3D-NDT. J. Field Robot. **24**, 803–827 (2007)
9. Mitteta, M.A., Nouira, H., Roynard, X., Goulette, F., Deschaud, J.E.: Experimental assessment of the quanergy M8 LIDAR sensor. In: ISPRS - International Archives of the Photogrammetry, Remote Sensing and Spatial Information Sciences, vol. 41B5, pp. 527–531, June 2016
10. Myronenko, A., Song, X.: Point set registration: coherent point drift. IEEE Trans. Pattern Anal. Mach. Intell. **32**(12), 2262–2275 (2010)
11. Pomerleau, F., Colas, F., Siegwart, R., Magnenat, S.: Comparing ICP variants on real-world data sets. Autonom. Robots **34**(3), 133–148 (2013)
12. Wu, D., Meng, Y., Zhan, K., Ma, F.: A LIDAR slam based on point-line features for underground mining vehicle. In: 2018 Chinese Automation Congress (CAC), pp. 2879–2883, November 2018

13. Xuexi, Z., Guokun, L., Genping, F., Dongliang, X., Shiliu, L.: Slam algorithm analysis of mobile robot based on LIDAR. In: 2019 Chinese Control Conference (CCC), pp. 4739–4745, July 2019
14. Yagfarov, R., Ivanou, M., Afanasyev, I.: Map comparison of LIDAR-based 2D slam algorithms using precise ground truth. In: 2018 15th International Conference on Control, Automation, Robotics and Vision (ICARCV), pp. 1979–1983, November 2018

An Application of Laser Measurement to On-Line Metal Strip Flatness Measurement

Marcos Alonso[1], Alberto Izaguirre[1], Imanol Andonegui[1], and Manuel Graña[2(✉)]

[1] ECS Department, Robotics and Automation Group, Mondragon University, Mondragon, Spain

[2] Computational Intelligence Group, UPV/EHU, San Sebastian, Spain
manuel.grana@ehu.es

Abstract. In this article we discuss the need for metal strip flatness and the state of the art for its measurement, which is of top importance for the metal processing industry. There is a strong pressure for quality that demands on-line measurement that is robust to the perturbations introduced by further processing down the line. We sketch the design of an innovative on-line metal strip flatness measurement device based on the recovery of depth information from two parallel laser projected lines. Preliminary results show its robustness on simulated and real data.

Keywords: Laser measurement · Steel sheet · Rolling mills

1 Introduction

The requirements on the surface quality of rolled sheet metal products are continuously increasing. Figure 1 shows an schematic representation of such large machineries, where a rolled sheet metal is processed to cut pieces for further process. The rolled sheet is unfolded and fed into a chain of rolling mills that flatten it. Further the sheet is feed to cutting station that produces the final sheets. Sheet flatness defects greatly decrease the value of the final product for markets such as architecture panels or the automotive industry. Flatness is the surface evenness of the metal sheet in the unstressed state. The American Society for Testing and Materials (ASTM) provides definitions and procedures for measuring flatness characteristics of steel sheet products so that purchasers and suppliers have common definitions and measuring procedures for flatness characteristics in order to provide common procedure(s) for quantifying flatness anomalies. Specifically, the ASTM defines two methods to standardize flatness measurement in rolled sheet metal products, namely the Steepness Index and the I-Unit [2]. Manual metal sheet flatness measurement methods demands skilled operators to locate flatness deviations and adjust rolling mill settings manually to

Á. Herrero et al. (Eds.): SOCO 2020, AISC 1268, pp. 835–842, 2021.
https://doi.org/10.1007/978-3-030-57802-2_80

correct those deviations. These methods have been replaced by automatic shape measuring devices, which allow for closed control loops. In the late 60s of the past century, an on-line flatness measuring system known as stressometer were introduced [16] measuring the transversal stress distribution in a strip using pressure transducers. Afterwards, shape measurement rolls based on piezoelectric load sensors and air-bearing rotors were developed [1,4,5]. These sensors allow to use the flatness measurements to control roll levelling process and the strip shape with reference to a target profile. However, their use in hot, very thick rolled products or sheets cutting lines is problematic or impossible. Moreover, the sensors could be damaged if the force applied on them exceed the hardware limits otherwise the shape measurement could be incorrect, and they are not suitable for high quality products because they can cause surface scratchs. In the 80s, optical gauges where introduced [8]. These systems are able to measure manifest flatness, i.e. flatness not hidden by tensions, whereas a shape roll relies in the determination of tensile stress, being capable of measure latent flatness. Most commonly used optical flatness inspection systems are usually based on laser triangulation principle. The triangulation principle enables distance measurement on a broad range of different material surfaces. Depending on whether a laser point or a laser line is projected onto the object surface, a one-dimensional or two-dimensional output signal is possible. Other types of optical flatness measuring devices are based in ultra diffuse light or moire pattern projection [6,12,17]. In this work, we present preliminary results of an innovative sensor based on synchronized measurement of two laser markers, we show surface reconstruction results, though the details of the system must be withdraw due to ongoing patent process.

The paper is organized as follows. The optical flatness measurement system is described first. Second, the numerical and experimental methods are presented. Third the results of representative simulations and experimental tests and are described. Finally, the conlcussions of this work are stated.

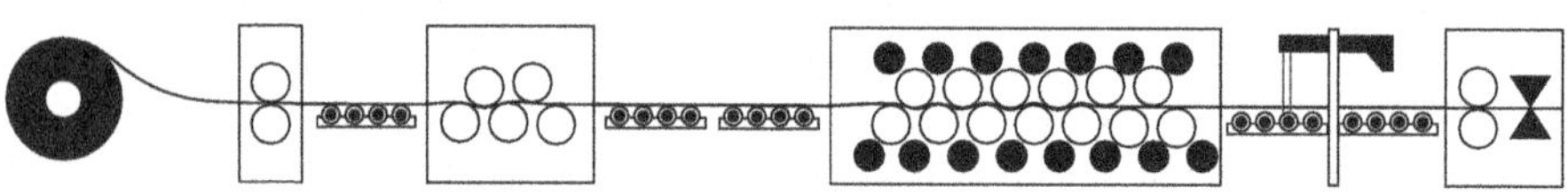

Fig. 1. Industrial levelling and strip cutting process line scheme.

2 Optical Flatness Measurement System

In this section we describe the overall structure of the measurement system with some emphasis on the computational aspects withdrawing details that could impede the patenting process. The rough idea of the sensor is the projection of two parallel lines on the running sheet that may allow to compute the instantaneously gradient on the surface of the metal sheet, i.e. without delay and additional uncertainties due to the motion of the metal sheet in the machine. Figure 2 shown a rough draft of the disposition of the laser projectors and the camera. The inset shows how the detection of the two projected points in the sheet allows the computation of the gradient on the longitudinal direction of the sheet, i.e. the direction of the sheet motion in the machine. We withdraw the operational geometrical and photogrammetric details of the system. The extraction of the laser illuminated points in the image captured by the camera is done by application of straighforward thresholding methods. For increased detection precision, we applied a Savitzky-Golay [14] finite impulse response (FIR) differential filter to the laser intensity profile, and we computed the zero-crossing with sub-pixel accuracy. In essence, the measurement of the two metal sheet height points is simultaneous, so the gradient computation is not affected by vibration and other sources of noise that would affect the gradient computation on the basis of a single laser line and the computational reconstruction from motion parameters. For the experiments the devised sensor has been installed in a industrial levelling and strip cutting process line. This sensor has been placed at the output of a rolling leveller and near the cutting station. The sensor outputs the 3D profile of the metal sheet for each laser using an encoder placed over the metal strip as trigger source. Depth measurements computed from laser-triangulation are synchronized with the motion of the metal strip using an incremental encoder located after the roll leveller stage. This ensures a uniform data acquisition and detection of small jitter in motion, as well as acceleration or de-acceleration.

2.1 3D Sheet Measurement

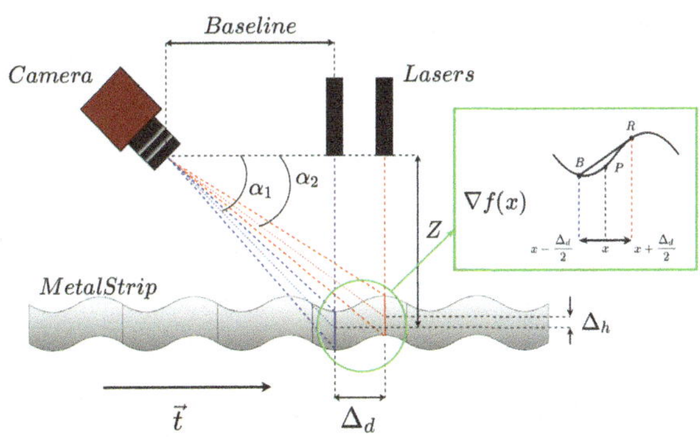

Fig. 2. Local longitudinal direction gradient computation.

Surface estimation from noisy gradient data has been investigated for several years [3, 7, 9, 10]. Several studies such as non-destructive measurements of three-dimensional specular object geometries, ground model surface reconstruction from terrestrial point clouds, optical testing based on phase-measuring deflectometry sensors etc. have take advantage of these research [11, 13]. In the case of

2D data, there exist mainly two different approaches to solve the stated problem [15]. On the one hand, there are "local methods", they integrate along predetermined paths, they are simple, fast, and reconstruct small local height variations quite well. However, they propagate both the measurement error and the discretization error along the path introducing a global shape deviation. On the other hand, there are "global methods", the advantage of global methods is that there is no propagation of the error. In general, it is crucial to note that this reconstruction methods depends on the slope measuring sensor and the properties of the acquired data. We have used a method based on piece-wise cubic Hermite spline interpolation that allows filtering these undesirable noise sources using both surface gradient and height information, details will be provided elsewhere.

3 Results

Simulated Data Results. First we report some simulation based results of our surface reconstruction approach. Figure 3(up) shows a synthetic flat surface with low frequency perturbations that simulate the most common defects in a roll leveller processing line, i.e. center buckles and wavy edges respectively. The middle image shows the effect of noisy detection of the laser projected lines and/or mechanical noise. The bottom image shows the strong impulse due to a cutting operation at the end of the line. This characteristic is very common in final processing lines, where after flattening, the sheet is cut into different sizes for transport and subsequent manufacture in sectors such as the automotive industry. Figure 4 shows some results of the Hermite polynomical based filtering and surface reconstruction. The left images correspond to the filtering of the impulsive noise, while the right images correspond to the filtering of the cutting noise,

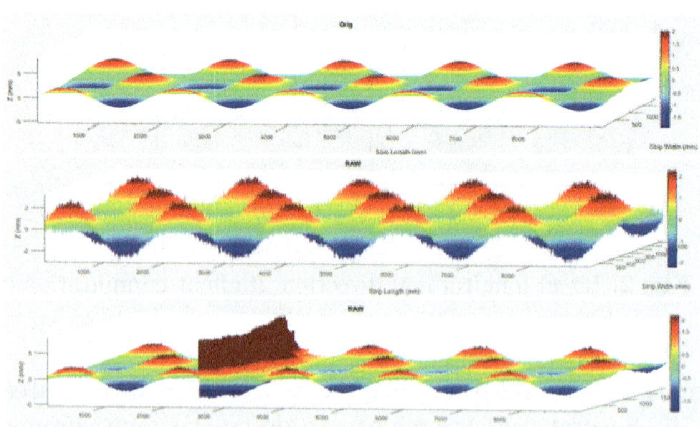

Fig. 3. Synthetic surface (up) Noise free, (middle) corrupted by impulsive noise (bottom) noise due to a cutting impulse.

with different parameter settings. The strongest filtering (below) removes noise but also the low frequency effects that we want to detect. So a fine tuning of the parameters is required for real experimentation and application.

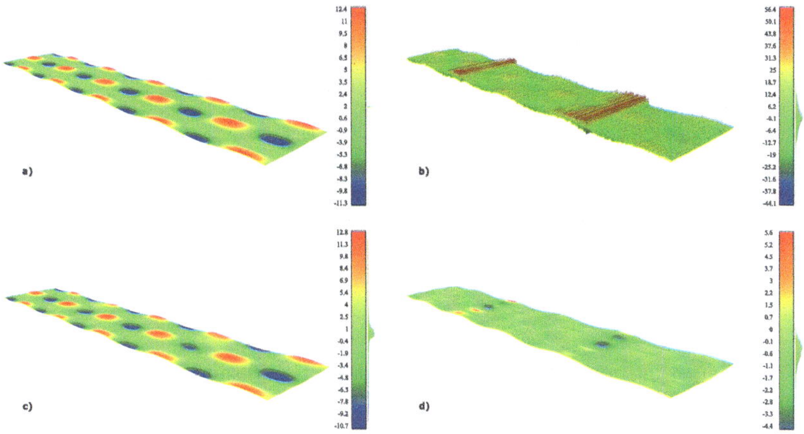

Fig. 4. Synthetic surface results 3D. (left) the removal of the impulsive noise with different cutt-off parameters. (right) the removal of the cutting induced noise.

Real Life Experimental Results. For the real life validation experiments the sensor has been installed in a industrial levelling and strip cutting process line at Fagor site. This sensor has been placed at the output of a rolling leveller and previous to the cutting stage shown in Fig. 1. Figure 5 shows the lines in the actual experimental deployment. Samples of the actual results are shown in Figs. 6 and 8. Figure 6 shows 2D intensity images of the raw sensor data (up) and the reconstructed surface after Hermite polynomial filtering. Some ghostly lines can still be appreciated after removal of the rippling effect produced by the cutting event. Similarly, Figure 8 provides a 3D representation of the raw sensor data (up) and

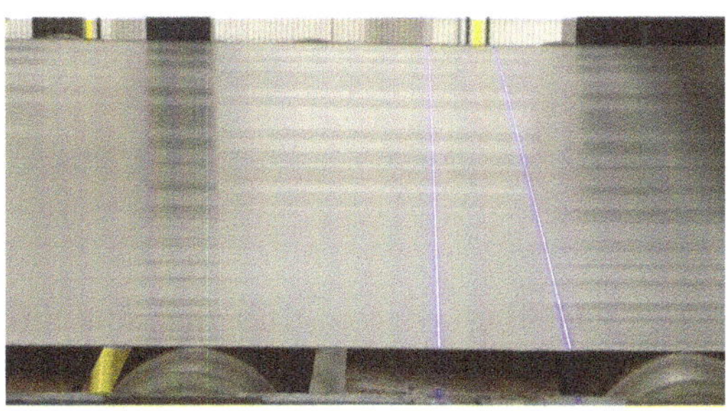

Fig. 5. a) Laser lines over metal strip. b) Flatness defects.

the results after Hermite polynomial filtering (bottom). Main rippling surface features are effectively removed, while the low frequency effects of interest are preserved for further quality control (Fig. 7).

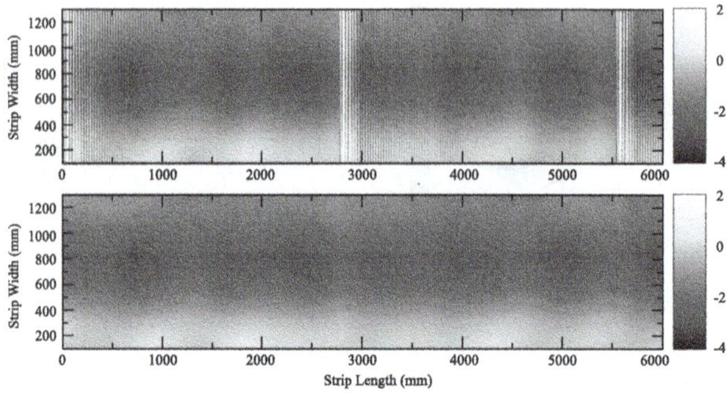

Fig. 6. 2D representation of filtering and reconstruction results. (up) Sensor raw data. (bottom) Surface reconstruction results.

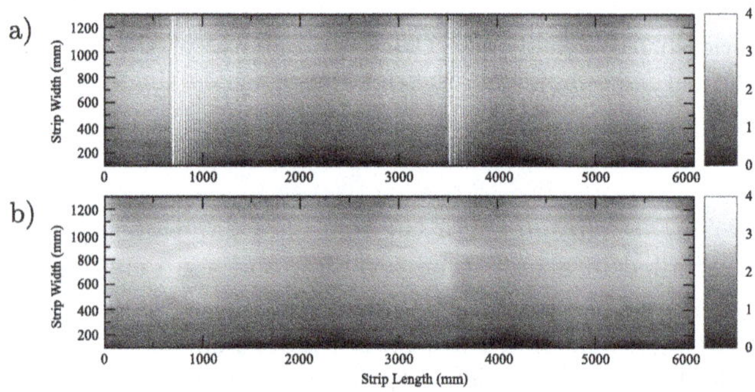

Fig. 7. a) Sensor raw data. b) Surface reconstruction results.

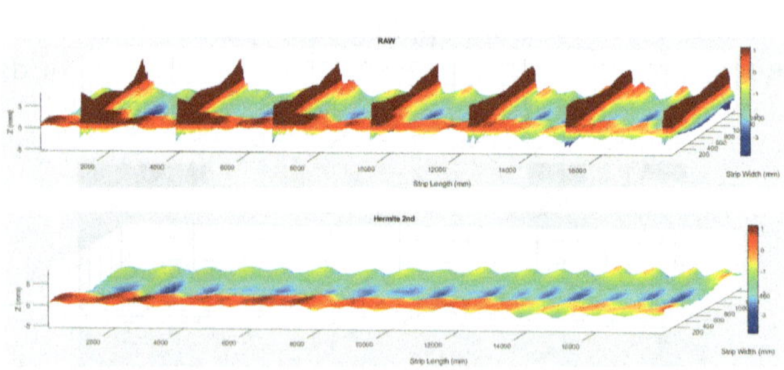

Fig. 8. 3D representation of filtering and reconstruction results. (up) Real noisy surface data. (bottom) Estimated filtered surface data.

4 Conclusions

In this paper we present preliminary results of an artificial vision system consisting of a sensor and two laser lines for the inspection of rolled sheet metal products in an industrial processing line. This system allows to retrieve an accurate and real-time estimation of the metal sheet surface allowing flatness defects detection, i.e. wavy edges, center buckles, bow. In addition, we propose a computational method of signal processing that enables the isolation of the actual surface measurements from the vibrations induced in the metal sheet by the different mechanical elements of the processing line. This method, based on cubic Hermite spline interpolation, is particularly robust even in the situation where a local high frequency and high amplitude noise produced by a cutting station located near the scanning area distort the measurement signal. This method is computationally efficient, so it does not require high cost computing resources.

Through of simulations we have verified that this method allows to analyse and filter the information even when the flatness information is extremely hindered. In particular, a noise source produced by the cutting stage near the machine vision system has been added. This source of noise has a sporadic character and excites vibrations in harmonics that propagate back-and-forth in the metal sheet; thus generating measurement errors that have not been considered by other authors up to date. The results obtained by means of these simulations demonstrate that the Hermite method proposed in this article allows us to accurately compute the flatness measurement of the simulated metal sheets. We have also tested the suitability of the method in a real production environment. The experimental measurements once again confirm the accuracy, robustness and reliability of the machine vision system and the surface estimation method presented in this article. In fact, we have set side by side the experimental results with a CMM (coordinate measurement machine) by including in our measurement experiments patterns whose known geometry consists of characteristic surface defects in this type of material. The proposed method can pave the way to closed-loop systems, low cost real-time flatness quality inspection, and high efficiency and quality rolled products production.

We intend to concentrate our future research in two areas. First, for surface reconstruction planned future work includes improving the performance of the proposed method and investigate a method based on compactly supported radial basis functions (CSRBFs) for Hermite surface interpolation and Hermite Radial Basis Functions(HRBF) Implicit with least squares for the surface reconstruction of scattered points. Interpolating incomplete meshes (hole-filling) and reconstructing surfaces from point-clouds derived from noisy 3D range scanners are important problems. The functional nature of the RBF representation offers new possibilities for surface registration algorithms, mesh simplification, compression and smoothing algorithms. Secondly, regarding sheet flatness error detection we will investigate a flatness anomaly detection approach based on deep convolutional neural networks (CNNs). The flatness defects of steel strips are classified according to various features, but it is hard for traditional methods to extract all these features and use them effectively.

Acknowlegment. This work has been partially supported by FEDER funds through MINECO project TIN2017-85827-P, RFCS EU funded project FLATBEND with grant number 800730, and grant IT1284-19 as university research group of excellence from the Basque Government.

References

1. Air bearing shapemeter, shapemeter for the rolling industry. White Paper (2015). https://www.primetals.com/fileadmin/user_upload/Air_bearing_shapemeter.pdf
2. Astm a568/a568m-17a, standard specification for steel, sheet, carbon, structural, and high-strength, low-alloy, hot-rolled and cold-rolled, general requirements for (2017)
3. Agrawal, A., Chellappa, R., Raskar, R.: An algebraic approach to surface reconstruction from gradient fields. In: Tenth IEEE International Conference on Computer Vision (ICCV 2005), vol. 1, pp. 174–181 (2005)
4. Tsuzuki, S., et al.: Flatness control system of cold rolling process with pneumatic bearing type shape roll. IHI Eng. Rev. **42**, 54–60 (2009). IHI, Tokyo, Japan
5. Bergman, G., Enneking, A., Thies, K.h: Displacement-type shape sensor for multiroll leveler (2005)
6. Classon, P.K.L.: A new generation optical flatness measurement systems (2015)
7. Frankot, R.T., Chellappa, R.: A method for enforcing integrability in shape from shading algorithms. IEEE Trans. Pattern Anal. Mach. Intell. **10**(4), 439–451 (1988)
8. Jouet, J., Francois, G., Tourscher, G., de Lamberterie, B.: Automatic flatness control at solmer hot strip mill using the lasershape sensor. Iron Steel Eng. **65**(8), 50–56 (1988)
9. Karaçali, B., Snyder, W.: Reconstructing discontinuous surfaces from a given gradient field using partial integrability. Comput. Vis. Image Underst. **92**(1), 78–111 (2003)
10. Klette, R., Schluens, K.: Height data from gradient maps. In: Solomon, S.S., Batchelor, B.G., Waltz, F.M. (eds.) Machine Vision Applications, Architectures, and Systems Integration, vol. 2908, pp. 204–215. International Society for Optics and Photonics, SPIE (1996)
11. Knauer, M.C., Kaminski, J., Hausler, G.: Phase measuring deflectometry: a new approach to measure specular free-form surfaces. In: Osten, W., Takeda, M. (eds.) Optical Metrology in Production Engineering, vol. 5457, pp. 366–376. International Society for Optics and Photonics, SPIE (2004)
12. Paakkari, J.: On-line flatness measurement of large steel plates using moiré topography (1998)
13. Rychkov, I.: Locally controlled globally smooth ground surface reconstruction from terrestrial point clouds (2012)
14. Savitzky, A., Golay, M.J.: Smoothing and differentiation of data by simplified least squares procedures. Anal. Chem. **36**, 1627–1639 (1964)
15. Schlüns, K., Klette, R.: Local and global integration of discrete vector fields, pp. 149–158 (1997)
16. Sivilotti, O., GuiseppePervi, C.: Arrangement in strip rolling mills for measuring the distribution of the strip tension over the strip width (1966)
17. Vollmer, F.: Vip08 flatness measurement system (2010). https://vollmeramerica.com/vip-08-flatness-measurement-system

Efficiency of Public Wireless Sensors Applied to Spatial Crowd Monitoring in Buildings

Anna Kamińska-Chuchmała(✉)

Department of Computer Science and Systems Engineering, Wrocław University of Science and Technology, Wybrzeże Wyspiańskiego 27, 50-370 Wrocław, Poland
anna.kaminska-chuchmala@pwr.edu.pl

Abstract. Contemporary world rest on wireless connections via networks developed rapidly. Majority of people using wireless networks i.e. WiFi at: work, home, and also when they are outside. A central issue in wireless networks is efficiency and stable connection to Internet. Additional challenge is to create and maintain reliable WiFi network with full open access, videlicet every user could connect with that network if only have a range. Exemplary aforesaid network is university network where users are very specific group. An innovative approach is to treat and use Access Points (APs) as wireless sensors network to crowd monitoring in buildings where such network is located. Thanks to this we receive accurate information about the number of users (people) without interfering with their sensitive data. The aim of this research is to use geostatistical methods to prepare spatial models and predict behaviour of crowd connected with wireless sensors on full of range considered area. The investigation has tended to focus on efficiency of wireless sensors belong to open WiFi network named PWR-WiFi located in building at the main campus of Wrocław University of Science and Technology (WUST) in Poland. The data gathered for analysis are acquired from three consecutive years 2014–2016 to better show the dynamic growth of number of PWR-WiFi network users it means number of people crowded in building. Parameter, which most reflecting behaviour of PWR-WiFi network is the number of users, obtained from APs, that was investigated during research. Preliminary and structural analysis with approximation of variogram models were made and as the next step spatial prediction models of wireless sensors network efficiency were performed by Turning Bands geostatistical simulation method. Three models of spatial prediction were prepared for three subsequent years 2014, 2015 and 2016. Following, the results were compared with spatial prediction models created previously by ordinary kriging estimation method.

Keywords: Wireless sensors network · Crowd monitoring · Spatial prediction models · Turning Bands method · Geostatistical methods

Á. Herrero et al. (Eds.): SOCO 2020, AISC 1268, pp. 843–852, 2021.
https://doi.org/10.1007/978-3-030-57802-2_81

1 Introduction

Many researchers have extensively studied prediction of wireless network in recent years, e.g. in [1] was presented the issues of location and request prediction in wireless networks characterizing them as a discrete sequence prediction problems, and surveyed the major Markovian prediction methods.

Analysis of WiFi performance data for a WiFi throughput prediction approach was made in [2]. Author implemented a WiFi parameter visualization tool to show users' WiFi performance in a graphic way. In this tool, machine learning method is used for WiFi performance analysis to predict WiFi throughput. A SVM-based classification model is proposed to work as a prediction function which takes WiFi parameters both for target AP and nearby interference APs as input, and output is categorized as WiFi throughput: good, medium, poor or very poor.

Authors in [3] propose a simple traffic prediction mechanism using the Recursive Least Squares (RLS) algorithm a which does not make any stationarity assumptions on the underlying time series and hence is able to operate on the raw data. Results prepared on real data performance evaluation show that RLS algorithm is capable of delivering accurate predictions and shows good adaptive behaviour at the same time being intuitively simple and lightweight from an implementation perspective.

The study of [4] Generalized Regression Neural Networks (GRNNs) has become an important aspect of predict the output, packets dropped of a sample DMesh network simulation. Authors observed that some of considered parameters e.g. traffic density and number of channels used, have a direct impact on error rate of the regression model. As result the high variance proved that GRNN approach can represent real characteristics of DMesh architecture.

[5] proposed a generic framework to approach the problem of mobility prediction using Hidden Markov Models (HMM). Authors used a real dataset with information regarding APs, users and derived mobility information from it. The data mined from the traces was useful in predicting the users movement and may be used to allocate resources in the network.

[6] in their work presented a survey on mobility prediction schemes proposed for wireless networks such as: prediction used in routing protocol, mobility prediction based on mobile user's behaviour, Markov based prediction scheme, Mobility Prediction Algorithm Based on Dividing Sensitive Ranges, Autoregressive Hello protocol, Mobility prediction using Neural Network and Bayesian network.

On basis of review of literature it could be claimed that no one till now adapt geostatistical methods to prediction efficiency of wireless networks especially in context to wireless sensors adapt to crowd monitoring. The issue of crowd monitoring was rather solving by using LiDAR (Light Detection and Ranging) and treat as camera sensor for pedestrian detection [7]. In consequence, the purpose of this research is using Turning Bands method (TBM) to spatial prediction of efficiency of WiFi sensors network. First Author's results with similar approach was published in [8], and the previous investigations had concentrated on applying geostatistical estimation and simulation methods to spatial prediction of performance but wired not wireless network (e.g. [9–12]).

2 Turning Bands Method

One of the most popular method from geostatistical simulations is Turning Bands method. It was used for the first time by [13] in the special case of Brownian random function. Next, it was developed by [14] and used for simulation by [15]. TBM is a stereologic tool used for reduction of multidimensional simulation to a one-dimensional one. The main idea of Turning Bands is to adding up a large number of independent simulations defined on lines scanning the plane [16]. The TBM consists in the reduction of a Gaussian random function of covariance C to the simulation of an independent stochastic process of covariance C_θ. According to [17] let $(\theta_n,\ n \in \mathbb{N})$ be a sequence of directions $\mathbb{S}_d^+$, and let $(X_n,\ n \in \mathbb{N})$ be a sequence of independent stochastic processes of covariance C_{θ_n}, then random function:

$$Y^{(n)}(x) = \frac{1}{\sqrt{n}} \sum_{k=1}^{n} X_k(< x, \theta_k >), \qquad x \in \mathbb{R}^d \tag{1}$$

assumes covariance equal to:

$$C^{(n)}(h) = \frac{1}{n} \sum_{k=1}^{n} C_{\theta_k}(< h, \theta_k >). \tag{2}$$

Turning Bands algorithm is given below:

1. Transform input data using Gaussian anamorphosis.
2. Choose a set of directions $\theta_1, ..., \theta_n$ such that $\frac{1}{n}\sum_{k=1}^{n} \delta_{\theta_k} \approx \varpi$.
3. Generate independent standard stochastic processes $X_1, ..., X_n$ with covariance functions $C_{\theta_1}, ..., C_{\theta_n}$.
4. Compute $Y^{(n)}(x) = \frac{1}{\sqrt{n}}\sum_{k=1}^{n} X_k(< x, \theta_k >)$ for any $x \in D$.
5. Make kriged estimate $y^*(x) = \sum_c \lambda_c(x) y(c)$ for each $x \in D$.
6. Simulate a Gaussian random function with mean 0 and covariance C in domain $\mathbb{D}$ on condition points. Let $(z(x), x \in D)$ and $(z(c), c \in C)$ be the obtained results.
7. Make kriged estimate $z^*(x) = \sum_c \lambda_c(x) z(c)$ for each $x \in D$.
8. Obtain the result $(y^*(x) + z(x) - z^*(x), x \in D)$.
9. Perform a Gaussian back transformation to return to the original data.

3 Experiment Background

The data considered in presented research were collected from open WiFi network named PWr-WiFi. This wireless network is located in the main campus at (WUST). PWR-WiFi network using the standard IEEE802.11 of wireless infrastructure. The data collected to this research are obtained from eleven sensors (AP's) given in five-storey building (Fig. 1), named B4. APs work with using frequency 2.4 GHz in IEEE 802.11b/g/n standards and 5 GHz in IEEE 802.11a/n standards. APs contained in PWR-WiFi network are wireless connected to switch and configured to get IP address from network and connecting to WiFi controller by LWAPP (Light Weight Access Point Protocol) protocol.

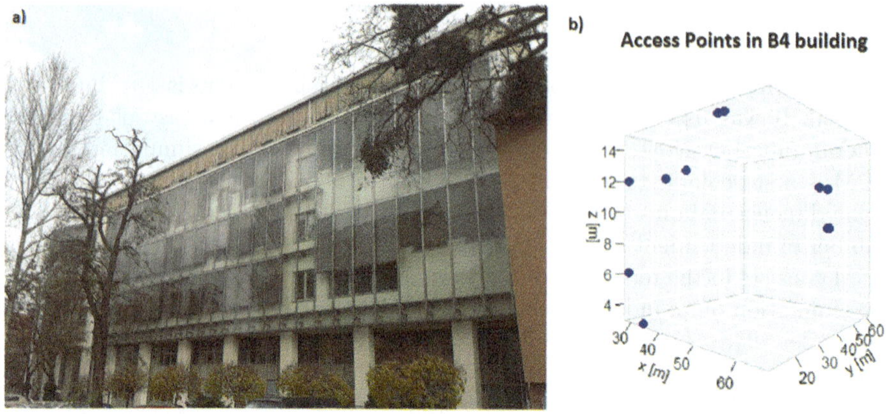

Fig. 1. (a) B4 building located in main campus of WUST (b) projection of localization sensors (APs) in B4

The analysed data were obtained from passive experiment (real data) which were taken from 14th - 29th April over three consecutive years: 2014, 2015 and 2016 collected every hour between 7:00 AM and 9:00 PM. Examined wireless sensors (APs) are installed in B4 building as follows: one on first floor, one on second floor, two on third floor, five on fourth floor, and two on fifth floor. All analysis and prediction presented in this paper were performed in R language under R environment in version 3.4.4 which is available as Free Software on GNU licence [18]. Moreover, prediction with geostatistical TBM was made by using RGeostats package [19].

4 Preliminary and Structural Analysis Data from Sensors

Examined PWR-WiFi sensors network is an open university wireless network where most of users are students and employees of WUST like lecturers or administrative workers. In considered B4 building there are two deaneries, few lecture halls, many laboratories, two libraries and also administrative and researcher's offices. Three databases from years 2014–2016 which are examined during research contain every hour measurement from 16 days between 7:00 AM and 9:00 PM each day. This part of day was chosen because of B4 building is closed by night and main traffic in network is during office hours and lectures. Predominantly, the first classes are started at 7:30 AM and the last one finished at 8:30 PM. Generally, during the day, classes are started quarter after an odd hour. This specific character of PWR-WiFi network could be seen in Fig. 2, where data of number of users are presented for whole examined period of 2015. This figure shows periodical behaviour of network users. It have to be mentioned that this considered part of month April is a time in semester where students coming regularly on classes thus regularly need to access to the wireless sensors.

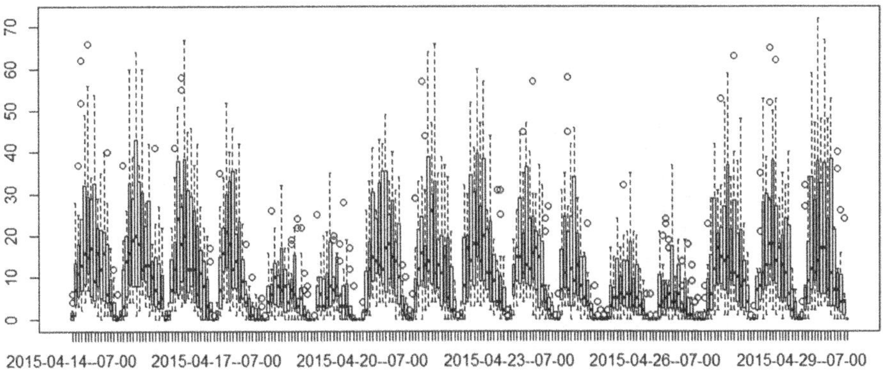

Fig. 2. Number of users served by 11 APs in B4 building between 14th and 29th April 2015

Basics statistics are presented in Table 1. Its cover three years 2014–2016. The maximum number of users is equal 90 in 2016 and the minimum equals 54 in 2014. It could be noticed growing trend of number of users. Mean value also confirms this trend, because every year mean number of users is higher by about 4 to 6 users. Variance of data is also growing up what showing variability of this process and significant data differentiation. Furthermore standard deviation is also the largest in 2016 and equals 14.69.

Table 1. Basic statistics of number of users for all considered sensors located in building within three years

Parameters	14-29.04.2014	14-29.04.2015	14-29.04.2016
Min number of users	0	0	0
Max number of users	54	72	90
Mean number of users	4	10	14
Variance	31.99	158.94	215.80
Standard deviation	5.66	12.61	14.69

Histogram of number of users in 2014 served by 11 Access Points is presented in Fig. 3. The highest frequency of users is in between 0 to 10 and also in the other bin between 10 to 20. The histogram is skewed left.

First step in structural analysis is performing of Gaussian anamorphosis (Fig. 4). For all datasets it was made by using 100 Hermite polynomials.

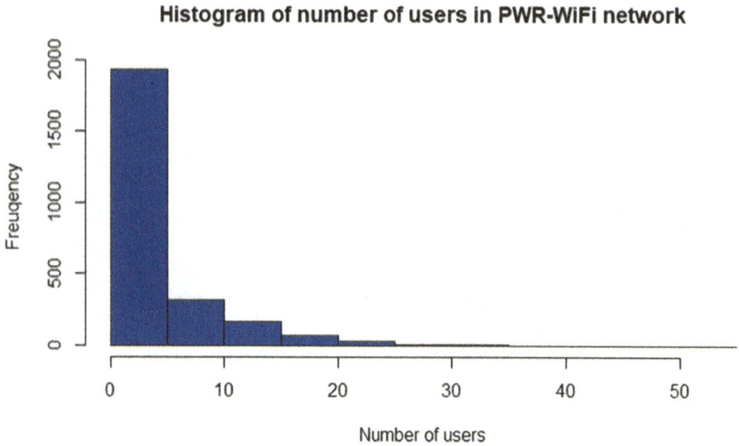

Fig. 3. Histogram of number of users in PWR-WiFi network in B4 building in 2014

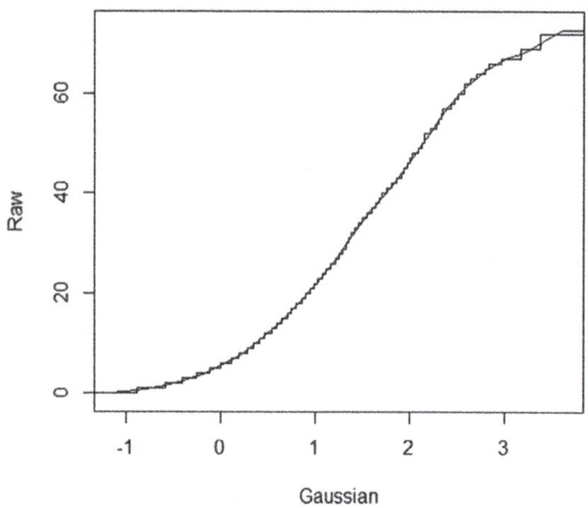

Fig. 4. Gaussian anamorphosis for number of users in PWR-WiFi network in 2015

The second step in structural analysis in geostatistical approach is to calculating variograms and approximate them with appropriate theoretical model or models. For each of databases from three years 2014–2016 variograms were calculated in four different direction: 0, 45, 90, 135 degrees. As example variogram of number of users in 2015 is presented in Fig. 5. Distance lags for each calculation direction equals 5. The number of lags for each calculation direction is 15. For all directions it could be seen the nugget effect. The range of variogram function for two directions was about 30 m and for two others direction almost 60 m. In the next step variogram function was approximated by theoretical functions: nugget effect, exponential, and spherical.

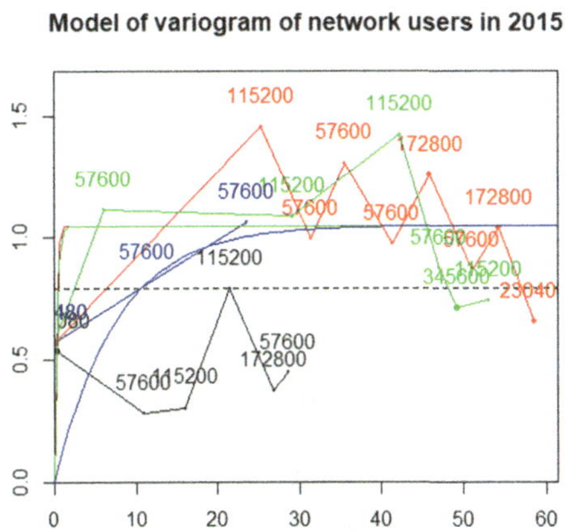

Fig. 5. Models of variograms calculated in four directions for number of users connected to the PWR-WiFi sensors in 2015

5 Spatial Prediction Models of PWR-WiFi Network Efficiency Simulated by TBM

Models have 3 dimensions (x and y geographical coordinates and z the altitude coordinates) and prediction cover the space where spreads the range of signal from APs belonging to PWR-WiFi wireless network in building B4. Three models of spatial prediction were prepared for three consecutive years 2014, 2015 and 2016 and contain: Gaussian anamorphosis models, theoretical models of variogram approximation and moving neighbourhood. The moving neighbourhood search is performed by angular sectors and the neighbourhood ellipsoid is anisotropic. The search ellipsoid has a three dimensions. Results of spatial predictions by Turning Bands simulation method are presented in Table 2.

Table 2. Statistics of simulated by TBM number of users of PWR-WiFi sensors

Parameters	14-29.04.2014	14-29.04.2015	14-29.04.2016
Min number of users	0	0	0
Max number of users	54	72	90
Mean number of users	10	12	13
Variance	92.77	139.01	224.55
Standard deviation	9.63	11.79	14.98

Maximum number of users is the highest in 2016 and equals 90 as well as variance and standard deviation. Growing trend in prediction models for whole considered area (all building B4) is also visible.

In Fig. 6 there are presented raster maps of mean number of PWR-WiFi users simulated by TBM in years 2014–2016. Localization of more concentrations of users such as students, lecturers or administrative workers are similar for all years. Probably it is related to the fact that it is nearby (depends of floor in the building): lecture hall, library or deanery. Unfortunately in 2015 and 2016 3 sensors were disabled thus in some areas on maps there are less users.

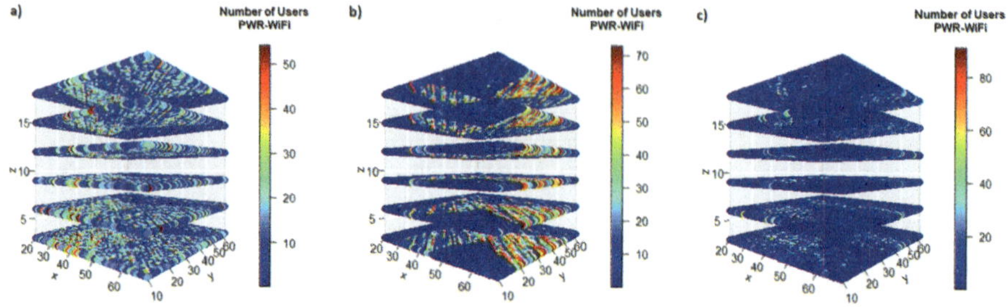

Fig. 6. Scatter 3D plot of number of users connected to the PWR-WiFi sensors in B4 building at WUST in (a) April of 2014 (b) April of 2015 (c) April of 2016

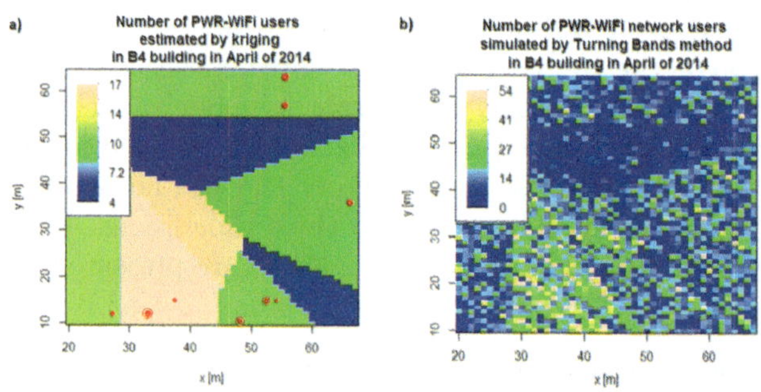

Fig. 7. Comparison of raster maps of number of users connected to the PWR-WiFi sensors in B4 building in April 2014 (a) estimated by kriging (b) simulated by Turning Bands

Previously, Author prepared also spatial predictions on these considered databases, albeit by using kriging estimation method [20]. Comparison of results from spatial prediction models made by different geostatistical methods is presented in Fig. 7. Obtained values confirmed characteristic feature of estimation method, it means smoothing. In these cases, in wireless sensors prediction of efficiency models, more realistic data are desirable and reflect the better character of the phenomenon being studied.

6 Conclusions

To bring the paper to a close, the summary of the main points is given here: preliminary and structural analysis of data obtained from sensors were conducted, and spatial (3D) prediction models of PWR-WiFi wireless sensors efficiency within three years: 2014, 2015, and 2016 with using Turning Bands geostatistical simulation method were presented. In conclusion, it seems that these kind of spatial predictions, especially obtained raster maps, could be very helpful for localization of people in buildings. Moreover, WiFi sensors may increasingly be used for surveillance and crowd monitoring in public places because their intrinsic respect for personal data. Additionally such approach could be alternative use to popular LiDAR systems.

Further research in this area will be include performing space-time (4D) prediction of PWR-WiFi wireless sensors with using more parameters like channel utilization. Models of prediction will be performed with using not only geostatistical estimation (Kriging) methods and simulation methods (Turning Bands), but also with other geostatistical methods like Sequential Gaussian Simulation.

References

1. Katsaros, D., Manolopoulos, Y.: Prediction in wireless networks by Markov chains. IEEE Wirel. Commun. **16**(2), 2–9 (2009)
2. Pan, D.: Analysis of Wi-Fi performance data for a Wi-Fi throughput prediction approach. MSc Thesis. KTH Royal Institute of Technology School of Information and Communication Technology, Stockholm (2017)
3. Kulkarni, P., Lewis, T., Fan, Z.: Simple traffic prediction mechanism and its applications in wireless networks. Wirel. Pers. Commun. **59**, 261–274 (2011)
4. Odabasi, S.D., Gumus, E.: A prediction model for performance analysis in wireless mesh networks. Int. J. Electron. Mech. Mechatr. Eng. **6**(3), 1241–1250 (2016)
5. Prasad, P.S., Agrawal, P.: Movement prediction in wireless networks using mobility traces. In: 2010 7th IEEE Consumer Communications and Networking Conference (CCNC) (2010). https://doi.org/10.1109/CCNC.2010.5421613
6. Ananthi, J., Ranganathan, V.: Review: on mobility prediction for wireless networks. Int. J. Emerg. Technol. Adv. Eng. **3**(4), 891–902 (2013)
7. Wu, T., Tsai, C., Guo, J.: LiDAR/camera sensor fusion technology for pedestrian detection. In: Asia-Pacific Signal and Information Processing Association Annual Summit and Conference (APSIPA ASC), Kuala Lumpur, pp. 1675–1678 (2017)
8. Kamińska-Chuchmała, A., Graña, M.: Indoor crowd 3D localization in big buildings from Wi-Fi access anonymous data. Sensors **19**(19), 4211 (2019). https://doi.org/10.3390/s19194211
9. Borzemski, L., Kamińska-Chuchmała, A.: Client-perceived web performance knowledge discovery through turning bands method. Cybern. Syst. Int. J. **43**(4), 354–368 (2012)
10. Borzemski, L., Kamińska-Chuchmała, A.: Knowledge engineering relating to spatial web performance forecasting with sequential Gaussian simulation method. In: Advances in Knowledge-Based and Intelligent Information and Engineering Systems. FAIA, vol. 243, pp. 1439–1448. IOS Press, Amsterdam (2012)

11. Borzemski, L., Kamińska-Chuchmała, A.: Distributed web systems performance forecasting using turning bands method. IEEE Trans. Industr. Inf. **9**(1), 254–261 (2013)
12. Kamińska-Chuchmała, A.: Spatial Internet traffic load forecasting with using estimation method. Procedia Comput. Sci. **35**, 290–298 (2014)
13. Chentsov, N.N.: Lévy Brownian motion for serveral parameters and generalized white noise. Theor. Prob. Appl. **2**(2), 265–266 (1957)
14. Matern, B.: Spatial variation – stochastic models and their application to some problems in forests surveys and other sampling investigations. In: Meddelanden fran Statens Skogsforskningsinstitut, vol. 48(5). Almaenna Foerlaget, Stockholm (1960)
15. Matheron, G.: The intrinsic random functions and their applications. Adv. Appl. Prob. **5**, 439–468 (1973)
16. Chiles, J.-P., Delfiner, P.: Geostatistics: Modeling Spatial Uncertainty. Wiley, Hoboken (2012)
17. Lantuejoul, C.: Geostatistical Simulation. Models and Algorithms. Springer, Heidelberg (2002)
18. R Core Team: R: A Language and Environment for Statistical Computing. R Foundation for Statistical Computing, Vienna (2018). https://www.R-project.org
19. Renard, D., Bez, N., Desassis, N., Beucher, H., Ors, F., Freulon, X.: RGeostats: The Geostatistical R package 11.2.2 MINES ParisTech/ARMINES (2018). http://cg.ensmp.fr/rgeostats
20. Kamińska-Chuchmała, A.: Spatial prediction models of wireless network efficiency estimated by kriging method. Rynek Energii **135**(2), 89–94 (2018)

Machine-Learning Techniques Applied to Biomass Estimation Using LiDAR Data

Leyre Torre-Tojal[1(✉)] and Jose Manuel Lopez-Guede[2,3]

[1] Department of Mining and Metallurgical Engineering and Materials Science, Faculty of Engineering, University of the Basque Country (UPV/EHU), Vitoria-Gasteiz, Spain
leyre.torre@ehu.es

[2] Department of Systems Engineering and Automatic Control, Faculty of Engineering, University of the Basque Country (UPV/EHU), Vitoria-Gasteiz, Spain

[3] Computational Intelligence Group, University of the Basque Country (UPV/EHU), Leioa, Spain

Abstract. With the development of artificial intelligence, alternative advanced machine learning approaches have allowed the training of increasingly sophisticated models via the available data. The light detection and ranging (LiDAR) remote sensing technique is being increasingly applied to obtain informative terrain maps, due to its ability to collect large amounts of data with satisfactory accuracy. Forest ecosystem management needs a multi-faceted approach, combining forest mapping and inventory in order to provide comprehensive knowledge on the current state and future trends of forest resources. Estimation of forestry aboveground biomass (AGB) by means of LiDAR data uses high-density point sampling data obtained in dedicated flights, which are often too costly for available research budgets. In this paper, we exploit already existing public low-density LiDAR data obtained for other purposes, such as cartography. This paper focuses on the application of machine-learning-based predictive systems for the extraction of biomass information from low-density LiDAR data (0.5 points/m^2) taking into account the Pinus radiata species in the Arratia-Nervión region (Spain).

Keywords: Biomass · LiDAR · Regression · Remote sensing

1 Introduction

Expert systems consist of two main components: a knowledge base and an inference engine. Expert systems are applicable to various scopes that involve human ideas, deductions and reasoning, which implies that any field that requires human expertise can use them to minimize risks associated with the issue to deal with [1].

Remotely acquired data (land, airborne or satellite based) have been successfully used for the assessment of tree characteristics such as average height, dominant height, or mean diameter [2]. Regarding the forest management applications, Light Detection and Ranging (LiDAR) stands out among the available remote sensing methods because it allows the acquisition of data in large areas and provides measures of variables describing

Á. Herrero et al. (Eds.): SOCO 2020, AISC 1268, pp. 853–861, 2021.
https://doi.org/10.1007/978-3-030-57802-2_82

the structure of the forest canopy [3], even allowing the discrimination between tree species [4]. Based on the previous variables, aboveground biomass estimation is easily manageable [5].

Specific LiDAR data capture campaigns for biomass measurement are very expensive, hindering the general application of the technology in forestry management. However, this obstacle can be overcome thanks to some institutions, which carry out periodic LiDAR data capture campaigns to build digital terrain and surface models, mostly for cartographic purposes.

In this study, we focus on the application of LiDAR for forest biomass estimation, which has been traditionally carried out using two main families of approaches, namely, destructive and non-destructive methods [6], applying machine-learning techniques. The ability of LiDAR to collect a large amount of densely sampled elevation data promises a more efficient and inexpensive tool—developed by training data-driven expert systems—for forest biomass management [7].

The paper is organized as follows. Section 2 provides the data and methods applied in this study. Section 3 presents the results of the applied methodologies for biomass estimation in the study area. Section 4 includes a discussion of existing literature and a comparison of the obtained results using the three specific methodologies: Multiple Linear Regression (MLR), Random Forest (RF) and Support Vector Regression (SVR). Finally, Sect. 5 presents the conclusions and proposals for future work.

2 Materials and Methods

2.1 Study Area

The Arratia-Nervión region encompasses 14 municipalities, covering a total area of 400 km^2. The average altitude of the region is 465 m, with an average slope of 18.6°. High slope (30–45°) areas are frequent across the entire region. Pine forests of P. radiata are the most important land cover in the Basque Country, 125,000 ha, accounting for 32% of the forested area in the Basque Country, equivalent to 49% of the area covered by this species in Spain.

2.2 Ground Truth Data

118 sample plots of the Fourth National Forestal Inventory are located inside the area of the study. We selected plots in which P. radiata occupied more than 80% of the area of the plot; 55 plots fulfilled this condition. The biomass of the sample plots was measured using the methodology based on nested subplots applied by the Spanish National Institute for Nature Conservation (ICONA after the Spanish name; [8]). In this methodology, each sample plot is subdivided into four circular sample subplots of variable radius, specifically, 5, 10, 15, and 25 m. Subplots with the largest radius cover an approximate area of 0.2 ha.

The nested plot method is suitable when there is a considerable variability in the trees diameters. In these data gathering campaigns, the minimum, mean, and maximum values for the trees diameters and heights were 10.5, 33.94, and 78.30 cm and 4.30, 22.91, and 42.20 m, respectively.

Fig. 1. Location of the study area (Arratia-Nerviόn, Biscay, Spain)

The volume per tree was calculated using an allometric model developed by the HAZI institute of the Basque Government. The model uses the diameter at breast height (d mm) and total tree height (h m) as independent variables according to the following equation:

$$VCC(dm^3) = 0.0006785d^{1.86004} \cdot h^{1.01378} \quad (1)$$

Once the reference values of the volume for each tree were calculated, these values were extrapolated to an extension of one hectare. The biomass was calculated by adding a correction of 4% of the volume because tree branches and the thinnest part of the tree trunk were not taken into account in the field measurements, due to the wood production processes.

2.3 LiDAR Data

The Basque Autonomous Community area was flown over using a Lite Mapper 6800 Airborne Laser Scanner with the flying parameters of Table 1.

The reference system of the data is the European Terrestrial Reference System 89 (ETRS89) and the coordinate system is UTM for the thirtieth time zone north. The dataset was divided into sheets of 2 × 2 km of extension, classified into eight classes: Unclassified, Ground, Low Vegetation, Medium Vegetation, High Vegetation, Building, Low Points, and Reserved. The data are publicly available at: ftp://ftp.geo.euskadi.eus/lidar/LIDAR_2012_ETRS89/LAS/

Table 1. Flying parameter

Parameter	Value
Average altitude aboveground	1,100 m
Average speed	67 m/s
Pulse Repetition Frequency	100 kHz
Scan Frequency	70 kHz
Maximum scan angle	60°
Beam divergence	<0.5 mrad
Average point density	0.5 points/m^2

2.4 Orthophotos

The orthopohotos used in this study were gathered in the flight campaign carried out by the Basque Government from 23 July to 28 August 2012 with a spatial resolution of 25 cm/pixel, which were used to detect possible defects in the NFI4 data, and contradictions between NFI4 and LiDAR data. These orthophotos were downloaded from the Spatial Data Infrastructure (SDI) of the Basque Country Government from the following site.

2.5 Methods

Biomass estimation using LiDAR data has been widely addressed in previous studies, empirical modelling of the biomass has been carried out using different statistical approaches [9, 10]. Despite of Multiple Linear Regression (MLR) being the most frequently used method; more sophisticated machine learning regression techniques have gained ground in biomass estimation [11, 12]. In the present study, we apply and compare three predictive machine-learning approaches: MLR, Random Forest (RF) and Support Vector Regression (SVR) using the caret package of R statistical software for model training. A five-fold cross-validation process was carried out for each approach, splitting the dataset into five data folders. The reported results are the averages over the five test datasets.

For the performance evaluation of these modelling approaches, we have considered the coefficient of determination (R^2) and the Root Mean Square Error (RMSE).

For the extraction of the LiDAR features, point cloud was treated to clip it to the area occupied by the parcels of the NFI4 (Fig. 2). Then the generated cloud points have been treated to extract LiDAR height related and density related metrics.

Multiple Linear Regression. Multiple Linear Regression (MLR) computes a prediction of the variable under study as a linear combination of a set of regressor variables, often called features or input factors:

$$Y = b_0 + b_1 x_1 + \ldots + b_k x_k \tag{2}$$

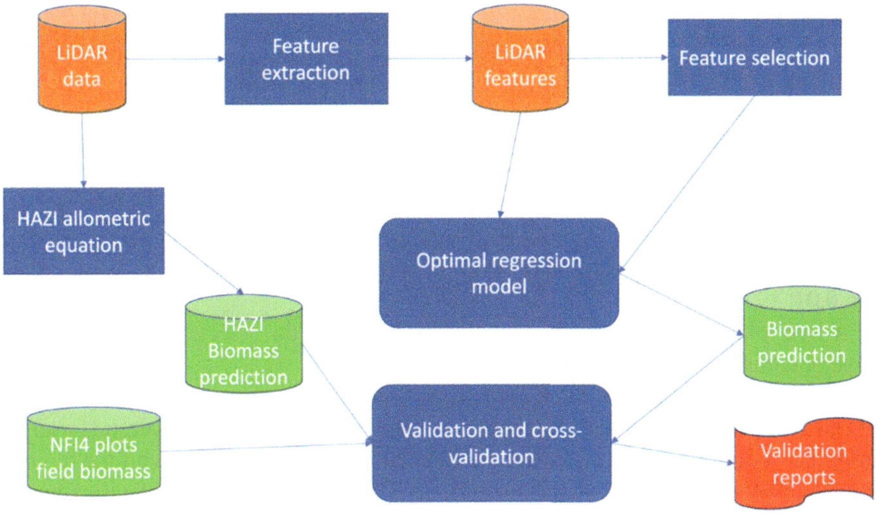

Fig. 2. Overall process carried out in the study.

where k is the model order and $b_0, b_1, \ldots, b_k$ are the coefficients of the linear combination.

After that extraction of the features of the LiDAR data, LiDAR height related variables and density metrics were calculated. For the density metrics the point cloud was divided into 10 vertical layers, then, the fraction of points falling inside each layer was counted. In that way, 10 canopy densities were computed (denoted as *tr_1,…, tr_10*).

For selecting the variables to be used in the model regression, computational experiments with all single variables and all possible combinations of two and three variables were carried out to select the best performing variable for further cross-validation experimentation.

To guarantee the underlying hypothesis of linear regression, Variance Inflation factor (VIF), Shapiro-Wilk test (SW), Breutsch-Pagan test (BP), Durbin-Watson test (DW), Ramsey's RESET linearity (RES) test and Bonferroni (BON) test were applied to the fitted models [13].

Random Forest. The Random Forest (RF) has gained acceptance in forestry applications due to its robustness and modelling flexibility in predicting/imputing the values of new unknown samples. By definition, RF is a non-parametric technique based on randomly growing decision trees, randomly deciding at each tree node which variables will be tested and which value will be the decision parameter [13]. This method first grows several decision trees and later combines the predictions from all the trees to produce the ensemble response.

Support Vector Machines. Support Vector Machines (SVM) have been increasingly used in land cover studies. The SVM training algorithm aims to find a hyperplane that separates the samples into two classes maximizing its margin, i.e., the distance between the discriminant hyperplane and the samples at the class boundaries [14].

3 Results

Regarding to the MLR, the 5 best fitting two-variables MLR models produced very similar results, as is shown in Table 2. No three-variable model exhibiting significant improvement over these two-variable models was found in our computational exploration. The first three models showed identical R^2 (0.80) and RMSE (0.25 ton/ha in logarithmic units) values. Their scores on the statistical tests were quite similar, including the detection of outliers according to Bonferroni's test. In all the entries of Table 2, the hypotheses of linear regression were guaranteed because all p values were greater than 0.01.

A five-fold cross-validation was also carried out on the selected model using five folds, obtaining an average RMSE = 0.07 ton/ha.

Finally, we computed the MLR with the two selected variables (95th percentile of the LiDAR heights, p95) and (density metric corresponding to the third layer, tr3) over the entire dataset to obtain the regression model that could be compared with the biomass estimation results published by the institutions. The fitted regression model is expressed in Eq. 3:

$$\ln \mathit{Biomass} = 3.77418 + (0.06729p_{95}) + (0.54792tr_3) \tag{3}$$

Table 2. Values obtained for the ten best two-variable models using MLR. P95,99 = 95th,99th LiDAR height percentile, abovemean = proportion of first all) returns above the mean, allabovemean (all returns above mean height) /(total returns), tri = percentage of points above the i-th layer from the total number of returns

Variables	R^2	RMSE	SW	BP	DW	VIF	RES	BON
(p99, abovemean)	0.80	0.25	0.76	0.13	0.40	1.28	0.75	0.02
(p99, allabovemean)	0.80	0.25	0.81	0.11	0.43	1.17	0.79	0.02
(p95, tr3)	0.80	0.25	0.48	0.19	0.35	1.11	0.83	0.02
(p95, tr2)	0.79	0.25	0.53	0.18	0.41	1.10	0.76	0.96
(p95, tr4)	0.79	0.25	0.49	0.19	0.30	1.14	0.68	0.02

It is necessary to invert the logarithmic transformation to obtain the actual biomass values [15], using the standard error of the estimation of Eq. (4):

$$\mathit{Biomass}(ton/ha) = 1.0342\exp(3.77418 + (0.06729p_{95}) + (0.54792tr_3)) \tag{4}$$

A five-fold cross-validation of RF was run 20 times with the same data set. No variable selection was carried out previously. The RMSE and R^2 performance measures of the five best RF models can be seen in Table 3. The number of independent trees was set to 500 (Ntree), and the number of variables randomly sampled as candidates at each split was set to 2 (Mtry). Even if the table only appear the 5 best models, the mean value when 20 iterations were carried out were 0.30 and 0.73, respectively. The minimum R^2 value was 0.69. The results are shown in Table 3.

Table 3. Best performance results corresponding to RF models

Ntree	Mtry	RMSE	R^2
500	2	0.28	0.78
500	2	0.29	0.76
500	2	0.29	0.75
500	2	0.29	0.75
500	2	0.29	0.75

Again, five-fold cross-validation of SVR was run 20 times over the same data set. No variable selection was carried out previously (Table 4). The results of the five best models are presented in Table 4. Even if in the table only appear the 5 best models, the mean value when 20 iterations were carried out was 0.35 for the RMSE and 0.55 for the R^2 value, with the minimum value of R^2 for the worst model being 0.51.

Table 4. Best performance results corresponding to SVR

C	Sigma	RMSE	R^2
1	0.01	0.37	0.63
1	0.01	0.36	0.58
1	0.01	0.37	0.58
1	0.02	0.38	0.57
1	0.02	0.38	0.57

The values of the R^2 and RMSE obtained by the three different approaches are shown in Table 5. MLR provided the best fit with $R^2 = 0.80$ and RMSE = 0.25. In terms of R^2, MLR achieved the best fitting, closely followed by RF. Regarding to the RMSE values, MLR had the lowest error and SVR indicated the highest one. For additional comparison, we computed the variance ratio (ratio of the standard deviations of the predicted and the observed biomass) and the bias (difference between the means of the predicted biomass to that of observed biomass). The values of the variance ratio for the three approaches were very similar, falling in the interval (0.8, 0.9). Regarding bias, the MLR obtained slightly better results than the other two approaches, with SVR being the most biased one. MLR and SVR had a positive bias, although RF was negatively biased. Taking into account all the accuracy measures, MLR was noted as the best performing methodology in this case study.

4 Discussion

The modelling performance results obtained in this study are comparable with those of other studies concerning plot-level biomass estimations, which have generally reported

Table 5. Comparison of the three applied methodologies

Method	RMSE	R^2
MLR	0.25	0.80
RF	0.28	0.78
SVR	0.37	0.63

R^2 values lower than the ones obtained in our study even with highest density point [16]. For instance, a study [5] carried out in the Canadian boreal zone combining LiDAR and Landsat surface reflectance composites to estimate the biomass. They applied RF technique in forestlands with both deciduous and coniferous tree species. They reported a validation measure $R^2 = 0.52$. Another study [15], in Scotland, focused on the biomass estimation for years 2002 and 2006, being most of the forest area covered by plantations of Sitka spruce, using MLR and RF. They concluded that MLR provided better models to capture the true empirical relationship between the biomass and LiDAR observations, as noted in the present study as well. In contrast, other studies, such as one carried out considering data from the New York state [12], taking into account deciduous trees and coniferous species, concluded that SVR performs better than RF in terms of the ratio of RMSE to the mean input biomass (RRMSE). MLR obtained the worst results. Another study in Canada [11], estimated the stand-level canopy cover and other forest structural parameters fusing the information from the LiDAR data and Landsat imagery. In this case, the authors noted that RF provided better results than MLR, with values of $R^2 = 0.72$ and $R^2 = 0.64$ respectively, being the relative RMSE values of RF and MLR 0.07 and 0.09 for mature forest stands. Similar results were reported for canopy height estimations, in which RF models yielded substantially lower RMSE than MLR.

Hence, our results, in general, agree with the range of results reported in the literature for biomass estimation, though there is no consensus on the best modelling approach. An advantage of the MLR linear approach is that it is well known and accepted by all communities, whereas machine learning approaches are still seen as non-linear black boxes by some research communities.

5 Conclusions

Our study demonstrated automated biomass estimation of P. radiata using data driven machine-learning approaches over public LiDAR data obtained from a low point density flight (0.5 points/m^2). In the present study, MLR has obtained better prediction performance than RF and SVR, with a coefficient of determination (R^2) of 0.8 and a RMSE of 0.25 expressed in logarithmic units. These results are comparable with results reported by other studies in the literature.

The incorporation of data from additional sensors could help improve the model-based results. The European Copernicus program could be a reasonable option to improve model predictive performance because the data set includes satellite-borne earth observation and in situ data.

References

1. Darlington, K.: The Essence of Expert Systems. Prentice Hall, Pearson Education, London (2000)
2. Guo, Z., Chi, H., Sun, G.: Estimating forest aboveground biomass using HJ-1 satellite CCD and ICESat GLAS waveform data. Sci. China Earth Sci. **53**, 16–25 (2010)
3. Nelson, R., Oderwald, R., Gregoire, T.G.: Separating the ground and airborne laser sampling phases to estimate tropical forest basal area, volume, and biomass. Remote Sens. Environ. **60**, 311–326 (1997)
4. Shi, Y., Wang, T., Skidmore, A.K., Heurich, M.: Important LiDAR metrics for discriminating forest tree species in central Europe. ISPRS J. Photogramm. Remote Sens. **137**, 163–174 (2018)
5. Matasci, G., Hermosilla, T., Wulder, M., White, J., Coops, N., Hobart, G., Zald, H.: Large-area mapping of Canadian boreal forest cover, height, biomass and other structural attributes using Landsat composites and LiDAR plots. Remote Sens. Environ. **209**, 90–106 (2018)
6. Parresol, B.: Assessing tree and stand biomass: a review with examples and critical comparisons. Forest Sci. **45**(4), 573–593 (1999)
7. Shao, G., Shao, G., Gallion, J., Saunders, M., Frankenberger, J., Songlin, F.: Improving LiDAR-based aboveground biomass estimation of temperate hardwood forests with varying site productivity. Remote Sens. Environ. **204**, 872–882 (2018)
8. ICONA: methods for the second national forest inventory (Segundo inventario forestal nacional. explicaciones y métodos. 1986–1995). ICONA, Madrid, Spain (1990)
9. Gobakken, T., Næsset, E., Nelson, R., Bollandsås, O., Gregoire, T., Ståhl, G., Astrup, R.: Estimating biomass in Hedmark county, Norway using national forest inventory field plots and airborne laser scanning. Remote Sens. Environ. **123**, 443 (2012)
10. Goldbergs, G., Levick, S., Lawes, M., Edwards, A.: Hierarchical integration of individual tree and area-based approaches for savanna biomass uncertainty estimation from airborne LiDAR. Remote Sens. Environ. **205**, 141–150 (2018)
11. Ahmed, O., Franklin, S., Wulder, M., White, J.: Characterizing stand-level forest canopy cover and height using Landsat time series, samples of airborne LiDAR, and the random forest algorithm. ISPRS J. Photogram. Remote Sens. **101**, 89–101 (2015)
12. Gleason, C., Im, J.: Forest biomass estimation from airborne LiDAR data using machine learning approaches. Remote Sens. Environ. **125**, 80–91 (2012)
13. Breiman, L.: Random forest. Mach. Learn. **45**, 5–32 (2001)
14. Mountrakis, G., Im, J., Ogole, C.: Support vector machines in remote sensing: a review. ISPRS J. Photogram. Remote Sens. **66**, 247–259 (2011)
15. Zhao, K., Suarez, J., Garcia, M., Hu, T., Wang, C., Londo, A.: Utility of multitemporal LiDAR for forest and carbon monitoring: tree growth, biomass dynamics, and carbon flux. Remote Sens. Environ. **204**, 883–897 (2018)
16. Hall, S.A., Burke, I.C., Box, D.O., Kaufmann, M.R., Stoker, J.M.: Estimating stand structure using discrete-return LiDAR: an example from low density, fire prone ponderosa pine forests. Forest Ecol. Manage. **208**(1–3), 189–209 (2005)

Active Learning for Road Lane Landmark Inventory with Random Forest in Highly Uncontrolled LiDAR Intensity Based Image

Asier Izquierdo[1] and Jose Manuel Lopez-Guede[2,3](✉)

[1] Airestudio Geoinformation Technologies Scoop, Albert Einstein Kalea, 44, E6, Oficina 8, 01510 Vitoria-Gasteiz, Spain
[2] Department of Systems Engineering and Automatic Control, Faculty of Engineering Vitoria-Gasteiz, Basque Country University (UPV/EHU), Nieves Cano 12, 01006 Vitoria-Gasteiz, Spain
jm.lopez@ehu.es
[3] Computational Intelligence Group, University of the Basque Country (UPV/EHU), San Sebastián, Spain

Abstract. Road landmark inventory is becoming an important industry for the maintenance of transport infrastructures among others. Several commercial sensors are available witch include LiDAR sensors allowing to capture up to 1.5 million data point per second. We obtain an intensity based image from the LiDAR point cloud intensity. The landmark detection is posed as a two class classification problem that may be solved by some standard approaches, for example, Random Forest (RF). Besides model parameter selection, a central problem is the construction of the labeled dataset due to human labor cost and the highly uncontrolled conditions of the data capture. We propose an open ended Active Learning approach with a human operator in the loop who can start the Active Learning process when detection quality is degraded by the change in data condition in order to achieve adaptation to them. As an additional contribution, we have assessed the ability of Active Learning to overcome the issues raised by highly class imbalanced dataset, reaching a True Pixel Ratio value of 0.98.

1 Introduction

Road landmark inventory is a flourishing industry around the world, as the traffic becomes more dense and the drivers must rely on a well maintained infrastructure. Specifically, horizontal signals and lane landmarks, such as lines, arrows or other drawings on the asphalt, are of great public concern. In this section we present the problem definition and motivation, an introductory review of Active Learning, the description of the proposed approach and finally, the paper contributions and structure.

Á. Herrero et al. (Eds.): SOCO 2020, AISC 1268, pp. 862–871, 2021.
https://doi.org/10.1007/978-3-030-57802-2_83

1.1 Problem Definition and Motivation

The task tackled in this paper is driven by the industrial exploitation of a car mounted sensor nicknamed "ladybug". More precisely, the sensor is the IP-S3 HD1 product of Topcon (Japan). It is composed of a positioning system (wheel encoder, GPS receiver, Inertial Measurement Unit), five cameras pointing at regular arc intervals of the circumference, and a sixth one pointing up and a LiDAR sensor. The task is to create an inventory of the road signals and landmarks using both LiDAR and image data. All images are tagged spatially with coordinates provided by on board GPS. In short, in this paper we deal with the extraction of the horizontal signals drawn on the road. The image capture conditions, specially the illumination, is wildly changing from one traveling capture to another, or during the same trip due to the changes in the position of the sun, the time of the year, the weather, etc. Besides, the road maintenance is often in bad condition, so the lines may be fading or interrupted. Finally, for supervised classification approaches the construction of the labeled dataset would be costly so we are extremely interested in exploiting other alternatives, such as the Active Learning strategy [1].

1.2 Active Learning

Classification approaches need careful selection and labeling of training data samples from the available data. In response to this issue, Active Learning [2] tries to achieve the most accurate classification using the smallest possible training set, minimizing the user interaction needed to label the training samples. Active Learning starts with a minimal training sample, adding new labeled samples in an iterative process. Aiming to provide the greatest increase in classifier accuracy [3], the additional samples are selected according to some classification uncertainty measures which does not require knowledge of the actual data label. Besides its benefits in economy of computation and data labeling, Active Learning is also useful when the underlying data statistics are non stationary, so that the classifier built at one time instant may not be optimal later on. Active Learning has been successfully applied to classification of remote sensing images [1,4,5], and image retrieval based on semisupervised Support Vector Machines [6]. Active Learning inspiration for the selection of a minimal collection of training images is proposed in [7] for the development of combined generative and discriminative models in the segmentation of CT scans. An active feedback approach is used in [8] to improve the classification based annotation of radiographs.

1.3 Proposed Approach

In this paper we formulate the road landmark segmentation problem [9] as a pixel classification into road landmark and background classes. We collect the input LiDAR data provided from the Mobile Mapping System (MMS) sensor. We only focus on intensity values from the LiDAR point cloud. From the intensity values, we generate an orthorectified image, and then we compute the Gabor features over those images collecting all the image features in a unique pool for the training of the classifiers and their validation. We apply an Active Learning strategy in order select the optimal training dataset. The classifier trained with the optimal training dataset is validated over the entire images, producing the performance report for the specific classifier. We repeat the validation for the different classifiers and classifier parameters explored. The Active Learning oracle providing sample labels in the reported experiments is the ground truth provided by manual segmentation.

For pixel classification we explore the results of Random Forest (RF) [10,11] classifiers based on texture features computed at pixel level. Specifically we apply a bank of Gabor filters, so that the feature vector of each pixel is composed of the Gabor coefficients plus some spatial localization information. We report performance results over a collection of road images in order to assess the most adequate classifier and parameter settings.

1.4 Paper Contribution and Content

Some specific contributions of the approach proposed in this paper relative to the state of the art of road image segmentation algorithms are: (1) Active Learning reduces the human intervention to the minimum in the process of training data selection and labeling, (2) we test an efficient and fast classifier approaches, namely RF, which allow quick adaptation to incremental training datasets, (3) the approach does not required *a priori* information or geometric models, (4) feature extraction is based on an specific systematic approach, i.e. Gabor filters (5) if we need to transfer the trained classifier to new data stream we only need to pick new training samples according to the Active Leaning approach, i.e. the process is an open ended learning process with a human in the loop. (6) In our experimental exploration we have found that Active Learning may provide an alternative avenue to tackle the issues raised by heavily class imbalanced datasets.

The structure of the paper is as follows: Sect. 2 describes the machine learning approaches, the Active Learning framework, and the image feature generation method. Section 3 describes the experimental setup, while Sect. 4 provides the experimental results. Finally, Sect. 5 provides our conclusions and some hints for future work.

2 Methods

In this section we first provide a short review of the machine learning approach used in this paper to tackle the classification problem. Next we present the Active Learning strategy for training dataset selection, including a discussion of its role dealing with highly imbalanced datasets. Finally, we comment on the feature extraction method by Gabor filter bank.

2.1 Random Forest Classifiers

Random Forest (RF) algorithm is a classifier [11] that encompasses bagging [12] and random decision forests [13,14], whose performance has been demonstrated in a variety of applications [10,15]. RF became popular due to its simplicity of training and tuning while offering a competitive performance to other machine learning approaches, such as support vector machines. Consider a RF as a collection of decision tree predictors, built so that they are as much decorrelated as possible, and denoted by Eq. (1):

$$\{h(\mathbf{x}; \psi_t); t = 1, ..., T\} \tag{1}$$

where $\mathbf{x}$ is a d-dimensional random sample of random vector X and ψ_t are independent identically distributed random vectors modeling the stochastic nature of the tree building process. Each tree $h(\mathbf{x}; \psi_t)$ casts a unit vote in order to decide the class assignment of $\mathbf{x}$. RF captures complex interaction structures in data, and are proposed [11] to be resistant to both over-fitting of data when individual trees are very deep and no pruned, and under-fitting when individual trees are too shallow.

Given a dataset of N samples, a bootstrapped training dataset is used to grow a tree $h(\mathbf{x}; \psi_t)$ on a randomly selected subset of data dimensions $\hat{d}$ such that $\hat{d} \ll d$. Decision tree growing recursively picks the best data split of each node based on these information measure of each dimension. In RF pruning is not required [11]. The RF training process picks randomly the dimension and the dataset bootstrapping according to independent identically distributed random vectors ψ_t. This randomness is the source of RF individual tree diversity ensuring the decorrelation of their outputs.

Classification of a new input $\mathbf{x}$ is done by majority voting over the responses of the trees in the RF $C_u(x)$. The critical parameters of the RF classifier for the experiments reported below are the number of trees in the forest, the dimension of the random subspace and the maximum tree depth. Later we report experiments assessing the effect of these parameters in our specific study.

2.2 Active Learning for Image Segmentation

We want to classify image pixels into two classes, the target and the background [15]. Target in our case are the pixels corresponding to the lane marks and other landmarks in the road. In a nutshell, an Active Learning system returns to the user an image whose intensity value corresponds to the degree of uncertainty in the classification of the pixel. Upon this image, the user, in its role as the oracle will pick some of the pixels with greatest intensity labeling them for insertion in the training dataset. Then, a new instance of the classifier is trained [3]. The features of each pixel are the result of the application of a bank of Gabor filters, the pixel intensity and its coordinates. Though the feature vector dimensionality is relatively high, we do not carry out any feature selection procedure because we prefer to leave open the possibility that a certain orientation or scale may be meaningful in future images. We assume that the stream of images may produce images that are quite different from the ones used for training. Hence, the final implementation will allow to restart the Active Learning process when the human operator detects deviations from optimal segmentation. It is an open-end learning process with a human in the loop. For the computational exploration reported in this paper, we do not resort to a human oracle. Instead, we use a hand delineated ground truth of a collection of images. Hence, the labeling process consists in the selection of the pixels with maximal uncertainty, applying random selection to solve ties.

2.3 Active Learning and Class Imbalance

Class imbalance is a widely know issue in machine learning [16] that is gaining increased attention due to its very strong bias effect in classifier training by almost any modeling approach. In the case of digital mapping, the effect of class imbalance has been also recognized [17]. The data dealt with in this paper is highly class imbalanced, hence we must take into account this fact in the use of performance metrics, and in the way we select the samples that are added to the training set. We have carried out comparative experiments ensuring that the added samples are class balanced and not ensuring this class balance. Usually, using subsampled balanced datasets for training does not achieve competitive test results. However, we hypothesized that the optimal selection of the samples followed by the Active Learning may compensate for the subsampling.

2.4 Gabor Texture Features

In order to have a systematic characterization of the surroundings of each pixel we use a bank of Gabor filters. The magnitude of the responses of the Gabor filters are used as the feature vector for classification. In other words, we use the local texture descriptor of the image as features [18,19] for classification. Formally, a Gabor filter is defined by the product of a sinusoidal wave, a plane wave in 2D, and a Gaussian function. The Gaussian component modulates the scale of the filter, while the wave component acts as a selector of the orientation and spatial frequency of the detected objects. In many implementations, Gaussian scale and wavelength are linked, so only the wavelength is specified. The Gabor filter provides a complex valued response, so its magnitude and phase can be used as features. In this paper we use only the phase. Formally, the impulse response of single filter is defined by Eq. (2):

$$g\left(x,y\right)=\left(\frac{1}{2\pi\sigma_x\sigma_y}\right)\exp\left[-\frac{1}{2}\left(\frac{x'^2}{\sigma_x^2}+\frac{y'^2}{\sigma_y^2}\right)\right]\exp\left[2\pi i\left(Ux+Vy\right)\right] \tag{2}$$

where we rotate the Euclidean coordinates by θ such that $x' = x\cos(\theta)+y\sin(\theta)$, and $y' = x\sin(\theta) + y\cos(\theta)$. The parameters σ_x, σ_y define the spatial support and bandwidth of the filter. The complex exponential factor is a 2D sinusoidal wave of frequency $F = \sqrt{U^2 + V^2}$ and orientation $\gamma = tan^{-1}\left(\frac{V}{U}\right)$.

3 Experimental Setup

In this section we introduce the actual dataset used for computational experiments, the design for model parameters exploration, and the performance measures used for validation and comparison among model results.

3.1 Dataset

From the intensity data obtained from the LiDAR point cloud, which contained 10,103,405 points, a set of 10 orthoimages have been generated. Figure 1 shows one of the experimental images (left) and their manually delineated ground truth. The actual imbalance ratio of the data is 1:117, the target minority class accounts 0.85% of the dataset, the remaining 99.15% corresponds to the background, including the road and the environment.

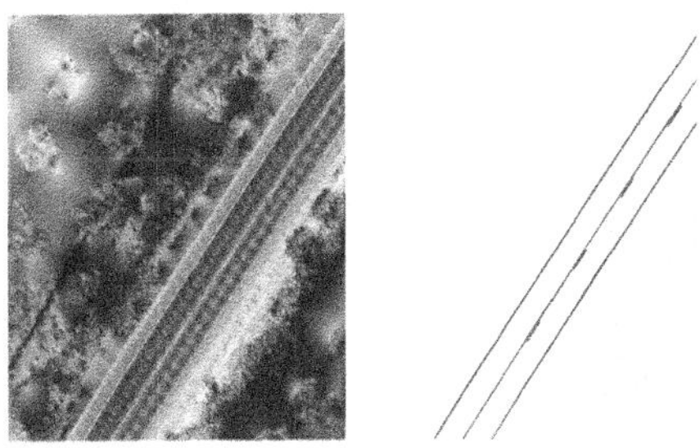

Fig. 1. Left: one of the experimental images, Right: its corresponding manually delineated ground truth (white is the background)

3.2 Model Parameter Exploration

We have carried out an exhaustive exploration of the model parameter settings in order to assess robustness of the approach and to find out which parameter is more influential. As a conclusion we can provide some guidance on the best parameter setting and the most promising model. Regarding RF computational experiments, we have explored the influence of the number of trees (NT), the number of variables taken into account at node split (NVS), and the size of the sample (NS) that is added to the current training dataset. We have also made the comparative between ensuring that the dataset subsample added to the current training dataset is class balanced and not ensuring class balance.

3.3 Validation

In order to evaluate the quality of the results [20], we report the sensitivity (SEN), specificity (SP), accuracy (AC) and true positive ratio (TPR) of the pixel-wise classification of the entire images using the classifiers built upon the selected training datasets at the end of the Active Learning process. The most valuable metric is the TPR because of the strong class imbalance of the dataset.

The pool of pixels used for the selection of the training dataset is composed of pixels of all labeled images, so the selection tries to have representatives from all images, in order to avoid overtraining on one image. Hence, at each Active Learning iteration we compute the classification uncertainty over all images. However we do not ensure that the selection is fair, in the sense of picking the same amount of pixels from each image to be added to the training dataset. Regarding the issue of the separation of training and test data for validation, it is ensured as far as we are reporting the performance measures over the pixels not in the training set. Active Learning is *per se* safe in this regard, because never uses the labeling information of data outside the training dataset [1].

4 Experimental Results

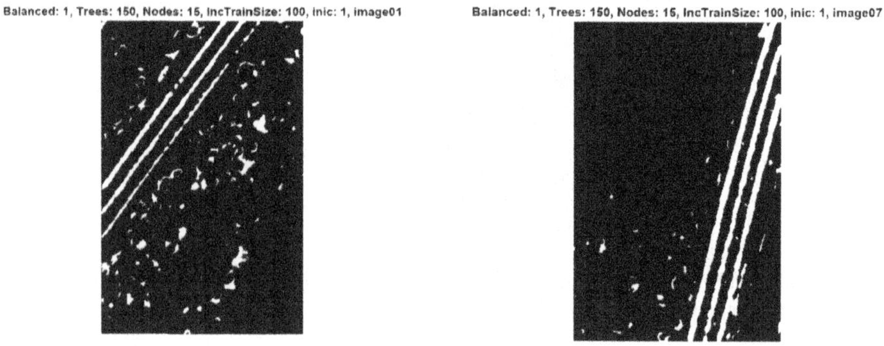

Fig. 2. Some visual results of the trained RF ensemble classifiers using balanced training sample increments.

For a qualitative appraisal of the results we show in Fig. 2 two images of the detections achieved by RF classifiers trained on the final training dataset of Active Learning processes. On the other hand, for a quantitative assessment, we also compare the performance achieved by the RF classifiers when the training dataset is increased by candidate samples with or without ensuring class balancing, as shown in Table 1.

If we examine the AC, SEN, SP and TPR performance, where imbalanced training dataset growth lead to high AC and SEN values but very poor SP and TPR, balanced training dataset growth lead to high AC, SEN and TPR values but also poor SP values. In fact, the mean TPR in Table 1(b) is 0.074 while studying balanced dataset in Table 1(a) it is 0.954.

The size of the sample (NS) that were added to the training dataset at each iteration is a global parameter of the Active Learning process. In the RF experiments, we find moderate but significant effect of the sample size 100 versus 50 as

Table 1. Active Learning using RF classifiers. Performance results measured by sensitivity (SEN), specificity (SP), accuracy (AC), True Positive Ratio (TPR) of RF varying the number of trees (NT), the number of variables considered for the split at each node (NVS), the number of samples added in each iteration of the active learning algorithm (NS). The added set of samples is class balanced (a) and unmbalanced (b)

NS	NVS	NT	SEN	SP	AC	TPR
50	5	50	0.890	0.174	0.883	0.968
50	5	100	0.851	0.194	0.844	0.951
50	5	150	0.870	0.181	0.863	0.949
50	5	200	0.878	0.179	0.871	0.930
50	10	50	0.884	0.176	0.877	0.961
50	10	100	0.870	0.186	0.863	0.956
50	10	150	0.856	0.196	0.850	0.963
50	10	200	0.870	0.185	0.863	0.955
50	15	50	0.855	0.199	0.849	0.973
50	15	100	0.870	0.186	0.863	0.955
50	15	150	0.907	0.141	0.900	0.890
50	15	200	0.869	0.185	0.862	0.969
50	20	50	0.887	0.170	0.880	0.948
50	20	100	0.892	0.165	0.885	0.863
50	20	150	0.877	0.184	0.870	0.941
50	20	200	0.869	0.185	0.862	0.969
100	5	50	0.906	0.167	0.899	0.966
100	5	100	0.901	0.172	0.894	0.981
100	5	150	0.919	0.150	0.912	0.957
100	5	200	0.898	0.168	0.891	0.969
100	10	50	0.917	0.158	0.909	0.909
100	10	100	0.922	0.153	0.914	0.962
100	10	150	0.910	0.161	0.902	0.948
100	10	200	0.914	0.159	0.907	0.964
100	15	50	0.922	0.150	0.915	0.953
100	15	100	0.924	0.151	0.917	0.957
100	15	150	0.927	0.154	0.919	0.950
100	15	200	0.910	0.159	0.903	0.952
100	20	50	0.912	0.156	0.905	0.951
100	20	100	0.904	0.165	0.896	0.964
100	20	150	0.889	0.181	0.882	0.966
100	20	200	0.914	0.159	0.906	0.958

(a)

NS	NVS	NT	SEN	SP	AC	TPR
50	5	50	0.999	0.003	0.989	0.068
50	5	100	0.999	0.001	0.990	0.027
50	5	150	0.999	0.001	0.989	0.076
50	5	200	0.999	0.002	0.989	0.063
50	10	50	0.999	0.005	0.990	0.026
50	10	100	0.999	0.001	0.989	0.033
50	10	150	0.998	0.002	0.988	0.082
50	10	200	0.999	0.001	0.989	0.075
50	15	50	0.996	0.005	0.987	0.119
50	15	100	0.998	0.006	0.988	0.052
50	15	150	0.999	0.005	0.989	0.060
50	15	200	0.998	0.005	0.988	0.107
50	20	50	0.997	0.004	0.987	0.125
50	20	100	0.997	0.003	0.987	0.108
50	20	150	0.998	0.002	0.989	0.088
50	20	200	0.998	0.005	0.988	0.105
100	5	50	0.999	0.001	0.989	0.060
100	5	100	0.999	0.002	0.990	0.041
100	5	150	0.998	0.002	0.988	0.094
100	5	200	0.998	0.003	0.988	0.116
100	10	50	0.999	0.002	0.990	0.040
100	10	100	1.000	0.002	0.990	0.020
100	10	150	0.999	0.003	0.989	0.073
100	10	200	0.999	0.001	0.989	0.085
100	15	50	0.998	0.007	0.988	0.106
100	15	100	0.998	0.002	0.988	0.128
100	15	150	0.999	0.002	0.989	0.083
100	15	200	0.999	0.004	0.989	0.066
100	20	50	0.998	0.003	0.988	0.078
100	20	100	0.998	0.002	0.989	0.068
100	20	150	0.999	0.002	0.989	0.075
100	20	200	1.000	0.001	0.990	0.023

(b)

we can see in TPR results from balanced dataset in Table 1(a) and unbalanced dataset in Table 1(b). It looks like that the greater sample increment is preferable, for instance, in the case of balanced dataset the mean value of TPR for NS = 100 is 0.97, while 0.95 for NS = 50.

5 Conclusion and Future Works

In this paper we introduce an Active Learning approach to deal with the labeling of road landmarks in intensity prepocessed images obtained by an on board sensor that includes LiDAR as well as positioning sensors for the purpose of detailed road signaling inventory. The underlying problem is a two class classification problem with strong class imbalance, and potentially large volume of images taken under very diverse light and atmospheric conditions, as well as road conditions. The proposed solution is an open ended process segmentation with a human in the loop that may start the adaptation to new images at any moment. Due to the cost of image labeling, the adaptation follows an Active Learning approach, where the training set is built incrementally with the most informative image samples. We have explored the performance of random forest (RF). Our computational experiments have found great results applying RF in terms of True Positive Ratio (TPR), a performance measure more appropriate than Accuracy for strong class imbalanced dataset. For instance, with an initial set up of 100 trees, 50 samples added in each iteration of the active learning algorithm and 5 variables considered for the split of each node, we have achieved a TPR of 0.98. Additionally we found a novel way to deal with class imbalance through Active Learning selection of optimal balanced training dataset. We think that the approach deserves further exhaustive study, as it has not been previously proposed in the literature. Future works would be addressed to the exploitation of the fused image and LiDAR information in order to enhance the road landmark recognition.

Acknowledgments. The work in this paper has been partially supported by Airestudio Geoinformation Technologies Scoop and Basque Government's BIKAINTEK grant. The work has also been supported by FEDER funds for the MINECO project TIN2017-85827-P, the grant IT1284-19 as university research group of excellence from the Basque Government and project 7-AA-3091-EG of the *Consejería de Fomento, Infraestructuras y Ordenación del Territorio. Dirección General de Infraestructuras de la Junta de Andalucía.*

References

1. Tuia, D., Volpi, M., Copa, L., Kanevski, M., Munoz-Mari, J.: A survey of active learning algorithms for supervised remote sensing image classification. IEEE J. Sel. Topics Signal Process. **5**(3), 606–617 (2011)
2. Cohn, D., Atlas, L., Ladner, R.: Improving generalization with active learning. Mach. Learn. **15**, 201–221 (1994)
3. Settles, B.: Active learning literature survey. Sciences **15**(2), 1–67 (2010)
4. Mitra, P., Shankar, B.U., Pal, S.K.: Segmentation of multispectral remote sensing images using active support vector machines. Pattern Recogn. Lett. **25**(9), 1067–1074 (2004)
5. Tuia, D., Pasolli, E., Emery, W.: Using active learning to adapt remote sensing image classifiers. Remote Sens. Environ. **115**(9), 2232–2242 (2011)
6. Hoi, S.C.H., Jin, R., Zhu, J., Lyu, M.R.: Semisupervised SVM batch mode active learning with applications to image retrieval. ACM Trans. Inf. Syst. **27**(3), 1–29 (2009)
7. Iglesias, J., Konukoglu, E., Montillo, A., Tu, Z., Criminisi, A.: Combining generative and discriminative models for semantic segmentation of CT scans via active learning. In: Information Processing in Medical Imaging, pp. 25–36. Springer, Heidelberg (2011)

8. Tao, Y., Peng, Z., Jian, B., Xuan, J., Krishnan, A., Sean Zhou, X.: Robust learning-based annotation of medical radiographs. In: Medical Content-Based Retrieval for Clinical Decision Support. Lecture Notes in Computer Science, vol. 5853, pp. 77–88. Springer, Berlin/Heidelberg (2010)
9. Izquierdo, A., Lopez-Guede, J.M., Graña, M.: Road lane landmark extraction: a state-of-the-art review. In: Pérez García, H., Sánchez González, L., Castejón Limas, M., Quintián Pardo, H., Corchado Rodríguez, E., (eds.) Hybrid Artificial Intelligent Systems, pp. 625–635. Springer International Publishing, Cham(2019)
10. Barandiaran, I., Paloc, C., Grana, M.: Real-time optical markerless tracking for augmented reality applications. J. Real Time Image Process. **5**, 129–138 (2010)
11. Breiman, L.: Random forests. Mach. Learn. **45**(1), 5–32 (2001)
12. Breiman, L.: Bagging predictors. Mach. Learn. **24**(2), 123–140 (1996)
13. Amit, Y., Geman, D.: Shape quantization and recognition with randomized trees. Neural Comput. **9**(7), 1545–1588 (1997)
14. Ho, T.: The random subspace method for constructing decision forests. IEEE Trans. Pattern Anal. Mach. Intell. **20**(8), 832–844 (1998)
15. Maiora, J., Ayerdi, B., Graña, M.: Random forest active learning for aaa thrombus segmentation in computed tomography angiography images. Neurocomputing **126**, 71–77 (2014)
16. Haixiang, G., Yijing, L., Shang, J., Mingyun, G., Yuanyue, H., Bing, G.: Learning from class-imbalanced data: review of methods and applications. Expert Syst. Appl. **73**, 220–239 (2017)
17. Sharififar, A., Sarmadian, F., Malone, B.P., Minasny, B.: Addressing the issue of digital mapping of soil classes with imbalanced class observations. Geoderma **350**, 84–92 (2019)
18. Fogel, I., Sagi, D.: Gabor filters as texture discriminator. Biol. Cybern. **61**(2), 103–113 (1989)
19. Maldonado, J.O., Graña, M.: Recycled paper visual indexing for quality control. Expert Syst. Appl. **36**(5), 8807–8815 (2009)
20. Ruiz-Santaquiteria, J., Bueno, G., Deniz, O., Vallez, N., Cristobal, G.: Semantic versus instance segmentation in microscopic algae detection. In: Engineering Applications of Artificial Intelligence, vol. 87, p. UNSP 103271, January 2020

Author Index

Á. Herrero et al. (Eds.): SOCO 2020, AISC 1268, pp. 873–876, 2021.
https://doi.org/10.1007/978-3-030-57802-2

The manufacturer's authorised representative in the EU is Springer Nature Customer Service Centre GmbH, Europaplatz 3, 69115 Heidelberg, Germany. If you have any concerns regarding our products, please contact ProductSafety@springernature.com

Printed and bound by CPI Group (UK) Ltd, Croydon, CR0 4YY
30/09/2026
02225509-0019